Astrophysics and Space Science Proceedings

Volume 61

The Astrophysics and Space Science Proceedings is a series of high-level proceedings volumes covering a broad range of subjects in Astrophysics, Astronomy, Cosmology, and Space Science. The contributors are distinguished specialists with international reputations in their fields of expertise. The subjects are placed in the broader context of related disciplines such as Engineering, Computer Science, Environmental Science, and Nuclear and Particle Physics.

Shubhrangshu Ghosh • A. R. Rao
Editors

The Relativistic Universe: From Classical to Quantum

Proceedings of the International Symposium on Recent Developments in Relativistic Astrophysics: ISRA 2023, SRMUS, Gangtok

Springer

Editors
Shubhrangshu Ghosh
Center for Astrophysics, Gravitation and Cosmology & Department of Physics
Shri Ramasamy Memorial (SRM) University Sikkim
Gangtok, Sikkim, India

A. R. Rao
Department of Astronomy and Astrophysics
Tata Institute for Fundamental Research
Mumbai, Maharashtra, India

International Symposium on Recent Developments in Relativistic Astrophysics isra isra2023 ISRA Gangtok India 2023 12 10 2023 12 12 https://srmus.ac.in/ISRA2023

ISSN 1570-6591 ISSN 1570-6605 (electronic)
Astrophysics and Space Science Proceedings
ISBN 978-3-031-90185-0 ISBN 978-3-031-90186-7 (eBook)
https://doi.org/10.1007/978-3-031-90186-7

This Springer imprint is published by the registered company Springer Nature Switzerland AG
The registered company address is: Gewerbestrasse 11, 6330 Cham, Switzerland

Foreword

In the present Indian scenario, very few Astrophysicists are pursuing diverse research topics with equal dexterity. Banibrata Mukhopadhyay, though just crossed the "tender" age of 50, is most certainly one of them. He has created a school of thought by infusing a deep sense of passion among his colleagues, and graduate and post-graduate students. Anyone interacting with Bani would immediately sense his genuine attachment to whatever physical problem he targets. His interest ranges from accretion phenomena around black holes, or neutron stars, equation of states of neutron stars, field theory and particle physics in curved space-time, astronomical data analysis, etc. His super-Chandrasekhar mass limit is certainly a great finding and could open up newer possibilities. His balance of enthusiasm, competence, and passion attracted a galaxy of students and coworkers around—a testimony of which was displayed in this very fruitful conference to celebrate his 50th Birthday in a beautiful setting of Sikkim.

The enthusiastic attendance, deliberations, and discussions in the meeting among scientists and scholars of diverse interest made the conference a very successful one. Particular emphasis was on the effects of magnetic fields on dynamics of inflow and outflows around compact objects. Discussions on observational effects of magnetic fields in and around these objects are presented in some of the articles. Observational signatures of shocks in these flows were discussed in other set of papers. State-of-the-art numerical simulations with or without magnetic fields provided food for thought to those wondering about the stabilities of theoretical solutions. A significant part of the conference was dedicated to Active galaxies, gamma-ray bursts, and ultra-luminous X-ray sources which resonated some of the thoughts of Banibrata.

Experts in completely different topics, namely, dark matter, cosmology, cosmic rays, astrobiology, modified theories of gravity and their effects on observations felt equally comfortable and found significant audience, thanks to magnetic attraction of Bani. Their contributions to this Volume enriched the Volume, and expectedly reflected overall sentiment of the conference. In all these topics Bani contributed significantly. This proceedings is thus an example of "*Astrofisica senza confini*" among the nextGen astrophysicists and astronomers of India. When such workers

Professor Banibrata Mukhopadhyay in a scientific discussion

from diverse interest converse, newer branches open up and that is what would be expected from this school of Bani in near future. I am sure Bani will remain as productive and passionate as ever in the next 10 to 20 or more years and would enrich the subject and the work force at a greater pace.

Indian Centre for Space Physics, Kolkata, India Prof. Sandip Kumar Chakrabarti
November 27, 2024

Banibrata with Professor Sandip Kumar Chakrabarti during a discusion at ICSP

Banibrata with Professor Ramesh Narayan of Harvard

Banibrata with Professor Rashid Sunyaev in Pescara

Felicitation of Professor Mukhopadhyay on his 50th birth anniversary at SRM University Sikkim (SRMUS), Gangtok. Left to right: Shubhrangshu Ghosh, Banibrata Mukhopdhyay, Professor Ch Satish Kumar, Hon'ble Vice Chancellor of SRMUS, and Kiranmayi Kothapalli

Message from Vice Chancellor

It is a matter of honor and immense pleasure that the first International Symposium on Relativistic Astrophysics (ISRA 2023) with the theme: "*International Symposium on Recent Developments in Relativistic Astrophysics*" was organized by Center for Astrophysics, Gravitation and Cosmology (CAGC), SRM University Sikkim (SRMUS), from December 11 to 13, 2023. The Symposium took an in-depth review on recent progresses in various aspects of Astrophysics and Space Science, gathering leading minds in this field from India and abroad. The event provided a key platform for distinguished scientists, researchers, and scholars in this field to disseminate knowledge and discuss current challenges and future perspectives in these areas. The exuberant participation, deliberations, and thoughtful discussions in the meeting among scientists, scholars, and a galaxy of young researchers of assorted interest, held in the picturesque setting of Sikkim, made this event a very fruitful and an enriching one; a one of its kind in the North-Eastern region of our country, and Sikkim in particular. One of the hallmark of this event is the felicitation of Prof. Banibrata Mukhopadhyay of Indian Institute of Science, on his 50th birth anniversary. It gave me immense pleasure to felicitate Prof. Mukhopadhyay on behalf of SRMUS, recognizing his significant contributions to this field.

At SRMUS, we take immense pride in our commitment to advancing the frontiers of knowledge in the field Astronomy & Astrophysics, Space Science, Gravitation, Cosmology, and related fields. In this regard, we have recently established a dedicated Center for Astrophysics, Gravitation, and Cosmology (CAGC), aimed at catalyzing research, innovation, and education in these specialized domains. Through our dedicated efforts, we have played a pivotal role in fostering a vibrant scientific community and made significant strides in shaping the scientific landscape of North Eastern India. Our institution continues to be a driving force, providing researchers and students with a conducive environment to push the boundaries of knowledge. This symposium stood as a testament to our institute's ongoing commitment to the advancement of Astrophysical and Space Sciences. It is really a matter of honor for me to extend my heartfelt gratitude to CAGC, Department of Physics, School of Basic Sciences, and the organizing committee for their commitment and superb drive in organizing this successful event.

This volume, which is more of a Festschrift in honor of Prof. Mukhopadhyay, is a collection of original research findings and reviews on recent advances in Astrophysics and Space Sciences, contributed by most of the speakers/delegates participated in the said Symposium, having made an overwhelming support in regard to this proceedings. The diverse nature of articles in this volume on the contemporary issues of Astrophysics and Space Science, reflecting upon broader interest of Mukhopadhyay, enriched this Festschrift significantly. I am sure that this volume would substantially benefit next generation of scientists and scholars, particularly young graduates and researchers working in this field, and a broader populace pursuing fundamental research in science. I take this opportunity to thank distinguished scientists and scholars who participated in this event, and for their invaluable contribution to this volume. I specially wish Prof. Banibrata Mukhopadhyay for his remarkable scientific journey ahead. My heartfelt thanks to the editors Dr. Shubhrangshu Ghosh and Prof. A. R. Rao, for nicely organizing and crafting this volume, especially Shubhrangshu (the convener of this Symposium), whose single-minded determination and painstaking effort made this symposium and proceedings a success.

Shri Ramasamy Memorial (SRM) University
Sikkim, Gangtok, India
December 6, 2024

Prof. Ch Satish Kumar

Preface

A *Festschrift*, literally meaning "celebration writing," alternatively in Latin *liber amicorum*, is a collection of *essays in honor* to mark a special occasion or honoring special achievements or celebrating notable career journey of a prominent academic or scholar. This Festschrift is dedicated to Professor Banibrata Mukhopadhyay, a leading Indian astrophysicist of our times, commemorating his 50th birth anniversary. The present volume of work is a collection of original research reports and reviews on contemporary issues on relativistic astrophysics and space science, and related topics, based on the record of the complete proceedings of the first *International Symposium on Recent Developments in Relativistic Astrophysics (ISRA 2023)*, convened in December 11–13 at Shri Ramasamy Memorial (SRM) University Sikkim (SRMUS), Gangtok, India. This is a tribute to Prof. Banibrata Mukhopadhyay in light of Mukhopadhyay's (BM hereinafter) illustrious academic journey. One of us (SG), the first de facto research student of BM, conceived of organizing this symposium celebrating BM's 50th birth anniversary and over 25 years of his academic journey, recognizing his significant contributions to this field. The event witnessed overwhelming response not only from his own doctoral students, post-doctoral fellows, senior undergraduate students, and his research collaborators (present and past), but even fellow distinguished scientists, researchers, and scholars alike—thanks to BM's alluring admirability and his scientific ingenuity. Another of us (ARR), almost served as his mentor, witnessed his growth from young hardworking innovative and intelligent student to a matured professor who himself is mentoring many students.

Although BM's formal training was under the stewardship of Prof. Sandip Kumar Chakrabarti, a pioneer in black hole astrophysics, creating an "institution of himself," BM is largely self-exploratory, surpassing topical boundaries. In just over 25 years of his active research, one observes phenomenal growth in his research capabilities, with his research evolving significantly and diversifying into several interconnecting areas. BM began his research career working simultaneously on nucleosynthesis in hot accretion disks around black holes and solutions of Dirac equation in various curved backgrounds. While they are apparently disconnected pieces of projects, actually there is a connection as the accreting matter consists of

Fermions: ions and electrons. This shows his versatile and innovative mind, evident as early as during PhD time itself. Subsequently, he worked in a broad range of fields, encompassing, relativistic astrophysics, astrophysical fluids, compact objects in general, nuclear astrophysics, high energy astrophysics, to very early cosmology, field theory in curved spacetime, modified gravity, quantum mechanics in gravitational background, etc. He, in fact, appears to have initiated a new research window, namely, super-Chandrasekhar white dwarfs, with a large number of followers among scientists and scholars of all age group. Further, his proposed methodology to derive a pseudo-Newtonian potential to describe general relativistic properties of an accretion disk in the Newtonian framework is a unique contribution. Such diversification not only remained confined to BM himself, but subsequent evolution and diversification of research happened with his 30 odd doctoral students and postdoctoral fellows (till date), starting with a specific subject under Mukhopadhyay and then transitioning to a newer field with equal dexterity. BM, indeed, has created a school of thought.

ISRA 2023 was a testimony of such diversification. The symposium witnessed the participation of a galaxy of scientists, scholars, and young researchers from diverse research backgrounds (however, interconnected) working on contemporary topics in astronomy and space science. The attendees not only comprised students, coworkers, research collaborators, and mentors or mentees of Mukhopadhyay, but across the community, reflecting upon broader research domain and interest of BM. The present volume gathers together 50 original research articles and/or reviews contributed by most of the speakers/attendees of the said Symposium, who are well-known scientists and scholars in their own right. It covers a wide spectrum of topics that includes physics of accretion and relativistic jets, consequence of supermassive black hole (SMBH) activities, X-ray, gamma-ray astronomy, and active galactic nucleus (AGN)-related observations, astrophysical fluids and plasma, general relativistic magnetohydrodynamic (GRMHD) simulations, physics of compact objects, gravitational wave astrophysics, cosmic-ray physics, dark matter problem and cosmological issues, modified theories of gravity, gravity-quantum interface including topics related to holography, quantum physics in space, astrochemistry and origin of life, and even space instrumentation. A majority of the authors were/are Mukhopadhyay's PhD students, postdoctoral fellows, and research collaborators. A major part of this document revolves around evolution and diversification of Mukhopadhyay's own research and his accomplishment, and subsequent research evolution of his 30 odd doctoral students and postdoctoral fellows into diverse fields of astronomy and space science. The book has been compiled in such a manner so as to present an overview of the contemporary research in relativistic astrophysics and related fields in space science, pointing out future perspectives, and updating with the latest and most impactful developments in the field. Such diverse nature of articles not only reflected broad sentiment of the meeting but also Mukhopadhyay's own diversification of research, which has enriched this Festschrift significantly.

For a better streamlining and minute detailing, we have broadly classified and segregated 50 odd articles into 13 sections, with each section focusing more on some specific areas, although many sections are profusely interconnected. Many of the sections have significant contributions of BM's past and present students with a strong component of Mukhopadhyay's own contribution. The number of chapters in each sections varies depending upon the number of contributions in each field; however, we have made efforts to give all sections an equal importance. At the end, a "special contributory section" is deliberately created to encapsulate Mukhopadhyay's own research journey, emphasizing on his contribution to science.

In Part I of this volume, we depict the recent advances in the theory of accretion flows and relativistic jets, their observational implications, such as on different aspects of two component advective flow (TCAF), magnetically arrested disk (MAD), two-mode accretion flows, radiatively driven jets, explaining observational properties of BH sources in the context of Comptonization process and state transitions in BH X-ray binaries (BHXRBs) due to accretion shocks, even on aspects of hydrodynamic turbulence in accretion disks, and analogue gravity in the context of Bondi flow. Part II deals with phenomenological aspects related to SMBH activities in AGNs, such as on aspects of episodic AGN jets and morphological dichotomy in AGN sources. In Part III, we mostly deal with general aspects of astrophysical fluids and plasma, GRHD/GRMHD simulations and their astrophysical implications, such as on aspects of mean field dynamo, hydromagnetic turbulence. This also contains GRHD simulations around accreting BHs, explaining observations such as hard state ultra-luminous X-ray sources (ULXs) through GRMHD simulations.

In Parts IV, V, and VI, we have taken up various aspects of high energy observational astronomy. Part IV primarily deals with various aspects of observations and modeling of XRBs and future developments in this field, such as on X-ray observations of BHXRBs, highlighting X-ray polarization measurements using AstroSat, recurrent BHXRB sources, on aspects of modeling of quasi-periodic oscillations (QPOs), estimating spin of stellar mass BH/compact sources, even on aspects of chaotic nature of BHXRB sources. In Part V, we have covered recent advances on the theoretical and observational aspects of gamma-ray bursts (GRBs), including modeling of GRB sources. Part VI deals with diverse topics related to AGNs (although fewer articles), which includes observations related to BL Lacertae objects, studying of star formation properties in spiral galaxies in the UV regions using UV space telescopes including AstroSat-UVIT highlighting AGN feedback issues, and very interesting studies related to relics of AGN jet feedback through citizen science astronomy.

Part VII devotes to a relatively newly opened window of astrophysics, namely, Gravitational Wave (GW) Astronomy. It captures articles from discussing merging compact binaries, which is indeed related to first direct discovery of GW due to black holes' coalescence, to how the presence of accretion disk could influence the GW emission. It also includes a work constraining neutron star properties by GW observation. Part VIII, however, concentrates more on the compact object per se. Its chapters enlighten the new window initiated by Mukhopadhyay, namely, super-Chandrasekhar white dwarf. The chapters cover how a massive white dwarf violating the Chandrasekhar-limit can be identified by GW astronomy, simulating the time evolution of magnetized main sequence stars leading to super-Chandrasekhar white dwarfs. Another chapter centers around EHT images and attempts to pinpoint the fundamental properties of the underlying BH.

In Part IX, we have covered a wide spectrum of topics related to astroparticle physics and cosmology, like ultralight dark matter (DM), origin of cosmic rays, and fast radio bursts (FRBs), highlighting their use as possible tools for cosmological investigations, topics related to cosmic glitch in gravity and modification to dark energy density, novel tools on measuring Hubble constant, and even exploring phase transitions in very early Universe through computer-aided simulations.

Part X covers the quantum aspects of various fundamental physics from space. As neutrino is an important ingredient in astrophysics cosmology, the chapters herein explores quantum correlations and effects of gravity in neutrino oscillations, the underlying quantum speed limit in gravitational and varying magnetic fields and their consequences. Part XI captures works in the interface of gravity and quantum physics. It includes works from holography, quantum gravity, and quantum physics in gravitational background to Hawking radiation and black hole thermodynamics in modified Einstein gravity. Hence, it includes a bit of the field theory in curved spacetime.

In Part XII, fascinating and emerging field of astrochemistry and astrobiology related to origin of life and complex biophysical systems have been covered. At last, in Part XIII, in the "special contribution" section, BM has very beautifully and systematically summarized his own research journey over his past two and a half decades. It is a well-written account of his scientific contribution, touching upon variety of interconnecting subjects he worked on or pursuing presently. As beautifully paraphrased by Prof. Sandip Kumar Chakrabarti in his foreword to this volume that this volume in true sense "is thus an example of *Astrofisica senza confini*," not only among future generation of scientists and researchers in astronomy and space community, but also would prove to be beneficial to potential graduates, doctoral students, young and professional researchers, and academicians alike, across the scientific community, encapsulating fascinating area of research.

A mentor also has mentors. In view of this, we have decided to depict a few photographs of BM in this volume among those who mentored and influenced him in his scientific journey till date. We have included pictures with Prof. Sandip Kumar Chakrabarti (his formal PhD supervisor), Prof. Ramesh Narayan (his postdoctoral mentor at CfA, Harvard), with Prof. R. A. Sunyaev who significantly influenced his scientific thoughts. Apart from them and many others, Mukhopadhyay was strongly inspired and influenced by many senior distinguished Indian scientists, like Prof. A. R. Rao (of Tata Institute of Fundamental Research; one of us here), Prof. Vinod Krishan (of Indian Institute of Astrophysics), Prof. T R Govindarajan (from The Institute of Mathematical Sciences, now at Krea University), Prof. Naresh Dadhich (of Inter-University Centre for Astronomy and Astrophysics), etc. We could not depict their pictures with BM here, due to page limit in this volume. We have also decided to attach formally one of the photographs of Mukhopadhyay's felicitation at SRMUS by Hon'ble Vice Chancellor of the institute.

We thank all our distinguished colleagues and friends who were present and enthusiastically participated in the international symposium in Gangtok in December 2023, and their overwhelming support in contributing to this volume. We are also thankful to the editors of all the journals for giving their permission to reuse or reproduce many figures in different chapters. We are also grateful to Science and Engineering Research Board (SERB), Department of Science and Technology (DST), Government of India, and Board of Research in Nuclear Sciences (BRNS), Department of Atomic Energy (DAE), Government of India, who substantially funded our said symposium in honor of BM. One of us (SG) would take this opportunity to show earnest gratitude to Hon'ble Vice Chancellor of SRMUS, Prof. Ch. Satish Kumar, without whose continued support and appreciation this symposium and this subsequent proceedings would not have come to fruition. Also, all SG's colleagues, friends, and staffs at SRMUS, and SRM family at large, deserve special gratitude. Our special thanks to Mayank Pathak, Zenia Zuraiq, Srishty Aggarwal, Abhishek Jha, Mriganka Dutta, Rohan Raha, Mayusree Das, Panchajanya Dey, and Anindya Guria for carefully proofreading this volume. Lastly, we thank, Dr. Ramon Khanna (Executive Editor at Springer) Sujatha Chakkala, Srivithya Santhanam, and the entire team of Springer for showing diligence and perseverance, meticulously going through every step, to make this book possible.

Gangtok, India
Mumbai, India
December 2, 2024

Shubhrangshu Ghosh
A. R. Rao

Contents

Part I Theory of Accretion and Relativistic Jets

1 **Black Hole Accretion Is All About Sub-Keplerian Flows** 3
Sandip K. Chakrabarti

2 **Radiative Acceleration of Relativistic Jets from Accretion Disks Around Black Holes** 15
Indranil Chattopadhyay, Raj Kishor Joshi, Sanjit Debnath, Priyesh Kumar Tripathi, and M. Saleem Khan

3 **Noisy Accretion Flows** 25
Amit K. Chattopadhyay

4 **Bondi Flow from Various Perspectives** 37
Souvik Ghose and Tapas K. Das

5 **Dissipative Shocks in Accretion Flows: A Theoretical Forecast of the Spectral State Transitions in Black Hole X-ray Binaries** 53
Soumen Mondal and Prasad Basu

6 **Extreme Magnetic Fields Around Black Holes** 67
Koushik Chatterjee, Matthew Liska, Alexander Tchekhovskoy, Sera Markoff, and Ramesh Narayan

7 **Study of Accretion Disks Around Black Holes with Two Types of Gas Inflows** 81
Rajiv Kumar

8 **Universality of Coronal Properties in Accreting Black Holes** 93
Sudip Chakraborty, Ajay Ratheesh, Francesco Tombesi, Rodrigo Nemmen, and Srimanta Banerjee

Part II Supermassive Black Hole Activities

9 Mode of Accretion and Jet Power in Episodic AGN Jets 103
Chiranjib Konar, Shobha Kumari, Sabyasachi Pal, and Ananda Hota

10 Can Morphological Dichotomy in Radio Galaxies Be Resolved Through Galactic Potential–Induced Shocks? 115
Shubhrangshu Ghosh

Part III Astrophysical Fluids and GRMHD Simulations

11 Mean Field Dynamo: Beyond MHD 129
Vinod Krishan

12 General Relativistic Hydrodynamic Simulations Around Accreting Black Holes 139
Sudip K. Garain and Pranayjit Dey

13 Towards a Theory of Hydromagnetic Turbulence with Higher Order Fluid Moments 151
Arindam Saikia and Rupak Mukherjee

14 Accretion Disk-Outflow/Jet and Hard State ULXs 165
Mayank Pathak and Banibrata Mukhopadhyay

15 Magnetically Arrested Advective Accretion Flows and Jets/Outflows Around Stellar Mass Black Holes: Explaining Hard State ULXs with GRMHD Simulations 177
Rohan Raha, Banibrata Mukhopadhyay, and Koushik Chatterjee

Part IV Observations/Modeling of X-ray Binaries

16 X-ray Observations of Black Hole Sources 191
A. R. Rao

17 Light Curves and HID Signatures of Recurrent Black Hole X-ray Binary Sources 205
M. C. Ramadevi and V. P. Shyam Prakash

18 QPOs in Compact Sources as a Non-linear Hydrodynamical Resonance: Determining Spin of Compact Objects..................... 219
Arghya Ranjan Das and Banibrata Mukhopadhyay

19 Estimation of Spin Parameters of Stellar Mass Black Holes: LMC X-1, LMC X-3 and GX 339-4....................................... 233
Meghamani Haldar, Tamal Sarkar, Shubrangshu Ghosh, and Rajat K. Dey

20 Decoding the Deterministic Nature of Black Hole IGR J17091–3624... 255
Anindya Guria and Banibrata Mukhopadhyay

Part V Gamma-Ray Bursts

21 Recent Observational and Theoretical Progress in the Study of Gamma-Ray Bursts 269
Asaf Pe'er

22 Multi-Temperature Blackbody Emission in GRB 170409A 283
Soumya Gupta and Sunder Sahayanathan

Part VI Active Galactic Nucleus Observations

23 Ten Years of Searching for Relics of AGN Jet Feedback Through RAD@home Citizen Science 293
Ananda Hota, Pratik Dabhade, Prasun Machado, Avinash Kumar, Ck. Avinash, Ninisha Manaswini, Joydeep Das, Sagar Sethi, Sumanta Sahoo, Shilpa Dubal, Sai Arun Dharmik Bhoga, P. K. Navaneeth, Chiranjib Konar, Sabyasachi Pal, Sravani Vaddi, Prakash Apoorva, Megha Rajoria, and Arundhati Purohit

24 UV Star Formation Properties of NGC 1097 307
Bannanje Ananthamoorthy and Debbijoy Bhattacharya

25 Simultaneous Optical to X-Ray Spectra of OJ 287 and X-Ray Spectral Changes 319
Pankaj Kushwaha

Part VII Gravitational Wave Astrophysics

26 On the Origin of Merging Compact Object Binaries 331
Tomasz Bulik

27 Observational Imprints of Accretion Disk on Observed Gravitational Wave Data from LISA 341
Prasad Basu, Soumen Mondal, and Sangita Chatterjee

28 Can We Detect Super-Chandrasekhar White Dwarfs via Continuous Gravitational Waves? 355
Mayusree Das and Banibrata Mukhopadhyay

Part VIII Compact Objects

29 Was it Really a Kerr Black Hole That EHT Imaged in 2019? 369
Chandrachur Chakraborty

30 Exploring Magnetic Field Limits in Anisotropic Neutron Stars through GW190814 Tidal Deformability 377
Debabrata Deb, Banibrata Mukhopadhyay, and Fridolin Weber

31 Observational Mass-Radius Relation for Magnetized White Dwarfs 391
Drisya Karinkuzhi and Banibrata Mukhopadhyay

32 **Simulating Super-Chandrasekhar White Dwarfs** 397
Zenia Zuraiq, Achal Kumar, Alexander J. Hackett, Surendra Bhattarai, Christopher A. Tout, and Banibrata Mukhopadhyay

Part IX Astroparticle Physics and Cosmological Issues

33 **Ultralight Dark Matter: A Novel proposal** 411
T. R. Govindarajan

34 **The Origin of Cosmic Rays: An Enduring Mystery in Physics** 423
Arunava Bhadra

35 **A Glitch in Gravity: Cosmic Lorentz-Violation from Fiery Big Bang to Glacial Heat Death** 435
Robin Y. Wen, Lukas T. Hergt, Niayesh Afshordi, and Douglas Scott

36 **Inference of Hubble Constant using Standard Sirens and Reconstructed Matter Density Field** 447
Supranta S. Boruah, Ghazal Geshnizjani, and Guilhem Lavaux

37 **Novel Understanding of Cosmological Phenomena Using Fast Radio Bursts** 459
Surajit Kalita, Shruti Bhatporia, and Amanda Weltman

38 **Cosmic Phase Transitions in a Supercomputer?** 471
Soumya Kanti Ganguly

Part X Quantum Physics in Space

39 **Quantum Correlations in Neutrino and Neutral Meson Oscillations** 489
Subhashish Banerjee

40 **Influence of Gravity on the Quantum Speed Limit in Neutrino Oscillations** 503
Abhishek Kumar Jha, Mriganka Dutta, Subhashish Banerjee, and Banibrata Mukhopadhyay

41 **Increasing Quantum Speed Limit of Relativistic Electron via Non-uniform Magnetic Field** 515
Srishty Aggarwal, Banibrata Mukhopadhyay, Subhashish Banerjee, Arindam Ghosh, and Gianluca Gregori

Part XI Gravity and Quantum Physics Interface

42 **Affine Connection and Quantum Theory** 527
Kaushik Ghosh

43 Holographic Approach to Anomalous Transport in a Massive $U(1)$ Gauge Theory 539
Nishal Rai and Eugenio Megías

44 Holographic Fermions in Nonlinear Arcsin Electrodynamics 549
Ashis Tamang, Shubhrangshu Ghosh, and Nishal Rai

45 A Covariant Approach to the Berry Phase Arising due to Einstein Gravity 557
Achal Kumar and Banibrata Mukhopadhyay

46 Black Hole Thermodynamics in Modified Gravity 565
Mriganka Dutta and Banibrata Mukhopadhyay

47 Faster Rate of Hawking Radiation in Modified Gravity Constraining Dark Matter 573
Panchajanya Dey and Banibrata Mukhopadhyay

Part XII Astrochemistry and Astrobiology

48 Deciphering Interstellar Ice Morphology: Atomistic Simulations Reveal the Complex Behavior of Ethanethiol 583
Jeet Majumdar, Shubhadeep Nag, Tejender S. Thakur, Subramanian Yashonath, Bhalamurugan Sivaraman, and Prabal K. Maiti

49 Representation of Continuum Equations in Physical Components for Arbitrary Curved Surfaces Appearing in Biophysical Modeling 597
Sujit Kumar Nath

Part XIII Special Contribution

50 Black Holes, White Dwarfs, Neutron Stars, Related Fluid Mechanics and Nuclear Astrophysics: My Research Journey from 1997 to 613
Banibrata Mukhopadhyay

Part I
Theory of Accretion and Relativistic Jets

Chapter 1
Black Hole Accretion Is All About Sub-Keplerian Flows

Sandip K. Chakrabarti

Abstract We review the advantages of fitting with a Two Component Advective Flow (TCAF) which uses only four physical parameters. We then present the results of hydrodynamic simulations to highlight the fact that the primary component of a black hole accretion remains the sub-Keplerian or the low angular momentum flow independent of whether we have a high, intermediate or low mass X-ray binary. Every aspect of spectral and timing properties, including the disk-jet connection could be understood well only if such a component is present along with a Keplerian component of variable size and accretion rate.

1.1 Introduction

Black holes are the simplest of all celestial bodies. It is no wonder that the behavior of matter would also be very simple. The inner boundary condition of any flow onto black holes of any mass is universal, i.e., the flow enters through the horizon with the velocity of light. This supersonic flow at the inner boundary, forces it to pass through a sonic surface outside the horizon [1–3]. If the angular momentum is high enough so that the barrier $\sim 1/r^3$ becomes comparable to the gravitational force $\propto 1/r^2$, the centrifugal barrier resists the nearly freely falling matter to slow it down, forcing it to pass through a standing or oscillating shock wave, puffing it up and then entering through the horizon supersonically after passing through a second sonic surface. The sonic surface farther from the black hole is of Bondi type, while the one just outside of the horizon is the result of a strong gravity created by the general relativistic effects. This rapidly falling transonic, sub-Keplerian flow does not find time to cool down, carries all the specific energy and specific angular momentum (but not the total angular momentum) to its 'grave', i.e., the black hole, and is generally known as the advective accretion flow [2, 4]. A part of the incoming flow is pushed in the

S. K. Chakrabarti (✉)
Indian Centre for Space Physics, Kolkata, India
e-mail: sandip@csp.res.in

S. Ghosh, A. R. Rao (eds.), *The Relativistic Universe: From Classical to Quantum*, Astrophysics and Space Science Proceedings 61,
https://doi.org/10.1007/978-3-031-90186-7_1

direction perpendicular to the disk plane to create the outflow, the core of which could become fast moving, collimated jets in presence of toroidal magnetic fields. The jet could be accelerated beyond hydrodynamic values in presence of radiative and magnetic fields. These are the exact solutions of radiatively inefficient advective flows, or advection dominated flows etc. Though originally these latter flows did not include any shocks and thus missed the most exciting component of the flow, slowly, they started appreciating the role of the shocks [2] in shaping the observed spectra as well [5].

A more complex situation occurs when the viscosity and radiative transfer are included. Depending on the viscosity, which transports angular momentum and weakens the centrifugal barrier, or the residual flow energy remained after radiating, the shock may or may not form, even though the flow remains transonic and enters the horizon supersonically. These transonic, advective, radiative, viscous disks are examples of complete disk models and are capable to explaining observational aspect of black hole accretion, except when strong magnetic fields are present which may produce an extra power-law component [2, 6–8].

In contrast, the oft-used [9] viscous, black body radiating, standard Keplerian disk hardly moves towards the black hole. It is chopped off at the marginally stable radius (3 Schwarzschild radii in case of a non-rotating black hole). Being optically thick and geometrically thin, it radiates multi-color black-body emission. It has no length scale of importance (like the shocks or sonic points in a transonic flow) apart from that due to optical depth depending on accretion rates. Strictly speaking, they are subsonic everywhere. The accretion rate $\dot{M} = 3\pi \Sigma \nu [1 - (\frac{r}{r_i})^{-\frac{1}{2}}]^{-1}$ depends on the kinematic viscosity coefficient ν and not directly on velocity. Any non-zero ν would make the flow Keplerian in this model. Thus a Keplerian disk formation is always guaranteed.

In our complete transonic/advective flow scenario, the flow above a critical viscosity parameter α, is shock-less, makes the disk Keplerian *and* forces the inner disk to become supersonic at a point between the marginally bound and marginally stable orbits. This is what a standard disk should have been. Its temperature distribution would always be modified black body type, independent of accretion rates at each disk annulus, as it could 'reflect' (absorb and re-emit) radiation of the Compton cloud (discussed below).

Similarly, a flow with high 'accretion rate' and strong radiative pressure OR a flow with low 'accretion rate' and strong ion pressure has angular momentum distribution which deviates from a Keplerian distribution [10, 11]. These doughnut shaped isolated rotating bodies produce the so-called 'Thick accretion disk' which are geometrically thick and optically 'thick' or 'thin' depending on the accretion rate but have no solution connecting to the black hole horizon, nor have a solution through which matter is supplied. In our complete transonic/advective flow scenario, their isolation is remedied by identifying them as the post-shock flow of a complete solution [12, 13].

1.2 Two Component Advective Flow or TCAF

In our approach, a simple flow geometry is used for illustration (Fig. 1.1). The exact shape is unimportant as we use only the gross characteristics of the flow. A thick sub-Keplerian flow will initially have a height-dependent viscosity and will have higher viscosity on the equatorial plane due to higher temperature, turbulence etc. So, by definition, a disk is segregated into two components by an imaginary boundary away from the equatorial plane and it becomes a two component flow. It is irrelevant how the actual viscosity is sourced or is distributed or where the boundary lies [6, 7]. The boundary will not exist if the viscosity is very low or very high, as the viscosity parameter does not cross critical value which differentiates between shock and no-shock solutions. The region with high viscosity will behave like a Standard Keplerian disk, and the region surrounding it will behave as a sub-Keplerian transonic flow with a shock [14]. The Keplerian disk maintains the high viscosity, not because of high temperature, but because of high pressure (due to high density). The post-shock region will behave as a thick disk with proper advection as decided by the transonic flow solution. We need to concentrate only on four physical quantities to understand this flow: The accretion rates of the two components (i.e., exact viscosity or net accretion rate is not needed), the location of the shock (i.e., exact viscosity, energy/angular momentum etc. not needed) and most importantly, the strength of the shock. With these choice the parameters which 'embarrass' scientists, such as the exact viscosity, opacity of the torus not needed,

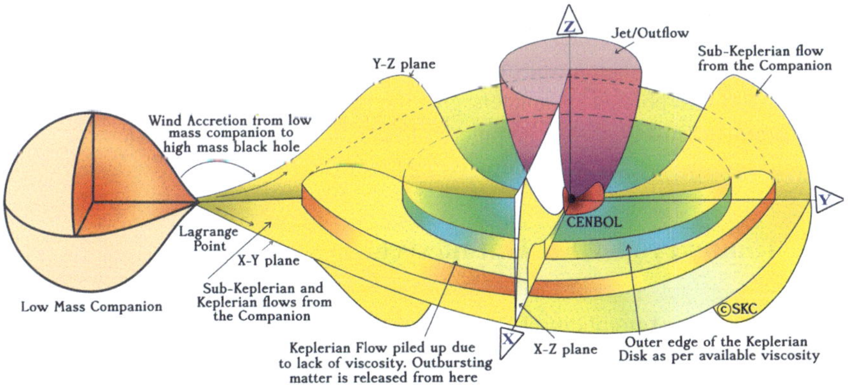

Fig. 1.1 Cartoon diagram (roughly in log scale) of a Two Component Advective Flow which seems to successfully fit X-ray data with only four physical flow parameters. From the normal companion both low and high angular momentum flow enters the Roche lobe of the compact object. In outbursting sources, high angular momentum matter halts at piling radius due to lack of viscosity. At outbursts, all or part of this stored matter is released due to enhanced viscosity, and a Keplerian disk moves in surrounded by the sub-Keplerian flow. The latter is slowed down at the shock front while the inner edge of the former is truncated there. The locations of successive slowing down, the strengths of the centrifugal barrier at these places, and the accretion rates in these components, decide all the spectral and timing properties

net accretion rate, specific energy or angular momentum etc. are not needed. The compression ratio or the strength of the shock decides the post-shock density and thus the opacity is determined. Furthermore, the post-shock flow being hot, it is assumed to evaporate the inner region of the Keplerian disk inside the shock location and the truncated inner edge location becomes identical to the shock location. The post-shock hot torus is also identified as the Compton cloud. So, computing the degree of interception of soft photons, reprocessing of the soft photons from the Keplerian disk to re-emit as hard photons become simpler. Step-by-step fitting spectral generation procedure was presented in [7] and this is used widely while fitting the spectra of stellar mass and supermassive black holes. This torus is called the CENtrifugal barrier supported BOundary Layer or the CENBOL for a good reason. Like a star with a hard surface, CENBOL also dissipates energy of the flow and is the primary source of high energy radiation of the accretion flow. Similarly, as a boundary layer, CENBOL is also the base of the outflow [15–17]. If the CENBOL is absent for excessive cooling, the hard radiation and the outflow are quenched [18, 19]. Most interestingly, the size of the CENBOL alone does not decide the outflow rate. For very large CENBOL, the outflow base area is high, but the temperature is low. For very small CENBOL, the base is smaller though the temperature could be higher. In both the cases, the outflow rate is low. Only when the CENBOL has an intermediate size, the outflow rate is high.

Combination of these four parameters produces the complete spectrum including effects of reflection at a given time. Generally speaking, as the disk rate is raised the spectrum becomes softer. If the sub-Keplerian or halo rate is raised, the spectrum becomes harder. If the shock location is reduced, the spectrum becomes softer. If the shock strength is enhanced, the spectrum becomes harder. Fitting spectrum has been extensively done either directly running the code [8] several times [20, 21] or making a model where results of running the code thousands of times are saved [22, 23] and interpolated parameters give the best fits. One can also use the code directly and iteratively change parameters to obtain the best fit [23, 24].

1.3 Timing Properties, Outbursts etc.

In the same spirit of simplicity, TCAF produces Quasi-Periodic Oscillations (QPOs) in radiation 'free of cost' without invoking any extra component [25–27]. The CENBOL itself oscillates when the cooling time scale roughly agrees with the infall time scale (Fig. 1.2). However, the CENBOL has a finite size and different part oscillates at a slightly different frequency. This produces 'quasi' periodicity and often higher harmonics. Depending on the mass of the black hole, radiation from CENBOL will be emitted at different wavelength. TCAF is the only solution which directly correlates the disk accretion rate with QPO frequency. Higher disk rate cools the CENBOL faster and increases the QPO frequency [26–28]. These are typically the type-C QPOs. This is observed in all the outbursts in X-ray binaries [29–32]. When the Rankine-Hugoniot relations at the shock is not satisfied, shock

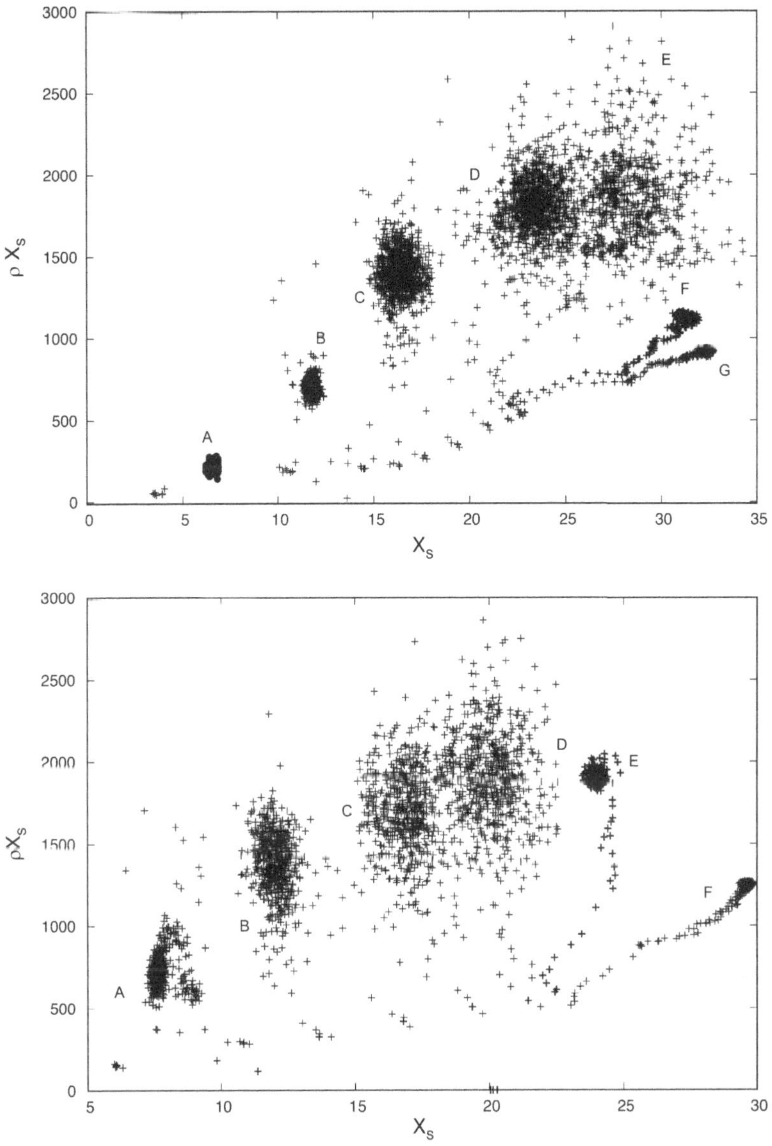

Fig. 1.2 The 'optical depth' variation with shock location for a set of injected parameters when cooling is decreasing (from A to F) for the stellar mass black holes (top) and supermassive black holes (bottom) in two dimensional SPH simulations. The scatter in shock location and optical depth is lowest when there is very low cooling (F) or very high cooling (A) and highest when resonance happens. The non-monotonicity of the optical depth is important as it rightly explains the non-monotonicity of the time lag behavior of QPOs [32]

starts oscillating 'searching for' the solution. This type of shocks does not follow direct relation with accretion rate or viscosity and are responsible for Type A and Type B QPOs.

Outbursts themselves are easy to understand in the TCAF scenario. In between two outbursts, much of the time, some matter with low-angular momentum will continuously reach the central black hole and the spectrum will show quiescence since the Keplerian disk is non-existent or exists only very far away and few soft photons are present for reprocessing by CENBOL. Assuming the net mass loss rate of the companion to be constant, clearly, most of the matter would be stuck at a 'pile-up' radius, accumulating steadily for months or years in this period (Fig. 1.1). The piled up matter, fully or partly, will eventually be driven inward triggered by enhanced viscosity on the hot equatorial plane. Both Keplerian and sub-Keplerian flow rates will suddenly go up but the sub-Keplerian component (halo) reaches the central black hole earlier than the Keplerian component (disk), causing sudden hardening of the spectrum before an outburst ensues [33–35]. This is typically the Hard state where type-C QPOs may form. The time delay of the soft peak with respect to hard peak is regularly observed and is proportional to the size of the Keplerian disk formed. The Keplerian disk will have its own outer boundary, since the supply of Keplerian matter is stopped with the drop of viscosity at the piling radius (after the convective viscosity is reduced, for example. As the Keplerian disk rate is increased, the spectrum softens from the hard state to the hard intermediate state with increasing QPO frequency (Fig. 1.2). In case the entire matter at piled up radius is evacuated in one shot, the outburst will last longer, and the next one will be expected with a large time gap. In case the piled up matter is partly evacuated, there could be secondary outbursts or several outbursts of smaller peaks.At some point of time, with a rise in Keplerian to sub-Keplerian flow rate ratio, the resonance condition is not fulfilled and only Type A/B QPOs are seen. This is the so-called Soft Intermediate State. Here the outflow rate is the highest. When the complete flow becomes Keplerian, the soft state is formed. Once the Keplerian supply from the pileup radius is stopped, the Keplerian disk can no longer sustain its identity or its angular momentum distribution. Its matter will slowly be entrained by the fast moving sub-Keplerian flow sandwiching it. In the rising stage of the outburst, the Keplerian disk is formed rapidly due to enhanced viscosity, but during the declining stage, the process of disappearance of the Keplerian disk is slower. This is the cause of hysteresis [36] effect. If one plots the dynamic hardness ratio or Comptonizing efficiency, the effect of hysteresis is seen clearly and is independent of the mass of the black hole [33, 37] unlike other diagrams (such as hardness ratio plots) which use rigid definition of hardness ratio.

Apart from the five spectral states (including the quiescent state with mainly a sub-Keplerian flow) mentioned above, there could be a super-soft state, when the Keplerian rate is very high and may produce radiation pressure supported thick disk at the inner edge from where radiation pressure driven outflow may form. In presence of magnetic field which could be brought by the flow, collimation and acceleration of the jet may occur when CENBOL is present through effects similar to Blandford-Payne (diverging fields) or Chakrabarti-Bhaskaran effects (diverging-

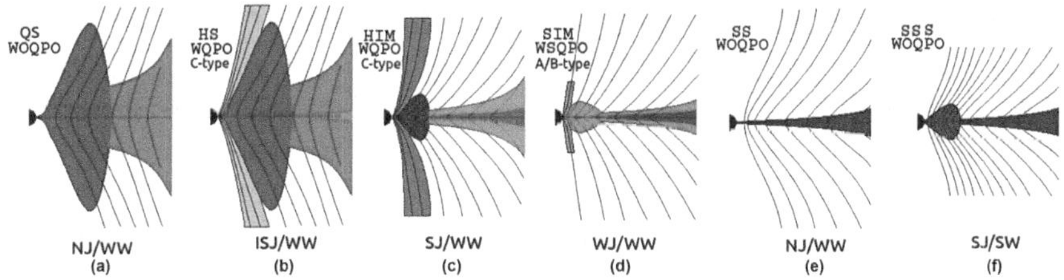

Fig. 1.3 All possible configurations of TCAF (**a**) Quiescent State (QS) with only very weak outflow, no jets, and no Keplerian component and no QPOs; (**b**) Hard State (HS) with a low-disk rate Keplerian component and high halo-rate sub-Keplerian component with Type-C QPOS, moderate or intermediate strong jets and weak winds (ISJ/WW); (**c**) Hard Intermediate (HIM) state having Type-C QPO, stronger jets from the CENBOL, stronger magnetically driven outflows; Strongest jets at (**c**) to (**d**) transitions; (**d**) Soft Intermediate (SIM) states have sporadic Type A/B QPOs, jets and outflows; (**e**) Soft States (SS) have only magnetically driven weak outflows but no jets and no QPOs and finally (**f**) Super-soft states (SSS) with very high accretion rates where inner-jet is blocked by the puffed-up matter. Collimated radiation and magnetically driven strong jets at the core of a strong wind from the whole disk. No QPOs are expected

converging fields) [38, 39]. In quiescent or soft states the kinematic luminosity of the jets would be negligible while it would be the highest in the soft-intermediate states. In the super-soft states, a part of the field lines would be blocked by puffed up thick disk, and thus the polarization is reduced and observed jets would emit at lower X-ray energies. Similar effects are expected in neutron stars and ULX sources with very high accretion rates. In Fig. 1.3, we show all the Spectral states of an outbursting X-ray binary.

One added advantage of fitting a spectrum with TCAF, be it in the context of a stellar mass black hole or a supermassive black hole, is that the mass of the central black hole is derived from the fitting procedure 'free of cost' from each data across spectral states. This is because, all the physical quantities derived from the four physical fitted parameters are functions of the mass of the black hole in a non-trivial way. Similarly, the normalization constant, which is the ratio of the photon flux emitted by the theoretical TCAF solution at the object frame, and the photon flux detected by the instruments (after correction due to the absorption) is also obtained after fitting. This should give us certain combination of the distance and the inclination angle of the source which can be cross-checked with those obtained from other methods [24, 29–31].

Since TCAF is the quasi-exact solution (in the sense, it uses average flow properties while fitting) for the disk accretion, any X-ray generated at the outflows and jets other than its base, namely, the CENBOL (which is an integral part of TCAF), would be in excess of what TCAF predicts at the object frame. This excess should be correlated with the radio/IR data of the jets [40]. Since the jets necessarily originate from the accretion, disk and jet are connected across the spectral states. A transonic flow solution connects the inflow and outflow properties as a function of the physical flow parameters.

1.4 Where Would Sub-Keplerian Flows Come From?

Since spectral fits by TCAF heavily depends on the existence of a low angular momentum flow (halo) component, it is pertinent to ask: where does the sub-Keplerian flow come from? In the case of an active galaxy with a super-massive black hole at the center, stellar winds from a large number of stars contribute. Ideally, if there are infinite number of stars, the tangential velocities will cancel totally and only the radial flow would dominate accretion. In reality, presence of a few stars would create a flow which will have very low net angular momentum as compared to the Keplerian value. In the worst case, the matter falling back from jets and outflows at larger radius, will have low angular momentum. In the case of a high mass X-ray binary, where the normal star is known to have profuse amount of winds, the net angular momentum would be lower than the Keplerian value at the Roche lobe. Matter may be accreted from both sides of the Roche-lobe also. But the compact object being lighter, high energy winds will not be gravitationally attracted from all directions. In both the cases, the flow will face the barrier due to centrifugal force close to the black hole and the size of this disk would be much smaller as compared to the accretion radius or Roche lobe distance. In the case of a low mass X-ray binary (LMXRB), it is *believed* without proof that disk is totally Keplerian (and the community is more worried about how to get rid of the excess angular momentum to create such a disk!). Surprisingly, the numerical simulations show that exactly the opposite situation prevails! We see that the average angular momentum still remains highly sub-Keplerian, though larger than that of a high mass X-ray binary system (Figs. 1.4 and 1.5). Because of this, the Keplerian disk is of larger size since the compact object is more massive and the L_1 point is farther from the compact primary, and the average angular momentum of the injected matter is higher. The sub-Keplerian nature in LMXRB is because the low mass stars also have winds and high mass compact captures most of this wind from all directions!

Figure 1.5 shows the average specific angular momentum distribution across the mass ratio of compact binaries. We observe that in all the cases, this is below the Keplerian value. In other words, a Keplerian disk could be formed only if sufficient viscosity transports angular momentum internally. This explains why low angular momentum flow is important in shaping the spectrum. Without this, fitting of a spectrum would require larger number of parameters.

1.5 Caveats in Fitting Spectral Data

In an evolving system, such as in an outburst, the matter approaches a black hole in a viscous time scale inside the Keplerian disk. It may take a few days to cross, say, 50,000–100,000 Schwarzschild radii. What this implies is that in the rising phase, at a given instant when the observation is made, the entire disk does not have a constant accretion rate. A propagation front of gradually changing accretion rate in the disk

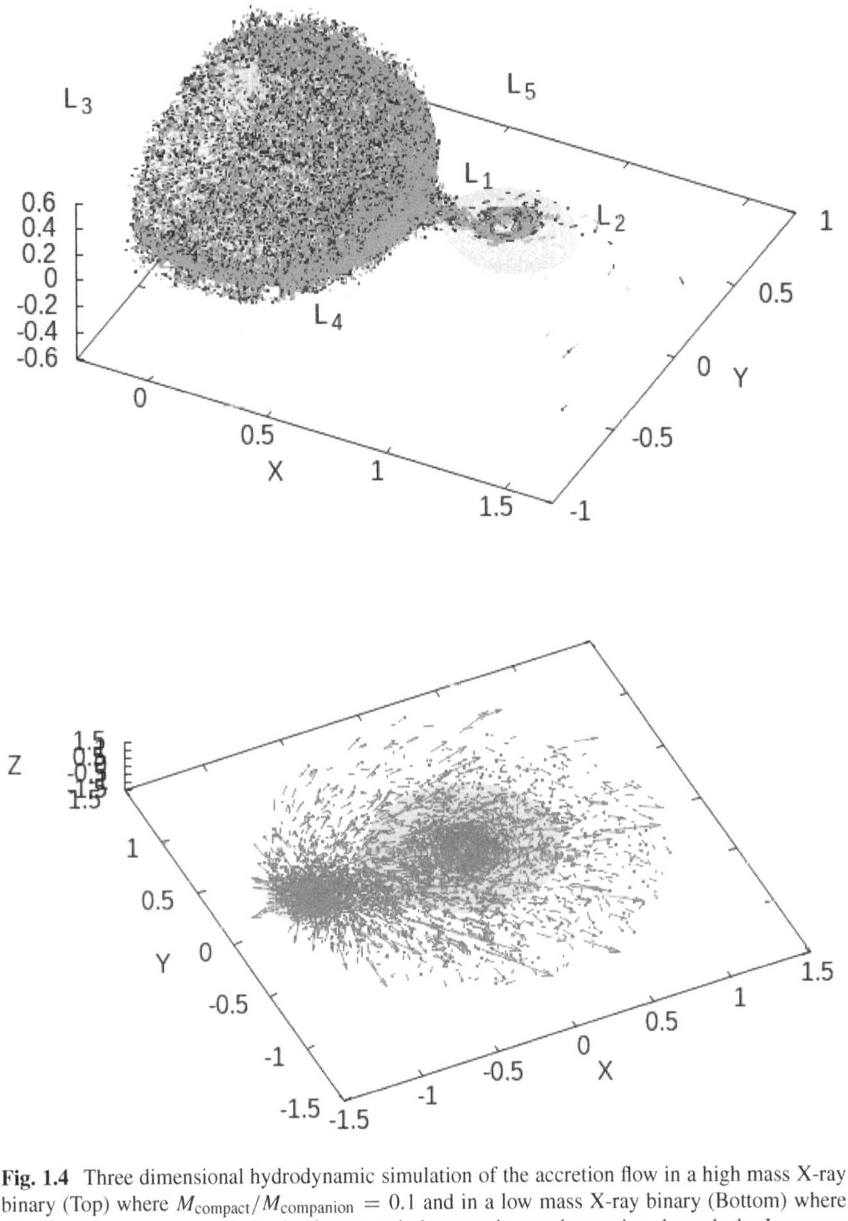

Fig. 1.4 Three dimensional hydrodynamic simulation of the accretion flow in a high mass X-ray binary (Top) where $M_{\rm compact}/M_{\rm companion} = 0.1$ and in a low mass X-ray binary (Bottom) where $M_{\rm compact}/M_{\rm companion} = 10$. In the former, wind matter is mostly passing through the Lagrange point as the light compact object is unable to attract winds from all directions. In the latter, even a little wind is attracted from all directions by the heavier compact resulting in a substantial amount of sub-Keplerian flow

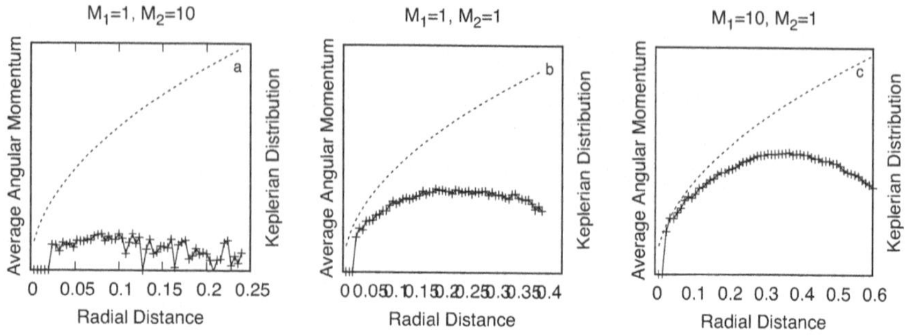

Fig. 1.5 Average angular momentum of matter accreting from Roche lobe to the compact as obtained by three dimensional simulations of inviscid flow. For reference, Keplerian distribution is also plotted. Note that across the mass-ratio the average flow is always sub-Keplerian. Viscosity transports angular momentum of this matter and creates a standard Keplerian disk as a bi-product. This is the main reason why TCAF requires a sub-Keplerian component to fit the spectra

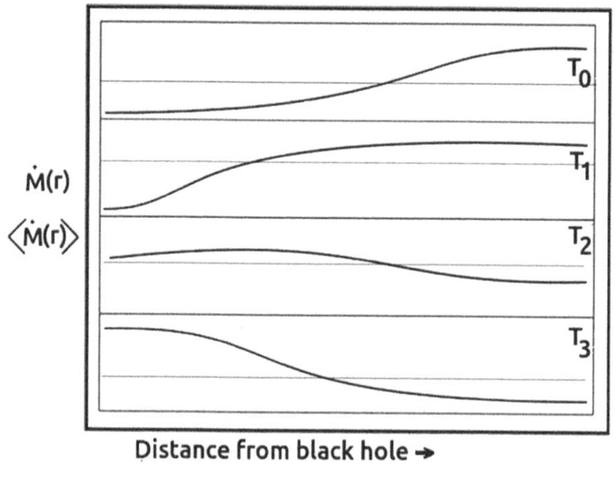

Fig. 1.6 A cartoon diagram showing propagation of matter in a Keplerian disk in viscous timescale towards the black hole in the rising phase of an outburst for a period of time (T) when the viscosity was high. As the viscosity is turned off, supply of Keplerian matter is also reduced, and this front continues to move in creating the declining state. $T_0 < T_1 < T_2 < T_3$. The true accretion rate ($\dot{M}(r)$) is shown in thick curve and the average ($< \dot{M}(r) >$) is shown in thin straight line for illustration

component brings in the matter. The true range of variation of $\dot{M}(r)$ is much higher than the average $< \dot{M}(r) >$. Hence fitted variation of $< \dot{M}(r) >$ may be much lower than the true variation (Fig. 1.6). In some occasions when violent changes in accretion rate takes place, such fitting with an average rate is not proper and would be misleading and one adds more components or parameters to get a good fit. This is true for any model which takes accretion rate or temperature as the parameter. In these cases spatially dependent mass flow has to be used to fit with TCAF and the code has to be re-written accordingly. Since the infall timescale of a sub-Keplerian flow is smaller, this problem is not so severe for the halo rate.

1.6 Conclusions

TCAF provides a four parameter family of spectra which appear to fit most of the observed data in X-ray binaries or in AGNs. Its fundamental premise is that the black hole accretion always (except in soft or supersoft states) has the sub-Keplerian flow and Keplerian disks form only when the viscosity rises. The fact that a high viscosity is difficult to have, this suits the TCAF scenario. Since rising and falling of viscosity will affect the formation and disappearance of the Keplerian component differently, hysteresis effect is seen. TCAF also provides the QPO frequency from spectral fits itself, provide a resonance like condition [27] between the infall time from the shock and the cooling time scale inside the CENBOL prevails. We showed through numerical simulations that the TCAF can address the spectral and timing properties very squarely. Large scale magnetic fields do not appear to be dynamically important in fitting the gross properties of the spectra except adding a power in certain states. However for the formation of collimated jets, it could be important. Power-law in soft states can be understood from the contribution by the bulk motion inside the inner sonic point of TCAF [7].

References

1. Chakrabarti, S. K. (1989). *The Astrophysical Journal, 347*, 365.
2. Chakrabarti, S. K. (1990). *Theory of Transonic Astrophysical Flows*. World Scientific.
3. Chakrabarti, S. K. (1996). *The Astrophysical Journal, 464*, 664.
4. Chakrabarti, S. K. (1996). *Physics Reports, 266*(5 & 6), 229–392.
5. Truong, L., Wood, K. S., Wolff, M. T., et al. (2016). *The Astrophysical Journal, 819*, 112.
6. Chakrabarti, S. K. (1995). In *Proceedings of 17th Texas Symposium (C94-12-12)* (p. 546). New York Academy of Sciences.
7. Chakrabarti, S. K., & Titarchuk, L. G. (1995). *The Astrophysical Journal, 455*, 623.
8. Chakrabarti, S. K. (1997). *The Astrophysical Journal, 484*, 313.
9. Shakura, N. I., & Sunyaev, R. A. (1973). *Astronomy & Astrophysics, 24*, 337.
10. Paczyńsky, B., & Wiita, P. J. (1980). *Astronomy & Astrophysics, 88*, 23.
11. Rees, M. J., Begelman, M. C., Blandford, R. D., & Phinney, E. S. (1982). *Nature, 295*, 17.
12. Molteni, D., Lanzafame, G., & Chakrabarti, S. K. (1994). *The Astrophysical Journal, 425*, 161.
13. Chakrabarti, S. K. (1996). *The Astrophysical Journal, 471*, 237.
14. Giri, K., & Chakrabarti, S. K. (2013). *Monthly Notices of the Royal Astronomical Society, 430*, 2836.
15. Chakrabarti, S. K. (1985). *The Astrophysical Journal, 288*, 1.
16. Chakrabarti, S. K. (1985). *The Astrophysical Journal, 288*, 7.
17. Chakrabarti, S. K. (1986). *The Astrophysical Journal, 303*, 582.
18. Chakrabarti, S. K. (1999). *Astronomy & Astrophysics, 351*, 185.
19. Garain, S. K., Ghosh, H., & Chakrabarti, S. K. (2012). *The Astrophysical Journal, 758*, 114.

20. Chakrabarti, S. K., Dutta, B., & Pal, P. S. (2009). *Monthly Notices of the Royal Astronomical Society, 394*, 1463.
21. Mandal, S., & Chakrabarti, S. K. (2008). *The Astrophysical Journal, 689*, 17.
22. Debnath, D., Chakrabarti, S. K., & Mondal, S. (2014). *Monthly Notices of the Royal Astronomical Society, 440*, 121.
23. Mondal, S., Chakrabarti, S. K., & Debnath, D. (2015). *The Astrophysical Journal, 798*, 57.
24. Nandi, P., Chatterjee, A., Chakrabarti, S. K., & Dutta, B. G. (2021). *Monthly Notices of the Royal Astronomical Society, 506*, 3111.
25. Molteni, D., Sponholz, H., & Chakrabarti, S. K. (1996). *The Astrophysical Journal, 457*, 805.
26. Chakrabarti, S. K., & Manickam, S. G. (2000). *The Astrophysical Journal Letters, 531*, L41.
27. Chakrabarti, S. K., Mondal, S., & Debnath, D. (2015). *The Astrophysical Journal, 452*, 345.
28. Garain, S. K., Ghosh, H., & Chakrabarti, S. K. (2014). *Monthly Notices of the Royal Astronomical Society, 437*, 1329.
29. Debnath, D., Molla, A. A., Chakrabarti, S. K., & Mondal, S. (2015). *The Astrophysical Journal, 803*, 59.
30. Molla, A. A., Debnath, D., Chakrabarti, S. K., Mondal, S., & Jana, A. (2016). *Monthly Notices of the Royal Astronomical Society, 460*, 3163.
31. Chatterjee, D., Debnath, D., Chakrabarti, S. K., Mondal, S., & Jana, A. (2016). *The Astrophysical Journal, 827*, 88.
32. Dutta, B. G., & Chakrabarti, S. K. (2016). *The Astrophysical Journal, 828*, 101.
33. Ghosh, A., & Chakrabarti, S. K. (2018). *Monthly Notices of the Royal Astronomical Society, 479*, 1210.
34. Chakrabarti, S. K., Debnath, D., & Nagarkoti, S. (1989). *Advances in Space Research, 63*, 3749.
35. Bhowmick, R., Debnath, D., Chatterjee, K., et al. (2012). *The Astrophysical Journal, 910*, 138.
36. Roy, A., & Chakrabarti, S. K. (2017). *Monthly Notices of the Royal Astronomical Society, 472*, 4689.
37. Pal, P. S., Chakrabarti, S. K., & Nandi, A. (2011). *International Journal of Modern Physics D, 20*, 1597.
38. Blandford, R. D., & Payne, D. G. (1982). *Monthly Notices of the Royal Astronomical Society, 199*, 883.
39. Chakrabarti, S. K., & Bhaskaran, P. (1992). *Monthly Notices of the Royal Astronomical Society, 255*, 255.
40. Jana, A., Chakrabarti, S. K., & Debnath, D. (2017). *The Astrophysical Journal, 850*, 91.

Chapter 2
Radiative Acceleration of Relativistic Jets from Accretion Disks Around Black Holes

Indranil Chattopadhyay, Raj Kishor Joshi, Sanjit Debnath, Priyesh Kumar Tripathi, and M. Saleem Khan

Abstract Matter falling onto black holes, also called accretion disks, emits intense high-energy radiation. Accretion disks during hard to hard intermediate spectral states also emit bipolar outflows. Radiation drag was supposed to impose the upper limit on the terminal speed. It was later shown that a radiation field around an advective accretion disk imposes no upper limit on speed, about a few hundred of Schwarzschild radius from the disk surface. We study radiatively driven electron-proton and electron-positron jets, for geometrically thick and slim transonic disks by using numerical simulation. We show that pair-dominated jets can reach ultra-relativistic speeds by radiation driving. We also discuss at what limits radiative acceleration may fail.

2.1 Introduction

Matter accreting onto black holes emits intense radiation in a wide range of the electromagnetic spectrum. Active galactic nuclei and microquasars are supposed to harbor black holes, and they are also observed to be associated with bipolar jets of varying energetics. Jets also originate from the accretion disk around the central black hole. So, it is natural to investigate the possibility of radiative acceleration of

I. Chattopadhyay (✉)
Aryabhatta Research Institute of Observational Sciences (ARIES), Nainital, India
e-mail: indra@aries.res.in

R. K. Joshi
Department of Astronomy, Astrophysics and Space Engineering, Indian Institute of Technology Indore, Simrol, India

S. Debnath · P. K. Tripathi
Aryabhatta Research Institute of Observational Sciences (ARIES), Nainital, India

Department of Applied Physics, Mahatma Jyotiba Phule Rohilkhand University, Bareilly, India

M. S. Khan
Department of Applied Physics, Mahatma Jyotiba Phule Rohilkhand University, Bareilly, India

S. Ghosh, A. R. Rao (eds.), *The Relativistic Universe: From Classical to Quantum*, Astrophysics and Space Science Proceedings 61,
https://doi.org/10.1007/978-3-031-90186-7_2

ejected matter into relativistic jets. The minimal criterion for radiative acceleration is that the jet should be optically thin. It was, however, soon realized that for optically thin medium and extended sources, i.e., jet medium close to the accretion disk, the radiation field penetrates the medium. So matter plying through this radiation field will experience a decelerating force proportional to various components of the local three-velocity of the jet [1–3]. This force is called radiation drag. Fukue and his collaborators investigated this aspect of the interaction of disk radiation with the outflow jets both in the fluid as well as in the particle limit [4–6]. The general consensus on the radiative driving of jets around different disk models is that for certain disk models (such as an infinitely extended Keplerian disk), the jet terminal speed is mildly relativistic [2]. For other models (like a thick disk or a combination of ADAF and Keplerian disk), the jets are relativistic with a Lorentz factor of about a few [1, 6]. It is accepted that ultra-relativistic jets cannot be produced by radiation driving. More recently, a number of numerical simulations of super-critical accretion disks showed that outflows driven by disk radiation can reach up to speeds $0.3-0.4c$, where c is the speed of light in vacuum [7–9].

Parallelly, a group of researchers investigated the issue of radiative driving of outflowing matter by the radiation field around an advective accretion disk [10–16]. In these analytical investigations, the authors showed that the radiation field from advective disks does not impose any upper limit on terminal speed. The radiation from advective disks mostly originates from the inner part of the disk, and at a few hundred Schwarzschild radii above, the disk ceases to be an extended object. In such cases, the upper limit imposed on jet speed is the speed of light, in other words, no upper limit on speed. That is, if the disk is sufficiently bright, jets can be accelerated to high speeds. Numerical simulations showed that not only can one accelerate jets to high speeds, but for given disk geometry, disk radiation can accelerate pair-dominated subsonic jets to ultra-relativistic speeds [17, 18]. In this chapter, we discuss the basics of radiative acceleration and why this process is more efficient around an advective disk than around other disk models.

In this chapter we are not studying the generation mechanism of the jet. Advective disks in a significant region of the energy-angular momentum parameter space harbors accretion shocks. Various numerical simulations have shown that thermal gradient force in the post-shock region drives bipolar jets in an orthogonal direction to the equatorial plane of the accretion disk [19–22]. In this chapter, we assume the jet has been launched by the mechanism of thermal driving of the post-shock disk (PSD), and the jet is accelerated by the radiation of the accretion disk. The accretion disk is auxiliary and, apart from being the source of radiation, plays no part in the simulations.

In this chapter, we solve time-dependent equations of motion of a relativistic gas impacted by the radiation of the accretion disk. We use a TVD scheme based on relativistic hydrodynamics [23], but incorporating gravity only through the time-time component of the metric tensor, the so-called 'weak field limit' [18]. We consider all the ten independent components of the radiation tensor from an advective disk [14]. The radiation field has two extended sources: (1) the post-shock part of the accretion disk or PSD and (2) the sub-Keplerian pre-shock disk or SKD.

The two sources produce a very interesting radiation field. In this chapter, we inject the jets with very low and subsonic velocity and study how much the jets can be accelerated.

In the next section, we will present the assumptions and equations of motion, and then, we will present the results and draw concluding remarks.

2.2 Assumptions and Equations of Motion

We have incorporated gravity through the weak field approximation in a special relativistic metric written in cylindrical coordinates $(r,\ \theta,\ z)$. If the metric is generally represented as $ds^2 = g_{\mu\nu}d^{\mu}dx^{\nu}$, where $\mu \equiv (t,\ r,\ \theta,\ z)$, then the weak field assumption implies, that the gravitation enters through g_{tt} or, the time-time component of the metric tensor. Therefore,

$$ds^2 = -(1+2\Phi)c^2dt^2 + dr^2 + r^2d\theta^2 + dz^2; \quad \Phi = \frac{GM}{R - r_{\rm g}} \tag{2.1}$$

Here, $R = \sqrt{r^2+z^2}$ is the radial distance from the central object, (r, θ, z) are the usual coordinates in cylindrical geometry, and $r_{\rm g} = 2GM/c^2$ is the Schwarzschild radius. The gravitational force is expressed via a pseudo-Newtonian potential Φ and is called Paczyńsky-Wiita potential [24]. The equations of motion are $T^{\mu\nu}_{;\nu} = 0$, and $(nu^{\nu})_{;\nu} = 0$, in terms of components they are:

$$\frac{\partial D}{\partial t} + \frac{1}{r}\frac{\partial}{\partial r}\left[r\alpha Dv^r\right] + \frac{\partial}{\partial z}\left[\alpha Dv^z\right] = 0 \tag{2.2a}$$

$$\frac{\partial M^r}{\partial t} + \frac{1}{r}\frac{\partial}{\partial r}\left[r\alpha\left(M^r v^r + p\right)\right] + \frac{\partial}{\partial z}\left[\alpha M^r v^z\right] = \frac{\alpha p}{r} + \frac{\alpha M^\theta v^\theta}{r} - E\frac{\partial\alpha}{\partial r} + G^r \tag{2.2b}$$

$$\frac{\partial M^\theta}{\partial t} + \frac{1}{r}\frac{\partial}{\partial r}\left[r\alpha M^\theta v^r\right] + \frac{\partial}{\partial z}\left[\alpha M^\theta v^z\right] = -\frac{\alpha M^\theta v^r}{r} + G^\theta \tag{2.2c}$$

$$\frac{\partial M^z}{\partial t} + \frac{1}{r}\frac{\partial}{\partial r}\left[r\alpha M^z v^r\right] + \frac{\partial}{\partial z}\left[\alpha(M^z v^z + p)\right] = -E\frac{\partial\alpha}{\partial z} + G^z \tag{2.2d}$$

$$\frac{\partial E}{\partial t} + \frac{1}{r}\frac{\partial}{\partial r}\left[r\alpha(E+p)v^r\right] + \frac{\partial}{\partial z}\left[\alpha(E+p)v^z\right] = -M^r\frac{\partial\alpha}{\partial r} - M^z\frac{\partial\alpha}{\partial z} + G^t. \tag{2.2e}$$

In Eqs. (2.2a–2.2e), $D,\ M^r,\ M^z,\ E$ is the mass density, the radial and axial component of momentum density, and the total energy density of the fluid, respectively.

Here v^r, v^θ, v^z are the components of three-velocity and $\alpha = \sqrt{(1+2\Phi)}$. It may be noted that the time and space components of 4-velocities are given as $u^t = \gamma/\sqrt{\alpha}$, $u^i = \gamma v^i/\sqrt{g_{ii}}$, respectively, here g_{ii} are the space components of the metric tensor components and $\gamma = 1/\sqrt{(1-v_i v^i)}$ is the Lorentz factor. G^μ are the components of radiation four-force density given as

$$G^r = G^{\hat{r}}_{co} + \frac{\gamma - 1}{v^2} v^r v_i G^{\hat{i}}_{co}, \tag{2.3a}$$

$$G^\theta = \frac{1}{r}\left[G^{\hat{\theta}}_{co} + \frac{\gamma - 1}{v^2} v^\theta v_i G^{\hat{i}}_{co}\right], \tag{2.3b}$$

$$G^z = G^{\hat{z}}_{co} + \frac{\gamma - 1}{v^2} v^z v_i G^{\hat{i}}_{co}, \tag{2.3c}$$

$$G^t = \frac{\gamma}{\alpha} v_i G^{\hat{i}}_{co}, \tag{2.3d}$$

where,

$$G^{\hat{i}}_{co} = \rho_e F^i \quad (i = r, \theta, z), \tag{2.3e}$$

F^i are the components of the flux measured in the comoving frame, and the general form of which is given as,

$$\begin{aligned} F^i = & -\gamma^2 v^i \mathcal{E} + \gamma\left[\delta^i_j + \left(\gamma + \frac{\gamma^2}{\gamma + 1}\right) v^i v_j\right]\mathcal{F}^j \\ & - \gamma v_j \left[\delta^i_k + \frac{\gamma^2}{\gamma + 1} v^i v_k\right]\mathcal{P}^{jk}. \end{aligned} \tag{2.3f}$$

In Eq. (2.3f), $\mathcal{E} = \frac{\sigma_T}{m_{el}c} E_{rd}$, $\mathcal{F}^i = \frac{\sigma_T}{m_{el}c} F^i_{rd}$, $\mathcal{P}^{jk} = \frac{\sigma_T}{m_{el}c} P^{jk}_{rd}$. Here, E_{rd}, F^i_{rd}, P^{ij}_{rd}, represent the radiation energy density, three components of radiation flux, and six independent components of radiation pressure [14, 16, 18]. ρ_e is the mass density of leptons and m_{el} is the mass of electron. σ_T is the Thomson scattering cross-section.

We solve these equations using a TVD scheme. For a detailed understanding of the eigenstructure and the code, one may refer to [23]. The thermodynamics is governed by a relativistic equation of state known as CR EoS [25]. CR EoS is given as

$$e = \rho c^2 f(\Theta, \xi); \text{ where,} \tag{2.4}$$

$$f(\Theta,\, \xi) = 1 + (2 - \xi)\Theta\left[\frac{9\Theta + 6/\eta}{6\Theta + 8/\eta}\right] + \xi\Theta\left[\frac{9\Theta + 6/\eta\tau}{6\Theta + 8/\eta\tau}\right] \tag{2.5}$$

Here, ρ is the mass density, $\Theta = p/\rho c^2$, p is the gas pressure, $\eta = m_{el}/m_p$, m_p is the mass of the proton, $\xi = n_p/n_{el}$ is the ratio of the number density of protons to the electron and $\tau = 2 - \xi + \xi/\eta$. The charge balance is maintained by the protons, electrons, and positrons. If $\xi = 1$, it implies electron-proton plasma; on the other, $\xi = 0$ implies electron-positron pair plasma. Any other ξ between $0 - 1$, would imply electron-positron-proton plasma.

2.3 Results

Numerical simulation of radiative acceleration of jets above an advective disk has been carried out in one (1D) and two spatial dimensions (2D), starting with subsonic, sub-relativistic jets to supersonic, relativistic jets in detail by [17, 18]. The 1D jet structure is way simpler, and multidimensional simulations create a lot more complicated structures, primarily due to various instabilities like Kelvin-Helmholtz type instabilities. As a result, in 1D simulations, one could get impressive (> 50) terminal Lorentz factors for pair-dominated flow [17], but for 2D simulations, jets reach Lorentz factors up to ~30 [18].

In this chapter, we focus on comparing the radiatively driven jets from the advective disk, the inner part of which are geometrically thick, with that of disks that were geometrically slim or thin. It may be noted that an advective-transonic disk [26–28] consists of matter falling onto a black hole through one or more sonic/critical points. In case matter becomes transonic at a larger distance from the black hole, it can go through a standing or a time-dependent shock before falling onto the black hole through the inner sonic point. It has been shown through various numerical simulations [19–21, 29] that such disks, when undergoing shocks, generate bipolar outflows from the post-shock region, which were proposed as precursors of astrophysical jets. Therefore, one can consider the shocks as the base of the jet. In Fig. 2.1, we present a schematic diagram of the cross-section of disk-jet geometry. The post-shock disk (PSD), the sub-Keplerian disk (SKD), and the jet are shown. The arrows represent radiations from PSD and SKD interacting with the jets, which are flowing perpendicular to the disk plane.

The vertical height of PSD (h_s) can be geometrically thick ($h_s/x_s > 1$) or can be geometrically thin ($h_s/x_s < 1$), where x_s is the radial extent of the PSD on the

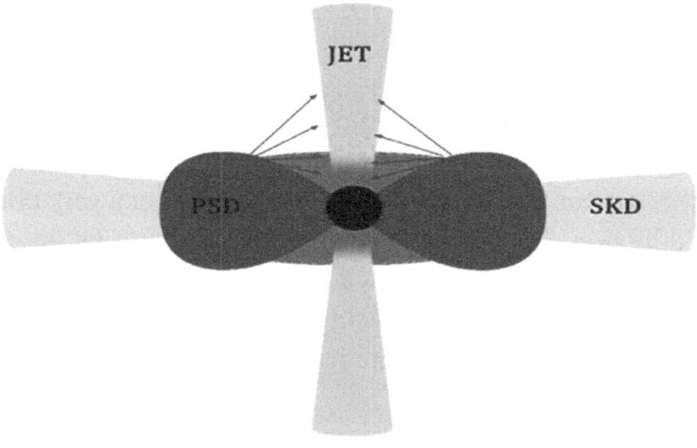

Fig. 2.1 Cartoon diagram of the cross-section of the advective accretion disk and associated jets. The post-shock disk, or PSD, and the pre-shock sub-Keplerian disk, or SKD, are shown. The arrows represent the radiation impinging on the jet

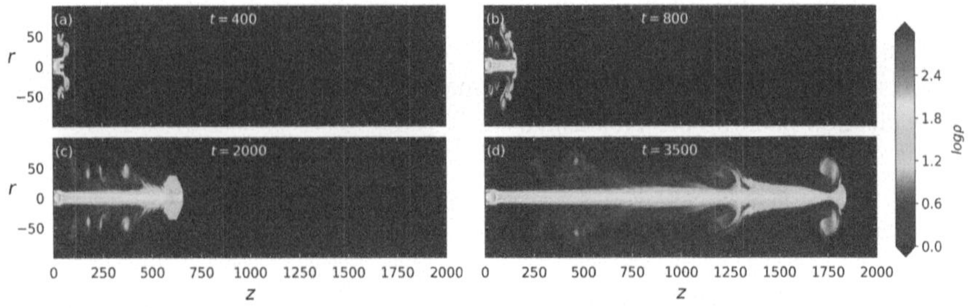

Fig. 2.2 Contours of $log\ \rho$ for an electron-positron jet driven by the radiation field of an accretion disk with $x_s = 10.0$, $h_s = 2.5x_s$. The disk luminosity is $l = 0.2L_{Edd}$ and $\dot{m} = 7.5\dot{M}_{Edd}$. The injection parameters for the jet beam are taken as $v_{\rm inj} = 0.001$, $\Theta = 2.0$ at $z = 3.0$. The jet beam is 1000 times heavier than the ambient medium and $p_{\rm a} = p_{\rm j}/10.0$

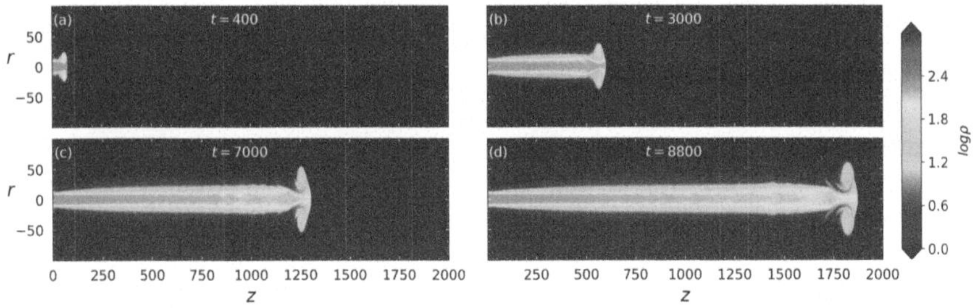

Fig. 2.3 Contours of $log\ \rho$ for an electron-proton jet. The accretion disk parameters and injection parameters at the jet base are the same as mentioned in Fig. 2.2

equatorial plane. In Fig. 2.2, we plot the evolution of a jet of fluid composed of purely electron and positron (i.e., $\xi = 0$), powered by radiation from an accretion disk, in which the dimensions of PSD are given by $h_s/x_s = 2.5$ and $x_s = 10r_{\rm g}$. The initial ratio of the ambient ($p_{\rm a}$) to the jet pressure ($p_{\rm j}$) is 0.1. The jet material remains denser than the ambient medium, and the backflow creates significant Kelvin-Helmhotlz plumes. The radiation field is such that there is a significant back flow at around $z \sim 6r_{\rm g}$; however, picks up the speed and is accelerated by radiation to about $0.8c$, followed by a strong shock at around $40r_{\rm g}$ (bluish region on the jet spine) and a compression ratio of about 4. The terminal speed of this jet along the spine is $\sim 0.6c$.

In Fig. 2.3, we plot the evolution of the density profile of an electron-proton jet but with the same jet injection and disk parameters of Fig. 2.2. Radiative acceleration of electron-proton medium is not a very efficient process. With the sub-Eddington disk luminosity, an electron-proton jet has less proportion of electrons, and therefore, radiation has a limited effect on the jet. Near the base, the jet expands due to thermal driving, reaching up to relatively high speeds $v \sim 0.45c$. Thereafter, the radiation drag slows the jet down to speeds up to $v \sim 0.25c$.

The traditional thick disk or the Polish doughnut [24], is without advection and is unstable. A geometrically thick, advective disk is stable, but the radiation field produced is such that, the radiative acceleration is moderate. However, for an advective disk that is geometrically slim, the radiation field looks away from the disk and accelerates the jet. In Fig. 2.4, we plot the contours of $log\ \rho$ (left panel) and $log\ P$ (right panel) of a pair dominated jet, and is acted on by the radiation field around a slim advective disk (the injection values and other factors are mentioned in the caption). Since the jets are accelerated, there are many shocks in the jet spine, which causes the braided structure of shocks and rarefaction regions. The shocks also show up as enhancements in the pressure plot.

In Fig. 2.5 (left), we plot the bulk Lorentz factor along the axis of the jet for the case presented in the previous figure, and in the right panel, we plot the jet kinetic luminosity. The presence of many shocks implies that the jet speed profile along its

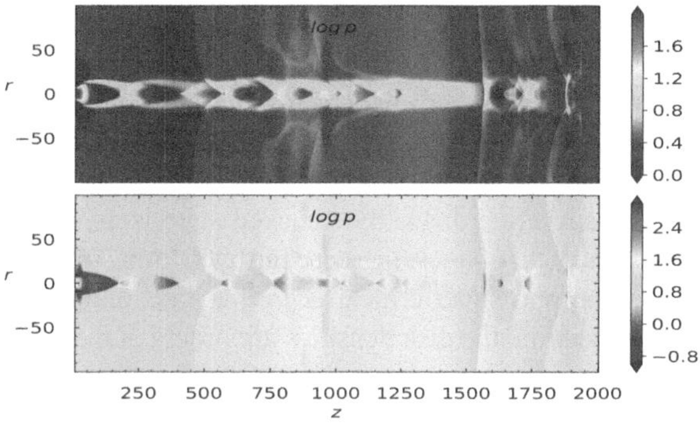

Fig. 2.4 Contours of $log\ \rho$ (top) and $log\ P$ (bottom) in the $r - z$ plane for a pair dominated jet $\xi = 0.001$. For this case $h_s = 0.6(x_s - 1.0)$. The disk luminosity is $l = 0.35L_{\rm Edd}$ and $\dot{m} = 9.5\dot{M}_{\rm Edd}$. The injection parameters for the jet beam are taken as $v_{\rm inj} = 0.001$, $\Theta = 0.1$ at $z = 3.0$

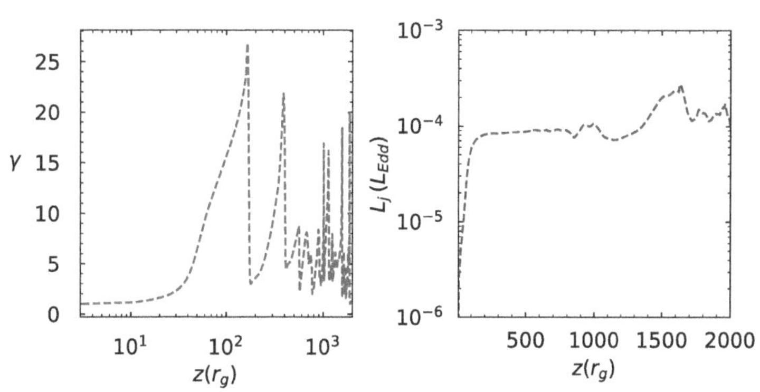

Fig. 2.5 Variation of Lorentz factor γ (left) and jet kinetic luminosity L_j (right) as a function of z for jet model presented in Fig. 2.4

spine will not be smooth. All the vertical jumps in the left panel represent shocks, and the spikes in γ coincide with the rarefaction regions (dark blue regions in the braided structure of the jet) in the left panel of Fig. 2.4. In these rarefaction regions, the Lorentz factor reaches $\gamma \gtrsim 20$!

2.4 Discussions

Radiative acceleration of jets, as a process, has never been favored in the community, mainly on account of radiation drag effects, as has been discussed in the introduction [1–3]. Radiation drag is effective when the source is extended, so all the arguments made in favor of dominant radiation drag involved either infinitely extended disks or the jet being studied in the funnel of a very thick torus. Later, a few simulations were done with jet generation from an initial thick tori and very high accretion rates [8, 9]. The medium above the accreting tori becomes optically thick to slim. Therefore, photons emerging from the accreting torus of a classical thick disk undergoes many scatterings, thereby increasing the percentage of back scattered photons. This limits the scope of acceleration of the outflowing jets, resulting in mildly relativistic speeds.

In the case of the advective disks, the shocked state is associated with the jet state. It has been shown that shock in accretion can form, even in the presence of significant advection and super-critical accretion rate, and yet because of the presence of strong advection, the disk densities are orders of magnitude lower, than that in super critical accretion tori. In Kumar and Chattopadhyay [30], we presented shocked accretion solutions with super critical accretion rates and we reproduce one such solution in Fig. 2.6. The most interesting part is that even for such high values of accretion rate, the maximum density in an accretion disk around a ten solar mass black hole, turns out to be $\sim 10^{-6}$g cm^{-3}, which is orders of magnitude less than typical thick disk solutions of similar accretion rates. As mentioned before, the presence of advection makes the advective disks more stable and of lower density. The presence of advection also ensures that from the post-shock region, only a fraction of inflowing matter can be redirected as outflows. This fact too, makes the outflows optically thin. As a result, the radiation can accelerate the plasma above an advective disk.

2.5 Summary and Conclusion

In this chapter, we explored the acceleration scenario of jets by radiation from the advective disk. Simulations show that the advective disks can be geometrically thick or thin, and that creates different radiation fields. We show that for thick disks, even electron-positron jets can be accelerated up to Lorentz factors of few. Electron proton jets are not significantly affected by radiation. The pair-dominated flows

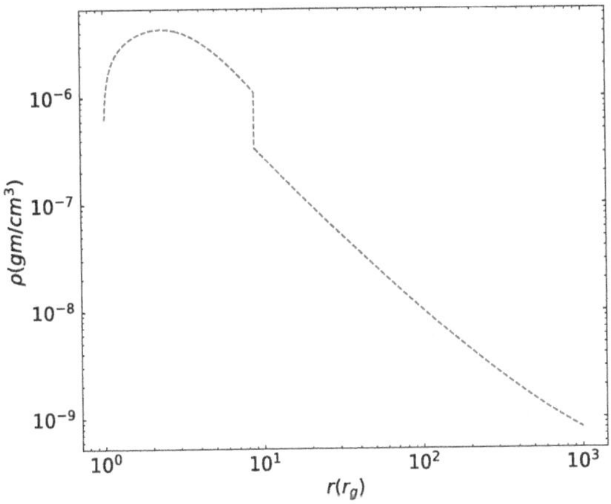

Fig. 2.6 Variation of mass density along the equatorial plane of the accretion disk for accretion rate $\dot{m} = 10\dot{M}_{Edd}$, generalized specific energy $\epsilon = 1.0001$ and specific angular momentum at the horizon $\lambda_0 = 1.6755$, for a viscosity parameter $\alpha_v = 0.005$, in the presence of bremsstrahlung and synchrotron and a simplified inverse-Compton processes. Adopted from [30]. All have been computed assuming a black hole mass $10M_\odot$

from thin, advective disks can be accelerated to ultra-relativistic speeds, and the jets can produce a large number of shocks as well. If the accretion rate is more than a hundred Eddington rate, then the jet will become optically thick, and this method of radiative acceleration will become ineffective.

Acknowledgments We thank the organizers of ISRA 2023. Our special thanks to Shubhrangshu Ghosh and Banibrata Mukhopadhyay.

References

1. Sikora, M., & Wilson, D. B. (1981). *Monthly Notices of the Royal Astronomical Society, 197*, 529.
2. Icke, V. (1989). *Astronomy & Astrophysics, 216*, 294.
3. Fukue, J. (1996). *Publications of the Astronomical Society of Japan, 48*, 631.
4. Fukue, J. (1999). *Publications of the Astronomical Society of Japan, 51*, 425.
5. Fukue, J. (2005). *Publications of the Astronomical Society of Japan, 57*, 691.
6. Fukue, J., Tojyo, M., & Hirai, Y. (2001). *Publications of the Astronomical Society of Japan, 53*, 555.
7. Eggum, G. E., & Katz, J. I. (1985). *The Astrophysical Journal, 298*, L41.
8. Sadowski, A., & Narayan, R. (2015). *Monthly Notices of the Royal Astronomical Society, 453*, 3213.
9. Oshuga, K., Mineshige, S., Mori, M., & Kato, Y. (2009). *Publications of the Astronomical Society of Japan, 61*, L7.
10. Chattopadhyay, I., & Chakrabarti, S. K. (2000). *International Journal of Modern Physics D, 9*, 57.
11. Chattopadhyay, I., & Chakrabarti, S. K. (2000). *International Journal of Modern Physics D, 9*, 717.
12. Chattopadhyay, I., & Chakrabarti, S. K. (2002). *Monthly Notices of the Royal Astronomical Society, 333*, 454.
13. Chattopadhyay, I., Das, S., & Chakrabarti, S. K. (2004). *Monthly Notices of the Royal Astronomical Society, 348*, 846.

14. Chattopadhyay, I. (2005). *Monthly Notices of the Royal Astronomical Society*, *356*, 145.
15. Vyas, M. K., Kumar, R., Mandal, S., & Chattopadhyay, I. (2015). *Monthly Notices of the Royal Astronomical Society*, *453*, 2992.
16. Vyas, M. K., & Chattopadhyay, I. (2019). *Monthly Notices of the Royal Astronomical Society*, *482*, 4203.
17. Joshi, R. K., Debnath, S., & Chattopadhyay, I. (2022). *The Astrophysical Journal*, *933*, 75.
18. Joshi, R. K., Chattopadhyay, I., Tsokaros, A., & Tripathi, P. K. (2024). *The Astrophysical Journal*, *971*, 13.
19. Molteni, D., Sponholz, H., & Chakrabarti, S. K. (1996). *The Astrophysical Journal*, *457*, 805.
20. Molteni, D., Ryu, D., & Chakrabarti, S. K. (1996). *The Astrophysical Journal*, *470*, 460.
21. Lee, S. J., Chattopadhyay, I., Kumar, R., & Ryu, D. (2016). *The Astrophysical Journal*, *831*, 33.
22. Das, S., Chattopadhyaym, I., Nandi, A., & Molteni, D. (2014). *Monthly Notices of the Royal Astronomical Society*, *442*, 251.
23. Ryu, D., Chattopadhyay, I., & Choi, E. (2006). *The Astrophysical Journal Supplement Series*, *166*, 410.
24. Paczýnsky, B., & Wiita, P. J. (1980). *Astronomy & Astrophysics*, *88*, 23.
25. Chattopadhyay, I., & Ryu, D. (2009). *The Astrophysical Journal*, *694*, 492.
26. Fukue, J. (1987). *Publications of the Astronomical Society of Japan*, *39*, 309.
27. Chakrabarti, S. K. (1989). *The Astrophysical Journal*, *347*, 365.
28. Chakrabarti, S. K. (1996). *The Astrophysical Journal*, *464*, 664.
29. Garain, S., & Kim, J. (2023). *Monthly Notices of the Royal Astronomical Society*, *519*, 4550.
30. Kumar, R., & Chattopadhyay, I. (2014). *Monthly Notices of the Royal Astronomical Society*, *443*, 3444.

Chapter 3
Noisy Accretion Flows

Amit K. Chattopadhyay

Abstract In this chapter, the origin of hydrodynamic turbulence in accretion disks is analyzed in the presence of temporally correlated noise to understand the origin of non-magnetic instability in Rayleigh-stable cold protoplanetary objects. In such flows, hydromagnetic/magnetohydrodynamic instability and the corresponding turbulent viscosity are expected to be sluggish or may even be ruled out. Previously, such flows were found to exhibit large transient hydrodynamic growth in two-dimension under linear perturbations, but insignificant growth in the presence of vertical columnar growth. To understand the origin of such instabilities that were not accounted for by conventional hydrodynamic (long-ranged) modes, in a series of articles between 2013 and 2014, we proposed thermal fluctuations as drivers of such instability and showed that an Orr Sommerfeld-Squire model with spatially correlated noise can indeed explain this conundrum. In this chapter, we extend this analysis to temporally correlated noise, focusing on a regime where the Flory-type scaling in spatially correlated noise breaks down. The target is to understand how non-spatially correlated non-magnetic accretion may still amount to instability while being Rayleigh stable. A range of angular velocity profiles (Ω) as functions of radial coordinate (r) have been explored: $\Omega \propto r^{-q}$ $(1 \leq q < 2)$. The model leads to a key understanding of non-magnetic accretion flows as a temporal noise-controlled instability. Additionally, we find remarkable differences between spatial and temporally correlated noise spectra, also with varying Reynolds numbers, that exhibit distinct universality classes.

3.1 Introduction

The origin of turbulence and associated transport in "cold" astrophysical flows, particularly accretion disks, is still an unsolved issue. The challenge of explaining

---

A. K. Chattopadhyay (✉)
School of Business, National College of Ireland, Dublin, Ireland
e-mail: amit.chattopadhyay@ncirl.ie

S. Ghosh, A. R. Rao (eds.), *The Relativistic Universe: From Classical to Quantum*, Astrophysics and Space Science Proceedings 61,
https://doi.org/10.1007/978-3-031-90186-7_3

linearly unstable modes in flows governed by low viscosity has baffled the community for a long time with potential answers given that include a combination of "bypass mechanics" [1–6], skewed-varicose-type instabilities [7] and nonnormal eigenmodes [8]. But none of these was completely successful in addressing the origin of non-magnetic instability.

From the perspective of statistical physics, it was obvious that the key anomaly lay in the "cold flow" description. While 3000 K is cold from an astrophysical description, it is very "hot" from the parlance of statistical physics. To incorporate such thermal energy, we construed a model around the paradigmatic work of Forster et al. [9] which studied large non-equilibrium fluctuations in unbounded flows. This together with a seminal follow-up by De Dominicis and Martin [10] provided our groundwork to address perturbing random forces. These two studies combined with the one by Medina et al. [11] inspired the well-known Kardar-Parisi-Zhang (KPZ) model [12] that studied hydrodynamic instabilities related to interface growth. The nature of coarsening predicted in the KPZ interface growth dynamics also sheds light on a varying range of non-equilibrium phenomena, including fluid flows, bacterial growth, paper burning front, etc. [13]. The article by Medina et al. [11] also sheds light on the importance of varying spatiotemporal correlations on low-dimensional Navier-Stokes models. These studies lay the foundation of the first non-trivial boundary-constrained flow involving a three-dimensional *bounded* incompressible Navier-Stokes fluid perturbed by boundary layer shear. Chattopadhyay and Bhattacharjee [14] applied the now well-established randomly stirred model of Navier-Stokes flows constrained by geometry.

The above studies lay the backbone of a series of collaborative publications with Mukhopadhyay et al. where we addressed non-magnetic instability as a statistical process that has its origin in thermal fluctuations [15–17]. These were the first known applications of stochastically perturbed Orr-Sommerfeld-Squire dynamics in analyzing accretion flows. Modeled on spatially correlated thermal noise, the results provided key leads to the origin of subcritical turbulence in Rayleigh flows.

The present study analyzes the impact of temporally correlated noise on rotating shear flows modeled within the Orr-Sommerfeld and Squire framework [15]. In Sect. 3.2, the stochastically perturbed Orr-Sommerfeld-Squire model is presented for a quasi-2-dimensional flow (x direction is assumed as a thin slice). Section 3.3 discusses two-point spatiotemporal correlations, focusing on the temporally correlated part of the noise and concludes with future directions.

3.2 Stochastically Perturbed Orr-Sommerfeld-Squire Flow

In the absence of magnetic coupling with the perturbed fluid, we can analyze thermally triggered cold accretion flows driven by a Coriolis force based on the Orr-Sommerfeld and Squire equations [15]. The model is a linearized version of the original 3-dimensional (non-magnetic) Navier-Stokes with a background plane shear $(0, -x, 0)$ $(-L/2 < x < L/2$ defined on a cube of size $L)$ and angular velocity $\Omega \sim r^{-q}$, where the Rayleigh stable parameter q has a strict bound for accretion flows: $1 < q < 2$. Note $q \to \infty$ defines a plane Couette flow. We focus on 3 key domains of q, respectively $q \to 2$ which is a constant angular momentum flow (we consistently use $q = 1.9$), $q = 1.5$ represents a Keplerian flow and $q = 1$ depicts a constant angular velocity flow. When triggered by thermal noise, for an incompressible fluid, the model equations are given by

$$\left(\frac{\partial}{\partial t} - x\frac{\partial}{\partial y}\right)\nabla^2 u + \frac{2}{q}\frac{\partial \zeta}{\partial z} = \frac{1}{R_e}\nabla^4 u + \eta_1(x,t), \tag{3.1}$$

$$\left(\frac{\partial}{\partial t} - x\frac{\partial}{\partial y}\right)\zeta + \frac{\partial u}{\partial z} - \frac{2}{q}\frac{\partial u}{\partial z} = \frac{1}{R_e}\nabla^2 \zeta + \eta_2(x,t), \tag{3.2}$$

where u, ζ represent the x-components of the velocity and vorticity perturbations, R_e is the Reynolds number, and $\eta_{1,2}$ are the stochastic thermal noise perturbing the linearised system defined by $< \eta_i(\mathbf{x},t)\eta_j(\mathbf{x}',t') >\sim \delta_{ij}|\mathbf{x}-\mathbf{x}'|^{2\rho-2}\,|t-t'|^{2\theta-1}$. Note that Eqs. (3.1) and (3.2) depict noise-driven Orr-Sommerfeld and Squire equations. In the Fourier-transformed space, the long-ranged noise correlations can be represented through the two-point structure function $< \eta_i(\mathbf{k},\omega)\eta_j(\mathbf{k}',\omega') >= 2D(k,\omega)\delta^2(\mathbf{k}+\mathbf{k}')\delta(\omega+\omega')\delta_{ij}$ where the spatiotemporally correlated noise strength is given by $D(k,\omega) \sim |\mathbf{k}|^{-2\rho}\,\omega^{-2\theta}$. Corresponding Fourier series expansions of all relevant variables $u(\mathbf{x},t)$, $\zeta(\mathbf{x},t)$, $\eta_i(\mathbf{x},t)$ can be presented as

$$\xi(\mathbf{x},t) = \int \tilde{\xi}_{\mathbf{k},\omega}\, e^{i(\mathbf{k}.\mathbf{x}-\omega t)} d^3k\, d\omega, \tag{3.3}$$

where $\xi(\mathbf{x},t) = u(\mathbf{x},t),\ \zeta(\mathbf{x},t),\ \eta_i(\mathbf{x},t)$. Imposing similar multifractality constraints as expected from a dynamically evolving interface [13], we can write

$$< |h(\mathbf{x},t) - h(\mathbf{x}',t')|^2 >\sim |\mathbf{x}-\mathbf{x}'|^{2\alpha} f\left(\frac{|t-t'|}{|\mathbf{x}-\mathbf{x}'|^z}\right), \tag{3.4}$$

where α and z are respectively the "roughening" and "dynamic" exponents that quantify the impact of the stochastic noise on the emergent dynamics [12]. In absence of any nonlinearity, simple scaling leads to $z = 2$ and $\alpha = \rho+2\theta$. However, nonlinearity renormalizes the exponents leading to a Flory-type scaling given by $z_F = (6-2\rho)/3$ and $\alpha_F = 2\rho/3$ in 3 spatial dimensions. Our focal regime for this study will be $\rho = 0$ which is the strong-coupling limit defined by $\theta \neq 0$. In particular, we will focus on the critical threshold $\theta_c = 1/2$ below which nonlinearity driven instabilities are generated and above which noise produces singularity.

A linear solution of Eq. (3.3) in the barrow-gap limit $x \to L$ leads to

$$\begin{pmatrix} \tilde{u}_{\mathbf{k},\omega} \\ \tilde{\zeta}_{\mathbf{k},\omega} \end{pmatrix} = \mathcal{M}\begin{pmatrix} \tilde{\eta}_{1\mathbf{k},\omega} \\ \tilde{\eta}_{2\mathbf{k},\omega} \end{pmatrix}, \tag{3.5}$$

where

$$\mathcal{M} = \begin{pmatrix} \mathcal{M}_1 & \mathcal{M}_2 \\ \mathcal{M}_3 & \mathcal{M}_4 \end{pmatrix}, \tag{3.6}$$

with $\mathcal{M}_1 = [q^2 R_e(k^2 - i R_e\omega - i L R_e k_y)]/\text{den}$, $\mathcal{M}_2 = [2iqR_e^2 k_z]/\text{den}$, $\mathcal{M}_3 = [i(q-2)qR_e^2 k_z]/\text{den}$ and $\mathcal{M}_4 = [k^2 q^2 R_e(k^2 - i R_e\omega - i L R_e k_y)]/\text{den}$, where den $= -k^2 q^2 (k^2 - i R_e\omega - i L R_e k_y)^2 + 2(q-2)R_e^2 k_z^2$. Here $k^2 = {k_x}^2 + {k_y}^2 + {k_z}^2$; in the following analysis, we will assume isotropicity, i.e. $k_x = k_y = k_z = k/\sqrt{3}$.

3.2.1 Temporal Correlation

From the perspective of statistical physics, temporal correlations are supposed to quantify both "roughening" and "dynamic" exponents of a developing front. For this study, the magnitude and gradients (scalings) of the u and ζ autocorrelation functions would quantify noise-induced turbulence in accretion disks. Assuming $< \tilde{\eta}_{1\mathbf{k},\omega} \tilde{\eta}_{2\mathbf{k},\omega} >= 0$, we obtain temporal correlations of velocity, vorticity and gradient of velocity given below as

$$< u(\mathbf{x}, t)\, u(\mathbf{x}, t+\tau) >= C_u(\tau) = \int d^3k \int d\omega\, e^{-i\omega\tau} < \tilde{u}_{\mathbf{k},\omega}\, \tilde{u}_{-\mathbf{k},-\omega} >, \tag{3.7a}$$

$$< \zeta(\mathbf{x}, t)\, \zeta(\mathbf{x}, t+\tau) >= C_\zeta(\tau) = \int d^3k \int d\omega\, e^{-i\omega\tau} < \tilde{\zeta}_{\mathbf{k},\omega}\, \tilde{\zeta}_{-\mathbf{k},-\omega} > . \tag{3.7b}$$

Using Eqs. (3.5), (3.6), the velocity correlation can be written as follows

$$\begin{aligned} C_u(\tau) = &\int d^3k \int d\omega\, e^{-i\omega\tau}\, k^{-2\rho}\, \omega^{-2\theta}\, [2D_1 q^4 R_e^2 (k^4 + (R_e\omega + L R_e k_y)^2) \\ &+ 8D_2 q^2 R_e^4 k_z^2]/[-4(-2+q)^2 R_e^4 k_z^4 \\ &+ 4R_e^2 q^2 (-2+q) k^2 k_z^2 (k^4 - (R_e\omega + L R_e k_y)^2) \\ &- k^4 q^4 (k^4 + (R_e\omega + L R_e k_y)^2)^2]. \end{aligned} \tag{3.8}$$

To perform the *omega* integration, we first need to calculate the poles of the kernel. Then on rescaling the integrand, Eq. (3.8) reduces to

$$\begin{aligned} C_u(\tau) = &-\frac{4\pi}{R_e^4 q^4} \int \frac{dk}{k^2} \int d\omega\, e^{-i\omega\tau}\, k^{-2\rho}\, \omega^{-2\theta} \\ &\times [\frac{4D_1 q^2 R_e^4 (2-q) + 8D_2 q^2 R_e^4 k^2/3}{(\omega-\omega_1)(\omega-\omega_2)(\omega-\omega_3)(\omega-\omega_4)} \\ &+ \frac{2D_1 q^4 R_e^2 ((R_e\omega + L R_e k/\sqrt{3})^2 + k^4 - \frac{2R_e^2(2-q)}{3q^2})}{(\omega-\omega_1)(\omega-\omega_2)(\omega-\omega_3)(\omega-\omega_4)}], \end{aligned} \tag{3.9}$$

where the poles are given by

$$\omega_{1,2,3,4} = \pm\frac{1}{R_e}\left(\sqrt{\frac{2R_e^2(2-q)}{3q^2} - k^4 \pm \left(\frac{2\sqrt{2}R_e(2-q)}{\sqrt{3}q}\right)k^2} - \frac{LR_ek}{\sqrt{3}}\right). \tag{3.10}$$

$C_\zeta(\tau)$ can be similarly calculated from Eq. (3.7b).

Expressing Eq. (3.9) as $C_u(\tau) = 4\pi \int_{k_0}^{k_m} dk\, k^{2-2\rho}\, \mathrm{RES}_{C_u}(q, R_e, \tau, \theta)$, with

$$\begin{aligned}
&\mathrm{RES}_{C_u}(q, R_e, \tau, \theta)\\
&= \int d\omega\, e^{-i\omega\tau}\, \omega^{-2\theta}\, \Big[\frac{4D_1q^2R_e^4(2-q) + 8D_2q^2R_e^4k^2/3}{(\omega-\omega_1)(\omega-\omega_2)(\omega-\omega_3)(\omega-\omega_4)}\\
&+ \frac{2D_1q^4R_e^2\left((R_e\omega + LR_ek/\sqrt{3})^2 + k^4 - \frac{2R_e^2(2-q)}{3q^2}\right)}{(\omega-\omega_1)(\omega-\omega_2)(\omega-\omega_3)(\omega-\omega_4)}\Big].
\end{aligned} \tag{3.11}$$

The argument $\mathrm{RES}_{C_u}(q, R_e, \tau, \theta)$ can be expressed as a complex integral that can be solved using the Residue Theorem. We avoid including the rather long expression here which resembles (but does not equate to) equation (10) of [15]. For numerical calculations, we use $k_0 = 2\pi/L_{\max}$ and $k_m = 2\pi/L_{\min}$ are the limits of perturbation wavenumber k, when $L = L_{\max} - L_{\min}$ being the size of the chosen small section of the disk in the radial direction; we later choose $L = 1$. Also, as explained before, we focus on the temporally correlated part which means $\rho = 0$ in all future calculations.

3.2.1.1 Turbulent Limit: Scaling

The turbulent regime is characterized by $R_e \to \infty$. Focusing on the critical limit $\theta = 1/2$, this gives us

$$C_u(\tau)_{R_e\to\infty,\theta=1/2} \to \frac{72\pi\, q^{-17/2}}{\sqrt{2}L^3}\,\frac{1}{\tau^3}. \tag{3.12}$$

This $1/\tau^3$ dependence of the two-point temporal correlation function in velocity is an important scaling relation that was previously deducted by Phinney [18] and Cannizzo et al. [19].

3.2.2 *Temporal Correlations: Velocity and Vorticity*

Figure 3.1 shows the variation of the velocity-velocity temporal correlations for Reynolds number $R_e = 5772$ [Fig. 3.1a] against $R_e = 10{,}000$ [Fig. 3.1b]; clearly, the higher Reynolds number shows suppressed correlation. Figure 3.1c and d shows the respective trends for the real components of the velocity correlation function.

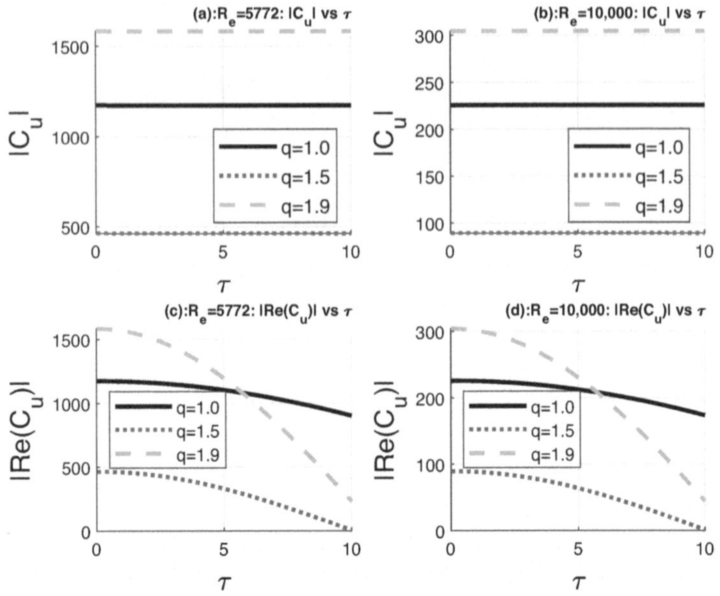

Fig. 3.1 Velocity temporal correlations for $q = 1$ (solid line), $q = 1.5$ (dotted line) and $q = 1.9$ (dashed line)

Note, that both Fig. 3.1c and d shows slowly decaying oscillatory profiles for both Reynolds numbers with the higher number showing hydrodynamically forced suppression as before.

Figure 3.2a and b shows similar comparisons between Reynolds numbers 5772 and 10,000 for the absolute values of vorticity (ζ) correlation while Fig. 3.2c and d compare the real vorticity correlations. With a higher flux rate than a velocity correlator, the real components of the vorticity correlator C_ζ demonstrate distinctly different trends compared to that of the velocity dynamics. While the absolute values of the correlator C_ζ follow a similar monotonic increase in correlations largely independent of the choice of the Reynolds number (Fig. 3.2a and b), the real components of the same structure-functions show oscillatory growth tending to finite-sized singularities for both Reynolds numbers (Fig. 3.2c and d). Clearly for the absolute values, Keplerian velocity ($q = 1.5$) dominates while the oscillatory components are dominated by constant angular velocity ($q = 1.0$) flows. The vorticity correlation features are in stark contrast to velocity correlation where the constant angular momentum ($q \to 2$) dominates Keplerian flows ($q = 1.5$).

3.2.3 Spatial Correlation

Like temporal correlation, here also we assume $<\tilde{\eta}_{1\mathbf{k},\omega}\tilde{\eta}_{2\mathbf{k},\omega}>= 0$ and obtain spatial correlations of velocity and vorticity, as given below

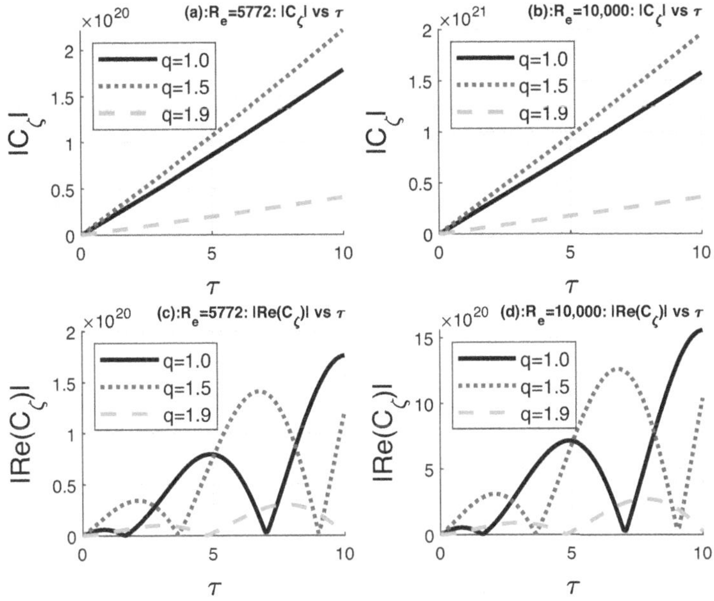

Fig. 3.2 Vorticity temporal correlations for $q = 1$ (solid line), $q = 1.5$ (dotted line) and $q = 1.9$ (dashed line)

$$< u(\mathbf{x}, t)\, u(\mathbf{x} + \mathbf{r}, t) >= S_u(r) = \int d^3k \int d\omega\, e^{i\mathbf{k}.\mathbf{r}} < \tilde{u}_{\mathbf{k},\omega}\, \tilde{u}_{-\mathbf{k},-\omega} >,$$

$$< \zeta(\mathbf{x}, t)\, \zeta(\mathbf{x} + \mathbf{r}, t) >= S_\zeta(r) = \int d^3k \int d\omega\, e^{i\mathbf{k}.\mathbf{r}} < \tilde{\zeta}_{\mathbf{k},\omega}\, \tilde{\zeta}_{-\mathbf{k},-\omega} > . \quad (3.13)$$

Now using Eqs. (3.5), (3.6), and (3.13) the velocity correlation $S_u(r)$ is explicitly given by

$$S_u(r) = 2\pi \int_{k_0}^{k_m} dk\, k^{2-2\rho} \int_0^{\pi} d\theta\, e^{ikr\cos\theta} \int d\omega\, \omega^{-2\theta}\, [2D_1 q^4 R_e^2$$
$$(k^4 + (R_e\omega + LR_e k_y)^2) + 8D_2 q^2 R_e^4 k_z^2]/[4(2-q)^2 R_e^4 k_z^4 +$$
$$4R_e^2 q^2 (2-q) k^2 k_z^2 (k^4 - (R_e\omega + LR_e k_y)^2) + k^4 q^4 (k^4 + (R_e\omega + LR_e k_y)^2)^2]$$
$$= 2\pi^2 \int_{k_0}^{k_m} dk\, k^{2-2\rho}\, J_0(kr) \int d\omega\, [2D_1 q^4 R_e^2 (k^4 + (R_e\omega + LR_e k_y)^2)$$
$$+8D_2 q^2 R_e^4 k_z^2]/[4(2-q)^2 R_e^4 k_z^4 + 4R_e^2 q^2 (2-q) k^2 k_z^2 (k^4 - (R_e$$
$$\omega + LR_e k_y)^2) + k^4 q^4 (k^4 + (R_e\omega + LRk_y)^2)^2]\, \omega^{-2\theta}, \quad (3.14)$$

where $J_0(kr)$ is the zeroth-order Bessel function. Similarly, one can obtain $S_\zeta(r)$ and $S_{du}(r)$ explicitly. The poles in Eq. (3.14) are identical to Eq. (3.8) given by (3.10). Similar profiles for vorticity correlations can also be extracted.

Comparing Figs. 3.1 and 3.3, we find that the spatiotemporal fluctuations get completely dominated by hydrodynamically stable modes for the absolute values of velocity-velocity and vorticity-vorticity correlations, reciprocating similar qualitative stability in Figs. 3.1a and 3.3a, and also in Figs. 3.1b and 3.3b. However, for $q = 1$, the real components from Figs. 3.1 and 3.3 (Fig. 3.1a versus Fig. 3.3a) show different characteristics compared to higher values of $q = 1.5$, 1.9 where the velocity and vorticity correlations show similar qualitative trends. This is not too difficult to understand though. For a constant angular velocity flow ($q = 1$), the vorticity flux rate dominates the velocity flux which explains the thermal noise-controlled oscillations in Fig. 3.3 for $q = 1$. For larger q-values, the flux rates start matching each other which explains the similar profiles in Figs. 3.1c:3.3c and Figs. 3.1d:3.3d. The point of interest here is that both temporal (Fig. 3.1) and spatial (Fig. 3.3) correlators are affected by the long-ranged noise correlation, a key outcome of this study.

Remarkably, the vorticity profiles as shown in Fig. 3.4a–d point to thermally controlled oscillations at short time scales that fast settle down into quiescent modes as time increases. This happens due to additional hydrodynamic flux affecting spatial components in vorticity correlations (Fig. 3.4), unlike temporal components (Fig. 3.3). As is expected, in the absence of external forcing (the noise here is

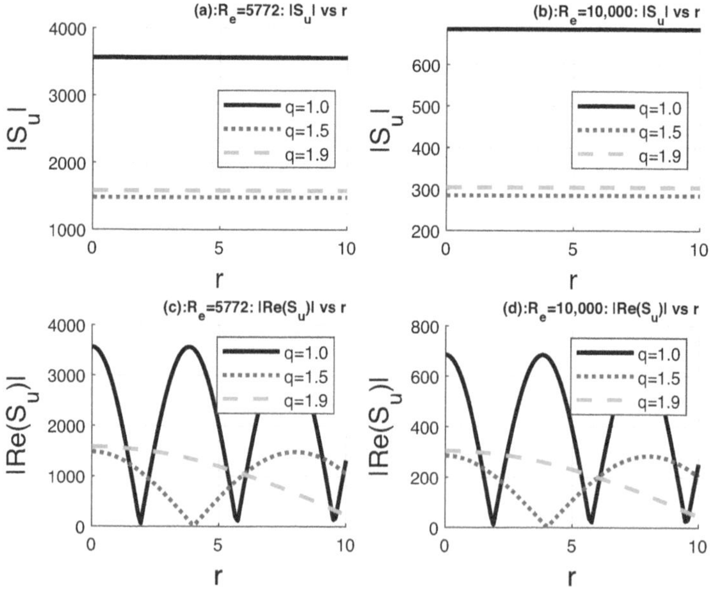

Fig. 3.3 Velocity spatial correlations for $q = 1$ (solid line), $q = 1.5$ (dotted line) and $q = 1.9$ (dashed line)

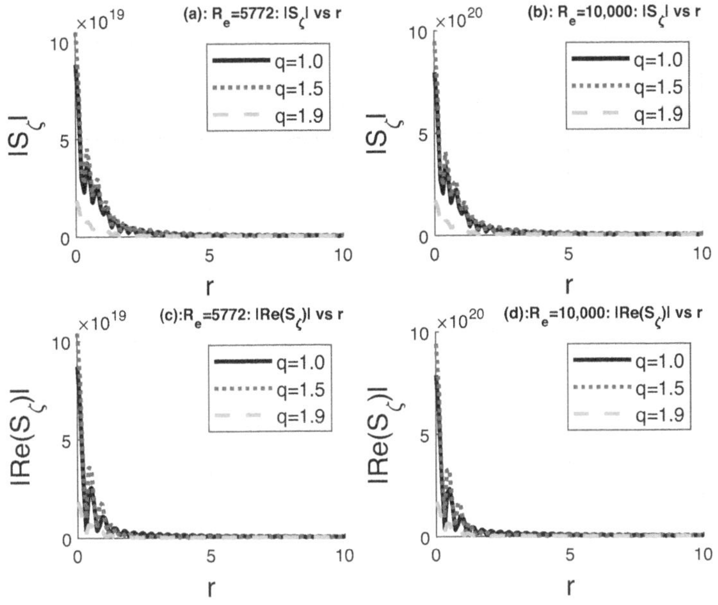

Fig. 3.4 Vorticity spatial correlation varying with time for $q = 1$ (solid line), $q = 1.5$ (dotted line) and $q = 1.9$ (dashed line)

only temporally correlated), spatial correlations quickly die out irrespective of the choice of q and R_e. Compared with our previous studies on spatially correlated noise-driven Orr-Sommerfeld-Squire model [15–17], the degree of oscillations with spatial correlations is decidedly more pronounced. To avoid repetition, we refrain from plotting the imaginary temporal or spatial correlations, both for velocity and vorticity, as they resemble their real counterparts.

3.3 Summary and Conclusion

Here we have analyzed temporally correlated noise-driven Orr-Sommerfeld-Squire flow to understand the origin of non-magnetic instability in accretion flow. As the molecular viscosity in accretion disks is negligible, any transport of matter therein would arise through turbulence. While these flows are Rayleigh stable in the absence of magnetic coupling, linearly unstable modes show up which could lead to turbulence. This has been observed in a multitude of "cold". Our starting point was the fact that while these are astrophysically cold flows, from the perspective of thermodynamics, these are thermal "hot" flows, an aspect that we previously studied [15–17]. These earlier studies focused on rotating and non-rotating flows driven by spatially correlated noise. The present study is the first in line to consider temporal noise.

Founded on statistical physics, we have shown that forced linearized rotating shear flows in a narrow gap limit (mimicking a small portion of accretion disks) reveal a large correlation between energy growth of perturbation in the presence of white noise, arising in the system due to stochastic perturbation. We have analyzed the evolutions of the temporal and spatial correlations, which are shown to be evolving to a large value in the flow time scale, triggering plausible instability and turbulence. While the correlations are directly proportional to the Reynolds number for nonrotating flows, it is independent of the Reynolds number for rotating flows when the flow is controlled by the dynamics and noise. While correlations decrease as the flow deviates from the type of constant angular momentum to the Keplerian and then of flat-rotation curve, still, they still appear large enough to trigger instability. Interestingly, while rotational effects reduce the amplitude of spatial correlations compared to that of a nonrotating case, the amplitude of temporal correlations increases in the presence of rotational effects and increases with the increase of Coriolis parameter, in the time scale of accretion flows. Moreover, on a smaller time scale, the amplitude of temporal correlation in non-rotating flows appears independent of Reynolds number. A key outcome of this study is the fast convergence of the spatial vorticity nodes in the absence of an external spatially correlated noise (Fig. 3.4). This is in contrast to sustained temporal growth (Figs. 3.1 and 3.2) which explains the original premise, that of temporal noise-driven non-magnetic instability.

The properties of temporal and spatial correlations together define the system of rotating shear flows and accretion disks to follow a single universality class. Our previous studies [15–17] on spatiotemporally correlated noise-driven Orr-Sommerfeld-Squire models have opened up a new chapter in magnetohydrodynamic turbulence. The key takeaway from the present study is that temporally correlated noise, which originates from the curse of higher dimensionality, can dominate spatial correlation and even hydrodynamic modes in the asymptotic limit. More studies in this line, comparing features against Kolmogorov scaling are expected to shed new light on the origin of nonlinear instability.

Acknowledgments I am grateful to my student Aimee Pascaline N Unkundiye for her help with the latex editing.

References

1. Chagelishvili, G. D., Zahn, J.-P., Tevzadze, A. G., & Lominadze, J. G. (2003). *Astronomy & Astrophysics*, *402*, 401.
2. Tevzadze, A. G., Chagelishvili, G. D., Zahn, J.-P., Chanishvili, R. G., & Lominadze, J. G. (2003). *Astronomy & Astrophysics*, *407*, 779.
3. Yecko, P. A. (2004). *Astronomy & Astrophysics*, *425*, 385.
4. Butler, K. M., & Farrell, B. F. (1992). *Physics of Fluids*, *4*, 1637.

5. Reddy, S. C., & Henningson, D. S. (1993). *Journal of Fluid Mechanics, 252*, 209.
6. Trefethen, L. N., Trefethen, A. E., Reddy, S. C., & Driscoll, T. A. (1993). *Science, 261*, 578.
7. Weishaar, E., Busse, F. H., & Nagata, M. (1991). *Journal of Fluid Mechanics, 226*, 549–564.
8. Mukhopadhyay, B., Afshordi, N., & Narayan, R. (2005). *The Astrophysical Journal, 629*, 383.
9. Forster, D., Nelson, D. R., & Stephen, M. J. (1977). *Physical Review A, 16*, 732.
10. De Dominicis, C., & Martin, P. C. (1979). *Physical Review A, 19*, 419.
11. Medina, E., Hwa, T., Kardar, M., & Zhang, Y.-C. (1989). *Physical Review A, 39*, 3053.
12. Kardar, M., Parisi, G., & Zhang, Y.-C. (1986). *Physical Review Letters, 56*, 889.
13. Barabási, A.-L., & Stanley, H. E. (1995). *Fractal concepts in surface growth*. Cambridge University Press.
14. Chattopadhyay, A. K., & Bhattacharjee, J. K. (2000). *Physical Review E, 63*, 016306.
15. Mukhopadhyay, B., & Chattopadhyay, A. K. (2013). *Journal of Physics A, 46*, 035501.
16. Nath, S. K., Mukhopadhyay, B., & Chattopadhyay, A. K. (2013). *Physical Review E, 88*, 013010.
17. Nath, S. K., & Chattopadhyay, A. K. (2014). *Physical Review E, 90*, 063014.
18. Phinney, E. S. (1989). *Manifestations of a massive black hole in the galactic center*.
19. Cannizzo, J. K., Lee, H. M., & Goodman, J. (1990). *Astrophysical Journal, 351*, 38–46.

Chapter 4
Bondi Flow from Various Perspectives

Souvik Ghose and Tapas K. Das

Abstract Realization of the stationary integral solutions of steady-state transonic accretion flow in spherical symmetry (so called Bondi flow) is crucial, since it helps to understand accretion phenomena on various astrophysical objects. Such flows are generally studied in the literature, in general, only from astrophysical contexts. In recent years, however, attempts have been made to study accreting black hole systems as an example of autonomous dynamical systems. The fixed-point solution scheme has been borrowed from the theory of dynamical systems to study the transonic properties of accretion flow onto an astrophysical black hole, and it has been demonstrated that the nature of the phase orbit of transonic flow solutions can be understood even without constructing integral solutions. Because a large-scale astrophysical fluid flow is vulnerable to external perturbations, it is necessary to ensure that the stationary accretion solutions are stable under such perturbations. Such a task can be accomplished by adopting a time-dependent stability analysis scheme for the accretion flow, to demonstrate, under which condition the perturbation will not diverge. While performing such stability analysis, it has also been observed that a space time metric, dubbed as the sonic (or acoustic) metric can be constructed to describe the propagation of perturbations embedded within the accreting fluid. Such a sonic metric is similar in nature to certain representation of the Schwarzschild metric. Thus, the acoustic metric mimics a black hole like spacetime within the accreting fluid, and the transonic surface can be identified with a black hole like horizon. Such identification is accomplished using the theory of causal structure, by constructing Carter-Penrose diagrams. An accreting black hole system, thus, can be perceived as an classical analogue gravity model naturally found in the universe. Hence, accretion phenomena onto astrophysical black holes can be looked upon from three apparently non overlapping perspectives—astrophysical processes, theory of dynamical systems, and emergent

---

S. Ghose (✉) · T. K. Das
Harish-Chandra Research Institute, HBNI, Prayagraj, India
e-mail: souvikghose@hri.res.in; tapas@hri.res.in

S. Ghosh, A. R. Rao (eds.), *The Relativistic Universe: From Classical to Quantum*, Astrophysics and Space Science Proceedings 61,
https://doi.org/10.1007/978-3-031-90186-7_4

gravity (alternatively, the analogue gravity) phenomena, respectively. The present chapter illustrates, by taking the simplest possible accretion flow model, how one can study astrophysical accretion processes from three aforementioned perspectives.

4.1 Introduction

Accretion refers to the process by which matter through its gravitational attraction. Accretion is ubiquitous in astrophysics and is responsible for the formation of most astronomical objects, including galaxies, stars, planets and compact astrophysical objects. The importance of accretion of interstellar gas has been recognized since the late 1930s, primarily through the pioneering work of Hoyle, Lyttleton, and Bondi [1–4]. However, these early studies overlooked the impact of pressure in the in-falling matter, assuming that any generated heat would be rapidly radiated away and that the gas temperature remained low, thereby justifying the neglect of pressure effects compared to dynamical effects. In 1952, Bondi's seminal paper [5] marked a significant advancement in this field by exploring gas accretion in an adiabatic scenario, incorporating pressure effects for the first time. He considered a simplified model where an accretor of mass M was placed in an infinite, gas cloud with uniform density ρ_∞ and pressure p_∞ at an infinitely large distance from the accretor. The accretion of gas was assumed to be spherically symmetric and steady, with the mass of the accretor remaining constant during the accretion process.

In the context of black holes, accretion processes are particularly critical. Black hole accretion involves the accumulation of surrounding compressible astrophysical fluid into an accretion disk, where intense gravitational forces heat the material to extreme temperatures, causing it to emit vast amounts of radiation. This process not only influences the growth and evolution of black holes but also makes impacts on their surroundings, contributing to phenomena such as the emergence of relativistic jets and energetic outflows. The study of black hole accretion is thus essential for understanding both the astrophysics of black holes and their broader astrophysical environments. In a simplified set up, spherically symmetric accretion onto astrophysical black holes has also been studied. In Bondi's original treatment the pressure and the density of the gas was always connected by the adiabatic equation of state of an ideal gas:

$$\frac{p}{p_\infty} = \left(\frac{\rho}{\rho_\infty}\right)^\gamma, \tag{4.1}$$

γ being the ratio of the specific heats at the constant pressure and the constant volume and $1 < \gamma < 5/3$. Bondi's analysis is based upon two basic equations. If r be the radial coordinate and v be the inward velocity of the gas, stationary integral solution of the continuity equation is given by:

$$4\pi r^2 v\rho = \text{constant}. \tag{4.2}$$

In steady state, the accreting fluid also satisfies the stationary integral solution of the Bernoulli's equation:

$$\frac{v^2}{2} + \int_{p_\infty}^{p} \frac{dp}{\rho} - \frac{GM_{BH}}{r} = \text{constant} \, (= 0), \tag{4.3}$$

where M_{BH} is the mass of the accretor (black hole). The right-hand side of Eq. (4.3) vanishes due to the boundary conditions [5], and the second term on the left-hand side can be integrated using the equation of state (Eq. (4.1)). Bondi investigated the steady-state solution through an algebraic analysis of Eqs. (4.2) and (4.3). Such analysis was facilitated by introducing a variable, the Mach number, which is the ratio of the dynamical velocity (v) of the gas to the local sound speed ($c_s = \sqrt{\frac{\gamma p}{\rho}}$).

Through his algebraic approach, Bondi identified various types of accretion solutions, including those that make transition from subsonic to supersonic regimes (or vice versa), referred to as transonic solutions hereafter. Bondi's work constitutes the simplest model of accretion within the framework of the Newtonian gravity framework, fully exploiting the spherical symmetry and by neglecting any rotational energy component of the in-falling material. The term Bondi Flow (or Bondi Accretion) is now synonymous with spherical, inviscid, steady accretion in Newtonian gravity.

However, rather than using an algebraic approach, it is often more practical to solve a system of differential equations to obtain the stationary, transonic accretion solution in the spherically symmetric flow. Additionally, instead of applying boundary conditions at infinity, it is computationally more efficient to consider boundary conditions at the critical point of the flow, which will be explained in subsequent sections (see [6] for details).

4.2 Phase Portrait for the Bondi Flow

Stationary integral solutions of the fluid dynamic equations governing the steady state the Bondi flow provide two first integrals of the motion, and it can be shown that the aforementioned two conserved quantities are the specific energy and the mass accretion rate, respectively with their explicit expressions as shown below:

$$\mathcal{E} = \frac{1}{2}v^2 + \frac{1}{\gamma - 1}c_s^2 - \frac{1}{r} = \frac{1}{\gamma - 1}c_{s_\infty}^2, \tag{4.4}$$

$$\dot{M} = \rho v r^2. \tag{4.5}$$

where c_s^2 represents the local adiabatic sound speed, $c_s^2 = \gamma p/\rho$, utilizing the adiabatic equation of state $p\rho^{-\gamma} = \text{constant}$. Here, c_{s_∞} denotes the sound speed at a considerable distance from the accretor, and $\mathcal{E}$ is identified as the conserved specific energy. We use natural units, i.e., $G = c = 1$. All masses are measured in the unit of the black hole mass (M_{BH}). We consider $M_{BH} = 1$ for algebraic convenience. Moreover, radial distance is scaled by GM_{BH}/c^2 and any speed is scaled by light-speed (c). Radial velocity profile of the accreting matter can be obtained by integrating the following equation:

$$\frac{dv}{dr} = \frac{\frac{1}{r^2} - \frac{2c_s^2}{r}}{\frac{c_s^2}{v} - v} = \frac{N}{D}. \tag{4.6}$$

Given the continuous nature of physical flow, if D vanishes at any r, then N must also vanish at that r. These locations (r_c) are termed *critical points* or *fixed points*, borrowing the term from the dynamical systems theory. At this point, $v_c = c_{s_c}$ and $r_c = \frac{1}{2c_{s_c}^2}$, and hence the critical point is actually the sonic point of the flow. These conditions are known as the *sonic point conditions*. Throughout this chapter, a subscript 'c' denotes values at the *critical points*. The Mach number at the critical point is $M = 1$. Substituting the sonic point conditions into Eq. (4.4) yields:

$$r_c = \frac{5 - 3\gamma}{4(\gamma - 1)\mathcal{E}}. \tag{4.7}$$

For any γ, the values of all flow variables at sonic points are fully determined by the conserved energy, which is in turn determined by the boundary conditions at infinity. Integrating Eq. (4.6) is simplified with boundary conditions expressed at the flow's critical point. The local sound speed $a(r)$ can be obtained from Eq. (4.4) once $v_c = v(r_c)$ is known. Alternatively, eliminating dv/dr from Eqs. (4.4) and (4.5) provides a differential equation for da/dr analogous to Eq. (4.6):

$$\frac{dc_{s_c}}{dr} = \frac{(\gamma - 1)\left(\frac{v_c^2}{r} - \frac{1}{2r_c^2}\right)}{\left(c_{s_c} - \frac{v^2}{c_{sc}}\right)}. \tag{4.8}$$

This system of two differential equations ((4.6) and (4.8)) can be numerically integrated to obtain $v(r)$ and $c_s(r)$ at various r. A compact representation of this information is the phase portrait, which plots the Mach number $M(r) = v(r)/c_s(r)$ versus r. The phase portrait of the Bondi flow is shown in Fig. 4.1. Figure 4.1 illustrates that only two transonic solutions connect the accretor and infinity. The accreting solution, where the Mach number increases as matter approaches the accretor, becomes supersonic in the region $r < r_c$. The spherical surface at $r = r_c$ is often termed the *sound horizon* or *acoustic horizon*, as acoustic disturbances generated within this region cannot cross the $r = r_c$ surface and must move towards the accretor. Thus, the sound horizon acts like an *event horizon* for acoustic perturbations. We will later explore the non-trivial nature of this analogy.

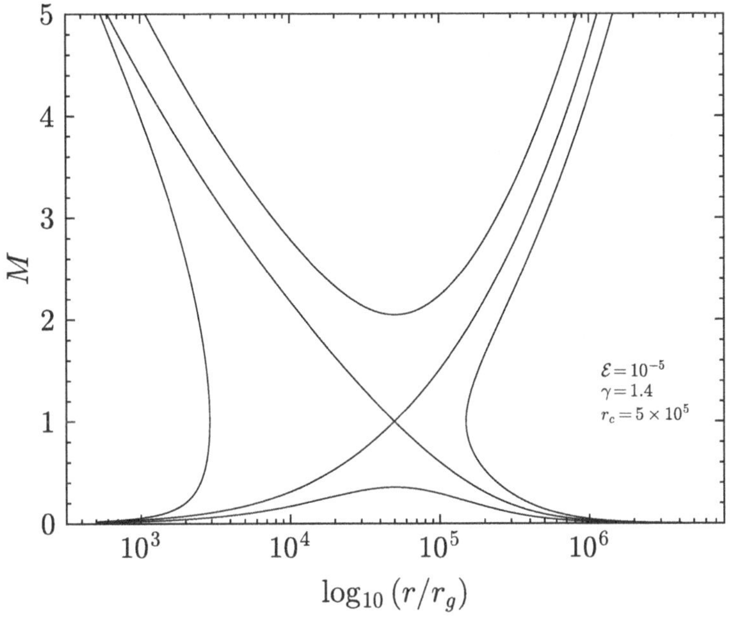

Fig. 4.1 Mach number (M) vs $\log_{10} r$ plot for Bondi type accretion. The radial distance is scaled by GM_{BH}/c^2. Apart from the transonic branches, supersonic, subsonic and unphysical solutions are also plotted for demonstration

4.3 A Dynamical Systems Perspective

In steady state, any quantitative information about any of the variables of interest must be obtained from the phase portrait of the system. However, Eq. (4.7) may not retain its simple form for more general scenario such as axially symmetric accretion. Depending on the system, this equation can be a higher order polynomial or even a transcendental equation with multiple solution. In such scenarios, where multiple critical points (and consequently multiple sound horizons) appear, obtaining the phase portrait at the first place might become challenging. Fortunately, a simple analogy with the autonomous dynamical systems enables us to at least have some qualitative insight about the transonic accretion profiles without drawing the corresponding phase portrait (for a detailed account of the methodology see [7, 8]).

One can introduce a dummy variable (say τ) and write Eq. (4.6) as:

$$\frac{dv}{d\tau} = u\left(2c_s^2 - \frac{1}{r}\right) = N, \tag{4.9}$$

$$\frac{dr}{d\tau} = r(v^2 - c_s^2) = D.$$

Since the critical points (equilibrium points in the terminology of dynamical systems) are fixed by the flow constants, it is simple to use a perturbative scheme around these points to extract information about the phase orbits (or separatrics) in the neighborhood of critical points. Thus, introducing $v = v_c + \delta v$, $c_s = c_{s_c} + \delta c_s$ and $r = r_c + \delta r$, where δ terms are small first order perturbations, one can linearize the system and obtain from Eq. (4.9):

$$\frac{d(\delta v)}{d\tau} = -v_c^2(\gamma - 1)\delta v + \frac{v_c(3 - 2\gamma)}{r_c^2}\delta r, \tag{4.10}$$

$$\frac{d(\delta r)}{d\tau} = r_c v_c(1 + \gamma)\delta v + 2(\gamma - 1)v_c^2 \delta r.$$

This kind of parametrization has been explored in fluid dynamics (e.g. see [9]). In astrophysical context such parametrization was used in [10, 11]. The above set of autonomous system can be written in a compact form:

$$\begin{pmatrix} \frac{d(\delta v)}{d\tau} \\ \frac{d(\delta r)}{d\tau} \end{pmatrix} = \begin{pmatrix} -v_c^2(\gamma - 1) & \frac{v_c(3-2\gamma)}{r_c^2} \\ r_c v_c(1 + \gamma) & 2(\gamma - 1)v_c^2 \end{pmatrix} \begin{pmatrix} \delta v \\ \delta r \end{pmatrix}, \tag{4.11}$$

or symbolically $\mathbf{Y} = A\mathbf{X}$. Using the solution of the type $\delta v \sim exp(\Omega\tau)$ and $\delta r \sim exp(\Omega\tau)$, one can obtain an eigenvalue equation for the above as $A\mathbf{X} = \Omega\mathbf{X}$, and the eigenvalue itself is obtained from the characteristic equation $\det\{(A - \Omega I)\} = 0$. This leads to a quadratic equation:

$$\Omega^2 + \mathbf{Tr}A + \Delta = 0, \tag{4.12}$$

where $\Delta = \det\{A\}$. Since, all the components of A are functions of various flow variables as defined at critical points, all are completely specified by the flow constants. Once Eq. (4.12) are obtain they also yield the nature of the critical points themselves. Clearly, the nature of the roots of Eq. (4.12) depend on the numerical values of Δ and the discriminant $D = (\mathbf{Tr}A)^2 - 4\Delta$. Various possible conditions are summarized in Table 4.1. For the present case, there is just one critical point and that is always of saddle type:

$$\Omega^2 = \frac{2\mathcal{E}^2(5 - 3\gamma)}{\left(\frac{1}{\gamma - 1} - \frac{3}{2}\right)^2}. \tag{4.13}$$

Table 4.1 Types of roots and nature of critical points under different conditions

Conditions	Types of roots (Ω)	Nature of critical points
$\Delta < 0$	Real, opposite sign	Saddle
$\Delta > 0$ and $D > 0$	Real, same sign	Node
$\Delta > 0$ and $D = 0$	Two equal roots	Node
$\Delta > 0$ and $\mathbf{Tr}A = 0$	Purely imaginary roots	Center
$\Delta > 0$ and $D < 0$	Complex roots	Spiral

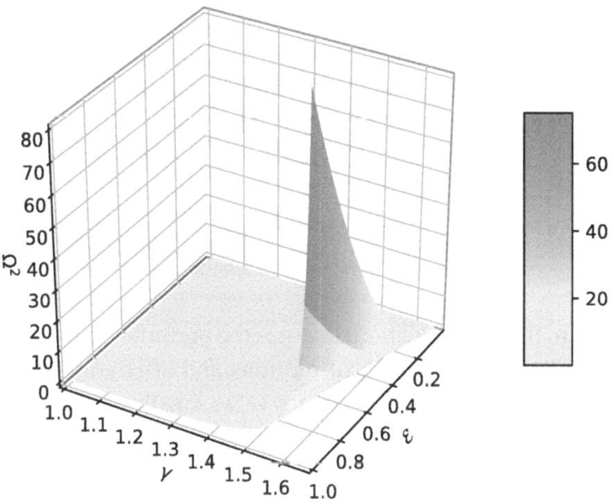

Fig. 4.2 Phase portrait of the Bondi flow

This is true for any viable parameter value, as is evident from Fig. 4.2. The scheme, however, is quite general and in more complicated scenario its real worth is appreciated [12, 13].

4.4 Stability of the Steady State Solution

In all the discussions above, we relied on the steady state version of the Euler and continuity equations. In any large-scale astrophysical fluid flow, such a state may be subjected to external perturbations. A star might pass by the accretor itself and it's gravitational influence might jeopardize the steady state configuration at least for some duration. It is thus imperative to check for the stability of the steady state solution against such cases. A standard way to look into such effects is to study the evolution of the time dependent perturbations [13]. As long as the perturbations do not diverge, the steady state solutions remain meaningful. The general form of the continuity equation is:

$$\frac{\partial \rho}{\partial t} + \nabla \cdot (\rho \mathbf{v}) = 0. \tag{4.14}$$

For an irrotational flow ($\nabla \times \mathbf{v} = 0$), satisfies the Euler's equation becomes:

$$\frac{\partial v}{\partial t} + \frac{1}{2}\nabla(v^2) = -\frac{1}{\rho}\nabla(p) - \nabla\phi, \tag{4.15}$$

ϕ being the potential of the body force, e.g. gravity. Since, the flow is irrotational, one can introduce a velocity potential ψ such that $\mathbf{v} = \nabla\psi$. Consider some arbitrary but exact stationary solutions of the equations of motion $[\rho_0, v_0, \psi_0]$. For a barotropic fluid ($p = p(\rho)$), perturbation in ρ induces a perturbation in p as well around some background value p_0. One can then study the linearized perturbations around this background solution by writing:

$$\begin{aligned} \rho(r,t) &= \rho_0(r,t) + \epsilon\tilde{\rho}(r,t), \\ p(r,t) &= p_0(r,t) + \epsilon\tilde{p}(r,t), \\ \psi(r,t) &= \psi_0(r,t) + \epsilon\tilde{\psi}(r,t), \end{aligned} \tag{4.16}$$

where the overhead *tilde* denotes the perturbed quantity and $0 < \epsilon \ll 1$. Substituting these into the equation of motions and after some tedious algebra, it is possible to write down the equation that $\tilde{\psi}(r,t)$ satisfies [14]:

$$\begin{aligned} &\frac{\partial}{\partial t}\left(c_s^{-2}\rho_0\frac{\partial\tilde{\psi}(r,t)}{\partial t}\right) + \frac{\partial}{\partial t}\left(c_s^{-2}\rho_0\mathbf{v}_0\cdot\nabla\tilde{\psi}(r,t)\right)\nabla\cdot\left(\mathbf{v}_0\rho_0 c_s^{-2}\frac{\partial\tilde{\psi}(r,t)}{\partial t}\right) \\ &+\nabla\cdot\left(c_s^{-2}\mathbf{v}_0\rho_0(\mathbf{v}_0\cdot\nabla\tilde{\psi}(r,t))\right) - \nabla\cdot\left(\rho_0\nabla\tilde{\psi}(r,t)\right) = 0. \end{aligned} \tag{4.17}$$

Equation (4.17) can be written in a much compact form:

$$\partial_\mu\left(f^{\mu\nu}\partial_\nu\right)\tilde{\psi}(r,t) = 0, \tag{4.18}$$

where $\mu, \nu = 0, 1, 2, 3$ and $f^{\mu\nu}$ is given by:

$$f^{\mu\nu} = \left(\frac{\rho_0}{c_s^2}\right)\begin{pmatrix} 1 & v^1 & v^2 & v^3 \\ v^1 & v^1v^1 - c_s^2 & v^1v^2 & v^1v^3 \\ v^2 & v^2v^1 & v^2v^2 - c_s^2 & v^2v^3 \\ v^3 & v^3v^1 & v^3v^2 & v^3v^3 - c_s^2 \end{pmatrix}, \tag{4.19}$$

where (v^1, v^2, v^3) are the components of the 3-velocity $\mathbf{v}$. For the spherically symmetric Bondi flow, of course, the only velocity component available is the radial velocity which we denote simply with v. The dummy indices in Eq. (4.18) then runs for (0, 1) only and the $f^{\mu\nu}$ in this case is given by:

$$f^{\mu\nu} = \left(\frac{\rho_0}{c_s^2}\right)\begin{pmatrix} 1 & v \\ v & (|v|^2 - c_s^2) \end{pmatrix}. \tag{4.20}$$

An wave equation such as Eq. (4.18) opens up new possibilities to explore which we shall do in the next section. For the moment let us concentrate on the stability of the stationary solution which we sought to investigate. Consider a trial acoustic wave function for wave equation for the Bondi flow:

$$\tilde{\Psi} = g_\omega(r)\exp\{(-i\omega t)\}. \tag{4.21}$$

Then the spatial part $g_\omega(r)$ satisfies

$$-\omega^2 f^{tt} g_\omega - i\omega(f^{tr}\partial_r g_\omega + g_\omega \partial_r f^{rt} + f^{rt}\partial_r g_\omega) + (g_\omega \partial_r f^{rr} + f^{rr}\partial_{rr} g_\omega) = 0, \tag{4.22}$$

where $f^{tt} = \frac{v_0}{\Psi_0}$, $f^{tr} = f^{rt} = \frac{v_0^2}{\Psi_0}$ and $f^{rr} = \frac{v_0}{\Psi_0}(v_0^2 - c_{s0}^2)$. The solution can represent a standing wave as well as a travelling wave.

4.4.1 *Standing Wave Analysis*

If the accretor is an astrophysical object with a well defined physical boundary, one can assume the perturbations to vanish at the two extremes, far away from the accretor and on its surface. A standing wave solution is useful in such scenarios. However, as a transonic flow becomes supersonic at the sound horizon the only way it can become subsonic again is through a discontinuous shock transition. But a standing wave solution is continuous. So, for the flow to have a standing wave solution it has to be *subsonic throughout*. Let the two boundaries be r_1 and $r_2(> r_1)$ such that $g_\omega(r_1) = g_\omega(r_2) = 0$. Now multiplying Eq. (4.22) and integrating in the range $r_1 < r < r_2$ we get

$$\omega^2 \int g_\omega^2 f^{tt} dr + i\omega \int \partial_r (g_\omega^2 f^{rt}) dr + \int f^{rr} (\partial_r g_\omega)^2 dr = 0. \tag{4.23}$$

After substituting the values of metric elements $f^{\mu\nu}$ we have

$$\omega^2 = -\frac{\int v_0(v_0^2 - c_{s0}^2)(\partial_r g_\omega)^2 dr}{\int v_0 g_\omega^2 dr}. \tag{4.24}$$

As the flow is subsonic everywhere, $v_0 < c_{s0}$, then ω is real and we get oscillatory solution for standing wave and the perturbations never diverge.

4.4.2 *Travelling Wave Analysis*

If the accretor happens to be a black-hole, there is no hard surface and any flow that reaches the event horizon is supersonic. In such a scenario the analysis described in Sect. 4.4.1 is not valid. The wave in this scenario is a traveling one. The wavelength of such a wave though is small compared to the radius of the accretor. Hence, employing *WKB approximation method* we can approximate the solution in power series of ω

$$g_\omega(r) = \exp\left[\sum_{n=-1}^{n=\infty} \frac{k_n(r)}{\omega^n}\right]. \tag{4.25}$$

Substituting this in Eq. (4.23) and equating the coefficients of ω's we get the following three equations

$$-f^{tt} - 2if^{tr}\frac{dk_{-1}}{dr} + f^{rr}\left(\frac{dk_{-1}}{dr}\right)^2 = 0 \tag{4.26}$$

$$-2if^{tr}\frac{dk_0}{dr} - i\frac{df^{tr}}{dr} + \frac{df^{rr}}{dr}\frac{dk_{-1}}{dr} + f^{rr}\left[2\frac{dk_{-1}}{dr}\frac{dk_0}{dr} + \frac{d^2k_{-1}}{dr^2}\right] = 0 \tag{4.27}$$

$$-2if^{tr}\frac{dk_1}{dr} + \frac{df^{rr}}{dr}\frac{dk_0}{dr} + f^{rr}\left[2\frac{dk_{-1}}{dr}\frac{dk_1}{dr} + \left(\frac{dk_0}{dr}\right)^2 + \frac{d^2k_0}{dr^2}\right] = 0. \tag{4.28}$$

Solving for $\frac{dk_{-1}}{dr}$, $\frac{dk_0}{dr}$ and $\frac{dk_1}{dr}$ and then integrating using the values of metric elements we obtain:

$$k_{-1} = i\int \frac{dr}{v_0 \mp c_{s0}}, \tag{4.29}$$

$$k_0 = -\frac{1}{2}\ln\left(\frac{v_0 c_{s0}}{\Psi_0}\right). \tag{4.30}$$

Thus for $\omega \gg 1$ the first three terms of $g_\omega(r)$ expression (Eq. (4.25)) are $\sim \omega r$, $\ln r$ and $1/(\omega r)$ respectively. For large values of r:

$$\omega r \ll \ln r \ll \frac{1}{\omega r} \tag{4.31}$$

$$\Longrightarrow \quad \omega \mid k_{-1} \mid \ll \mid k_0 \mid \ll \frac{1}{\omega} \mid k_1 \mid . \tag{4.32}$$

Hence, power series in $g_\omega(r)$ does not diverge even as n increases, that is,

$$\omega^{-n} \mid k_n(r) \mid \ll \omega^{-(n+1)} \mid k_{n+1}(r) \mid,$$

and one obtains a traveling wave solution with finite amplitude.

Thus, in either case, i.e., standing wave analysis as described in Sect. 4.4.1 and traveling wave analysis as described in Sect. 4.4.2, the stability of the steady state solution is ensured.

4.5 Analogue Spacetime

It is interesting that Eq. (4.20) looks similar to the equation satisfied by a massless scalar field (φ) propagating in a curved spacetime:

$$\partial_\mu \left(\sqrt{-g} g^{\mu\nu} \partial_\nu \varphi\right) = 0. \tag{4.33}$$

A direct comparison of Eq. (4.33) with Eq. (4.20) and some algebraic manipulation lead to an expression for the space-time metric governing the propagation of the linear perturbations embedded within the fluid:

$$g_{\mu\nu} = -\frac{1}{c_{s_0}} \begin{bmatrix} v_0^2 - c_{s_0}^2 & -v_0 \\ -v_0 & 1 \end{bmatrix}, \tag{4.34}$$

and the corresponding line element (apart from the conformal factor):

$$ds^2 = \frac{1}{c_{s_0}}[(v_0^2 - c_{s_0}^2)dt^2 + dr^2 - 2v_0 dr dt]. \tag{4.35}$$

The line element is similar to the Schwarzschild line element in Painlevé-Gullstrand coordinates and has a nonzero curvature. However, since Eq. (4.20) holds for acoustic perturbations, the curved spacetime is visible only to these perturbations, while the background spacetime remains Newtonian. The emergence of such an effective metric in the context of an inviscid, irrotational, classical fluid (known as the "acoustic metric") has been recognized since Unruh's pioneering work in 1981 [15], and has further been formalized by Visser [14]. Models using this concept are called 'analogue gravity' models (for detailed discussion, see [16]). In astrophysical accretion, analogue gravity models are unique, as they may involve both a *gravitational* event horizon and an *acoustic* horizon (if the accretor is a black hole). Once the analogy is established for a given configuration, certain methodologies as developed in the theory of General Relativity can be used to study the causal structure of the spacetime. Earlier, a similarity between the sound horizon and a black hole event horizon was noted. It is particularly interesting to establish that the sound horizon is indeed a null horizon in the causal structure of the acoustic spacetime. With a steady-state solution (and a phase portrait such as Fig. 4.1), we can calculate the acoustic spacetime at every r. Coordinate transformations can then be used to draw the Carter-Penrose diagram, which visualizes the causal structure. The procedure is discussed in the next section.

4.5.1 Carter-Penrose Diagram for the Bondi Flow

We start by choosing null (sonic) coordinates to write down the line element (4.35). We note that for null curves (soundlike in case of analogue space-time) $ds^2 = 0$, which yields

$$(dt - A_+(r)dr)\,(dt - A_-(r)dr) = 0. \tag{4.36}$$

where,

$$A_\pm = \frac{-g_{tr} \pm \sqrt{g_{tr}^2 - g_{rr}g_{tt}}}{g_{tt}}. \tag{4.37}$$

So instead of co-ordinates (t, r), we choose new co-ordinates to be null co-ordinates (χ, ω) such that

$$d\omega = dt - A_+(r)dr, \tag{4.38}$$

$$d\chi = dt - A_-(r)dr. \tag{4.39}$$

Using coordinate transformation introduced in (4.38), the line element (4.35) can be written as

$$ds^2 = g_{tt}d\chi d\omega. \tag{4.40}$$

We now expand $A_-(r)$ and $A_+(r)$ up to first order of $(r - r_c)$. Thus by expanding v_0 near r_c as

$$v_0(r) = -v_c + \left|\frac{dv}{dr}\right|_{r_c} (r - r_c) + O\left((r - r_c)^2\right). \tag{4.41}$$

Expanding c_{s_0} in a similar manner we can show that:

$$v_0^2 - a_0^2 \approx -2(v_0)_c\left(v_0' - a_0'\right)(r - r_c). \tag{4.42}$$

considering up to the first order terms, where $'$ denotes derivative with respect to r.

Now we expand $A_-(r)$ and $A_+(r)$ up to linear order of $(r - r_c)$. For that we first note that $g_{tt} \propto (u_0 - c_{eff})^2$ is very small near r_c which implies $|g_{tt}g_{rr}/g_{tr}^2| \ll 1$. Thus we obtain

$$A_+(r) = \frac{-g_{tr} + g_{tr}\left(1 - \frac{g_{tt}g_{rr}}{g_{tr}^2}\right)^{1/2}}{g_{tt}} \tag{4.43}$$

$$\approx -\frac{g_{rr}}{2g_{tr}} \tag{4.44}$$

and

$$A_-(r) = \frac{-g_{tr} - g_{tr}\left(1 - \frac{g_{tt}g_{rr}}{g_{tr}^2}\right)^{1/2}}{g_{tt}} \tag{4.45}$$

$$\approx -\frac{2g_{tr}}{g_{tt}} \tag{4.46}$$

$$= \frac{1}{\kappa}\frac{1}{r - r_c}, \tag{4.47}$$

where

$$\kappa = (v_0')_c - (c_{s0}')_c. \tag{4.48}$$

Even though

$$\chi \approx t - \frac{1}{\kappa}\ln|r - r_c| \tag{4.49}$$

shows a logarithmic divergence at $r = r_c$, the form of g_{rr} and g_{tr} ensure that

$$\omega = t + \int \frac{g_{rr}}{2g_{tr}} dr \tag{4.50}$$

is regular there.

Thus, near critical points

$$e^{-\kappa\chi} \propto e^{-\kappa t}\,|r - r_c| \propto e^{-\kappa t}(v_0^2 - c_{s0}^2). \tag{4.51}$$

In this case, one can compare the acoustic null coordinates with those of the Schwarzschild metric and deduce a coordinate transformation that removes the singularity of the metric element at the critical point. The transformation equations are given by:

$$\begin{aligned} U(\chi) &= -e^{-\kappa\chi}, \\ W(\omega) &= e^{\kappa\omega}. \end{aligned} \tag{4.52}$$

This is analogous to transforming to a coordinate system for the Schwarzschild metric, which behaves regularly at $r = r_c$. Finally, to compactify the infinite space into a finite patch, we use coordinates (T, R) such that:

$$T = \tan^{-1}(W) + \tan^{-1}(U), \tag{4.53}$$

$$R = \tan^{-1}(W) - \tan^{-1}(U). \tag{4.54}$$

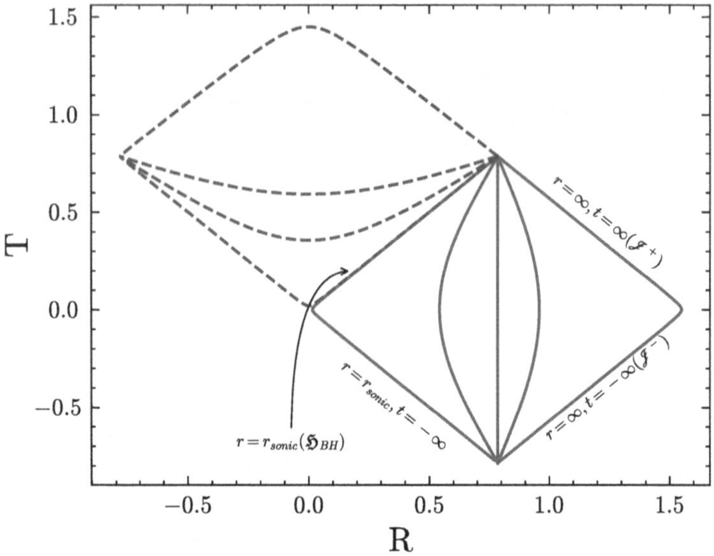

Fig. 4.3 Carter-Penrose diagram for the analogue spacetime associated with the transonic, Bondi accretion branch. The blue (dashed) region is where the flow is supersonic and the red (solid) region is the region where the flow is subsonic

In (T, R) coordinates, lines where r = constant can be drawn, allowing the resulting diagram to represent the causal structure of the original spacetime in a compactified region. This diagram, shown in Fig. 4.3, is known as the Carter-Penrose diagram (hereafter referred to as CP). For brevity, we will not delve into the algebra or detailed methodology for drawing the CP diagram. Readers interested in the usefulness of CP diagrams for exploring black-hole spacetimes are referred to Townsend's excellent review [17] or [18] for more mathematical rigor. The detailed methodology for obtaining CP diagrams in the context of analogue gravity is discussed in [19] and [20]. Here, we will highlight some features of CP, but first, it is helpful to quote a lemma from differential geometry [21].

If two metrices G and g on the same manifold $\mathscr{M}$ are conformally related, then the null geodesics with respect to metric G are null geodesics also with respect to the metric g and vice-versa. Thus any perturbation propagating with the sound-speed is null-like in the acoustic space-time. One can define the boundary of the mapping ψ of the entire analogue space-time $\mathscr{M}$ as

$$\partial\psi\left(\mathscr{M}\right) = i^0 \bigcup \mathscr{I}^+ \bigcup \mathscr{I}^-, \tag{4.55}$$

where

1. i_0, known as *Spatial Infinity* is the endpoint of the ψ image of all space-like curves in $(\mathscr{M}, g)$.

2. $\mathscr{J}^+$, known as *Future Causal Infinity* is the endpoint of the ψ image of all future directed causal curves in $(\mathscr{M}, g)$.
3. $\mathscr{J}^-$, known as *Past Causal Infinity* is the endpoint of the ψ image of all past directed causal curves in $(\mathscr{M}, g)$.

The sound horizon is immediately identified as a null hypersurface from the Fig. 4.3. We have already shown that no acoustic perturbation, created within the sound horizon (blue region in the Fig. 4.3 can cross the sound horizon and escape to a large distance away from the accretor (blue region in the Fig. 4.3. That along with the fact that the sound horizon is a null horizon in the acoustic space-time shows that sound horizon behaves like an event horizon for acoustic perturbations.

4.6 Conclusion

The procedures discussed in the present chapter is fairly general and are applicable to any inviscid, irrotational fluid flow. In the context of accretion, Bondi flow is the simplest possible configuration and are discussed merely for a demonstrative purpose. The methodology has been consistently applied to various accretion disk models both in Newtonian and in General relativistic frameworks [13, 20]. One immediate application of the Carter-Penrose diagram is worth noting. For axially symmetric accretion disks in hydrostatic equilibrium along vertical directions, the *sonic points* and *critical points* do not coincide. In such scenarios the Carter-Penrose diagrams conclusively show that the horizons lie on the *critical points* of the flow and not on the *sonic points* [20].

References

1. Hoyle, F., & Lyttleton, R. A. (1939). *Proceedings of the Cambridge Philosophical Society*, *35*, 405.
2. Hoyle, F., & Lyttleton, R. A. (1940). *Proceedings of the Cambridge Philosophical Society*, *36*(3), 325. https://doi.org/10.1017/S0305004100017369
3. Hoyle, F., & Lyttleton, R.A. (1940). *Proceedings of the Cambridge Philosophical Society*, *36*(4), 424. https://doi.org/10.1017/S0305004100017461
4. Bondi, H., & Hoyle, F. (1944). *Monthly Notices of the Royal Astronomical Society*, *104*, 273. https://doi.org/10.1093/mnras/104.5.273
5. Bondi, H. (1952). *Monthly Notices of the Royal Astronomical Society*, *112*(2), 195.
6. Chakrabarti, S. (1996). *Physics Reports*, *266*(5), 229. https://doi.org/https://doi.org/10.1016/0370-1573(95)00057-7
7. Strogatz, S. H. (2018). *Nonlinear dynamics and chaos: With applications to physics, biology, chemistry, and engineering*. CRC Press.
8. Jordan, D., & Smith, P. (2007). *Nonlinear ordinary differential equations: An introduction for scientists and engineers*. OUP Oxford.
9. Bohr, T., Dimon, P., & Putkaradze, V. (1993). *Journal of Fluid Mechanics*, *254*, 635.
10. Ray, A. K., & Bhattacharjee, J. (2002). *Physical Review E*, *66*(6), 066303.

11. Afshordi, N., & Paczyński, B. (2003). *The Astrophysical Journal*, *592*(1), 354.
12. Mandal, I., Ray, A. K., & Das, T. K. (2007). *Monthly Notices of the Royal Astronomical Society*, *378*(4), 1400.
13. Chaudhury, S., Ray, A. K., & Das, T. K. (2006). *Monthly Notices of the Royal Astronomical Society*, *373*(1), 146.
14. Visser, M. (1998). *Classical and Quantum Gravity*, *15*(6), 1767. https://doi.org/10.1088/0264-9381/15/6/024
15. Unruh, W. G. (1981). *Physical Review Letters*, *46*(21), 1351.
16. Barcelo, C., Liberati, S., & Visser, M. (2011). *Living Reviews in Relativity*, *14*, 1.
17. Hawking, S. W., & Ellis, G. F. (2023). *The large scale structure of space-time*. Cambridge University Press.
18. Townsend, P. K. (1997). *Black holes*.
19. Barceló, C., Liberati, S., Sonego, S., & Visser, M. (2004). *New Journal of Physics*, *6*(1), 186. https://doi.org/10.1088/1367-2630/6/1/186
20. Maity, S., Shaikh, M. A., Tarafdar, P., & Das, T. K. (2022). *Physical Review D*, *106*(4), 044062.
21. Frè, P. G., et al. (2013). *Gravity, a geometrical course: Volume 2: Black holes, cosmology and introduction to supergravity*. Springer.

Chapter 5
Dissipative Shocks in Accretion Flows: A Theoretical Forecast of the Spectral State Transitions in Black Hole X-ray Binaries

Soumen Mondal and Prasad Basu

Abstract In this chapter, we forecast the spectral state transition (hard to soft state) theoretically for the transient sources at a constant accretion rate. We first demonstrate a systematic evolution of the relative size of the Comptonizing medium, supposed to be an independent second parameter (accretion rate be the first) for the harder to softer spectral transition. The second parameter is also essential in obtaining the hysteresis of the hardness ratio during outburst and to procure the "q" diagram theoretically. We therefore study extensively the variation of the emitted post-shock radiation in the relativistic accretion flow harboring a dissipative shock. With the dissipation, as the shock moves inward, the inner edge of the disk reduces progressively and attains a minimum for maximum dissipation where transition occurs. The extreme shock location determines the minimum corona size which is obtained around 10 r_g $(= \frac{GM}{c^2})$ for moderate spin. The present chapter is therefore significantly important for the black hole X-ray binary (BHXRBs) sources to examine the origin of spectral transitions, and QPO evolution etc.

5.1 Introduction

The spectro-temporal behavior of the emitted radiation of transient black hole binaries are well studied in literature. In most of the transient binaries with the emission of the radiation the corona shrinks progressively on successive days as the spectral state changes from hard state (HS) $\rightarrow$ hard-intermediate (HIM) $\rightarrow$ soft-intermediate (SIM) $\rightarrow$ soft states (SS) [1, 2]. The luminosity changes greatly on these days. Several dissipative processes namely the Comptonization removes

S. Mondal (✉)
Department of Physics, Jadavpur University, Kolkata, India

P. Basu
Department of Physics, Cotton University, Guwahati, Assam, India

S. Ghosh, A. R. Rao (eds.), *The Relativistic Universe: From Classical to Quantum*, Astrophysics and Space Science Proceedings 61,
https://doi.org/10.1007/978-3-031-90186-7_5

the significant part of the bulk energy from the post-shock corona [3] and thus emits hard X-rays from the hot Compton cloud [4, 5] of a stellar-mass black hole. As a result, pressure drops and the corona shrinks gradually, and thereby changes the inner geometry of the accretion disk. This causes the physical state of the system to evolve with dissipation, and the end exhibits spectral transitions. The frequency of quasi-periodic oscillations (QPOs) also rises monotonically during the hard (HS) and hard-intermediate spectral states (HIM) and decreases in the declining phase and vanishes in soft state [6, 7].

Therefore, the study of the systematic evolution of the corona size with the emission of radiation is important to understand the physics of the accretion disk. In the present communication, we try to explain these evolution systematically with the help of radiative shock (also known as dissipative shock). The most interesting part of the radiative shock is that a significant portion of the post-shock energy get released from the shock transition. The dissipation gradually enhances and the shock propagates inward. Thus, we first investigate the progressive increases of the dissipation with shock location and find out the maximum amount of total bulk energy (kinetic and thermal) which could be removed from the post shock matter so that accretion shock becomes marginally weak. With the dissipation, post-shock entropy gain reduces monotonically and shock becomes weaker. It is thus restricted by the entropy change in the post-shock corona. We finally obtain the critical flow parameters for which the dissipation becomes extremum (almost close to 100%). This happens when shock is closest indicating that the size of the corona or the radius of the inner edge of the accretion disk is minimum. Once the corona reaches minimum size, the dissipation reduces greatly just crossing the maxima, indicating a spectral transition. We examined this feature in detail for all possible flow parameters namely, energy and angular momentum), and spin of the black hole. The high spin increases the magnitude of energy release greatly, reducing the corona further. The radial and vertical oscillations of the post shock boundary layers, induced due to these propagating shock, may excite the QPOs in the emitted X-rays. Thus the radiative shocks have been well studied in literature and a number of applications have already been investigated in black hole accretion physics [8–17] as well as in study of the Interstellar medium, Supernova and accretion processes on to the protostars [18–22]. Thus, we study the properties of the radiative shocks for fully relativistic accretion flows around spinning black holes. In compare to the series of previous study [8, 23] in Newtonian framework, we generalize it in the relativistic framework [24, 25] and come up with the significantly new results that can predict spectral transition from the systemic evolution of post-shock corona, inner edge of disk and QPOs evolution, etc. The spin elevated dissipation further and therefore we also discuss the influences of the spin parameter [25]. However the variation of energy dissipated from the post-shock corona as shock moves inward is correlated with the entropy change in the post-shock corona during shock transition which is not explored yet. This is the key parameter, we study here, regulates the geometry and emitted radiation simultaneously. Therefore, the variation of the corona size with the emission is important to study in explaining the observed phenomenon such as change in luminosity, spectral state transitions and systematic evolution of QPOs etc.

We choose $G = c = M = 1$. Here, G, c and M are the universal gravitational constant, velocity of light and M is the mass of the black hole respectively. This chapter is organized as follows. In the relativistic accretion flows Sect. 5.2, we first discuss the entropy change and energy loss from the post-shock corona. Using this, we identify the parameters for maximum energy dissipation from the post-shock corona in Sect. 5.3. We finally discuss the possibility of spectral transitions for transient sources in Sect. 5.4 and end up with discussion in Sect. 5.5

5.2 The Entropy Change and Energy Loss

Accretion flows may harbor shocks and shock location is determined from Rankine-Hugoniot shock condition [3, 26]. In radiative shocks, the energy balance condition does not hold. The relation modifies to $\mathcal{E}_- = \mathcal{E}_+ + \Delta\mathcal{E}$, where $\Delta\mathcal{E}$ be the amount of energy release released during shock [24]. During the shock transition, flow gains entropy. Post-shock entropy is thus higher than pre-shock flow, that means if (ΔS) be the entropy gains in the post-shock flow then

$$S_+ - S_- = \Delta S, \quad \text{and,} \ 0 \leq \Delta S \leq \Delta S_{max}.$$

The subscripts '$-$' and '+' denote the pre-shock and post shock quantities, respectively. There is a one to one relation between the Rankine-Hugoniot shock conditions and entropy generation in the post-shock flows: (1) Post-shock entropy ΔS decreases with the release of post-shock energy $\Delta\mathcal{E}$. At some point of energy release, if there is no entropy-gain, no accretion shock forms in the flow. (2) For $\Delta S > S_{max}$, the Rankine-Hugoniot shock conditions do not match and flow does not have shock. The ΔS_{max} is marked as boundary between the 'Shock in Accretion (SA)' and 'No Shock in Accretion (NSA)'- regions in the parameters space of energy and angular momentum [27]. (3) The $\Delta S = 0$ provides another boundary between the 'Shock in Accretion (SA)' and 'Shock in Wind (SW)' regions. Thus $\Delta S = 0$ and ΔS_{max}, provide upper and lower boundaries of the Radiative 'Shock in Accretion (SA)' region. Hence we get

$$\mathcal{E}_{max} \geq \mathcal{E} \geq \mathcal{E}_{min},$$

for a fixed angular momentum (l). Both $\mathcal{E}_{max}$ and $\mathcal{E}_{min}$ vary with l and spin a of the black hole. For a given a, the parameters namely $\mathcal{E}, l$ taken from SA-region, the flow dissipates energy $\Delta\mathcal{E}$ during shock transition.

As $\Delta\mathcal{E}$ releases, pressure drops in the post shock corona. To maintain pressure balance, and shock propagates inward. With the decrease of shock location, energy $\Delta\mathcal{E}$ increases successively, but entropy gain ΔS also reduces monotonically. Hence the strength of the shock reduces gradually as it progresses inward. The variation

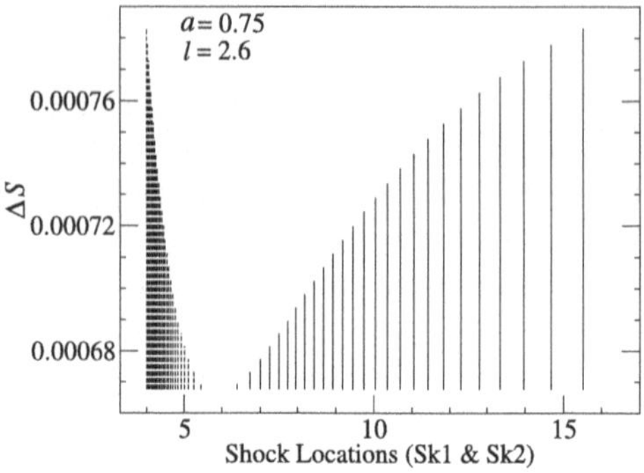

Fig. 5.1 In this figure, we plot the change in entropy accretion rate (S) at different shock locations. The shock locations Sk1 (dashed) and Sk2 (solid) are presented. The length of the vertical line decreases gradually, the change in entropy also decreases, hence shocks become weaker. Both shocks tend to merge towards each other

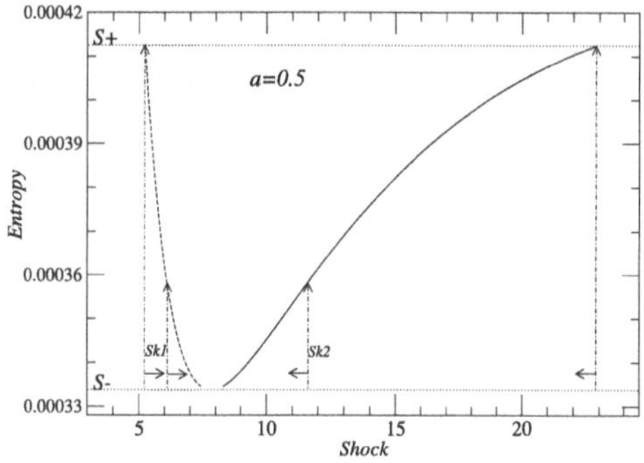

Fig. 5.2 In this figure, we plot the envelope of entropy change to show how the strength of the shock decreases as shock Sk2 moves inside and Sk1 moves outside, as a result, $\Delta\mathcal{E}$ increases. The energy loss from corona is regulated by the entropy change during shock transition

of the entropy gain (the strength) during shock transition is presented in Fig. 5.1 at various shock locations. In Fig. 5.2, we plot the envelope of entropy change to show how the strength of the shock decreases as shock moves. The solid vertical line represents the change in entropy accretion rate (S) as the shock Sk2 moves. The magnitude of this change decreases monotonically as a result of increasing $\Delta\mathcal{E}$ indicating that the entropy gain ΔS decreases. The top- and bottom points of the straight lines correspond to pre-shock and post-shock entropies respectively. The pre-shock energy is the same therefore all the bottom points of the vertical lines remain on the same horizontal line. With the propagation, the shocks become more and more weaker. The weakest shock forms (smallest vertical line on left) when $\Delta\mathcal{E}$ becomes maximum and after that no further shock appears in the flow. The position where the extreme shock forms, ΔS may or may not be zero. Figures 5.1 and 5.2 show a pair of shocks appearing symmetrically on the opposite ends of the middle sonic point, one is stable (Sk2), another is unstable (Sk1). As energy dissipates from the post-shock corona, both shocks Sk1 and Sk2 move towards each other (see Fig. 5.2), and finally merge together at the middle sonic point. Considering

Sk1, a systematic evaluation of the shock Sk2 and it's extreme location is obtained accurately [25, 26].

Further, as seen the $\Delta\mathcal{E}$ depends on the amount of entropy gain $\Delta S > 0$. As ΔS decreases, $\Delta\mathcal{E}$ increases. The maximum $\Delta\mathcal{E}$ occurs at ΔS minimum value. If this minimum value $\Delta S = 0$, then $\Delta\mathcal{E}$ becomes maximum at that angular momentum. For the other inflow energies, the change in entropy ΔS is minimum and energy loss reaches a local maxima at that inflow energy. We will see later that the local maxima resides at the top most points on each individual energy dissipated curve whereas $\Delta S = 0$ is the tallest point among all such curves (see in Fig. 5.5).

As discussed, the energy dissipation and the change of post-shock entropy are deeply co-related but the relation is not linear and simple. We also need to observe the location of the closest shock in order to study the energy dissipation systematically with the entropy gain. At this closest location, even if $\Delta S \neq 0$, no further dissipation occurs. This part of ΔS is no longer useful to release energy from the post-shock region. This is because extreme shock locations can not be compressed further. This is clearly shown in Fig. 5.3. Here we plot the

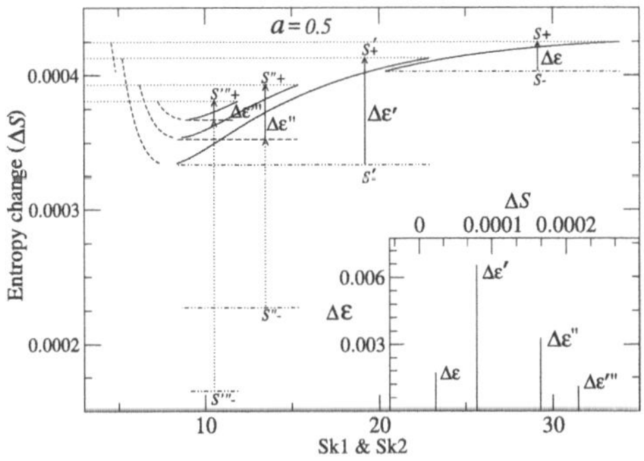

Fig. 5.3 In this figure, we study the pre-shock and post-shock entropies with the variation of the shock location Sk1 (dashed) and Sk2 (solid) at 4-different inflow energies $\mathcal{E}_{in} = 1.0075, 1.006555, 1.005, 1.004$ for $l = 2.934$. The pre-shock entropies are marked by horizontal dotted-dashed line, shown by S_-, S'_-, S''_-, S'''_-, and the post-shock entropies are marked by dotted horizontal line, shown by S_+, S'_+, S''_+, S'''_+. At $\mathcal{E}_{in} = 1.0075, 1.006555$, $S_+ - S_- = \Delta S$ and $S'_+ - S'_- = \Delta S'$ are shown by a solid vertical arrow line only. For $\mathcal{E}_{in} = 1.005$, and 1.004, $S''_+ - S''_- = \Delta S''$ and $S'''_+ - S'''_- = \Delta S'''$ are shown by a dotted vertical line along with a solid vertical arrow line. As Sk2 moves inside, the post-shock entropy decreases along the solid curve and reaches the lower point of the curve, denoting the minimum value of the post-shock entropy. The dotted vertical arrow lines indicate difference between the pre- to the post-shock entropy gain at minimum and solid vertical arrow lines indicate the decrease in entropy gain when shock moves inward. When shock reaches its closest position (minimum of the curve), energy dissipation becomes maximum and then it falls rapidly. The entropy gain for $\mathcal{E}_{in} = 1.005$, and 1.004 is not effective to release the energy because shock can not move inside further. The minimum value of post-shock entropy merges with pre-shock entropy at $\mathcal{E}_{in} = 1.006555$, thus dissipation reaches global maximum. The entropy change and the energy dissipations are not linear. This non-linear behavior is shown in the sub-panel of the figure. For the variation of inflow energies from the higher to the lower value, $\Delta\mathcal{E}$ increases initially with the increases of ΔS, and reaches a maximum value when $\Delta S = 0$ at $\mathcal{E}_{in} = 1.006555$. After that $\Delta\mathcal{E}$ decreases with increases of ΔS

pre- and post-shock entropy difference with the variation of the shock location Sk1 (dashed) and Sk2 (solid). We choose four-different pre-shock ($\mathcal{E}_{in}$ = 1.0075, 1.006555, 1.005, 1.004) at l = 2.934 (chosen). The pre-shock entropies are labeled by S_-, S'_-, S''_-, S'''_- and denoted by horizontal dotted-dashed line. The corresponding post-shock entropies are S_+, S'_+, S''_+, S'''_+ and denoted by dotted horizontal line at those energies. At $\mathcal{E}_{in}$ = 1.0075, 1.006555, we denote $\Delta S(= S_+ - S_-)$ and $\Delta S'(= S'_+ - S'_-)$ respectively (represented by solid vertical arrow line). For the other two $\mathcal{E}_{in}$ = 1.005, and 1.004, $\Delta S''(= S''_+ - S''_-)$ and $\Delta S'''(= S'''_+ - S'''_-)$ respectively and represented by dotted vertical line along with a solid vertical arrow line.

In all four cases, the solid curves of the respective inflow energies show the decreases in the post-shock entropy when shock moves inward. The minimum value of the post-shock entropy is thus the lower point of the curves. This minimum value of the post-shock entropy generally does not equal to the pre-shock entropy, i.e., $\Delta S \neq 0$, except for $\mathcal{E}_{in}$ = 1.006555 for which $\Delta S' = 0$. The difference between the pre- to the minimum of the post-shock entropy is represented by dotted vertical arrow lines and the amount of allowed entropy decrease is represented by solid vertical arrow lines.

In Fig. 5.3, we see that when $\mathcal{E}_{in}$ = 1.0075 is high, the entropy gain ΔS reduces and becomes zero. But shock is not the closest in this case. The curves are truncated at S_-, and below this point no shock form. The energy dissipation increases initially as Sk2 propagates inside, however the $\Delta\mathcal{E}$ is small. An interesting case is discussed at a moderate value of $\mathcal{E}_{in}$ = 1.006555. Here $\Delta S'$ is also moderate and can reduce to $\Delta S' = 0$. We see that this occurs when (1) the closest shock forms, both shocks coincide at the minimum of the curve and (2) pre- and post-shock entropies become closer and finally coincide. For $\mathcal{E}_{in}$ = 1.006555, these two conditions are guaranteed, the dissipated energy $\Delta\mathcal{E}'$ becomes highest. For other $\mathcal{E}_{in}$ dissipations are less. Thus, this highest dissipation at a given angular momentum can be obtained at an intermediate energy when inflow energies are varied. At $\mathcal{E}_{in}$ = 1.005, and 1.004, the $\Delta S''$ and $\Delta S'''$ is higher, however the dissipated energy $\Delta\mathcal{E}''$ and $\Delta\mathcal{E}'''$ is lower and reduce gradually. In this case, the pre- and post-shock entropies do not merge to the same value even though both Sk1 and Sk2 coincide together (form closest/weakest shock). Hence $\Delta S'' \neq 0$ or $\Delta S''' \neq 0$. The minimum value of the post-shock entropy is represented by new dotted horizontal lines, below this no dissipation occurs during shock transition. The solid vertical arrow lines represent the effective entropy loss. The magnitude of the energy loss $\Delta\mathcal{E}''$ and $\Delta\mathcal{E}'''$ however reduces with the decrease of inflow energy successively although the total entropy change increases.

To summarize it, we see that for the variation of inflow energies from the higher to the lower value, the dissipation $\Delta\mathcal{E}$ increases initially with the increases of ΔS, and reaches a maximum in between (where $\Delta S = 0$ and exhausted fully). Thereafter $\Delta\mathcal{E}$ decreases however ΔS increases. This entropy increase is not effective in energy dissipations and remains as a residual. This is also shown in the sub-panel of Fig. 5.3 for the same set of parameters. Thus, the entropy change and the energy dissipations are not linearly correlated.

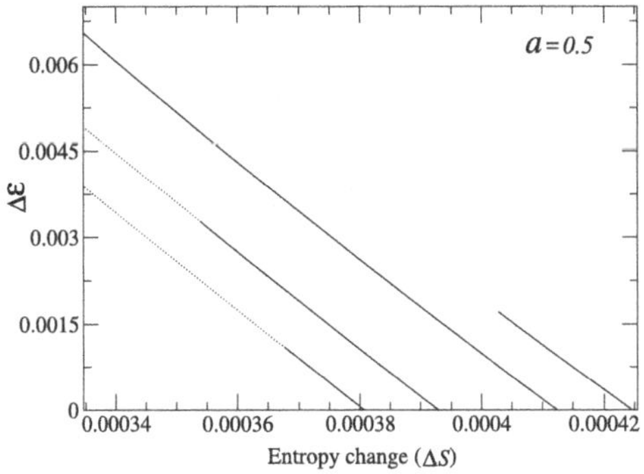

Fig. 5.4 In this figure, we also present the variation of the dissipated energy $\Delta\mathcal{E}$ with the entropy change at different inflow energies. For a fixed inflow energy, the energy emission increases linearly as the entropy change decreases and is maximum at the closest shock. The dotted lines indicate that no emission generated after the closest approach of shock. The top most points on the lines represent pre-shock entropy S_- and bottom points on the x-axis represent post-shock entropy S_+. We see that energy dissipation $\Delta\mathcal{E}$ is maximum at an intermediate value of ΔS (at $\mathcal{E}_{in} = 1.006555$)

For more clarity, we draw Fig. 5.4 where we plot the energy dissipations vs. the entropy change in the four different curves for the same set of inflow energies. We see that for a given inflow energy, the $\Delta\mathcal{E}$ increase linearly as ΔS decreases. For the second curve (from right), it becomes maximum when shock is closest. Thereafter $\Delta\mathcal{E}$ decreases for the other two curves. Here the dotted lines are drawn to indicate that no emission is possible even though ΔS increases successively. The top most points on the lines represent pre-shock entropy S_- and bottom points on the x-axis represent post-shock entropy S_+. From both the figures, we see that $\Delta\mathcal{E}$ is maximum at an intermediate value of ΔS i.e., it occurs at an intermediate value of inflow energy (here $\mathcal{E}_{in} = 1.006555$, $l = 2.934$, and $a = 0.5$) where the ΔS gradually reduces to zero. This parameter ($l - 2.934$, $\mathcal{E}_{in} = 1.006555$) is therefore identified as a critical parameter in the flow for which energy dissipation $\Delta\mathcal{E}$ is maximum. Therefore, for a fixed angular momentum, there exists a critical point for which flow dissipates maximum energy. The value of $\mathcal{E}_{in}$ and the magnitude of $\Delta\mathcal{E}$ maxima at the critical point changes with the inflow angular momentum and spin of the black hole.

5.3 The Parameters for Maximum Dissipation from Corona

The corona emits a part of the post-shock bulk energy. With the emission the post-shock flow entropy reduces gradually. In case of a radiative shock, this emitted energy $\Delta\mathcal{E}$ depends on $(\mathcal{E}_{in}, l, a)$. The values of $\mathcal{E}_{in}$ at the boundaries of SA-region $\Delta\mathcal{E} = 0$ (both for lower and upper boundaries) and exhibits a maximum $\Delta\mathcal{E} = \Delta\mathcal{E}_l$ in between two boundaries. To demonstrate it clearly, we plot the emission curves

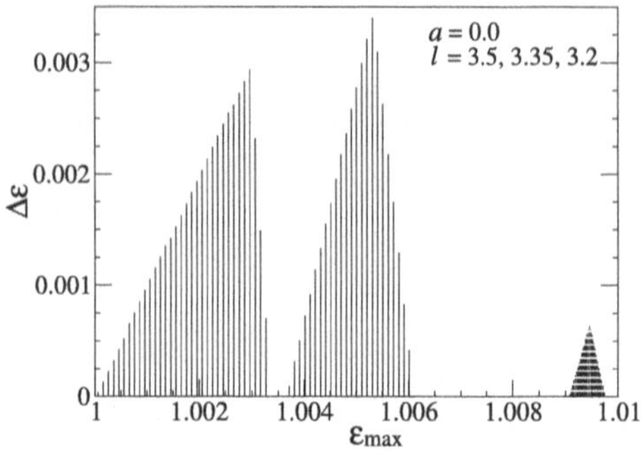

Fig. 5.5 In this figure, the variation of $\Delta\mathcal{E}$ is shown in terms of the inflow energy $\mathcal{E}_{in}$ in the energy range between the upper and lower boundaries of the SA-region at three different values of l. The $\Delta\mathcal{E}$ is zero at both ends of the upper and lower boundaries of the SA-region. The energy loss is elevated with the decrease of energy $\mathcal{E}_{in}$ initially. It reaches a maximum value and then again decreases. The maximum energy loss generated at intermediate l and depends on the flow parameters

at different inflow energies in Fig. 5.5 at a given angular momentum shown in the figure. At both ends $\Delta\mathcal{E} = 0$ and the tallest line in-between represents maximum dissipation $\Delta\mathcal{E}_l$ at that angular momentum. The parameters for which this maximum dissipation happens are the parameters of 'max-emission' $(\mathcal{E}_l, l)$. Therefore, we determine the all possible 'max-emission' parameters $(\mathcal{E}_l, l)$ and variation of $\Delta\mathcal{E}_l$ for a given spin a, by exploring the all inflow energies $\mathcal{E}_{in}$ from the upper to the lower boundary of the SA-region in the entire range of l. To see the variation of $\Delta\mathcal{E}_l$, we observe the peaks of these tallest curves in Fig. 5.5. At low angular momentum, peak is low and is highest at intermediate l values. At high l and low energy region, $\Delta\mathcal{E}_{in}$ tends to ($\mathcal{E}_{in}-$ rest mass energy) which means that the dissipation $\Delta\mathcal{E}_{in}$ tends to 100%, i.e., the post-shock flow dissipates all of its thermal and kinetic energy. Thus, the percentage of dissipation is maximum for high angular momentum and low energy regions, and the magnitude of dissipation $\Delta\mathcal{E}$ becomes maximum at an intermediate value of angular momentum.

We then identify all the 'max-emission parameters' $(\mathcal{E}_l, l)$ for different l values at which energy dissipation is maximum and find a 'max-emission curve' for the entire range of flow parameters in the SA-region. The values of these 'max-emission parameters' $(\mathcal{E}_l, l)$ are shown in Fig. 5.6. This is also calculated by varying the spin of the black hole and obtaining the same curve for different spins $a = 0.999, 0.9, 0.75, 0.5, 0, -0.5, -0.999$ [25]. In the figure, we see that maximum energy dissipation occurs when the flow parameters close to the upper boundary of the SA-region. A vertical line is drawn where the maximum dissipation curve shows a deep. The point at which this deep occurs is known as 'critical point' of the curve. The parameters at the critical point are known as 'critical flow parameters' $(\mathcal{E}_{cr}, l_{cr})$. The critical flow parameters vary only with the spin of the black hole. It moves toward the lower angular momentum and high energy region, as the SA-regions or upper boundary also follow the same shift [25].

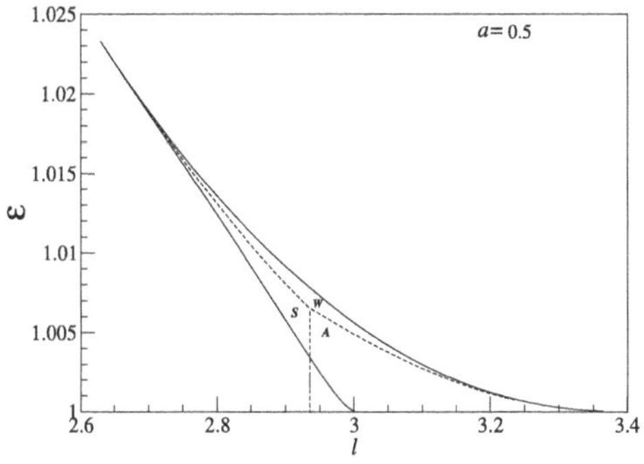

Fig. 5.6 In this figure, we plot the maximum energy dissipation curve in the SA-region for $a = 0.5$. The curve gives the value of the inflow energy $\mathcal{E}_{in}$, l at which dissipations become maximum. At both ends (low and high l) the curve almost merges with the upper boundary and the deviation increases at the middle from where we draw a vertical line. This triple point is known as the critical point which lies almost at the center of geometry (CG) of the SA-region. The parameters at the critical point is denoted as critical flow parameters

We draw Fig. 5.7 to show the variation of magnitude of the maximum dissipation of energy at various inflow energies $\mathcal{E}_{in} = \mathcal{E}_l$ taken from the maximum emission curve for $a = 0.5$. We see that in region I, $\Delta\mathcal{E}$ increases monotonically with the increase of $\mathcal{E}_{in}$, becomes maximum at the critical point and then in region III, it falls rapidly at the high energy region. This is very similar to the variation of fast rise and exponential decay (FRED) curves. In both the high and low energy the dissipation is low and at the critical point the dissipation becomes maximum, denoted by a dotted vertical line. In referring to Fig. 5.5, we see that the dissipation is insignificant indicating that high inflow energy and low angular momentum region does not emit much of its total bulk energy. This is because the shock location moved to the closest point at the critical point i.e., the size of the corona is minimum. In the low energy and high angular momentum region, shock initially forms at a large radius. As we move towards the high inflow energy and low angular momentum region along the maximum dissipation curve, shock moves inside and dissipation increases monotonically. Shock reaches the closest point at the critical parameters, the dissipation becomes maximum. After that the dissipation decreases gradually at the high energy and low angular momentum region because the corona becomes minimum in size as the shock remains in the closest location. This variation of the shock location is presented in the sub-panel of Fig. 5.7 for the entire range of 'max-emission' inflow parameters. We here see that initially energy dissipation increases gradually with the decrease of shock location, becomes maximum at the extreme shock location at $10\,r_g$ for $a = 0.5$ and finally stays in the same location from where the dissipation falls exponentially. For the moderate spin, the minimum corona radii are close to the $10\,r_g$ and depend on the Kerr parameter a only [25].

Generally during outbursts, with the emission of the radiation the corona shrinks progressively on successive days as the spectral state changes from hard, hard-intermediate, soft-intermediate to soft states. This study will help us to find out

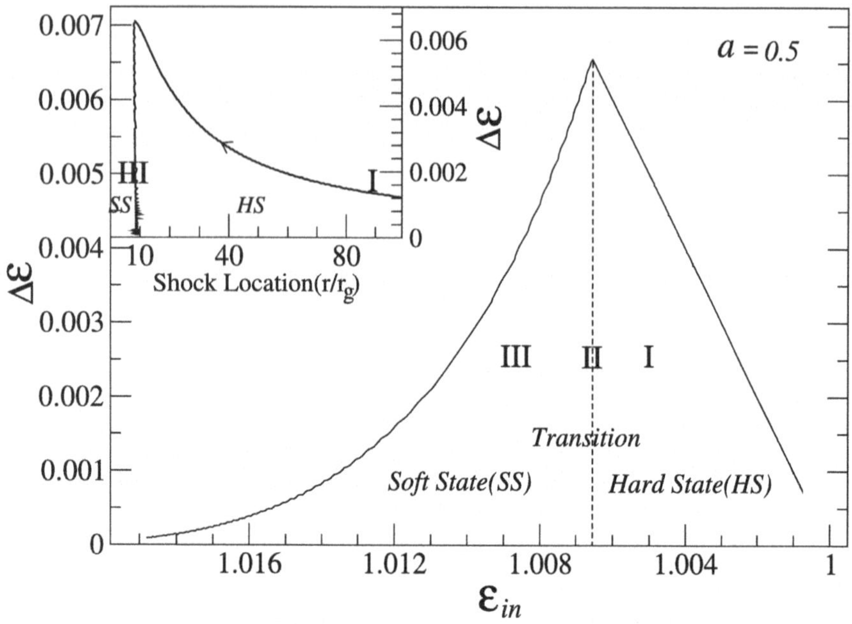

Fig. 5.7 Here, we plot how the maximum energy dissipation varies with the inflow energy $\mathcal{E}_{in}$. We see that with the increase of $\mathcal{E}_{in}$, $\Delta\mathcal{E}$ increases monotonically, becomes maximum at the critical point and then falls exponentially. At low (region-I) and high (region-III) dissipated energy is low and becomes high at the intermediate values of $\mathcal{E}_{in}$. The dissipation maxima is shown by a vertical dotted line at the critical point (region-II). At the critical point the energy dissipation reaches maximum. It is a global maximum for the entire SA-region for a given spin ($a = 0.5$). The variation of the shock location is shown in the sub-panel of the figure. The shock location becomes closest at the critical point of maximum dissipation. At the high energy and low angular momentum region dissipation falls exponentially because shock can not move further inside from $10\, r_g$

the variation of corona geometry with the emitted radiation from post-shock corona and to predict the spectral state transition, in particular for the FRED sources.

5.4 Possibility of Spectral State Transition

We now focused on the spectral transition from harder to relatively softer states. The present study opens up certain feasibility on this issue. Generally, harder (Hard and Hard Intermediate) to softer (Soft Intermediate and Soft) state transition is explained by the accretion rate when it crosses a critical value [28]. With the increasing observational evidence, this one-parameter canonical scheme faces a number of challenges and discrepancies. For example, it is not clear what is the critical value of the accretion rate for a source to trigger a transition. Further despite of having substantial brightness or enough accretion rate, a number of *"failed outbursts"* in BHBs (e.g., SwJ1745, GRS 1716-249 during its 2016-17, Swift J1753.5-0127 during its 2015 outburst, 2005 outburst of swift J1753.5-0127 etc.) is seen which never exhibited soft state. In addition, it is also not clearly understood why for a few transient sources having longer orbital periods (e.g. XTE J1550-564), the accumulated mass is relatively large, however it produces small outbursts. On the contrary, the transitions have been also seen for the sources with a more or less constant mass accretion rate (GS 2023-338 [29–31] and Cyg X-1 [32]). Furthermore, even for the complete outbursting sources, the size of the Comptonizing medium and its correlation with the spectral hardness, has not been established in order to explain the q-diagram theoretically. All this evidence indicates that mass accretion rate can not be a sole parameter to tiger a transition. The relative size of the corona or the geometry of varying corona could possibly be the second parameter to trigger a transition [33]. Both of these physical parameters may vary to a large extent independently, however the change in one parameter may influence the change in other. The nonlinear correlation between them however may explain this two-dimensional hysteresis behavior (hardness ratio 'q'-diagram).

From our results we see that as the day progresses, the shock moves in and its strength decreases. The relative size of the hot Comptonized region shrinks progressively as spectral state changes on successive days from hard, hard-intermediate, soft-intermediate state to soft states [4, 34, 35]. Such scenarios are possible to explain using dissipative shock and one can determine the physical parameters, namely the critical parameters for which such transition occurs. From Fig. 5.7, we see that dissipation/luminosity elevated monotonically decreasing corona size. This is similar to the rising phase of outburst seen in the *harder state*. The dissipation becomes maximum and then starts to fall when the size of the corona reduces to minimum. The transition starts here. The spectral state transition from harder state to the softer state occurs resembling the rising to decline phase of the outburst [36, 37]. Once the decline phase starts, luminosity decays rapidly in the *Softer state* [38–41] since during that period the smallest corona already formed. The shock reached the closest location and stayed there. At the low angular momentum region, dissipation decreases rapidly because shock does not compress the corona further. Thus a large corona gradually becomes smaller in size. Thus the hot corona exhausted gradually by emitting a substantial part of available bulk energy from where the emission of hard flux started to decline and thus cooled down significantly. An increase of accretion rate accelerates the cooling processes faster. Thus, the transition to a softer state is favorable for the sources [25].

However, in order to exhibit spectral transitions, the flow parameters must have varied. The viscosity causes the flow parameters to evolve. For the outbursting sources viscosity varies at the outer edge of the disk and if the viscosity is higher than a critical value then a flow injected from far away would become a low-angular momentum inward by transporting flow angular momentum outside which indeed are favorable for the transition at the critical point. When the parameters evolve and reach to the critical point, the transition occurs. The spin further enhances the dissipation, reducing the corona size faster [25].

As shock moves inside, compressing the corona, the disk extends largely up to the interior minima which then predominantly produces a multi-colored black body spectra ([38] and ref. therein). In such a situation if the accretion rate is enhanced, the flux of soft photons goes up rapidly. This results in softening the spectrum faster and thus has achieved a softer state satisfactorily.

5.5 Discussions

Transient sources have a few interesting properties: the luminosity variation in rising and decline phase, the spectral state transitions and the QPOs evolution etc. are predominantly observed. So far we try to explain the state transition from harder to softer state. The intermediate states (Hard-Intermediate or and Soft-Intermediate) or the hardness intensity 'q'-shape diagram could also possibly be obtained by simultaneously relaxing the energy condition along with the mass conservation relation. Since most of the sources are often seen to produce outflow/jets in the intermediate state.

Moreover, for transient sources, QPOs evolve monotonically on a daily basis [6, 7], frequency increases during rising phase and decreases in the decline phase of outburst. Possibly there is a one to one correlation between the shock location and QPO frequency [36, 40, 42]. The propagating dissipative shock discussed here can be helpful to understand a systematic evolution of these QPOs [24]. Once outburst is set in, the size of the Compton cloud decreases gradually and the shock strength also decreases. The propagation of the shock excites oscillations in the boundary layer of discontinuity [26]. The radial and vertical oscillations of the post shock boundary layers may excite the quasi-periodic oscillations (QPOs) in the emitted X-rays radiation [43–47]. The systematic evolution of the Compton cloud with the variation of emitted radiation (first rise and then falls) can demonstrate the origin of QPOs evolution satisfactorily in both the rising and declining phases of outburst [24, 25].

Furthermore, conventionally the models, such as diskbb+PL, the disk inner edge is estimated from the normalization, or uses inner-most stable circular orbit (ISCO) radius for fitting the spectral data or taken as a free parameter. In the proposed mechanism, the boundary of the post-shock region is identified as the inner radius of the disk which varies in progression with the cooling of the corona. The minimum value of the inner disk radius can be obtained consistently [25] and dissipative shock provides an independent estimation of this value.

Finally, implementation of dissipative shock over the non-dissipative propagating shock perhaps gives a better description for TCAF, particularly in spectral studies, QPO evolution, inner disk radius etc. Since it is coupled with the variation of the emitted flux and thus can explain the evolution of the spectral parameters namely the size of the Compton cloud, shock location, compression ratio which decides the optical depth and hence the hardness-ratio etc. more accurately [25]. Our new description presented in the relativistic framework, one can extract the physical parameters independently. The estimated values could accurately provide the compatibility and consistency of two-component advective accretion flow (TCAF) in demonstrating the spectral analysis for a large number of sources [7, 36–42, 48] (and reference therein).

Thus the present work is significantly important for the transient black hole binaries to demonstrate the one to one correlation between the radiative loss, the change in the spectral states, QPOs evolution, etc. with the progressive evolution of corona and to find the physical parameters. The dissipative shock dispenses a unique mechanism to infer the inference of the spin and to demonstrate the precise nature of the accretion dynamics of the outbursting sources.

Acknowledgments Authors acknowledge IUCAA, Pune for their support through the 'Visiting Associateship Program'.

References

1. Klis, M. V. D. (2005). *Astronomische Nachrichten*, *326*, 798.
2. Homan, J., & Belloni, T. (2005). *Astrophysics and Space Science*, *300*, 107.
3. Chakrabarti, S. K., & Titarchuk L. G. (1995). *Astrophysics Journal*, *455*, 623.
4. Narayan, R., & McClintock, J. E. (2008). *New Astronomy Reviews*, *51*(10–12), 733–751.
5. Brandt, W. N., & Alexander, D. M. (2015). *Astronomy and Astrophysics Review*, *23*, 1–93.
6. Dutta, B. G., & Chakrabarti, S. K. (2016). *Astrophys. J.*, *828*, 101.
7. Dutta, B. G., Pal, P. S., & Chakrabarti, S. K. (2018). *Monthly Notices of the Royal Astronomical Society*, *479*, 2183.
8. Das, S., Chakrabarti S. K., & Mondal S. (2010). *Monthly Notices of the Royal Astronomical Society*, *401*, 2053.
9. Singh, C. B., & Chakrabarti S. K. (2011). *Monthly Notices of the Royal Astronomical Society*, *410*, 2414.
10. Singh, C. B., & Chakrabarti S. K. (2011). *International Journal of Modern Physics D*, *20*, 2507.
11. Singh, C. B., & Chakrabarti S. K. (2012). *Monthly Notices of the Royal Astronomical Society*, *421*, 1666.
12. Fragile, P. C., Gillespie A., Monahan T., Rodriguez M., Anninos P. (2012). *Astrophysics Journal Suuplement Series*, *201*, 9.
13. Sarkar, B., & Das S. (2013). In *Astronomical Society of India Conf. Seri. 8*, Recent Trends in the Study of Compact Objects: Theroy and Observations, ed. S. Das, A. Nandi, & I. Chattopadhyay, 143.
14. Mondal, S., Chakrabarti, S. K., & Debnath, D. (2014). *Astrophysics and Space Science*, *353*, 223.
15. Nagarkoti, S., & Chakrabarti S. K. (2016). *Monthly Notices of the Royal Astronomical Society*, *462*, 850.
16. Fukue, J. (2019). *Monthly Notices of the Royal Astronomical Society*, *483*, 3839–3850.
17. Fukue, J. (2019). *Monthly Notices of the Royal Astronomical Society*, *483*, 2538–2546.
18. Zeldovich, Y. B., & Raizer, Y. P. (1967). In W. D. Hayes & R. F. Probstein (Eds.), Academic Press. 1966/1967. Physics of Shock Waves and High-Temperature Hydrodynamic Phenomena.
19. Shull, J. M., & Draine, B. T. (1987). In *Interstellar Processes; Proceedings of the Symposium, Grand Teton National Park, WY, July 1-7, 1986 (A88-14501 03-90)* (pp. 283–319). D. Reidel Publishing Co.

20. Drake, R. P. (2007). *Physics of Plasmas*, *14*, 043301.
21. Tolstov, A., Blinnikov, S., Nagataki, S., & Nomoto, K. (2015). *Astrophysics Journal*, *811*, 47.
22. Ferguson, J. M., Morel J. E., & Lowrie R. B. (2017). *Journal of Quantitative Spectroscopy and Radiative Transfer*, *202*, 176.
23. Lu, J. F., & Feng, Y. (1997). *Publications of the Astronomical Society of Japan*, *49*, 525–533.
24. Mondal, S., & Basu, P. (2020). *Monthly Notices of the Royal Astronomical Society*, *497*, 21192132.
25. Mondal, S. (2023). *Astrophysics Journal*, *950*, 26.
26. Mondal, S. (2010). *Astrophysics Journal*, *708*, 1442.
27. Mondal, S., & Chakrabarti S. K. (2006). *Monthly Notices of the Royal Astronomical Society*, *371*, 1418.
28. Abramowicz, M. A., Chen, X., Kato, S., Lasota, J. P., & Regev, O. (1995). *The Astrophysical Journal*, *438*, 37.
29. Sunyaev, R. A., et al. (1991). *Soviet Astronomy Letters*, *17*, 123.
30. Terada, K., Miyamoto, S., Kitamoto, S., Tsunemi, H., & Hayashida, K. (1992). In Y. Tanaka (Ed.), Frontiers of X-ray Astronomy (p. 323). Universal Academy.
31. Miyamoto, S., Kitamoto, S., Iga, S., Negoro, H., & Terada, K. (1992). *Astrophysics Journal*, *391*, 21.
32. Zhang, S. N., Cui, W., Harmon, B. A., Paciesas, W. S., Remillard, R. E., & van Paradijs, J. (1997). *Astrophysics Journal*, *477*, L95.
33. Homan J., Wijnands R., van der Klis M., & Belloni T. (2001). *Astrophysics Journal Supplement Series*, *132*, 377.
34. Manmoto, T. (2000). *The Astrophysical Journal*, *534*, 734.
35. Harris, D. E., & Krawczynski H. (2006). *Annual Review of Astronomy and Astrophysics*, *44*, 463–506.
36. Debnath, D., Mondal, S., & Chakrabarti, S. K. (2015). *Monthly Notices of the Royal Astronomical Society*, *447*, 1984.
37. Jana, A., Debnath, D., Chatterjee, D., Chatterjee, K., Chakrabarti, S. K., Naik. S., Bhowmick, R., & Kumari, N. (2020). *The Astrophysical Journal*, *897*, 3.
38. Chakrabarti, S. K. (2018). In M. Bianchi, R. T. Jansen, & R. Ruffini (Eds.), *Proc. of The Fourteenth Marcel Grossmann Meeting 2015* (pp. 369–384). World Scientific Publishing, ISBN 9789813226609.
39. Mondal, S., Debnath, D., & Chakrabarti, S. K., *The Astrophysical Journal*, 786, 4
40. Debnath, D., Chakrabarti, S. K. & Nandi, A. (2013). *Advances in Space Research*, 52, 2143.
41. Nandi, A., Debnath, D., & Mandal, S., et al. (2012). *Astronomy & Astrophysics*, 542, 56.
42. Debnath, D., Chakrabarti, S. K. & Nandi, A. (2010). *Astronomy & Astrophysics*, 520, 98.
43. Kato, S., & Fukue, J. 1980, PASJ, 32, 377.
44. Nowak, M., & Lehr D. (1999). Theory of Black Hole Accretion Disks, ed. M. A. Abramowicz, G. Bjornsson, J. E. Pringle (Cambridge: Cambridge Univ. Press).
45. Abramowicz, M. A., Karas, V., Kluźniak, W., Lee, W., & Rebusco, P. (2003). PASJ, 55, 46.
46. Fragile, C., Straub O & Blaes O. (2016). *Monthly Notices of the Royal Astronomical Society*, 461, 1356.
47. Dewberry, J. W., Latter H. N., & Ogilvie G. I. (2018). *Monthly Notices of the Royal Astronomical Society*, 476, 4085.
48. Mondal, S., Chakrabarti, S. K., Nagarkoti, S., & Arevalo, P. (2017). *The Astrophysical Journal*, *850*, 47.

Chapter 6
Extreme Magnetic Fields Around Black Holes

Koushik Chatterjee, Matthew Liska, Alexander Tchekhovskoy, Sera Markoff, and Ramesh Narayan

Abstract Recent results of the event horizon-scale images of M87* and Sagittarius A* from the Event Horizon Telescope Collaboration show that strong magnetic fields are likely present around the central black holes (BHs) in these sources. Magnetically arrested disks (MADs), the end stage of magnetic flux saturation around BHs, are especially rich in horizon-scale physics due to the presence of powerful jets and magnetic flux eruptions that provide significant feedback on the accretion mechanism. Here, we present an overview of our current knowledge about the magnetic field evolution in numerical simulations of accreting BHs, focusing on jet launching, black hole-interstellar medium feedback, and black hole imaging of MADs. We find that misaligned MAD accretion flows seemingly exhibit jet ejection cycles that could produce flaring states in radio-quiet active galactic nuclei. Further, we show that advances in horizon-scale interferometric telescopes could identify disk misalignment by imaging the disk-jet connection region.

K. Chatterjee (✉)
Institute for Research in Electronics and Applied Physics, University of Maryland, College Park, MD, USA
e-mail: kchatt@umd.edu

M. Liska
Center for Relativistic Astrophysics, Georgia Institute of Technology, Atlanta, GA, USA
e-mail: mliska3@gatech.edu

A. Tchekhovskoy
Center for Interdisciplinary Exploration & Research in Astrophysics (CIERA), Physics & Astronomy, Northwestern University, Evanston, IL, USA
e-mail: atchekho@northwestern.edu

S. Markoff
Anton Pannekoek Institute for Astronomy, University of Amsterdam, Amsterdam, Netherlands
e-mail: S.B.Markoff@uva.nl

R. Narayan
Black Hole Initiative at Harvard University, Cambridge, MA, USA
e-mail: rnarayan@cfa.harvard.edu

S. Ghosh, A. R. Rao (eds.), *The Relativistic Universe: From Classical to Quantum*, Astrophysics and Space Science Proceedings 61,
https://doi.org/10.1007/978-3-031-90186-7_6

6.1 Introduction

While the idea of magnetically dominated accretion flows has been around for at least 50 years [1], this mode of accretion, dubbed magnetically arrested disks (MADs), truly became a subject of interest over the past 2 decades or so [2, 3]. Part of the interest stemmed from advances in computational development of 3D general relativistic magnetohydrodynamic (GRMHD) codes that enabled self-consistent simulations of black hole (BH) accretion. On the observational front, the multitude of Sagittarius A* (or Sgr A*) X-ray flares captured by the Chandra observatory [4–6] provided indications of particle acceleration, possibly driven by reconnection of magnetic fieldlines close to the BH [7, 8]. Figure 6.1 shows that our standard picture of magnetorotational instability (MRI)-driven accretion produces X-ray and near-infrared (nIR) flares too weak to explain some of Sgr A*'s brightest flares [6, 9], suggesting the need for stronger magnetic fields. More recently, the polarized horizon-scale EHT images of M87 and Sgr A* also favor strong vertical magnetic fields [10, 11]. Indeed, the MAD state could be a natural outcome of starved BHs (i.e., accreting at highly sub-Eddington rates), which is the case for both Sgr A* and M87. There is seemingly an abundant supply of poloidal magnetic field either by advecting μGauss-level magnetic fields from the interstellar medium (ISM) [3, 12] or from magnetized stellar winds (as thought to occur in Sgr A*; [13]), or by in-situ generation of poloidal flux via dynamo processes in the disk [14], favoring the emergence of the MAD state. Motivated by these advances in theory, numerics and observations, we discuss the many features of MAD accretion flows and their

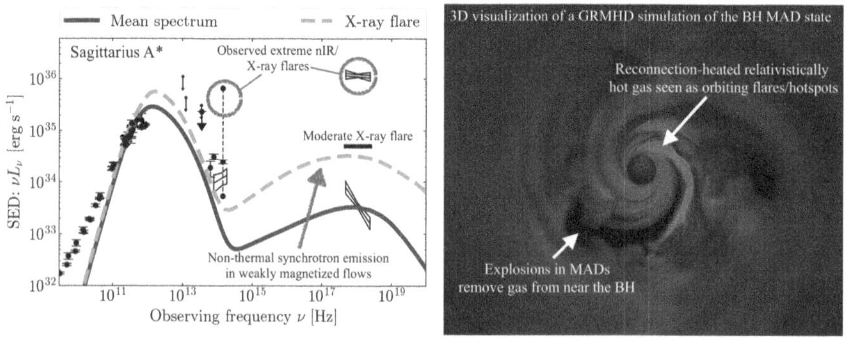

Fig. 6.1 (*Left*) Magnetized gas turbulence triggers particle acceleration that leads to simultaneous moderate nIR/X-ray flares, but cannot explain the most energetic flares [19]. We show a multi-wavelength SED (solid and dashed lines) of Sagittarius A* from a GRMHD simulation of a weakly magnetized accretion flow around a spinning BH assuming synchrotron emission from both thermal and non-thermal (i.e. accelerated) electron populations. Sgr A* data is shown in black (see [19] for more details). Note that we ignore synchrotron self-Compton calculations in this SED. (*Right*) Magnetically arrested disks (MADs) exhibit magnetic explosions that heat up gas (red) in the disk (green/blue), a mechanism favorable for producing extreme high-energy flares (Color figure online)

outflows, with a focus on recent results of misaligned MAD flows from Chatterjee et al. [15].

6.2 Revisiting Magnetically Arrested Accretion Flows

In this section, we introduce some salient features of the MAD state. We divide our discussion into two parts: time-averaged and time-dependent disk/jet properties. The former category of analysis allows us to understand broad features of MADs and cross-compare between multiple simulations, while the latter involves tracking the gas and magnetic field dynamics over time with the aim of studying the evolution of magnetic explosions often seen in MADs, and their effect on jet morphology. For our discussion, we make use of results from MAD simulations, performed using the GRMHD code `H-AMR` [16], from a number of published papers [15, 17, 18]. These results are largely consistent with simulations from other codes.

We focus on three quantities: the accretion rate $\dot{M} = -\iint \rho u^r dA_{\theta\varphi}$, the poloidal magnetic flux $\Phi_{\rm BH} = (1/2)\iint |B^r| dA_{\theta\varphi}$ calculated at the event horizon, and the total outflow power $P = \dot{M}c^2 - \dot{E}$. For these quantities, we require the gas density ρ, the radial component of the gas 4-velocity u^r, the radial component of the 3-magnetic field vector B^r and the total energy flux in the radial direction $\dot{E} = \iint T^r_t dA_{\theta\varphi}$, defined using the stress-energy tensor T^μ_ν and the area element $dA_{\theta\varphi} = \sqrt{-det(g_{\mu\nu})} d\theta d\varphi$. We assume natural units $G = c = 1$, which implies that our characteristic length-scale, the gravitational radius $r_{\rm g} = GM/c^2$, and the corresponding light-crossing time-scale $t_{\rm g} = r_{\rm g}/c$, are normalised to the BH mass M. For our simulations, we can set $M = 1$. Further, we can normalise the magnetic flux and the outflow power with the time-averaged accretion rate $\langle \dot{M} \rangle$, giving us the dimensionless magnetic flux $\phi_{\rm BH} = \Phi_{\rm BH}/\sqrt{\langle \dot{M} \rangle}$ and the outflow efficiency $\eta = P/\langle \dot{M} \rangle$ (note that $r_{\rm g}$ and the speed of light c are unity according to our units). We refer the reader to Porth et al. [20] for more details about these quantities.

6.2.1 Time-Averaged Properties of MADs: Magnetic Flux Saturation and Relativistic Jets

MADs deal with magnetically-saturated BH environments where forces due to magnetic fields are roughly in radial balance with the ram pressure from the accreting gas. It thus follows that such a system should exhibit the maximum horizon-scale poloidal magnetic flux possible given sufficiently large supply from the disk. Any amount of extra magnetic flux accumulated near the horizon results in expulsion of magnetic fluxtubes, or “magnetic flux eruptions”, creating temporal oscillations of the (largely radial) magnetic flux $\phi_{\rm BH}$ at the BH horizon. We will discuss these eruptions in the next subsection. Figure 6.2 shows the time-averaged

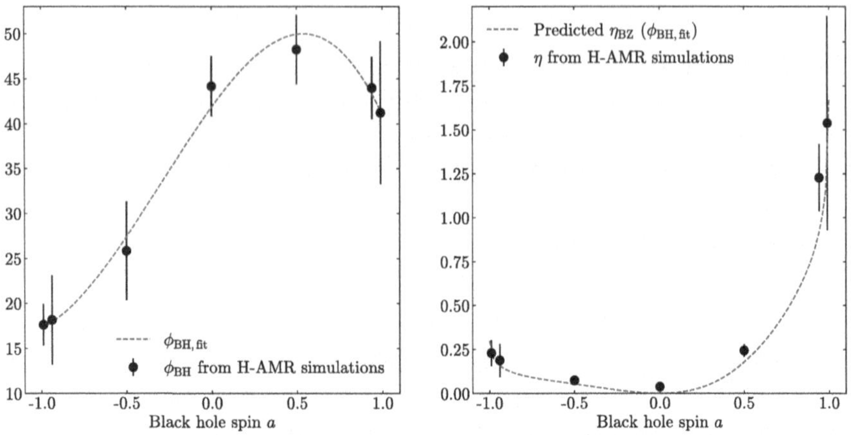

Fig. 6.2 (*Left*) We show the accretion-rate normalised magnetic flux at the BH horizon for GRMHD simulations of MADs around a variety of BH spins [18]. We also show the 1σ standard deviation for each model to indicate the amount of variability in $\phi_{\rm BH}$. The dashed line shows a cubic fit to the numerical data, using the `scipy` function `curvefit`. (*Right*) We show the corresponding outflow efficiency η along with the 1σ standard deviation for each model. The dashed line shows the expected Blandford-Znajek (BZ) jet efficiency $\eta_{\rm BZ} \propto \phi^2_{\rm BH,fit}\Omega^2_{\rm H}$ [21–23], where $\Omega_{\rm H} = a/2r_{\rm H}$ is the horizon angular frequency and $r_{\rm H} = (1+\sqrt{1-a^2})r_{\rm g}$ is the event horizon radius. We find that for BHs with a large prograde spin, (1) the BZ power matches the simulation results as jets dominate the total outflow power, and (2) both $\phi_{\rm BH}$ and η exhibit the largest variability among the models, presumably due to powerful and frequent magnetic flux eruptions

$\phi_{\rm BH}$ as a function of BH spin, calculated over $5000\,t_{\rm g}$ using `H-AMR` simulations [18, 24]. We see that $\phi_{\rm BH}$ has a cubic dependence of the BH spin parameter a: $\phi_{\rm BH,fit} = 41.84 + 26.06a - 12.68a^2 - 14.23a^3$, which is roughly consistent with previous work using GRMHD simulations [15, 22, 25]. While this cubic dependence on spin has not yet been worked out from first principles, it is reasonable to assume that the BH spin should strongly affect the MAD magnetic flux since (1) the magnitude of $\phi_{\rm BH}$ depends on the horizon area (which, in turn, is spin-dependent), and (2) the accretion disk close to the BH is subject to BH frame-dragging. For example, Narayan et al. [25] showed that the disk scale-height appears to be closely tied to the BH spin. Additionally, for BHs accreting at or close to their Eddington rate, radiation physics would also strongly affect disk structure [26–28].

In MADs, the strong magnetic fields near the BH not only modulate disk dynamics, but also power highly efficient relativistic jets. GRMHD simulations show that, in most cases, there is always an outflow of energy (and sometimes mass) in the form of jets and disk winds, provided that the BH environment is sparse enough to allow jet propagation. Figure 6.2 shows one of the most interesting properties of MADs: rapidly spinning prograde BHs can launch jets with efficiencies η exceeding 100% in the MAD state, first shown by Tchekhovskoy et al. [29], which provides direct evidence of spin energy extraction from BHs, a direct consequence of the Blandford-Znajek (BZ) mechanism [21]. Indeed, it has been shown that jets from MADs spindown their BHs, with the equilibrium BH spin close to zero [18, 22, 25, 26, 28], which could have exciting implications for spindown of BHs accreting at super-Eddington rates, e.g., in gamma-ray bursts and ultra-luminous X-ray binaries [30].

6.2.2 How MADs Evolve over Time: The Role of Magnetic Flux Eruptions

Numerical simulations show that the BH magnetosphere can continue accumulating magnetic flux well past the time-averaged $\phi_{\rm BH}$ shown in Fig. 6.2, slowly forming a current sheet layer, typically in the equatorial plane. At the point of magnetic flux over-saturation, magnetic reconnection occurs close to the BH, and nearly-cylindrical fluxtubes filled with vertical magnetic fields are ejected out into the disk. Vital questions about the ejection mechanism remain due to the complex 3D evolution of the reconnection event, such as:

(1) Is the formation of the current sheet only due to reconnection of the azimuthal magnetic field? (2) Does reconnection of the radial magnetic field trigger the eruption event, or does the azimuthal field also play a role? (3) What determines the limited azimuthal range covered by the expelled fluxtubes? (4) When are fluxtubes ejected and at what frequency in time? (5) Does the disk structure and dynamics also affect the ejection mechanism apart from shearing out the fluxtube once ejected into the disk?

Figure 6.3 shows an example of how an ejected fluxtube propagates through the accretion disk in a GRMHD simulation. Additionally we show the time-evolution of the magnetic flux $\phi_{\rm BH}$ and the outflow efficiency η. We clearly see temporal oscillations in the time-series plots where there is a slow rise as magnetic flux is advected along with the gas towards the BH, thereby strengthening the outflow, and a fast decay due to fluxtube ejections. High resolution ideal and resistive GRMHD as well as GR-Particle-in-Cell simulations are actively investigating the ejection mechanism with a particular focus on the decay rate of the magnetic flux and its relation to the local reconnection rate [31–33]. Further, once ejected, magnetic flux eruptions push against gas in the disk and can power disk winds, effectively regulating the disk angular momentum [18, 34], possibly alongside Maxwell stresses due to large-scale poloidal fields [35]. In summary, recent advances in simulations present an exciting opportunity to study magnetic flux eruptions in great detail, which could lead to new insights in the triggering, frequency and energetics of flares in accreting BHs (e.g., see Fig. 6.1).

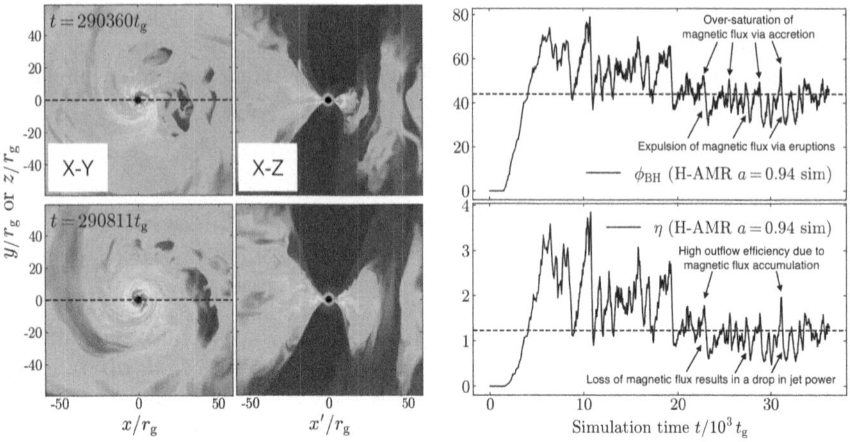

Fig. 6.3 (*Left*) We show the x-y and x-z planes of two representative snapshots of a MAD flow around a Schwarzschild BH from Chatterjee et al. [18], tracking the movement of a magnetic fluxtube as it propagates into the disk. From the figures, we see that a bundle of fieldlines get separated from the jet and form a fluxtube. The accretion flow gradually shears out the fluxtube over time, giving rise to Rayleigh-Taylor-like (RT) and Kelvin-Helmholtz (KH) instabilities at the fluxtube boundary that could explain nIR flares seen in Sgr A* [36]. These fluxtubes push out disk material, powering short-term winds, thus influencing angular momentum transport [18]. (*Right*) Time evolution of ϕ_{BH} and η for the $a = 0.94$ simulation from Fig. 6.2. We also indicate the time-averaged values from Fig. 6.2 with dashed lines. Flux eruptions repeatedly take away magnetic flux from the BH magnetosphere when magnetic forces overcome the accreting gas

6.3 Jet-ISM Collisions in MAD Accretion Flows

Jets travelling through dense ISM environments, such as that found in quasi-spherical accretion flows, could be subject to large-scale kink instabilities, resulting in jet collapse. Gas infall in the polar regions of the BH could also inhibit jet production. Thus jet propagation is intimately connected to both the intrinsic jet power via the BZ process as well as environment properties. Powerful jets can burrow through the environment, creating a cocoon of magnetized gas that forms backflows and helps in jet collimation [37]. For jets with smaller ϕ_{BH}, the jets can quickly lose stability and collapse, at times creating shocked cavities. Recent advances in computational resources have enabled GRMHD simulators to expand to larger length-scales and experiment with different types of accretion flows in order to self-consistently capture the complex dynamics of jet-ISM interactions [13, 37–41].

Figure 6.4 shows an interesting case when the incoming gas and magnetic fields are so misaligned from the BH spin vector such that the jet in the MAD state collides directly with the disk bulk, turns almost 90° and eventually falls apart [15]. In this scenario, the jets actively regulate the advection of magnetic flux onto the BH horizon in a manner similar to flux eruptions (but with a lot more power!), creating

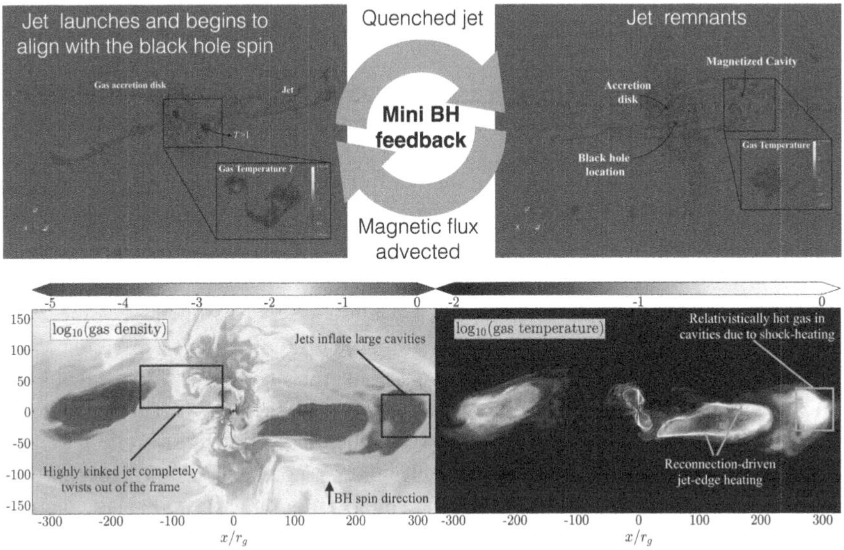

Fig. 6.4 (*Top*) Chatterjee et al. [15] finds that a small-scale BH feedback loop can form for highly misaligned MAD accretion flows. As the newly formed jet strengthens over time due to the accumulation of magnetic flux at the horizon. When the jet begins to align with the BH spin, it collides with the accreting gas, disrupting the supply of magnetic flux to the horizon, leading to jet quenching. Once the jet subsides, new magnetic flux can be advected and brought to the BH, thereby restarting jet launching, and completing the feedback loop. In this particular simulation, the timescale for the feedback loop is around $5 \times 10^4 t_g$, or around 3 days to 8 years for a $10^{6-9} M_\odot$ BH, which presents an interesting possibility for flaring states in radio-quiet AGN and quiescent X-ray Binaries. (*Bottom left*) Vertical cross-section plot of gas density plot of the same simulation as above showing an extremely warped jet (blue) launched out of a spinning BH which is accreting from a 90° misaligned disk (red). (*Bottom right*) The distorted jet inflates shocked cavities in the ISM. The shocked gas shows relativistic temperatures ($T > 1$ in relativistic units) suggesting significant non-thermal X-ray flaring. The jet also shows limb-brightened features as magnetic fields become squeezed between the jet and the ISM, which triggers reconnection and heats the gas. See this link for a movie of this simulation: https://youtu.be/mTLbZgd5CtY, https://youtu.be/mTLbZgd5CtY (Color figure online)

a feedback loop as a result. The figure shows a plethora of interesting features such as jet remnants, magnetized filaments, cavities during the "quenched jet phase" and a twisted or bent jet during the "jetted MAD phase". The gas temperature (T) plot in Fig. 6.4 shows that such systems could exhibit relativistically-hot gas that can produce X-ray emission alongside small-scale radio jets. We require more in-depth analysis of the radiative emission of these systems to see if such feedback loops can explain radio and X-ray flaring in radio-quiet AGN as an alternative mechanism to binary BH-disk interactions. See [15] for details on the simulations and jet/disk evolution for such systems.

6.4 Outlook

Magnetically arrested disks present a uniquely rich laboratory for exploring magnetized plasmas in extreme conditions, from the micro-scale physics of reconnection, hotspots and shocks, to the meso-scales of BH accretion, disk warping in misaligned inflows and magnetic flux eruptions, and all the way to the large-scales of twisted and kink-unstable jets, magnetized backflows and cocoons. Figure 6.5 shows one such example. We show raytraced horizon-scale images of 3 misaligned near-MAD models at sub-millimetre wavelengths (using the GR-radiative transfer code BHOSS [46]) and their reconstructions assuming the telescope array properties of the 2017 EHT array [43], and a future large interferometric array (the next-generation EHT, or ngEHT [44, 45]), including only thermal noise for simplicity. We find that even after accounting for the systematic errors in imaging, it is possible to capture 3 primary features with the ngEHT array that indicate misaligned accretion: (1) the shape of the BH shadow region (i.e., the brightness depression in the center), given our knowledge of the source inclination, (2) the ring asymmetry, and (3) the warped disk-jet just outside the ring. We note that while these reconstructions are quite simplistic and more careful work needs to be done (much like that seen in the ngEHT Analysis Challenge papers; [47, 48]), it is encouraging that observing the horizon-scale accretion flow with such high resolution is possible either with ground-based arrays and potentially future space missions.

Capturing the tell-tale signs of magnetic flux eruptions would be crucial to identify the MAD state. The equatorial plane reconnection event that triggers the eruption as well as the disk-fluxtube shearing/RT/KH instabilities could produce high-energy flares [32], while the expulsion of accretion material from the BH magnetosphere could produce a dip in the submillimetre flux [49] in AGNs like M87 and Sgr A* in addition to affecting polarimetric features in the horizon-scale image [50] as well as the jet boundary [51]. Further, the development of the MAD state in misaligned accretion often forces the inner part of the disk to "magneto-spin align" with the BH spin [15, 52], which have crucial consequences for interpreting the hard state of X-ray Binaries [53], where some sources show quasi-periodic oscillations in the X-ray lightcurve. It is possible that flux eruptions may explain some quasi-periodic activity such as in AGNs [54], but power-spectral-density calculations of the MAD $\dot{M}$ time-series have not yet revealed promising results, primarily showing red noise. Evidently further investigations of the MAD state in standard torii need to be done, such as the dependence of accretion properties on simulation grid resolutions [55, 56], similar to analysis done for weakly-magnetized, MRI-dominated accretion flows [20]. Similarly, understanding the ejection mechanism of fluxtubes holds important consequences for the interpretation of high-energy flares. Guided by the plethora of multi-wavelength observations of accreting BHs and jets, the future looks bright for numerical simulations dealing with accreting BHs with plenty of exciting scientific achievements to look forward to.

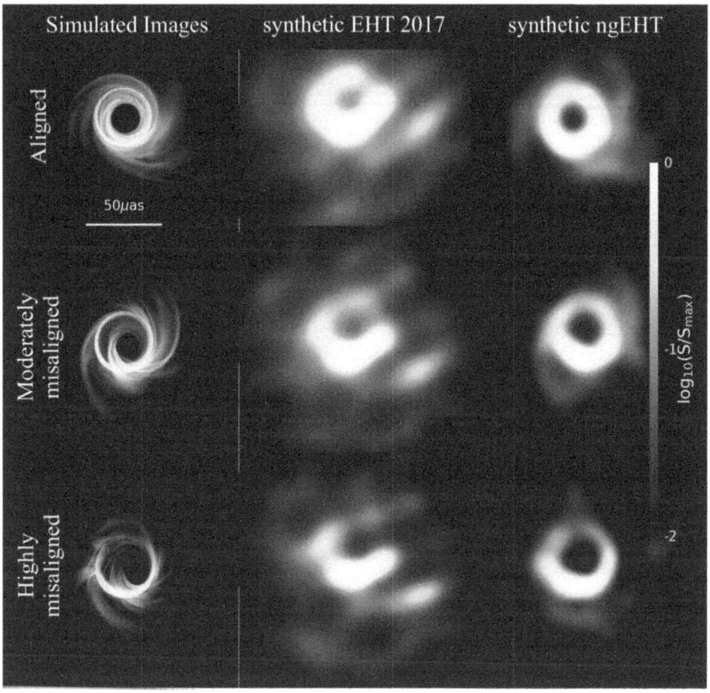

Fig. 6.5 Misaligned accretion onto spinning BHs result in bent jets that do not follow the BH spin direction. Misalignment significantly affects the evolution of the horizon-scale structure in these systems, altering the BH shadow size and shape as well as the extended jet structure. We show a snapshot of raytraced images for disks with misalignment angles of 0°, 30° and 60° [17], measured at large distances, and their corresponding synthetic reconstructions at 230 GHz (using the `eht-imaging` toolkit [42]) with the 2017 EHT array [43] and the future ngEHT array [44, 45], assuming the M87 BH mass and distance. The 2017 EHT array fails to properly capture the differences between these disks, demonstrating the need to resolve the disk-jet connection region by including more short-baseline arrays

Acknowledgments KC thanks Alex Raymond and Daniel Palumbo for their help with scripts to generate and reconstruct synthetic data from GRMHD simulations using `eht-imaging`. KC and RN were supported in part by grants from the Gordon and Betty Moore Foundation and the John Templeton Foundation to the Black Hole Initiative at Harvard University, and by NSF award OISE-1743747. ML was supported by the John Harvard, ITC and NASA Hubble Fellowship Program fellowships. AT acknowledges support from the NSF AST-2009884 and NASA 80NSSC21K1746 grants. AT was supported by BSF grant 2020747 and NSF grants AST-2107839, AST-1815304, AST-1911080, AST-2206471, OAC-2031997. This research was enabled by support provided by an INCITE program award PHY129, using resources from the Oak Ridge Leadership Computing Facility, Summit, which is a US Department of Energy office of Science User Facility supported under contract DE-AC05- 00OR22725, as well as Calcul Quebec (http://www.calculquebec.ca) and Compute Canada (http://www.computecanada.ca). This work has made use of NASA's Astrophysics Data System (ADS).

References

1. Bisnovatyi-Kogan, G. S., & Ruzmaikin, A. A. (1974). The accretion of matter by a collapsing star in the presence of a magnetic field. *Astrophysics and Space Science*, *28*, 45–59. https://doi.org/10.1007/BF00642237
2. Igumenshchev, I. V., Narayan, R., & Abramowicz, M. A. (2003). Three-dimensional magnetohydrodynamic simulations of radiatively inefficient accretion flows. *The Astrophysical Journal*, *592*, 1042–1059. https://doi.org/10.1086/375769
3. Narayan, R., Igumenshchev, I. V., & Abramowicz, M. A. (2003). Magnetically arrested disk: an energetically efficient accretion flow. *Publications of the Astronomical Society of Japan*, *55*, L69–L72. https://doi.org/10.1093/pasj/55.6.L69
4. Baganoff, F. K., et al. (2001). Rapid X-ray flaring from the direction of the supermassive black hole at the Galactic Centre. *Nature*, *413*, 45–48. https://doi.org/10.1038/35092510
5. Nowak, M. A., et al. (2012). Chandra/HETGS observations of the brightest flare seen from Sgr A*. *The Astrophysical Journal*, *759*. https://doi.org/10.1088/0004-637X/759/2/95
6. Haggard, D., et al. (2019). Chandra spectral and timing analysis of Sgr A*'s brightest X-ray flares. *The Astrophysical Journal*, *886*. https://doi.org/10.3847/1538-4357/ab4a7f
7. Markoff, S., Falcke, H., Yuan, F., & Biermann, P. L. (2001). The Nature of the 10 kilosecond X-ray flare in Sgr A*. *Astronomy and Astrophysics*, 379, L13–L16. https://doi.org/10.1051/0004-6361:20011346
8. Yuan, F., Quataert, E., & Narayan, R. (2003). Nonthermal electrons in radiatively inefficient accretion flow models of Sagittarius A*. *The Astrophysical Journal*, *598*, 301–312. https://doi.org/10.1086/378716
9. Do, T., et al. (2019). Unprecedented near-infrared brightness and variability of Sgr A*. *The Astrophysical Journal*, *882*. https://doi.org/10.3847/2041-8213/ab38c3
10. Event Horizon Telescope Collaboration et al. (2021). First M87 event horizon telescope results. VIII. Magnetic field structure near the event horizon. *The Astrophysical Journal*, *910*. https://doi.org/10.3847/2041-8213/abe4de
11. Event Horizon Telescope Collaboration and 286 colleagues 2024. *First Sagittarius A* event horizon telescope results. VIII. Physical interpretation of the polarized ring*. *The Astrophysical Journal*, *964*. https://doi.org/10.3847/2041-8213/ad2df1
12. Haverkorn, M., & Spangler, S. R. (2013). Plasma diagnostics of the interstellar medium with radio astronomy. *Space Science Reviews*, *178*, 483–511. https://doi.org/10.1007/s11214-013-0014-6
13. Ressler, S. M., White, C. J., Quataert, E., & Stone, J. M. (2020). Ab Initio horizon-scale simulations of magnetically arrested accretion in Sagittarius A* fed by stellar winds. *The Astrophysical Journal*, *896*. https://doi.org/10.3847/2041-8213/ab9532
14. Liska, M., Tchekhovskoy, A., & Quataert, E. (2020). Large-scale poloidal magnetic field dynamo leads to powerful jets in GRMHD simulations of black hole accretion with toroidal field. *Monthly Notices of the Royal Astronomical Society*, *494*, 3656–3662. https://doi.org/10.1093/mnras/staa955
15. Chatterjee, K., Liska, M., Tchekhovskoy, A., & Markoff, S. (2023). Misaligned magnetized accretion flows onto spinning black holes: magneto-spin alignment, outflow power and intermittent jets. arXiv e-prints. https://doi.org/10.48550/arXiv.2311.00432
16. Liska, M. T. P., et al. (2022). H-AMR: A new GPU-accelerated GRMHD code for exascale computing with 3D adaptive mesh refinement and local adaptive time stepping. *The Astrophysical Journal Supplement Series*, *263*. https://doi.org/10.3847/1538-4365/ac9966
17. Chatterjee, K., et al. (2020). Observational signatures of disc and jet misalignment in images of accreting black holes. *Monthly Notices of the Royal Astronomical Society*, *499*, 362–378. https://doi.org/10.1093/mnras/staa2718
18. Chatterjee, K., & Narayan, R. (2022). Flux eruption events drive angular momentum transport in magnetically arrested accretion flows. *The Astrophysical Journal*, *941*. https://doi.org/10.3847/1538-4357/ac9d97

19. Chatterjee, K., et al. (2021). General relativistic MHD simulations of non-thermal flaring in Sagittarius A*. *Monthly Notices of the Royal Astronomical Society, 507*, 5281–5302. https://doi.org/10.1093/mnras/stab2466
20. Porth, O., et al. (2019). The event horizon general relativistic magnetohydrodynamic code comparison project. *The Astrophysical Journal Supplement Series, 243*. https://doi.org/10.3847/1538-4365/ab29fd
21. Blandford, R. D., & Znajek, R. L. (1977). Electromagnetic extraction of energy from Kerr black holes. *Monthly Notices of the Royal Astronomical Society, 179*, 433–456. https://doi.org/10.1093/mnras/179.3.433
22. Tchekhovskoy, A., McKinney, J. C., & Narayan, R. (2012). General relativistic modeling of magnetized jets from accreting black holes. *Journal of Physics Conference Series, 372*. https://doi.org/10.1088/1742-6596/372/1/012040
23. Chatterjee, K., Younsi, Z., Kocherlakota, P., & Narayan, R. (2023). On the universality of energy extraction from black hole spacetimes. arXiv e-prints. https://doi.org/10.48550/arXiv.2310.20043
24. Event Horizon Telescope Collaboration et al. (2022). First Sagittarius A* event horizon telescope results. V. Testing astrophysical models of the galactic center black hole. *The Astrophysical Journal, 930*. https://doi.org/10.3847/2041-8213/ac6672
25. Narayan, R., Chael, A., Chatterjee, K., Ricarte, A., & Curd, B. (2022). Jets in magnetically arrested hot accretion flows: geometry, power, and black hole spin-down. *Monthly Notices of the Royal Astronomical Society, 511*, 3795–3813. https://doi.org/10.1093/mnras/stac285
26. Ricarte, A., Narayan, R., & Curd, B. (2023). Recipes for jet feedback and spin evolution of black holes with strongly magnetized super-eddington accretion disks. *The Astrophysical Journal, 954*. https://doi.org/10.3847/2041-8213/aceda5
27. Liska, M. T. P., Kaaz, N., Chatterjee, K., Emami, R., & Musoke, G. (2024). Magnetic flux plays an important role during a black hole X-ray binary outburst in radiative two-temperature general relativistic magnetohydrodynamic simulations. *The Astrophysical Journal, 966*. https://doi.org/10.3847/1538-4357/ad344a
28. Lowell, B., Jacquemin-Ide, J., Tchekhovskoy, A., & Duncan, A. (2024). Rapid black hole spin-down by thick magnetically arrested disks. *The Astrophysical Journal, 960*. https://doi.org/10.3847/1538-4357/ad09af
29. Tchekhovskoy, A., Narayan, R., & McKinney, J. C. (2011). Efficient generation of jets from magnetically arrested accretion on a rapidly spinning black hole. *Monthly Notices of the Royal Astronomical Society, 418*, L79–L83. https://doi.org/10.1111/j.1745-3933.2011.01147.x
30. Narayan, R., Sadowski, A., & Soria, R. (2017). Spectra of black hole accretion models of ultraluminous X-ray sources. *Monthly Notices of the Royal Astronomical Society, 469*, 2997–3014. https://doi.org/10.1093/mnras/stx1027
31. Bransgrove, A., Ripperda, B., & Philippov, A. (2021). Magnetic hair and reconnection in black hole magnetospheres. *Physical Review Letters, 127*. https://doi.org/10.1103/PhysRevLett.127.055101
32. Ripperda, B., et al. (2022). Black hole flares: ejection of accreted magnetic flux through 3D plasmoid-mediated reconnection. *The Astrophysical Journal, 924*. https://doi.org/10.3847/2041-8213/ac46a1
33. Galishnikova, A., et al. (2023). Collisionless accretion onto black holes: dynamics and flares. *Physical Review Letters, 130*. https://doi.org/10.1103/PhysRevLett.130.115201
34. Manikantan, V., et al. (2024). Winds and disk turbulence exert equal torques on thick magnetically arrested disks. *The Astrophysical Journal, 965*. https://doi.org/10.3847/1538-4357/ad323d
35. Narayan, R., Sadowski, A., Penna, R. F., & Kulkarni, A. K. (2012). GRMHD simulations of magnetized advection-dominated accretion on a non-spinning black hole: role of outflows. *Monthly Notices of the Royal Astronomical Society, 426*, 3241–3259. https://doi.org/10.1111/j.1365-2966.2012.22002.x

36. Zhdankin, V., Ripperda, B., & Philippov, A. A. (2023). Particle acceleration by magnetic Rayleigh-Taylor instability: Mechanism for flares in black hole accretion flows. *Physical Review Research*, *5*. https://doi.org/10.1103/PhysRevResearch.5.043023
37. Rohoza, V., et al. (2024). How to turn jets into cylinders near supermassive black holes in 3D general relativistic magnetohydrodynamic simulations. *The Astrophysical Journal*, *963*. https://doi.org/10.3847/2041-8213/ad24fc
38. Lalakos, A., et al. (2022). Bridging the bondi and event horizon scales: 3D GRMHD simulations reveal X-shaped radio galaxy morphology. *The Astrophysical Journal*, *936*. https://doi.org/10.3847/2041-8213/ac7bed
39. Cho, H., et al. (2023). Bridging scales in black hole accretion and feedback: magnetized bondi accretion in 3D GRMHD. *The Astrophysical Journal*, *959*. https://doi.org/10.3847/2041-8213/ad1048
40. Kaaz, N., Murguia-Berthier, A., Chatterjee, K., Liska, M. T. P., & Tchekhovskoy, A. (2023). Jet formation in 3D GRMHD simulations of Bondi-Hoyle-Lyttleton accretion. *The Astrophysical Journal*, *950*. https://doi.org/10.3847/1538-4357/acc7a1
41. Ressler, S. M., White, C. J., & Quataert, E. (2023). Wind-fed GRMHD simulations of Sagittarius A*: tilt and alignment of jets and accretion discs, electron thermodynamics, and multiscale modelling of the rotation measure. *Monthly Notices of the Royal Astronomical Society*, *521*, 4277–4298. https://doi.org/10.1093/mnras/stad837
42. Chael, A. A., et al. (2019). ehtim: Imaging, analysis, and simulation software for radio interferometry. Astrophysics Source Code Library. ascl:1904.004
43. Event Horizon Telescope Collaboration et al. (2019). First M87 event horizon telescope results. I. The shadow of the supermassive black hole. *The Astrophysical Journal*, *875*. https://doi.org/10.3847/2041-8213/ab0ec7
44. Raymond, A. W., et al. (2021). Evaluation of new submillimeter VLBI sites for the event horizon telescope. *The Astrophysical Journal Supplement Series*, *253*. https://doi.org/10.3847/1538-3881/abc3c3
45. Johnson, M. D., et al. (2023). Key science goals for the next-generation event horizon telescope. *Galaxies*, *11*. https://doi.org/10.3390/galaxies11030061
46. Younsi, Z., et al. (2020). Modelling the polarised emission from black holes on event horizon-scales. In *Perseus in Sicily: From Black Hole to Cluster Outskirts* (Vol. 342, pp. 9–12). https://doi.org/10.1017/S1743921318007263
47. Roelofs, F., et al. (2023). The ngEHT analysis challenges. *Galaxies*, *11*. https://doi.org/10.3390/galaxies11010012
48. Chatterjee, K., et al. (2023). Accretion flow morphology in numerical simulations of black holes from the ngEHT Model library: The impact of radiation physics. *Galaxies*, *11*. https://doi.org/10.3390/galaxies11020038
49. Gelles, Z., Chatterjee, K., Johnson, M., Ripperda, B., & Liska, M. (2022). Relativistic signatures of flux eruption events near black holes. *Galaxies*, *10*. https://doi.org/10.3390/galaxies10060107
50. Jia, H., et al. (2023). Millimeter observational signatures of flares in magnetically arrested black hole accretion models. *Monthly Notices of the Royal Astronomical Society*, *526*, 2924–2941. https://doi.org/10.1093/mnras/stad2935
51. Davelaar, J., et al. (2023). Synchrotron polarization signatures of surface waves in supermassive black hole jets. *The Astrophysical Journal*, *959*. https://doi.org/10.3847/2041-8213/ad0b79
52. McKinney, J. C., Tchekhovskoy, A., & Blandford, R. D. (2013). Alignment of magnetized accretion disks and relativistic jets with spinning black holes. *Science*, *339*, 49. https://doi.org/10.1126/science.1230811
53. Fragile, P. C., Chatterjee, K., Ingram, A., & Middleton, M. (2023). The luminous, hard state can't be MAD. *Monthly Notices of the Royal Astronomical Society*, *525*, L82–L86. https://doi.org/10.1093/mnrasl/slad099
54. Cui, Y., et al. (2023). Precessing jet nozzle connecting to a spinning black hole in M87. *Nature*, *621*, 711–715. https://doi.org/10.1038/s41586-023-06479-6

55. White, C. J., Stone, J. M., & Quataert, E. (2019). A resolution study of magnetically arrested disks. *The Astrophysical Journal*, *874*. https://doi.org/10.3847/1538-4357/ab0c0c
56. Sosapanta Salas, L., et al. (2024). Resolution analysis of magnetically arrested disk simulations. arXiv e prints. https://doi.org/10.48550/arXiv.2405.00564

Chapter 7
Study of Accretion Disks Around Black Holes with Two Types of Gas Inflows

Rajiv Kumar

Abstract We utilized a steady-state, axisymmetric, viscous hydrodynamic fluid equations around a black hole in Schwarzschild geometry background. Here, the relativistic Schwarzschild geometry is mimicked by the Paczyński-Wiita potential. We investigated two types of inflowing gases that can generate different kinds of accretion flows around the central objects. The inflowing gases are presented on the local energies (B_{ob}) of the gases versus the outermost accretion boundary locations (r_{ob}) plane we named it the outermost boundary condition (OBC)-plane. Based on the energies of the inflowing gases we found two types of inflowing gas classified as cold-mode and hot-mode inflowing gases in the OBC-plane. Doing so we have found the initial temperature of the inflowing gases can be a parameter for the study of the accretion process. As it can affect the disk structure and optical depth of the accretion flow, which in turn can impact the radiative emissions observed in many accreting sources.

7.1 Introduction

Accretion is a fundamental process in astrophysics that describes the accumulation of matter, typically gas or dust, onto a massive object due to gravity [1]. As material falls onto the central object, it forms a rotating disk known as an accretion disk. These disks are common in various astrophysical systems, including young stellar objects, black holes, and active galactic nuclei. Accretion disks can be extremely hot and luminous, emitting radiation across the electromagnetic spectrum. The physics governing the evolution of accretion disks involves processes such as viscosity and angular momentum transport [2]. Viscosity within the disk causes the material to lose angular momentum and spiral inward towards the central object. The exact

R. Kumar (✉)
Harish-Chandra Research Institute, Prayagraj, India
e-mail: rajivkumar@hri.res.in

S. Ghosh, A. R. Rao (eds.), *The Relativistic Universe: From Classical to Quantum*, Astrophysics and Space Science Proceedings 61,
https://doi.org/10.1007/978-3-031-90186-7_7

mechanism responsible for viscosity in accretion disks remains an area of active research, with turbulence, magnetic fields, and other effects playing potential roles.

The accretion rate at which material accretes onto the central object can vary significantly depending on factors such as the mass of the central object, the properties of the surrounding material, and the efficiency of angular momentum transport within the accretion disk [3, 4]. Usually, the accretion rate is a free parameter, however, in extreme cases, it can be limited by the Eddington luminosity limit which depends on the mass of the central object. Accretion rates are crucial for understanding the growth of stars, black holes, and other celestial bodies. In the presented study, we will study the qualitative aspects of the accreting gases which are defined at the outer-accretion boundary location (OBL), such as the initial temperature, compositions, magnetization level, ionization level, and initial angular momentum of the gases. Due to the limitations of the present model equations, we will focus only on the initial temperature of the inflowing gases around the black hole. With the changing temperature of inflowing gas, we have investigated many kinds of accretion solutions. This chapter is organized as follows: in the next Sect. 7.2 governing equations and assumptions are given, classification of inflowing gases is described in Sect. 7.3, results are discussed in Sect. 7.4, and finally we conclude in Sect. 7.5.

7.2 Governing Equations and Assumptions

In this study, we focused only on the fluid properties of the accretion disk and corresponding OBCs. We assumed the axis-symmetric, steady state, and viscous hydrodynamic fluid flow around a non-rotating black hole (BH). The space-time properties around the non-rotating BH are described by the pseudo-Newtonian potential [5]. We used the geometric unit system to represent the fluid equations of motion and flow quantities where $2G = M = c = 1$. Here, M is the mass of the BH, c is the speed of light, and G is the Gravitational constant. So, the units of length, mass, speed and time are the Schwarzschild radius or $R_s = 2GM/c^2$, M, c, and $R_s/c = 2GM/c^3$, respectively.

7.2.1 *Equations of Motion for Accretion Flow*

The equations of motion for viscous hydrodynamic flow on the equatorial plane in cylindrical coordinates (r, ϕ, z) are described by,

the radial-momentum equation:

$$v\frac{dv}{dr} + \frac{1}{\rho}\frac{dp}{dr} - \frac{\lambda^2}{r^3} + \frac{1}{2(r-1)^2} = 0. \tag{7.1}$$

The azimuthal-momentum equation:

$$v\frac{d\lambda}{dr} + \frac{1}{\Sigma r}\frac{d(r^2 W_{r\phi})}{dr} = 0. \tag{7.2}$$

The z-momentum equation with considering the disk in hydrostatic equilibrium in z-direction, which gives us the local disk half-height expression:

$$h = \sqrt{\frac{2}{\gamma}} a r^{1/2}(r-1). \tag{7.3}$$

The mass-accretion rate equation:

$$\dot{M} = 2\pi \Sigma v r, \tag{7.4}$$

here, $\dot{M}$ is a constant of motion.
The entropy generation equation:

$$\Sigma v T\frac{ds}{dr} = f_{\mathrm{adv}}\, Q^{+}. \tag{7.5}$$

The local variables v, a, p, ρ and λ in the above equations are the radial bulk velocity, sound speed, isotropic pressure, gas density, and specific angular momentum (AM) of the flow, respectively. The local viscous heating therm is given by $Q^+ = W^2_{r\phi}/\eta$, and we assumed radiative inefficient flow, so advection factor $f_{\mathrm{adv}} = 1$ in this study. Here, $\Sigma = 2\rho h$ and $W_{r\phi} = \eta r\frac{d\Omega}{dr}$ are the vertically integrated density and the viscous stress tensor. $\eta = \rho \nu h$ is the dynamic viscosity coefficient, $\nu = \alpha a^2/(\gamma\Omega_k)$ is the kinematic viscosity, α is the viscosity parameter, Ω and Ω_k are the local angular velocity and local Keplerian angular velocity, respectively. Other quantities, s, T, and h are denoted for the entropy density, the local temperature in the physical unit, and the local half-height of the disk, respectively.

The energy per unit mass can be obtained by integrating Eq. (7.1) with the help of Eqs. (7.2), (7.4), and (7.5), which yields the constant of motion of the flow.

$$E = \frac{v^2}{2} + \frac{a^2}{\gamma-1} - \frac{\lambda^2}{2r^2} + \frac{\lambda_0\lambda}{r^2} - \frac{0.5}{r-1}, \tag{7.6}$$

where, E is called the global specific energy of the flow, and the constant throughout the accretion flow even in the presence of viscosity [6, 7]. The local specific energy of the accretion flow can be given by the canonical Bernoulli parameter (B) that is a constant of motion for inviscid flow and defined as,

$$B = \frac{v^2}{2} + \frac{a^2}{\gamma-1} + \frac{\lambda^2}{2r^2} - \frac{0.5}{r-1}. \tag{7.7}$$

The adiabatic sound speed is defined as

$$a = \sqrt{\frac{\gamma p}{\rho}} = \sqrt{\frac{2\gamma\Theta}{\tilde{t}}}, \tag{7.8}$$

where γ is an adiabatic index, Θ is a dimensionless temperature of the gas, and $\tilde{t} = 1 + m_{p^+}/m_{e^-}$.

Integrating Eq. (7.2) with the help of Eq. (7.4) and the expression of angular velocity gradient is obtained,

$$\frac{d\Omega}{dr} = -\frac{\gamma v \Omega_k (\lambda - \lambda_0)}{\alpha a^2 r^2}, \tag{7.9}$$

where λ_0 is specific AM at the BH event horizon obtained by considering vanishing torque at the horizon [6]. Moreover, Ω_k denotes the Keplerian angular velocity. Since $\lambda = r^2\Omega$, the radial derivative of λ can be written by

$$\frac{d\lambda}{dr} = 2r\Omega + r^2\frac{d\Omega}{dr}. \tag{7.10}$$

Simplifying Eqs. (7.1), (7.2), and (7.5), with the help of Eqs. (7.3) and (7.9) we obtain,

$$\frac{dv}{dr} = \frac{N}{D}, \tag{7.11}$$

where,

$$N = \frac{2}{\gamma+1}\frac{(5r-3)v}{2r(r-1)} + \frac{(\lambda^2 - \lambda_k^2)v}{a^2 r^3} + \gamma^2\left(\frac{\gamma-1}{\gamma+1}\right)\frac{v^2\lambda_k(\lambda-\lambda_0)^2}{\alpha a^4 r^4},$$

and

$$D = \frac{v^2}{a^2} - \frac{2}{\gamma+1}.$$

The gradient of sound speed is obtained,

$$\frac{da}{dr} = \left(\frac{a}{v} - \frac{\gamma v}{a}\right)\frac{dv}{dr} + \frac{(5r-3)a}{2r(r-1)} + \frac{\gamma(\lambda^2 - \lambda_k^2)}{ar^3}. \tag{7.12}$$

The accretion disk solutions in vertical equilibrium are solved by simultaneously integrating Eqs. (7.10), (7.11), and (7.12) using Runge-Kutta 4th order numerical method. We are following the same methodology as used in these studies [3, 8]. In doing so, firstly, we have investigated a critical point (CP) of the flow corresponding to the given parameters (E, λ_0, α) by the iteration method with the help of the CP conditions as described in those studies. The CP conditions are obtained from Eq. (7.11) by setting $N = 0$ and $D = 0$. As we know the accretion flow can have multiple critical points (MCPs), which can have shocks. So we have also investigated the adiabatic shock conditions in the accretion flows which can give us the shock location and outer CP of the accretion flow if it occurs.

The Rankine-Hugoniot shock conditions are obtained by conservation of the momentum flux, mass flux, and energy flux across the shock. The shock conditions are given by, the mass flux conservation,

$$\dot{M}_+ = \dot{M}_-, \tag{7.13}$$

The momentum conservation,

$$W_+ + \Sigma_+ v_+^2 = W_- + \Sigma_- v_-^2, \tag{7.14}$$

and the third shock condition is the energy conservation,

$$E_+ = E_-. \tag{7.15}$$

Here, subscripts minus $(-)$ and plus $(+)$ of the flow variables denote the quantities before and after the shock, respectively. W is the vertically integrated pressure. We used shock conditions Eqs. (7.13)–(7.15) and found out the shock locations with outer CP in the global accretion solutions.

7.3 OBC-Plane

This study has focused on the investigations of the nature of accretion flows by changing the initial temperature of the inflowing gases at the outermost boundary locations (OBLs) around the black hole. Recently, we have defined outer boundary conditions (OBCs) for the accretion flows which can represent the external gas environment around the accreting object [4]. Those OBCs can be presented on the plane of local specific energy (B_{ob}) of the inflowing gases and OBLs (r_{ob}) and we named this plane as OBC-plane. The details of the OBC-plane are presented in Fig. 7.1. The B_{ob} of the gas is defined by using the canonical Bernoulli parameter in Eq. (7.7). The definition of B_{ob} at r_{ob} is that the initial inflow velocity $v \sim 0$, initial temperature Θ_{ob} varies $\Theta_{\text{vir}} >> \Theta_{\text{ob}} \sim \Theta_{\text{vir}}$, and initial specific AM $\lambda_{\text{ob}} = \lambda_{\text{K}}$ of the gases. Here, λ_{K}, and Θ_{vir} are Keplerian AM, and dimensionless Virial temperature at r_{ob}, respectively. The virial temperature is used to show the variation of the B_{ob} in the OBC-plane. However, the initial temperature of inflowing

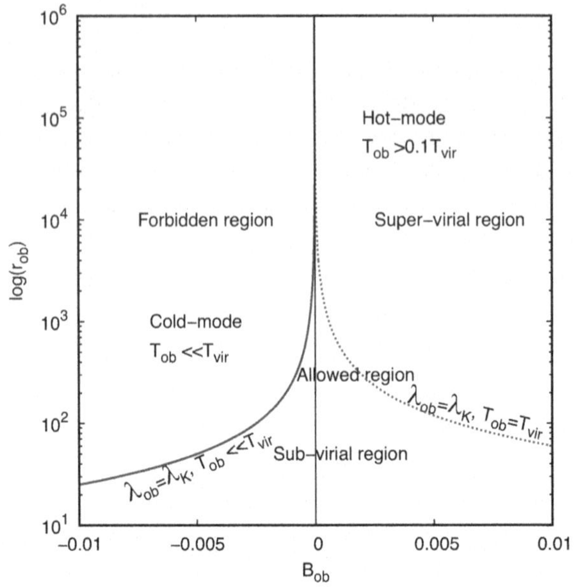

Fig. 7.1 OBC-plane: Represents variation of B_{ob} with r_{ob}. The region left side of the black thin line ($B_{\text{ob}} = 0$) represents cold-mode inflow gases and the solid-violet line represents $\lambda_{\text{ob}} = \lambda_{\text{K}}$ with $\Theta_{\text{ob}} << \Theta_{\text{vir}}$. The region's right side of $B_{\text{ob}} = 0$ represents hot-mode inflow gases and the dotted-red line represents $\lambda_{\text{ob}} = \lambda_{\text{K}}$ with $\Theta_{\text{ob}} = \Theta_{\text{vir}}$. An upper left region of the solid-violet line is named a forbidden region because these energies can not be possible for inflowing gases. The rest of the region, the right side of the solid-violet line is named an allowed region which can be further divided into sub- and super-virial regions (Color figure online)

gas at the OBL is also a free parameter. The Virial temperature of a gas at OBL is defined by following the Virial Theorem, and energy equipartition theorem for gases, so Θ_{vir} is expressed as

$$\frac{3}{2\mu}kT_{\text{vir}} = \frac{1}{2}\frac{GMm_H}{R - R_s} \Rightarrow \Theta_{\text{vir}} = \frac{\mu}{6}\frac{(1 + m_p/m_e)}{r - 1}. \tag{7.16}$$

Here, $1.5kT_{\text{vir}}$ is the average kinetic energy of the particle, $R = rR_s$, and $\Theta_{\text{vir}} = kT_{\text{vir}}/(m_{e^-}c^2)$. k, μ, and T_{vir} are the Boltzmann constant, mean molecular weight of the gas, and Virial temperature in the physical unit. The Virial sound speed can be defined using Eq. (7.8) as

$$a_{vir} = \sqrt{\frac{2\gamma\Theta_{\text{vir}}}{\tilde{t}}} = \sqrt{\frac{\mu\gamma}{3(r-1)}}. \tag{7.17}$$

Here we used $\mu = 1$, $\gamma = 5/3$ and $r = r_{\text{ob}}$. So the B_{ob} versus r_{ob} is plotted in Fig. 7.1.

Primarily, the whole OBC-plane can be divided into forbidden and allowed energy regions of the inflowing gases at the OBLs as shown in Fig. 7.1. A solid violet line in the cold-mode region is drawn with conditions $\lambda_{ob} = \lambda_K$ and $\Theta_{ob} << \Theta_{vir}$ at r_{ob} and it separates the forbidden and allowed energy regions in the OBC-plane. A region on the left side of this line is the forbidden energy region, and the right side region is the allowed energy region for the inflowing gases. In the forbidden region, the energies are not possible for the inflowing gases. The allowed energy region can be further divided into cold-mode and hot-mode inflowing gases/regions, and both regions are separated by the $B_{ob} = 0$ solid black thin line. The hot-mode region is again divided into sub- and super-virial regions, and it is separated by the dotted red line which is generated by using conditions $\lambda_{ob} = \lambda_K$ and $\Theta_{ob} = \Theta_{vir}$. This dotted red curve not only separates virial and super virial regions is also provides the qualitatively the same variation of the B_{ob} corresponding to the types of accretion solutions for a particular viscosity parameter. This we will show in Fig. 7.4. for the shocked flows only. Moreover, the OBCs corresponding to the three CP solutions are mostly found in the super virial region and this kind of detailed study will be reported elsewhere in the future. The cold-mode region has only sub-virial temperature.

7.4 Results and Discussions

We scanned the whole OBC-plane and found five types of global accretion solutions that are represented by some of their properties in the following Figs. 7.2–7.5. Interestingly, the cold-mode gas with $B_{ob} < 0$ gives only global advection-dominated accretion flow (ADAF) [9, 10]. However, the hot-mode gas with $B_{ob} > 0$ gives other four types of global accretion solutions with different initial temperatures or B_{ob} at the r_{ob}. Figure 7.2 represents λ distributions for the five global accretion solutions with different initial temperatures or B_{ob} of inflowing gases at the same OBL as mentioned in the caption. A solid violet line represents ADAF. A dashed green line represents ADAF-thick flow [11]. Both solutions have a single inner CP, but the ADAF-thick flow is geometrically thicker than the ADAF. Dotted, dash-dotted, and dash-double dotted curves represent W-type (3-CP but passes only through inner CP), shocked (3-CP), and Bondi-types (single outer CP) accretion solutions, respectively [12, 13]. Interestingly, the ADAF-thick, W-type, shocked and Bondi-type accretion solutions are only generated by the hot-mode gas $B_{ob} > 0$ at OBL. Moreover, it is seen that the AM transportation due to viscosity is more efficient with higher temperatures of inflowing gases at the OBL. Thus, the ADAF solution with the lowest temperature (cold-mode gas) at OBL has the highest AM distribution, and the Bondi-type with the highest temperature (hot-mode gas) at OBL has the lowest AM distribution. Here, it can be noted that the low AM distribution allows the formation of outer CP in the solutions, such as in the shocked and Bondi-type flows. So, the initial temperature of inflowing gases should be high enough because $W_{r\phi} \propto a$ or Θ so that the outward AM transportation by the viscosity can become

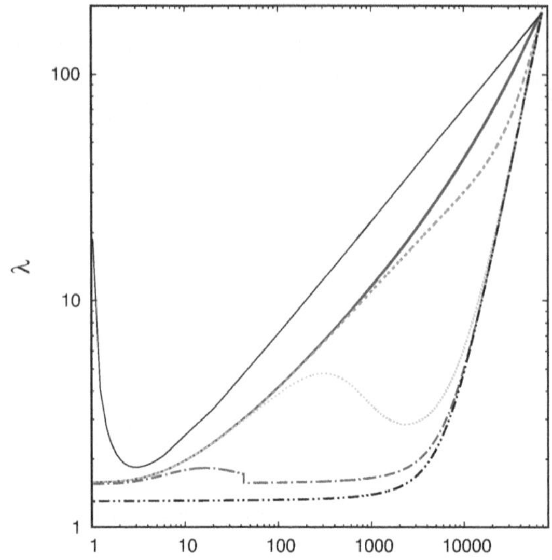

Fig. 7.2 Variations λ versus r for five types of accretion solutions with same $r_{\rm ob} \sim 6.92e4$, $\lambda_{\rm ob} = \lambda_{\rm K}$ and $\alpha = 0.05$ but different initial temperature or $B_{\rm ob} = -2.70e{-}6$ (solid-violet line, ADAF), $1.23e{-}5$ (dashed-green, ADAF-thick), $1.07e - 4$ (dotted-blue, W-type), $2.43e{-}4$ (dashed-dotted red, Shock) and $2.65e{-}4$ (dashed-double dotted black, Bondi-type) at the OBL. A solid black curve represents Keplerian AM distribution (Color figure online)

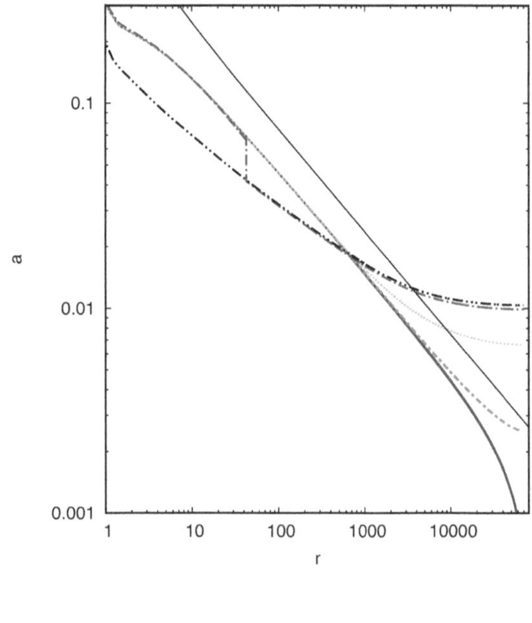

Fig. 7.3 Variations of a versus r with same five solutions as in Fig. 7.2. A solid black curve represents the variation of the Virial sound speed (a_{vir})

efficient to reduce the AM distribution in the disk to make the outer CP if the inflowing gases at $r_{\rm ob}$ has Keplerian AM.

Figure 7.3 shows the variation of the sound speed (a) with radial distance (r) corresponding to the same solutions of Fig. 7.2 with the same line styles. A solid black curve represents the variation of the viral sound speed (a_{vir}), which is expressed in Eq. (7.17). The cases of Fig. 7.3 show that the a of W-type,

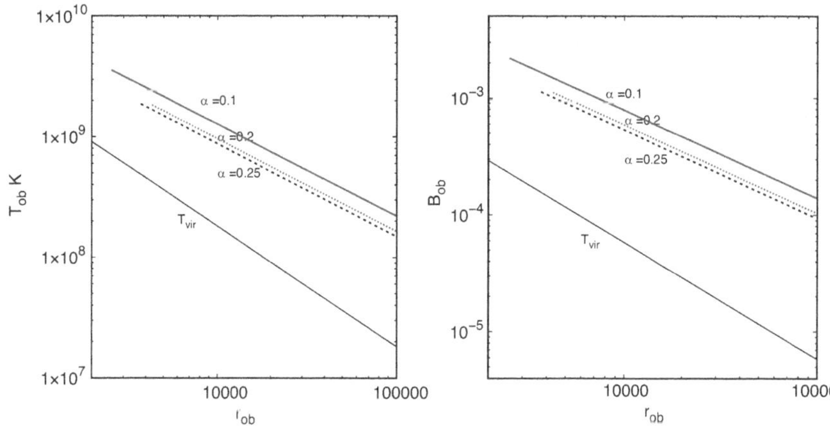

Fig. 7.4 Variation of $T_{\rm ob}$ (in physical unit) with $r_{\rm ob}$ for global shocked accretion solutions for different $\alpha = 0.1$ (solid-violet line), 0.2 (dotted-red line), and 0.25 (dashed-black line). A solid thin black line represents the variation of $T_{\rm vir}$. The $r_{\rm ob}$ is calculated when $\lambda_{\rm ob} = \lambda_{\rm K}$. In the right panel, we plotted $B_{\rm ob}$ corresponding to the same solutions of the left panel. The right panel represents a part of the hot-mode region of the OBC-plane (Color figure online)

shocked, and Bondi-types solutions at OBL exceed the a_{vir}, however the ADAF-thick solution has sub-virial sound speed in hot-mode inflow gases. Usually, we consider the Virial temperature or corresponding sound speed as the maximum temperature/sound speed of the gases in the gravitationally bound system. However, the super-virial temperature of the gases can be achieved through external heating, such as radiative heating. The sub- and super-virial temperature positions can be changed for particular solutions with changing flow parameters, such as E, λ_0, and α. Here, particularly we are interested in shocked solutions because of two reasons, (1) The parameter space of shocked flow is small compared to other solutions, and it shrinks with increasing α [4, 8], and (2) Shocked solutions can have many importance, such as high energy nonthermal radiations, outflow, and QPOs generation [14–16]. So, the OBC of the shocked flow is explored with different viscosity parameters and presented in Fig. 7.4. In this figure, we plotted the initial temperature ($T_{\rm ob}$ in the unit of K) of the inflowing gases at OBL in the left panel. Here, $T_{\rm ob}$ is the proton temperature of the gas. We plotted it because electron and proton temperatures are almost the same around the OBL in the two-temperature accretion flows [17]. Here, we plotted three curves for $\alpha = 0.1,\ 0.2,\ 0.25$ and shocked parameter space for higher $\alpha > 0.25$ is almost vanished. This figure shows that the $T_{\rm ob}$ for the shocked flow is much higher than the $T_{\rm vir}$. So this can raise questions about the existence of the steady-state shocks in accretion flows with $\lambda_{\rm ob} = \lambda_{\rm K}$ at $r_{\rm ob}$ if the super-virial temperature of the inflowing gases at OBL is not found around the BHs. For example, the shocked flow at $r_{\rm ob} \sim 10^5 R_{\rm s}$ in the left panel of Fig. 7.4 has temperature $T_{\rm ob} > 10^8$K and this temperature at such radius is not found yet in observational studies in the BH accretion system.

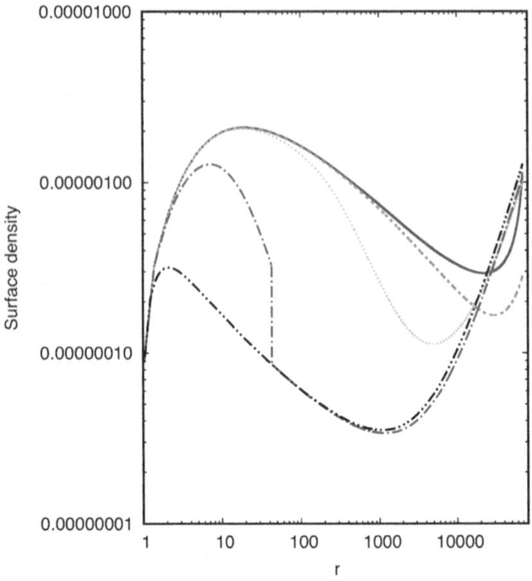

Fig. 7.5 Variations of the surface density (Σ) with radial distance (r) for five different accretion solutions of Fig. 7.2

Interestingly, in Fig. 7.4, the variations of the T_{ob} for shocked flow follow almost the same trends as the variation of T_{vir} with distance in the gravitational system. However, the slope of the decreasing T_{ob} of the shocked flows with r_{ob} is flatter than the variation of the T_{vir} in especially large radii. Now we checked the variation of the B_{ob} corresponding to the shocked solutions in the OBC-plane. So we have plotted r_{ob} versus B_{ob} in the right panel of Fig. 7.4 with B_{ob} corresponding to the T_{vir}. This panel of Fig. 7.4 represents only a part of the hot-mode region of the OBC plane that focuses on the OBCs corresponding to the shocked flows. We again found that the variation of B_{ob} for shocked flow is almost the same as the variation of the B_{ob} corresponding to the T_{vir}. Thus, the dotted red curve in Fig. 7.1 gives a qualitative understanding of the variation of the B_{ob} corresponding to the OBCs of the global accretion solutions for a particular viscosity parameter with $\lambda_{\text{ob}} = \lambda_{\text{K}}$ in the OBC plane. Here, we have plotted B_{ob} for only the shocked flows. Interestingly the variation of the B_{ob} with r_{ob} is also the same for the other kinds of accretion solutions, however, the location of those curves is different in the OBC plane.

Figure 7.5 represents the variation of the surface density ($\Sigma = 2\rho h$) for the same five accretion solutions. The variation of the surface density can represent the optical depth of the disk with a multiplication of Thompson scattering cross-section. The change in optical depth can affect the radiative emission of the disk. This figure shows that the surface density of the disk can change more than one order of magnitude by only changing the initial temperature of the inflowing gases at OBL. So we can use the T_{ob} at the OBL as a parameter as the mass accretion rate is used in the spectral modeling studies.

7.5 Conclusions

In this study, we have investigated the OBC-plane (Fig. 7.1) which has shown two types of the behavior of variations of the r_{ob} with B_{ob} and gives cold-mode and hot-mode inflowing gases at the OBLs. Both types of gases have given us different types of accretion solutions. Interestingly, $B_{ob} < 0$ gives only global ADAF solutions [10]. $B_{ob} > 0$ gives global ADAF-thick [11], W-types, shocked and Bondi-type solutions [12]. This study also explains why some researchers find accretion solutions with multiple sonic points [8, 12] while others find single sonic point ADAF solutions [9, 10]. We have investigated these solutions by keeping r_{ob} and α fixed but only changing the initial temperature of the inflowing gases at r_{ob} as shown in Fig. 7.2. Here, we have found that the ADAF has the highest AM distribution and Bondi-type has the lowest AM distribution. Similarly, these solutions have different distributions of the surface density (Fig. 7.5) which can affect the radiative emissivities of the disk. So we proposed that the initial temperature of the inflowing gases at OBLs can be a parameter for the spectral modeling studies of the accreting sources. In this study, we have also found that the global shocked and Bondi-type solutions have been generated from the super-virial temperature of the inflowing gases at the OBLs. So this result can be a demerit for the having of the steady state shocks in the BH accretion systems unless they have super-virial hot gas in their outer environment. However, it can be possible that the super-virial hot gas can be generated in the outer environment by radiative heating.

Acknowledgments I would like to express my gratitude to the organizers of the International Symposium on Recent Developments in Relativistic Astrophysics at SRM University, Gangtok, India.

Competing Interests Author has no competing interest.

References

1. Bondi, H. (1952). *Monthly Notices of the Royal Astronomical Society*, *112*, 195.
2. Shakura, N. I., & Sunyaev, R. A. (1973). *Astronomy & Astrophysics*, *24*, 337.
3. Kumar, R., & Gu, W.-M. (2019). *The Astrophysical Journal*, *876*, 114.
4. Kumar, R., & Yuan, Y.-F. (2021). *The Astrophysical Journal*, *910*, 9.
5. Paczyńsky, B., & Wiita, P. J. (1980). *Astronomy & Astrophysics*, *88*, 23.
6. Becker, P. A., & Le, T. (2003). *The Astrophysical Journal*, *588*, 408.
7. Gu, W.-M., & Lu, J.-F. (2004). *Chinese Physics Letters*, *21*, 2551.
8. Kumar, R., & Chattopadhyay, I. (2013). *Monthly Notices of the Royal Astronomical Society*, *430*, 386.
9. Narayan, R., & Yi, I. (1994). *The Astrophysical Journal Letters*, *428*, L13.
10. Narayan, R., Kato, S., & Honma, F. (1997). *The Astrophysical Journal*, *476*, 49.
11. Lu, J.-F., Gu, W.-M., & Yuan, F. (1999). *The Astrophysical Journal*, *523*, 340.
12. Chakrabarti, S. K. (1989). *The Astrophysical Journal*, *347*, 365.
13. Chakrabarti, S. K. (1996). *The Astrophysical Journal*, *464*, 664.
14. Chakrabarti, S., & Titarchuk, L. G. (1995). *The Astrophysical Journal*, *455*, 623.

15. Mandal, S., & Chakrabarti, S. K. (2010). *The Astrophysical Journal Letters*, *710*, L147.
16. Lee, S.-J., Chattopadhyay, I., Kumar, R., Hyung, S., & Ryu, D. (2016). *The Astrophysical Journal*, *831*, 33.
17. Sarkar, S., Chattopadhyay, I., & Laurent, P. (2020). *Astronomy & Astrophysics*, *642*, A209.

Chapter 8
Universality of Coronal Properties in Accreting Black Holes

Sudip Chakraborty, Ajay Ratheesh, Francesco Tombesi, Rodrigo Nemmen, and Srimanta Banerjee

Abstract Low-luminosity active galactic nuclei (LLAGN) share characteristics with the hard state of stellar-mass black hole X-ray binaries (BHBs), particularly in their radio loudness, lack of thermal UV emission from an accretion disk, and spectra dominated by power-law emission. In this chapter, we present a detailed hard X-ray spectral analysis of a carefully selected sample of unobscured LLAGN using archival NuSTAR data, aiming to investigate the properties of their central engines in the low-accretion regime. By applying physically motivated Comptonization models to the continuum emission, we discover a pronounced anti-correlation between the optical depth and electron temperature in the corona. This trend, similar to what has been observed in more luminous Seyferts, appears to hold across various coronal geometries. The slope of the anti-correlation in logarithmic space

S. Chakraborty (✉)
Université Paris Saclay, Université Paris Cité, CEA, CNRS, AIM, Gif-sur-Yvette, France
e-mail: sudip.chakraborty@cea.fr

A. Ratheesh
INAF-IAPS, Rome, Italy

F. Tombesi
Department of Physics, University of Rome 'Tor Vergata', Rome, Italy

INAF - Astronomical Observatory of Rome, Monte Porzio Catone (Rome), Italy

Department of Astronomy, University of Maryland, College Park, MD, USA

NASA/Goddard Space Flight Center, Greenbelt, MD, USA

INFN - Rome Tor Vergata, Rome, Italy

R. Nemmen
Universidade de São Paulo, Instituto de Astronomia, Geofísica e Ciências Atmosféricas, Departamento de Astronomia, São Paulo, Brazil

Kavli Institute for Particle Astrophysics and Cosmology (KIPAC), Stanford University, Stanford, CA, USA

S. Banerjee
Inter University Centre for Astronomy and Astrophysics, Pune, India

S. Ghosh, A. R. Rao (eds.), *The Relativistic Universe: From Classical to Quantum*, Astrophysics and Space Science Proceedings 61,
https://doi.org/10.1007/978-3-031-90186-7_8

for LLAGN (ranging from 0.68 to 1.06) is remarkably consistent with that seen in Seyferts (0.55–1.11) and the hard state of BHBs (around 0.87). This behavior likely stems from the Compton thermostat mechanism and also suggests deviations from a static disk-corona radiative configuration. Our results imply a potential universality in the Comptonization process across black hole systems spanning a wide range of masses and accretion rates. We also explore the implications of the Compton thermostat theory to account for this anti-correlation in LLAGN and discuss open questions that remain.

8.1 Introduction

Observational studies of black hole X-ray binaries (BHBs) and active galactic nuclei (AGN) suggest that accretion flows near black holes likely feature a hot, thick corona situated above a cooler, thin accretion disk. In AGN, the corona manifests in the X-ray spectrum as a dominant power-law component, where optical and ultraviolet photons from the accretion disk are upscattered to the X-ray band through Comptonization. A defining feature of this process is the presence of a high-energy cut-off in the X-ray spectrum, indicative of the corona's temperature [1]. X-ray spectra also display reflection features, such as the iron line complex near 6.4 keV and occasionally a "Compton hump" around 30 keV.

Low-luminosity AGN (LLAGN) are systems with supermassive black holes (SMBHs) that exhibit relatively low intrinsic brightness, with bolometric luminosities in the range of $L_{\rm bol} \sim 10^{38-43}$ erg s^{-1}. Unlike brighter AGN, which typically have Eddington ratios $\lambda_{\rm Edd}$ between 0.1 and 1, LLAGN operate in a significantly sub-Eddington regime with $\lambda_{\rm Edd}$ ranging from about 10^{-2} to 10^{-5} [2]. It is believed that in the nearby universe, most SMBHs are in this low-accretion state, which suggests LLAGN could represent a significant phase in the lifecycle of SMBHs [3]. A large fraction of LLAGN are categorized as Low-Ionization Nuclear Emission-line Regions (LINERs).

LLAGNs differ notably from brighter AGN. Their broadband spectral energy distributions (SEDs) lack the characteristic "big blue bump" in the UV and show weak or absent narrow Fe Kα lines [4], indicating the possible absence of a standard, geometrically thin, optically thick accretion disk [5]. Instead, LLAGN accretion is thought to be dominated by advection. The X-ray emission in these systems is often explained by radiatively inefficient accretion flows (RIAFs), a subset of Advection-Dominated Accretion Flows (ADAFs) [6]. In ADAFs, the energy produced in the disk is only partially radiated away, with a substantial fraction being advected towards the black hole. In RIAFs, this accretion mode is characterized by low radiative efficiency ($L \ll 0.1\dot{M}c^2$) and low accretion rates ($\dot{M} \lesssim 0.01\dot{M}_{\rm Edd}$). Throughout this chapter, the terms ADAF and RIAF are used interchangeably.

Despite low accretion rates, LLAGNs still produce power-law X-ray emission similar to that seen in higher-accretion AGN and BHBs. However, unlike the Comptonization of thermal disk photons in more luminous AGN, LLAGN X-ray power-law emission may result from inverse Compton scattering of synchrotron or soft blackbody photons emitted by a truncated thin disk ($\gtrsim 100GM/c^2$) [7]. Power-law emission in LLAGNs also differs slightly from that of bright Seyferts, particularly in the inverse correlation between the power-law index (Γ) and Eddington ratio [8]. However, LLAGNs bear resemblance to the low/hard states of BHBs and tend to occupy a similar region in the $\lambda_{\rm Edd}$ vs. disk fraction plot (akin to the 'q'-diagram for BHB evolution), contrasting with the soft state usually seen in Seyferts [9]. The Seyfert galaxies have been observed extensively in the hard X-ray band (above 10 keV) by instruments such as BeppoSAX [10], INTEGRAL [11], and *Swift*/BAT [12]. The 2013 launch of NuSTAR, the first focusing hard X-ray telescope, significantly advanced these observations, offering exceptional spectral resolution and sensitivity across 3–79 keV range. NuSTAR has enabled precise measurements of high-energy cut-offs and detailed studies of reflection features, including the Fe lines around 6–7 keV and the Compton hump at 20–50 keV. [13] utilized NuSTAR to study Seyfert galaxies and found a geometry-independent anti-correlation between the optical depth (τ) and electron temperature ($kT_{\rm e}$) of the corona, posing challenges for fixed disk-corona radiative models (see Sect. 8.3).

No large-scale study of LLAGN had been conducted until recently. Our work [14] aims to address this gap. In this chapter, we present our results and discuss their broader implications.

8.2 Methods and Results

In our analysis, we utilized data from the BASS DR2 survey[1] [15] to select low-luminosity ($L_{\rm bol} \lesssim 10^{43}$ erg/s), Compton-thin ($N_{\rm H} < 10^{24}\,{\rm cm}^{-2}$) AGN. We ensured that the selected AGN had a sufficient signal-to-noise ratio in NuSTAR observations (3–50 keV count rate $\geq$0.04 counts/s) for reliable spectral analysis. For each of the 16 LLAGNs in our sample, we conducted detailed spectral modeling using the `compTT` Comptonization model and included additional models (`gauss/APEC/xillver/pexmon`) to account for reflection features.

A key result of our study was the discovery of an anti-correlation between the optical depth (τ) and electron temperature ($kT_{\rm e}$) in LLAGN, similar to the trend observed in more luminous AGN. This anti-correlation was found to be consistent across different coronal geometries. In the slab geometry, we found a Spearman's rank coefficient of $\rho = -0.81 \pm 0.11$, while for the spherical geometry, the coefficient was -0.75 ± 0.14.

We applied the bivariate correlated errors and intrinsic scatter (BCES) regression method[2] to further explore this relationship, fitting the anti-correlation in logarithmic form as $\log_{10}(kT_{\rm e}) = a \log_{10}(\tau) + b$. For the slab and spherical geometries, the best-fit parameters were:

[1] https://www.bass-survey.com/dr2.html.

[2] https://github.com/rsnemmen/BCES.

$$a = -0.77 \pm 0.09 \text{ and } b = 1.62 \pm 0.01 \text{ (slab)},$$

and

$$a = -0.98 \pm 0.08 \text{ and } b = 2.04 \pm 0.03 \text{ (sphere)},$$

respectively. These values align closely with those of Seyferts:

$$a = -0.97 \pm 0.14 \text{ and } b = 1.53 \pm 0.15 \text{ (slab)},$$

and

$$a = -0.70 \pm 0.15 \text{ and } b = 1.81 \pm 0.05 \text{ (sphere)}.$$

We show these anti-correlations in Fig. 8.1. Remarkably, when we compared the kT_e and τ values from a set of 438 hard-state RXTE (PCA+HEXTE) observations of 11 BHBs [16], we found a similarly strong anti-correlation, with a Spearman's

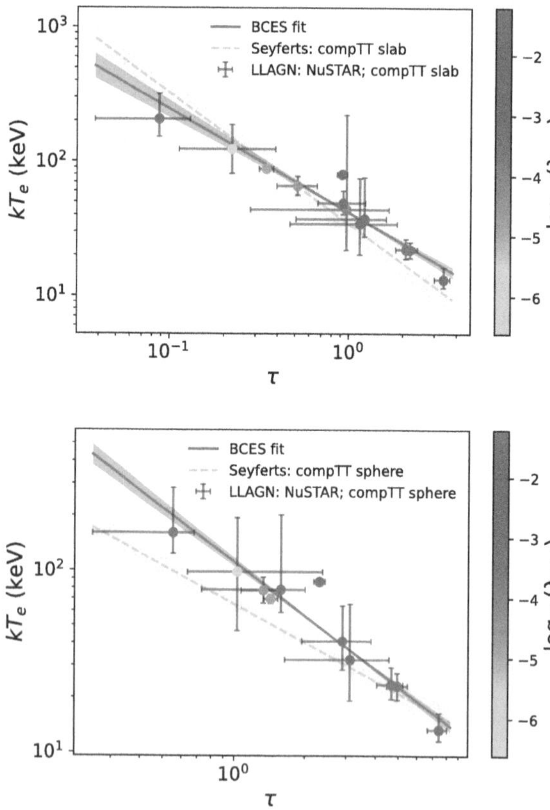

Fig. 8.1 The figure displays the anti-correlation between electron temperature (kT_e) and optical depth (τ) for different coronal geometries [14]. The top panel represents a slab corona, while the bottom panel shows a spherical corona. Data points are color-coded according to the corresponding Eddington ratio (λ_{Edd}). In both panels, the solid lines denote the BCES best-fit lines for our LLAGN sample, with the shaded grey regions representing their confidence intervals. For comparison, the best-fit BCES lines and confidence regions for Seyferts [13] are shown with dashed and light blue lines, respectively (Color figure online)

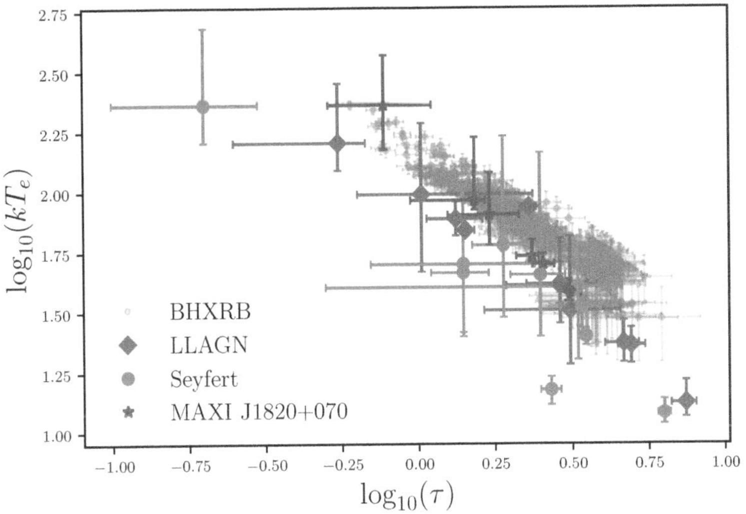

Fig. 8.2 The plot shows the relationship between electron temperature (kT_e) and optical depth (τ) for spherical coronal geometry across various classes of objects. LLAGN from [14] are indicated by blue diamond markers, Seyferts from [13] are represented by orange solid circles, and hard-state data from 11 BHBs from [16] are depicted as light grey hollow circles. This plot illustrates the notable universality in Comptonization properties among different black hole sources discussed in this proceeding. Additionally, hard-state data points for MAXI J1820+070 [17] are shown with red star markers. MAXI J1820+070 is observed to move along this anti-correlation line from right to left as its corona contracts [18] (Color figure online)

rank coefficient of $\rho = -0.84 \pm 0.01$, and best-fit parameters $a = -0.87 \pm 0.02$ and $b = 2.17 \pm 0.01$ (see Fig. 8.2).

These findings suggest a striking universality in the Comptonization behavior of accreting black holes, spanning several orders of magnitude in both black hole mass and accretion rates (Fig. 8.2).

It is important to note that assuming a single τ or kT_e for the entire corona may be an oversimplification. Due to the low fluxes of these sources, we lack the data quality necessary to detect regions with varying τ. While such an approach might work for sources like MAXI J1820+070, which require two distinct τ/kT_e components to model the corona [17], it is impractical for low-luminosity AGNs. Additionally, the high coronal temperatures observed in some LLAGN may arise from inefficient Compton cooling, possibly due to highly truncated accretion disks.

8.3 Discussions and Work in Progress

The anti-correlation can be theoretically interpreted through the concept of the "Compton thermostat" [19, 20]. Consider a slab-like corona heated at a rate F_h

per unit area by thermal photons from the disk, with the photon flux F_s. If the coronal electrons follow a Maxwell-Boltzmann distribution at temperature θ (where $\theta = kT_e/m_ec^2$) and cool mainly through inverse Compton scattering with seed photons, the energy balance equation can be written as:

$$F_h = \tau[A(\Theta) - 1]4\pi J. \tag{8.1}$$

Here, J represents the mean photon intensity in the corona, and A is the Compton amplification factor. In the optically thin, small Compton parameter (y) regime, we can assume $J \propto F_s/c$, with a proportionality factor dependent on both temperature and optical depth. Thus, in this scenario, J becomes $f_C(\Theta, \tau)F_s/c$, leading to the energy conservation equation:

$$\frac{F_h}{F_s} = \frac{l_h}{l_s} = \tau[A(\Theta) - 1]f_C(\Theta, \tau). \tag{8.2}$$

Here, $l = \frac{L}{R}\frac{\sigma_T}{m_ec^3}$ represents the compactness, where the luminosity L and the size R are normalized by the Eddington luminosity and gravitational radius, respectively. This equation implies that, for fixed τ and l_h/l_s, the coronal temperature remains independent of the heating rate. If the heating increases, the constant l_h/l_s ensures that the corona receives more seed photons, cooling it via Comptonization and maintaining the temperature at equilibrium. This equilibrium state can be well described by the scaling law [21]:

$$\tau\Theta \approx a(l_h/l_s)^{1/4}, \tag{8.3}$$

where a depends on geometry, with $a \sim 0.06$ for slab coronas and $a \sim 0.1$ for spherical ones. This equilibrium confirms that the Compton y parameter is regulated to around 1 when a sufficient number of seed photons is available. Thus, this y-dependent Compton thermostat can account for the observed anti-correlation.

However, this straightforward explanation, which assumes a fixed disk-corona configuration in radiative balance, faces challenges. In bright Seyferts, within the $l_h - \Theta$ framework, the system approaches the "pair thermostat" region [22, 23], where l_h should constrain Θ to a narrow parameter range (see [24] for detailed analysis). The observed anti-correlation between kT_e and τ suggests a deviation from the fixed disk-corona configuration assumption [13]. One potential reason for this deviation could be changes in the coronal geometry. Observations from the hard state of the black hole binary MAXI J1820+070 [17] show that it follows this anti-correlation, demonstrated with red star markers in Fig. 8.2, as its corona contracts over time [18]. Another possibility is that Seyferts may exhibit this behavior due to hybrid plasma in their corona, as proposed by [25], allowing them to remain below the pair-thermostat boundary.

For LLAGN, this raises a different question: where do the necessary thermal seed photons for the Compton thermostat come from? In the RIAF scenario, the standard thin disk is either absent or significantly truncated, challenging the foundation of the Compton thermostat hypothesis. Furthermore, in LLAGN with RIAF, τ is proportional to the thickness and density of the accretion flow, implying that τ is roughly connected to the mass accretion rate $\dot{m}$, as $\tau \approx 0.03(\dot{m}/10^{-3})$ [6]. Yet, our data do not reveal any significant correlation between τ and $\dot{m}$. Using the Eddington ratio ($\lambda_{\rm Edd}$) from BASS DR2 as a proxy for $\dot{m}$, we found weak correlations, with Spearman's rank coefficients of 0.20 ± 0.16 for spherical corona and 0.24 ± 0.14 for slab corona. This suggests that the Comptonizing region in LLAGN may be distinct from the RIAF zone and more closely resembles that of Seyfert corona.

To address this, we are currently exploring alternative models, such as the TCAF framework [26], to explain the anti-correlation. Additionally, we are conducting a comprehensive multi-wavelength spectro-polarimetric campaign on select LLAGN to investigate the properties of the Comptonizing region and to resolve the degeneracy between coronal geometries in the $\tau - kT_{\rm e}$ space. With better simulations and more extensive data, we hope to gain a clearer understanding of coronal properties across a wide range of black hole masses and accretion rates.

Lastly, bremsstrahlung emission might also produce spectra similar to those we observe, as these photons could undergo Comptonization. In our current study, we assume the seed photons originate from the disk, following the standard approach. In future work, we aim to examine the effects of Comptonizing bremsstrahlung photons more thoroughly.

Acknowledgments This work was supported by funding from the European Union's Horizon 2020 research and innovation program through grant agreement no. 101004168, as part of the XMM2ATHENA project. The authors extend their gratitude to Prof. A. R. Rao and Prof. Sudip Bhattacharyya for their valuable feedback and constructive suggestions. RN acknowledges financial support from FAPESP (São Paulo Research Foundation) under grant numbers 2017/01461-2 and 2022/10460-8. We also thank Prof. Julian Krolik for his critical insights, and the anonymous referee for their helpful comments that enhanced this chapter.

References

1. Brenneman, L. W., Madejski, G., Fuerst, F., et al. (2014). *The Astrophysical Journal, 788*, 61.
2. Ho, L. C. (2009). *The Astrophysical Journal, 699*, 626.
3. Shin, M.-S., Ostriker, J. P., & Ciotti, L. (2010). *The Astrophysical Journal, 711*, 268.
4. Terashima, Y., Iyomoto, N., Ho, L. C., & Ptak, A. F. (2002). *The Astrophysical Journal Supplement Series, 139*, 1.
5. Nemmen, R. S., Storchi-Bergmann, T., Yuan, F., et al. (2006). *The Astrophysical Journal, 643*, 652.
6. Narayan, R. & Yi, I. (1995). *The Astrophysical Journal, 452*, 710.
7. Nemmen, R. S., Storchi-Bergmann, T., & Eracleous, M. (2014). *Monthly Notices of the Royal Astronomical Society, 438*, 2804.
8. Diaz, Y., Hernàndez-García, L., Arévalo, P., et al. (2023). *Astronomy & Astrophysics, 669*, A114.
9. Fernández-Ontiveros, J. A., López-López, X., & Prieto, A. (2023). *Astronomy & Astrophysics, 670*, A22.
10. Dadina, M. (2007). *Astronomy & Astrophysics, 461*, 1209.

11. Molina, M., Bassani, L., Malizia, A., et al. (2013). *Monthly Notices of the Royal Astronomical Society*, *433*, 1687.
12. Ricci, C., Trakhtenbrot, B., Koss, M. J., et al. (2017). *The Astrophysical Journal Supplement Series*, *233*, 17.
13. Tortosa, A., Bianchi, S., Marinucci, A., Matt, G., & Petrucci, P. O. (2018). *Astronomy & Astrophysics*, *614*, A37.
14. Chakraborty, S., Ratheesh, A., Tombesi, F., Nemmen, R., & Banerjee, S. (2023). *Astronomy & Astrophysics*, *676*, L13.
15. Koss, M. J., Ricci, C., Trakhtenbrot, B., et al. (2022). *The Astrophysical Journal Supplement Series*, *261*, 2.
16. Banerjee, S., Gilfanov, M., Bhattacharyya, S., & Sunyaev, R. (2020). *Monthly Notices of the Royal Astronomical Society*, *498*, 5353.
17. Buisson, D. J. K., Fabian, A. C., Barret, D., et al. (2019). *Monthly Notices of the Royal Astronomical Society*, *490*, 1350.
18. Kara, E., Steiner, J. F., Fabian, A. C., et al. (2019). *Nature*, *565*, 198.
19. Liang, E. P. T. (1979). *The Astrophysical Journal Letters*, *231*, L111.
20. Haardt, F. & Maraschi, L. (1991). *The Astrophysical Journal Letters*, *380*, L51.
21. Pietrini, P. & Krolik, J. H. (1995). *The Astrophysical Journal*, *447*, 526.
22. Svensson, R. (1982). *The Astrophysical Journal*, *258*, 335.
23. Zdziarski, A. A. (1985). *The Astrophysical Journal*, *289*, 514.
24. Fabian, A. C., Lohfink, A., Belmont, R., Malzac, J., & Coppi, P. (2017). *Monthly Notices of the Royal Astronomical Society*, *467*, 2566.
25. Fabian, A. C., Lohfink, A., Belmont, R., Malzac, J., & Coppi, P. (2017). *Monthly Notices of the Royal Astronomical Society*, *467*, 2566.
26. Chakrabarti, S. & Titarchuk, L. G. (1995). *The Astrophysical Journal*, *455*, 623.

Part II
Supermassive Black Hole Activities

Chapter 9
Mode of Accretion and Jet Power in Episodic AGN Jets

Chiranjib Konar, Shobha Kumari, Sabyasachi Pal, and Ananda Hota

Abstract In this chapter, we would like to discuss some of the results from the observations of radio loud active galactic nuclei (AGNs) showing episodic jet activity. We, for the first time, constrained the length of time during which the jet activity remained dormant. Moreover, we will discuss about strong evidence that the jet powers in two different consecutive episodes remain constant. We will furnish possible explanations for this. We will also show how we attempted to explore any possible connection between the episodic jet activity and mode of accretion in AGNs. Recent analysis of some of our GMRT observations show strong evidence of some AGNs being in the dormant state and having not yet started the new episode of jet activity.

9.1 Introduction

Radio galaxies are the radio loud active galactic nuclei (AGNs). All AGNs don't have jets. When they have very long jets (a few 100 kpc to a few Mpc, where 1 pc = 3.0857×10^{18} cm), then they inflate what are called lobes. Morphologically the jet-lobe structures are of two types, namely Fanaroff–Riley type I (FR I) and

C. Konar (✉)
Amity Institute of Applied Sciences, Amity University Uttar Pradesh, Noida, India

RAD@home Astronomy Collaboratory, Kharghar, Navi Mumbai, India

S. Kumari
Midnapore City College, Midnapore, WB, India

S. Pal
Midnapore City College, Midnapore, WB, India

RAD@home Astronomy Collaboratory, Kharghar, Navi Mumbai, India

A. Hota
UM-DAE Centre for Excellence in Basic Sciences, University of Mumbai, Kalina, Mumbai, India

RAD@home Astronomy Collaboratory, Kharghar, Navi Mumbai, India

S. Ghosh, A. R. Rao (eds.), *The Relativistic Universe: From Classical to Quantum*, Astrophysics and Space Science Proceedings 61,
https://doi.org/10.1007/978-3-031-90186-7_9

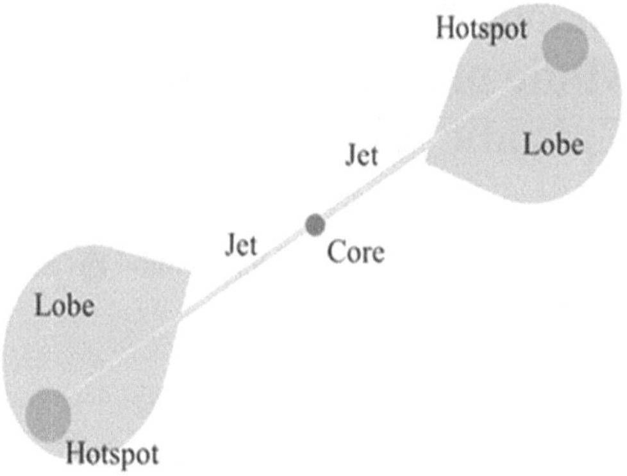

Fig. 9.1 A schematic picture of an FR II radio galaxy as observed in radio band

Fanaroff–Riley type II (FR II) [1]. The FR IIs are the ones having hotspots at the end of the jets where the flow of jet fluid creates termination shocks and the post-shock jet fluid, without being able to mix with the thermal inter-galactic medium, comes back towards the core in a backflow. Thus, the lobes are edge-brightened for FR II jets. For this piece of work, we confine ourselves to FR II radio galaxies in which multiple episodes of the jet ejection activity are observationally well detectable because of the edge-brightened structure of the lobes in each episode. A schematic picture of a typical radio galaxy is shown in Fig. 9.1. The composite structure has been drawn. Radio jets are not always visible in the radio images. If jet is visible, then often this is more prominent on one side than on the other side. That is because of the relativistic boosting effect due to the ultra-relativistic bulk motion of the jet fluid. At the end of the jets, there are prominent and compact radio emission called hotspots. It is known that radio galaxies can be episodic. Radio image of an episodic radio galaxy (ERG) can be seen in Fig. 9.2. In this image what is seen as a core is actually resolved into a mini double radio sources in high resolution image (for details, see [2]). By episodic we mean that the jet activity is episodic, or in other words, the jet activity is recurrent in nature. The causes of episodic behavior of the jet activity are not yet fully understood. As more and more observations in various bands will be done, hopefully more clues will be obtained to unravel the mystery. In this chapter, we will discuss the canonical model of AGNs, accretion states of AGNs, constraining the quiescent phase of jet activity, similarity/dissimilarity of jet power in different episodes, and finally our attempt to find a radio galaxy in the quiescent phase of its jet activity.

9.2 Episodic Radio Galaxies

Often an FR II radio galaxy stops its jet activity for a period of a few to a few tens of Myr, and then again starts its jet activity. The new jets plough through the non-thermal plasma dumped in the previous jet activity. These new jets also have

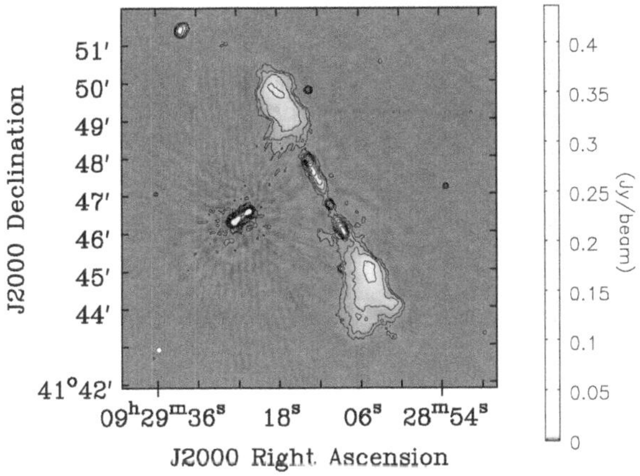

Fig. 9.2 A radio image of an ERG B0925+420 at 144 MHz frequency. This image is made with the image-data of LOFAR Two-metre Sky Survey (LoTSS) from the Data Release-2 (i.e., DR2) database. False color image is overlayed with contour image of itself. Color bar of the false color image is given at the bottom. Contour levels are 0.0005, 0.00111111, 0.00294444, 0.006, 0.0102778, 0.0157778, 0.0225, 0.0304444, 0.0396111, 0.05 Jy/beam (at the multiple of $\sqrt{2}$ from 0.0005 to 0.05). See [2] for science details about the source (Color figure online)

edge brightened structure with the hotspots at the end of the jets, and backflow structure in the lobes. When imaged in radio frequencies, inner jets, which are the current jets, are clearly visible within the non-thermal plasma dumped in the previous cycle(s). Hotspots of the outer lobes have become warm spots, i.e., they have diffused out. These radio galaxies with multiple episodes of jet activities are called episodic radio galaxies (ERGs). ERGs can be double-double radio galaxies (DDRGs) or triple double radio galaxies (TDRGs). If, in a radio continuum image, we detect structural signature of two episodes in a radio galaxy, it is called a DDRG (see Fig. 9.3). Similarly, if there is structural signature of three episodes in radio continuum image of a radio galaxy, it is called a TDRG. It is to be borne in mind that DDRGs or TDRGs can be radio loud quasars also. Since radio galaxies and radio quasars are unified through "Unification Scheme" [3], in this chapter we refer to all such sources as radio galaxies irrespective of whether they are radio galaxies or radio quasars. An FR II type ERG (namely J1158+2621) was first discovered by Owen et al. in 1993 [4] in their 20 cm VLA survey of Abell clusters of galaxies. However, the authors could not realize its episodic nature. It was Schoenmakers who in his Ph.D. thesis in 1999 reported several DDRGs and interpreted their double-double structure to be due to episodic jet activity [5–7]. Afterwards, the astronomers with their hindsight came to know that the DDRG nature of the radio galaxy J1158+2621 was first detected in 1993.

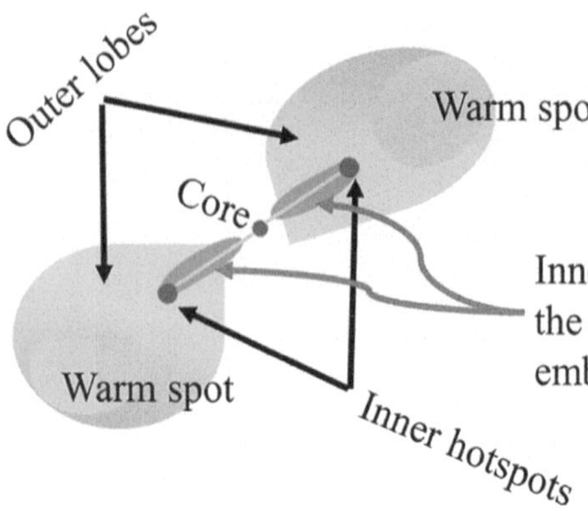

Fig. 9.3 A schematic picture of an ERG which has two episodes of jet activity. It is called a double-double radio galaxy. Various parts of such a radio galaxy are labelled in the picture

9.2.1 Formation Scenario of Episodic Radio Galaxies

The formation scenario of DDRGs, which are ERGs with only two episodes, has been depicted through diagram in Fig. 9.4. We need to bear in mind that we confine ourselves to FR II radio galaxies.

Stage-1 At first the episode-1 of jet activity starts by the dynamics of Active Galactic Nuclei (AGNs) at the core regions of galaxies. These jets propagate through the ambient thermal media. These thermal media can be either Intra-Cluster Media (ICM), or Intra-Group Media (IGrM), or Inter-Galactic Media (IGM), depending upon where the host galaxies of the AGNs are situated. Since the jets carry jet-fluid with ultra-relativistic bulk speed, they impinge on the ambient thermal media and the jet-fluid suddenly slows down in its speed; thereby causing a jet termination shock. These jet termination shocks appear as hotspot in radio images. The jet flow continues for a few Myr to a few hundred Myr. This is the Stage−1 as depicted in Fig. 9.4.

Stage-2 Jet activity stops at some point. The reasons behind episodic activity in various kinds of AGNs can be different and they cannot always be known source by source. However, we can guess some plausible models for the cessation of jet activity. Those plausible causes will be discussed later in this chapter. When jets stop, as we understand from our conventional wisdom, the radio core which is the base of the jets will/should disappear. However, the last jet matter ejected from the central AGN will take about typically 0.1 Myr to travel to the hotspots. During that period, the hotspots will appear as strong and compact point source, as there is replenishment of energetic particles continuously. This is depicted in Stage-2 in Fig. 9.4. In this stage, core will be invisible, however hotspots and outer side of the jets will be visible for some time which will be very small compared to cosmic time, but long enough compared to human life span.

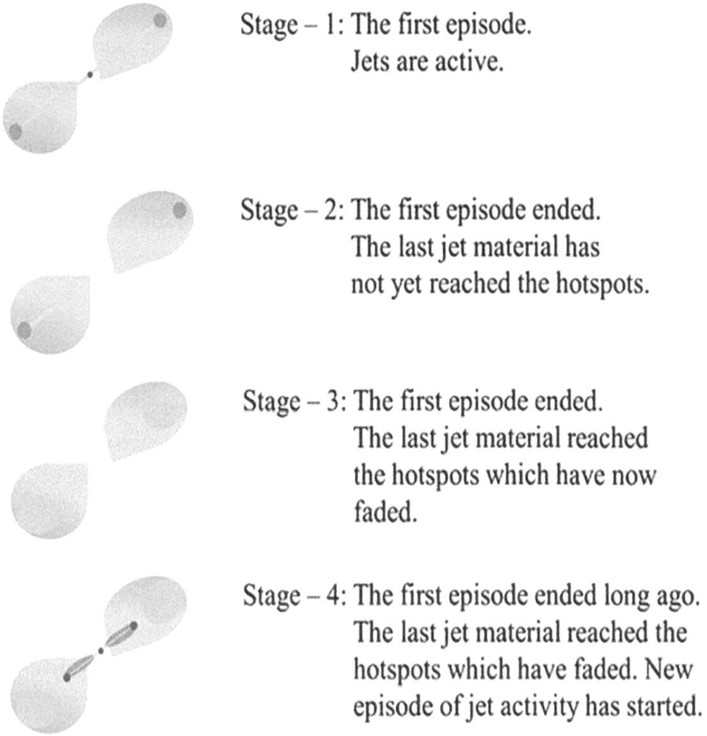

Fig. 9.4 Schematic diagrams showing various stages of DDRGs (which are ERGs) in which two episodes are visible in radio images

Stage-3 After the last jet matter reaches the hotspots, they (hotspots) will start losing compactness and radio intensity. They will fade and become diffuse emission embedded in even more diffuse emission at the edge of the lobes. At this stage, neither core, nor compact hotspots and jets will be visible in radio images. This is the dormant stage of jet activity. The plasma dumped in the lobes in the first episode will remain visible for up to billion years before they go below the sensitivity level of radio telescopes. This stage is depicted in Stage-3 of Fig. 9.4.

Stage-4 After a long lull, a new episode (second episode) of jet activity starts. The new jets plough through the non-thermal plasma dumped in the previous episode. Though the matter density in the non-thermal plasma through which the new jets are propagating is extremely tenuous, jet termination shock is likely to form. Prima facie, it seems that the jets will be moving ballistically and there will be no hotspots at the end of the jets. However, Konar and Hardcastle [8] argued that since the non-thermal plasma is ultra-relativistic and jet heads are propagating with relativistic speed, the jet termination shock can form. The reasons are that (a) the internal energy density and rest mass energy density of the non-thermal plasma yields higher relativistic mass density, and (b) the relativistic mass density gets further enhanced by the square of Lorentz factor of the hotspot as observed in the hotspot frame. This physical aspect may help formation of the hotspots, even if there is no entrainment of thermal matter into the older lobes created in the previous episode. This is depicted in Stage-4 in Fig. 9.4.

This is the way we can explain the DDRG or TDRG structures in radio galaxies. We further studied to investigate if there is any preference for ERGs to be either high excitation radio galaxies (HERGs) or low excitation radio galaxies (LERGs). We discuss on that issue in Sect. 9.4.

9.3 Jet Properties in Different Episodes

Study of jet properties is an intriguing aspect for ERGs. Since what causes episodic activity is still a mystery, we can understand some constraints from the observations of jets and lobes. From our detailed spectral ageing analysis in our previous work [8] we found that jet powers are strongly correlated with the injection spectral index of the lobes. Constraining the duration of quiescent phase is very important for such a study. We devised a noble method with the help of spectral ageing analysis to constrain the duration of quiescent phase in our work in 2013 [9]. The working formula used to calculate the duration of quiescent phase (t_{quies}) is given by (see [9] for explanation)

$$t_{quies} = (t_{ws} + t_{jet}) - t_{innd}, \tag{9.1}$$

where t_{quies} is the duration of quiescent phase, t_{ws} is the spectral age of the warm spot of the outer lobes, t_{jet} is the travel time of an element of jet matter (moving with speed of light) from the core to the warm spot, and t_{innd} is the age of the inner double. It is to bear in mind that the outer hotspots will start ageing (in the sense of spectral age) after the last jet matter reaches the outer hotspots which will gradually become warm spots.

From a statistical study of 8 sample sources, we found that t_{quies} can be between 0.01 and 102 Myr. That's good constrain to model the cause of episodic activity. Whatever may be the cause of episodic jet activity, while modeling this, the timescales of quiescent phase and active phase need to be within the observed ballpark.

Related to jet properties, in 2013 two important results were obtained in our work by Konar and Hardcastle [8] who found that the injection spectral indices of inner and outer doubles were similar, and the injection spectral indices of lobes (similar for outer and inner) were strongly correlated with the jet powers of outer doubles for the 8 sample sources in the same paper. Since the injection spectral index is dependent upon the strength of the jet termination shock, the jet powers and the jet Lorentz factors are expected to remain similar in two episodes to produce the similar injection spectral indices in two consecutive episodes of jet activity [8]. Therefore, we can draw an inference that the jet powers in every episode should be similar for the ERGs. We think that this is a very good constraint to build a model to explain the episodic behavior of radio galaxies. Hopefully, our ongoing work will throw some light in that direction.

9.4 Accretion States of Radio Galaxies and Episodic Radio Galaxies

Before we discuss about the accretion states of radio galaxies, we need to discuss the canonical model of an AGN. The apparatus responsible for strong optical activity and jet activity are furnished well by Urry and Padovani [3] in 1995. In Fig. 9.5 we furnish the pictorial depiction of the canonical model of AGNs.

Though it may seem from the canonical model that every AGN has all the components shown, it is not so. The model in Fig. 9.5 is a composite model. From the work of Laing and Peacock [10], Hardcastle et al. [11], Best and Heckman [12] and Konar et al. [13] it is quite clear that the radio loud AGNs are of two types as per the accretion states. They are called HERGs and LERGs. HERGs are likely to have a Radiatively Efficient Accretion Disk (READ). This accretion disk might be a Shakura-Sunyaev type (See [14]) accretion disk; hereafter this will be called standard accretion disk). That's because the accretion flow is cold. As a result, geometrically thin and optically thick accretion disk can in principle form. This mode of accretion is called cold mode accretion. We do not know the disk geometry for sure. Even a moderately advective (sub-Keplerian) flow with strong magnetic field may be quite radiatively efficient [15]. It is more likely that in HERGs, inner disk dynam-

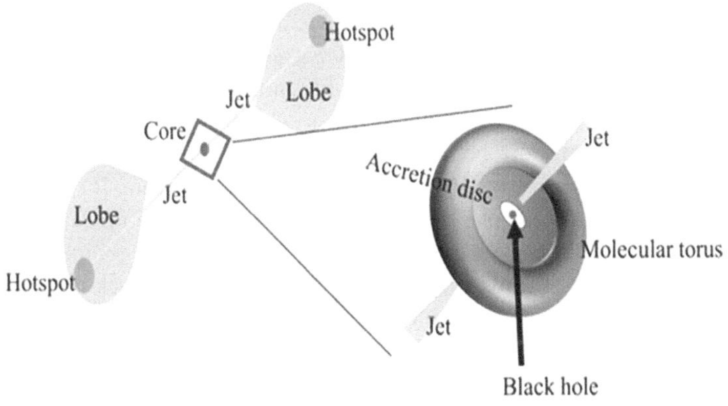

Fig. 9.5 Picturizing the canonical model of AGNs. It is a composite model. What we see as radio core in radio image of a radio galaxy is a very compact component. Its actual size is less than a parsec. Within the core hidden a molecular torus (brown component in picture) and an accretion disk (blue component in picture). At the middle of the disk, there is supermassive black hole (Color figure online)

ics is moderately advective, surrounded by standard Shakura & Sunyaev type disk. Although this requires further investigation. The READs of HERGs are supposed to radiate strongly in UV and optical band of electromagnetic spectrum. The strong UV radiation doubly ionise the oxygen atoms which shows up in [O III] lines in the optical spectrum. Moreover, the modified blackbody spectrum from the standard accretion disk becomes modified into an X-ray power law spectrum, as seed photons from standard disk are up-scattered into the hard part of 0.3–10 keV X-ray band by the electrons of hot corona around the AGN. Jet electrons up-scatter jet radio photons and Cosmic Microwave Background (CMB) photons in the soft part of 0.3–10 keV X-ray band to form a soft X-ray power law spectrum. That's why HERGs show double power law in 0.1–10 keV X-ray band. The hard power law is heavily absorbed, which is interpreted to be due to torus material. The presence of standard accretion disk means the presence of molecular torus around the black hole. The host galaxies of HERGs are invariably visible in the band-4 image in the all-sky survey by Wide-field Infrared Survey Explorer (WISE). Hereafter we will refer to this survey as WISE survey. Band-4 WISE survey is at 22 μm which is in the mid-infrared regime. The molecules in the molecular torus radiate around this wavelength.

On the other hand, LERGs do not have tori. Hence not much cold matter is available to form READ. They are surrounded by hot gas that flows into the black hole. That's why the accretion is very inefficient. This mode of accretion is called hot mode accretion (see [11, 13]). Therefore, the LERGs are not detected in WISE band-4 survey images, they don't have strong [O III] lines; because due to the lack of standard accretion disk, there is no UV light to ionise the oxygen atoms. Moreover, LERGs show single power law spectra in the soft part of 0.1–10 keV X-ray band, because there is no standard disks in them (see [11, 13]).

In our study in 2019 [13], we investigated which mode of accretion (hot or cold) is prevalent in our small sample of DDRGs. Our sample consisted of three DDRGs, namely J0116−4722, J1158+2621 and J1548−3216. All these three DDRGs are LERGs, so in all three there must be hot mode accretion on. Therefore, feedback between supermassive black holes of the AGNs and the ambient media of the host galaxies of these DDRGs might be a plausible model for the episodic jet activity in these DDRGs. However, it is not obvious whether this feedback model can universally explain episodic behavior in all DDRGs. It is well known from the literature that the radio galaxy 3C 219 is an ERG and has strong [O III] lines (see Fig. 9.6). Therefore, DDRG 3C 219 is a HERG which has cold mode accretion. In a HERG, for the feedback model to work, torus material has to completely vanish, and then the supermassive black hole has to come under feedback operation with the ambient medium. That should mean that during quiescent phase, if we observe the AGNs in all DDRGs, we must not find any accretion disk and molecular torus signature in X-ray (double power law in 0.1–10 keV band), optical ([O III] lines) and mid infra-red band (detection in WISE band-4 which is at around 22 μm). Therefore, to test the hypothesis of feedback between supermassive black hole and ambient medium as the universal cause of episodic behavior we must find a sample of radio galaxies in the quiescent phase of jet activities. To identify a radio galaxy in its quiescent phase is not an easy task.

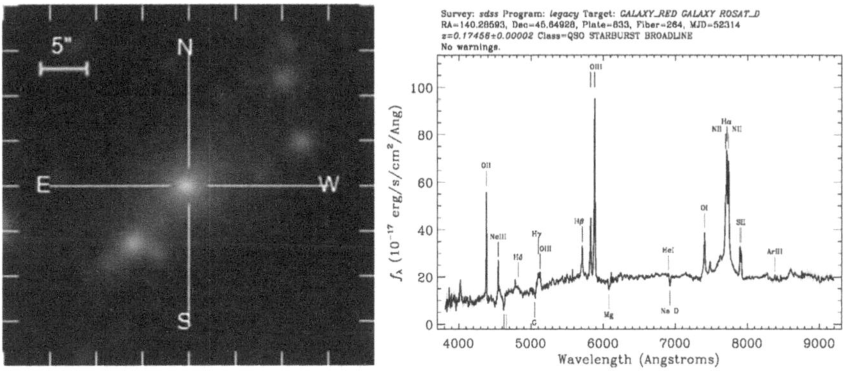

Fig. 9.6 Left panel: SDSS DR12 image of host galaxy of 3C219. Right panel: Optical spectrum of 3C219 obtained from SDSS DR12 database

9.5 Our Ongoing Work

Our ongoing work is to find and then study a sample of radio galaxies which are in quiescent phase. Through our search in the NRAO VLA sky survey (NVSS), we found about half a dozen candidate sources which are thought to be in the quiescent phase of jet activity. Prima facie evidence of the sources to be in quiescent phase is to have the criteria which are (1) the sources should be hotspot less, the sources should not be FR I type, i.e., should not have jets visible near the core, (2) the source images should not have any detectable radio core, and (3) the lobes should be very diffuse. We have multifrequency GMRT data of these sources. Before the conference was held, we could finish making a GMRT image of the radio galaxy J073318.47−365453.30 at 612.651 MHz (see Fig. 9.7). NVSS image (at 1400 MHz) of this source exists (see Fig. 9.8). In both the images, we have not detected any radio core up to the rms noise of 0.05 mJy/beam (beam size $\sim 7.27'' \times 3.79''$) and 0.5 mJy/beam (beam size $\sim 45'' \times 45''$) in GMRT 612.651 MHz and NVSS images respectively. We have not yet done the high frequency radio imaging to check if radio core is visible at C-band and X-band data of VLA. Though we cannot say for sure, it seems this radio galaxy does not have any core even at C- and X- bands. We have usually seen radio cores at 1400 MHz image of similar sensitivity of most of the radio galaxies with active jets. Since the lobes of these sources seem to be very diffuse with no trace of being fed by the jets, and we don't see any core of this radio galaxy at both 612.651 and 1400 MHz, we can expect it to be in quiescent phase of its jet activity.

The radio galaxy J073318.47−365453.30 is most probably a LERG as is not detected in WISE band-4 (see Fig. 9.9). We understand that we don't have X-ray data and optical spectrum of this source to cross check its LERG nature. Our ongoing work will verify the truthfulness of our conjecture that this source is a LERG.

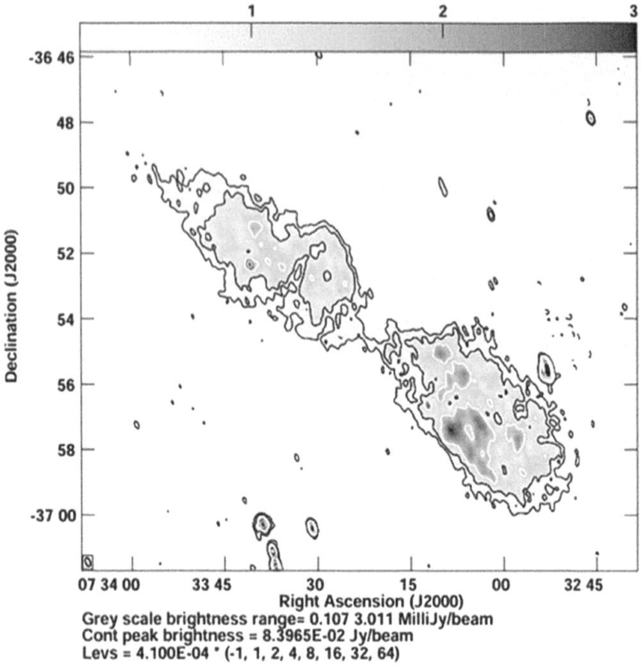

Fig. 9.7 Radio continuum image of J073318.47−365453.30. The image was made with the GMRT at 612.651 MHz. Contour levels are given at the bottom of the image. Beam size ∼ 7.27″ × 3.79″, and rms noise ∼0.05 mJy/beam

Fig. 9.8 Radio continuum image of J073318.47−365453.30 at 1400 MHz from NVSS survey. North-Eastern (NE) lobe and South-Western (SW) lobes are clearly visible. There is an unrelated source at the north of SW lobe. Beam size ∼ 45″ × 45″, and rms noise ∼0.5 mJy/beam

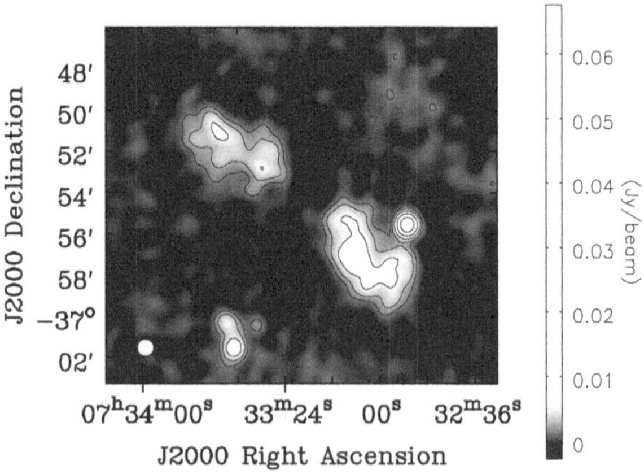

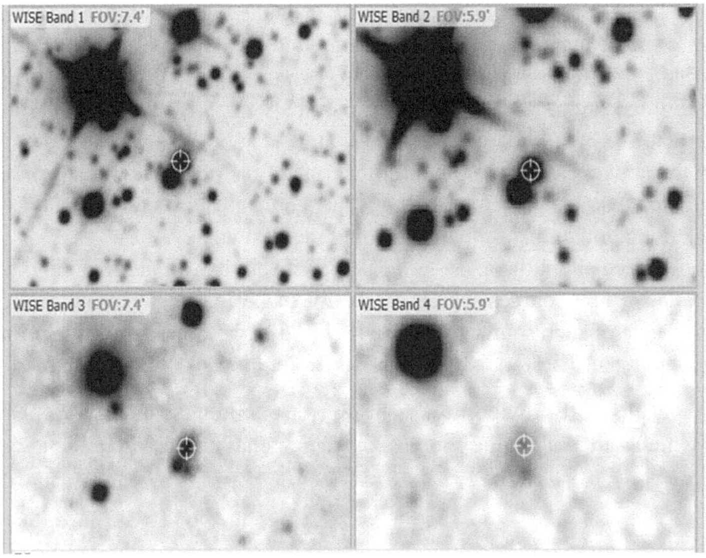

Fig. 9.9 The image of the host galaxy of J073318.47−365453.30 at 4 WISE bands which are at 3.4 μm (top left panel), 4.6 μm (top right panel) 12 μm (bottom left panel) and 22 μm (bottom right panel) respectively

In our ongoing work, we will study the X-ray and optical spectra to verify the accretion state of this AGN. From the radio data at C and X band, we will check the detectability of core of this source to correctly infer if this source is a LERG or a HERG. If it turns out to be in the quiescent state of its jet activity and it is a LERG, then we know the accretion state of this source at the quiescent state, which will provide a constraint in building a model to explain the cause of episodic jet activity. In future we need a statistical study to unravel the mystery of episodic jet activity.

9.6 Results and Conclusion

In this chapter, we have discussed some of our observational results that were already published in various papers. Some are preliminary results obtained through our ongoing work discussed in Sect. 9.5. Those results are itemised as follows.

1. The injection spectral indices of two lobes of a given episodes are similar.
2. While determining the injection spectral indices from the combined spectrum of the double-lobe structure, we found that the injection spectral indices of inner double and outer double are similar.
3. Since injection spectral indices (α_{inj}) are set by the shock acceleration process, it depends on the strength of the shock. That means, the α_{inj} is likely to depend on the jet power, jet Lorentz factor and the ratio of magnetic energy to particle

energy. Therefore, similar injection index is tantamount to similar jet power with similar ratio of magnetic to particle energy and similar Lorentz factor of the jets.

4. We found from our work [13] that ERGs can be both HERGs and LERGs. In fact, all three sample sources in our paper [13] were LERGs. From literature we know that ERGs can be HERGs (see [13]).
5. From the above results, we can infer that one of the plausible models of episodic activity in DDRGs which are LERGs is the feedback between supermassive black hole and the hot ambient medium around the host galaxy.
6. To know more about the cause of episodic jet activity, we are in a ongoing work and found a few sources which are likely to be in the quiescent phase of jet activity. One of such sources is J073318.47−365453.30. This source prima facie appears to be a LERG, i.e., this source is in hot mode accretion state.

All these results will provide important clues to building a model of episodic jet activity which will explain the cause of episodic behavior in radio galaxies.

References

1. Fanaroff, B. L., & Riley, J. M. (1974). *Monthly Notices of the Royal Astronomical Society, 167*, 31.
2. Brocksopp, C., Kaiser, C. R., Schoenmakers, A. P., & de Bruyn, A. G. (2007). *Monthly Notices of the Royal Astronomical Society, 382*, 1019.
3. Urry, C. M., & Padovani, P. (1995). *Publications of the Astronomical Society of the Pacific, 107*, 803.
4. Owen, F. N., White, R. A., & Ge, J. P. (1993). *The Astrophysical Journal Supplement Series, 87*, 1350.
5. Schoenmakers, A. P. (1999). PhD thesis, Universiteit Utrecht.
6. Schoenmakers, A. P., de Bruyn, A. G., Rottgering, H. J. A., van der Laan, H., Kaiser, C. R. (2000). *Monthly Notices of the Royal Astronomical Society, 315*, 371.
7. Schoenmakers, A. P., Mack K.-H., de Bruyn, A. G., Röttgering, H. J. A., Klein, U., van der Laan, H. (2000). *Astronomy and Astrophysics Supplement Series, 146*, 293.
8. Konar, C., & Hardcastle, M. J. (2013). *Monthly Notices of the Royal Astronomical Society, 436*, 1595.
9. Konar, C., Hardcastle, M. J., Jamrozy, M., & Croston, J. H. (2013). *Monthly Notices of the Royal Astronomical Society, 430*, 2137.
10. Laing R. A., & Peacock J. A. (1980). *Monthly Notices of the Royal Astronomical Society, 190*, 903.
11. Hardcastle, M. J., Evans, D. A., & Croston J. H. (2007). *Monthly Notices of the Royal Astronomical Society, 376*, 1849.
12. Best, P. N., & Heckman, T. M. (2012). *Monthly Notices of the Royal Astronomical Society, 421*, 1569.
13. Konar, C., Hardcastle, M. J., Croston, J. H., Jamrozy, M., Hota, A., & Das Tapas, K. (2019). *Monthly Notices of the Royal Astronomical Society, 486*, 3975.
14. Shakura, N. I., & Sunyaev, R. A. (1973). *Astronomy & Astrophysics, 24*, 337.
15. Yuan, F., & Narayan, R. (2014). *Annual Review of Astronomy and Astrophysics, 52*, 529.

Chapter 10
Can Morphological Dichotomy in Radio Galaxies Be Resolved Through Galactic Potential–Induced Shocks?

Shubhrangshu Ghosh

Abstract Based on their extended radio morphologies, radio-loud (RL) active galactic nuclei (AGNs) are principally separated into two classes: the low-radio-power Fanaroff-Riley class I (FRI) and the more radio-powerful FRIIs. This FR-dichotomy has been commonly perceived to arise either (a) owing to the impact of external ambient medium on outflow/jet propagation, or (b) due to intrinsic differences in black hole (BH) properties and/or jet production mechanism. Alternatively, based on the characteristics of their optical emission-line spectra, RL population can be classified into two distinct classes, namely low- and high-excitation radio galaxies (RGs) (i.e., the LERGs and HERGs). LERGs, a majority of them harboring low-radio-power FRI-type jets, while HERGs and/or RL quasars, mostly associated with more radio powerful FRII jets, could be viewed as possible counterparts of low-hard state (LHS) BH X-ray binaries (BHXRBS), and transition state [during LHS-to-high-soft state (HSS) at hard-intermediate state] XRBs, respectively. An intriguing aspect, however being observed, is that the significant population of LERGS are found to possess FRII-type jets, while some HERGs are found to be associated with FRI-type jets. Within the paradigm, in which different faces of AGNs appear to be the broad analogues of BHXRB states, it is expected that almost all LERGs should harbor FRI-type jets, while HERGs should harbor FRII-type jets. The existence of sub-population of FRI/FRII type jets in LERG/HERGs and the nature of connection between accretion modes and jet-morphology (FR-class) is a subject of long-standing debate, given the fact that LERGs/HERGs have distinct galactic environments, while accretion modes in AGNs are distinctly related to galactic environments. Here I attempt to address this issue from a different approach, in the context of 'shock' induced by galactic gravitational potential of the host elliptical galaxy on the outflowing matter, considering a five-component

S. Ghosh (✉)
Center for Astrophysics, Gravitation and Cosmology, Shri Ramasamy Memorial (SRM) University Sikkim, Gangtok, India

Department of Physics, Shri Ramasamy Memorial (SRM) University Sikkim, Gangtok, India
e-mail: shubhrangshughosh.r@srmus.edu.in

S. Ghosh, A. R. Rao (eds.), *The Relativistic Universe: From Classical to Quantum*, Astrophysics and Space Science Proceedings 61,
https://doi.org/10.1007/978-3-031-90186-7_10

galactic system (SMBH + stellar + dark matter + diffuse hot gas + Λ); as a first approximation, confining to the simplest scenario of wind-type outflows. For most of the relevant range of galactic mass-to-light ratio (Υ_B), shock formation occurs in the central-to-outer regions of wind-type outflows, beyond few kiloparsecs. I discuss about the plausible implication of these shocks in the context of realistic outflow/jets, and comment on the observed FR-dichotomy in the radio sources.

10.1 Introduction

One of the remarkable phenomena associated with black hole (BH) accretion is the ejection of powerful, relativistic, well-collimated, radio emitting twin jets of plasma, often observed both from black hole X-ray binaries (BHXRBs), as well as from a small fraction of active galactic nuclei (AGNs). The AGNs dominating in radio-luminosity, radio loud (RL) AGNs [the radio galaxies (RGs) and RL QSOs], are predominantly hosted by large elliptical galaxies. Based on their extended radio morphologies, RL AGNs are principally separated into two classes: the low-radio-power Fanaroff-Riley class I (FRI) and the more radio-powerful FRIIs. The morphological (FR) dichotomy reveals the way of the propagation of outflow/jet through the host galaxy and the manner in which the energy transport occurs in the radio source. FRI sources are less collimated, inefficient in transporting energy, and gets decelerated to sub-relativistic speeds at a comparatively lesser distance from central nucleus [on the scales of a few to tens of kiloparsec (kpc)]. On the other hand, FRII jets remain accelerated, collimated and largely relativistic till the end of the radio-lobes, out to the scales of hundreds of kpcs, or even more [1–7].

This FR dichotomy has been suggested to arise either due to the impact of the external ambient medium on outflow/jet propagation, or owing to the intrinsic differences in BH properties (like BH spin) and/or jet production mechanism, or due to a combination of both (e.g., [8–12]). Alternatively, based on the characteristics of their optical emission-line spectra, RL population can be partitioned again into two distinct classes, namely the low- and high-excitation RGs (i.e., the LERGs and the HERGs), with the HERGs being the high-luminosity sources displaying prominent optical emission lines, believed to be powered by surrounding cold molecular torus through radiatively efficient cold mode accretion. In contrast, LERGs are low-luminosity sources devoid of any strong nuclear emission lines, are thought to be powered by ambient hot X-ray emitting gaseous medium through an advection dominated (quasi-spherical) accretion, in a radiatively inefficient mode (or hot mode accretion). The HERG and LERG populations are believed to represent intrinsically different types of objects.

Over recent years, attempts have been made to draw broad analogy between different classes of AGNs and corresponding BHXRB states (e.g., [13–19]). This is to be expected given the very simple scalings with mass for BHs in general relativity (GR), in which case the AGNs can be viewed as scaled-up BHXRBs. The existence of the BHXRB jets is being closely tied to accretion modes/accretion states. Low-hard state (LHS) of BHXRBs, in which the X-ray emission is characterized by a low-luminosity and hard power-law spectral continuum, is usually associated with low-power continuous, steady radio jets, similarly like low-power FRI-type jets in RL AGNs. LERGs/Low luminous RL AGNs have been suggested to be possible counterparts of BHXRBs in LHS (see, e.g., [13, 15, 20]). On the other hand, during LHS-to-high-soft state (HSS) transition, as mass accretion rate ($\dot{M}$) increases [$\dot{M} \sim (0.01 - 0.1)\dot{M}_{\rm Edd}$] (e.g., [21, 22]), at high X-ray luminosity (particularly at hard-intermediate state), a different type of high-powered relativistic transient jet phenomenon is observed, similar like FRII-type jets in RL AGNs. RL radiative-mode AGNs like HERGs and/or RL quasars, have properties that seem to make them viewed as possible counterparts of transition-phase BHXRBs (at hard-intermediate state) (e.g., [14, 15, 18, 19, 23]).

In this paradigm, in which different faces of AGNs could be viewed as broad analogues of of BHXRB states, it is expected that almost all LERGs should harbor FRI-type jets, while HERGs should harbor FRII-type jets. However, quite intriguingly, it is being observed that a significant population of the LERGs are found to possess radio-powerful FRII jets, while some HERGs are found to be associated with low radio-power FRI-type jets (e.g., [6, 7, 24–28]). The existence of sub-population of FRI/FRII-type jets in LERG/HERGs and the nature of connection between accretion modes and jet-morphology (FR-class) is a subject of long-standing debate, given the fact that LERGs/HERGs have distinct galactic environments, while accretion modes in AGNs are distinctly related to galactic environments. Here it is to be noted that the primary reason behind the observed morphological (FR) difference is due to shocks arising via some mediation of the ambient medium on the outflowing matter. However, the primary explanation for FR division in LERGs or HERGs (sub-population of FRI/FRII jets in LERG/HERGs), perhaps, could be less likely framed, solely in the context of direct interaction of the outflowing matter with the ambient medium through which the outflows/jets propagate. Alternatively, attempts have been made to understand this FR division in the context of spin paradigm (e.g., [8–11]). The role of BH spin in triggering jets, however, remains unclear. Moreover, in the context of supermassive BHs (SMBHs), it is more uncertain to constrain the SMBH spin (e.g., [29]). Also, there lies the possibility of spinning-down of BH through mergers (e.g. [30–32]).

Here, I will attempt to tentatively address this issue from a different perspective, by exploring a possible viable alternative, in the context of 'shock' induced by galactic gravitational potential of the host elliptical galaxy on the outflowing matter. Here I wish to point out that, as the outflow/jet propagates through the galactic medium, it is conceivable that the flow would be expected to be influenced by the gravitational field of the host elliptical galaxy, owing to different galactic mass components, like, stellar, dark matter (DM), hot gas, etc. One then needs to consider the galactic contribution to gravitational potential in a more realistic scenario. For our purpose, as a first approximation, we confine ourselves to the 'wind-type' outflows. We examine, how far such host galaxy potential could impact the outflow dynamics.

10.2 Spherical-Type Flow in Five-Component Elliptical Galaxy

In recent times, in two of our previous works [33, 34], we undertook a detailed study of giant (Bondi-type) spherical accretion (inflow)-outflow, in the context of massive elliptical galaxies, by considering the effect of entire host galaxy gravitational potential, in the back drop of five-component galaxy system (central SMBH + stellar + DM + hot gas + repulsive cosmological constant Λ). Such giant Bondi-type accretion flow towards central SMBH with the accretion length-scale well exceeding hundreds of pc to kpc (i.e., on the scales of interstellar or even intergalactic), is quite relevant in the context of massive/giant elliptical galaxies (see [33, 34], and references therein). An important feature of Bondi-type flow is that, the solution inherently comprises of both accretion (inflow) branch, as well as wind-type outflow component. In an actual scenario, in the context of giant elliptical galaxies, the outflow/jet lengths exceed well beyond kpc scales even reaching Mpc scales. In addition to the flow being influenced by host galaxy potential comprising of galactic mass components, the effect of dark energy (DE) on the flow can not be ignored, particularly in the outer regions (e.g. [35, 36]). For which we considered the simplest and the most attractive DE candidate: repulsive cosmological constant Λ in accordance with the current paradigm (ΛCDM model) of cosmology. We modeled the mass-density distribution profiles for different mass components of the galaxy following, [37]. The generalized version of Navarro, Frenk & White (1995,1996) (NFW) [38, 39] DM mass-density profile, prescribed by Jing & Suto (2000) (JS) [40], having an outer slope of $\simeq -3$ and an arbitrary inner slope of $-\gamma$ has been adopted in our study, that provides considerably better fit to simulated DM halos. For stellar mass distribution, and diffuse hot gas, we followed standard Sérsic profile and isothermal β-model, respectively (for more details see [33, 37]). One of the defining parameters in our study is the galactic mass-to-light ratio (Υ_B); expressed in units of $M_\odot/L_\odot$), that is intrinsically related to host galaxy potential through virial mass of the galaxy. For our purpose and comprehensive analysis, we have adopted a wide range in the value of $\Upsilon_B \sim (14–390)$, with $\Upsilon_B \sim 14$ corresponding to no DM case, and $\Upsilon_B \sim 390$, mass-to-light ratio of the observed Universe [33, 34, 37].

Here, as a first approximation, we focus on a simplistic scenario, the spherical 'wind-type' flow, which one can envisage as some 'spherical outflow of matter', propagating through the galaxy. Although in actual reality, outflows/jets are collimated structures, however, understanding the effect of galactic potential on spherical wind-type outflowing solution will provide deeper insights about the actual influence of galactic potential on realistic outflow/jet dynamics. If indeed, host galaxy potential significantly impacts the wind-type outflow dynamics, it could lead to interesting ramifications in regard to realistic outflows and jets. It is to be noted that spherical wind-type outflow can be realized as a wind-branch solution in the context of Bondi-type flow. We do not describe the hydrodynamical (Bondi-type) spherical flow model which is extensively discussed in the literature (for details see [33, 34] and references therein). For our purpose, we considered a generic class of polytropic flows (both adiabatic and isothermal class of flows), considering adiabatic index Γ in the range 4/3 (for relativistic flow) $\lesssim \Gamma \lesssim 5/3$ (for a non-relativistic flow). For a flow towards a point gravitating mass, spherically symmetric (inviscid) polytropic flow is usually characterized by a single 'saddle-type' sonic/critical point (i.e., what is called classical Bondi solution), through which the subsonic to supersonic flow transition occurs. Although, in the literature, a few instances of deviation from this 'criticality-condition' has been discussed, if the polytropic condition in the flow is found to be strictly violated (see [33] and references therein). Transonic nature of the flow is usually determined through 'sonic energy (E_c) – sonic radius (r_c)' or 'sonic entropy ($\dot{\mu}_c$) $- r_c$' profiles in case of adiabatic flows, and 'temperature (T) $- r_c$' profile for isothermal case.

It is being found that owing to the inclusion of galactic contribution to the gravitational potential, in the context of our five-component galaxy system, the spherical flow displays a remarkable behavior, with the occurrence of multi-transonicity in the flow, irrespective in the value of Υ_B, implying a notable departure from classical Bondi case (see [33, 34] for more details). We depict $E_c - r_c$ (Figs. 10.1 and 10.2) and $T - r_c$ profiles (Fig. 10.3) for adiabatic and isothermal flows, for a few sample cases. For demonstrative purpose, here, I show the adiabatic cases for only $\Gamma = 4/3$, and correspond to JS-3/2 DM profile (i.e., for inner slope value $-\gamma = -3/2$). The profiles clearly show multi-transonicity in the flow.

Figure 10.1 shows that within a certain range of E_c, i.e. from some $E_{\rm min}$ to some $E_{\rm max}$, the flow consists of three sonic points for any specific choice of E_c. To illustrate, a representative line of constant energy $E_c = 2.24 \times 10^{-6}$ has been marked on the figure, intersecting at three sonic points. The inner and outer sonic radii in the curve having negative slopes are the 'saddle (or X)-type' sonic points through which the subsonic to supersonic transition occurs. One with the positive slope between two X-type points is the 'center (or O)-type'. Note that at some radial location specified as static radius ($r_{\rm stat}$; the location where the gravitational attraction owing to galactic mass components is offsetted by the effect of Λ); truncation of profile curve occurs. $E_{\rm min}$ corresponds to this static radius ($r_{\rm stat}$). Figure 10.1 also show the relative contributions of various galactic components to the transonic behavior of the flow. Figure indicates that stellar component is dominating in the regions of central pc to kpc, the DM and the hot gas dominate the regions from a few to few hundreds of kpc, while the influence of Λ becomes significant beyond 100 kpc scale. Figure 10.2 is similar to that of Fig. 10.1, however depicts the comparison with different values of Υ_B. Rest of the figure is interpreted in the same way as that in Fig. 10.1. With the enhancement of Υ_B, the range in E_c for which one obtains multi-transonicity in the flow enhances. Similarly, for isothermal case (as depicted in Fig. 10.3), for some specific range in T (i.e. from some $T_{\rm min}$ to some $T_{\rm max}$), three sonic points emerge in the flow [Fig. 10.3a]. For comparatively higher Υ_B, five sonic points may even occur [Fig. 10.3b]. For clarity, constant temperature representative lines (short dotted-dashed lines; *pink online*) has been marked for both three and five sonic point cases.

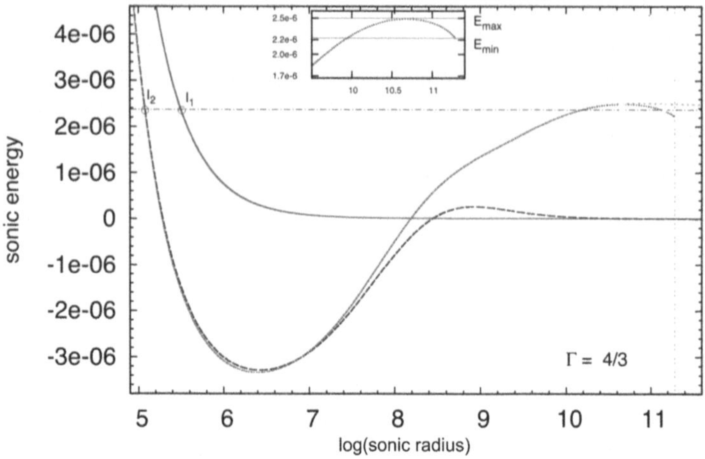

Fig. 10.1 $E_c - r_c$ profiles (r_c (sonic radius) in units of $r_g = GM_{\rm BH}/c^2$, $M_{\rm BH}$ is the BH mass) for spherical-type flow in 5-component elliptical galaxy. For demonstrative purpose, I choose standard value of $\Upsilon_B = 100$, corresponding to JS-3/2 DM case, and $\Gamma = 4/3$. Solid (*red*) correspond to classical Bondi case. Dotted (*dark-green*) curve correspond to full five-component case. To show the relative contributions of various galactic components to the transonic behavior of the flow, as an example, I show $E_c - r_c$ profiles for (BH + stellar) case [long-dashed (*blue*) curves] as well as for (BH + stellar + DM + hot gas) case [short-dashed (*light-green*) curves]. Note that the truncation of dotted curve at $r_c = 1.9168 \times 10^{11} r_g$ implies 'static radius' ($r_{\rm stat}$, marked by vertical double-dashed line). The horizontal dotted-dashed line in is a representative line of constant energy $E_c = 2.24 \times 10^{-6}$ that intersects the dotted curve at three sonic locations, with two 'X-type' sonic points with negative slopes, and in-between two X-type sonic points one with the positive slope of the curve is the 'O-type' sonic point. I_2 denotes the inner X-type sonic point for five-component case, whereas I_1 corresponds to the sonic point for classical Bondi case. The inset image in the figure depicts the zoomed view corresponding to the five-component case in the outer regions, denoting minimum energy $E_{\rm min}$ and maximum energy $E_{\rm max}$, the energy range for which the flow will have three sonic points. Here I consider $M_{\rm BH} = 10^8 M_\odot$, for which 1 parsec $\sim 10^5 r_g$. See Figure 1 in [33] (Color figure online)

The usual reason for emergence of multi-sonicity in accretion/wind flow is due to occurrence of some centrifugal potential barrier (as in the case of high angular momentum advective flows close to the BH/compact object in the context of point gravitating source) (e.g., Chakrabarti and collaborators). For our spherical flow scenario, instead of angular momentum, the requisite potential barrier is provided by galactic gravitational potential in the central-to-outer regions, mostly where it dominates (see [33, 34]). Very significantly, the occurrences of multi-sonicity in the flow indicates the possibility of having 'shocks' in the fluid flow.

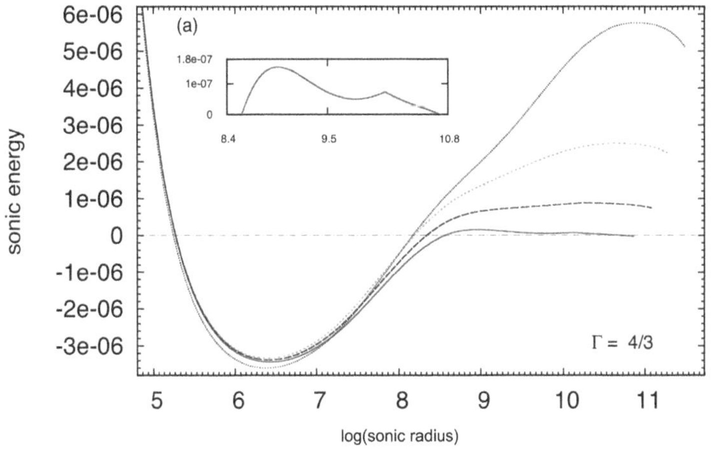

Fig. 10.2 Similar to that of Fig. 10.1, however depicts the comparison with different values of Υ_B. The curves in the plot from bottom to top, correspond to $\Upsilon_B \equiv (14, 33, 100, 390)$. Rest of the figure is interpreted in the same way as that in Fig. 10.1. r_c (sonic radius) is in units of r_g. $M_{\rm BH} = 10^8 M_{\odot}$. See Figure 2 in [33]

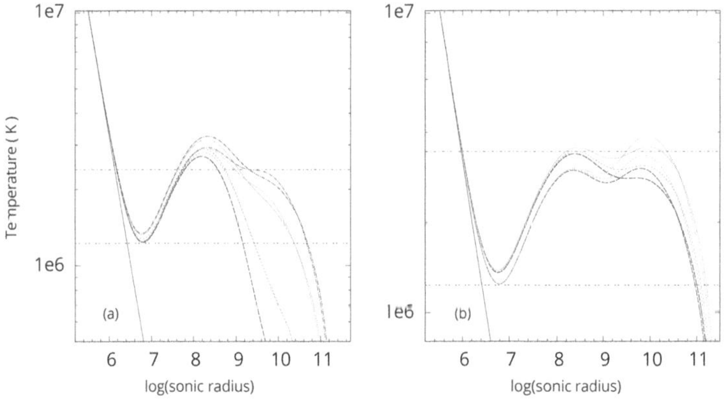

Fig. 10.3 Variation of T as a function of r_c (sonic radius; in units of r_g) for various values of Υ_B, showing multi-transonicity in isothermal spherical flow. Solid line (red) in both the figures correspond to classical Bondi case. In figure (**a**), thick curves correspond to JS-3/2 DM, while thin curves for NFW case. Short-dashed (green), dotted (cyan), and long dotted-dashed (brown), are for $\Upsilon_B \equiv (33, 11, 150)$. Long-dashed (*blue*) curve is for no DM case with $\Upsilon_B = 14$. Figure (**b**) is alike panel (**a**), however for those Υ_B yielding 5-sonic points in the flow. Thick curves are for JS-3/2 DM case with long-dashed (*blue*), short-dashed (*green*), and dotted (*cyan*) correspond to $\Upsilon_B \equiv (250, 300, 390)$, while thin curves identically correspond to NFW DM case. The double-dashed (lower horizontal) lines represent $T_{\rm min}$. Rest of the figure is to be interpreted in the same way as that in Fig. 10.1. $M_{\rm BH} = 10^8 M_{\odot}$. See Figure 1 in [34] (Color figure online)

10.3 Flow Topology and Shocks

As emphasized in the previous section, it is possible for generation of 'shocks' in the flow, within the multi-transonic parameter space region. For adiabatic flows, the global parameter space region is defined by quantities $\langle \Upsilon_B, E_c, \Gamma \rangle$, while for isothermal flows global parameter space is defined by $\langle \Upsilon_B, T \rangle$. In general, shocks can only possibly occur in the multi-transonic regime consisting of at least 3-sonic points in the flow, with inner 'X-type' and outer 'X-type' points, and a 'O-type' sonic point in-between. For shocks to occur in the adiabatic flows, one needs to satisfy (non-dissipative) Rankine–Hugoniot shock conditions, whereas for isothermal shock solutions, isothermal shock conditions need to be satisfied (see e.g., [41–43]). The details about Rankine–Hugoniot and isothermal shocks in accretion and wind solutions have been widely discussed in the literature, and I do not elaborate it here (for more details see [33, 34], and references therein). For adiabatic case, it is being found that for $\Upsilon_B \gtrsim 100$ (relevant for our scenario), and for range of $4/3 \lesssim \Gamma \lesssim 3/2$, corresponding to both JS-3/2 and NFW DM distribution profiles, possible Rankine–Hugoniot shocks occurs in the flow, in the 'wind-type' outflowing solutions, within the multi-transonic domain. Similarly, isothermal shocks occur in the wind flow branch corresponding to both JS-3/2 and NFW profiles, for the entire gamut of $\Upsilon_B(\sim 14–390)$. However, no shocks are found to be generated in accretion flow branch. It is significant to note that these shocks in 'wind-type' outflowing solutions (in which I am primarily interested in) mostly occur in the central-to-outer radii where the gravitational potential due to host elliptical galaxy dominates.

In Figs. 10.4 and 10.5, I depict a few sample cases of flow topologies (or phase portraits), for which one obtains shock transition in the flow, corresponding to adiabatic and isothermal cases, respectively. The vertical lines in the figures indicate

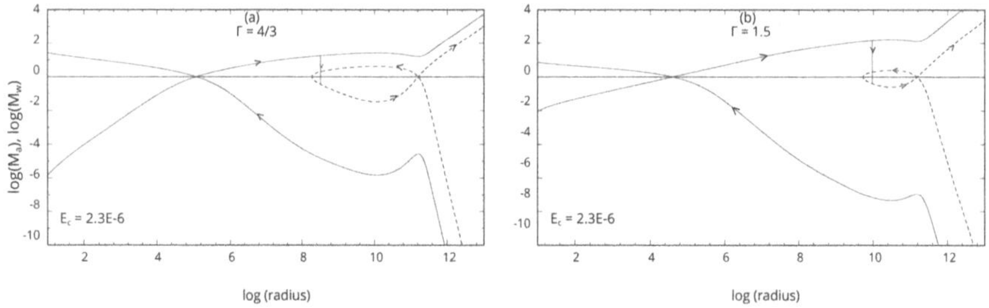

Fig. 10.4 Phase portraits and shock transitions, corresponding to adiabatic case, depicting '$X-\alpha$' type flow topologies. For this sample case, I choose standard value of $\Upsilon_B = 100$. Left and right panels correspond to $\Gamma = 4/3$ and $3/2$, respectively, correspond to JS-3/2 DM case. Solid (*red*) and long-dashed (*blue*) curves in both panels symbolize accretion/wind solutions passing through the inner and outer 'X (saddle)-type' sonic points, respectively. Note that the vertical (arrowhead) lines in both the panels denote shock transitions along the wind branch solutions. The radial location r (in 'X-axis') is in units of r_g. $M_{\rm BH} = 10^8 M_\odot$. See Figure 9 in [33] (Color figure online)

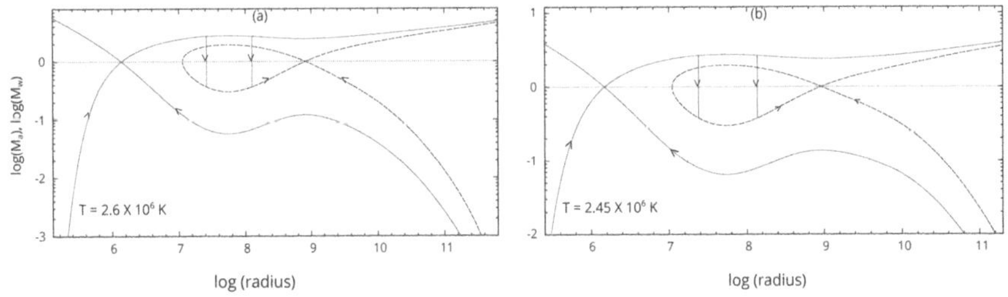

Fig. 10.5 Same as that of Fig. 10.4, but for isothermal case. Here left and right panels correspond to JS-3/2, and NFW DM cases, respectively. The radial location r (in 'X-axis') is in units of r_g. Rest of the figure is to be interpreted in the same way as that in Fig. 10.4. $M_{\rm BH} = 10^8 M_{\odot}$. See Figure 4 in [34]

the possible shock-transitions in the wind-type outflow branch. Here note that, the flow topologies (or phase portraits) as shown in the figures are of '$X - \alpha$'-types, which are typically associated with 'shocks' in wind-type solutions; '$X - \alpha$' profile is improbable to give rise to a shock in accretion flow branch (e.g., [42]). Although one do obtain '$\alpha - X$'-type flow topologies, but no shock transitions occur in the flow (which are meant to have shocks in accretion branch). Figure 10.5 reveals that, for isothermal case, one obtains two physical shock locations, identified by two vertical lines. Following the stability analysis (e.g., [43–45]), outer shock is always found to be stable (which is only considered). An important aspect to be noted here is that the flow topology and the shocks that are being obtained in the spherical flow in the context of five-component galactic system, are similar to that one obtains in advective flows close to the BHs/compact objects. More importantly, the strength of these wind shocks in the central-to-outer radii are comparable with the shocks generated in advective flows close to the BHs/compact objects, with magnitudes even >10 for some parameter space region (see [33, 34]). The galactic gravitational potential (uniquely represented through Υ_B) renders the requisite centrifugal barrier to generate shocks in spherical wind-type flows.

In Fig. 10.6, for demonstrative purpose, I show the variation of shock location ($r_{\rm sck}$) for different values of Υ_B, in the wind-type outflowing solutions, in the context of isothermal shocks. $r_{\rm sck}$ is found to have strong dependence on parameter Υ_B; with the increase in the value of Υ_B, wind shock location shifts considerable outwards. Interestingly, these shocks in the 'wind-type' outflows are found to occur in the scales of kiloparsecs and beyond. Notably, these type of shocks are terminal-type shocks which could lead to abrupt deceleration of outflows propagating in the ambient medium. This could arguably have interesting ramifications in regard to realistic outflows/jets, which I endeavor to comment in the next section.

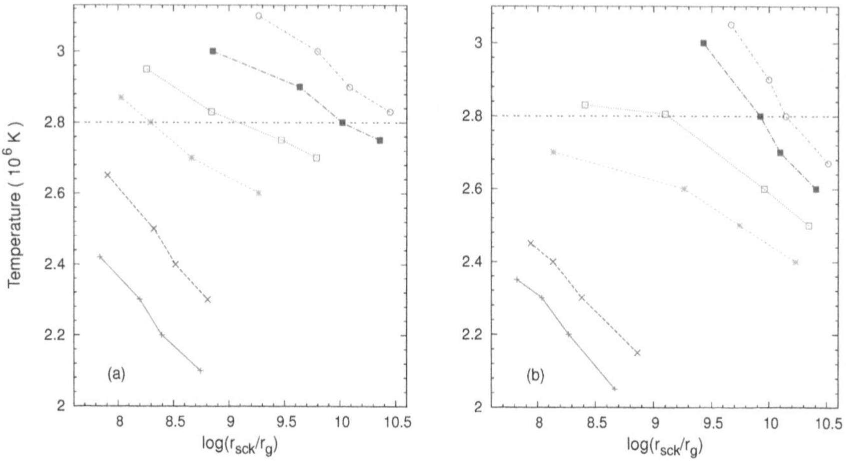

Fig. 10.6 Variation of r_{sck} with T for various Υ_B in wind-type outflows. The left and right panels correspond to JS-3/2, and NFW DM cases, respectively. The lines in each panel from bottom to top, correspond to $\Upsilon_B \equiv (33, 100, 250, 300, 350, 390)$. Double-dashed (horizontal) lines, in both the figures denote $T = 2.8 \times 10^6 K$. $M_{\rm BH} = 10^8 M_\odot$. See Figure 6 in [34]

10.4 Discussion and Concluding Remarks

Here, I emphasized on the aspect of shock formation in wind-type outflows owing to the effect of host galaxy gravitational potential on the propagating outflowing matter in massive elliptical galaxies, emanating from the central active nucleus. The host galaxy potential provides the requisite centrifugal potential barrier for the emergence of such shocks in wind-type outflows. I speculate a potential scenario where this mode of shock (solely due to the contribution of galactic gravitational potential) could possibly occur in realistic outflows and jets propagating through the ambient galactic medium. Three major attributes of such type of shock would be following: (a) they would most likely to be generated in central-to-outer radii in the outflows on the scales of kiloparsecs and beyond, (b) they are terminal-type shocks which would lead to abrupt deceleration of outflowing matter, affecting the outflow/jet speed, and (c) depending on the parameter Υ_B (likely to be distinct for different galaxies), these terminal-type shocks in the outflowing matter would occur at different radii for different galaxies.

The distinct jet morphologies in radio galaxies (i.e. FR-division) reveal the way the outflow/jet propagates in the ambient galactic medium, with FRI morphology indicating an abrupt deceleration of jet to a sub-relativistic speeds at a relatively lesser distance from the central nucleus (on the scales of a few-to-10 kpc), whereas FRII morphology indicates that the jets remain accelerated and largely relativistic till the end of the radio-lobes, out to the scales of hundreds of kpcs, or even more. The primary reason behind this morphological difference has been envisaged to be related to the direct interaction of outflow/jet matter with the external ambient galactic medium, with the morphological division closely tied to the galactic environment. However, quite intriguingly, RL galaxies: LERGS and HERGs, although having distinct galactic environments, are found to harbor sub-population of FR1/FR2-type jets, more so in LERGs. This suggests lesser inevitability of this aspect of *outflow/jet interaction with external ambient medium*, as the primary cause for FR discrepancy. A few authors, have suggested an alternative 'spin paradigm' to account for this FR division, however, the precise role of BH spin in triggering jets, remains unclear, and still a subject of long-standing debate (see Sect. 10.1 for more details).

Here I hypothesize a following alternative scenario to address this issue in the context of 'shock' induced by galactic gravitational potential of the host elliptical galaxy on the outflowing matter: Such mode of shocks, if indeed occurs in outflows/jets, would substantially impact the dynamics and outflow/jet propagation, in a similar fashion like other scenarios, mentioned above. If such galactic shocks occur at a relatively lesser distance from the active nucleus (say on the scales of a few-to-10 kpc), depending upon 'galactic mass-to-light ratio', one can relate it to FRI-type morphology, otherwise, would eventually lead to FRII-type jets. Galactic potential induced shocks, thus, could be conceived as one of the 'contributing factors' to explain the observed morphological (FR) division in the radio sources. Nonetheless, one needs to examine thoroughly, whether such multi-transonicity and/or shocks can actually occur in collimated outflows/jets due to the influence of host galaxy potential. Such a work will be pursued in the near future.

Acknowledgments This chapter is based on the scientific talk in 'International Symposium on Recent Developments in Relativistic Astrophysics (ISRA 2023)' held during December 11–13, 2023, in Gangtok, India, commemorating 50th birth anniversary of Prof. Banibrata Mukhopadhyay of Indian Institute of Science, Bangalore. The author acknowledges IUCAA associateship program for their kind support.

References

1. Fanaroff, B. L., & Riley, J. M. (1974). *Monthly Notices of the Royal Astronomical Society*, *167*, 31.
2. Laing, R. A., & Bridle, A. H. (2002). *Monthly Notices of the Royal Astronomical Society*, *336*, 328.
3. Hardcastle, M. (2008). Astronomical society of the pacific conference series. In T. A. Rector & D. S. De Yong (Eds.), *Extragalactic jets: Theory and observation from radio to gamma ray* (Vol. 386, p. 46).
4. Best, P. N., & Heckman, T. M. (2012). *Monthly Notices of the Royal Astronomical Society*, *421*, 1569.
5. Mingo, B. et al. (2014). *Monthly Notices of the Royal Astronomical Society*, *440*, 269.
6. Heckman, T. M., & Best, P. N. (2014). *Annual Review of Astronomy & Astrophysics*, *52*, 589.
7. Tadhunter, C. (2016). *The Astronomy & Astrophysics Review*, *24*, 10.
8. Meier, D. L. (1999). *The Astrophysical Journal*, *522*, 753.
9. Meier, D. L. (2001). *The Astrophysical Journal*, *548*, L9.
10. Tchekhovsko, A., Narayan, R., & McKinney, J. C. (2010). *The Astrophysical Journal*, *711*, 50.

11. Garofalo, D., Evans, D. A., & Sambruna, R. M. (2010). *Monthly Notices of the Royal Astronomical Society*, *406*, 975.
12. Turner, R. J., & Shabala, S. S. (2015). *The Astrophysical Journal*, *806*, 59.
13. Merloni, A., Heinz, S., & di Matteo, T. (2003). *Monthly Notices of the Royal Astronomical Society*, *345*, 1057.
14. Nipoti, C., Blundell, K. M., & Binney, J. (2005). *Monthly Notices of the Royal Astronomical Society*, *361*,, 633.
15. Körding, E. G., Jester, S., & Fender, R. (2006). *Monthly Notices of the Royal Astronomical Society*, *372*, 1366.
16. Sobolewska, M. A., Siemiginowska, A., & Gierliński, M. (2011). *Monthly Notices of the Royal Astronomical Society*, *413*, 2259.
17. Svoboda, J., Guainazzi, M., & Merloni, A. (2017). *Astronomy & Astrophysics*, *603*, A127.
18. Fernández-Ontiveros, J. A., & Muñoz-Darias, T. (2021). *Monthly Notices of the Royal Astronomical Society*, *504*, 5726.
19. Moravec, E., et al. (2022). *Astronomy & Astrophysics*, *662*, A28.
20. Bariuan, L. G. C., et al. (2022). *Monthly Notices of the Royal Astronomical Society*, *513*, 4673.
21. Churazov, E., et al. (2005). *Monthly Notices of the Royal Astronomical Society*, *363*, L91.
22. Yuan, F., & Narayan, R. (2014). *Annual Review of Astronomy & Astrophysics*, *52*, 529.
23. Zhu, S. F., et al. (2020). *Monthly Notices of the Royal Astronomical Society*, *496*, 245.
24. Buttiglione, S., et al. (2010). *Astronomy & Astrophysics*, *509*, A6.
25. Tadhunter, C. N., et al. (1998). *Monthly Notices of the Royal Astronomical Society*, *298*, 1035.
26. Torresi, E., Grandi, P., Costantini, E., & Palumbo, G. C. C. (2012). *Monthly Notices of the Royal Astronomical Society*, *419*, 321.
27. Grandi, P., et al. (2021). *The Astrophysical Journal*, *911*, 17.
28. Mingo, B., et al. (2022). *Monthly Notices of the Royal Astronomical Society*, *511*, 3250.
29. Bonson, K., & Gallo, L. C. (2016). *Monthly Notices of the Royal Astronomical Society*, *458*, 1927.
30. Gammmie, C. F., Shapiro, S. L., & McKinney, J. C. (2004). *The Astrophysical Journal*, *602*, 312.
31. King, A. R., & Pringle, J. E. (2006). *Monthly Notices of the Royal Astronomical Society*, *373*, L90.
32. King, A. R., & Pringle, J. E. (2007). *Monthly Notices of the Royal Astronomical Society*, *377*, L25.
33. Raychaudhuri, S., Ghosh, S., & Joarder, P. S. (2018). *Monthly Notices of the Royal Astronomical Society*, *479*, 3011.
34. Raychaudhuri, S., Ghosh, S., & Joarder, P. S. (2021). *Journal of Computational Physics*, *2021*, 025.
35. Sarkar, T., Ghosh, S., & Bhadra, A. (2014). *Physical Review D*, *90*, 3008.
36. Ghosh, S., & Banik, P. (2015). *International Journal of Modern Physics D*, *24*, 50084.
37. Mamon, G. A., & Lokas, E. L. (2005). *Monthly Notices of the Royal Astronomical Society*, *363*, 705.
38. Navarro, J. F., Frenk, C. S., & White, S. D. M. (1995). *Monthly Notices of the Royal Astronomical Society*, *275*, 720.
39. Navarro, J. F., Frenk, C. S., & White, S. D. M. (1996). *The Astrophysical Journal*, *462*, 563.
40. Jing, Y. P., & Suto, Y. (2000). *The Astrophysical Journal*, *529*, L69.
41. Chakrabarti, S. K. (1989). *The Astrophysical Journal*, *240*, 7.
42. Abramowicz, M. A., & Chakrabarti, S. K. (1990). *The Astrophysical Journal*, *350*, 281.
43. Das, T. K., Pendharkar, J. K., & Mitra, S. (2003). *The Astrophysical Journal*, *592*, 1078.
44. Chakrabarti, S. K., & Molteni, D. (1993). *The Astrophysical Journal*, *417*, 671.
45. Yang, R., & Kafatos, M. (1995). *Astronomy & Astrophysics*, *295*, 238.

Part III
Astrophysical Fluids and GRMHD Simulations

Chapter 11
Mean Field Dynamo: Beyond MHD

Vinod Krishan

Abstract Magnetic fields pervade the entire universe, including even extremely weakly ionized regions. The generation and sustenance of the magnetic fields is an attendant important issue in such systems. In a multi-fluid system such as a partially ionized plasma, the magnetic induction is subjected to the ambipolar diffusion and the Hall effect in addition to the usual advective and resistive dissipation processes. The Hall effect arises from the treatment of the electrons and the ions as two separate fluids. The ambipolar diffusion arises due to the inclusion of neutrals as the third fluid with consequent ion-neutral drag. Mean field dynamo is one of the most acceptable mechanisms of the generation of large-scale field from small scale field fluctuations in a turbulent plasma. In this talk I shall discuss the mean field dynamo mechanism in a weakly ionized turbulent plasma. It is shown that these non-ideal effects modify the so-called coefficients α, β of the mean field dynamo in a rather substantial way. The Hall effect may enhance or quench the dynamo action altogether. The ambipolar diffusion brings in a contribution to α which depends on the mean magnetic field. The new correlations embodying the coupling of the charged fluids and the neutral fluid appear in a decisive manner. The turbulence is necessarily magnetohydrodynamic with new spatial and time-scales. The nature of the new correlations is demonstrated by taking the Alfvénic turbulence as an example. An illustrative example of magnetic field configuration resulting due to these beyond MHD effects is given.

11.1 Introduction

Plasmas may contain more species of particles than the commonly understood to be the electrons and the ions. Neutrals, charged dust grains and ions with different charged states could be some of the other constituents of a plasma. A

V. Krishan (✉)
Indian Institute of Astrophysics, Bangalore, India
e-mail: vinod@iiap.res.in

S. Ghosh, A. R. Rao (eds.), *The Relativistic Universe: From Classical to Quantum*, Astrophysics and Space Science Proceedings 61,
https://doi.org/10.1007/978-3-031-90186-7_11

solar prominence, for example, has been modeled with 5 species of particles. This necessitates different levels of description of plasmas. More often than not, we treat plasmas as a single fluid, the magnetohydrodynamic (MHD) model wherein the dynamics of all the species of particles are combined in an appropriate manner to construct a single fluid with one density, one pressure and one velocity. The MHD model has no characteristic spatial and temporal time scales. One can also treat each species of particles as a separate fluid. Thus an electron-ion plasma consists of two fluids, the electron fluid and the ion fluid. Inclusion of multi-fluid effects brings out additional physical features at different spatial and time scales. Multi-fluid plasmas exist everywhere: the early universe, neutralization and reionization stages, molecular clouds [1], stellar atmospheres, in particular the solar photosphere and the chromosphere, neutron stars, cold accretion disks and planetary atmospheres to name a few [2]. Here, I consider a partially ionized plasma consisting of three species of particles, electrons, ions and neutrals. A simpler variant of a partially ionized plasma is a weakly ionized plasma described in Sect. 11.2. The Faraday law of magnetic induction, in a weakly ionized plasma, acquires novel, nonlinear and non-ideal effects which are not included in a single fluid MHD formulation. In Sect. 11.3, a brief tutorial on the mean-field dynamo in MHD description is given [3]. I shall discuss the mean-field dynamo mechanism in a weakly ionized plasma in Sect. 11.4. The multifluid effects appear in the spatial and time dependence of the magnetic field. The nature of the plasma turbulence manifests itself in the growth and form of the magnetic field thus generated. A few directions for future investigations in this area are given in the conclusion section.

11.2 Weakly Ionized Plasmas

Partially ionized plasmas are generally cold and collisional. The collisional processes, in a three-component plasma, constitute the electron-neutral and the ion-neutral collisions in addition to the Coulomb collisions between electrons and ions [4]. The elastic collisions result in particle scattering and energy exchange. The scattering cross-sections are typically of the order of geometrical dimensions of atoms or molecules. The inelastic collisions cause excitation, ionization, charge exchange and polarizability. The electron-neutral collision frequency is given as [5],

$$\nu_{\rm en} \sim n_n < \sigma v >_{en} \sim 10^{-15} n_{\rm n} \sqrt{8K_{\rm B}T/(\pi m_{\rm en})}\, \sec^{-1}, \tag{11.1}$$

where the symbols have their usual meaning. The ion-neutral collision frequency is given by

$$\nu_{\rm in} \sim n_n < \sigma v >_{in} \sim 5 \times 10^{-15} n_{\rm n} \sqrt{8K_{\rm B}T/(\pi m_{\rm in})}\, \sec^{-1}. \tag{11.2}$$

Thus, the electrical resistivity of a partially ionized plasma is predominantly due to the electron-neutral collisions in contrast to a fully ionized plasma where electron-ion Coulomb collisions are the main contributors. The dynamics of the three component partially ionized plasma (e,i,n) is governed by the usual laws of mass and momentum conservation for each fluid along with the conservation of energy. The weakly ionized plasma is a simplified form resulting from the neglect of electron and ion inertial terms in their momentum transfer equations. The electron momentum equation yields Ohm's law as

$$\mathbf{E} = -\frac{\mathbf{V}_e \times \mathbf{B}}{c} - \frac{m_e}{e}\nu_{\mathrm{en}}(\mathbf{V}_\mathrm{e} - \mathbf{V}_\mathrm{n}); \mathbf{J} = en_\mathrm{e}(\mathbf{V}_\mathrm{i} - \mathbf{V}_\mathrm{e}). \tag{11.3}$$

where $\mathbf{E}$ is the electric field and $\mathbf{J}$ is the current density. Here pressure forces and the Coulomb collisional forces have been neglected. The inertia-less ion momentum equation, on substituting for $\mathbf{E}$, furnishes the ion-neutral velocity difference:

$$\mathbf{V}_\mathrm{n} - \mathbf{V}_\mathrm{i} = -\frac{m_e \nu_{\mathrm{en}}\mathbf{J}}{e\nu_{\mathrm{in}}\rho_\mathrm{i}} - \frac{\mathbf{J}\times\mathbf{B}}{c\nu_{\mathrm{in}}\rho_\mathrm{i}}, \tag{11.4}$$

and the electron-neutral velocity difference:

$$\mathbf{V}_\mathrm{n} - \mathbf{V}_\mathrm{e} = \mathbf{V}_\mathrm{n} - \mathbf{V}_\mathrm{i} + \frac{\mathbf{J}}{en_\mathrm{e}}. \tag{11.5}$$

The dynamics of the neutral fluid is described as

$$\rho_\mathrm{n}\left[\frac{\partial \mathbf{V}_\mathrm{n}}{\partial t} + (\mathbf{V}_\mathrm{n}\cdot\nabla)\mathbf{V}_\mathrm{n}\right] = -\nabla p_\mathrm{n} - \nu_{\mathrm{ni}}\rho_\mathrm{n}(\mathbf{V}_\mathrm{n} - \mathbf{V}_\mathrm{i}) - \nu_{\mathrm{ne}}\rho_\mathrm{n}(\mathbf{V}_\mathrm{n} - \mathbf{V}_\mathrm{e}), \tag{11.6}$$

which, on substituting for the velocity differences, becomes

$$\rho_\mathrm{n}\left[\frac{\partial \mathbf{V}_\mathrm{n}}{\partial t} + (\mathbf{V}_\mathrm{n}\cdot\nabla)\mathbf{V}_\mathrm{n}\right] = -\nabla p + \frac{\mathbf{J}\times\mathbf{B}}{c}. \tag{11.7}$$

The neutral fluid feels the Lorentz force and behaves like a conducting fluid due to its coupling with electrons and ions. The Faraday law of magnetic induction, on substituting for the electric field and the velocity differences, takes the form [6]:

$$\frac{\partial \mathbf{B}}{\partial t} = \nabla\times\left[\left(\mathbf{V}_\mathrm{n} - \frac{\mathbf{J}}{en_\mathrm{e}} + \frac{\mathbf{J}\times\mathbf{B}}{c\nu_{\mathrm{in}}\rho_\mathrm{i}}\right)\times\mathbf{B}\right] + \eta\nabla^2\mathbf{B}. \tag{11.8}$$

Here, on the RHS, one can easily recognize the first term to be the advective term which shows that the field is being advected by the neutral fluid, the second term represents the Hall electric field which is perpendicular to the current density, the third term is the ambipolar electric field generated due to ion-neutral drag and the last term is the resistive electric field wherein the electrical resistivity is now predominantly due to electron-neutral collisions. Thus, the Hall and the ambipolar terms describe the beyond MHD effects and contain the signatures of the three fluids. It is clear that the ambipolar effect dominates at large values of the magnetic field. In fact the relative importance of the Hall and the ambipolar term can be described in terms of the ratio $\frac{\nu_{in}}{\omega_{ic}}$ where ω_{ic} is the ion-cyclotron frequency. In this system the magnetic field is not frozen to any of the fluids. One can put Eq. (11.8) in a dimensionless form to bring out the new spatial and time scales introduced by the multi-fluid effects.

11.3 Mean Field Dynamo in MHD

In this section, salient features of the mean field dynamo in the single fluid or MHD approximation are described [3]. One begins with a fully ionized two component plasma consisting of electrons and ions in a state of turbulence. The turbulence consists of velocity and magnetic field fluctuations such that there exists a scale separation. One can then split the velocity and the magnetic fields into two components, one on small scales and the other on large scales, the microscopic and the macroscopic as:

$$\mathbf{V} = \mathbf{V}_0 + \mathbf{v}'; \mathbf{B} = \mathbf{B}_0 + \mathbf{b}', \tag{11.9}$$

such that $< \mathbf{v}' >= 0$, $< \mathbf{b}' >= 0$ where the angular brackets stand for ensemble average and $\mathbf{V}_0$ and $\mathbf{B}_0$ are the average large-scale quantities. The mean field dynamo mechanism addresses the question: can the small-scale fluctuations $\mathbf{v}'$ and $\mathbf{b}'$ co-operate to produce large scale $\mathbf{B}_0$ even in the absence of the mean flow $\mathbf{V}_0$? The MHD magnetic induction equation reads:

$$\frac{\partial \mathbf{B}}{\partial t} = \nabla \times [\mathbf{V} \times \mathbf{B}] + \eta \nabla^2 \mathbf{B}. \tag{11.10}$$

Substitute Eq. (11.9) into Eq. (11.10) to get

$$\frac{\partial [\mathbf{B} + \mathbf{b}']}{\partial t} = \nabla \times [(\mathbf{V} + \mathbf{v}') \times (\mathbf{B} + \mathbf{b}')] + \eta \nabla^2 [\mathbf{B} + \mathbf{b}']. \tag{11.11}$$

Taking the ensemble average gives the evolution of the large-scale field:

$$\frac{\partial \mathbf{B}_0}{\partial t} = \nabla \times [\mathbf{V}_0 \times \mathbf{B}_0 + < \mathbf{v}' \times \mathbf{b}' >] + \eta \nabla^2 \mathbf{B}_0. \tag{11.12}$$

The new term

$$\mathbf{E}_F =< \mathbf{v}' \times \mathbf{b}' >, \tag{11.13}$$

quadratic in fluctuations and presumably nonzero acts as a generating term, in addition to the advective term, for the large-scale magnetic field. Subtraction of Eqs. (11.11) and (11.12) gives the evolution of the small-scale magnetic field to the first order in fluctuations

$$\frac{\partial \mathbf{b}'}{\partial t} = \nabla \times [\mathbf{V}_0 \times \mathbf{b}' + \mathbf{v}' \times \mathbf{B}_0] + \eta \nabla^2 \mathbf{b}'. \tag{11.14}$$

Integration of Eq. (11.14) from sometime in the past up to the present time t gives $\mathbf{b}'$ using which one can determine the generating electric field in the absence of the large-scale flow ($\mathbf{V}_0 = 0$) as

$$\mathbf{E}_F = \alpha_v \mathbf{B}_0 - \beta_v \nabla \times \mathbf{B}_0, \tag{11.15}$$

The details of the procedure to go from Eq. (11.14) to (11.15) can be seen in [3] and [2]. Here, the kinetic helicity α_v responsible for the growth of the magnetic field is defined as

$$\alpha_v = -\frac{\tau_{cor}}{3}(< \mathbf{v}' \cdot \mathbf{\Omega}' >), \mathbf{\Omega}' = \nabla \times \mathbf{v}', \tag{11.16}$$

where $\mathbf{\Omega}'(r, t)$ is the fluid vorticity and τ_{cor} is the time interval over which the fluctuations are correlated. The second dynamo quantity β_v represents dissipation of the magnetic field and it is defined as

$$\beta_v = \frac{\tau_{cor}}{3} < v'^2 > . \tag{11.17}$$

The large-scale field can now be determined from

$$\frac{\partial \mathbf{B}_0}{\partial t} = \nabla \times [\alpha_v \mathbf{B}_0] + (\eta + \beta_v)\nabla^2 \mathbf{B}_0, \tag{11.18}$$

where the dynamo quantities α_v, β_v are constants. Since Eq. (11.18) is a linear equation, one can easily check that the magnetic field grows exponentially as $e^{ikz - i\omega t}$ for

$$k < \alpha_v(\eta + \beta_v)^{-1} \tag{11.19}$$

or for large spatial scales. This is the mechanism of the mean field dynamo in the single fluid which will now be carried over to a multifluid system.

11.4 Mean Field Dynamo in a Weakly Ionized Plasma

The magnetic induction equation in a weakly ionized plasma, Eq. (11.8), can be rewritten as

$$\frac{\partial \mathbf{B}}{\partial t} = \nabla \times [\mathbf{V}_E \times \mathbf{B}] + \eta \nabla^2 \mathbf{B}. \tag{11.20}$$

The effective velocity$\mathbf{V}_E$ is defined as

$$\mathbf{V}_E = \mathbf{V}_n + \mathbf{V}_H + \mathbf{V}_A, \tag{11.21}$$

with the Hall velocity

$$\mathbf{V}_H = -\frac{\mathbf{J}}{en_e}, \tag{11.22}$$

and the ambipolar velocity

$$\mathbf{V}_A = \frac{\mathbf{J} \times \mathbf{B}}{c\rho_i \nu_{in}}. \tag{11.23}$$

Following the procedure outlined in the previous section we split V_E into its large scale and small-scale parts as

$$\mathbf{V}_E = \mathbf{V}_{E0} + \mathbf{V}'_E, \tag{11.24}$$

with

$$\mathbf{V}_{E0} = \mathbf{V}_{n0} - \frac{\mathbf{J}_0}{en_e} + \frac{\mathbf{J}_0 \times \mathbf{B}_0}{c\rho_i \nu_{in}}, \tag{11.25}$$

and

$$\mathbf{V}'_E = \mathbf{V}'_n - \frac{\mathbf{J}'}{en_e} + \frac{\mathbf{J}' \times \mathbf{B}_0}{c\rho_i \nu_{in}} + \frac{\mathbf{J}_0 \times \mathbf{b}'}{c\rho_i \nu_{in}}. \tag{11.26}$$

The effective helicity, α_E, has now three contributions, the kinetic helicity, the Hall helicity and the ambipolar helicity:

$$\alpha_E = -\frac{\tau_{cor}}{3}[< \mathbf{V}'_E.(\nabla \times \mathbf{V}'_E) >] = \alpha_v + \alpha_H + \alpha_A.\mathbf{B}_0, \tag{11.27}$$

with

$$\alpha_v = -\frac{\tau_{cor}}{3}[< \mathbf{V}'_n.(\nabla \times \mathbf{V}'_n) >], \alpha_H = \frac{2\tau_{cor}}{3en_e}(< \mathbf{J}'.\boldsymbol{\Omega}'_n >), \tag{11.28}$$

and

$$\alpha_A = \frac{2\tau_{cor}}{3c\rho_i \nu_{in}}(< \mathbf{j}' \times \mathbf{\Omega}'_n >). \tag{11.29}$$

The coupling of the neutral fluid and the charged components appears through the correlation between the current density fluctuations and the neutral fluid vorticity fluctuations. This is in essence the manifestations of the MHD turbulence in contrast to the case where only α_v, a purely hydrodynamic quantity, is the contributor to the generation of the magnetic field. The contribution of the ambipolar term is essentially nonlinear since it depends on the large-scale field $\mathbf{B}_0$ which may provide a saturation mechanism for the growing large-scale field. One also notices the difference in the nature of correlations in α_H and α_A. While the Hall effect needs a component of the current density fluctuations parallel to the velocity fluctuations, the ambipolar effect thrives if the current density fluctuations have a component perpendicular to the velocity fluctuations. The other dynamo quantity β_E corresponding to β_v can be similarly defined as

$$\beta_E = \frac{\tau_{cor}}{3}[< \mathbf{V}_E'^2 >] = \beta_v + \beta_H + \beta_A.\mathbf{B}_0, \tag{11.30}$$

with

$$\beta_v = \frac{\tau_{cor}}{3}[< \mathbf{V}_n'^{\,2} >], \beta_H = -\frac{2\tau_{cor}}{3en_e} < (\mathbf{J}'.\mathbf{V}_n') >, \tag{11.31}$$

and

$$\beta_A = \frac{2\tau_{cor}}{3c\rho_i \nu_{in}} < \mathbf{J}' \times \mathbf{\Omega}'_n >, \tag{11.32}$$

keeping terms only first order in $\mathbf{J}'$. It is instructive to examine the new correlations for the case, say, of Alfvénic turbulence. In a weakly ionized plasma, the relation between the velocity and the magnetic fluctuations for the Alfven waves is found to be [7]

$$V_n' = \pm\frac{b'}{\sqrt(4\pi\rho_i)}[\frac{\rho_i}{\rho_n}], \tag{11.33}$$

where ρ_i and ρ_n are respectively the mass densities of the ionic and the neutral fluids. Substitution for $\mathbf{J}'$ (through $\mathbf{b}'$) in α_H fetches

$$\alpha_H = \pm 2\frac{\rho_n}{\rho_i}\frac{\lambda_H}{\lambda_n}\alpha_v, \tag{11.34}$$

where $\lambda_H = \frac{(c\sqrt{(m_i)}}{\sqrt{(4\pi n_i e^2)}}$ is the ion inertial length and $\lambda_n = \frac{<(\mathbf{V}'_n.\mathbf{\Omega}'_n)>}{<\Omega'^2_n)>}$ is the ratio of the kinetic helicity and the enstrophy of the neutral fluid turbulence and represents the typical spatial scale of the turbulence. The Hall effect can enhance or quench the dynamo mechanism for the Alfvénic turbulence. It is noteworthy that the ambipolar effect vanishes for the Alfvénic turbulence. The dynamo equation in a weakly ionized plasma takes the form

$$\frac{\partial \mathbf{B}_0}{\partial t} = \nabla \times [\alpha_E \mathbf{B}_0] + (\eta + \beta_E)\nabla^2 \mathbf{B}_0. \tag{11.35}$$

An illustrative example can be shown by taking $\mathbf{B}_0 = (0, B, \partial A/\partial x)$ in Cartesian coordinates (x, y, z) which become (θ, ϕ, r) in the corresponding spherical coordinates. Here A is the vector potential. Thus $(B, \partial A/\partial x)$ represent respectively the toroidal and the poloidal components of the field. Following [8], the boundary conditions then become the vanishing of (B, A) at the end points of a finite interval $x = (0, \pi R)$ corresponding to the poles of the sphere. The details of the solution procedure of Eq. (11.35) can be seen in [6]. A couple of field plots are reproduced here for a glimpse of the qualitative appreciation of the beyond MHD effects in the mean field dynamo mechanism. The induction equation has been put in a dimensionless form by using a spatial scale R, a normalizing field b and a time scale (R^2/η). The initial conditions are those taken in [8] viz $A'(x, 0) = 0$; $B(x, 0) = \sin x$. A representative solution of the dynamo equation is presented in Fig. 11.1. One notices a sharp growth of both the components of the magnetic field after a nearly steady state. Formation of sharp magnetic structures or current sheets due to the Hall and the ambipolar effects is quite evident. The range of variation of the field components needs to be studied in detail for the emergence of other potential configurations.

11.5 Conclusion

The mechanism of the mean field dynamo operates in a turbulent conducting fluid. It can magnify an initial seed magnetic field and is believed to be responsible for the maintenance of cosmic magnetic fields in different locales. First studied in a single conducting fluid in the limit of MHD, it has been later extended to include beyond MHD effects. Here we have demonstrated the action of the mean field dynamo in a weakly ionized plasma wherein the multi-fluid effects contribute in a decisive and significant manner to the growth of the magnetic field. The Hall effect arises from treating the electron and the ion fluids as two separate fluids and the ambipolar effect due to the drag between the ions and the neutrals. These effects result not only in a modified growth rate but generates specific magnetic configurations such as current sheets and filaments which may be prone to processes such as magnetic reconnection, nonlinear dissipation and saturation. The beyond MHD effects need

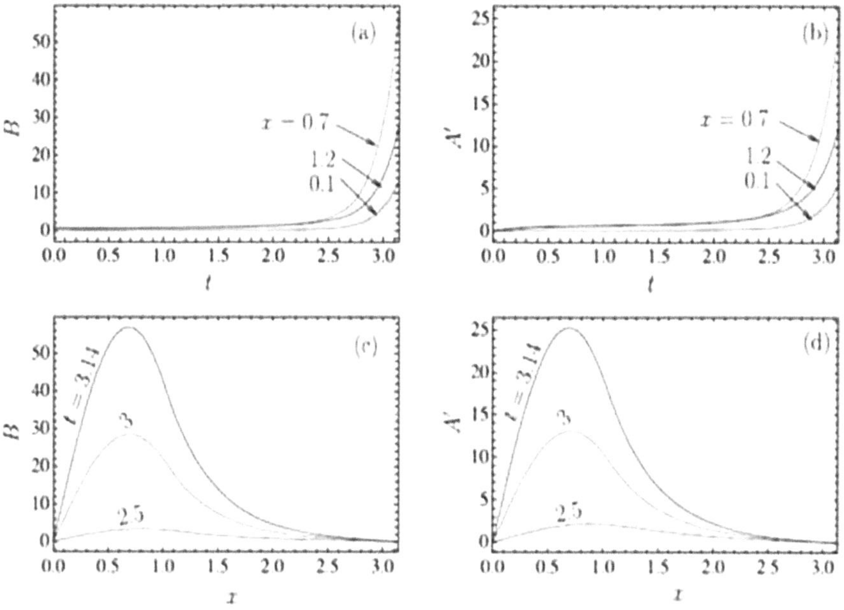

Fig. 11.1 Plots of the toroidal and poloidal components of the magnetic field evolved through the dynamo mechanism contained in Eq. (11.35) (re-produced from [6])

to be explored in several different astrophysical objects and sites as for example in neutron stars and the epochs of neutralization and reionization of the early universe. In addition, the multi fluid effects including finite ion inertia and higher order turbulent fluctuations may reveal more novel features of a beyond MHD description of a mean field dynamo.

Acknowledgments The author thanks the organizers for giving her an opportunity to share her thoughts on the topic of the magnetic dynamos.

References

1. Brandenburg, A., Zweibel, E. G. (1994). *The Astrophysical Journal, 427*, L91.
2. Krishan, V. (2016). Physics of partially ionized plasmas. Cambridge University Press.
3. Krause, F., & Radler, K. H. (1980). *Mean field magnetohydrodynamics and dynamo theory*. Pergamon Press.
4. Alfvén, H., & Falthammer, C-G. (1962). *Cosmic electrodynamics*. Clarendon Press.
5. Krishan, V. (2022). *Journal of Astronomy & Astrophysics, 43*.
6. Krishan, V., & Gangadhara, R. T. (2008). *Monthly Notices of the Royal Astronomical Society, 85*, 849.
7. Krishan, V., & Varghese, B. A. (2008). *Solar Physics, 247*, 343.
8. Stix, M. (1972). *Astronomy & Astrophysics, 20*, 9.

Chapter 12
General Relativistic Hydrodynamic Simulations Around Accreting Black Holes

Sudip K. Garain and Pranayjit Dey

Abstract Strong gravity in the immediate vicinity of compact objects (e.g., black holes, neutron stars) necessitates inclusion of general relativistic effects. Traditionally, pseudo-Newtonian potential representations of gravity were favored to simulate the fluid motion in this region since that reduced the calculation complexity. However, with the advent of easily implementable, reliable numerical algorithms and computer hardware, more and more research groups are shifting towards the numerical solutions of general relativistic fluid dynamics equations. In this chapter, we report our progress on the development of such simulation tool and present results of sub-Keplerian accretion flow onto black holes.

12.1 Introduction

Fluid configuration around the black holes determines the spectro-temporal and polarimetric signatures for stellar mass black hole X-ray binaries (BHXRBs) as well as the active galactic nuclei (AGNs) [1, 2]. Strong gravity around the black holes mostly dictates the fluid behavior. Therefore, solution of general relativistic fluid dynamics equations are frequently used to infer the spectro-temporal as well as polarimetric properties.

Majority of the general relativistic simulations of accretion disk start from an initially equilibrium torus threaded with a seed magnetic field ([3] and references therein). However a more realistic simulation set up may be constructed by letting the matter ideally come from far out and enter the simulation domain situated at a

S. K. Garain (✉)
Department of Physical Sciences and Center of Excellence in Space Sciences India, Indian Institute of Science Education and Research, Kolkata, India
e-mail: sgarain@iiserkol.ac.in

P. Dey
Department of Physical Sciences, Indian Institute of Science Education and Research, Kolkata, India

S. Ghosh, A. R. Rao (eds.), *The Relativistic Universe: From Classical to Quantum*, Astrophysics and Space Science Proceedings 61,
https://doi.org/10.1007/978-3-031-90186-7_12

finite distance. Model based analytical solution for velocity components, pressure, density etc. can be supplied as the inflow boundary condition at the outer boundary of the simulation domain and the time-dependent fluid dynamical equations, with the help of this boundary condition, should determine the dynamical solution inside the domain. Accretion simulations of collapsing plasma with initially Bondi type spherically symmetric matter distribution, but with latitude-dependent low angular momentum (sub-Keplerian), has been done earlier using non-GR codes ([4–6]) and recently using GRMHD simulations ([7–10]). We are, rather, interested in accretion simulations of axisymmetric, thick disk type configuration. Such set-up has been used to simulate sub-Keplerian and Keplerian matter accretion onto black holes [11–16]. However, all these simulations are done using pseudo-Newtonian potential proposed by [17].

Our aim is to extend such simulations using general relativistic fluid dynamics solvers. A few general relativistic hydrodynamics (GRHD) simulations of transonic, sub-Keplerian accretion disk with this type of set-up have been conducted earlier [18, 19]. However, further extensions are not reported. We are in process of developing a three dimensional GRHD code, designed specifically to implement the above set-up, and in this chapter, report the initial results using that code.

Our chapter is organized as follows: In Sect. 12.2, we provide a very brief overview of the general relativistic analytical solution of sub-Keplerian accretion flow. In Sect. 12.3, we introduce the GRHD equations and our numerical solution methodology. In Sect. 12.4, we present the results. Finally, in Sect. 12.5, we provide concluding remarks.

In our following calculations, we use $r_g = GM_{bh}/c^2$ as unit of distance, r_g/c as unit of time and $r_g c$ as unit of specific (i.e., per unit mass) angular momentum. Specific energy is measured in the unit of c^2. Here, G is the gravitational constant, M_{bh} is the mass of the black hole and c is the speed of light in vacuum.

12.2 Sub-Keplerian Flow: Analytical Solutions

The general relativistic analytical solutions for transonic, sub-Keplerian accretion flow onto compact objects are discussed in [20–22] with sufficient details. A non-dissipative, transonic accretion flow is usually characterized by the conserved parameters specific energy ($\epsilon = u_t/(1 - na_s^2)$) and specific angular momentum ($l = -u_\phi/u_t$). Here, u_t and u_ϕ are the t and ϕ components of the four-velocity u_μ, n is the polytropic index and a_s is the local sound speed.

The analytical solution aims to find the radial variation of the fluid variables under the stationary and axisymmetric conditions. The calculation starts by considering the equations of the conserved specific energy $\epsilon = u_t/(1 - na_s^2)$ and the mass accretion rate $\dot{m} = Au^r\rho$. Here, A is a geometric quantity representing the surface area through which mass flux is considered, u^r is the r component of the four-velocity u^ν and ρ is the rest mass density. Further, using the adiabatic equation of state $P = K\rho^{\Gamma}$, with P being pressure and Γ being the adiabatic index, and the relation between enthalpy h and a_s as $h = 1/(1 - na_s^2)$, ϵ and $\dot{m}$ are expressed in terms of a_s and the radial velocity $\mathcal{V}$ in the rotating frame. Using the fact that ϵ and $\dot{m}$ are conserved, by differentiating these equations w.r.t. r and eliminating da_s/dr, one can obtain an equation for $d\mathcal{V}/dr$ [18, 19]. Numerical integration of this equation will result in $\mathcal{V}(r)$ and subsequently all other fluid variables as a function of r.

12.3 General Relativistic Hydrodynamics: Basic Equations and Solution Procedure

For numerical simulations, we solve the following conservation equations:

$$\nabla_{\mu}\left(\rho u^{\mu}\right) = 0\,, \tag{12.1}$$

$$\nabla_{\mu} T^{\mu\nu} = 0\,. \tag{12.2}$$

Here, ∇_{μ} represents the covariant derivative, u^{μ} is the four-velocity and $T^{\mu\nu}$ is the stress-energy tensor. $T^{\mu\nu} = \rho h u^{\mu} u^{\nu} + P g^{\mu\nu}$ for ideal fluid with h as the specific enthalpy given by $h = 1 + \frac{\Gamma}{\Gamma-1}\frac{P}{\rho}$, $\Gamma = 4/3$ being the adiabatic index. Following {3+1} formalism [23, 24], we write the space-time metric $g_{\mu\nu}$ in terms of lapse (α), shift vector (β^i) and the spatial metric (γ_{ij}). After some algebraic manipulations, this set of equations can be written as a set of five partial differential equations (PDEs):

$$\frac{1}{\sqrt{-g}}\left[\frac{\partial\sqrt{\gamma}D}{\partial t} + \frac{\partial}{\partial x^i}\left(\sqrt{-g}D\left(v^i - \frac{\beta^i}{\alpha}\right)\right)\right] = 0\,, \tag{12.3}$$

$$\frac{1}{\sqrt{-g}}\left[\frac{\partial\sqrt{\gamma}S_j}{\partial t} + \frac{\partial}{\partial x^i}\left(\sqrt{-g}\left(S_j\left(v^i - \frac{\beta^i}{\alpha}\right) + P\delta^i_j\right)\right)\right]$$
$$= T^{\mu\nu}\left(\frac{\partial g_{\nu j}}{\partial x^{\mu}} - \Gamma^{\lambda}_{\nu\mu} g_{\lambda j}\right)\,, \tag{12.4}$$

and

$$\frac{1}{\sqrt{-g}}\left[\frac{\partial\sqrt{\gamma}\tau}{\partial t} + \frac{\partial}{\partial x^i}\left(\sqrt{-g}\left(\tau\left(v^i - \frac{\beta^i}{\alpha}\right) + Pv^i\right)\right)\right]$$
$$= \alpha\left(T^{\mu 0}\frac{\partial \ln\alpha}{\partial x^{\mu}} - T^{\mu\nu}\Gamma^{0}_{\mu\nu}\right)\,. \tag{12.5}$$

Here, $\sqrt{-g} \equiv det(g_{\mu\nu})$ and $\sqrt{\gamma} \equiv det(\gamma_{ij})$, and these are connected by $\sqrt{-g} = \alpha\sqrt{\gamma}$. We denote the set of five-component vector $U = \left(D, S_j, \tau\right)$ as vector of conserved variables and can be expressed in terms of vector of primitive variables $V = \left(\rho, v^i, P\right)$ as follows:

$$D = \rho W, \quad S_j = \rho h W^2 v_j, \quad \tau = \rho h W^2 - P - D.$$

Here, W is the Lorentz factor given by $W = 1/\sqrt{1 - v^i v_i} = \alpha u^t$. v^i are the components of three-velocity given as $v^i = \frac{u^i}{\alpha u^t} + \frac{\beta^i}{\alpha}$ and the co-variant counter part can be calculated as $v_j = \gamma_{ij} v^i$.

The above set of PDEs is further written in integral form and subsequently discretized on a given mesh [23, 24]. For our present calculations, we use Schwarzschild space-time metric for which $\alpha = \sqrt{(1 - 2/r)}$, $\beta^i = 0$, $\sqrt{-g} = r^2 \sin\theta$ and $\sqrt{\gamma} = r^2 \sin\theta\,(1 - 2/r)^{-1/2}$. The resulting discretized equations on a spherical mesh constructed using Boyer–Lindquist coordinates (t, r, θ, ϕ) [25] are solved using finite volume method. A better choice may be to use horizon penetrating Kerr-Schild coordinate representation of Schwarzschild metric, which we may adopt in future works.

For spatial reconstruction, we have used second order accurate van Leer slope limiter following [26]. We perform reconstruction on vector $\left(\rho, W v^i, P\right)$, instead of primitive variable vector V, since the reconstruction on $W v^i$ ensures sub-luminal reconstructed profile of v^i inside a zone [27]. We have provisions for HLL and LLF Riemann solvers for calculating the interfacial fluxes. Second-order accurate strong stability preserving Runge-Kutta(RK) time integration is used for time advancement. One of the non-trivial step in general relativistic hydrodynamics is conserved-to-primitive conversion as it requires a non-linear equation solution employing a root solver (e.g., Newton-Raphson). We have implemented two methods following [28] and [29]. For our calculations, we prefer the method of [29]. It may happen that the root solver does not converge for a few zones after the time-update step and for such pathetic zones, we use the previous time-step solution as it is already saved in a RK type time-update. The timestep dt is calculated following standard Courant-Friedrichs-Lewy (CFL) condition ([30, 31])

$$dt = C_{\text{CFL}} \frac{1}{\frac{\lambda^r}{dr} + \frac{\lambda^\theta}{r d\theta} + \frac{\lambda^\phi}{r \sin\theta d\phi}},$$

where, λ^i is maximum speed in i^{th} direction and C_{CFL} is the CFL number. For all the runs, we use $C_{CFL} = 0.9$. For one- or two-dimensional simulations, contribution from the corresponding inactive dimension(s) is switched off.

12.4 Results

In this section, we present results of a few standard test problems validating our implementation. Later in this section, we present results for couple of production runs for sub-Keplerian accretion disk.

12.4.1 One-dimensional Test Problems

In this sub-section, we present results of a couple of test problems to demonstrate the achievement of global accuracy and correctness of our code.

12.4.1.1 Stationary Torus

In this benchmarking test problem, we initialize a one-dimensional (radial direction) computation domain [4:40] using a constant specific angular momentum ($l = 3.9$), stationary torus solution [32–34] and evolve the solution for one-full rotation period $t = 100$ at the density maximum. Next, we subtract the numerical solution from the stationary solution and compute the errors. For a globally (i.e., spatially and temporally) second order accurate code, the error should converge with second order accuracy. For this test problem, we run simulations with zones ranging from 64 to 4096. The results are shown in Table 12.1. Both the L_1 and L_{inf} accuracy columns show achievement of second order accuracy in the asymptotic limit.

12.4.1.2 Standing Accretion and Wind Shock Solutions

In these one-dimensional benchmarking test problems, we demonstrate our code's capability to capture discontinuous solutions that are predicted in [35, 36]. For the black hole accretion (wind) solution having multiple sonic points, it is possible that the solution branch passing through the outer (inner) sonic point is connected to the solution branch passing through the inner (outer) sonic point via a shock

Table 12.1 Second order convergence for static torus problem

Zones	L_1 error	Accuracy	L_{inf} error	Accuracy
64	5.11×10^{-5}		7.14×10^{-3}	
128	1.58×10^{-5}	1.69	1.99×10^{-3}	1.84
256	3.72×10^{-6}	2.09	1.66×10^{-3}	0.26
512	5.05×10^{-7}	2.88	2.79×10^{-4}	2.57
1024	1.06×10^{-7}	2.26	1.71×10^{-5}	4.03
2048	2.80×10^{-8}	1.92	2.84×10^{-6}	2.56
4096	7.36×10^{-9}	1.93	7.46×10^{-7}	1.93

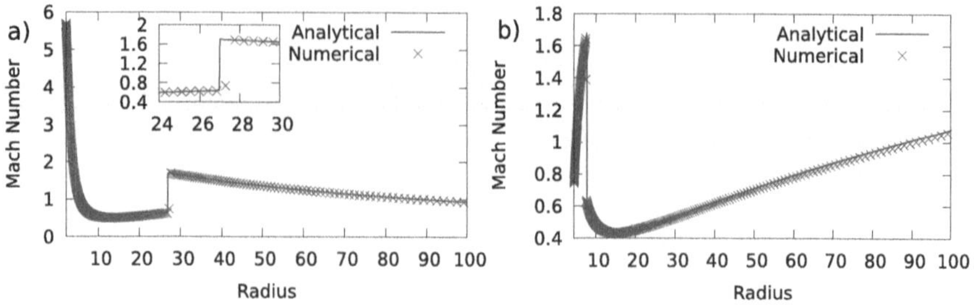

Fig. 12.1 (**a**) shows the comparison between the analytical (*solid line*) and the numerical solutions (*crosses*) for the shock in accretion solution whereas (**b**) shows the same for shock in wind solution. Both the plots show a very good matching between the analytical and numerical results

transition. Analytically, one needs to perform the Rankine-Hugoniot analysis to find the shock location. Using numerical simulation, we are able to detect the shock at the analytically predicted location with a satisfactory level of accuracy.

Figure 12.1a shows an example of the shock in accretion solution. The solid line shows the analytical solution and the crosses show the numerical solution. The numerical solution is shown at a time $t = 20,000$. This simulation is run on a one-dimensional radial domain [2.2:100] using 300 ratioed grid (with common ratio = 1.018). We use inflow type boundary condition at the outer radial boundary $r = 100$ and outflow type boundary condition at the inner radial boundary $r = 2.2$. This solution is specified by the specific energy ($\epsilon = 1.007$) and specific angular momentum ($l = 3.4$). The outer sonic point, shock and the inner sonic points are located at 89.6, 27 and 5.4, respectively. The numerical solution clearly captured all this locations. We notice from Fig. 12.1a that flow is highly supersonic when it crosses the inner boundary. Therefore, acoustic waves do not propagate any feedback upstream from the inner boundary.

Figure 12.1b shows an example of shock in wind solution. The line-point styles are same as in Fig. 12.1a. Here, the numerical solution is shown at a time $t = 30,000$. This simulation is run on a one-dimensional radial domain [4.5:100] using 300 ratioed grid (with common ratio = 1.018). We use inflow type boundary condition at the inner radial boundary $r = 4.5$ and outflow type boundary condition at the outer radial boundary $r = 100$. This solution is specified by $\epsilon = 1.007$ and $l = 3.48$. The inner sonic point, shock and the outer sonic points are located at 5.04, 7.7 and 88.23, respectively. Here again, the numerical solution captured all the locations.

12.4.2 Two-dimensional Test Problems

In this sub-section, we present results of a couple of test problems to demonstrate the operability of our code in multi-dimensions.

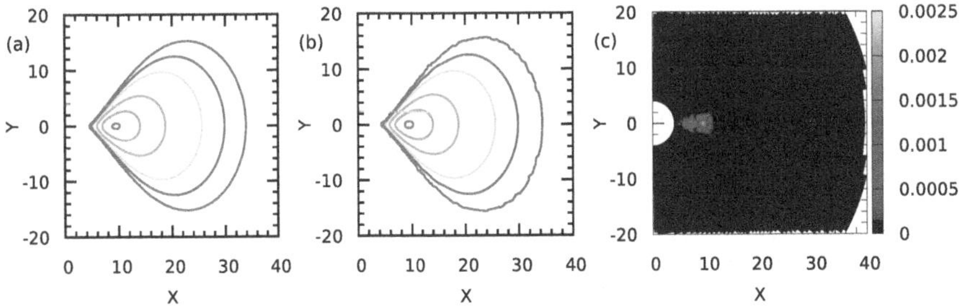

Fig. 12.2 (**a**) and (**b**) show the contours of constant rest mass densities on log scale at the initial and the final times, respectively. The torus structure is found to be well maintained except towards the outer layers of the torus. (**c**) shows the plot of grid-by-grid difference between the density values. We see difference is highest around the location of density maximum

12.4.2.1 Stationary Torus

This is a multi-dimensional extension of the test problem presented in Sect. 12.4.1.1. We initialize a constant specific angular momentum $l = 3.9$ stationary torus on the $r - \theta$ domain [4:40] $\times$ [0 : π] using 128 logarithmically binned radial grids and 180 uniform angular grids. The initial condition is evolved till $t = 500$ GM/c^3 using outflow boundary conditions on radial boundaries and reflective boundary conditions on polar boundaries. Since the torus solution is a result of hydrostatic equilibrium, it is expected that the torus structure will be well-maintained except the numerical dissipation errors. Figure 12.2a and b show the contours of constant rest mass densities on log scale at the initial and the final times, respectively. The inner-most contour corresponds to $\log_{10}$(density) = -1.25 and successive contours correspond to -1.5, -2, -3, -4 and 7. The contours maintain their overall structure unchanged except a few wiggles on the outer most contour. This is due to the numerical dissipation of our second-order accurate code. Figure 12.2c shows the grid-by-grid difference of the rest mass density values and we notice that the maximum difference arises at the center of the torus, which is the location of density maximum.

12.4.2.2 Bondi Accretion

In this test problem, we simulate the spherically symmetric Bondi accretion [37] onto a black hole. The simulation is performed on the $r - \theta$ domain [2.2:100] $\times$ [0 : π] using 150 logarithmically binned radial grids and 180 uniform angular grids. The simulation domain is initially filled with a background matter having density 10^{-8} times lower than the incoming matter density. We use inflow boundary condition at the outer radial boundary and outflow boundary condition at the inner radial boundary. Thus, at all the ghost zones of the outer radial boundary, we maintain the

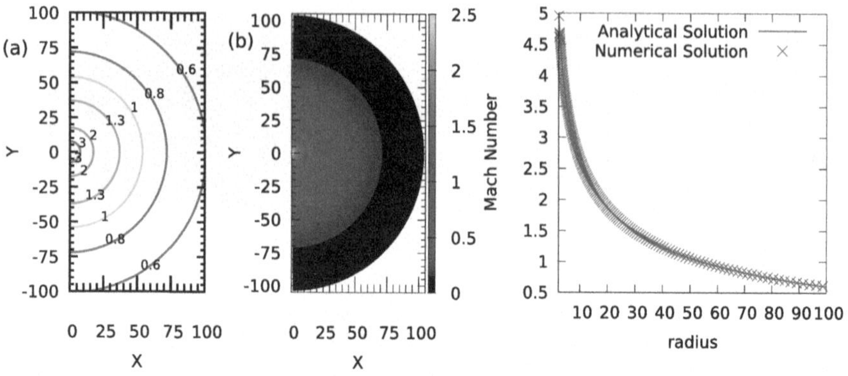

Fig. 12.3 (**a**) shows the contours of constant Mach numbers whereas (**b**) shows the rest mass densities on log scale at the final time. Both these figures demonstrate the spherical symmetry of the solution. (**c**) shows the comparison of radial Mach number variation between the analytical (*solid line*) and the numerical (*crosses*) results

vector of primitive variables $V = (1.0, -0.05233, 0, 0, 0.005967)$ corresponding to $\epsilon = 1.015$ during the entire simulation.

The simulation is run till the time of $t = 10,000$. By this time, a steady state solution is achieved. Figure 12.3a shows the contours of constant Mach numbers whereas b shows the rest mass density distribution on log scale at the final time. Both these figures demonstrate the spherical symmetry of the final solution. Figure 12.3c shows the comparison of radial Mach number variation between the analytical (solid line) and the numerical (crosses) results. We find slight mismatch very close to the inner boundary. Here, the gradients of fluid variables are very steep and we believe, to capture the correct solution, we need to use either high order accurate reconstruction schemes or finer resolution if we continue to use second order accurate schemes.

12.4.3 Sub-Keplerian Accretion Disk

In this sub-section, we present results for couple of multi-dimensional simulations of sub-Keplerian accretion disk. We run two cases with two different l values : for run R1, $l = 3.5$ and run R2, $l = 3.56$. $\epsilon = 1.002$ for both the runs. Analytically, these parameters allow shock formation in the sub-Keplerian accretion flow. Because of the higher l value for R2, the average shock location is expected to be at higher radial distance.

For both the cases, the simulations are performed on a $r - \theta$ domain $[2.1{:}200] \times [0 : \pi]$ using 300 logarithmically binned radial grids and 180 uniform angular grids. The simulation domain is initially filled with a static matter having density and pressure as corresponding floor values (10^{-8} and 10^{-11} respectively). Matter enters

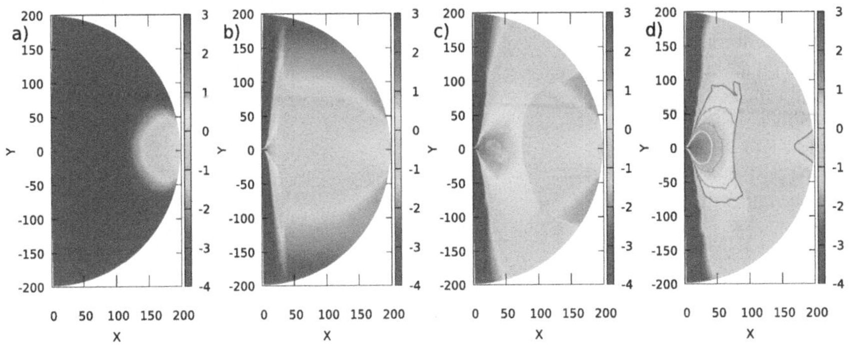

Fig. 12.4 shows the time evolution of the sub-Keplerian accretion disk structure. Colors show the rest mass density distribution. Snapshots are shown at times (**a**) 400, (**b**) 1600, (**c**) 6000 and (**d**) 18,000. Density isocontours corresponding to density values 10, 5, 2 and 1 as we move from inner-most contour to the outer one, are over-plotted in (**d**). The isocontours clearly show the formation of density torus in the post-shock region

the simulation domain at a constant rate through the outer radial boundary. Velocity components and sound speed of the incoming matter are calculated following the analytical solution provided in Sect. 12.2 for a given (ϵ, l) pair. Since we don't have radiative cooling or viscous dissipations, incoming matter density ρ_{out} is normalized to 1.0. The sound speed and ρ_{out} together allow one to calculate the pressure of the incoming matter for a given Γ which is 4/3 for these runs. Thus, the vector of primitive variables V for the incoming matter is calculated using (ϵ, l) pair and this V is maintained at the ghost zones of the outer radial boundary throughout the simulation to mimic constant matter supply to the black hole. This inflow boundary condition is maintained for zones having $80° \leq \theta \leq 110°$. Other than this, we use outflow boundary condition at all other outer radial grids to allow outflow from the accretion disk. We also use outflow boundary condition at the inner radial boundary to mimic free-flow of matter towards the horizon. Simulations are run till a stopping time of $t = 18,000$.

Figure 12.4 shows the time-evolution for run R1. Colors show the density distribution at times (a) 400, (b) 1600, (c) 6000 and (d) 18,000. Time evolution for run R2 follows similar pattern. The simulations achieve a nearly steady state around time $t = 6000$ (i.e., state corresponding to Fig. 12.4c). Rest mass density isocontours corresponding to density values 10, 5, 2 and 1 as we move from inner-most contour to the outer one, are over-plotted in Fig. 12.4d. The isocontours clearly show the formation of density torus in the post-shock region. The density torus resembles the thick torus which are constructed using hydrostatic equilibrium assumption (e.g., Fig. 12.2). However, our simulated torus has advection included and is highly dynamic. The former is a result of balance between the inward gravitational force and outward combined effect of centrifugal and pressure gradient forces. The flow is purely azimuthal with radial component of four velocity set to

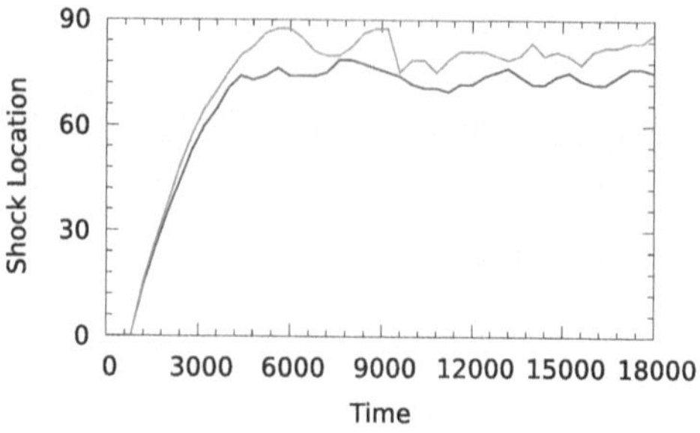

Fig. 12.5 shows the time variation of the shock location on the equatorial plane for the two different runs (green—R2, purple—R1). This plot shows that the post-shock torus is dynamic rather than being static

zero. However, in our case, the flow has both azimuthal and radial non-zero velocity components. Because of this radial component, energy-momentum is advected towards the black hole. The outer boundary of the torus coincides with the location of shock in the accreting sub-Keplerian matter. Figure 12.5 shows the time variation of the shock location on the equatorial plane for the two different runs (green—R2, purple—R1). This plot shows that the post-shock torus is dynamic rather than being static. Also, the torus size is larger for higher l. Such post-shock dynamical tori are used to explain the observed spectral and temporal properties of accreting black holes [38–41].

12.5 Summary and Conclusions

In this chapter, we present our progress of developing a general relativistic fluid dynamics solver. Our aim is to use the said solver for simulating an accretion disk configuration that mimics mass inflow from far out rather than starting from an equilibrium torus. We have demonstrated that our presently developed GRHD code works for Schwarzschild spacetime, is globally second order accurate and performs robustly in multi-dimensions. Finally, using this code, we simulate geometrically thick sub-Keplerian accretion disks. At the time of writing this chapter, we have extended the code's operability in three-dimensions and the performance is being tested. We'll report the results in future publications.

Acknowledgments We acknowledge the usage of Kepler cluster of DPS, IISER Kolkata and Pegasus cluster of IUCAA, Pune for running a few simulations. SKG also acknowledges the support of start-up research grant provided by IISER Kolkata. The authors thank the anonymous referee for constructive suggestions for improving the chapter.

Competing Interests The authors have no conflicts of interest to declare that are relevant to the content of this chapter.

References

1. Belloni, T. M., & Motta, S.E. (2016). Astrophysics of black holes: From fundamental aspects to latest developments. In: C. Bambi (Ed.), *Astrophysics and space science library* (Vol. 440, p. 61). https://doi.org/10.1007/978-3-662-52859-4-2.
2. Netzer, H. (2015). *Annual Review of Astronomy and Astrophysics*, *53*, 365. https://doi.org/10.1146/annurev-astro-082214-122302.
3. Porth, O., Chatterjee, K., Narayan, R., Gammie, C. F., Mizuno, Y., Anninos, P., Baker, J. G., Bugli, M., Chan, C. K., Davelaar, J., et al. (2019). *The Astrophysical Journal Supplement*, *243*(2), 26. https://doi.org/10.3847/1538-4365/ab29fd.
4. Proga, D., & Begelman, M. C. (2003). *The Astrophysical Journal*, *592*(2), 767. https://doi.org/10.1086/375773.
5. Janiuk, A., Proga, D., & Kurosawa, R. (2008). *The Astrophysical Journal*, *681*(1), 58. https://doi.org/10.1086/588375.
6. Li, J., Ostriker, J., & Sunyaev, R. (2013). *The Astrophysical Journal*, *767*(2), 105. https://doi.org/10.1088/0004-637X/767/2/105.
7. Ressler, S. M., Quataert, E., White, C. J., & Blaes, O. (2021). *Monthly Notices of the Royal Astronomical Society* *504*(4), 6076. https://doi.org/10.1093/mnras/stab311.
8. Lalakos, A., Gottlieb, O., Kaaz, N., Chatterjee, K., Liska, M., Christie, I. M., Tchekhovskoy, A., Zhuravleva, I., & Nokhrina, E. (2022). *The Astrophysical Journal Letter*, *936*(1), L5. https://doi.org/10.3847/2041-8213/ac7bed.
9. Kaaz, N., Murguia-Berthier, A., Chatterjee, K., Liska, M. T. P., & Tchekhovskoy, A. (2023). *The Astrophysical Journal*, *950*(1), 31. https://doi.org/10.3847/1538-4357/acc7a1.
10. Cho, H., Prather, B. S., Narayan, R., Natarajan, P., Su, K. Y., Ricarte, A., & Chatterjee, K. (2023). The Astrophysical Journal Letter, *959*(2), L22. https://doi.org/10.3847/2041-8213/ad1048.
11. Chakrabarti, S. K., & Molteni, D. (1993). *The Astrophysical Journal*, *417*, 671. https://doi.org/10.1086/173345.
12. Molteni, D., Ryu, D., & Chakrabarti, S. K. (1996). *The Astrophysical Journal*, *470*, 460. https://doi.org/10.1086/177877.
13. Ryu, D., Chakrabarti, S. K., & Molteni, D. (1997). *The Astrophysical Journal*, *474*(1), 378. https://doi.org/10.1086/303461.
14. Giri, K., & Chakrabarti, S. K. (2013). *Monthly Notices of the Royal Astronomical Society*, *430*, 2836. https://doi.org/10.1093/mnras/stt087.
15. Giri, K., Garain, S. K., & Chakrabarti, S. K. (2015). *Monthly Notices of the Royal Astronomical Society*, *448*, 3221. https://doi.org/10.1093/mnras/stv223.
16. Garain, S. K., & Kim, J. (2023). *Monthly Notices of the Royal Astronomical Society*, *519*(3), 4550. https://doi.org/10.1093/mnras/stac3736.
17. Paczyńsky, B., & Wiita, P. J. (1980). *Astronomy & Astrophysics*, *88*, 23.
18. Kim, J., Garain, S. K., Balsara, D. S., & Chakrabarti, S. K. (2017). *Monthly Notices of the Royal Astronomical Society*, *472*(1), 542. https://doi.org/10.1093/mnras/stx1986.
19. Kim, J., Garain, S. K., Chakrabarti, S. K., & Balsara, D. S. (2019). *Monthly Notices of the Royal Astronomical Society*, *482*(3), 3636. https://doi.org/10.1093/mnras/sty2953.
20. Chakrabarti, S. K. (1990). *Theory of transonic astrophysical flows*. World Scientific Publishing Co. https://doi.org/10.1142/1091.
21. Chakrabarti, S. K. (1996). *Monthly Notices of the Royal Astronomical Society*, *283*, 325. https://doi.org/10.1093/mnras/283.1.325.
22. Chakrabarti, S. K. (1996). *The Astrophysical Journal*, *471*, 237. https://doi.org/10.1086/177965.
23. Banyuls, F., Font, J. A., Ibáñez, J. M., Martí, J. M., & Miralles, J. A. (1997). *The Astrophysical Journal*, *476*, 221. https://doi.org/10.1086/303604.
24. Font, J. A. (2008). *Living Reviews in Relativity*, *11*(1), 7. https://doi.org/10.12942/lrr-2008-7.

25. Boyer, R. H., & Lindquist, R.W. (1967). *Journal of Mathematical Physics*, *8*, 265. https://doi.org/10.1063/1.1705193.
26. Mignone, A. (2014). *Journal of Computational Physics*, *270*, 784. https://doi.org/10.1016/j.jcp.2014.04.001.
27. Balsara, D. S., & Kim, J. (2016). *Journal of Computational Physics*, *312*, 357. https://doi.org/10.1016/j.jcp.2016.02.001.
28. Mignone, A., & Bodo, G. (2005). *Monthly Notices of the Royal Astronomical Society*, *364*(1), 126. https://doi.org/10.1111/j.1365-2966.2005.09546.x.
29. Del Zanna, L., & Bucciantini, N. (2002). *Astronomy & Astrophysics*, *390*, 1177. https://doi.org/10.1051/0004-6361:20020776.
30. LeVeque, R. (2002). *Finite volume methods for hyperbolic problems*. Cambridge texts in applied mathematics. Cambridge University Press. https://books.google.co.in/books?id=QazcnD7GUoUC.
31. Toro, E. (2009). *Riemann solvers and numerical methods for fluid dynamics: A practical introduction*. Springer Berlin Heidelberg. https://books.google.co.in/books?id=SqEjX0um8o0C.
32. Abramowicz, M., Jaroszynski, M., & Sikora, M. (1978). *Astronomy & Astrophysics*, *63*, 221.
33. Chakrabarti, S. K. (1985). *The Astrophysical Journal*, *288*, 1. https://doi.org/10.1086/162755.
34. Font, J. A., & Daigne, F. (2002). *Monthly Notices of the Royal Astronomical Society*, *334*(2), 383. https://doi.org/10.1046/j.1365-8711.2002.05515.x.
35. Fukue, J. (1987). *Publications of the Astronomical Society of Japan*, *39*(2), 309.
36. Chakrabarti, S. K. (1989). *The Astrophysical Journal Letter*, *337*, L89. https://doi.org/10.1086/185385.
37. Bondi, H. (1952). *Monthly Notices of the Royal Astronomical Society*, *112*, 195. https://doi.org/10.1093/mnras/112.2.195.
38. Cui, W., Zhang, S. N., Focke, W., & Swank, J. H. (1997). *The Astrophysical Journal*, *484*(1), 383. https://doi.org/10.1086/304341.
39. Chakrabarti, S. K., & Manickam, S. G. (2000). *The Astrophysical Journal Letter*, *531*(1), L41. https://doi.org/10.1086/312512.
40. Radhika, D., Nandi, A., Agrawal, V. K., & Mandal, S. (2016). *Monthly Notices of the Royal Astronomical Society*, *462*(2), 1834. https://doi.org/10.1093/mnras/stw1755.
41. Shang, J. R., Debnath, D., Chatterjee, D., Jana, A., Chakrabarti, S. K., Chang, H. K., Yap, Y. X., & Chiu, C. L. (2019). *The Astrophysical Journal*, *875*(1), 4. https://doi.org/10.3847/1538-4357/ab0c1e.

Chapter 13
Towards a Theory of Hydromagnetic Turbulence with Higher Order Fluid Moments

Arindam Saikia and Rupak Mukherjee

Abstract In this chapter, Chandrasekhar's deductive theory of turbulence is extended to the case of hydrodynamics and hydromagnetics, with higher order fluid moments. We include the contributions from correlation tensors at two different points of space and two different points in time to account for the energy transfer between Fourier modes with different k values. We start with three governing equations using the first three moments of Vlasov equation, which gives us particle, momentum and energy conservation relations, respectively. We obtain the scalar representations of correlation tensors for hydrodynamics as well as hydromagnetics. Finally we pick an example from literature where the energy spectra alone was not capable of explaining the observations reported by the authors.

13.1 Introduction

Turbulence is a challenging topic to articulate mathematically due to the complexity in deriving precise equations that encapsulate the chaotic nature of fluid flows involved. Kolmogorov's theory provides a heuristic model to understand turbulence that is able to capture the energy cascade process across different scales. However a rigorous phenomenological theory remains necessary, as Kolmogorov's model provides statistical insight rather than a deterministic picture that fully describe turbulent motions.

An alternative path to model such a system is Subrahmanyan Chandrasekhar's approach where he made significant strides in taking a more rigorous route, advancing our understanding, especially in magneto-hydrodynamics. His approach involved formulating equations solely containing correlation scalars. His basic idea was to introduce correlations in the velocity components (u_i) at two different points (x and x') and at two different times (t and t') and deduce a single equation

A. Saikia · R. Mukherjee (✉)
Department of Physics, Sikkim University, Gangtok, Sikkim, India
e-mail: rmukherjee@cus.ac.in

S. Ghosh, A. R. Rao (eds.), *The Relativistic Universe: From Classical to Quantum*, Astrophysics and Space Science Proceedings 61,
https://doi.org/10.1007/978-3-031-90186-7_13

through his framework to describe the turbulent dynamics [1]. He also showed that at $t - t' = 0$, i.e., when the time difference in taken to be zero, the differential equations take the form one supposes on Kolmogorov's similarity principles. In a later communication, Chandrasekhar extended his theory for hydromagnetics as well, following the same methodology [2, 3].

One can consider Vlasov equation and take moments of different order. The zeroth order moment of Vlasov equation gives continuity equation. The first order moment gives momentum equation and so on. We know, each moment equation contains a higher order moment of the distribution function. For example, the continuity equation (zeroth order moment of Vlasov equation) contains density (zeroth order moment of the distribution function) as well as mean velocity (first order moment of the distribution function). Similarly the momentum equation (first order moment of Vlasov equation) contains density (zeroth order moment of the distribution function), mean velocity (first order moment of the distribution function) and pressure (second order moment of the distribution function), and so on. Thus the set of moment equations of Vlasov equation is never closed, and is known as the BBGKY hierarchy problem. To make the set of fluid equations well-posed, we often take resort to a closure equation, that is to express a higher order moment of the distribution function, in terms of the lower order moments. For example, under certain circumstances, pressure (second order moment of the distribution function) can be written in terms of density (zeroth order moment of the distribution function). It is known that the set of moment equations involving higher order moments describe the properties of Vlasov equation better than a set of moment equations involving lower order moments [4–10]. For example, the set of continuity, momentum and pressure equations describe a fluid medium better than a set of continuity and momentum equation only.

Our primary motivation in this chapter was to extend Chandrasekhar's work for higher order moments of the Vlasov equation. An example case for working with lower order moments would be the closure at the first order using Fick's law forms [11]. Taking higher order moments has been shown to take us closer to a kinetic picture of the system rather than a purely statistical fluid picture which, as has been shown in the later sections, fails to account for and predict a couple of expected behaviors from complex hydromagnetic systems. In our work, we have derived them for continuity equation, momentum conservation equation and energy conservation equation with a Hammett-Perkins type closure [11]. A further extension of the work in terms of including the third moment is in progress. In addition, following Chandrasekhar's work on hydromagnetics, we extend our study with higher order fluid moments for magnetized fluid as well.

13.2 Chandrasekhar's Theory of Hydrodynamics

Chandrasekhar's argument was that a description of turbulence in terms of its spectrum $F(k)$ alone cannot be a complete one as there exists phase relationship between the different Fourier components of the velocity field, which is necessary for the transfer of energy between the modes. Thus, he introduces correlations in the velocity components at two different points and two different times to account for the phase relationships. The theory was proposed under conditions of isotropy when a stationary state prevails.

Under the assumption that the correlations depend, apart from the vector $\xi(= r'' - r')$ joining the two points, only on the difference in the times, $|t'' - t'|$, the following correlation tensors were introduced:

$$Q_{ij} = \overline{u_i(r', t')u_j(r'', t'')}, \tag{13.1}$$

$$T_{ij;k} = \overline{u_i(r', t')u_j(r', t')u_k(r'', t'')}, \tag{13.2}$$

$$P_{ij} = \overline{\overline{\omega}(r', t')u_i(r'', t'')u_j(r'', t'')}, \tag{13.3}$$

$$Q_{ij;kl} = \overline{u_i(r', t')u_j(r', t')u_k(r'', t'')u_l(r'', t'')}, \tag{13.4}$$

where $\overline{\omega} = p/\rho$ and u_i denote the components of turbulent velocity. Here, the semi-colon notation on the tensor subscripts serves as a convention for separating the indices measured at t' or t'' coordinates. In the later case, the indices are kept on the right side of the semi-colon. This notation doesn't represent a covariant derivative or contraction in this use case.

In addition to these correlation tensors, an additional statistical hypothesis had to be put forward in order to reduce the final relation into a single equation in terms of Q alone (here Q is the scalar representation of the tensor Q_{ij}):

$$Q_{ij;kl} = Q_{ik}Q_{jl} + Q_{il}Q_{jk} + Q_{ij}(0, 0)Q_{kl}(0, 0). \tag{13.5}$$

This hypothesis states that the quadruple moment $Q_{ij;kl}$ is related to the second-order moment Q_{ij} as in a *normal distribution*.

Now we are equipped to tackle the equation of motion for the fluid under study:

$$\frac{\partial u_i'}{\partial t'} + \frac{\partial}{\partial x_k'}u_i'u_k' = -\frac{\partial \overline{\omega}'}{\partial x_i'} + \nu\nabla^2 u_i'. \tag{13.6}$$

The above equation expressed in terms of correlation tensors takes the form:

$$\pm\frac{\partial Q_{ij}}{\partial t} = \frac{\partial}{\partial \xi_k}T_{ik;j} + \nu\nabla_\xi^2 Q_{ij}. \tag{13.7}$$

Now, using scalar representations for these tensor equations [1], the final equation is obtained,

$$\frac{\partial}{\partial r}(\frac{\partial^2}{\partial t^2} - \nu^2 D_5^2)Q = -2Q\frac{\partial}{\partial r}D_5 Q. \tag{13.8}$$

Here, D_n is the Laplacian Operator for a spherically symmetric function in an n-dimensional Euclidean space. Hence, D_5^2 represents the square of the Laplacian operator in a 5-dimensional Euclidean space.

13.3 Chandrasekhar's Theory of Hydromagnetics

Similar to the previous case of hydrodynamics by Chandrasekhar, a parallel approach is taken in the case of hydromagnetics. This time starting from the following sets of equations:

$$\frac{\partial u_i}{\partial t} + \frac{\partial}{\partial x_k}(u_i u_k - h_i h_k) = -\frac{\partial \overline{\omega}}{\partial x_i} + \nu \nabla^2 u_i, \tag{13.9}$$

and
$$\frac{\partial h_i}{\partial t} + \frac{\partial}{\partial x_k}(h_i u_k - u_i h_k) = \lambda \nabla^2 h_i, \tag{13.10}$$

again, u_i denotes the components of turbulent velocity, h_i the components of the magnetic field divided by $(4\pi\rho/\mu)^{\frac{1}{2}}$, ρ the density, P the pressure and μ,ν and σ are the coefficients of magnetic permeability, kinematic viscosity and electrical conductivity, respectively, and

$$\lambda = 1/4\pi\mu\sigma \quad \text{and} \quad \overline{\omega} = P/\rho + \frac{1}{2}|h|^2. \tag{13.11}$$

In addition, the following divergence conditions are used

$$\frac{\partial u_i}{\partial x_i} = 0 \quad \text{and} \quad \frac{\partial h_i}{\partial x_i} = 0. \tag{13.12}$$

This time around, the amount of correlation tensors required to represent the system are as follows:

$$Q_{ij} = \overline{u_i(r', t')u_j(r'', t'')}$$

$$H_{ij} = \overline{h_i(r', t')h_j(r'', t'')}$$

$$P_{ij} = \overline{\overline{\omega}(r', t')u_i(r'', t'')u_j(r'', t'')}$$

$$\Pi_{ij} = \overline{\overline{\omega}(r', t')h_i(r'', t'')h_j(r'', t'')}$$

$$T_{ij;k} = \overline{u_i(r', t')u_j(r', t')u_k(r'', t'')}$$

$$S_{ij;k} = \overline{h_i(r', t')h_j(r', t')u_k(r'', t'')}$$

$$F_{ij;k} = \overline{\left[h_i(r', t')u_j(r', t') - u_i(r', t')h_j(r', t')\right] h_k(r'', t'')}$$

$$Q_{ij;kl} = \overline{u_i(r', t')u_j(r', t')u_k(r'', t'')u_l(r'', t'')}$$

$$H_{ij;kl} = \overline{h_i(r',t')h_j(r',t')h_k(r'',t'')h_l(r'',t'')}$$

$$R_{ij;kl} = \overline{\left[h_i(r',t')u_j(r',t') - u_i(r',t')h_j(r',t')\right]\left[h_k(r'',t'')u_l(r'',t'') - u_k(r'',t'')h_l(r'',t'')\right]}$$

which allows us to reduce the dynamics into a set of two scalar equations:

$$\frac{\partial}{\partial r}\left(\frac{\partial^2}{\partial t^2} - \nu^2 D_5^2\right) Q = -2Q\frac{\partial}{\partial r}D_5 Q - 2H\frac{\partial}{\partial r}D_5 H, \tag{13.13}$$

and
$$\left(\frac{\partial^2}{\partial t^2} - \lambda^2 D_5^2\right) H = -2Q D_5 H - 2H D_5 Q - 2\frac{\partial Q}{\partial r}\frac{\partial H}{\partial r}. \tag{13.14}$$

Equations (13.13) and (13.14) are the required equations governing Q and H.

13.4 Basic Ideas of the Proposed Theory

As in the theory proposed by Chandrasekhar, we will incorporate the phase coupling by introducing correlation tensors that relate measurements at two different points of space and at two different points in time. This consideration alone serves as a manifestation of the underlying phase coupling between these two points.

To do so, we start with the Vlasov equation, which describes the simplest possible case of linear one-dimensional electrostatic waves given by:

$$\frac{\partial f}{\partial t} + v\frac{\partial f}{\partial x} + \frac{e}{m}E\frac{\partial f}{\partial x} = 0. \tag{13.15}$$

Proceeding with Eq. (13.15), we take moments of said equation, and arrive at the following three relations [11]:

$$\frac{\partial n}{\partial t} + \frac{\partial}{\partial x}(un) = 0, \tag{13.16}$$

$$\frac{\partial}{\partial t}(mnu) + \frac{\partial}{\partial u}(umnu) = -\frac{\partial P}{\partial x} + enE + \frac{\partial S}{\partial x}, \tag{13.17}$$

and
$$\frac{\partial P}{\partial t} + \frac{\partial}{\partial x}(uP) = -(\Gamma - 1)(P + S)\frac{\partial u}{\partial x} - \frac{\partial q}{\partial x}. \tag{13.18}$$

Equations (13.16), (13.17), and (13.18) represent particle, momentum, and energy conservation relations, respectively. The heat flux moment 'q' is considered as prescribed, effectively closing this set of coupled equations.

Here, $q = m\int dv\, f(v-u)^3$, and the dissipative momentum flux S is given by $S = mn\mu\frac{\partial u}{\partial x}$

In the following sections, hydrodynamics and hydromagnetics will be extended to include these moments while using a correlation tensor approach to represent the resulting dynamics.

13.5 Extension of Hydrodynamics for Higher Order Moments

Before we proceed with the derivations, we will define the following correlation tensors:

$$C_j = \overline{mn'u''_j} \qquad M_{jl;i} = \overline{P'_i u''_j u''_l}$$

$$F_{ik} = \overline{u'_i u'_k} \qquad \tilde{\Sigma}_{jl;i} = \overline{enE'_j u''_j u''_l}$$

$$N_{ij} = \overline{P'_i u''_j} \qquad \widetilde{M}_{ij} = \overline{u'_i P'_i u''_j}$$

$$\Sigma_{ij} = \overline{enR'_i u''_j} \qquad X_{ij} = \overline{q'_i u''_j}$$

$$D_{jl} = \overline{mn'u''_j u''_l} \qquad M^0_{jl;i} = \overline{u'_i P'_i u''_j u''_l}$$

Starting with Eq. (13.16), the continuity equation, and introducing the following sign conventions to describe measurements at two different points: $'$ representing measurements at spatial point r' and at time t', whereas, $''$ represents r'' and t'' points, we expand the derivative term and arrive at:

$$\frac{\partial n'}{\partial t'} + u'_i \frac{\partial}{\partial x'_i} n' + n' \frac{\partial}{\partial x'_i} u'_i = 0. \tag{13.19}$$

Multiplying the equation by u''_j and averaging all the terms across ensembles, transforms the operators as below (with $\xi = (r'' - r')$) and the $\pm$ sign depending on whether $t'' > t'$ or $t'' < t'$:

$$\pm \frac{\partial}{\partial t} C_j - \langle u \rangle \frac{\partial}{\partial \xi_i} C_j - \langle \rho \rangle \frac{\partial}{\partial \xi_i} Q_{ij} = 0. \tag{13.20}$$

Now, utilizing the fact that a scalar representation of the above tensor equation is possible in the case of homogeneous isotropic turbulence, where the tensors are invariant under arbitrary rotations and reflections of the reference axes [12], we obtain the following expression:

$$\pm \frac{\partial C}{\partial t} - \langle \rho \rangle \frac{\partial}{\partial r} Q = 0. \tag{13.21}$$

Here (in Eq. (13.21)), the second term goes to zero due to the implication of equation of continuity for an incompressible fluid.

Again, multiplying Eq. (13.21) with $u''_j u''_l$ and averaging all the terms across ensembles, we get

$$\pm\frac{\partial}{\partial t}D_{jl}-\langle u\rangle\frac{\partial}{\partial\xi_i}D_{jl}+\langle\rho\rangle\frac{\partial}{\partial\xi_i}T_{jl;i}=0. \tag{13.22}$$

We obtain the following scalar equation from Eq. (13.22)

$$\pm\frac{\partial}{\partial t}D-\langle u\rangle\frac{\partial}{\partial r}D+\langle\rho\rangle\left(r\frac{\partial}{\partial r}+5\right)T=0, \tag{13.23}$$

where, we have used the general form of a solenoidal isotropic tensor of second order [12], in order to find the corresponding scalar operators for D_{jl}.

Next, Using definition of S in Eq. (13.18) and expanding the derivative terms, we get the following form

$$mn'\frac{\partial u_i'}{\partial t}+u_i'\frac{\partial(mn')}{\partial t}+mn'\frac{\partial(u_i'u_k')}{\partial x_k}+u_i'u_k'\frac{\partial(mn')}{\partial x_k'}=-\frac{\partial P_i'}{\partial x_i'}+enE_i'+mn'\mu\frac{\partial^2u_i'}{\partial x_i'^2}. \tag{13.24}$$

As before, we multiply with u_j'' and take ensemble average,

$$\pm\langle\rho\rangle\frac{\partial}{\partial t}Q_{ij}\pm\langle u\rangle\frac{\partial}{\partial t}C_j-\langle\rho\rangle\frac{\partial}{\partial\xi_k}T_{ik;j}-F_{ik}\frac{\partial}{\partial\xi_k}C_j=\frac{\partial}{\partial\xi_i}N_{ij}+\Sigma_{ij}+\mu\langle\rho\rangle\frac{\partial^2}{\partial\xi_i^2}Q_{ij}. \tag{13.25}$$

In scalar representation, taking C_j and N_{ij} terms to be equal to zero from equation of continuity,

$$\pm\langle\rho\rangle\frac{\partial}{\partial t}Q\pm\langle u\rangle\frac{\partial}{\partial t}C-\langle\rho\rangle(r\frac{\partial}{\partial r}+5)T=\Sigma+\mu\langle\rho\rangle D_7Q. \tag{13.26}$$

Similarly, multiplying with $u_j''u_l''$ and averaging, we get,

$$\begin{aligned}&\pm\langle\rho\rangle\frac{\partial}{\partial t}(-T_{ij;l})\pm\langle u\rangle\frac{\partial}{\partial t}D_{jl}-\langle\rho\rangle\frac{\partial}{\partial\xi_k}Q_{ik;jl}-F_{ik}\frac{\partial}{\partial\xi_k}D_{jl}\\&=\frac{\partial}{\partial\xi_i}M_{jl;i}+\widetilde{\Sigma}_{jl;i}+\mu\langle\rho\rangle\frac{\partial^2}{\partial\xi_i^2}(-T_{jl;i}).\end{aligned} \tag{13.27}$$

And in scalar representation,

$$\mp\langle\rho\rangle\frac{\partial}{\partial t}T\mp\langle u\rangle\frac{\partial}{\partial t}D-\langle\rho\rangle\hat{O}_3Q-F\cdot\frac{\partial}{\partial r}D=(r\frac{\partial}{\partial r}+5)M+\widetilde{\Sigma}-\mu\langle\rho\rangle D_7T. \tag{13.28}$$

Lastly, we perform multiplication with u_j'' on equation (13.19) in the form:

$$\frac{\partial P_i'}{\partial t'} + \frac{\partial}{\partial x_i'}(u_i' P_i') = -\Gamma P_i' \frac{\partial u_i'}{\partial x_i'} - \Gamma m n' \mu \frac{\partial^2 u_i'}{\partial x_i'^2} + P_i' \frac{\partial u_i'}{\partial x_i'} + m n' \mu \frac{\partial^2 u_i'}{\partial x_i'^2} - \frac{\partial q_i'}{\partial x_i'}. \tag{13.29}$$

to get,

$$\pm \frac{\partial}{\partial t}[N_{ij}] - \frac{\partial}{\partial \xi_i}\widetilde{M}_{ij} = \Gamma \langle P \rangle \frac{\partial}{\partial \xi_i} Q_{ij} - \Gamma \langle \rho \rangle \mu \frac{\partial^2}{\partial \xi_i^2} Q_{ij} - \langle P \rangle \frac{\partial}{\partial \xi_i} Q_{ij} + \langle \rho \rangle \mu \frac{\partial^2}{\partial \xi_i^2} Q_{ij} + \frac{\partial}{\partial \xi_i} X_{ij}, \tag{13.30}$$

which gives us,

$$\pm \frac{\partial N}{\partial t} = \frac{\partial}{\partial r}(\widetilde{M} + \Gamma \langle P \rangle Q - \langle P \rangle Q + X) - (1 - \Gamma)\langle \rho \rangle \mu D_5 Q. \tag{13.31}$$

Multiplying Eq. (13.24) with $u_j'' u_l''$ and averaging

$$\pm \frac{\partial}{\partial t} M_{jl;i} - \frac{\partial}{\partial \xi_i} M_{jl;i}^0 = \Gamma \langle P \rangle \frac{\partial}{\partial \xi_i}(-T_{jl;i}) - \Gamma \langle \rho \rangle \mu \frac{\partial^2}{\partial \xi_i^2}(-T_{jl;i}) - \langle P \rangle \frac{\partial}{\partial \xi_i}(-T_{jl;i}) + \langle \rho \rangle \mu \frac{\partial^2}{\partial \xi_i^2}(-T_{jl;i}) + \frac{\partial}{\partial \xi_i} \tilde{X}_{jl;i}, \tag{13.32}$$

giving us,

$$\pm \frac{\partial M}{\partial t} = \left(r \frac{\partial}{\partial r} + 5 \right) \left(M^0 + (1 - \Gamma)\langle P \rangle T - \tilde{X} \right) + (\Gamma D_7 - D_5) \langle \rho \rangle \mu T. \tag{13.33}$$

We can now simplify equations (13.21), (13.23), (13.25), (13.28), (13.30) and (13.33) to get three concise equations that govern the dynamics, as

$$\pm \langle \rho \rangle \frac{\partial Q}{\partial t} + \langle u \rangle \langle \rho \rangle \frac{\partial}{\partial r} Q - \langle \rho \rangle \left(r \frac{\partial}{\partial r} + 5 \right) T - F \cdot \frac{\partial}{\partial r} C = \frac{\partial}{\partial r} N + \Sigma + \mu \langle \rho \rangle D_5 Q, \tag{13.34}$$

$$\begin{aligned} &\left(-\frac{\partial^2}{\partial t^2} \pm \mu \frac{\partial}{\partial t} D_7 \pm \langle u \rangle \left(r \frac{\partial}{\partial r} + 5 \right) \langle \rho \rangle \frac{\partial T}{\partial t} \pm (\langle u \rangle^2 + F) \frac{\partial}{\partial t} (\frac{\partial}{\partial r} D) \right) \\ &= \left(r \frac{\partial}{\partial r} + 5 \right)^2 \left(M^0 + (1 - \Gamma)\langle \rho \rangle T - \tilde{X} \right) \\ &\quad + \left(r \frac{\partial}{\partial r} + 5 \right) (\Gamma D_7 - D_5)\langle \rho \rangle \mu T \pm \frac{\partial \tilde{\Sigma}}{\partial t} \pm \langle \rho \rangle \frac{\partial}{\partial t} \hat{O}_3 Q, \end{aligned} \tag{13.35}$$

and

$$\pm \frac{\partial N}{\partial t} = \frac{\partial}{\partial r}\left(\widetilde{M} + \Gamma\langle P\rangle Q - \langle P\rangle Q + X\right) - (1-\Gamma)\langle\rho\rangle\mu D_5 Q. \tag{13.36}$$

13.6 Extension of Hydromagnetics for Higher Order Moments

Now that we have established a framework for hydrodynamics, with Eqs. (13.34), (13.35), and (13.36), and we have Chandrasekhar's relations for hydromagnetics (Sect. 13.3), with Eqs. (13.13) and (13.14), we can put forward a more accurate description of hydromagnetic turbulence by unifying these relations and obtaining a higher order representation for Chandrasekhar's hydromagnetics:

$$\frac{\partial}{\partial r}(\frac{\partial^2}{\partial t^2} - \nu^2 D_5^2)Q = -2Q\frac{\partial}{\partial r}D_5 Q - 2H\frac{\partial}{\partial r}D_5 H, \tag{13.37}$$

$$(\frac{\partial^2}{\partial t^2} - \lambda^2 D_5^2)H = -2QD_5 H - 2HD_5 Q - 2\frac{\partial Q}{\partial r}\frac{\partial H}{\partial r}, \tag{13.38}$$

$$\pm\langle\rho\rangle\frac{\partial Q}{\partial t} + \langle u\rangle\langle\rho\rangle\frac{\partial}{\partial r}Q - \langle\rho\rangle\left(r\frac{\partial}{\partial r}+5\right)T - F\cdot\frac{\partial}{\partial r}C = \frac{\partial}{\partial r}N + \Sigma + \mu\langle\rho\rangle D_5 Q, \tag{13.39}$$

$$\begin{aligned}&\left(-\frac{\partial^2}{\partial t^2} \pm \mu\frac{\partial}{\partial t}D_7 \pm \langle u\rangle\left(r\frac{\partial}{\partial r}+5\right)\langle\rho\rangle\frac{\partial T}{\partial t} \pm (\langle u\rangle^2 + F)\frac{\partial}{\partial t}\frac{\partial}{\partial r}D\right)\\ &= \left(r\frac{\partial}{\partial r}+5\right)^2\left(M^0 + (1-\Gamma)\langle\rho\rangle T - \tilde{X}\right)\\ &+ \left(r\frac{\partial}{\partial r}+5\right)(\Gamma D_7 - D_5)\langle\rho\rangle\mu T \pm \frac{\partial\tilde{\Sigma}}{\partial t} \pm \langle\rho\rangle\frac{\partial}{\partial t}\hat{O}_3 Q,\end{aligned} \tag{13.40}$$

and

$$\pm \frac{\partial N}{\partial t} = \frac{\partial}{\partial r}\left(\widetilde{M} + \Gamma\langle P\rangle Q - \langle P\rangle Q + X\right) - (1-\Gamma)\langle\rho\rangle\mu D_5 Q. \tag{13.41}$$

Now the above set of five equations describe higher order hydromagnetic turbulence.

13.7 Example

The relations shown above soon gets difficult to track, as noticed by Chandrasekhar as well, in his work on hydromagnetics [2, 3]. However, the need of such work has always been undermined as energy spectra alone has been found to cater to the need to explain hydrodynamic as well as hydromagnetic turbulence well.

Below we discuss at least one example [13] where, energy spectra alone has been found to fail to explain the numerical observation the authors performed.

A chaotic flow and uniform magnetic field is initialized. The initial condition is chosen such that, it is 3D MHD unstable. The flow and field variables are allowed to evolve numerically in a 3D periodic domain and the simulation is run enough for the system to get thermalized. Ideally, energy must have got equipartitioned once the system gets thermalized. However, the authors observe the following: even though, quite a number of the simulation results showed a thermalized energy spectrum [14], few have been found to reconstruct the initial flow as well as the field structures. In particular, an initial Taylor-Green flow with a uniform background magnetic field, is found to recur, while an initial Arnold-Beltrami-Childress flow with same uniform magnetic field, is found to thermalize.

The work was a first of its kind and has been now extended for several more initial flow and field variables [15]. For example, an initial Robert's flow with a uniform background magnetic field, is found to recur, while an initial cat's-eye flow with same uniform magnetic field, is found to thermalize [15]. Extension of such observation in terms of less restrictive conditions [16–18] have also been achieved. The authors have also reproduced the similar observations with different simulation packages, for example, finite-volume [19] as well as pseudo-spectral [20–22] algorithms for spatial discretization and explicit and implicit time marching algorithms.

The general assumptions in all the works performed so far in regard to above nonlinear oscillations are the following: so far, 'recurrence' phenomena have been observed for single-fluid plasmas, where, no large scale electric field is obtained due to charge separation between different species. In addition, it is assumed that a hydromagnetic description of plasma is good enough, such that, the individual fluid-elements can be assumed to posses a Maxwellian velocity distribution of the particles. This requires the plasma to be warm or cold, such that the plasma remains highly collisional within each fluid-element; such that any distortion of the parent Maxwellian distribution can get restored at a timescale faster than modes of any growing instability. Currently we are in pursuit of observing 'recurrence' with two-fluid and fully kinetic models as well.

13.8 Governing Equations

A typical recurrence problem solves the following equations:

Continuity equation:

$$\frac{\partial \rho}{\partial t} + \nabla \cdot (\rho \mathbf{u}) = 0. \tag{13.42}$$

Momentum equation:

$$\rho \frac{\partial \mathbf{u}}{\partial t} + \rho(\mathbf{u} \cdot \nabla)\mathbf{u} = \mathbf{j} \times \mathbf{B} - \nabla P + \mu \nabla^2 \mathbf{u} + \rho \mathbf{F}, \tag{13.43}$$

where the left hand side $\frac{D}{Dt} \equiv \frac{\partial}{\partial t} + \mathbf{u} \cdot \nabla$ is the convective derivative, $\mathbf{j} \times \mathbf{B}$ is the Lorentz force acting on the fluid element, ∇P represents the compressibility of the fluid and $\mu \nabla^2 \mathbf{u}$ is the viscous term with μ being the compressibility of the fluid. Here in this simulation, we concentrate on decaying turbulence, with no external force acting on the fluid.
Equation of state:

$$\frac{d}{dt}\left(\frac{P}{\rho^\gamma}\right) = 0. \tag{13.44}$$

This follows that pressure is proportional to the density of the fluid.
Ampere's law:

$$\nabla \times \mathbf{B} = 4\pi \mathbf{j}, \tag{13.45}$$

where we ignore the contribution from the displacement current being relativistically suppressed. Faraday's law of induction

$$\frac{\partial \mathbf{B}}{\partial t} = -c\nabla \times \mathbf{E}. \tag{13.46}$$

Ohm's law for conducting fluid

$$\mathbf{E} + (\mathbf{u} \times \mathbf{B})/c = \eta \mathbf{j}, \tag{13.47}$$

where, η is the coefficient of resistivity.

13.9 Initial Conditions

As mentioned in the earlier section, few of the initial conditions are found to recur. Two such examples are,

Taylor-Green Flow:

$$u_x = A\, U_0 \left[\cos(kx)\sin(ky)\cos(kz)\right],$$
$$u_y = -A\, U_0 \left[\sin(kx)\cos(ky)\cos(kz)\right],$$
$$u_z = 0.$$

Roberts Flow:

$$u_x = A\, U_0 \sin(kz),$$
$$u_y = B\, U_0 \sin(kx),$$
$$u_z = C\, U_0 \sin(ky).$$

For completeness, we took two more initial flow structures, that have not been found to recur:

Arnold-Beltrami-Childress Flow:

$$u_x = U_0 \left[A \sin(kz) + C \cos(ky)\right],$$
$$u_y = U_0 \left[B \sin(kx) + A \cos(kz)\right],$$
$$u_z = U_0 \left[C \sin(ky) + B \cos(kx)\right].$$

Cats-Eye Flow:

$$u_x = U_0\, B\, \sin(ky),$$
$$u_y = U_0\, A\, \sin(kx),$$
$$u_z = U_0 \left[A \cos(kx) - B \sin(ky)\right].$$

13.10 Simulation Parameters

We repeat the simulation reported earlier, and solve the single-fluid equations mentioned above [Eqs. (13.42)–(13.47)] using different initial conditions, with the plasma parameters provided in the table below. The numerical simulation is performed using the simulation package TARA [22] on the Sikkim University 'Brahmagupta' HPC system. We choose the following simulation parameters:

R_e	R_m	M_A	M_s
10^3	10^3	1	0.1

N	L	ρ_0	U_0	$A = B = C$	k
128	2π	1	0.1	1	1

As mentioned above, the results have been also reproduced at even higher resolution runs previously. In TARA, we use NVIDIA cuSolverRF with super-time-stepping algorithm that runs on our NVIDIA-A30 GPU card.

13.11 Analysis

As reported in all the recurrence related papers mentioned above, the energy spectra has been found to be inadequate in analyzing the origin of such phenomena. This raises a fundamental question: is the description of MHD system via energy-spectrum alone, complete? We find that this is something that Chandrasekhar realized precisely, but has been overlooked for decades [1]. Thus, development of an alternate theory that gives a more realistic depiction of the system is beneficial for application purposes and understanding complex phenomena. Chandrasekhar also observed that such description of turbulence in terms of phase-relations becomes even more complicated for hydromagnetic systems [2, 3]. But this approach remains a largely unexplored territory to describe such phenomena displayed by the MHD system. In this chapter, we have tried to extend Chandra's work upto third-moment (of kinetic Vlasov equation) for hydro and magneto-hydrodynamic models. However, an appropriate theory is still under-developed. This chapter doesn't claim to solve the limitations inherent in Kolmogorov's model; rather, it seeks to explore potential alternate frameworks that may offer deeper insights and represent the true behavior of complex hydromagnetic or hydrodynamic fluids.

13.12 Conclusion

In summary, we venture to re-think turbulence in the way Chandrasekhar described it in 1950s. Our findings extend Chandrasekhar's deductive theory of turbulence to encompass higher order moments for hydrodynamics as well as hydromagnetics by deriving three governing equations from the moments of Vlasov equation. This framework incorporates correlation tensors to account for energy transfer between Fourier modes with different wave numbers. Through rigorous mathematical analysis, we bridge the gap between heuristic models and detailed phenomenological descriptions, enhancing our understanding of turbulence in electrically influenced environments. This advancement not only builds upon the foundational work of Chandrasekhar and Heisenberg but also provides a robust basis for future research in turbulence theory, offering new insights and predictive capabilities.

As a relevant example of the need of such analysis we recall a well established phenomena called 'recurrence', where the energy spectra does not throw light to explain the effect. But by integrating a higher order model, we aim to be able to derive some useful insights on the complexities that traditional spectral methods might overlook. Our work, thus strengthens the motivation for exploring a more kinetic-based description of the dynamics of such a system.

Acknowledgments Both the authors are grateful to the referee(s) for their helpful comments. One of the authors Rupak Mukherjee acknowledges IUCAA visiting associateship program for their kind support. The simulation reported in this chapter is performed in Sikkim University's 'Brahmagupta' HPC Centre.

References

1. Chandrasekhar, S. (1955). Theory of turbulence. *Physical Review*, *102*(4), 941.
2. Chandrasekhar, S. (1955). Hydromagnetic turbulence. I. A deductive theory. *Proceedings of the Royal Society of London. Series A. Mathematical and Physical Sciences*, *233*(1194), 322–330.

3. Chandrasekhar, S. (1955). Hydromagnetic turbulence II. An elementary theory. *Proceedings of the Royal Society of London. Series A. Mathematical and Physical Sciences*, *233*(1194), 330–350.
4. Wang, L., Hakim, A., Bhattacharjee, A., & Germaschewski, K. (2015). Comparison of multi-fluid moment models with particle-in-cell simulations of collisionless magnetic reconnection. *Physics of Plasmas*, *22*(1), 012108.
5. Cagas, P., Hakim, A., Juno, J., & Srinivasan, B. (2017). Continuum kinetic and multi-fluid simulations of classical sheaths. *Physics of Plasmas*, *24*(2) 022118.
6. Wang, L., Germaschewski, K., Hakim, A., Dong, C., Raeder, J., & Bhattacharjee, A. (2018). Electron physics in 3-d two-fluid 10-moment modeling of ganymede's magnetosphere. *Journal of Geophysical Research: Space Physics*, *123*(4), 2815–2830.
7. Dong, C., Wang, L., Hakim, A., Bhattacharjee, A., Slavin, J. A., DiBraccio, G. A., & Germaschewski, K. (2019). A novel ten-moment multifluid model for mercury: From the planetary conducting core to the dynamic magnetosphere. Preprint. arXiv:1904.02695.
8. Ng, J., Hakim, A., Wang, L., & Bhattacharjee, A. (2020). An improved ten-moment closure for reconnection and instabilities. *Physics of Plasmas*, *27*(8) 082106.
9. Struchtrup, H., & Frezzotti, A. (2019). Grad's 13 moments approximation for enskog-vlasov equation. In *AIP Conference Proceedings* (Vol. 2132). AIP Publishing.
10. Struchtrup, H., & Frezzotti, A. (2022). Twenty-six moment equations for the Enskog–Vlasov equation. *Journal of Fluid Mechanics*, *940*, A40.
11. Hammett, G. W., & Perkins, F. W. (1990). Fluid moment models for landau damping with application to the ion-temperature-gradient instability. *Physical Review Letters*, *64*(25), 3019.
12. Chandrasekhar, S. (1953). Some aspects of the statistical theory of turbulence. In *Proceedings of the Symposium on Applied Mathematics* (Vol. 4, pp. 1–17). The American Mathematical Society.
13. Mukherjee, R., Ganesh, R., & Sen, A. (2019). Recurrence in three dimensional magnetohydrodynamic plasma. *Physics of Plasmas*, *26*(2) 022101.
14. Mukherjee, R., Ganesh, R., & Sen, A. (2019). Coherent nonlinear oscillations in magnetohydrodynamic plasma. *Physics of Plasmas*, *26*(4).
15. Mukherjee, R. (2019). *Turbulence, flows and magnetic field generation in plasmas using a magnetohydrodynamic model*. PhD thesis. Institute for Plasma Research, India, 2019.
16. Mukherjee, R., Ganesh, R., & Sen, A. (2018). Numerical study of driven 3d magnetohydrodynamics: dynamos and recurrences. In *APS Division of Plasma Physics Meeting Abstracts* (Vol. 2018, pp. BO8–004).
17. Mukherjee, R., Ganesh, R., & Sen, A. (2019). Nonlinear alfven waves and recurrences in 3d magnetohydrodynamics. In *APS Division of Plasma Physics Meeting Abstracts* (Vol. 2019, pp. JO4–013).
18. Biswas, S., Ganesh, R., Mukherjee, R., & Sen, A. (2021). Quasi-recurrence: a new novel feature observed in 3d-magnetohydrodynamic plasmas. In *5th Asia-Pacific Conference on Plasma Physics*.
19. Mignone, A., Bodo, G., Massaglia, S., Matsakos, T., Tesileanu, O., Zanni, C., & Ferrari, A. (2007). Pluto: a numerical code for computational astrophysics. *The Astrophysical Journal Supplement Series*, *170*(1), 228.
20. Mukherjee, R., Ganesh, R., Saini, V., Maurya, U., Vydyanathan, N., & Sharma, B. (2018). Three dimensional pseudo-spectral compressible magnetohydrodynamic gpu code for astrophysical plasma simulation. In *2018 IEEE 25th International Conference on High Performance Computing Workshops (HiPCW)* (pp. 46–55). IEEE.
21. Mukherjee, R., & Ganesh, R. (2018). Numerical relaxation of a 3D MHD Taylor-Woltjer state subject to abrupt expansion. *arXiv preprint arXiv:1811.09803*.
22. Mukherjee, R., Adhikari, S., & Biswas, S. (2021, April). Rupakmukherjee/tara. *Zenodo*, https://doi.org/10.5281/zenodo.4682188

Chapter 14
Accretion Disk-Outflow/Jet and Hard State ULXs

Mayank Pathak and Banibrata Mukhopadhyay

Abstract Ultraluminous X-ray sources (ULXs) have been objects of great interest for the past few decades due to their unusually high luminosities and spectral properties. A few of these sources exhibit super-Eddington luminosities assuming them to be centering around stellar mass objects, even in their hard state. It has been shown via numerical steady state calculations that ULXs in hard state can be interpreted as highly magnetized advective accretion sources around stellar mass black holes. We use general relativistic magnetohydrodynamic (GRMHD) framework to simulate highly magnetized advective accretion flows around a black hole and show that such systems can indeed produce high luminosities like ULXs. We also verify that the magnetic fields required for such high emissions is around 10^7 G, in accordance with previous numerical steady state calculations. We further present power profiles for zero angular momentum observer (ZAMO) frame. These profiles show interesting features which can be interpreted as effects of emission due to the Blandford-Znajek and Blandford-Payne mechanisms.

14.1 Introduction

Ultraluminous X-ray sources (ULXs) are known to exhibit very high X-ray luminosities. ULXs are point-like, non-nuclear sources which are very rare and most galaxies host either one or none at all. These sources cannot be stars, due to their very high luminosites, which would easily tear apart a star due to radiation pressure.

M. Pathak (✉)
Joint Astronomy Programme, Department of Physics, Indian Institute of Science, Bengaluru, India
e-mail: mayankpathak@iisc.ac.in

B. Mukhopadhyay
Department of Physics, Indian Institute of Science, Bengaluru, India
e-mail: bm@iisc.ac.in

S. Ghosh, A. R. Rao (eds.), *The Relativistic Universe: From Classical to Quantum*, Astrophysics and Space Science Proceedings 61,
https://doi.org/10.1007/978-3-031-90186-7_14

While bright X-ray sources are not uncommon with active galactic nuclei (AGN) having high X-ray luminosities, ULXs are neither as massive as AGNs nor located in galactic nuclei.

Luminosities of ULXs are in the range of $3 \times 10^{39} - 3 \times 10^{41}$ ergs/s. For a stellar mass source, these luminosities are more than their corresponding Eddington luminosity (L_{edd}), which is given by

$$L_{edd} = \frac{4\pi c G M}{\kappa_{es}}, \tag{14.1}$$

where G is the gravitation constant, M is the mass of the central gravitating object and κ_{es} is the electron scattering opacity.

To explain these high luminosities, ULXs have been modeled as accreting intermediate mass black holes (IMBH) with mass in the range of $10^2 - 10^4 M_{\odot}$, where $M_{\odot}$ is the solar mass. ULXs have also been modeled as stellar mass black holes with slim disks [1] or radiation pressure dominated super-Eddington accretion flows [2].

Another way to explain the super-Eddington emissions from ULXs is by modifying the Eddington limit itself. In the presence of high magnetic fields ($B \geq 10^{12} G$), κ_{es} reduces, which in-turn reduces the effect of radiation pressure on matter, thereby increasing the effective Eddington luminosity [3].

A certain number of ULXs also show pulsations which indicates the presence of a neutron star in the system [4–6]. However, a few ULXs show power-law spectra in their hard state. This behavior remains mysterious and quite counter-intuitive. Mondal and Mukhopadhyay (MM19 hereafter) considered numerical steady state calculations to show that this behavior can be explained by considering ULXs to be highly magnetized sub-Eddington advective accretion flows around a stellar mass black hole [7].

In this chapter, we have considered general relativistic magnetohydrodynamic (GRMHD) simulations to explore and verify the model developed in MM19. We have simulated an advective magnetized accretion flow around a stellar mass black hole with various initial conditions and have calculated outflow power from the system. This power comes out be in the range of the observed ULX luminosities. We also discuss outflow power profiles in the zero angular momentum observer (ZAMO) frame. These profiles show interesting behavior which can be used to determine/separate the Blanford-Znajek and the Blandford-Payne components of the outflow power.

14.2 Simulation Setup

We have used the publicly available GRMHD code, BHAC (Black Hole Accretion Code) [8] to simulate a system of an accretion disk around a Kerr black hole. The simulation uses horizon-penetrating modified Kerr-Schild (MKS) coordinates to evolve the system and data is output in Boyer-Lindquist (BL) coordinates. We adopt geometric units, i.e., $GM_{BH} = c = 1$, $r_g = GM_{BH}/c^2 = 1$ and the light crossing time, $t = GM_{BH}/c^3 = 1$, in our simulations. Here, M_{BH} is the mass of the black hole and c is the speed of light. The accretion disk is initiated by using the Fishbone Moncrief (FM) torus setup [9]. The black hole spin has been fixed at $a = 0.9375$. We have carried out 2.5-dimensional simulations, by exploiting the axisymmetry of the system. The computational domain extends from $1.22\, r_g$ to $2500\, r_g$ in the radial direction and 0 to 2π in the azimuthal direction. The simulations have been run at a resolution of $384 \times 192 \times 1$. All simulations are evolved to 3×10^4 timesteps. In a future work, we will elaborate these.

14.2.1 Magnetic Field Evolution

The simulation solves the induction equation to evolve the magnetic field in curved space-time. The field is initiated by defining the plasma-beta (β) and initial vector potential. In our simulations, we start with a poloidal magnetic field, initial $\beta = 100$ and have used the following vector potentials to initiate SANE (Standard and Normal Evolution) and MAD (Magnetically Arrested Disk) accretions [10], respectively:

1. $A_\phi = \max(\rho/\rho_0 - 0.2, 0)$,
2. $A_\phi = \exp(-r/r_o)(r/r_{in})^3 \sin^3\theta \max(\rho/\rho_0 - 0.01, 0)$,

where ρ_0 is maximum density, r_{in} is inner edge of the disk and $r_o = 400 r_g$. Note that GRMHD simulations with same parameters have also been carried out using the alternate publicly available code HARMPI by Raha et al. [11]. These results are also published in the present volume.

14.3 Results

Figures 14.1 and 14.2 show density contours with magnetic field streamlines at different time for SANE and MAD simulations respectively. These contours show the evolution of flow density and magnetic fields over the course of the simulations.

As evident from Fig. 14.1, the SANE accretion flow is uninterrupted throughout the evolution. At later times, the flow is accreted via a thin funnel onto the black hole. This is due to the formation of strong magnetic fields near the black hole, which oppose the accretion flow via magnetic pressure.

Figure 14.2 shows that the accretion is not continuous for MAD systems and the flow oscillates due to the formation of a barrier near the black hole. This is because MAD flows accumulate as much magnetic flux as possible in a short evolution time, leading to large magnetic pressure near the black hole which in-turn results in a

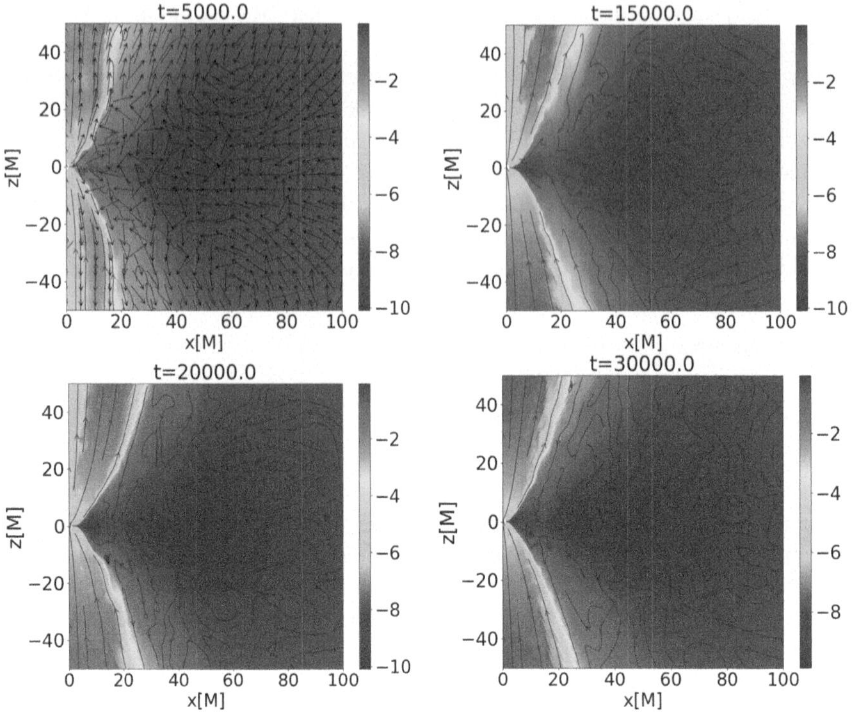

Fig. 14.1 Density contours for SANE with magnetic field streamlines. The top-left panel shows the velocity vectors

magnetic barrier. Due to the turbulent nature of the flow, the ram pressure of the accreting matter occasionally breaks through the magnetic barrier but due to high magnetic flux accumulation, the barrier forms again. This effect is exaggerated in 2-dimensional simulations, as the accretion flow is confined to only one azimuthal plane.

To explore the outflow power of the flow, we define the following two quantities:

1. Mass Accretion rate ($\dot{M}$):

$$\dot{M}(r) = -\int \sqrt{-g}\rho u^r d\theta d\phi. \tag{14.2}$$

2. Inward energy flux ($\dot{E}$):

$$\dot{E}(r) = \int \sqrt{-g}T_t^r d\theta d\phi, \tag{14.3}$$

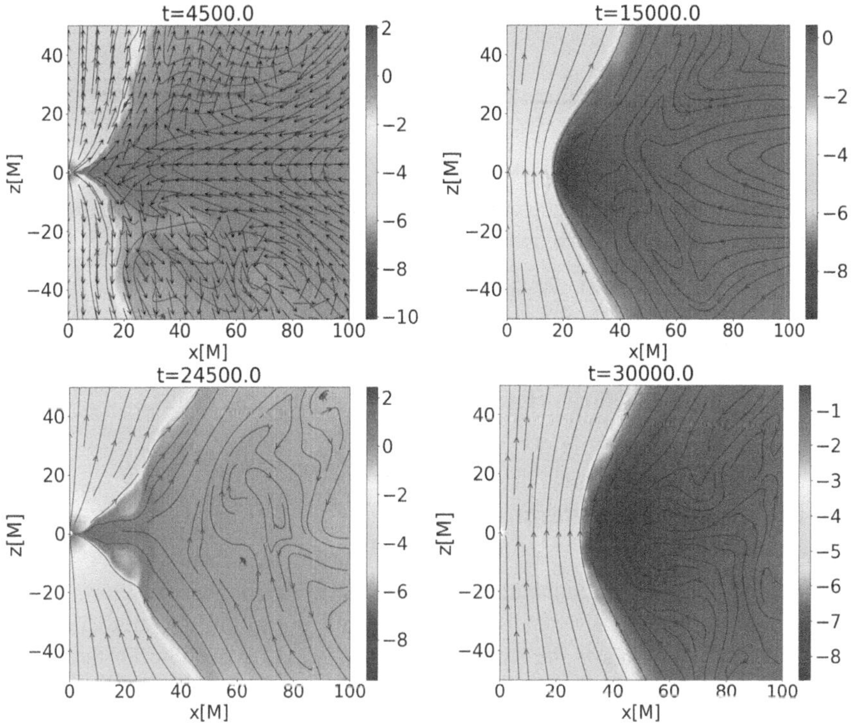

Fig. 14.2 Density contours for MAD with magnetic field streamlines. The top-left panel shows the velocity vectors

where g is the metric determinant and T^{μ}_{ν} is the stress energy tensor, given by

$$T^{\mu}_{\nu} = (\rho + \tilde{P} + \tilde{P}/(\gamma + 1) + b^2)u^{\mu}u_{\nu} - b^{\mu}b_{\nu}, \tag{14.4}$$

where ρ is the density, $\tilde{P}$ is the pressure of the flow and γ is the adiabatic constant; u^{μ} and b^{μ} are the four-velocity and four-magnetic field respectively and $b^2 = b^{\mu}b_{\mu}$.

Here the signs are chosen such that positive values mean the flow of the quantity into the black hole. The net output power is then defined as [10]:

$$P(r) = \dot{M}(r) - \dot{E}(r) = -\int \sqrt{-g}\rho u^r d\theta d\phi - \int \sqrt{-g}T^r_t d\theta d\phi\,. \tag{14.5}$$

The quantity $P(r)$ calculated following Eq. (14.5) gives the net outflow power in code units. To infer the physical power from $P(r)$, we need to multiply it with $\dot{M}_s c^2$, where $\dot{M}_s$ is the scale of accretion rate, defined by $\dot{M}_{phy} = \dot{M}_s\dot{M}$, with $\dot{M}_{phy}$ being the physical accretion rate. Since we are considering advective accretion flows, we have chosen $\dot{M}_{phy} = 0.05\dot{M}_{edd}$, where $\dot{M}_{edd}$ is the Eddington accretion rate, given by, $\dot{M}_{edd} = 1.39 \times 10^{18}(M_{BH}/M_{\odot})$ gm/s and $M_{BH} = 20M_{\odot}$.

Thus, the dimensional power is given by

$$P(r) = \left(\frac{\dot{M}(r) - \dot{E}(r)}{\dot{M}(r')} \right) \dot{M}_{phy} c^2, \tag{14.6}$$

where r' is the radius chosen for normalizing the outflow power.

The definition of time averaged power is ambiguous. Thus, we have considered three power definitions based on the types of time-averaging:

1. $P(r) =< \dot{M}(r) - \dot{E}(r) > / < \dot{M}(r) > \dot{M}_{phy} c^2$,
2. $P(r) =< \dot{M}(r) - \dot{E}(r) > / < \dot{M}(r_{eq}) > \dot{M}_{phy} c^2$,
3. $P(r) =< \dot{M}(r) - \dot{E}(r) > / < \dot{M}(r_h) > \dot{M}_{phy} c^2$,

where r_{eq} is the steady flow radius and r_h is the event horizon radius.

14.3.1 Steady Flow Radius

The accretion flow in our simulation domain undergoes inflow and outflow. As the simulation evolves, the flow reaches an inflow-outflow equilibrium out to a certain radius [12]. This radius is the steady flow radius (r_{eq}). Depending on the initial magnetic vector potential and time of evolution, r_{eq} changes. To determine r_{eq}, we investigate the time averaged accretion rate profiles for both SANE and MAD simulations.

As evident from Fig. 14.3, the radii r_{eq}-s for the SANE and MAD simulations are 10 r_g and 20 r_g respectively [13]. The lower value of r_{eq} for SANE indicates the slower evolution of the system compared to MAD. This is because MAD systems accumulate much more magnetic flux near the black hole in a smaller amount of time as compared to SANE. Thus MAD systems have more time to equilibrate with the conditions within the simulation domain. We compute the outflow power only from the inflow-outflow equilibrium region of the accretion flow, i.e., from the horizon of the black hole to r_{eq}. The region outside r_{eq} has not achieved inflow-outflow equilibrium and is thus unphysical from an astrophysical standpoint.

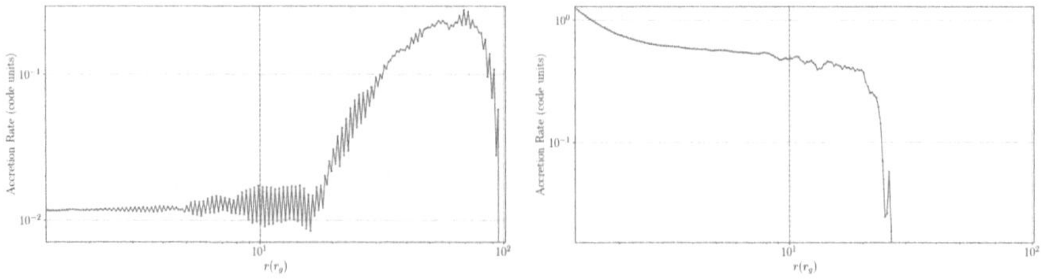

Fig. 14.3 Time averaged accretion rate profiles. *Red dashed line* indicates r_{eq}. (**a**) SANE. (**b**) MAD

14.3.2 Outflow Power Profiles

Using Eq. (14.6) and the above formalism for r_{eq}, we have explored the time averaged power profiles for the aforementioned three definitions. Figure 14.4 clearly shows that all the power definitions lie within the ULX range for both SANE and MAD simulations. However, the average power is higher for the MAD simulations than SANE. This shows that MAD systems are capable of producing higher outflow power. Nevertheless, the important point to note is that we do not include radiative cooling in the present chapter, hence the estimated outflow power is only the precursor to ULX luminosities or the upper bound.

Upon considering the same parameters as Raha et al. [11] used for their MAD simulation in HARMPI, i.e., the system domain going from 1.29 r_g to 10,000 r_g with resolution of 256×256, keeping the other MAD simulation parameters same

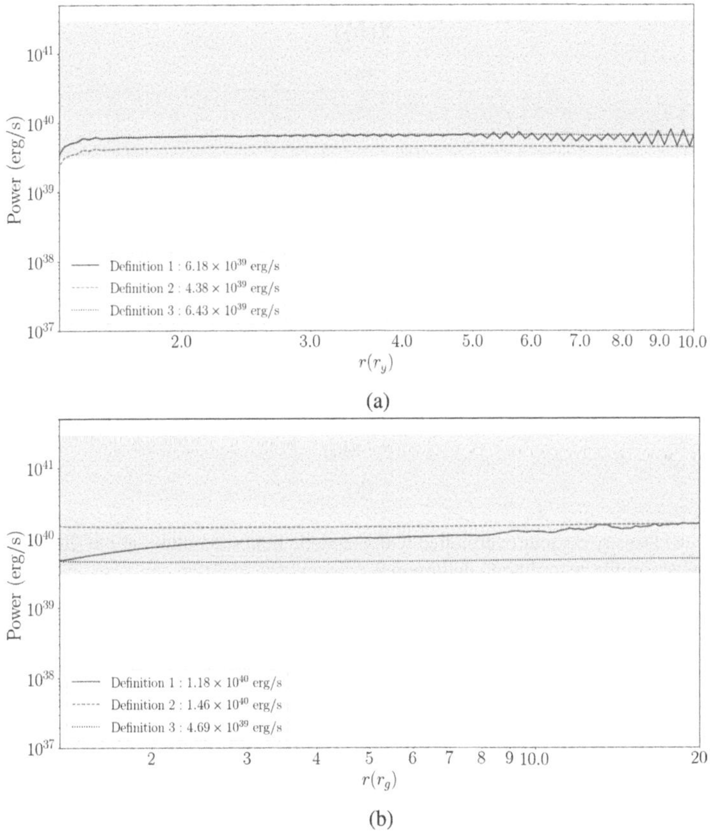

Fig. 14.4 Power profiles for all definitions of time averaging. Shaded region indicates the observed ULX luminosity range. Average power for each definition of power is also mentioned next to the indicated line-type. (**a**) SANE. (**b**) MAD

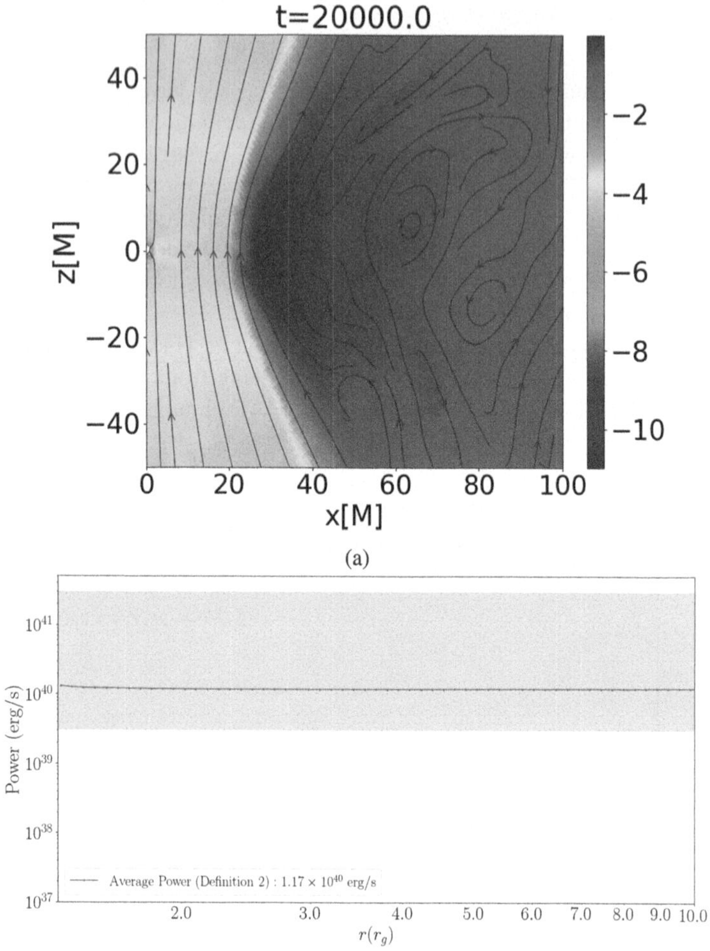

(a)

(b)

Fig. 14.5 (**a**) Density contour overplotted with magnetic field streamlines at $t = 20{,}000\ r_g/c$. (**b**) Outflow power profile according to definition 2

as mentioned above, the obtained density contour and power profile are given by Fig. 14.5. The time averaging for the power profile has been done from 15,000 r_g/c to 20,000 r_g/c, as Raha et al. [11]. The density contour shows strong poloidal magnetic fields forming a barrier, resulting in a MAD-like configuration. The power profile shows similar outflow power magnitude as obtained in Fig. 14.4. This shows that our outflow power remains similar irrespective of computational domain size, resolution and time averaging window, thus exhibiting the robustness of our results.

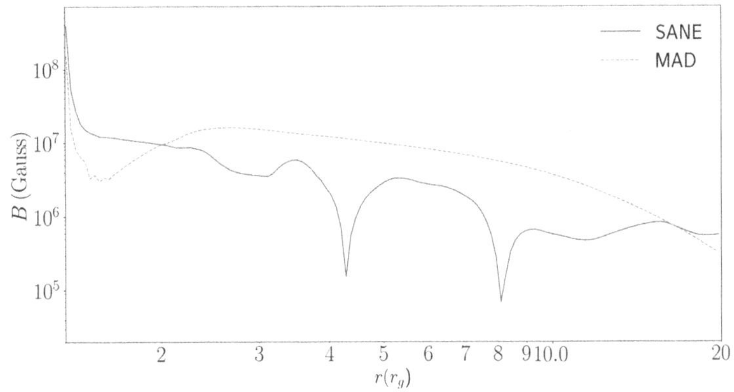

Fig. 14.6 Magnetic field profiles for SANE and MAD simulations

14.3.3 Magnetic Field Profile

MM19 showed that the fields required to produce ULX luminosities from advective flows are of the order of 10^7 G. However, to achieve such high fields, they had to consider super-Eddington magnetic fields far away from the black hole.

Figure 14.6 shows that the field strength obtained from our simulations is also $\sim 10^7$ G, without however any requirement for super-Eddington fields. It is also evident that although the peak magnetic field in the SANE simulations is slightly higher than the MAD simulations very close to the black hole, the higher MAD magnetic fields at other radii lead to higher luminosities compared to SANE. The differences in powers, whatever they be, between SANE and MAD is also reflected in their magnetic fields.

14.4 Discussion

MAD simulations are known to show more than 100% efficiency. This fact has been used to explain apparent more than 100% efficient accretion flows in AGNs [14]. In our simulations as well, we observe higher outflow power for MAD compared to SANE, even though SANE has higher peak magnetic field. This is because MAD flows are very efficient and are capable of producing strong outflows with lower fields.

This result can also be attributed to the fact that magnetic fields evolve differently for MAD and SANE simulations. Figure 14.7 shows the magnetic field components for SANE and MAD simulations. Unlike the SANE system, the MAD system exhibits a significant increase in B_θ at larger radii which results in sustained magnetic fields in MAD leading to higher outflow power.

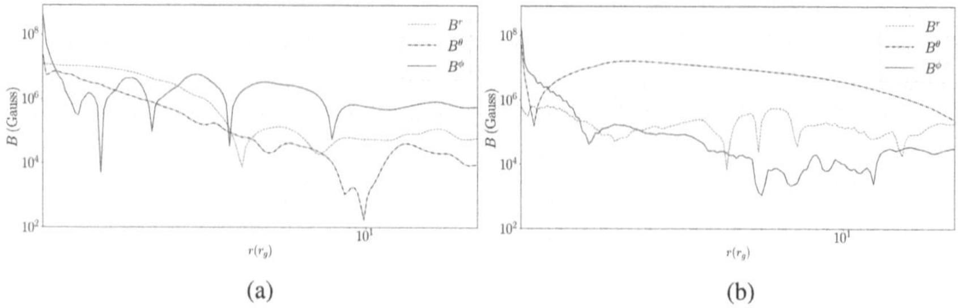

Fig. 14.7 Magnetic field components for (**a**) SANE and (**b**) MAD simulations

14.5 Outflow Power in the ZAMO Frame

The above analysis is valid for the observers at infinity and the obtained power profiles are flat. Due to this, the driving mechanism of the outflow cannot be attributed to the ergosphere (Blandford-Znajek (BZ) mechanism [15]), magnetocentrifugal effect of the disk (Blandford-Payne (BP) mechanism [16]) or simply disk wind. In our study, the combined outflow due to the magnetocentrifugal effect and disk wind is considered to be BP outflow.

There is no observer-independent way of defining energy of the gravitational field at a point. To study the interaction between the black hole and jet, we need to be in a local frame, which for the Kerr metric is fiducially the zero angular momentum observer (ZAMO) frame. To study the aforementioned outflow mechanisms, we investigate outflow power in the ZAMO frame.

14.5.1 Formalism

The outward energy flux is defined as:

$$\dot{E}(r) = -\int \sqrt{-g}T^{\mu}_{\nu}u^{\nu}e_{\mu}d\theta d\phi, \tag{14.7}$$

where u^{ν} and e_{μ} are the fluid four velocity and local four unit vector respectively, T^{μ}_{ν} is same as in Eq. (14.4).

The jet region is considered to be the part of the simulation domain in which $-T^{\mu}_{\nu} > 0$ [17]. We only consider outflow in the radial direction, which leads to $e_{\mu} = (0, 1/g^{rr}, 0, 0)$, where g^{rr} is the 'rr' component of the contravariant Kerr metric. We have run simulations for various values of black hole spin parameter, namely, $a = 0.9, 0.8, 0.6, 0.0$, to study the effect of a on the outflow power.

14.5.2 Power Profiles

Figure 14.8 shows the obtained power profiles for the different a values. It is clearly evident that the profile has two contributing components, a peak near the horizon that dies down by the end of the ergosphere and another smaller component around $r = 30r_g$.

The peak near the horizon decreases monotonically with a and reaches a minimum at $r = 2r_g$, i.e. the ergosphere. This can be interpreted as the Blandford-Znajek mechanism, which states that jets/outflows are powered by the spin of the black hole due to the strong inertial frame dragging effect. Particles entering the ergosphere tap the energy of the angular momentum reservoir of the black hole and are kicked out along the magnetic field lines. The field lines themselves are twisted due to the frame dragging which introduces magnetic tension force along the axis of rotation of the black hole. The magnetic tension force collimates the jet outflow.

This also explains the peak value decreasing with a. For slowly rotating black holes, there is not enough energy to tap from the black hole reservoir. This confirms that for non-rotating black holes, there will not be any emission due to the BZ mechanism. This is also evident from Fig. 14.8.

The second peak near $r = 30r_g$ is attributed to the BP process. This is because BP states that in the presence of strong magnetic fields, matter in the accretion disk starts to move along the magnetic field lines under the effect of magnetocentrifugal forces. Due to the acceleration of the particles, the field line can only support the particle flow till the Alfvén radius, after the magnetic tension force dominates over the magnetic pressure force and collimates the flow. Thus BP mechanism is valid for emissions from any part of the disk and is not strictly dependent on the central compact object.

However since the particles are not taping any energy reservoir in BP mechanism, the outflow power is lower than that of the BZ process. BZ on the other hand is an active process, in which the black hole acts as the initial energy source leading to stronger outflows.

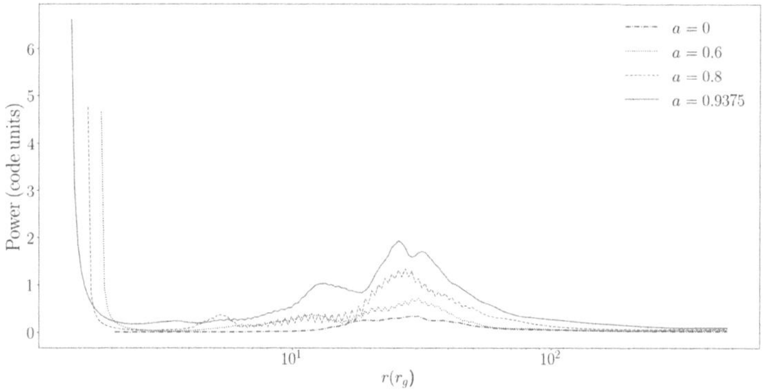

Fig. 14.8 Power profiles for ZAMO observer for different a values

14.6 Conclusion

Our simulations show that highly magnetized advective accretion flows can indeed produce high outflow power, well within the observed ULX luminosity range. This shows that the peculiar properties of hard state ULXs can be explained without invoking principles like modifying the Eddington luminosity by changing the electron scattering cross-section or considering ULXs to be intermediate mass black holes, only very few of which are currently known.

ZAMO frame results show that studying accretion-outflows by changing observer frames might give insights into the outflow generation mechanisms from accreting sources. Our results show two distinct peaks in the power profiles which can be attributed to BZ and BP mechanisms for outflow generation.

References

1. Ebisawa, K., Zycki, P., Kubota, A., Mizuno, T., & Watarai, K.-Y. (2003). *The Astrophysical Journal, 597*, 780.
2. Begelman, M. C. (2002). *The Astrophysical Journal, 568*, L97.
3. Herold, H. (1979). *Physical Review D, 19*, 2868.
4. Bachetti, M., et al. (2014). *Nature, 514*, 202.
5. Fürst, F., et al. (2016). *The Astrophysical Journal, 831*, L14.
6. Israel, G. L., et al. (2017). *Science, 355*, 817.
7. Mondal, T., & Mukhopadhyay, B. (2019). *Monthly Notices of the Royal Astronomical Society, 482*, L24.
8. Porth, O., Olivares, H., Mizuno, Y., et al. (2017). *Computational Astrophysics, 4*, 1.
9. Fishbone, L. G., & Moncrief, V. (1976). *The Astrophysical Journal, 207*, 962.
10. Chatterjee, K., & Narayan, R. (2022). *The Astrophysical Journal, 941*, 30.
11. Raha, R., Chatterjee, K., & Mukhopadhyay, B. (2024). Accretion disk-outflow/jet and hard state ULXs. *The relativistic universe: From classical to quantum*. Springer.
12. Narayan, R., Chael, A., Chatterjee, K., et al. (2022). *Monthly Notices of the Royal Astronomical Society, 511*, 3795.
13. Pathak, M., & Mukhopadhyay, B. (2025). *The Astrophysical Journal, 981*, 162.
14. Tchekhovskoy, A., Narayan, R., & Mckinney, J. C. (2011). *Monthly Notices of the Royal Astronomical Society, 418*, L79.
15. Blandford, R. D., & Znajek, R. L. (1977). *Monthly Notices of the Royal Astronomical Society, 179*, 433.
16. Blandford, R. D., & Payne, D. G. (1982). *Monthly Notices of the Royal Astronomical Society, 199*, 883.
17. Penna, R., Narayan, R., & Sądowski, A. (2013). *Monthly Notices of the Royal Astronomical Society, 436*, 3741.

Chapter 15
Magnetically Arrested Advective Accretion Flows and Jets/Outflows Around Stellar Mass Black Holes: Explaining Hard State ULXs with GRMHD Simulations

Rohan Raha, Banibrata Mukhopadhyay, and Koushik Chatterjee

Abstract An optically thin advective accretion disk is crucial for explaining the hard state of black hole sources. Using general relativistic magnetohydrodynamic (GRMHD) simulations, we investigate how a large-scale, strong magnetic field influences accretion and outflows/jets, depending on the field geometry, magnetic field strength, and the spin parameter of the black hole. We simulate a sub-Eddington, advective disk-outflow system in the presence of a strong magnetic field, which likely remains in the hard state. The model simulations based on HARMPI successfully explain ultra-luminous X-ray sources (ULXs) in the hard state, typically observed with luminosities ranging from 10^{39}–10^{40} ergs s^{-1}. Our simulations generally describe the bright, hard state of stellar-mass black hole sources without requiring a super-Eddington accretion rate. This chapter explores the characteristics of ULXs without invoking intermediate-mass black holes. The observed high luminosity is attributed to the energy stored in the strong magnetic fields, which can generate super-Eddington luminosity. The combined energy of the matter and magnetic field leads to such significant luminosity.

15.1 Introduction

Accretion disks surrounding black holes predominantly involve rotating gas, converting gravitational energy into radiation with high efficiency. This process is recognized as the key power source for various luminous astronomical systems,

R. Raha (✉) · B. Mukhopadhyay
Department of Physics, Indian Institute of Science, Bengaluru, India
e-mail: raharohan@iisc.ac.in; bm@iisc.ac.in

K. Chatterjee
University of Maryland, College Park, MD, USA
e-mail: kchatt@umd.edu

S. Ghosh, A. R. Rao (eds.), *The Relativistic Universe: From Classical to Quantum*, Astrophysics and Space Science Proceedings 61,
https://doi.org/10.1007/978-3-031-90186-7_15

from quasars and active galactic nuclei (AGNs) with supermassive black holes to X-ray binaries (XRBs) with stellar-mass black holes. It is widely acknowledged that magnetic fields are present in these accretion disks and play a significant role in altering their physical states. Therefore, it is expected that magnetic fields contribute to numerous dynamic processes in these disks. For instance, magnetic stress could replace viscous stress as explained by the standard model [1] proposed by Shakura and Sunyaev.

Accretion flows around black holes are generally divided into two categories: cold and hot flows. Cold flows consist of cool, optically thick gas, which typically occurs at higher mass accretion rates. This includes the standard thin disk producing multi-color blackbody radiation at sub-Eddington rates, as well as the slim disk at super-Eddington rates. Such flows are associated with bright objects like AGNs near the Eddington luminosity and soft-state black hole XRBs. On the other hand, hot flows can reach virial temperatures, are optically thin, and occur at lower accretion rates, featuring advection [2, 3]. These flows generally have lower radiative efficiency compared to thin disks, and their efficiency decreases as the accretion rate drops.

Black hole accretion disks transition through various X-ray spectral states, is largely based on their X-ray luminosity and spectral characteristics [4]. The most common states are the high/soft (HS) and low/hard (LH) states [4, 5], which have been the focus of extensive research. These states are believed to reflect different accretion geometries depending on the mass accretion rates and the density of matter.

However, ULXs in the hard state pose a significant scientific puzzle. These sources exhibit luminosities surpassing the Eddington limit for stellar-mass black holes, yet they persist in the hard state—a condition typically linked to lower accretion rates. This apparent contradiction challenges our current understanding. The Eddington luminosity, a crucial concept in this context, is mathematically expressed as:

$$L_{Edd} = \frac{4\pi c G M_{BH}}{\kappa_{es}} \approx 1.4 \times 10^{38} \left(\frac{M_{BH}}{M_{\odot}}\right) \text{erg s}^{-1} . \tag{15.1}$$

In this equation, M_{BH} represents the black hole's mass, G is Newton's gravitational constant, c denotes the speed of light, and κ_{es} stands for the electron scattering opacity.

A ULX in the hard state must be associated with an optically thin, advective accretion disk [6]. To explore ULXs in this state, we analyze how a strong, large-scale magnetic field in the accretion disk, outflows or jets, may transport angular momentum, primarily through magnetic shear rather than α-viscosity, depending on the magnetic field configuration and plasma-β parameter. Interestingly, while the disk becomes thermally unstable at higher accretion rates, it regains stability when stronger magnetic fields are present [7]. It was previously theorized [6, 8–11] that magnetically arrested advective accretion flow (MA-AAF) in an optically thin environment holds critical significance in explaining ULXs and other phenomena. This research confirms these behaviors through numerical simulations.

In this chapter and in a companion chapter [12], we simulate a sub-Eddington, disk-outflow symbiotic model in the advective regime (hence hard state; though no radiation physics included) in the presence of large-scale strong magnetic fields. Numerical simulations of a sub-Eddington, disk-outflow model in the advective regime with strong magnetic fields aim to explain hard-state ULXs typically observed with luminosities of $10^{39} - 10^{40}$ ergs s^{-1} (e.g., Antennae X-11, X-16, X-42, X-44 [13]; NGC 1365 [14]; M99 X1 [15]; M82 X42.3+59 [16]). Our model can also account for bright, hard states of stellar-mass black hole sources without requiring super-Eddington accretion rates or intermediate-mass black holes.

15.2 General Relativistic Magnetohydrodynamic (GRMHD) Framework

The behavior of accretion flows is largely governed by the interplay of non-Maxwellian plasma with intense gravitational and electromagnetic fields, and potentially significant radiation. We model this plasma as a fluid, necessitating the use of GRMHD equations. Our study employs HARMPI [17], a high-performance parallel GRMHD code, to analyze these flows. HARMPI utilizes a robust numerical approach, combining conservative methods, shock-capture capabilities, and stability, founded on the MUSCL scheme [18] and incorporating HLL fluxes [19]. We concentrate on resolving the GRMHD momentum conservation equations and the continuity equation within specific constraints, applying appropriate boundary and initial conditions for an accretion disk. This approach enables temporal evolution of the solution. Using HARMPI, we investigate GRMHD solutions for radiatively inefficient, magnetized accretion flows surrounding a rotating black hole, characterized by the Kerr metric in Kerr-Schild coordinates.

The first governing equation that describes the accretion flow is the conservation of particle number

$$(nu^{\mu})_{;\mu} = 0 \,, \tag{15.2}$$

where n is the particle number density and u^{μ} is the 4-velocity. By replacing n with mass density $\rho = mn$, when m being the mass of a particle, we can rewrite the above equation in a coordinate basis

$$\frac{1}{\sqrt{-g}}\partial_{\mu}(\sqrt{-g}\rho u^{\mu}) = 0 \,, \tag{15.3}$$

which reduces to

$$\partial_t(\sqrt{-g}\rho u^t) = -\partial_i(\sqrt{-g}\rho u^i) \,, \tag{15.4}$$

where i denotes spatial index. The next set of equations is the conservation of energy and momentum, given by

$$(T^{\mu}_{\nu})_{;\mu} = 0 , \tag{15.5}$$

where T^{μ}_{ν} is the stress-energy tensor. The equation in the coordinate basis reduces to

$$\partial_t(\sqrt{-g}T^t_{\nu}) = -\partial_i(\sqrt{-g}T^i_{\nu}) + \sqrt{-g}T^{\kappa}_{\lambda}\Gamma^{\lambda}_{\nu\kappa} , \tag{15.6}$$

where $\Gamma^{\lambda}_{\nu\kappa}$ is the affine connection.

The hydrodynamic part of stress-energy tensor for a perfect fluid with an electromagnetic field is given by

$$T^{\mu\nu}_{HD} = (\rho + u + p)u^{\mu}u^{\nu} + pg^{\mu\nu} , \tag{15.7}$$

(where u is the internal energy density of the gas and p is the gas pressure; for ideal MHD they are related by $p = (\gamma - 1)u$ with γ being polytropic constant) and the electromagnetic part is given by

$$T^{\mu\nu}_{EM} = F^{\mu\alpha}F^{\nu}_{\alpha} - \frac{1}{4}g^{\mu\nu}F^{\alpha\beta}F_{\alpha\beta} , \tag{15.8}$$

where $F^{\alpha\beta}$ is the electromagnetic tensor. Using the ideal MHD approximation, $\mathbf{E} + \mathbf{v} \times \mathbf{B} = 0$ as the fluid has very high conductivity. Hence, the Lorentz force on the plasma vanishes, which is equivalent to

$$u_{\mu}F^{\mu\nu} = 0 . \tag{15.9}$$

The magnetic field in the fluid frame is defined as

$$b^{\mu} = \frac{1}{2}\epsilon^{\mu\nu\kappa\lambda}u_{\nu}F_{\kappa\lambda} , \tag{15.10}$$

where ϵ is the Levi-Civita symbol. Using $b_{\mu}u^{\mu} = 0$ we get the electromagnetic part of the stress energy tensor as

$$T^{\mu\nu}_{EM} = b^2u^{\mu}u^{\nu} + \frac{1}{2}b^2g^{\mu\nu} - b^{\mu}b^{\nu} . \tag{15.11}$$

Therefore, the total stress-energy tensor is given by

$$T^{\mu\nu} = (\rho + u + p + b^2)u^{\mu}u^{\nu} + (p + \frac{1}{2}b^2)g^{\mu\nu} - b^{\mu}b^{\nu} . \tag{15.12}$$

The electromagnetic field evolution is given by the Maxwell equation

$$F_{\mu\nu,\lambda} + F_{\lambda\mu,\nu} + F_{\nu\lambda,\mu} = 0 \,, \tag{15.13}$$

and (15.14)

$$F^{*\mu\nu}_{;\nu} = 0 \,, \tag{15.15}$$

where $F^{*\mu\nu}$ is the dual of the electromagnetic field tensor given by

$$F^{*\mu\nu} = \frac{1}{2}\epsilon^{\mu\nu\kappa\lambda}F_{\kappa\lambda}. \tag{15.16}$$

15.3 Simulation Setup

We use modified Kerr–Schild coordinates (x_0, x_1, x_2, x_3) given by

$$t = x_0,\ \phi = x_3,\ r = e^{x_1},\ \theta = \pi x_2 + \frac{1}{2}(1-h)\sin(2\pi h x_2). \tag{15.17}$$

Our 2D simulation grid extends from $r = r_{in} = 0.6 + 0.4r_H$ to $10^3 r_g$ with r_H being the radius of event horizon and r_g is the gravitational radius given by $r_g = GM_{BH}/c^2$. The grid resolution used is $N_r \times N_\theta \times N_\phi \equiv 256 \times 256 \times 1$. We apply outflowing radial boundary conditions (BCs), transmissive polar BCs, and periodic BCs in the azimuthal direction.

Our simulation starts with a black hole surrounded by a standard Fishbone-Moncrief (FM) torus [20]. We use an ideal gas equation of state with the gas pressure $p = (\gamma - 1)u$, where $\gamma = 5/3$. We evolve the initial magnetized FM torus from time $t = 0$ to $t = 20,000 r_g/c$.

We initialize our model with a poloidal magnetic field by applying a toroidal magnetic vector potential given by, $A_\phi = max(q, 0)$ [21], where

$$q = \frac{\rho}{\rho_{max}} - 0.2 \,, \tag{15.18}$$

and ρ is the rest-mass gas density. The magnetic field strength in the initial setup is normalized by setting $max(p_{gas})/max(p_{mag}) = 0.1$, where $p_{mag} = b^2/2$, is the magnetic pressure. Here max denotes the maximum of the respective quantity.

All quantities are computed from the results produced in every $\sim 10r_g/c$. For quantities Q, averages over space ($< Q >$) and time ($[Q]_t$) are performed directly on Q.

The mass accretion rate as a function of radius (r) is calculated as

$$\dot{M}(r) = -\int\int \sqrt{-g}\rho u^r d\theta d\phi \,, \tag{15.19}$$

and the radial energy flux is given by

$$\dot{E}(r) = \int\int \sqrt{-g}T_t^r d\theta d\phi \, . \tag{15.20}$$

The net outflow efficiency is then given by

$$\eta = \frac{\dot{M} - \dot{E}}{\dot{M}} = \frac{P_{out}}{\dot{M}} \, . \tag{15.21}$$

This refers to the power of outflows that escape to infinity normalized by the mass accretion rate. Positive efficiency values correspond to energy extraction from the system through jets or outflows. We report the outflow power/efficiency time averaged from $t = 15,000 r_g/c$ to $t = 20,000 r_g/c$.

15.4 Results

We aim to investigate, through computational models, how the magnetic field and black hole spin influence the accretion process. We investigate the mechanism through which material is pulled into the black hole via magnetic fields associated with the accretion disk. To validate our findings, we have conducted simulations based on previously established parameters [21]. We have began with a magnetized torus, using a black hole with a spin parameter of $a = 0.9375$ and an initial plasma-β=100. We also compare our results with the GRMHD simulations carried out using the publicly available code BHAC (Black Hole Accretion Code) [22], for the same parameters, published in the present volume [12]. We evolve the disk in time till $t = 20,000 r_g/c$. We compute the power based on Eq. (15.21).

Throughout, when we report the results in CGS units, we consider a stellar mass black hole of mass $M_{BH} = 20 M_\odot$ with total mass accretion rate, $\dot{M} = 0.05 \dot{M}_{Edd}$, where $\dot{M}_{Edd} = L_{Edd}/\eta c^2 = 1.39 \times 10^{18} (M_{BH}/M)$ g s^{-1}, considering radiative efficiency $\eta = 0.1$.

Figure 15.1 shows the snapshot of density in log-scale at time $t = 20,000 r_g/c$ along with the outflow power. This shows an efficiency of around 1.2, as was demonstrated earlier [21]. A figure with similar parameters is presented in other work exploring BHAC [21] for comparison.

We further run simulations for two different spin parameters $a = 0.5$ and 0.998, i.e., low, and high spins, respectively, and start with an initial plasma-β=0.1. To capture the effect of the magnetic field, we also run simulations with $a = 0.998$ and initial plasma-β=1, which corresponds to a lower magnetic field case. Figures 15.2, 15.3 and 15.4 show snapshots of the simulation results of accretion disk for the three cases, respectively, where they show the inflow-outflow structure at an evolved time $t = 20,000 r_g/c$.

During the accretion process, the matter pulls the magnetic field lines inward because the magnetic flux is nearly frozen into the matter due to the ideal MHD

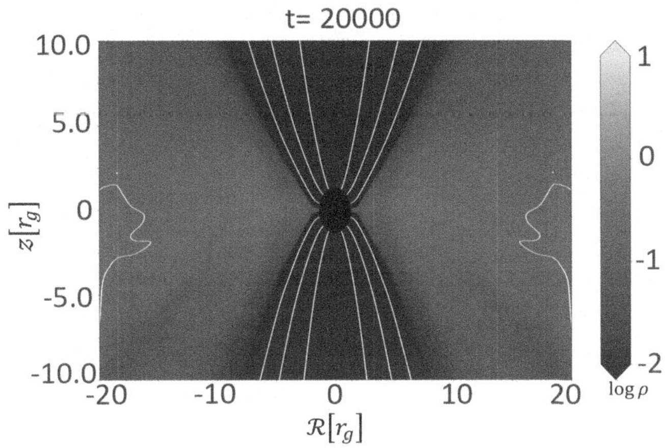

(a)

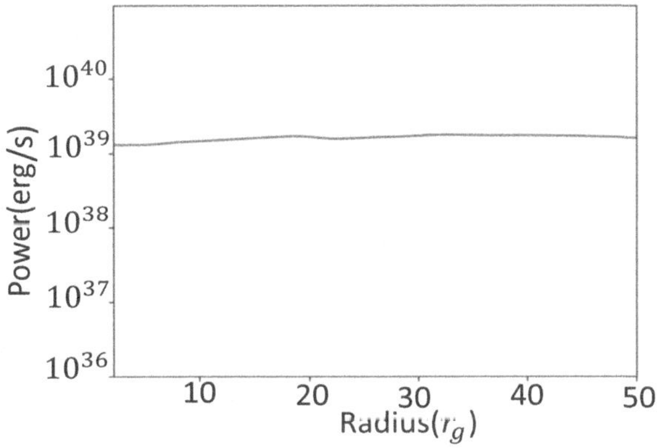

(b)

Fig. 15.1 (**a**) The density contour at the flow vertical plane in log-scale at time $t = 20,000 r_g/c$ with the white lines indicating magnetic field lines, for $a = 0.9375$, initial plasma-$\beta = 100$, (**b**) the corresponding outflow power in CGS units

condition. As the gas draws the magnetic field inward and interacts with the black hole's spin, the surrounding spacetime also gets dragged along. Moreover, the disk's rotation twists the magnetic field lines around the black hole [23]. Consequently, the magnetic field lines are carried along with the space around the black hole. This causes the field lines to form a helical shape around the black hole, moving outward in a wave-like fashion. This wave transports energy and angular momentum away from the black hole through outflows and jets [24]. Therefore, a black hole with a higher spin can more effectively drag both the matter and magnetic field, leading to faster inward motion of the material due to increased transport by magnetic shear. This mechanism generates more potent jets compared to scenarios involving lower spin rates. As compared in Figs. 15.2, 15.3 and 15.4, for the $a = 0.998$ and

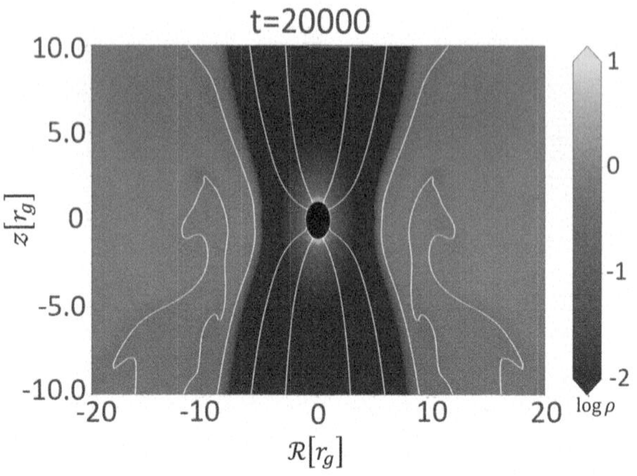

Fig. 15.2 Density contour in log-scale at time $t = 20,000 r_g/c$ with the white lines indicating magnetic field lines for the case $a = 0.998$, initial plasma-$\beta = 0.1$

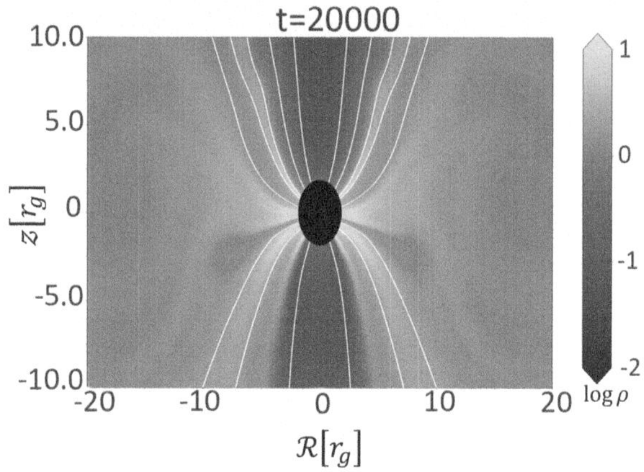

Fig. 15.3 Density contour at the flow vertical plane in log-scale at time $t = 20,000 r_g/c$ with the white lines indicating magnetic field lines for $a = 0.5$, initial plasma-$\beta = 0.1$

initial plasma-β=0.1 case, the magnetic field forms a barrier near the black hole, thus forming a magnetically arrested disk through which matter diffuses after some time. As the matter brings in more magnetic flux due to flux freezing, it forms a barrier again, and this process goes on. However, in the case of $a = 0.5$ and initial plasma-β=0.1, in Fig. 15.3, the spin is not high enough to drag the magnetic field as efficiently to form a barrier. In Fig. 15.4, for the case of $a = 0.998$ and initial plasma-β=1, although the spin is high, the initial magnetic field supply is not high enough to produce magnetic barriers. They produce outflows that are weaker than the case with initial plasma-β=0.1.

The possible explanations for these jets and outflows have been proposed by Blandford–Znajek (BZ) [25] and Blandford–Payne (BP) [26] mechanisms. It has been explained, how the toroidal magnetic field helps launch powerful outflows, which are then collimated by the poloidal magnetic field.

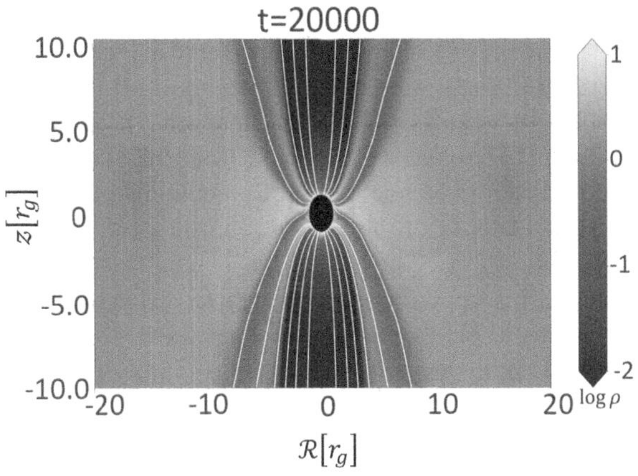

Fig. 15.4 Density contour at the flow vertical plane in log-scale at time $t = 20,000 r_g/c$ with the white lines indicating magnetic field lines for $a = 0.998$, initial plasma-$\beta = 1$

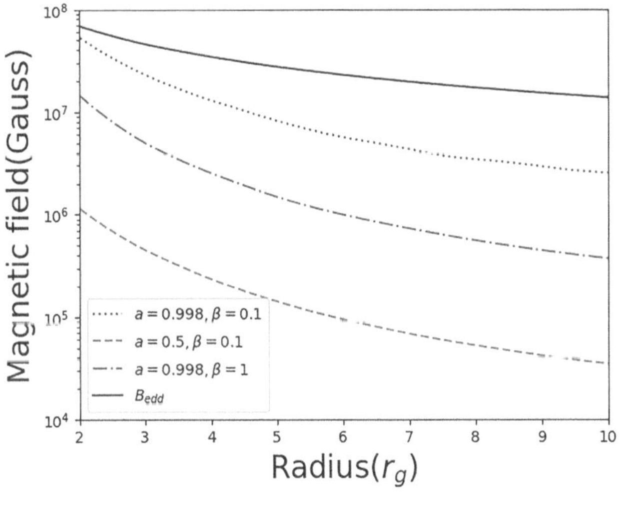

Fig. 15.5 Magnetic field in Gauss for (i) $a = 0.998$, initial plasma-$\beta = 0.1$, (ii) $a = 0.5$, initial plasma-$\beta = 0.1$, and (iii) $a = 0.998$, initial plasma-$\beta = 1$

The Blandford-Znajek model suggests that the power of jets is determined by the magnetic flux and the black hole's spin, expressed as $P \propto \phi_{BH}^2 a^2$, where ϕ_{BH} is the poloidal magnetic flux accumulated near the black hole and a is the spin of the black hole. Consequently, a higher black hole spin leads to matter falling inward at a greater velocity due to increased transport by magnetic shear, resulting in stronger outflows compared to a lower spin scenario.

Figure 15.5 shows the time averaged magnetic field for the three cases. It demonstrates how the magnetic field evolves for different spins. As discussed earlier, the higher spinning black hole can gather a larger magnetic field due to an increased frame dragging effect and hence has more magnetic flux at its disposal to launch more powerful outflows. High polarization in the hard state of Cyg X-1 with $M \sim 14M_{\odot}$ and related jet suggest magnetic field $B \sim 10^6 G$ at the base of jet (i.e. disk) [27], which is recovered from the simulations. Near black holes, the magnetic field's influence becomes predominant due to the process of flux freezing, which results from the inward movement of magnetic flux during accretion. However, there's a ceiling on the magnetic flux that can accumulate around a black hole. This upper bound on magnetic field intensity is established by equating the magnetic field's energy density to that of the accreting matter, which generates the Eddington luminosity in the vicinity of the black hole. This limit is referred to as the Eddington magnetic field [8] and can be described by the expression $B_{Edd} \sim 10^4 G(M/10^9 M_{\odot})^{-1/2}$. Due to the super-Eddington magnetic field very close to the black hole in the two cases, i.e., $a = 0.998$, initial plasma-$\beta = 0.1$, and $a = 0.998$, initial plasma-$\beta = 1$, these systems can produce very high energy outflows as the high magnetic field can throw out matter to produce energetic outflows.

We calculate the outflow power from the efficiency based on the definitions

$$\eta = \frac{[\dot{M}]_t - [\dot{E}]_t}{[\dot{M}]_t}. \tag{15.22}$$

This is equivalent to the definition in Eq. (15.21) but here we have normalized the power by time averaged mass accretion rate $[M]_t$ and time is averaged from $t = 15,000 r_g/c$ to $t = 20,000 r_g/c$.

Figure 15.6 shows the outflow power for the three cases discussed above. As expected, the outflow power increases with the spin of the black hole and also increases with the magnetic field strength (shown in Fig. 15.5). We see a direct cor-

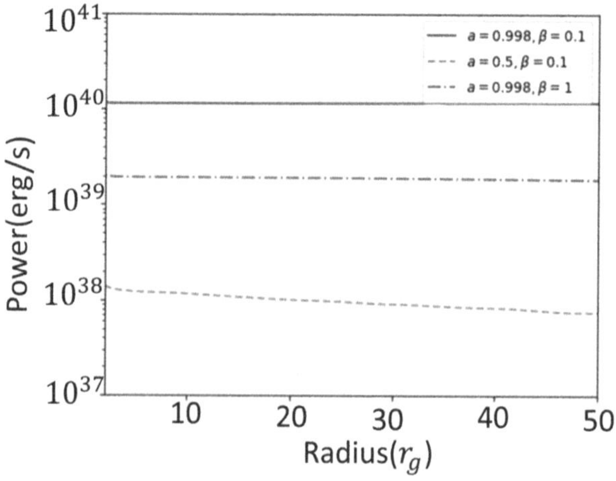

Fig. 15.6 Outflow Power in erg/s: (a) Power for (i) $a = 0.998$, initial plasma-$\beta = 0.1$, (ii) $a = 0.5$, initial plasma-$\beta = 0.1$, and (iii) $a = 0.998$, initial plasma-$\beta = 1$

relation between the magnetic field and outflow power, because the outflow power increases as we have a higher magnetic field very close to the black hole horizon. As predicted from Blandford-Znajek model discussed earlier, we observe a higher outflow power with increased magnetic field strength and higher black hole spin.

15.5 Summary

We have conducted simulations of a sub-Keplerian, magnetized, advective accretion disk, featuring both inflowing and outflowing matter. This disk, being sub-Keplerian and advective, is also optically thin, with inflow and outflow present. The presence of strong magnetic fields generates effective magnetic forces, which act similarly to viscous forces. These forces facilitate the transport of angular momentum both within the disk and through the outflow. To sustain hard X-rays, the accretion rate needs to have a strict upper limit, which is why the system remains non-radiative in nature with a lower accretion rate. However, radiative GRMHD simulations could provide a more accurate picture of the dynamics of accretion flows. We have computed the total outflow power from the system and have found that to generate highly energetic outflows, a high spin of the black hole, along with a substantial magnetic field, is required. The black hole can drag in the magnetic field, launching powerful jets.

The presence of strong magnetic fields leads to the formation of a highly energetic disk and strong outflows or jets. We observed magnetic field strengths of up to 10^6 Gauss at the bases of jets, consistent with observations of systems like Cyg X-1. Our simulations show a direct correlation between the magnetic field strength and outflow power, as predicted by the Blandford-Znajek model. Higher black hole spin and stronger magnetic fields result in more powerful outflows. For cases with high spin ($a = 0.998$) and strong initial magnetic fields (plasma-$\beta = 0.1$), we observed the formation of magnetically arrested disks, where magnetic barriers periodically form and dissipate.

The outflow power achieved in our simulations reaches levels consistent with observed ULXs, with luminosities in the range of $10^{39} - 10^{40}$ ergs/s, without requiring super-Eddington accretion rates or intermediate-mass black holes. This model can be applied to explain ULXs in their hard state. These ULXs can be described as magnetically dominated, advective, sub-Keplerian accretion flows, or MA-AAF, surrounding a stellar-mass black hole with a high spin. Our simulations demonstrate that the combination of strong magnetic fields and high black hole spin can produce the high luminosities observed in ULXs while maintaining the characteristics of the hard state.

Future work could include incorporating radiative GRMHD simulations to provide a more complete picture of the accretion flow dynamics, exploring a wider range of initial conditions, including different black hole masses and accretion rates, investigating the long-term stability of these magnetically arrested flows and their implications for ULX variability, and comparing simulated spectra with observed ULX spectra to further validate the model. These findings contribute to our understanding of ULXs and suggest that some of these enigmatic sources may be explained by extreme magnetic fields around rapidly spinning stellar-mass black holes, rather than requiring intermediate-mass black holes or super-Eddington accretion.

References

1. Shakura, N. I., & Sunyaev, R. A. (1973). *Astronomy & Astrophysics*, *24*, 337.
2. Chakrabarti, S., & Titarchuk, L. G. (1995). *The Astrophysical Journal*, *455*, 623.
3. Narayan, R., & Yi, I. (1995). *The Astrophysical Journal*, *452*, 710.
4. Das, A. R., & Mukhopadhyay, B. (2023). *The Astrophysical Journal*, *955*, 86.
5. Fender, R. P., Belloni, T. M., & Gallo, E. (2004). *Monthly Notices of the Royal Astronomical Society*, *355*, 1105.
6. Mondal, T., & Mukhopadhyay, B. (2020). *Monthly Notices of the Royal Astronomical Society*, *495*, 350.
7. Datta, S. R., Mondal, T., & Mukhopadhyay, B. (2022). *Monthly Notices of the Royal Astronomical Society*, *513*, 204.
8. Mondal, T., & Mukhopadhyay, B. (2019). *Monthly Notices of the Royal Astronomical Society Letter*, *482*, L24.
9. Mondal, T., & Mukhopadhyay, B. (2018). *Monthly Notices of the Royal Astronomical Society*, *476*, 2396.
10. Mondal, T., & Mukhopadhyay, B. (2019). *Monthly Notices of the Royal Astronomical Society*, *486*, 3465.
11. Mukhopadhyay, B., & Chatterjee, K. (2015). *The Astrophysical Journal*, *807*, 43.
12. Pathak, M., & Mukhopadhyay, B. (2024). Magnetically arrested advective accretion flows and jets/outflows around stellar mass black holes: Explaining hard state ULXs with GRMHD simulations. *The relativistic universe: From classical to quantum*. Springer.
13. Feng, H., & Kaaret, P. (2009). *The Astrophysical Journal*, *696*, 1712.
14. Soria, R., Risaliti, G., Elvis, M., Fabbiano, G., Bianchi, S., & Kuncic, Z. (2009). *The Astrophysical Journal*, *695*, 1614.
15. Soria, R., & Wong, D. S. (2006). *Monthly Notices of the Royal Astronomical Society*, *372*, 1531.
16. Feng, H., Rao, F., & Kaaret, P. (2010). *The Astrophysical Journal*, *710*, L137.
17. Gammie, C. F., McKinney, J. C., & Tóth, G. (2003). *The Astrophysical Journal*, *589*, 444.
18. van Leer, B. (1979). *Journal of Computational Physics*, *32*, 101.
19. Harten, A. (1983) *Journal of Computational Physics*, *49*, 357.
20. Fishbone, L. G., & Moncrief, V. (1976). *The Astrophysical Journal*, *207*, 962.
21. Chatterjee, K., & Narayan, R. (2022). *The Astrophysical Journal*, *941*, 30.
22. Porth, O., Olivares, H., Mizuno, Y., Younsi, Z., Rezzolla, L., Moscibrodzka, M., Falcke, H., & Kramer, M. (2017). *Computational Astrophysics and Cosmology*, *4*, 1.
23. Raha, R., Mukhopadhyay, B., Chatterjee, K., & Gopika, S. M. (2023). *Astronomy Reports*, *67*, S189.
24. Bisnovatyi-Kogan, G. S., & Ruzmaikin, A. A. (1974). *Astrophysics and Space Science*, *28*, 45.
25. Blandford, R. D., & Znajek, R. L. (1977). *Monthly Notices of the Royal Astronomical Society*, *179*, 433.
26. Blandford, R. D., & Payne, D. G. (1982). *Monthly Notices of the Royal Astronomical Society*, *199*, 883.
27. Zdziarski, A. A., Pjanka, P., Sikora, M., & Stawarz, L. (2014). *Monthly Notices of the Royal Astronomical Society*, *442*, 3243.

Part IV
Observations/Modeling of X-ray Binaries

Chapter 16
X-ray Observations of Black Hole Sources

A. R. Rao

Abstract X-ray astronomy is closely related to the study of black hole sources. The discovery that some unseen objects, more massive than any degenerate star, emit huge amounts of X-rays helped accept the concept that back holes are present in X-ray binaries. The detection of copious amounts of highly variable X-rays helped the emergence of the paradigm that all Active Galactic Nuclei harbored a supermassive black hole. Since the bulk of the emission in these sources are in X-rays and X-rays are thought to be originating from regions closest to the black holes, it was expected that X-ray observations would yield significant inputs for our understanding of the physical phenomena happening close to black holes like the disk-jet connection and help measure many important parameters like the mass and spin of the black holes. I will trace the developments in this area for the past several decades and, noting the relatively limited success, stress the need for more sensitive measurements. I will highlight the recent X-ray polarization measurements of the Galactic black hole candidate source Cygnus X-1 using the CZT Imager instrument of AstroSat and sketch possible future developments.

16.1 Introduction

X-ray astronomy started with the serendipitous discovery of the bright X-ray source Sco X-1 during the legendary rocket flight flown for the ostensible reason of measuring reflected X-rays from the moon [1]. Based on the then prevailing wisdom of stellar structure and evolution and the extrapolation from the measured X-rays from the Sun, bright X-ray sources were not expected in the sky. The detection of Sco X-1 and the subsequent detection of several X-ray sources near the Galactic center region tickled the imagination of the scientific community and nudged them to accept the existence of certain unusually bright X-ray sources [2, 3]. It was,

A. R. Rao (✉)
Department of Astronomy and Astrophysics, Tata Institute of Fundamental Research, Mumbai, India

S. Ghosh, A. R. Rao (eds.), *The Relativistic Universe: From Classical to Quantum*, Astrophysics and Space Science Proceedings 61,
https://doi.org/10.1007/978-3-031-90186-7_16

however, the launch of UHURU in 1970 that firmly established our understanding of X-ray binaries. Within a few years of UHURU's launch, it was understood and accepted that accretion onto compact objects is the prime energy source responsible for the copious emission of X-rays in these binary sources. In particular, the enigmatic X-ray source Cygnus X-1 was identified with a normal optical star based on its X-ray and radio positions, and the radial velocity measurements of this optical star indicated the presence of a massive unseen companion [4]. Cygnus X-1 was thought to be an accreting black hole source, though, at that time, it was usual to refer to it as a 'black hole candidate' because other possibilities could not be unequivocally eliminated [5]. It, however, took a while to extend these concepts to understand Active Galactic Nuclei (AGN) as accreting supermassive black holes [6, 7]. The unified model for the AGN emerged in the nineties with the basic tenet that the source of energy is accretion and attempts to understand AGN and black hole X-ray binaries (BH-XRB) as manifestations of the accretion phenomena across black hole masses of several orders of magnitude came to be universally accepted [8, 9].

Our understanding of accreting black hole sources is indeed intimately related to the development of X-ray observational capabilities. Over the past six decades, the sensitivity of X-ray detectors has leapfrogged several orders of magnitudes, and the diversity of observations in terms of cadence, energy range, and other features like polarization is truly breathtaking [10]. These developments, however, are mostly restricted to lower energy X-rays or soft X-rays below $\sim$10 keV and hence probe only some specific features of the X-ray emission mechanisms in these accreting sources. In this chapter, I will argue that the skewed development in the sensitivity of X-ray detectors favoring low-energy X-rays has hampered a deeper understanding of the accretion phenomena occurring near black holes.

In Sect. 16.2, I will give a historical perspective on the development of X-ray observations and their impact on our understanding of black hole sources. In Sect. 16.3, I will expand on the idea that the X-ray sensitivity improvements are heavily skewed towards low-energy X-rays. Section 16.3.2 describes the recent measurements of the hard X-ray polarization in the archetypical BH-XRB Cygnus X-1 using the CZT-Imager instrument onboard the AstroSat satellite. In the last section (Sect. 16.4), conclusions are given, emphasizing suggestions for further work.

16.2 Historical Perspectives

X-ray astronomy, the child of the space era, proliferated in tandem with the developments in space technology, culminating in the Great Observatory, *Chandra*. A large number of X-ray satellites have been flown (see [10] for a comprehensive list), and here I have selected a few of them to highlight the strategic developments at

Table 16.1 X-ray astronomy satellite missions[a]

Missions	Year	Mass[b]	Instruments[c]	Results[d]
UHURU	1970–1973	142	840; 1.7–18	X-ray binaries; BH-XRB Cygnus X-1
HEAO-A	1977–1979	2552	**A1:** 2000; 0.25–25 **A2:** 3000; 0.15–60	X-ray timing & spectra of BH-XRB
	.		**A3:** 450; 1–13 **A4:** 220; 15–10000	
HEAO-B	1978–1981	3130	**HRI:** 20; 0.15–3 **IPC** 10; 0.4–4	X-rays from all types (including AGN)
	.		**SSS:** 200; 0.5–45 **FPCS:** 1; 0.4–2.6	
RXTE	1995–2012	3200	**PCA:** 6500; 2–60 **HEXTE:** 1600; 15–250	QPOs; superluminal jets
Chandra	1999–	5000	**HRC:** 200; 0.06–10 **ACIS:** 600; 0.2–10	Deep imaging
XMM	1999–	3700	**EPIC:** 3325; 0.1–15	Spectra; grating spectra
Suzaku	2005–2015	1700	**XRS:** 600; 0.3–12; **XIS:** 1000; 0.2–12; **HXD:** 300; 10–600	Wide band spectra
Swift	2004–	1470	**XRT:** 135; 0.2–10 **BAT:** 5200; 15–150	Soft and hard X-ray monitoring
NuSTAR	2012–	350	847; 3–78	Spectra
AstroSat	2015–	1513	**SXT:** 90, 0.3–8 **LAXPC:** 6500; 3–80	Higher energy timing
			CZTI: 1000; 20–200	Polarization
HMXT	2017–	2800	**LE:** 384; 0.7–15 **ME:** 952; 5–30	Timing and spectra
			HE: 5100; 20–200	
NICER	2017–	372	1900; 0.2–12	Timing and spectra
IXPE	2021–	330	500; 0.3–10	Polarization

[a] Details of only a few important missions are given here
[b] Instrument mass in kg
[c] Nominal effective area (in cm^2), and energy rage (in keV) are given
[d] Prominent results pertaining to BH-XRBs for each era are given

different eras, and these are listed in Table 16.1. The Table segregates these satellites into different eras (separated by horizontal lines in the Table) and also gives certain highlights in our understanding of accreting black hole sources. The mass of the instruments, the nominal effective area, and the energy range are also given in the Table to give a sense of the complexity of the instruments.

16.2.1 The Early Effort

Though there was a flurry of activity in X-ray astronomy using rockets and balloons after the momentous rocket flight of 1962, the short duration of exposure in rocket-based experiments and the blockage due to the residual Earth's atmosphere limiting the energy range of observations in balloon-borne experiments resulted in a pretty poor sensitivity for observations [2, 3]. X-ray astronomy came of age only after the launch of the UHURU satellite.

UHURU, launched in 1970, though modest by current standards (142 kg), significantly changed our views of the X-ray sky due to the continuous observations possible in a satellite platform. UHURU operated in a scan mode and detected a large number of sources, and identified several of them, including X-ray pulsars, the BH-XRB Cyg X-1, and also identified a large number of extragalactic objects like AGN, QSOs, and clusters of galaxies. In the seventies, several small-scale experiments like Copernicus (OAO-3), ANS, Ariel-5, Ariel-6, and SAS-3 gave useful X-ray observations [10]. Masses of compact objects were estimated in about six X-ray binaries, and evidence for BH in XRBs was found for only three sources, viz., Cyg X-1, LMC X-3, and A0620-00 [11]. The very fact that copious X-ray emission, including in hard X-rays, is seen in these accreting objects was sufficient to formulate extensive and intricate accretion theories [12].

The next significant advance in X-ray observations came with the launch of the HEAO satellites. HEAO-A consisted of the most extensive suite of X-ray instruments, whereas HEAO-B (the Einstein Observatory) used X-ray focusing techniques for the first time, thus improving detection sensitivity by several orders of magnitude. X-ray emission was detected from all classes of objects ranging from stellar coronae to clusters of galaxies. Several innovative niche experiments like EXOSAT in 1983 (enabling long uninterrupted broadband observations), Ginga in 1987 (having an all-sky monitor), Rosat in 1990 (full sky survey at low energies), ASCA in 1993 (the first satellite to use an X-ray CCD), though relatively modest in scale as compared to the HEAO satellites, enriched X-ray astronomy with their unique contributions.

Due to the availability of all-sky monitors, many X-ray transients were discovered, and several of them were identified as possible black hole candidates. In the mid-nineties, among the 124 known LMXBs, there were 41 transients and 18 BH LMXBs. Radio outbursts were seen in several black hole sources, but the concept of jet emission from BH-XRBs emerged only in the late nineties [13]. QPOs were noticed in BH-XRBs in the EXOSAT era, but 'the full pattern, if there is one, has not yet emerged' [14]. Though Active Galactic Nuclei (AGN) as accreting supermassive black holes were not thought about in the eighties [6, 7], the unified model of AGN as accreting supermassive black holes was emerging in the nineties due to the extensive studies in the optical and radio wavelengths.

16.2.2 The Golden Era

The golden era of X-ray astronomy can be considered to be in the late nineties and early this century. The launch of the Great Observatory *Chandra* in 1999 and the equally sophisticated *XMM-Newton* in the same year established the maturity of X-ray astronomy. The very high-quality imaging by Chandra and very good spectroscopy (including grating) from both these observatories were nicely complemented by the timing studies by *RXTE* (1995–2012) and wide band spectral studies by *Suzaku* (2005–2015).

In this era, new phenomena at millisecond time scales were discovered in X-ray binaries. Though relativistic jets were discovered in the Galactic X-ray binary GRS 1915+105 (discovered by the *Granat* satellite) before the launch of *RXTE*, the all-sky monitoring ability, quick observation sampling and easy availability of data from *RXTE* enabled the discovery of many relativistic jet sources and their detailed understanding [15], including the fascinating object GRS 1915+105 [9]. There were 20 BH-XRBs with mass measurements, all monitored by RXTE, and almost half discovered by RXTE [16]. Such extensive studies helped connect the accretion phenomena in XRBs and AGNs by discovering fundamental relationships between X-ray and radio intensities and masses of black holes across black hole mass ranging from stellar mass to supermassive. The quick pointing ability of RXTE helped very long-term monitoring of bright AGNs, enabling the discovery of the dependence of the break time scale in the power spectral density to the mass and luminosity across mass scales [8]. It was established that the 'physics of disk-jet coupling in XRBs and AGN are closely linked' [9].

16.2.3 The Current Era

The legacy of X-ray astronomy continued in this century by operating several small-scale topically significant experiments, some of which are listed in Table 16.1. Swift, though a satellite meant for studying Gamma-ray bursts (GRB), had the wide field hard X-ray monitor, BAT, providing a continuous record of hard X-ray sources, in particular AGN, and the X-ray Telescope, XRT, due to the swift skewing ability of the satellite could provide snapshot fluxes of a large number of X-ray sources. NuSTAR is the first X-ray satellite with a hard X-ray focusing telescope, thus extending the sensitivity leap due to X-ray focusing to hard X-rays (up to 78 keV). AstroSat and HXMT provided sensitive wide-band measurement capabilities, and IXPE is a dedicated X-ray polarization measuring satellite.

The insight and information we have obtained on the concepts of accretion physics from recent studies can be described as 'more of the same thing' and incremental information gathering. The expectation that a detailed study of the spectral and timing properties of BH-XRBs would provide answers to the very fundamental astrophysical questions like measuring the parameters of black holes (mass and spin), and understanding the intricacies of the Physics in extreme conditions near these objects did not fully materialise.

The excellent spectral and timing data at low energies enabled the development of intricate models for the putative accretion disk and the Comptonising plasma. Using these models, efforts were made to measure the spin of the black holes in BH-XRBs, and the spin measurements of many sources have been reported [17]. When two methods of spin measurements, spectral fitting and timing (QPO) measurements, were applied to the same source, extremely divergent results were obtained [18]. In Cygnus X-1, it was found that the measured BH spin was strongly dependent on how the disk is modeled, mainly because the shape of the high-energy tail is relatively complex [19].

The number of mass measurements in BH-XRBs did not increase significantly. Most of the BH mass measurements, of course, came from radial velocity measurements in the optical band. There were attempts to measure the mass of the black holes in BH-XRBs from X-ray spectral fitting using physics-based accretion theories like the Two Component Accretion Flow (TCAF) model [20, 21]. These methods provided reasonable mass estimates for BH-XRBs [22] and intermediate-mass black holes [23] based on continuum fitting and also by studying the variation of the QPOs. The spectral models, however, are inconsistent with the models evolved over the years for the accretion disk and the Compton cloud (invoking high reflection), and, perhaps due to this, the mass measurements using TCAF have yet to gain wide acceptance by the community. Even the latest tool of X-ray polarimetric measurements using IXPE though indicated the presence of a combination of thermal disk, coronal, or reflected emission in BH sources, is yet to help in zooming into the correct accretion disk model for BH sources [24].

The current status of our understanding of the accretion phenomenon near black holes can be summarized as the one which can 'explain' new observations by tweaking the parameters of several existing 'models', but has not yet emerged as a widely accepted paradigm with clear predictive powers explaining possible outcomes in settings with widely different initial conditions (like understanding and explaining the disk-jet connection in widely different classes like TDEs and GRBs).

In the next section, I will highlight the limitations of the sensitivity of X-ray instruments for studying black holes.

16.3 X-ray Detection Sensitivity

To understand and evaluate the X-ray sensitivity of X-ray telescope/ detector systems, I have chosen data from four of the best telescope systems for X-ray observations. As an example of a bright source, particularly at higher energies ($>$ 50 keV), I have taken the data for Crab (nebula and pulsar, together). Near simultaneous data were taken from observations made by NuSTAR (the only operating hard X-ray focusing instrument) and CZT Imager (CZTI) of AstroSat

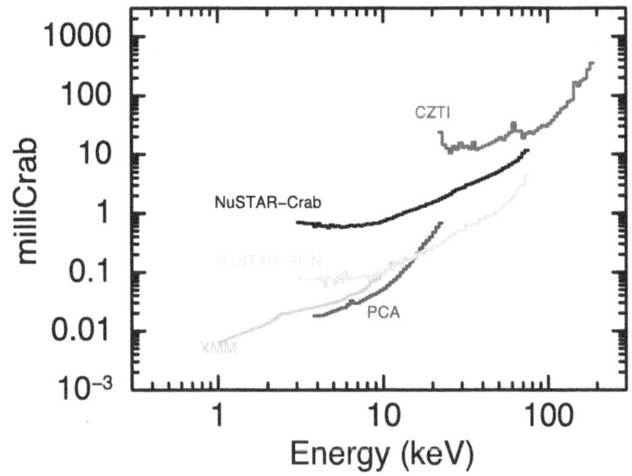

Fig. 16.1 X-ray spectral sensitivity is shown for four instruments. Data for NuSTAR and CZTI of AstroSat are taken from Crab observations, whereas for XMM-Newton-pn, RXTE-PCA and NuSTAR data are taken from observations of a nearby AGN, IC4329A. Sensitivity is taken as the errors in the data for a uniform logarithmic binning of 50 data channels per decade of energy, rescaled as the square root of time in the units of 10 ks, and renormalized to milliCrab units. The sensitivity of NuSTAR for Crab is inferior to that for the AGN because of the limitations of NuSTAR data storage while observing bright sources

[25], which has a box type Coded Aperture Mask (CAM) arrangement enabling simultaneous background measurement [26]. The NuSTAR and CZTI data are fit simultaneously with a power law in the 20–200 keV range. The sensitivity of each instrument, deemed as the errors in the data for the uniform logarithmic binning of 50 data channels per decade of energy, is calculated and rescaled as the square root of time in the units of 10 ks to take care of the fact that the durations of observation are different for each instrument. The sensitivity numbers, renormalised to milliCrab units, are shown in Fig. 16.1.

As an example of a fainter source, I have taken the data for the well-known AGN, IC 4329A. Data were taken from observations made by XMM-Newton and RXTE (simultaneously in 2003) and NuSTAR (non-simultaneous, observed in 2012). Data are fit using a model consisting of a power law and reflection component [27]. As was done earlier, the sensitivities of these three instruments, normalised to the Crab flux and scaled to 10 ks observation, are also shown in Fig. 16.1. It can be seen that the sensitivity of NuSTAR for Crab is inferior compared to that for the AGN because of the limitations of NuSTAR data storage while observing bright sources.

It can be seen that the huge improvement in sensitivity in X-ray astronomy, brought about by the use of focusing techniques, is confined to low energies (a few μCrab at one keV). In the intermediate energies (10–20 keV), the spectral sensitivities are modest and comparable to each other for focusing instruments and large-area collimated detectors (0.1–1 milliCrab). The sensitivity drops quite sharply above 20 keV, and beyond 50 keV, the sensitivity, in fact, is quite poor. It is doubtful that there will be drastic improvements in the spectral sensitivity of X-ray instruments in the future. The area drops drastically with energy for X-ray mirrors, and above 70 keV, it is very challenging to build them. For the collimated detectors, a high particle-induced background is a serious issue.

These sensitivity limitations seriously affect our understanding of the accretion phenomena near black holes. For example, for the archetypical black hole source Cygnus X-1, it is possible to look at the broad features of the spectral energy distribution from radio to GeV [28], and conclude that the observed fluxes at different wavelengths are consistent with a preferred jet model, but 'the available data do not provide unique constraints on the model free parameters' [28]. The bulk of the emission from the putative accretion disk is in the 10–500 keV region (in the hard state of the source): the region, as pointed at earlier, where the X-ray detection sensitivity is relatively poor.

If we accept that the improvement in the X-ray spectral sensitivity in the crucial hard X-ray region is very unlikely to improve dramatically in the near future, one avenue to exploit would be to explore the accretion phenomena with available tools, including the nascent polarimetry tools, albeit of limited sensitivity.

16.3.1 X-ray Polarization in Cygnus X-1

The IBIS and SPI instruments onboard the INTEGRAL satellite independently measured high polarization for Cygnus X-1 with a polarization degree (PD) of $\sim$ 65% with the polarization angle (PA) perpendicular to the jet direction, indicating a strong jet synchrotron component above 400 keV [29, 30]. The balloon-borne hard X-ray polarimeter POGO+, sensitive in the energy range 19–181 keV, however, placed an upper limit of 5.6% for PD in Cygnus X-1 and also placed an upper limit for polarization from the jet component of around 5–10% [31, 32]. The X-ray polarimetric observations of Cygnus X-1 in the hard state using IXPE (2–10 keV) reported a precise measurement of the polarization with a PD of $4.0\% \pm 0.2\%$ and an increasing trend of PD with energy [33]. The alignment of the X-ray polarization in soft X-rays, however, was with the radio jet (almost perpendicular to the polarization angle reported by the INTEGRAL data). The precise polarization data and its energy dependence could be 'explained' with the current ideas of accretion disks. The two variants of the models, viz., a hot corona sandwiching the accretion disk or a truncated accretion disk with an inner region of hot corona, could not be distinguished by the available data.

Recently, the high polarization at high energies has been independently confirmed by measurements using CZT Imager instruments of AstroSat [34], which will help us in resolving the tension between the INTEGRAL and the IXPE observations.

16.3.2 CZTI Measurements of the X-ray Polarization in Cygnus X-1

The Cadmium Zinc Telluride Imager (CZTI) instrument of the AstroSat satellite is a collimated hard X-ray detector operating in the 20–200 keV range [25]. The CZTI detectors are pixelated, and the information about any ionization in any of the pixels is recorded, thus making it sensitive to measure Compton scattered events in the detector plane. The scattered events modulate the azimuthal angle distribution if the incident radiation is polarized, and this polarization capability (in the 100–380 keV region) was demonstrated in the laboratory [35, 36]. Polarization measurements using CZTI have been reported for Crab [37] and a large sample of GRBs [38].

After the launch of AstroSat in 2015, Cygnus X-1 was observed several times. The initial polarization results were not very convincing due to the short exposures of each observation and the difficulty of adding data from different exposures. To alleviate these problems, three long ($>$ 200 ks) observations were made, triggered by the transition of the source from soft to hard state (AstroSat Observations ID2992, ID4646, and ID5146). Following the methodology of Lubinski et al. [39], these three observations were classified as intermediate hard state (HIMS), intermediate soft state (SIMS), and pure hard state (PH), respectively, by examining the variation of the flux with the 20–100 keV power-law spectral index (Fig. 16.2a).

The polarization analysis of these three observations shows exciting results. No evidence for polarization was seen in the PH and SIMS states, providing an upper limit of $\sim$10%. In the HIMS state, polarized emission was detected with very high ($\sim$6 σ) significance with a PD of 23$\pm$4%. An energy resolved analysis hints at resolving the earlier puzzle of low polarization at low energies and very high polarization seen at high energies: no polarization was detected below 175 keV, and very high polarization (PD $\sim$40%) is seen at energies above 175 keV, thus smoothly connecting the earlier low and higher energy observations.

These results are summarised in Fig. 16.2 (adapted from [34]). Figure 16.2a shows the segregation of the three observations into the three states. The azimuthal scattering angle distributions and the contour plots for PD and PA are shown for the individual observations (Fig. 16.2b) and for the energy-resolved measurements in the HIMS state (Fig. 16.2c). From all the available measurements, the PD and PA of Cygnus X-1 in different spectral states (pure hard, intermediate hard, intermediate soft, averaged hard, and soft states) are shown in Fig. 16.2d. We see an apparent increase in PD and a swing in PA at higher energies.

The CZTI measurements confirm the high polarization at high energies detected by the INTEGRAL satellite. Though this observation sort of resolves the tension between the IXPE and INTEGRAL measurements, understanding of such high polarization only confined to the higher energies within the premises of our current understanding of the steady state jet emission [28] still remains an issue. However, the current observations provide several hints at resolving this issue. In the spectral data, there is a signature of a separate spectral component above 100 keV[34].

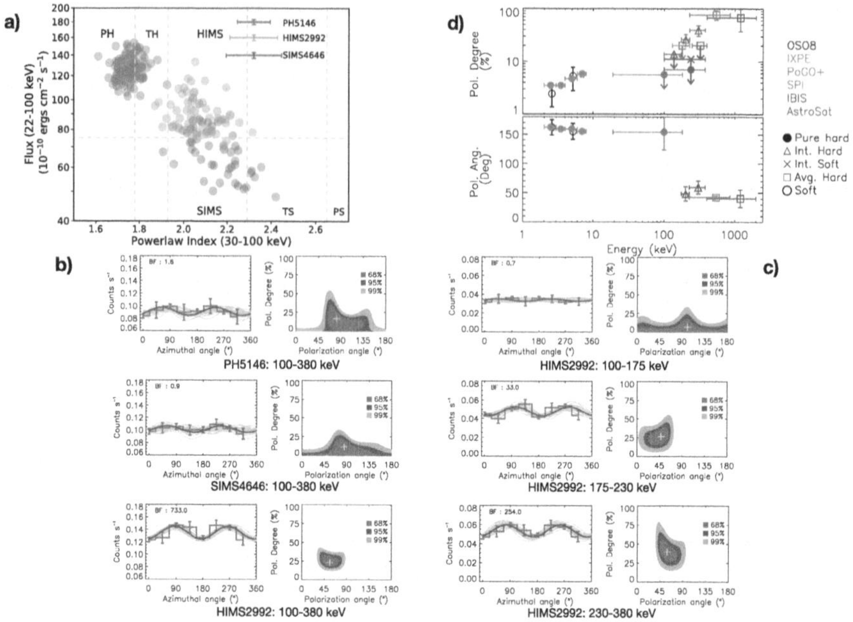

Fig. 16.2 Polarization measurements of Cygnus X-1. (**a**) Three long observations using AstroSat CZTI were segregated into three spectral states of PH, HIMS, and SIMS. (**b**) CZTI polarization measurements for the three states, the azimuthal distribution of scattered photons shown on the left and the confidence contour levels for the PD and PA, shown on the right. (**c**) Energy resolved polarization measurements for HIMS. (**d**) A collage of all polarization measurements of Cygnus X-1, plotted as a function of energy and segregated into spectral states, showing that the CZTI data smoothly joins the low-energy and high-energy measurements (Figure adapted from [34])

Further, the high polarization is seen only in the HIMS, a state known for its high and variable radio emission [39], indicative of a variable jet component. Hence, the steady-state jet models may not be applicable to the high polarization measurements seen in this state. Further detailed state-wise observation, simultaneously in broadband (using IXPE, AstroSat, INTEGRAL and Polix) along with radio observations and building good quality state-wise spectral energy distribution along with polarization information will undoubtedly help us in advancing our understanding of the accretion phenomena and the disk jet connection.

16.4 Conclusions

We have given a brief historical overview of the developments in X-ray astronomy, giving particular attention to the advancements in our understanding of the accretion phenomena near black holes. Despite the huge advancement in X-ray detection techniques, the behavior of the innermost region of the accretion disk in black holes has still remained a difficult area of research. It is noted that the technical developments in X-ray astronomy are confined mainly to low energies, thus making observations of limited impact on our understanding of the accretion phenomenon.

At low energies (2–10 keV), the region where the thermal and non-thermal emissions co-exist, multiple emission processes are possible and, more importantly, since the observational localization accuracy possible in X-rays (fraction of an arc-second, at best) spans a vast region close to the black hole where multiple hot regions are conceivably possible, any spectrally accurate data can be 'understood' in terms of these multiple regions (if all else fails, put a lamppost ! [40]). In hard X-rays, on the other hand, the emission processes are limited and the very fact that these high-energy photons mandate the existence of high-energy electrons with high emissivity demands the emission region very close to the energy generation region, which, in black hole sources, has to be within a few Schwarzschild radii. Thus, any features in hard X-rays necessarily have to come from very close to the black hole with limited emission mechanisms possible. Though the X-ray sensitivity in the crucial hard X-ray range is limited and unlikely to improve in the near future, X-ray polarization studies can give important hints about the working of the innermost regions of the accretion disk, including the disk-jet connection. Comprehensive spectral and polarization observations in an extremely wide band in X-rays, along with radio observations, will help us to get a firm grip on the accretion phenomenon.

Acknowledgments I thank Gulab Dewangan and N. P. S. Mithun for helping analyze data for the sensitivity calculations. I thank Tanmoy Chattopadhyay for assisting me in making this chapter's polarization Figure (Fig. 16.2).

Competing Interests The author has no conflicts of interest to declare that are relevant to the content of this chapter.

References

1. Giacconi, R., Gursky, H., Paolini, F. R., & Rossi, B. B. (1962). Evidence for x Rays From Sources Outside the Solar System. *Physical Review Letters, 9*(11), 439–443.
2. Morrison, P. (1967). Extrasolar X-Ray Sources. *Annual Review of Astronomy and Astrophysics, 5*, 325.
3. Giacconi, R., Gursky, H., & van Speybroeck, L. P. (1968). *Observational Techniques in X-Ray Astaraa, 6*, 373.
4. Bolton, C. T. (1975). Orbital elements and an analysis of models for HDE 226868 = Cygnus X-1. *The Astrophysical Journal, 200*, 269–277.
5. Blumenthal, G. R., & Tucker, W. H. (1974). Compact X-ray sources. *Annual Review of Astronomy and Astrophysics, 12*, 23–46.
6. Gursky, H., & Schwartz, D. A. (1977). Extragalactic X-ray sources. *Annual Review of Astronomy and Astrophysics, 15*, 541–568.
7. Weedman, D. W. (1977). Seyfert galaxies. *Annual Review of Astronomy and Astrophysics, 15*, 69–95.
8. McHardy, I. M., Koerding, E., Knigge, C., Uttley, P., & Fender, R. P. (2006). Active galactic nuclei as scaled-up Galactic black holes. *Nature, 444*(7120), 730–732.

9. Fender, R., & Belloni, T. (2004). GRS 1915+105 and the Disc-Jet Coupling in Accreting Black Hole Systems. *Annual Review of Astronomy and Astrophysics, 42*(1), 317–364.
10. Santangelo, A., Madonia, R., & Piraino, S. (2023). A Chronological History of X-Ray Astronomy Missions. arXiv e-prints, arXiv:2307.06652.
11. Bahcall, J. N. (1978). Masses of neutron stars and black holes in X-ray binaries. *Annual Review of Astronomy and Astrophysics, 16*, 241–264.
12. Shakura, N. I., & Sunyaev, R. A. (1973). Black holes in binary systems. Observational appearance. *Astronomy & Astrophysics, 24*, 337–355.
13. Tanaka, Y., & Shibazaki, N. (1996). X-ray Novae. *Annual Review of Astronomy and Astrophysics, 34*, 607–644.
14. van der Klis, M. (1989). Quasi-periodic oscillations and noise in low-mass X-ray binaries. *Annual Review of Astronomy and Astrophysics, 27*, 517–553.
15. Mirabel, I. F., & Rodríguez, L. F. (1999). Sources of Relativistic Jets in the Galaxy. *Annual Review of Astronomy and Astrophysics, 37*, 409–443.
16. Remillard, R. A., & McClintock, J. E. (2006). X-Ray Properties of Black-Hole Binaries. *Annual Review of Astronomy and Astrophysics, 44*(1), 49–92.
17. Draghis, P. A., Miller, J. M., Costantini, E., Gallo, L. C., Reynolds, M., Tomsick, J. A., & Zoghbi, A. (2023). A Systematic Review of 24 Previously Measured Black Hole Spins in X-ray Binaries. arXiv e-prints, arXiv:2311.16225.
18. Mall, G., Liu, H., Bambi, C., Steiner, J. F., & García, J. A. (2024). Measuring black hole spins through X-ray reflection spectroscopy and the relativistic precession model: the case of XTE J1859+226. *Monthly Notices of the Royal Astronomical Society, 527*(4), 12053–12064.
19. Zdziarski, A. A., et al. (2024). What Is the Black Hole Spin in Cyg X-1? arXiv e-prints, arXiv:2402.12325.
20. Chakrabarti, S. K., & Titarchuk, L. G. (1995). Spectral properties of accretion disks around galactic and extragalactic black holes. . *The Astrophysical Journal, 455*, 623.
21. Chakrabarti, S. K. (1997). Spectral properties of accretion disks around black holes. II. Sub-keplerian flows with and without shocks. *The Astrophysical Journal, 484*(1), 313–322.
22. Nath, S. K., Debnath, D., Chatterjee, K., Bhowmick, R., Chang, H.-K., & Chakrabarti, S. K. (2024). Accretion flow properties of EXO 1846-031 during its multi-peaked outburst after long quiescence. *The Astrophysical Journal, 960*(1), 5.
23. Mondal, S., Palit, B., & Chakrabarti, S. K. (2022). Study of accretion flows around an ultraluminous X-ray source M82 X-1 using NuSTAR data. *Journal Astronomy & Astrophysics, 43*(2), 90.
24. Cavero, N. R., & IXPE Collaboration. (2024). First year of stellar-mass black hole observations with the imaging X-ray polarimetry explorer. In *Multifrequency behaviour of high energy cosmic sources XIV*, 12–17 June 2023, Palermo, Italy. Published April 17, 2024. Online at https://pos.sissa.it/cgi-bin/reader/conf.cgi?confid=447,id.39
25. Bhalerao, V., et al. (2017). The cadmium zinc telluride Imager on AstroSat. *Journal Astronomy & Astrophysics, 38*(2), 31.
26. Mithun, N. P. S. (2024). CZTI calibration. *Private Communications*, 1–10.
27. Markowitz, A., Reeves, J. N., & Braito, V. (2006). Fe K emission and absorption in the XMM-EPIC spectrum of the Seyfert galaxy IC 4329a. *The Astrophysical Journal, 646*(2), 783–800.
28. Zdziarski, A. A., Pjanka, P., Sikora, M., & Stawarz, Ł. (2014). Jet contributions to the broad-band spectrum of Cyg X-1 in the hard state. *Monthly Notices of the Royal Astronomical Society, 442*(4), 3243–3255.
29. Laurent, P., Rodriguez, J., Wilms, J., Cadolle Bel, M., Pottschmidt, K., & Grinberg, V. (2011). Polarized gamma-ray emission from the galactic black hole cygnus X-1. *Science, 332*(6028), 438.
30. Jourdain, E., Roques, J. P., Chauvin, M., & Clark, D. J. (2012). Separation of two contributions to the high energy emission of cygnus X-1: polarization measurements with INTEGRAL SPI. *The Astrophysical Journal, 761*(1), 27.
31. Chauvin, M., et al. (2018). Accretion geometry of the black-hole binary Cygnus X-1 from X-ray polarimetry. *Nature Astronomy, 2*, 652–655.

32. Chauvin, M., et al. (2019). PoGO+ polarimetric constraint on the synchrotron jet emission of Cygnus X-1. *Monthly Notices of the Royal Astronomical Society, 483*(1), L138–L143.
33. Krawczynski, H., et al. (2022). Polarized x-rays constrain the disk-jet geometry in the black hole x-ray binary Cygnus X-1. *Science, 378*(6620), 650–654.
34. Chattopadhyay, T., et al. (2024). High hard X-Ray polarization in Cygnus X-1 confined to the intermediate hard state: Evidence for a variable jet component. *The Astrophysical Journal Letter, 960*(1), L2.
35. Chattopadhyay, T., Vadawale, S. V., Rao, A. R., Sreekumar, S., & Bhattacharya, D. (2014). Prospects of hard X-ray polarimetry with Astrosat-CZTI. *Experimental Astronomy, 37*(3), 555–577.
36. Vadawale, S. V., Chattopadhyay, T., Rao, A. R., Bhattacharya, D., Bhalerao, V. B., Vagshette, N., Pawar, P., & Sreekumar, S. (2015). Hard X-ray polarimetry with Astrosat-CZTI. *Astronomy & Astrophysics, 578*, A73.
37. Vadawale, S. V., et al. (2018). Phase-resolved X-ray polarimetry of the Crab pulsar with the AstroSat CZT Imager. *Nature Astronomy, 2*, 50–55.
38. Chattopadhyay, T., et al. (2022). Hard X-Ray polarization catalog for a five-year sample of Gamma-Ray bursts using AstroSat CZT imager. *The Astrophysical Journal, 936*(1), 12.
39. Lubiński, P., Filothodoros, A., Zdziarski, A. A. & Pooley, G. (2020). Distinct accretion modes of Cygnus X-1 revealed from hard X-Rays. *The Astrophysical Journal, 896*(2), 101.
40. Podgorný, J., Dovčiak, M., Goosmann, R., Marin, F., Matt, G., Różańska, A., & Karas, V. (2023). Spectral and polarization properties of reflected X-ray emission from black-hole accretion discs for a distant observer: the lamp-post model. *Monthly Notices of the Royal Astronomical Society, 524*(3), 3853–3876.

Chapter 17
Light Curves and HID Signatures of Recurrent Black Hole X-ray Binary Sources

M. C. Ramadevi and V. P. Shyam Prakash

Abstract Most of the Black Hole X-ray Binaries are transient in nature and these sources show outbursts with different recurrence timescales. The variability timescales seen in the light curves of these sources vary from few tens of milliseconds to few years. The profile of the light curves of the outbursts are also found to vary from one outburst to the other of the same source. The physics behind trigger of such outbursts are still unclear. Here in this chapter, we discuss recurrent outbursts of select black hole X-ray transient sources with the Hardness Intensity Diagram (HID) signatures in each of the outbursts shown by the source. The study of HIDs of the outbursts of these sources clearly indicates that not all outbursts of black hole X-ray binaries trace a 'q' pattern during an outburst. While in one of the BHB sources the light curve profile is found to repeat along with the 'q' pattern in all the HIDs of its outbursts, it has been observed that in other black hole binary systems, there are possibilities of varying light curve profiles with varying HID patterns which need not be a 'q'. We bring out the possible similarities and differences observed in the outbursts of these sources with reference to the profiles of the light curves and the respective HIDs traced in the outbursts.

17.1 Introduction

Black Hole X-ray Binaries (BHXRBs) are binary systems comprising a companion star and a black hole as the compact object [1]. There are a handful of confirmed BHXRBs with appropriate mass estimates for the stellar mass Black Holes in these systems [2, 3]. The black hole binaries in which the mass of the companion star is known from optical observations using which the mass of the black hole is estimated

M. C. Ramadevi (✉)
Space Astronomy Group, ISITE, URSC, Bangalore, India

V. P. Shyam Prakash
Space Astronomy Group, ISITE, URSC, Bangalore, India
Calicut University, Calicut, India

S. Ghosh, A. R. Rao (eds.), *The Relativistic Universe: From Classical to Quantum*, Astrophysics and Space Science Proceedings 61,
https://doi.org/10.1007/978-3-031-90186-7_17

Table 17.1 List of select BHXRBs and BHCs. The first three X-ray source names in bold text here in the Table are discussed in this chapter. The confirmed BHBs have the mass estimates for the black holes as the companion stars are observed in visible wave band and the mass of the black holes are estimated from the mass functions. The mass estimates of confirmed black hole binaries listed here are taken from [3]

Source name	Class	Mass ($M_\odot$)
4U 1630-472	BHC	3–9 [13]
GX 339-4	BHC	8.3–11.9 [14]
H1743-322	BHC	9–14.07 [15–18]
MAXI J1535-571	BHC	5.1–7.83 [19]
MAXI J1348-630	BHC	8–12 [20]
Swift J1753.5-0127	BHC	7.4–10 [21, 22]
4U 1543-475	BHB	8.4–10.4 [3]
XTE J1550-564	BHB	8.4–10.8 [3]
Cyg X-1	BHB	6.8–13.3 [3]
GRS 1915+105	BHB	10.0–18.0 [3]
LMC X-1	BHB	4.0–10.0 [3]
XTEJ 1118+480	BHB	6.5–7.2 [3]

are listed under confirmed black hole binaries. Other than these confirmed Black Hole Binaries, there are known X-ray binaries to host black holes which are called Black Hole Candidates that could have a stellar mass black hole as the compact object. Few examples of Black Hole Binaries and Black Hole Candidates are given in Table 17.1 for reference in which the mass of the confirmed black hole binaries are explicitly mentioned and are taken from [3]. The sources that are discussed here in this chapter are highlighted in the Table.

Black Hole Binaries are known to exhibit spectral and temporal variations during spectral state transitions [2–4]. Studying these sources in detail helps understand the physics and geometry of accretion in strong gravity regimes. These systems have been studied extensively in spectral and timing domain which has given a wealth of understanding of the physical processes involved in radiation emission in these systems. However there are still open questions on the accretion geometry which could be answered with X-ray polarimetry which is being explored now.

BHXRB sources show enhanced X-ray emission at different timescales. There are two different classes of BHXRBs: persistent and transient. Persistent BHXRBs show variabilities as low as few milliseconds and are always found to be bright enough to be detected all the time with any X-ray instrument. However, transient BHXRBs are in the quiescent state (very low intensity state) most of the time and undergo outbursts which show enhanced intensities of about few times to an order higher in magnitude than the quiescent phase of the source. The profile of the light curves are found to be varied like Fast Rise Exponential Decay (FRED), triangular, flat-topped plateaued, multiple short-recurrent outbursts etc. [5]. There are just a handful of confirmed BHXRBs which show different types of outbursts and amongst these, there are few sources that are known to show recurrence timescales in the outbursts they exhibit. The profile of each of the outburst shown by the same source need not follow the previous. There are repeated profiles observed in some of the recurrent outbursts while some are not. The cause for such behavior in the outbursts of these sources are not understood.

Typically during an outburst, the system traverses through different spectral states: low/hard state, high/soft state, steep power law state and back to low/hard state[2, 6]. Energy spectra of the source is modeled generally with two components: thermal and non-thermal. The thermal component is associated with the geometrically thin and optically thick Keplerian disk [7] and the non-thermal component is associated with the geometrically thick and optically thin corona around the inner regions of the accretion disk [8]. The intensity variations during the outburst can be mapped as a function of the hardness ratio which is the ratio of the fluxes in the hard energy band to that of the soft energy band called the "Hardness Intensity Diagram" (HID) [9, 10]. The Hardness Intensity Diagrams (HID) can be indicative of the spectral signature and evolution of the source during the outburst.

During the outburst and the spectral state transitions, the Black Hole system traverses the hardness intensity diagram (HID) in a manner to form a 'q' shaped pattern as discussed in [11, 12]. This HID representation also gives the possible disk-jet coupling in Black Hole Binaries [11]. There are few observations seen in Black Hole systems where the source does not traverse the 'q' pattern in the HID, but deviates to result in a different pattern.

In this chapter, we discuss the recurrent outbursts of select BH binary systems and the evolution of the sources in the HIDs during each of the outbursts. The select sources are 4U 1630-472, GX 339-4 and H1743-472. These sources show recurrent outbursts with various light curve profiles in every other outburst.

17.2 Data and Analysis

MAXI, all sky monitor onboard ISS has been continuously monitoring the X-ray sky since 2009 [23]. MAXI has wide field X-ray instruments to monitor the X-ray sky for transient outbursts from different class of X-ray sources. It has two types of instruments to monitor the sky: Gas Glit Camera (GSC) and Solid-state Slit Camera (SSC). Data from observations of X-ray sources by MAXI for certain standard energy bands as defined by MAXI can be accessed from the MAXI website, http://maxi.riken.jp/top/index.html. Data for any source for custom defined energy bands can be accessed via MAXI on-demand option available in the MAXI website.

We study the MAXI light curves of the outbursts of these select black hole sources along with the HID for each of the outbursts. One-day average fluxes of the sources for the energy bands 2–6, 6–20 and 2–20 keV are obtained via MAXI on-demand from the MAXI website. Hardness ratios are computed from the ratio of the fluxes in the energy bands 6–20 keV to 2–6 keV respectively. HID is the plot of intensity in the energy band 2–20 keV as a function of hardness ratio. We discuss the profile of the light curve along with the pattern traced in HID by the source in each of the outburst. It is observed that not all outbursts trace a 'q' pattern in the HID. The probable causes for this behavior is analyzed for the outbursts of each of these sources.

17.3 4U 1630-472

4U 1630-472 is a galactic black hole candidate (BHC) which shows recurrent outbursts since the time of discovery in 1969 [24, 25]. The light curve of the source from MAXI observations is shown in Fig. 17.1. The black hole candidate has undergone several outbursts over the past few decades and has been known to shown recurrent outbursts with a timescale of 600 days [26]. The outbursts of this black hole system has shown variations in various parameters like the duration of the burst, peak intensity, profile of the burst etc as can be seen in Fig. 17.1.

This source has been observed to show two types of bursts called the "mini" outbursts and the "super" outbursts. The mini outbursts that scale to peak intensities about < 0.5 Crab [27, 28] and have recurrence timescales of about 500 days lasts for about 100–200 days, while the super outbursts have a recurrence timescales of about 1.5–2.5 years with one or more mini outbursts between the recurrences and scale in peak intensities to about < 0.8 Crab [28]. This makes 4U 1630-472 a special source and it is postulated that this source may be a three body system with a compact object which is a black hole along with two companion stars [29]. This could possibly explain the mini and super outbursts observed in this system. A detailed spectral and timing analyses of two select outbursts of this source in 2016 and 2018 are discussed in [13]. It has been shown in [13] that the HID of the source is found to deviate from the canonical 'q' pattern seen in other sources.

Figure 17.1 shows the light curve of the source as observed by MAXI instrument onboard the ISS in the energy band 2–20 keV. The outbursts are numbered as OBn, where n is the index of the respective outburst. The MAXI flux in the energy band 2–6 and 6–20 keV are considered to obtain the hardness ratio. The light curve and HID of the respective outbursts are separately analysed and plotted in subsequent Figures.

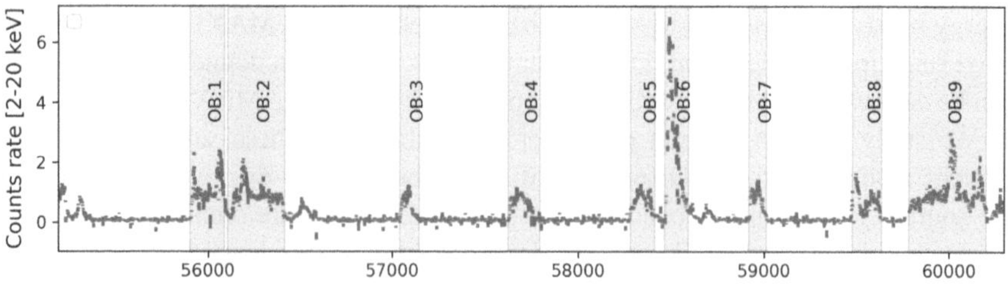

Fig. 17.1 MAXI Light curve of BHB source 4U 1630-472 showing various outbursts observed in the source

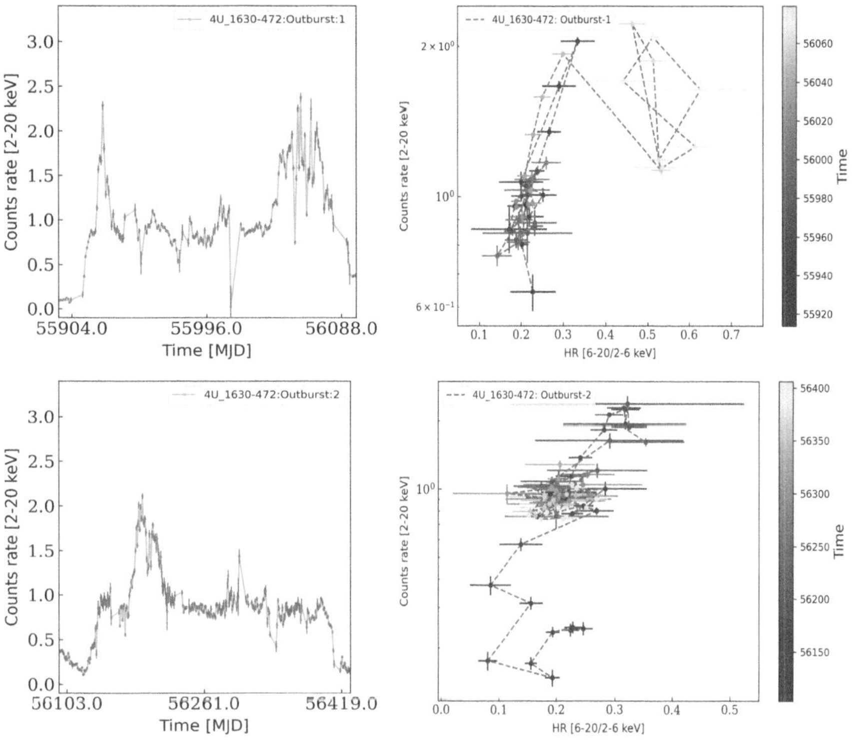

Fig. 17.2 Light curve (left) and HID (right) of outburst 1 (top) and 2 (bottom) as marked in Fig. 17.1 for source 4U 1630-472

Figure 17.2 shows the light curve and HID for the first two outbursts as marked in Fig. 17.1. The profile of the light curves of outburst 1 and 2 are not the typical Fast Rise Exponential Decay (FRED). The HID does not follow the typical 'q' shaped pattern as observed in previous outbursts during 2006, 2008 and 2010 as discussed in [29]. Here, for these two outbursts, the light curve profile is rather erratic as compared to any typical profile like the FRED or triangular shape in any Black Hole binaries [5]. Also, we find that the HID does not follow any systematic pattern.

Figures 17.3, 17.4 and 17.5 shows the light curves and HIDs for the 3rd to 7th of the outbursts as marked in Fig. 17.1. The profile of the light curves of the four outbursts from 3rd to 7th (except the outburst OB6 which replicates a FRED profile) mimic almost a near-triangular shape. The HIDs does not follow the typical 'q' shaped pattern, while it traces a knot shaped, distorted '&'. This pattern of the HID for two of the outbursts have been reported in [13] earlier.

Figure 17.6 shows the light curve and HID for the 8th and the 9th outbursts as marked in Fig. 17.1. The profile of the light curves of both these outbursts do not have a specific pattern and are similar to outbursts 1 and 2. Also, the HIDs do not follow the typical 'q' shaped pattern as observed in outbursts 1 and 2 and the

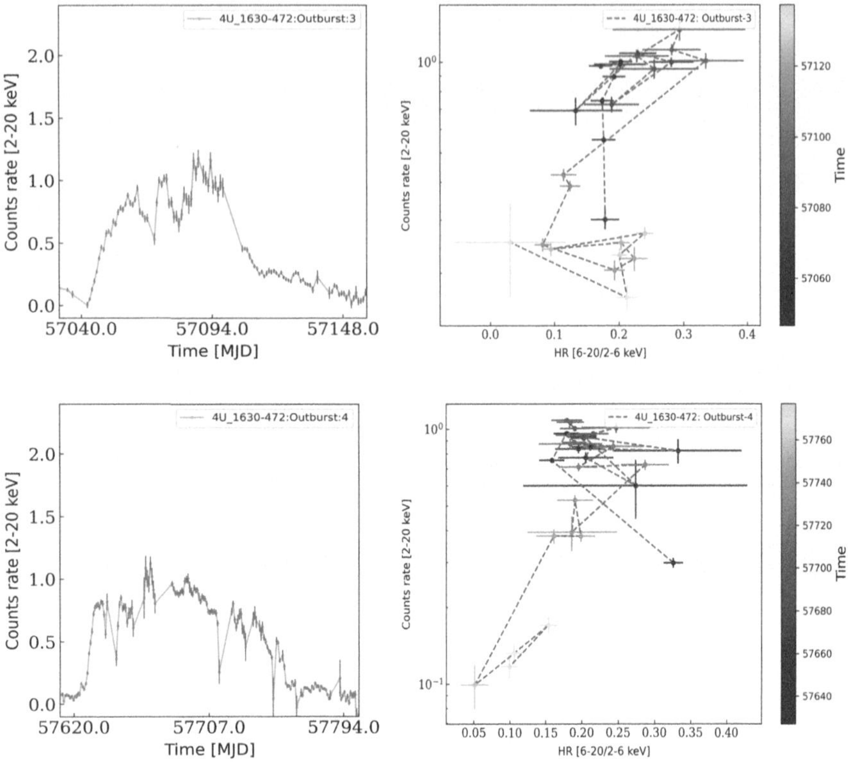

Fig. 17.3 Light curve (left) and HID (right) of outburst 3 (top) and 4 (bottom) as marked in Fig. 17.1 for source 4U 1630-472

HIDs does not follow any systematic pattern. It is interesting to note that when the light curve profile appears to be more erratic, the HID also does not seem to have any systematic pattern. This can indicate a complete different physical mechanism underlying the radiation during these outbursts as compared to the distorted knot pattern of the HID in the other outbursts observed in the source. With the possibility of this source being a three-body system with a compact star and two companion stars, it could be possible that the two discrete patterns of HID are observed in this system. More detailed study of these observations are underway in order to understand these two different HID patterns seen in the source.

17.4 GX 339-4

GX 339-4 is one of the most studied black hole transients during its several outbursts over the years. It was discovered by OSO-7 in 1972 [30]. GX 339-4 is a

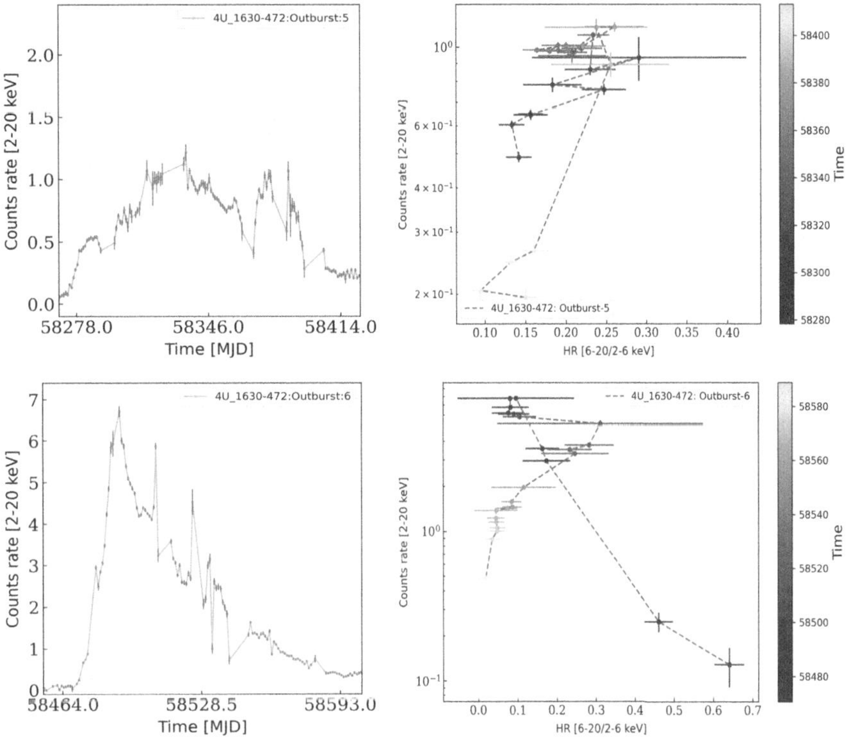

Fig. 17.4 Light curve (left) and HID (right) of outburst 5 (top) and 6 (bottom) as marked in Fig. 17.1 for source 4U 1630-472

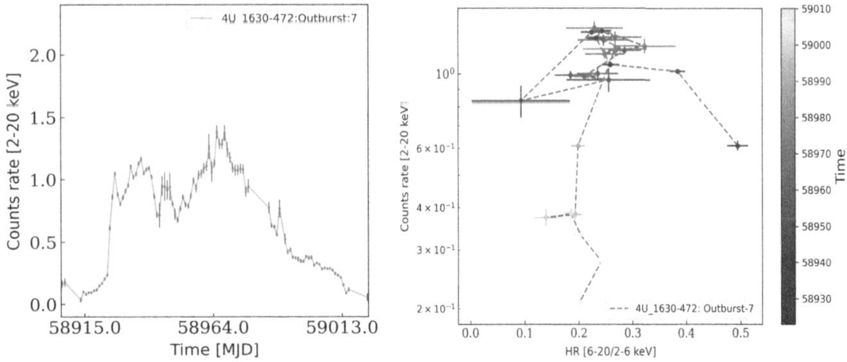

Fig. 17.5 Light curve (left) and HID (right) of outburst 7 as marked in Fig. 17.1 for source 4U 1630-472

recurrent transient source with outbursts occurring once in about few years. During an outburst, GX 339-4 transits through the canonical states as seen in a typical black hole binary system. This source has been seen to traverse a 'q' shaped pattern in

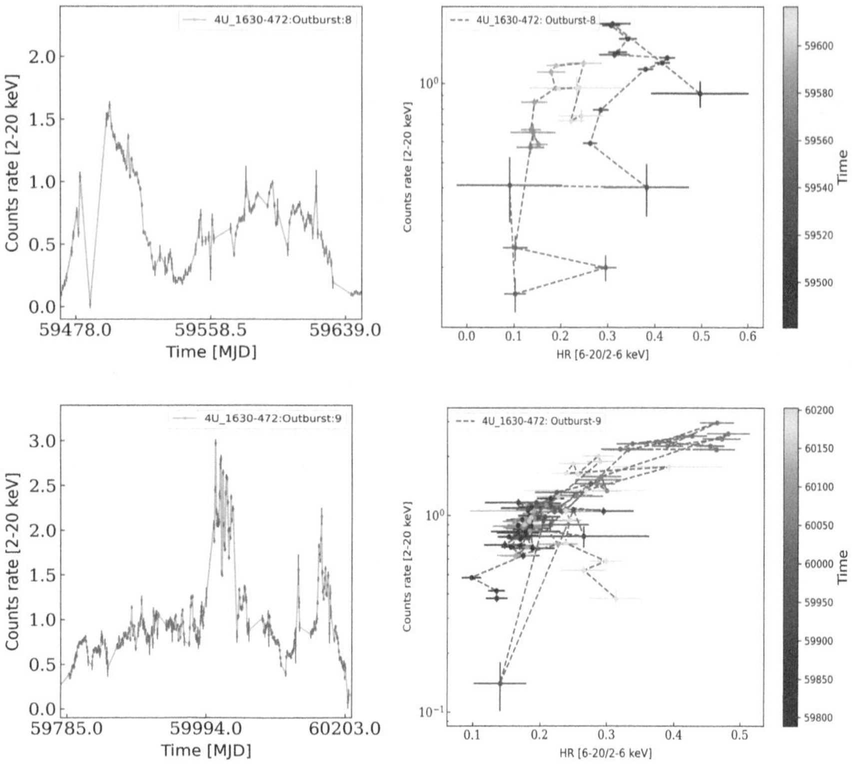

Fig. 17.6 Light curve (left) and HID (right) of outburst 8 (top) and 9 (bottom) as marked in Fig. 17.1 for source 4U 1630-472

the HID for almost all of its outbursts. GX 339-4 is the first source in which the 'q' pattern in the HID was identified along with the spectral evolution during its outburst [10]. The hysteresis behavior observed in spectral evolution of a typical black hole binary system in which the transition from hard-to-soft state during the rising phase is relatively faster and occurs at higher luminosity than the soft-to-hard transition during the decline was first identified in GX 339-4 [31].

Figure 17.7 shows the MAXI light curve of GX 339-4 with its several outbursts over few decades. The light curves of the four outbursts as marked in Fig. 17.7 are discussed further in detail along with the HID of the outbursts.

Figures 17.8 and 17.9 shows the light curves and HIDs of outbursts OB1, OB2, OB3 and OB4 of GX 339-4 as marked in Fig. 17.7. It can be seen that all the four outbursts show a triangular profile with the first one slightly deviating in the decaying phase of the outburst. However, the HIDs of all the four outbursts traces a 'q' pattern indicating that the source GX 339-4 follow the typical canonical HID profile of a black hole binary system. More detailed look of all the spectral and temporal signature of the source during its spectral evolution in each of these outbursts are being studied and compared with the other sources and will be discussed in a future work.

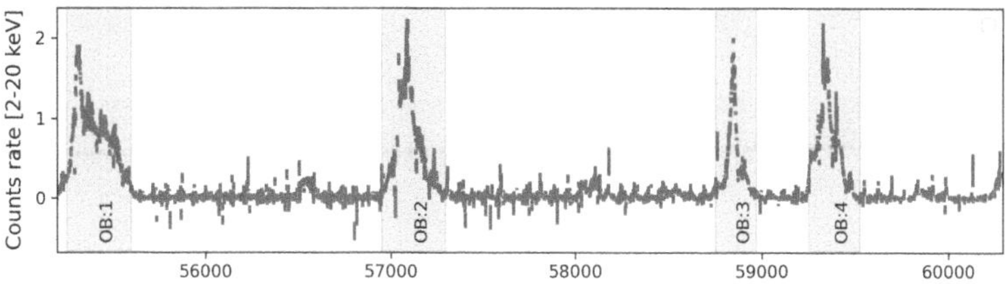

Fig. 17.7 MAXI Light curve of BHB source GX 339-4 showing various outbursts observed in the source

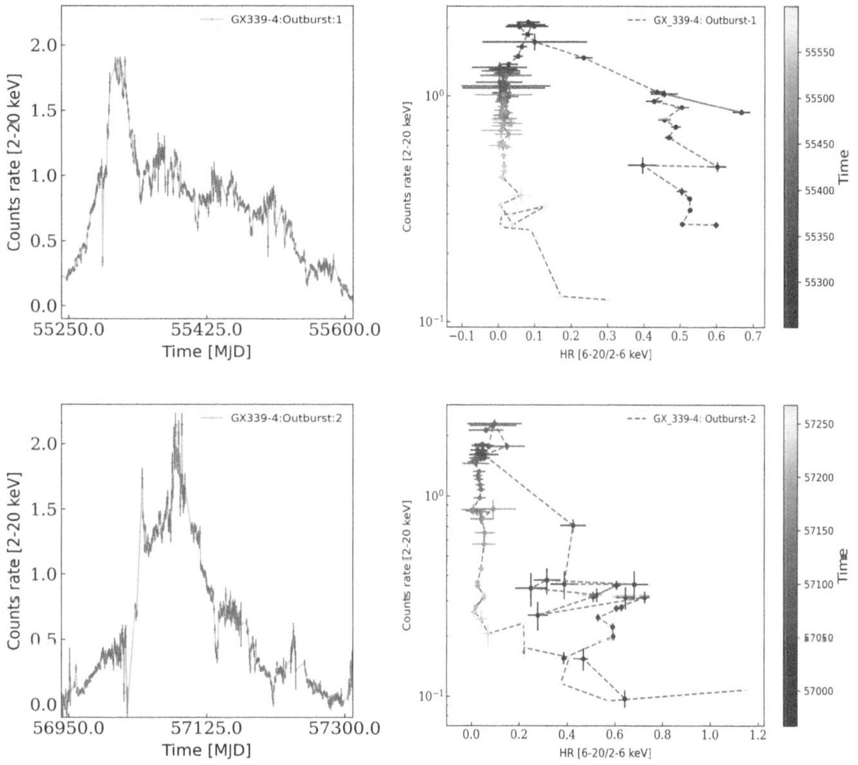

Fig. 17.8 Light curve (left) and HID (right) of outburst 1 (top) and 2 (bottom) as marked in Fig. 17.7 for source GX 339-4

17.5 H1743-322

H1743-322 is a Black Hole Candidate discovered by Ariel- 5 and HEAO-1 in 1977 [32]. This source has undergone several outbursts, with a recurrent period of about a year or less. In these outbursts of the source, few of them are tagged as "failed outbursts" [33, 34] where the source does not make it to the high/soft state during

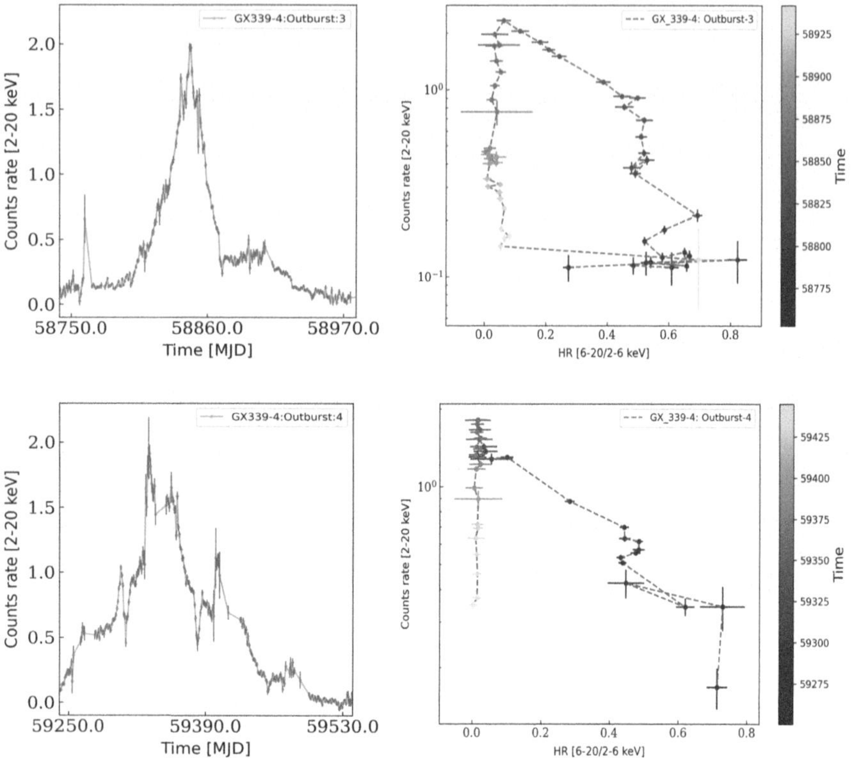

Fig. 17.9 Light curve (left) and HID (right) of outburst 3 (top) and 4 (bottom) as marked in Fig. 17.7 for source GX 339-4

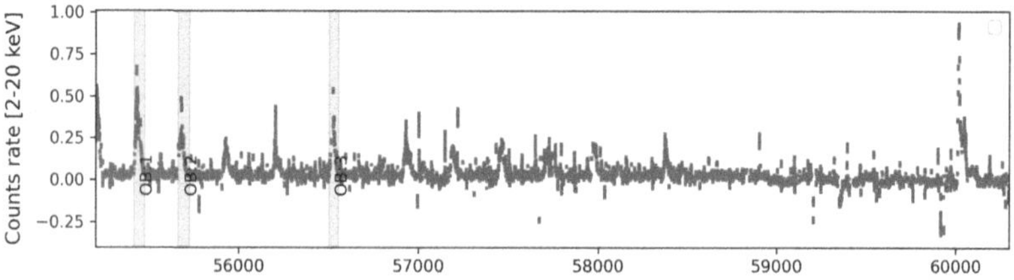

Fig. 17.10 MAXI Light curve of BHB source H1743-322 showing various outbursts observed in the source

an outburst and stays in the low/hard state throughout the outburst. However, the profile of the light curve would look like a triangular shape or a FRED, which is generally associated with a canonical outburst of a typical black hole binary system.

Figure 17.10 shows the MAXI light curve of the source H1743-322 with several outbursts. Three of the outbursts are highlighted for which the HIDs are studied in the following Figures. The individual light curves and the respective HIDs are shown in Figs. 17.11 and 17.12. It can be noted that most of the outbursts are having

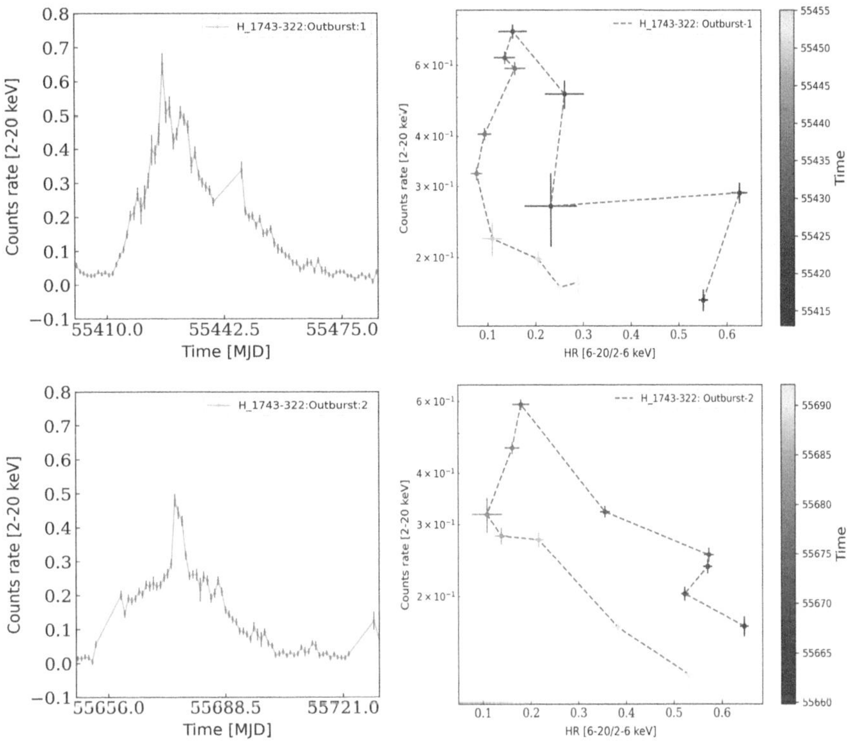

Fig. 17.11 Light curve (left) and HID (right) of outburst 1 and 2 as marked in Fig. 17.10 for source H 1743-322

a triangular profile in this source and the corresponding HIDs also appear similar with respect to each other. All these outbursts show spectral state transitions and hence the HID mimics a 'q' pattern. However, the pattern is rather skewed for all of the outbursts.

In two of its outbursts one in 2016 and the other in 2018, the source showed canonical state transitions in its outburst during 2016, while it was a failed outburst in 2018 as discussed in [34]. Figure 17.13 shows the MAXI light curve for the 2016 and the 2018 outbursts of the source. It can be noted that both the light curve profiles are very similar and are of triangular shape. The right plot in the Fig. 17.13 shows the HID for both the outbursts. As the source has undergone a canonical outburst in 2016 [35], the HID traces almost a 'q' shape. During its failed outburst, the 2018 outburst, the source does not show the 'q' pattern in the HID as shown in Fig. 17.13 discussed in [34]. A detailed study of all the outbursts of H1743-322 is underway and will be discussed in the upcoming work.

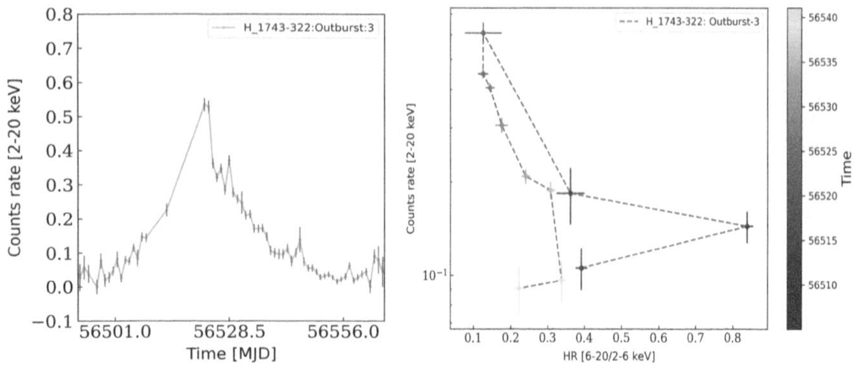

Fig. 17.12 Light curve (left) and HID (right) of outburst 3 as marked in Fig. 17.10 for source H 1743-322

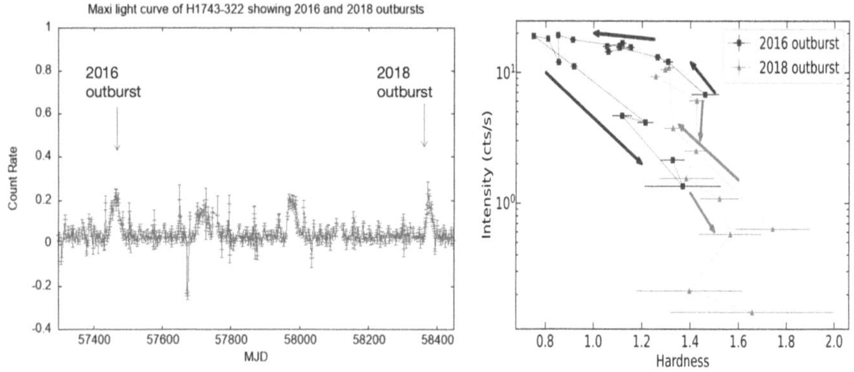

Fig. 17.13 Maxi light curve on the left showing the 2016 and 2018 outbursts of H1743-322 and on the right is the HID of 2016 (canonical outburst) and 2018 (failed outburst) outbursts of H1743-322 [34]

17.6 Conclusion

The three different sources 4U 1630-472, GX 339-4 and H1743-322 which are discussed here in this chapter are black hole binary systems. All the three fall under the recurrent transients sources which undergo outbursts over some intervals of time.

There are three scenarios observed here in this study: (1) varying light curve profiles along with varying HID patterns, (2) similar light curve profiles in subsequent outbursts, but with varying HID patterns, (3) light curve profiles repeat in subsequent outbursts along with 'q' pattern in HID of all the outbursts.

The light curves of the outbursts of 4U 1630-472 source clearly show that the profiles of the light curves of the outbursts of the same source could vary over different instances. Accordingly the HID also indicates a different pattern implicitly implying that the source 4U 1630-472 could be having different emission mechanisms during different outbursts or it could be a three body system as postulated form earlier studies.

Also, the light curves of a similar profile does not indicate that the emission mechanism driving the outburst could be the same as can be seen in H1743-322. As with the same triangular profile in the light curves of the outbursts of the source H1743-322, we find different HID patterns, some of them show canonical trends while some trace out a different pattern indicating failed outbursts.

However, all the outbursts seen in GX 339-4 show a similar light curve profile and HID pattern with a 'q' shape indicating a canonical spectral evolution as would be expected from a typical black hole binary system.

The study of HIDs of the outbursts of these sources clearly indicates that not all outbursts of black hole binaries trace a 'q' pattern during an outburst. More detailed study of these sources and the outbursts for both spectral and timing evolution of the systems along with HID signatures is underway.

References

1. Tanaka, Y. (2001). Black-hole binaries. In J. A. M. Bleeker, J. Geiss, & M. C. E. Huber (Eds.), *The century of space science*. Springer. https://doi.org/10.1007/978-94-010-0320-9_36.
2. McClintock, J. E., Remillard, R. A. (2006). Black hole binaries, "Black hole binaries." In W. Lewin, & M. van der Klis (Eds.), *Compact stellar X-ray sources*. Cambridge astrophysics series (Vol. 39, pp. 157–213). Cambridge University Press.
3. Remillard, R. A., McClintock, J. E. (2006). X-ray properties of black hole binaries. *Annual Review of Astronomy and Astrophysics, 44*, 49–92.
4. Belloni, T. M. (2010). *The jet paradigm*. Lecture notes in physics (Vol. 794, p. 53). Springer-Verlag. ISBN 978-3-540-76936-1.
5. Chen, W., Shrader, C. R., Livio, M. (1997). The properties of X-ray and optical light curves of X-ray novae. *The Astrophysical Journal, 491*, 312–338.
6. van der Klis, M. (2006). In W. H. G. Lewin, & M. van der Klis (Eds.), *Compact stellar X-Ray sources*. Cambridge astrophysics series (Vol. 39). Cambridge University Press. https://doi.org/10.2277/0521826594.
7. Shakura, N. I., & Sunyaev, R. A. (1973). *Astronomy & Astrophysics, 24*, 337.
8. Tanaka, Y., Lewin, W. H. G. (1995). In W. H. G. Lewin, J. van Paradijs, & E. P. J. van den Heuvel (Eds.), *X-ray binaries* (p. 126). Cambridge University Press.
9. Homan, J., Wijnands, R., van der Klis, Belloni, M., van Paradijs, T. M. J., Klein-Wolt, M., Fender, R., & Mendez, M. (2001). *The Astrophysical Journal Supplement, 132*, 377.
10. Belloni, T. M., Homan, J., Casella, P., van der Klis, M., Nespoli, E., Lewin, W. H. G., Miller, J. M., & Méndez, M. (2005). *Astronomy & Astrophysics, 440*, 207–222.
11. Fender, R. P., Belloni, T. M., & Gallo, E. (2004). *Monthly Notices of the Royal Astronomical Society, 355*, 1105–1118.
12. Homan, J., & Belloni, T. M. (2005). *The Astrophysical Journal Supplement Series, 300*, 107–117.
13. Baby, B. E., Agrawal, V. K., Ramadevi, M. C., Katoch, T., Antia, H. M., Mandal, S., & Nandi, A. (2020). *Monthly Notices of the Royal Astronomical Society, 497*, 1197.
14. Sreehari, H., Iyer, N., Radhika, D., Nandi, A., & Mandal, S. (2019). *Advances in Space Research, 63*, 1374.
15. Bhattacharjee, A., Banerjee, I., Banerjee, A., Debnath, D., & Chakrabarti, S. K. (2017). *Monthly Notices of the Royal Astronomical Society, 466*, 1372.

16. Aneesha, U., & Mandal, S. (2020). *Astronomy & Astrophysics, 637*, A47.
17. Connors, R. M. T., et al. (2019). *Monthly Notices of the Royal Astronomical Society, 485*, 3696.
18. Molla, A. A., Chakrabarti, S. K., Debnath, D., & Mondal, S. (2017). *The Astrophysical Journal, 834*, 88.
19. Sreehari, H., Ravishankar, B. T., Iyer, N., Agrawal, V. K., Katoch, T. B., Mandal, S., & Nandi, A. (2019). *Monthly Notices of the Royal Astronomical Society, 487*, 928–941.
20. Titarchuk, L., & Seifina, E. (2023). *Astronomy & Astrophysics, 669*, A57.
21. Ramadevi, M. C., & Seetha, S. (2007). *Monthly Notices of the Royal Astronomical Society, 378*, 182–188.
22. Shaw, A. W., et al. (2016). *Monthly Notices of the Royal Astronomical Society, 458*, 1636.
23. Tatehiro, M., Hiroshi, T., & Hitoshi, N. (2022). In C. Bambi, & A. Santangelo (Eds.), *Handbook of X-ray and Gamma-ray astrophysics*. Springer living reference work. ISBN: 978-981-16-4544-0, id.47. https://doi.org/10.1007/978-981-16-4544-0_38-1
24. Jones, C., Forman, W., Tanabaum, H., & Turner, M. J. L. (1976). *The Astrophysical Journal Letter, 210*, 9.
25. Priedhorsky, W. (1986). *The Astrophysical Journal Supplement Series, 126*, 89.
26. Parmar, A. N., Angelini, L., & White, N. E. (1995). *The Astrophysical Journal, 452*, L129.
27. Abe, Y., Fukazawa, Y., Kubota, A., Kasama, D., & Makishima, K. (2005). *Publications of the Astronomical Society of Japan, 57*, 629.
28. Tomsick, J. A., Corbel, S., Goldwurm, A., & Kaaret, P. (2005). *The Astrophysical Journal, 630*, 413.
29. Capitanio, F., Campana, R., De Cesare, G., & Ferrigno, C. (2015). *Monthly Notices of the Royal Astronomical Society, 450*, 3840–3854.
30. Markert, T. H., Canizares, C. R., Clark, G. W., Lewin, W. H. G., Schnopper, H. W., & Sprott, G. F. (1973). *The Astrophysical Journal, 184*, L67.
31. Nowak, M. A., Wilms, J., & Dove, J. B. (2002). *Monthly Notices of the Royal Astronomical Society, 332*, 856.
32. Doxsey, H., Bradt, G., & Fabbiano, R. (1977). *IAUC, 3113*.
33. Capitanio, F., Belloni, T. M., Del Santo, M., & Ubertini, P. (2010). *AIP Conference Proceedings, 1248*, 149–150.
34. Wang, P. J., et al. (2022). *Monthly Notices of the Royal Astronomical Society, 512*, 4541–4555.
35. Chand, S., Agrawal, V. K., Dewangan, G. C., Tripathi, P., & Thakur, P. (2020). *The Astrophysical Journal, 893*, 142.

Chapter 18
QPOs in Compact Sources as a Non-linear Hydrodynamical Resonance: Determining Spin of Compact Objects

Arghya Ranjan Das and Banibrata Mukhopadhyay

Abstract Quasi-periodic oscillations (QPOs) in compact astrophysical objects present a crucial window into the fundamental physics of very high-gravity environments. The present chapter proposes a comprehensive model that interprets QPOs across different frequency range from millihertz to kilohertz in different compact objects. Our model connects the observed QPO frequencies to the non-linear resonance of fundamental modes within the accretion disks, influenced by the central object's spin. This model specifically explains the high-frequency QPOs in black holes and neutron stars and suggests that the same underlying physics is present behind the generation of lower-frequency QPOs. Additionally, the model extends to make predictions about the spins, masses, and radii of these compact objects based on the properties of the observed QPOs.

18.1 Introduction

Quasi-periodic oscillations (QPOs) have been a subject of significant interest since their discovery in the 1980s [1, 2]. They have been observed across a variety of compact sources, including black holes, neutron stars, and even white dwarfs. These QPOs, which range in frequency from mHz to kHz, raise fundamental questions about their origins.

Despite extensive observations, the exact mechanism driving the origin of QPOs remains one of the most intriguing questions in the study of compact objects. In neutron star systems, it is commonly observed that in paired kHz QPOs, the difference between the frequencies stays roughly constant and equal to the stellar

A. R. Das (✉)
Department of Physics and Astronomy, Purdue University, West Lafayette, IN, USA
e-mail: das214@purdue.edu

B. Mukhopadhyay
Department of Physics, Indian Institute of Science, Bangalore, India
e-mail: bm@iisc.ac.in

S. Ghosh, A. R. Rao (eds.), *The Relativistic Universe: From Classical to Quantum*, Astrophysics and Space Science Proceedings 61,
https://doi.org/10.1007/978-3-031-90186-7_18

spin frequency or its half [3, 4], with a slightly decreasing trend as the frequency increases. On the other hand, for black holes, high frequency (HF) twin QPOs are often observed with frequencies in a 3:2 ratio [5–7]. Alternative viewpoints suggest that the ratios of high to low frequencies in QPO pairs can be 3/2, 4/3, 5/4, 7/5, or 9/7, for sources like Sco X-1, 4U 160852, 4U 1636-53, 4U 1728–34, and 4U 1820–30, with different ratios in different sources [8, 9].

In this chapter, we will delve into looking at possible origins of the kHz/HF QPOs, proposing a unified model that can explain all types of QPOs. In accretion disks, the epicyclic frequencies are the fundamental frequencies of oscillation. Earlier work [10] showed that the higher order non-linearity between these frequencies can explain the QPOs in some sources. Here, we extend this approach by showing that the fundamental frequencies, i.e., the epicyclic frequencies, get modified when we consider the motion of a fluid instead of a test particle around the compact object, leading to modified epicyclic frequencies. Resonance of these frequencies with the stellar spin frequency can lead to non-linear resonance and mode locking, resulting in the observed QPOs. See [11] for details along with the mentions of some previous works which are based on epicyclic frequencies.

The present model unifies all QPOs under a common framework involving modified epicyclic frequencies for fluids and nonlinear resonances. Understanding the physics behind QPOs is not just an academic exercise; it could serve as a powerful tool for estimating the radius and mass of neutron stars, as well as estimating the spins and constraining the mass of black holes.

In the sections that follow, we will explore the theoretical foundations of our model, derive the modified epicyclic frequencies for fluids in accretion disks, and examine the implications of nonlinear resonance in these systems. We will also apply our model to a range of observational data, demonstrating its robustness and predictive power, and bridging the gap between theory and observation to advance our understanding of QPOs.

18.2 Modified Epicyclic Frequency for Fluid

The epicyclic frequencies for test mass particles are often calculated by considering their perturbed motion around the compact object. A more general approach involves studying the perturbed motion of fluid annuli around the compact object, and thus epicyclic frequencies for fluids. Starting from the momentum balance equation we have

$$\frac{\partial \mathbf{u}}{\partial t} + (\mathbf{u} \cdot \nabla)\mathbf{u} + \nabla \left(\frac{P}{\rho_0} \right) = \mathbf{F}, \qquad (18.1)$$

where $\mathbf{u}$, P, and ρ_0 represent the fluids velocity, pressure, and density respectively, and $\mathbf{F}$ is the external force per unit mass including contributions from gravity.

Given the significant disturbances in motion around compact objects, it is reasonable to assume that the fluid annulus will deviate from an ideal streamline motion, experiencing continual perturbations. Considering a small perturbation $\delta\mathbf{r}$ to a fluid element's position $\mathbf{r}$ and assuming incompressibility and a locally polytropic equation of state, the perturbed momentum balance equation takes the form of

$$\frac{\partial \delta \mathbf{u}}{\partial t} + (\delta \mathbf{u} \cdot \nabla)\mathbf{u} + (\mathbf{u} \cdot \nabla)\delta \mathbf{u} = \delta \mathbf{F}, \tag{18.2}$$

where δ denotes the perturbation of the variables. Simplifying Eq. (18.2) leads to a formulation that closely resembles to forced damped oscillator equation as

$$\frac{d\delta \mathbf{u}}{dt} + (\delta \mathbf{u} \cdot \nabla)\mathbf{u} = \delta \mathbf{F}. \tag{18.3}$$

Assuming an equatorial axisymmetric disk around the compact object with circular orbits (hence $u_r = \dot{\Omega}_\phi \sim 0$), the components of the perturbed Eq. (18.3) turns out to be

$$\delta\ddot{r} + \gamma\delta\dot{r} - \Omega_\phi(r\delta\dot{\phi} - \Omega_\phi\delta r) = (\delta\mathbf{F})_r, \tag{18.4}$$

$$\delta\ddot{z} = (\delta\mathbf{F})_z, \tag{18.5}$$

$$r\delta\ddot{\phi} + \Omega_\phi\delta\dot{r} + \xi\delta\dot{r} = (\delta\mathbf{F})_\phi, \tag{18.6}$$

where $\xi = r\Omega_{\phi,r}$ and $\gamma = \partial u_r/\partial r$. These equations can now be seen to be very similar to the equations describing the forced damped harmonic oscillators with the first and/or the second order time derivatives of the position variables in the left hand side relating to the force term on the right side of the equations. We consider the perturbation to the force arising from two primary sources: orbital perturbations and an external forcing mechanism within the system; leading to the total perturbed force of the form as

$$\delta\mathbf{F} = \delta\mathbf{F}_{orbit} + \mathbf{F}_{forced}, \tag{18.7}$$

where the perturbation in the orbital force within an axisymmetric equatorial system is given by:

$$\delta\mathbf{F}_{orbit} = -\left(\Omega_r^2\delta r\right)\hat{r} - \left(\Omega_z^2\delta z\right)\hat{z}. \tag{18.8}$$

Here, Ω_j is defined as $\Omega_j = 2\pi\nu_j$, where ν_r and ν_z are the epicyclic frequencies, and ν_ϕ is the orbital frequency of a test particle in a circular orbit in the equatorial plane around the compact object.

In the Kerr metric, the epicyclic frequencies and the orbital frequency are given as [12]

$$\nu_\phi = \frac{1}{2\pi} \frac{1}{r^{3/2} + a} \frac{c^3}{GM}, \tag{18.9}$$

$$\nu_r = \frac{\nu_\phi}{r} \sqrt{\Delta - 4(\sqrt{r} - a)^2}, \tag{18.10}$$

$$\nu_z = \frac{\nu_\phi}{r} \sqrt{r^2 - 4a\sqrt{r} + 3a^2}, \tag{18.11}$$

where $\Delta = r^2 - 2r + a^2$, a is the Kerr parameter, M is the mass of central compact object, G is the Newton's gravitational constant and c is the speed of light.

Now the forcing term can be expressed as a combination of cosine functions with varying magnitude and frequency, hence we consider the form of the force to be

$$\mathbf{F}_{forced} = (F_{0r}\hat{r} + F_{0z}\hat{z} + F_{0\phi}\hat{\phi}) \cos \omega t. \tag{18.12}$$

Hence, the following equations of perturbation can be written from Eqs. (18.4), (18.5), (18.6), (18.8) and (18.12) as

$$\delta\ddot{r} + \gamma\delta\dot{r} - \Omega_\phi(r\delta\dot{\phi} - \Omega_\phi\delta r) = -\Omega_r^2\delta r + F_{0r} \cos \omega t, \tag{18.13}$$

$$\delta\ddot{z} = -\Omega_z^2\delta z + F_{0z} \cos \omega t, \tag{18.14}$$

$$r\delta\ddot{\phi} + \Omega_\phi\delta\dot{r} + \xi\delta\dot{r} = F_{0\phi} \cos \omega t. \tag{18.15}$$

These equations describe a coupled, forced, damped oscillator system, where the coupling occurs between the radial and angular components, while the vertical direction remains decoupled, owing to our confinement to the equatorial disk.

To solve these equations, we express the forcing terms $F_{0j} \cos \omega t$ as $F_{0j} e^{i\omega t}$ (considering the real part), where $j \in \{r, z, \phi\}$, and we make the ansatz for the perturbations as,

$$\delta r = \mathcal{R}(\omega)e^{i\omega t}, \quad \delta z = \mathcal{Z}(\omega)e^{i\omega t}, \quad \delta\phi = \Phi(\omega)e^{i\omega t}. \tag{18.16}$$

Substituting this ansatz of Eq. (18.16) into Eqs. (18.13), (18.14), and (18.15), we obtain three equations for $\mathcal{R}(\omega)$, $\mathcal{Z}(\omega)$ and $\Phi(\omega)$ which can be solved. As an initial step of solving, we assume $F_{0r} = F_{0z} = F_{0\phi} = F_0$ (for unequal force amplitudes, the method of multiple scales can be implemented to get the results as in Sect. 18.3). From this, the solutions for the absolute values of the amplitudes (i.e. up to a phase factor) are given by

$$|\mathcal{R}(\omega)| = \frac{|F_0|\sqrt{\omega^2 + \Omega_\phi^2}}{\omega\sqrt{\gamma^2\omega^2 + \left[\Omega_\phi\left(\xi + 2\Omega_\phi\right) + \Omega_r^2 - \omega^2\right]^2}}, \tag{18.17}$$

$$|\mathcal{Z}(\omega)| = \frac{|F_0|}{\omega^2 - \Omega_z^2}, \tag{18.18}$$

$$|\Phi(\omega)| = \frac{|F_0|\sqrt{\omega^2\left(\xi - \gamma + \Omega_\phi\right)^2 + \left(\Omega_r^2 - \omega^2 + \Omega_\phi^2\right)^2}}{\omega^2 r\sqrt{\gamma^2\omega^2 + \left[\Omega_\phi\left(\xi + 2\Omega_\phi\right) + \Omega_r^2 - \omega^2\right]^2}}. \tag{18.19}$$

Therefore, the perturbation δz reaches its maximum when

$$\omega = \Omega_z, \tag{18.20}$$

and δr and $\delta\phi$ are maximized when

$$f(\omega) = \gamma^2\omega^2 + \left[\Omega_\phi\left(\xi + 2\Omega_\phi\right) + \Omega_r^2 - \omega^2\right]^2 \tag{18.21}$$

is minimized. It is found that $f(\omega)$ is minimized for $\gamma = 0$ (circular orbit) when

$$\omega = \sqrt{r\Omega_{\phi,r}\Omega_\phi + \Omega_r^2 + 2\Omega_\phi^2}. \tag{18.22}$$

From these results, Eqs. (18.20) and (18.22) are the modified epicyclic frequencies for a fluid element in an accretion disk, moving in a circular orbit within the equatorial plane (i.e. a Keplerian accretion disk), denoted as ν_1 and ν_2, are given by

$$\nu_1 = \sqrt{\nu_r^2 + 2\nu_\phi^2 + r\nu_{\phi,r}\nu_\phi}, \tag{18.23}$$

$$\nu_2 = \nu_z. \tag{18.24}$$

As evident from the results, this modification introduces a coupling between the radial epicyclic and orbital frequencies, leading to a new epicyclic frequency ν_1 for the fluid. However, the vertical epicyclic frequency ν_2 remains unchanged, consistent with the frequency derived for a test particle.

18.3 QPO Model

In astrophysical systems, particularly around compact objects like neutron stars and black holes, QPOs provide critical insights into physics around these objects. This section delves into the mathematical modeling of these QPOs under both nonlinear and linear regimes, illustrating how the interplay between different modal frequencies leads to observable resonance effects.

When a compact object rotates, it can induce a new mode in the surrounding medium at the spin frequency ν_s, which interacts with existing disk modes ν_1 and ν_2. This interaction might stimulate generation of additional modes [13, 14] by combination of these frequencies as $\nu_{1,2} \pm p\nu_s$ (e.g. Eqs. (18.29) and (18.32)); with difference between two such modes in a system being approximately $p\nu_s$, where p is a fraction like $n/2$ with n being an integer (e.g., 1 or 2 in this case). Resonance occurs when $\nu_1 \approx \nu_2$ and the difference between ν_h and ν_l matches $\nu_s/2$ (nonlinear regime) or ν_s (linear regime), locking the frequency difference. In fact, observations from some neutron star systems reveal that the frequency gap between twin kHz QPOs remains around either the stellar spin frequency or half of it, despite variations in the QPO frequencies themselves [3].

In systems with j degrees of freedom, where each has a distinct natural frequency denoted as $\nu_1, \nu_2, \cdots, \nu_j$, strong frequency coupling can emerge if these frequencies are commensurable or nearly so, leading to internal resonance [15]. This is further emphasized when there is a harmonic external excitation frequency ν_s, exhibiting the resonance as

$$p\nu_s = \Sigma_{i=1}^{j} b_i \nu_i, \tag{18.25}$$

where p and b_i are integers, satisfying the relation

$$p + \Sigma_{i=1}^{j} |b_i| = k + 1, \tag{18.26}$$

where k is the order of non-linearity. For fundamental natural frequencies ν_1 and ν_2, Eq. (18.25) can be recast into

$$(n - m + 1)\nu_s = b_1\nu_1 + b_2\nu_2. \tag{18.27}$$

18.3.1 Nonlinear Regime

In the nonlinear region, setting $b_1 = -b_2 = 2$ leads to

$$\frac{\nu_s}{2}(n - m + 1) = \nu_1 - \nu_2, \tag{18.28}$$

with m and n being integers.

Resonance is expected to occur at an appropriate radius where $\nu_1 - \nu_2 \approx 0$, yielding $n = m - 1$ from Eq. (18.28) and hence $k = 3$ from Eq. (18.26), which suggests a cubic order of non-linearity. Now we propose the higher and lower QPO frequencies of a pair to be, respectively,

$$\nu_h = \nu_1 + \frac{m+1}{2}\nu_s, \quad \nu_l = \nu_2 + \frac{m}{2}\nu_s. \tag{18.29}$$

The rearrangement of the terms of Eq. (18.28) gives the resonance condition for the frequency difference

$$\left(\nu_1 + \frac{m+1}{2}\nu_s\right) - \left(\nu_2 + \frac{m}{2}\nu_s\right) = \frac{\nu_s}{2}. \tag{18.30}$$

This expression shows that the difference between the proposed QPO frequencies ν_h and ν_l, satisfies the condition for $\Delta\nu$ as

$$\Delta\nu = \nu_h - \nu_l \approx \frac{\nu_s}{2}. \tag{18.31}$$

This explains why the frequency gap between twin kHz QPOs in many neutron star systems consistently remains close to half of the stellar spin frequency, or sometimes the full spin frequency (as discussed later), despite variations in the QPO frequencies. This happens even as the excitation order m varies.

18.3.2 Linear Regime

For black holes or weakly magnetized neutron stars, due to the absence of a strong magnetosphere, the interaction between modes is likely to be linear rather than nonlinear. In these cases, resonance locking occurs when the difference between the frequencies ν_h and ν_l is approximately equal to ν_s. Under these circumstances, the first order excitation needs to be considered in Eq. (18.27) by setting $b_1 = -b_2 = 1$, which can be seen as linear regime since k becomes *unity* from Eq. (18.26). Now following a similar approach as in the nonlinear regime, the QPO frequencies are then given by

$$\nu_h = \nu_1 + (m+1)\nu_s, \qquad \nu_l = \nu_2 + m\nu_s. \tag{18.32}$$

It is evident from Eq. (18.32) that the resonance condition $\Delta\nu \approx \nu_s$ is satisfied.

Summarizing the QPO Frequency Models
Nonlinear Regime
Applicable for magnetized neutron stars

$$\nu_h = \nu_1 + \frac{m+1}{2}\nu_s, \quad \nu_l = \nu_2 + \frac{m}{2}\nu_s, \quad \Delta\nu \approx \frac{\nu_s}{2}.$$

Linear Regime
Applicable for black holes and weakly magnetized neutron stars

$$\nu_h = \nu_1 + (m+1)\nu_s, \quad \nu_l = \nu_2 + m\nu_s, \quad \Delta\nu \approx \nu_s$$

18.4 QPOs of Black Holes and Neutron Stars

QPOs are observed in both neutron stars and black holes, with slight differences in the resonance conditions for QPO formations (due to linear or non-linear mechanism). Although there is a slight difference, the underlying physical process still remains the same.

We can fit the observed QPO frequency for different neutron stars with our model to infer neutron stars' properties such as mass and radius. Now considering deformation to the neutron stars' shape from a perfect solid sphere to an ellipsoidal shape due to rapid rotation, as suggested in previous studies [10, 12, 16], the radius of the neutron star lies within the range

$$\frac{R_G}{\sqrt{0.35}} \leq R \leq \frac{R_G}{\sqrt{0.5}}, \tag{18.33}$$

where $R_G = \sqrt{GMa/2\pi\nu_s c}$, M is the mass of the star, and ν_s is the spin frequency.

For fast rotators like 4U 1636–53, with observed $\nu_s = 581.75$ Hz, we see $\Delta\nu \approx \nu_s/2$ which remains nearly constant exhibiting a slightly decreasing trend with increasing ν_h.

Fitting our model with the data we have a mass of $M = 1.5M_\odot$ corresponding to a radius in the range of 14.52 to 17.35 km (for $m = -2, -1$). Another set of solutions for $M = 1.65M_\odot$ yields a radius between 15.61 and 18.66 km (for $m = -2, 0$). For $m = -1, 1$ and $M = 1.23M_\odot$, the radius range is between 10.21 and 12.21 km, representing the best fit.

The point to note here is that different modes m can be excited corresponding to different frequencies but the same $\Delta\nu$.

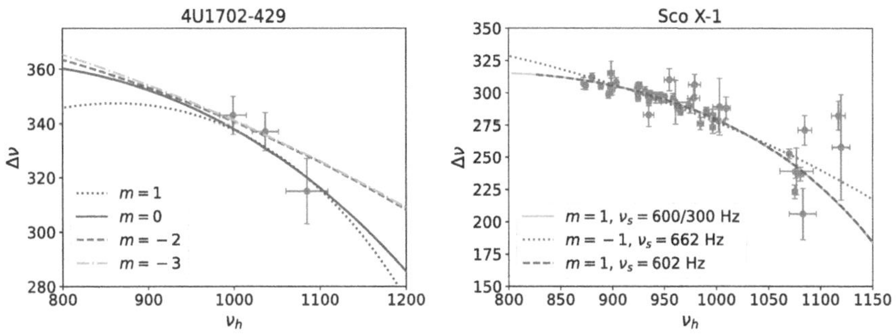

Fig. 18.1 Variation of QPO frequency difference with the change of higher QPO frequency and fitting the data with our QPO model, by varying its parameters and listing the best fits for (**a**) 4U 1702–429 and (**b**) Sco X-1. This also estimates the spin frequency of Sco X-1

For slow rotators, such as 4U 1702–429, with observed $\nu_s = 330$ Hz we see $\Delta\nu \approx \nu_s$ which remains almost constant and having a general decreasing trend, as can be seen in Fig. 18.1.

Fitting our model with the data we have a mass of $M = 1.67M_\odot$ giving the radius range of 9.25 to 11.06 km. Other solutions with lower mass $M < 1.4M_\odot$ also fits the observational data; including $M = 0.85M_\odot$, $1.2M_\odot$ and $0.97M_\odot$, with corresponding mode numbers as $m = 0, -2$ and -3 which give the radius as 7.62 to 9.1 km; 11.98 to 14.32 km; and 10.77 to 12.87 km, respectively. These fits are also illustrated in Fig. 18.1.

We can also estimate the spin frequency of neutron stars like Sco X-1, whose spin is not confirmed to date. By varying the spin frequency ν_s from 200 to 800 Hz and assuming a mass around $1.4M_\odot$ [17, 18], we obtain the best fit for $\nu_s = 600$ Hz in nonlinear regime (also 300 Hz in linear regime which estimates very large radius) for the radius range 15.91 to 19.08 km, a bit high. However, $\nu_s = 662$ Hz gives radius range of 10.21 to 12.2 km, with mass $M = 1.44M_\odot$. Nevertheless, if the mass is adjusted to a higher value, e.g., $M = 1.89M_\odot$, the best-fit radius in the nonlinear regime falls between 11.71 and 14 km with $\nu_s = 602$ Hz.

These fits are shown in Fig. 18.1.

18.4.1 Black Hole QPOs

For black holes, energy, and angular momentum are transferred to the surrounding accretion disk through magnetic field lines, which are typically weaker than those around neutron stars, particularly in soft states [19].

And unlike neutron stars, black holes lack a solid surface, so the black hole's spin primarily influences the accreting fluid via the frame-dragging effect. Therefore, we define the spin frequency for a black hole as

$$\nu_{BH} = \nu_s = \frac{\Omega_{BH}}{2\pi} \frac{c^3}{GM}, \tag{18.34}$$

Several black holes have been observed to have twin HF QPOs, such as GRO J1655-40 [20, 21], XTE J1550-564 [22], GRS 1915+105 [23, 24], H1743-322 [25] and IGR J17091-3624 [26] with their mass determined through independent measurement. However, the spin parameters of these black holes remain uncertain. By fitting observed QPO frequencies to our model, we can estimate black hole's spin and also give constraints on its mass.

For example, the twin HF QPOs of GRS 1915+105 have been reported at 67 and 40 Hz [27, 28], with additional twin peaks at approximately 168 and 113 Hz [29, 30]. A prominent QPO around 70 Hz and sporadic peaks at 41 and 34 Hz have also been observed [30–33]. The mass is varied in our model in the range $10.6 - 14.4M_{\odot}$ estimated by [24] and found our model fits the 168 and 113 Hz QPOs for mass $11.65 - 14.4M_{\odot}$ with spin parameters a from 0.99 to 0.8 along with $m = -2$; however the best fit for 69 and 41.1 Hz pair QPOs is for $m = -6$.

18.5 Explaining Other QPOs

The HF/kHz QPOs discussed in previous sections are not the only types observed in astrophysical phenomena. QPOs can also be detected without a corresponding pair, as shown in various studies (e.g., [34–36]). Additionally, numerous low-frequency QPOs are frequently observed. While currently no universally accepted model exists that explains all QPO types, the present analysis provides a unified explanation for both high and low-frequency QPOs and also single QPO frequencies, offering an explanation based on same fundamental physical mechanisms.

18.5.1 Single QPO Frequency

From Eq. (18.28) it follows that when $\Delta\nu = 0$, both frequencies coincide to form a single QPO frequency. By setting $b_1 = -b_2$ in Eq. (18.27), we have

$$\nu_1 - \nu_2 = \frac{n - m + 1}{b_1} \nu_s. \tag{18.35}$$

To generalize Eq. (18.29) for an arbitrary b_1, we modify it as

$$\nu_h = \nu_1 + \frac{m+1}{b_1} \nu_s, \quad \nu_l = \nu_2 + \frac{m}{b_1} \nu_s. \tag{18.36}$$

Thus, from Eqs. (18.35) and (18.36), we obtain

$$\nu_h - \nu_l = \Delta\nu = \frac{n - m + 2}{b_1}\nu_s. \tag{18.37}$$

Hence, $\Delta\nu = 0$, requires either a higher order nonlinearity, corresponding to a very large b_1 or the condition $n = m - 2$. In the latter case, the mode-locking is governed by $\nu_h - \nu_l = 0$, rather than $\nu_1 - \nu_2 = 0$.

18.5.2 *Low-Frequency QPOs*

For the case of hard accretion state, it is crucial to consider a nonzero radial velocity u_r and its radial gradient γ which makes the flow hot and radiatively inefficient (see, e.g., [37–39]). An example of such hard states of sources is GRS 1915+105 [30]. Thus with $u_r \neq 0$ and $\gamma \neq 0$, minimizing $f(\omega)$ yields

$$\omega^2 = X - \frac{\gamma^2}{2} \pm \frac{\sqrt{\gamma^4 - 4X\gamma^2}}{2}, \tag{18.38}$$

where $X = \Omega_\phi(\xi + 2\Omega_\phi) + \Omega_r^2$. In an advective sub-Keplerian flow where $\Omega_\phi \sim r^{-q}$ and $q > 1.5$, with $\gamma^2 >> \Omega_r^2$, the resonance frequency from Eq. (18.38) turns out to be

$$\omega \approx \frac{\sqrt{2}\Omega_r^2}{\gamma}, \tag{18.39}$$

for $q = 2$.

In an advective accretion disk with α-viscosity of 0.01 around a rotating black hole with a Kerr parameter of $a = 0.99$ [39], this model predicts a low-frequency QPO $\gtrsim$ 0.1 Hz; which is consistent with the observed frequencies in the hard state of GRS 1915+105.

18.6 Summary and Conclusion

In this chapter we have demonstrated that HF and kHz QPOs observed in black holes and neutron stars originate from the same underlying mechanism.

During the mathematical derivation of our model we have assumed locally polytropic equation of state and incompressibility of the fluid in the accretion disk. In reality, accretion disks are compressible and hence will have additional terms that depend on pressure and density. These terms could be explored to investigate more complex scenarios, such as including magnetic pressure or studying magnetohydrodynamic (MHD) effects. Future work could extend the model to incorporate these factors for a broader range of physical conditions. Nevertheless, in a local analysis, when the QPOs are expected to be originating from a local region, they are not odd assumptions.

We have shown how the epicyclic frequencies get modified when we consider the motion of fluid instead of the test particle in the accretion disk leading to the change in the fundamental frequencies of accretion disks. We have also shown that by relaxing the assumption of circular orbits, low-frequency QPOs can be explained by the same mechanism. A more general case would involve non-equatorial orbits, which would introduce coupling with the vertical direction. This would modify the vertical epicyclic frequency ν_z, similarly to how the radial frequency ν_r replaces by ν_1, and results in additional cross terms.

We further propose that the other types of QPOs are likely to arise from this same basic mechanism. Our analysis also shows that the fundamental frequencies of the accretion disks are modified compared to those for test particle. The spin of the compact object perturbs the system leading to the formation of new modes due to combination of fundamental and perturbation frequencies, resulting in resonance and mode locking at specific conditions leading to the observed QPO frequencies.

Additionally, our model also provides a tool to predict and/or constraint the mass and spin of black holes and neutron stars. It also offers predictions for the radius of neutron stars.

Acknowledgments The authors express their gratitude to Tomaso Belloni for sharing the latest data and for his careful review of the manuscript, comments, and suggestions. The work is partly dedicated to him, who we miss.

A.R.D. acknowledges the financial support from KVPY, DST, India.

References

1. Motch, C., Ricketts, M. J., Page, C. G., Ilovaisky, S. A., & Chevalier, C. (1983). Simultaneous x-ray/optical observations of gx339-4 during the may 1981 optically bright state. *Astronomy & Astrophysics, 119*, 171–176.
2. van der Klis, M., Jansen, F., van Paradijs, J., Lewin, W. H. G., van den Heuvel, E. P. J., Trümper, J. E., & Sztajno, M. (1985). Intensity-dependent quasi-periodic oscillations in the x-ray flux of gx5-1. *Nature, 316*(6025), 225–230.
3. van der Klis, M. (2000). Millisecond oscillations in x-ray binaries. *Annual Review of Astronomy and Astrophysics, 38*(1), 717–760.
4. Kluźniak, W., Abramowicz, M. A., Kato, S., Lee, W. H., & Stergioulas, N. (2004). Nonlinear resonance in the accretion disk of a millisecond pulsar. *The Astrophysical Journal, 603*(2), L89.
5. Rezzolla, L., Yoshida, S'i., Maccarone, T. J., & Zanotti, O. (2003). A new simple model for high-frequency quasi-periodic oscillations in black hole candidates. *Monthly Notices of the Royal Astronomical Society, 344*(3), L37–L41.
6. Remillard, R. A., & McClintock, J. E. (2006). X-ray properties of black-hole binaries. *Annual Review of Astronomy and Astrophysics, 44*(1), 49–92.
7. Török, G., Goluchová, K., Horák, J., Šrámková, E., Urbanec, M., Pecháček, T., & Bakala, P. (2015). Twin peak quasi-periodic oscillations as signature of oscillating cusp torus. *Monthly Notices of the Royal Astronomical Society: Letters, 457*(1), L19–L23.
8. Belloni, T., Méndez, M., & Homan, J. (2005). The distribution of khz QPO frequencies in bright low mass x-ray binaries. *Astronomy & Astrophysics, 437*(1), 209–216.

9. Belloni, T., Méndez, M., & Homan, J. (2007). On the kHz QPO frequency correlations in bright neutron star X-ray binaries. *Monthly Notices of the Royal Astronomical Society, 376*(3), 1133–1138.
10. Mukhopadhyay, B. (2009). Higher-order nonlinearity in accretion disks: Quasi-periodic oscillations of black hole and neutron star sources and their spin. *The Astrophysical Journal, 694*(1), 387.
11. Das, A. R., & Mukhopadhyay, B. (2023). QPOs in compact sources as a nonlinear hydrodynamical resonance: Determining spin of compact objects. *The Astrophysical Journal, 955*(2), 86.
12. Cook, G. B., Shapiro, S. L., & Teukolsky, S. A. (1994). Rapidly rotating neutron stars in general relativity: Realistic equations of state. *The Astrophysical Journal, 424*, 823.
13. Landau, L. D., & Lifshitz, E. M. (1976). *Mechanics*. Pergamon.
14. Abramowicz, M. A., Karas, V., Kluzniak, W., Lee, W. H., & Rebusco, P. (2003). Nonlinear resonance in nearly geodesic motion in low mass x-ray binaries. *Publications of the Astronomical Society of Japan, 55*, 466–467.
15. Nayfeh, A. H., & Mook, D. T. (1979). *Nonlinear oscillations*. Pure and Applied Mathematics: A Wiley Series of Texts, Monographs and Tracts. Wiley.
16. Bejger, M., & Haensel, P. (2002). Moments of inertia for neutron and strange stars: Limits derived for the crab pulsar. *Astronomy & Astrophysics , 396*(3), 917–921.
17. Steeghs, D., & Casares, J. (2002). The mass donor of scorpius x-1 revealed. *The Astrophysical Journal, 568*(1), 273.
18. Messenger, C., Bulten, H. J., Crowder, S. G., Dergachev, V., Galloway, D. K., Goetz, E., Jonker, R. J. G., Lasky, P. D., Meadors, G. D., Melatos, A., Premachandra, S., Riles, K., Sammut, L., Thrane, E. H., Whelan, J. T., & Zhang, Y. (2015). Gravitational waves from scorpius x-1: A comparison of search methods and prospects for detection with advanced detectors. *Phys. Rev. D, 92*, 023006.
19. Blandford, R. D., & Znajek, R. L. (1977). Electromagnetic extraction of energy from Kerr black holes. *Monthly Notices of the Royal Astronomical Society, 179*, 433–456.
20. Orosz, J. A., & Bailyn, C. D. (1997). Optical observations of GRO j1655-40 in quiescence. I. A precise mass for the black hole primary. *The Astrophysical Journal, 477*(2), 876–896.
21. Shahbaz, T., van der Hooft, F., Casares, J., Charles, P. A., & van Paradijs, J. (1999). The mass of X-ray Nova scorpii 1994 (=GRO J1655-40). *Monthly Notices of the Royal Astronomical Society, 306*(1), 89–94.
22. Orosz, J. A., Groot, P. J., van der Klis, M., McClintock, J. E., Garcia, M. R., Zhao, P., Jain, R. K., Bailyn, C. D., & Remillard, R. A. (2002). Dynamical evidence for a black hole in the microquasar XTE j1550-564. *The Astrophysical Journal, 568*(2), 845–861
23. Greiner, J., Cuby, J. G., & McCaughrean, M. J. (2001). An unusually massive stellar black hole in the galaxy. *Nature, 414*(6863), 522–525. 35107019.
24. Reid, M. J., McClintock, J. E., Steiner, J. F., Steeghs, D., Remillard, R. A., Dhawan, V., & Narayan, R. (2014). A parallax distance to the microquasar GRS 1915+105 and a revised estimate of its black hole mass. *The Astrophysical Journal, 796*(1), 2.
25. Miller, J. M., Raymond, J., Homan, J., Fabian, A. C., Steeghs, D., Wijnands, R., Rupen, M., Charles, P., van der Klis, M., & Lewin, W. H. G. (2006). Simultaneous chandra and RXTE spectroscopy of the microquasar h1743–322: Clues to disk wind and jet formation from a variable ionized outflow. *The Astrophysical Journal, 646*(1), 394.
26. Altamirano, D., & Belloni, T. (2012). Discovery of high-frequency quasi-periodic oscillations in the black hole candidate IGR J17091-3624. *The Astrophysical Journal Letters, 747*(1), L4.
27. Strohmayer, T. E. (2001). Discovery of a 450 HZ quasi-periodic oscillation from the microquasar GRO J1655–40 with the ROSSI x-ray timing explorer. *The Astrophysical Journal, 552*(1), L49.
28. Morgan, E. H., Remillard, R. A., & Greiner, J. (1997). RXTE observations of QPOs in the black hole candidate GRS 1915+105. *The Astrophysical Journal, 482*(2), 993.
29. McClintock, J. E., & Remillard, R. A. (2006). *Black hole binaries. Cambridge astrophysics* (chap. 4, pp. 157–206). Cambridge University Press.

30. Belloni, T., Soleri, P., Casella, P., Méndez, M., & Migliari, S. (2006). High-frequency quasi-periodic oscillations from GRS 1915+105 in its C state. *Monthly Notices of the Royal Astronomical Society, 369*(1), 305–310.
31. Belloni, T. M., Bhattacharya, D., Caccese, P., Bhalerao, V., Vadawale, S., & Yadav, J. S. (2019). A variable-frequency HFQPO in GRS 1915+105 as observed with AstroSat. *Monthly Notices of the Royal Astronomical Society, 489*(1), 1037–1043.
32. Majumder, S., Sreehari, H., Aftab, N., Katoch, T., Das, S., & Nandi, A. (2022). Wide-band view of high-frequency quasi-periodic oscillations of GRS 1915+105 in 'softer' variability classes observed with AstroSat. *Monthly Notices of the Royal Astronomical Society, 512*(2), 2508–2524.
33. Belloni, T. M., & Altamirano, D. (2013). Discovery of a 34Hz quasi-periodic oscillation in the X-ray emission of GRS 1915+105. *Monthly Notices of the Royal Astronomical Society, 432*(1), 19–22.
34. Homan, J., Klein-Wolt, M., Rossi, S., Miller, J. M., Wijnands, R., Belloni, T., van der Klis, M., & Lewin, W. H. G. (2003). High-frequency quasi-periodic oscillations in the black hole x-ray transient xte j1650–500. *The Astrophysical Journal, 586*(2), 1262.
35. Belloni, T. M., Sanna, A., & Méndez, M. (2012). High-frequency quasi-periodic oscillations in black hole binaries. *Monthly Notices of the Royal Astronomical Society, 426*(3), 1701–1709.
36. Remillard, R. A., & Morgan, E. H. (1999). X-ray oscillations at 184 GHz in the black hole candidate 4U1630-47. In *American Astronomical Society Meeting Abstracts* (vol. 195, p. 37).
37. Narayan, R., & Yi, I. (1995). Advection-dominated accretion: Underfed black holes and neutron stars. *The Astrophysical Journal, 452*, 710.
38. Chakrabarti, S. K. (1996). Grand unification of solutions of accretion and winds around black holes and neutron stars. *The Astrophysical Journal, 464*, 664.
39. Rajesh, S. R., & Mukhopadhyay, B. (2010). Two-temperature accretion around rotating black holes: A description of the general advective flow paradigm in the presence of various cooling processes to explain low to high luminous sources. *Monthly Notices of the Royal Astronomical Society, 402*(2), 961–984.

Chapter 19
Estimation of Spin Parameters of Stellar Mass Black Holes: LMC X-1, LMC X-3 and GX 339-4

Meghamani Haldar, Tamal Sarkar, Shubrangshu Ghosh, and Rajat K. Dey

Abstract Stellar mass black holes are essential laboratories for studying the extreme properties of matter and gravity. The spin parameter of a black hole, which describes the degree of its rotational angular momentum, is a crucial parameter that can provide insights into the formation and evolution of these compact objects. The continuum fitting method is a widely used technique to estimate the spin parameters of stellar mass black holes by analyzing their X-ray spectra. In this review, we present a comprehensive analysis of the spin parameter estimations for three well-studied stellar mass black holes, LMC X-1, LMC X-3, and GX 339-4, using data from the XMM-Newton, NICER, and NuSTAR X-ray astronomy missions. We discuss the underlying assumptions, data analysis techniques, and the resulting spin parameter values for LMC X-1 as 0.99 ± 0.08, for LMC X-3 as $0.24 - 0.30$ and for GX339-4 as $0.997 - 0.998$ with $T_{in} \sim 0.78 - 1.02$ keV, $0.95 - 1.21$ keV and $0.26 - 1.39$ keV, respectively. The implications of these spin measurements for our understanding of black hole formation, accretion, and astrophysics are also discussed.

M. Haldar · R. K. Dey
Department of Physics, University of North Bengal, Siliguri, India
e-mail: rs_meghamani@nbu.ac.in

T. Sarkar (✉)
HECRRC, University of North Bengal, Siliguri, India
e-mail: tsarkar@nbu.ac.in

S. Ghosh
Center for Astrophysics, Gravitation and Cosmology (CAGC), SRM University Sikkim, Gangtok, India

Department of Physics, SRM University Sikkim, Gangtok, India

S. Ghosh, A. R. Rao (eds.), *The Relativistic Universe: From Classical to Quantum*, Astrophysics and Space Science Proceedings 61,
https://doi.org/10.1007/978-3-031-90186-7_19

19.1 Introduction

Stellar mass black holes are remnants of massive stars that have undergone gravitational collapse. Their spin parameters hold crucial information about their formation, growth, and the dynamics of their surrounding accretion disks [1]. The continuum fitting method has emerged as a robust tool for estimating these spin values through the analysis of X-ray emission from the accreting material. The measurement of black hole spins has advanced significantly since the inception of X-ray astronomy. Measuring the spin parameter of a stellar mass black hole can provide valuable information about its formation, accretion history, and the underlying astrophysical processes at play. Shafee et al. [2] first proposed the continuum fitting method to estimate black hole spins effectively. The continuum fitting method is a widely used technique to estimate the spin parameters of stellar mass black holes. This method relies on the analysis of the X-ray continuum emission from the accretion disk surrounding the black hole, which is directly influenced by the spin parameter. By modeling the X-ray spectrum and its various components, it is possible to infer the spin parameter of the black hole [3]. Subsequent works have refined this approach by incorporating various models of accretion disks. Notably, [4] applied this method to LMC X-1, yielding high spin values that suggest rapid rotation. Similarly, [5] provided insights into the spin of GX 339-4, revealing its highly efficient accretion dynamics. Other studies, including those by Liu and Qiao [6] and Gúrpide et al. [7] addressed the meticulous relationships between spin values and accretion rates, emphasizing that spin measurements can provide insights into the evolutionary paths of black holes.

In this chapter, we focus on the spin parameter estimations of three well-studied stellar mass black holes, LMC X-1, LMC X-3, and GX 339-4, with the continuum fitting method using X-ray data from various instruments, including XMM-Newton, NICER, and NuSTAR [Table 19.1]. In Sect. 19.2, we have given an overview of each source and the reported spin parameter values for each source. Next, in the same section, we discuss data analysis techniques. In Sect. 19.3, we have presented the results obtained from spectral data analysis and fitting for each of the sources, and in the next section, i.e., Sect. 19.4 we have given our conclusion as well as the implications of these spin measurements for our understanding of black hole formation, accretion, and astrophysics are also explored.

Table 19.1 X-ray astronomy mission used for observation

Mission	Launch year	Instrument with energy
NICER	2017	X-ray timing instrument (XTI) with range 0.2–12 keV
XMM-Newton	1999	EPIC–MOS1, MOS2, PN with range 0.1–15 keV
NuSTAR	2012	Dual X-ray telescopes with range 3–79 keV

19.2 Methods

19.2.1 Overview of the X-Ray Sources

19.2.1.1 LMC X-1

LMC X-1 is a high-mass X-ray binary system located in the Large Magellanic Cloud, a satellite galaxy of the Milky Way. The black hole in this system has a mass of approximately 10 solar masses and is orbited by a massive, early-type companion star [8]. Several studies have attempted to measure the spin parameter of LMC X-1 using the continuum fitting method. One such study, conducted by Gou et al. [4] using XMM-Newton data, reported a spin parameter of a = 0.92 ± 0.01 [4]. The authors used a relativistic accretion disk model (KERRBB) to fit the X-ray spectrum and derive the spin parameter. They also accounted for the effects of the companion star's wind on the accretion disk, which can influence the X-ray continuum. In a later study, Steiner et al. [9] utilized data from the RXTE and Chandra X-ray observatories to estimate the spin parameter of LMC X-1 [9]. They found a spin parameter of a = 0.97 ± 0.01, consistent with the previous XMM-Newton-based estimate. The authors used an updated version of the KERRBB model, which incorporated improvements in the treatment of radiative transfer and relativistic effects. More recently, [10] and [11] analyzed data from the NICER X-ray observatory to refine the spin parameter estimation for LMC X-1 . They reported a spin parameter of a $\sim$ 0.92–0.95 and a $\sim$ 0.90 ± 0.01, in agreement with the earlier studies. The authors also discussed the importance of accounting for the effects of the companion star's wind and the limitations of the continuum fitting method in some cases.

19.2.1.2 LMC X-3

LMC X-3 is another high-mass X-ray binary system located in the Large Magellanic Cloud. The black hole in this system has a mass of approximately 7 solar masses and is orbited by a massive, early-type companion star [12]. Similar to LMC X-1, several studies have attempted to measure the spin parameter of LMC X-3 using the continuum fitting method. One notable study, conducted by Davis et al. [13] using XMM-Newton data, reported a spin parameter of a = 0.25 ± 0.16 for LMC X-3 [13]. The authors used the KERRBB model to fit the X-ray spectrum and derive the spin parameter, taking into account the effects of the companion star's wind. In a more recent study, Steiner et al. [14] analyzed data from the NICER X-ray observatory to provide a refined spin parameter estimate for LMC X-3 [14]. They reported a spin parameter of a = 0.30 ± 0.01, consistent with the earlier XMM-Newton-based estimate. The authors discussed the importance of using a robust model for the accretion disk and the potential influence of the companion star's wind on the observed X-ray continuum. More recently, using the data of the X-ray

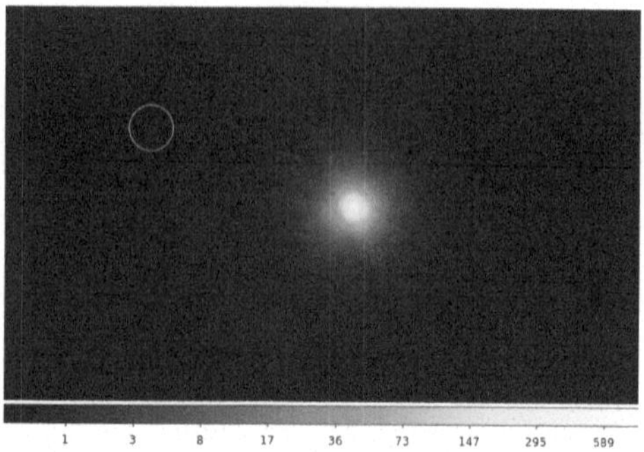

Fig. 19.1 LMC X-1 for Obs ID L7Nu 'source', 'background'

Mission NuStar and Swift with combined spectra in the 0.5–30 keV energy range, [10] reported spin of LMC X-3 a ~ 0.19 to 0.29 using KERRBB model with reflection component.

19.2.1.3 GX 339-4

GX 339-4 is a well-known X-ray binary system that has been extensively studied. The black hole in this system has a mass of approximately 9 solar masses and is believed to be in a low-mass X-ray binary system with a less massive companion star [15]. The spin parameter of GX 339-4 has been estimated using the continuum fitting method by several research groups. One study, conducted by Miller et al. (2004) using XMM-Newton data, reported a spin parameter of a = 0.94 $\pm$ 0.02 for GX 339-4 [16]. The authors used the KERRBB model to fit the X-ray spectrum and derive the spin parameter, taking into account the effects of the disk's inner radius. In a more recent study, Garcîa et al. [5] analyzed data from the NuSTAR X-ray observatory to provide a refined spin parameter estimate for GX 339-4 [5]. They reported a spin parameter of a = 0.95 $\pm$ 0.03, higher than the previous XMM-Newton-based estimate. The authors discussed the importance of using a more comprehensive spectral model that includes the reflection component and the potential influence of the companion star's properties on the X-ray continuum. Further, the works of [17] indicates reported the spin value of the $a \sim 0.998$ where they have used reflection model for their work.

Description NuSTAR images of the sources LMC X-1 [Figs. 19.1 and 19.2] and LMC X-3 [Figs. 19.3 and 19.4] obtained using FPMA module are shown respectively in the up and down panels. These two images represent observations during 2015-09-30 and 2015-12-29 respectively. We consider a circular source and background regions of 60 arcsec for both LMC X-1 and LMC X-3 as marked in the images [Figs. 19.1 and 19.3] (green one for source, yellow one for background) from which the spectrum is extracted.

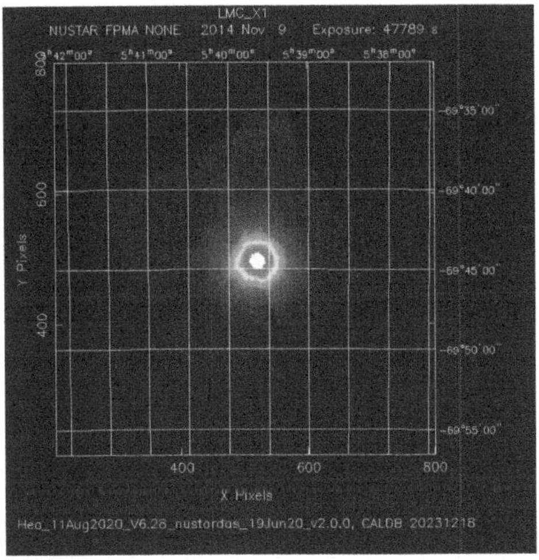

Fig. 19.2 LMC X-1 of Obs ID L7Nu RA, DEC information

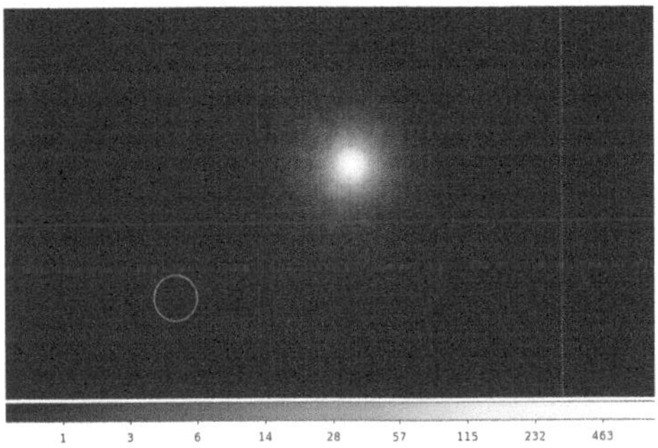

Fig. 19.3 LMC X-3 for Obs ID 15Nu 'source', 'background'

19.2.2 Observations and Data Reduction

Data acquisition was performed from the archives of XMM-Newton, NICER, and NuSTAR. We focused on observations with the highest quality and exposure times. For LMC X-1, data from XMM-Newton's EPIC cameras were utilized alongside NICER Timing Analysis and NuSTAR's hard X-ray observations. These multiband datasets allow for a comprehensive analysis of the spectral properties of the respective black holes.

LMC X-1 observations include three epochs with NICER (2018–2022), two epochs with XMM-Newton (2000–2002), and two epochs with NuSTAR (2014–2016). So, we have suitably chosen a total of 7 observations. LMC X-3 observations include two epochs with NICER (2017–2019), two epochs with XMM-Newton (2011), and one epochs with NuSTAR (2015). So, we have suitably chosen a total of 5 observations. Observations of GX339-4 consist of seven selected epochs: two from NICER (2021), three from XMM-Newton (2015), and two from NuSTAR (2015). The detailed of the observations are presented in Table 19.2. Once the raw

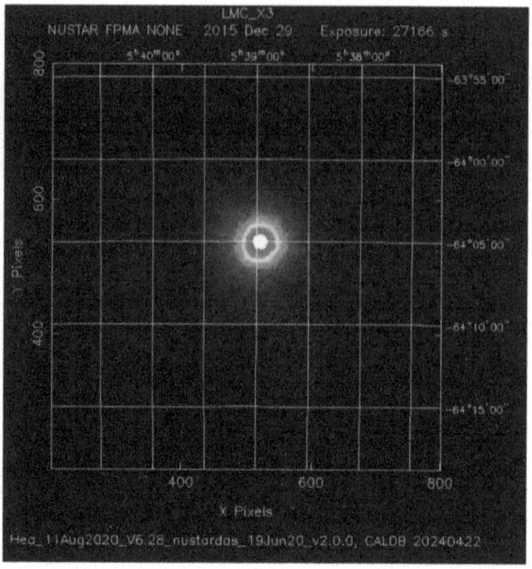

Fig. 19.4 LMC X-3 of Obs ID I5Nu RA, DEC information

Table 19.2 Log of observation of LMC X-1, LMC X-3 and GX339-4

SOURCES	Mission	ID	Obs ID	Date	Exposure time(s)
LMC X-1	NICER	L1Ni	1100070102	2018-01-11	3326
		L2Ni	5100070103	2022-10-21	3240
		L3Ni	5100070105	2022-10-23	3946
	XMM	L4X	0023940401	2002-09-26	39973
		L5X	0112900101	2000-10-21	7259
	NuSTAR	L6Nu	30201029002	2016-05-11	40223
		L7Nu	30001143002	2014-11-19	47789
LMC X-3	NICER	I1Ni	1101010107	2017-10-27	10761
		I2Ni	2101010101	2019-04-18	11978
	XMM	I3X	0671420301	2011-05-27	12913
		I4X	0671420501	2011-11-02	11913
	NuSTAR	I5Nu	30101052002	2015-12-29	27166
GX339-4	NICER	G1Ni	3558010903	2021-03-08	10998
		G2Ni	4133010110	2021-04-02	12205
	XMM	G3X	0760646401	2015-09-07	23700
		G4X	0760646501	2015-09-12	22114
		G5X	0760646701	2015-09-30	49200
	NuSTAR	G6Nu	80102011010	2015-09-17	38502
		G7Nu	80102011012	2015-09-30	41334

data sets are obtained, the data reduction was performed using HEASoft v.6.28. Below are the particulars of each mission.

Data reduction involved processing the raw data through standard pipelines specific to each mission. XMM-Newton data were analyzed using the Science Analysis Software (SAS) along with SAS-Thread "xmmextractor", where primary

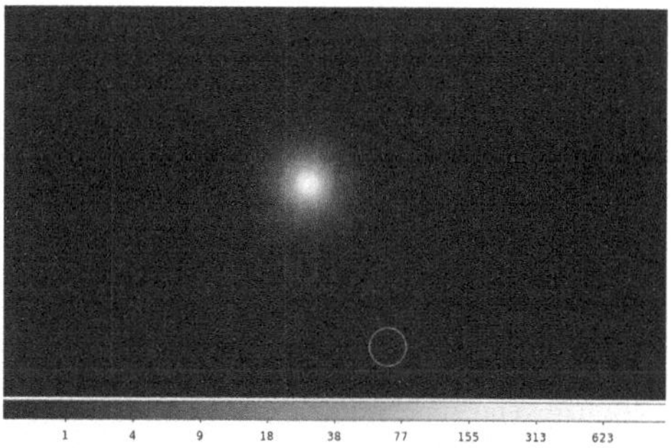

Fig. 19.5 GX339-4 for Obs ID G6Nu 'source', 'background'

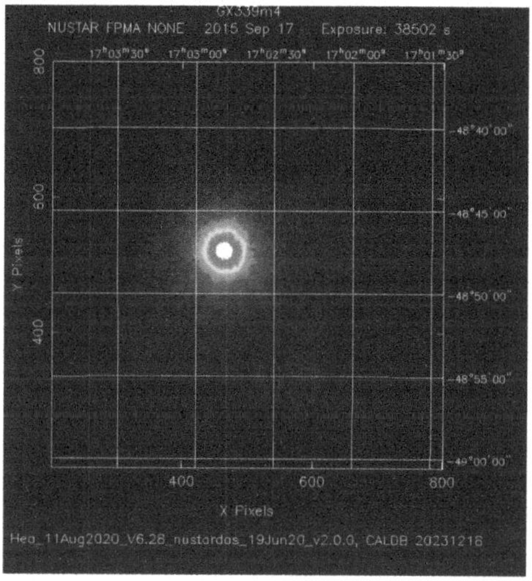

Fig. 19.6 GX339-4 of Obs ID G6Nu RA, DEC information

steps included filtering event lists to exclude bad pixels and source contamination, resulting in high-quality spectra. NICER data were processed with NICERDAS via the successive commands "*nicerl2*" to handle initial data reduction and calibration and "*nicerl3-spect*" with the specified object ID and 3C50 background model type, producing cleaned and calibrated spectral data. Data from NuSTAR were reduced using HEASOFT, ensuring proper calibration and background subtraction. We obtained source products by selecting circular regions with a radius of 60 arcseconds centered on the source coordinates. Additionally, a background region with the same radius was chosen at a distance from the source position Fig. 19.1. Then we extracted the spectra and light curves using the nuproduct task.

Description Figures 19.5 and 19.6: NuSTAR images of the sources GX339-4 obtained using FPMA module is shown in the above panels. This images represent observations during 2015-09-17. Here We also consider same circular source and background regions of 60 arcsec for this X-ray binary source as marked in the images (green one for source , yellow one for background) from which the spectrum and light curve are extracted.

19.2.3 Spectral Analysis and Data Fitting

We have conducted spectral analysis by fitting a model to the X-ray spectra of two extra galactic sources LMC X-1 and LMC X-3 and one galactic source GX339-4. To analyze the spectra, we primarily need four files: source, background, RMF and ARF files, the source and background files includes the count of photons detected in each energy channel and the response files (RMF and ARF generally with the extension .rmf and .arf), required to convolve the model with the detector's response. The spectral analysis is done using XSPEC v(12.11.1) which is a standard software for analysing the X-ray spectra, is a part of the High Energy Astrophysics (HEAsoft–Xanadu) software suite. Further, we have re-binned the data for better statistic and combine the files using GRPPHA, wherever required.

We carried out spectral analysis of the different energy range according to their background noise. The spectral analysis was done using HEASARC's spectral anaysis software package XSPEC v(12.11.1). In our analysis, we used **diskbb** [18] and **kerrbb** [19] models for the thermal component, **po** (powerlaw) model for the Comptonized component and occasionally **ga** (Gaussian) model for spectral line emission of different type.We also used **const** (constant) model only for NuSTAR data. The constant factor is included as the cross-normalization factor between the two modules FPMA and FPMB of NuSTAR. This method addresses any potential calibration discrepancies or systematic uncertainties between the two detectors. The cross-normalization factor for FPMA was fixed at 1 while it was allowed to vary for FPMB. We also used **Gaussian** model and **Tbabs** for the interstellar absorption in the spectral fitting.We carried out spectral analysis of the data in 0.5–10 keV range fot XMM and 0.3–10 keV for NICER observations of three sources. As LMC X-1 and LMC X-3 are weak in hard X-rays and typically beyond 30 keV background signals dominate, our spectral analysis is primarily constrained to the 3–30 keV range for NuStar and for GX339-4, it extends upto 45 keV.

Firstly, we employed a model that includes absorbed thermal and Comptonized emission components. The thermal component was modeled with the **diskbb** and the Comptonized emission was modeled with **po**. We refer this as Model-1. We continue the fitting of the data by changing the parameters and adding new components till we obtain physical acceptable model which is acceptable. Next, to estimate the black hole's spin, we have used the **kerrbb** model, which represents a relativistic thin accretion disk, for the thermal emission. Additionally, we combined it with the **po** model once more. We refer to this model as Model-2. We kept some of parameters like M_{Bh}(mass), i(inclination angle) and D_{Bh} (distance) of the black holes of **kerrbb** model fixed.

The continuum fitting process involved modeling the spectrum of black hole emissions. A multi-component model comprising cold disk emission, Comptonization etc was applied to the combined X-ray spectra. We utilized the Xspec software package to perform $\tilde{\chi}^2$ fitting, varying the spin parameters systematically to achieve a minimal $\tilde{\chi}^2$ statistic. The best-fit parameters were evaluated through Monte Carlo simulations to assess statistical uncertainties associated with the spin estimates.

The parameters obtained from fitting data with Model-1 and Model-2 are given in the Tables 19.3, 19.4, and 19.5 for LMC X-1, LMC X-3 and GX339-4, respectively. And, the folded spectrum of one of the Obs ID of each BHXB sources of each mission i.e. of LMC X-1 (L1Ni, L4X, L7Nu), LMC X-3 (l2Ni, l3X, l5Nu) and GX339-4 (G2Ni, G3X, G6Nu) are shown in Figs. 19.7, 19.8, 19.9, 19.10, 19.11, 19.12, 19.13, 19.14, and 19.15 for Model 1 and Figs. 19.16, 19.17, 19.18, 19.19, 19.20, 19.21, 19.22, 19.23, and 19.24 for Model 2, respectively.

19.3 Results and Discussion

LMC X-1 We obtained a satisfactory fit applying Model-1 to the spectrum of LMC X-1 across all observation epochs [Table 19.3]. The hydrogen column density was found to be $N_H = 0.74 - 1.3 \times 10^{22}$ cm^{-2}. The spectral analysis revealed that the inner disk temperature varies (T_{in}) in the range of 0.78–1.02 keV and the photon index, $\Gamma \sim 2.5$–4.5. ($T_{in} \sim 0.85$–0.90 keV for NICER, 0.78–0.88 keV for XMM, 0.88–1.02 keV for NuSTAR and $\Gamma \sim 2.48$–4.32 for NICER, 2.73–4.54 for XMM, 3.12–3.23 for NuSTAR). We detected a narrow Gaussian line at 0.96 keV in two of our observation Ids (L2Ni, L4X) haing a width of 0.04 and 0.17 keV respectively. The X-ray emission line at 0.96 keV corresponds to the emission from the hydrogen-like Ne X (neon) ion. By applying Model-2 to fit the data, we obtained accretion rate $\dot{M}$, spin of the black hole (a) .The mass accretion rate was found to be $\dot{M} \sim 0.28$–0.72×10^{18} gm s^{-1}. We found that LMC X-1 is a high spinning BH with spin a $\sim$ 0.99. We fixed spectral hardening factor at κ at 1.5 [10]. We plot the folded spectrum and delta chi(delchi) against energy in keV for Obs IDs L1Ni, L4X, L7Nu of LMC X-1 in Figs. 19.7, 19.8, and 19.9 for Model 1 and in Figs. 19.16, 19.17, and 19.18 for Model 2, respectively.

LMC X-3 We began the spectral analysis of LMC X-3 using Model-1. We used the hydrogen column density as $N_H = 0.045 \times 10^{22}$ cm^{-2}. (information collected from NED—NASA/IPAC Extragalactic Database). The spectral analysis showed that the inner disk temperature (T_{in}) ranges from 0.95 to 1.17 keV, and the photon index Γ is approximately between 1.81 and 2.85. ($T_{in} \sim 1.12$–1.21 keV for NICER, 0.95–1.17 keV for XMM, 1.15 keV for NuSTAR and $\Gamma \sim 1.81$–1.98 for NICER, 2.67–2.85 for XMM, 2.68 for NuSTAR). In two of our observation IDs (l3X, l4X), we identified a Gaussian line at 0.76 keV with widths of 0.31 and 0.85 keV with width of 0.31 keV respectively [Table 19.4]. The X-ray emission line at 0.76 keV corresponds to the emission from the hydrogen-like O VII (Oxygen VII) triplet and 0.85 keV corresponds to O VIII (Oxygen VIII) Ly- alpha transition. By using Model-2 to to fit the data, we determined the accretion rate $\dot{M}$, spin of the black hole (a). The mass accretion rate was found to be $\dot{M} \sim 3.47$–9.39×10^{18} gm s^{-1}. We found that LMC X-1 is a low spinning BH with spin 'a' $\sim$ 0.24–0.30. (a $\sim$ 0.29–0.30 for NICER, 0.24–0.30 for XMM, 0.29 for NuSTAR). We fixed spectral hardening factor at κ 1.7 [10]. We plot the folded spectrum and delta chi(delchi) against energy in keV

Table 19.3 Best-fitted spectral parameters of LMC X–1

*Model 1: [const*tbabs*(po+diskbb+ga)]*

ID	Const	N_H	T_{in}	N_{DBB}	Γ	N_{PO}	LineE	$\frac{\chi^2}{dof}$	
		(10^{22} cm^{-2})	(keV)						
L1Ni	–	1.08 ± 0.01	0.87 ± 0.008	52.50 ± 2.23	2.81 ± 0.03	0.24 ± 0.01	–	653.47/795	
L2Ni	–	1.30 ± 0.05	0.90 ± 0.009	59.14 ± 4.43	4.32 ± 0.27	0.35 ± 0.05	0.96 ± 0.01	152.18/143	
L3Ni	–	0.99 ± 0.01	0.85 ± 0.009	57.71 ± 3.30	2.48 ± 0.04	0.18 ± 0.01	–	627.55/964	
L4X	–	0.74 ± 0.007	0.78 ± 0.005	37.72 ± 1.67	2.73 ± 0.02	0.18 ± 0.007	0.96	2077.76/1837	
L5X	–	0.92 ± 0.01	0.88 ± 0.002	49.24 ± 0.98	4.54 ± 0.08	0.45 ± 0.02	–	2081.75/1839	
L6Nu	0.97 ± 0.003	1.1	0.88 ± 0.003	44.85 ± 1.70	3.12 ± 0.02	0.28 ± 0.01	–	1639.17/1343	
L7Nu	1.02 ± 0.003	1.1	1.02 ± 0.004	19.08 ± 0.62	3.23 ± 0.02	0.47 ± 0.02	–	1648.99/1343	

*Model 2: [const*tbabs*(po+kerrbb+ga)]*

ID	const	N_H	a	$\dot{M}$	N_{kerrbb}	Γ	N_{PO}	LineE	$\frac{\chi^2}{dof}$
		(10^{22} cm^{-2})		(10^{18} $gm\ s^{-1}$)					
L1Ni	–	1.03 ± 0.01	0.99 ± 0.04	0.49 ± 0.43	1.57 ± 0.57	2.73 ± 0.07	0.16 ± 0.01	–	661.65/794
L2Ni	–	1.38 ± 0.07	0.99 ± 0.05	0.49 ± 0.47	1.89 ± 0.69	4.99 ± 0.38	0.38 ± 0.07	0.96 ± 0.01	165.68/142
L3Ni	–	0.88 ± 0.02	0.99 ± 0.08	0.43 ± 0.75	1.98 ± 1.55	1.97 ± 0.18	0.05 ± 0.02	–	655.56/963
L4X	–	0.68 ± 0.02	0.99 ± 0.04	0.28 ± 0.21	1.40 ± 0.32	2.53 ± 0.08	0.12 ± 0.02	0.96	2082.78/1836
L5X	–	0.92 ± 0.01	0.99 ± 0.01	0.39 ± 0.06	2.07 ± 0.08	5.81 ± 0.16	0.79 ± 0.07	–	2080.68/1838
L6Nu	0.97 ± 0.003	1.1	0.99 ± 0.02	0.39 ± 0.12	1.7 ± 0.12	2.99 ± 0.05	0.20 ± 0.02	–	1574/1342
L7Nu	1.02 ± 0.003	1.1	0.99 ± 0.02	0.72 ± 0.31	0.77 ± 0.06	3.10 ± 0.06	0.33 ± 0.05	–	1617.44/1342

Kerr model fitting done taking i=36.38°, $M_{Bh} = 10.00M_{\odot}$, $D_{Bh} = 48.00$ kpc, hd=1.5 (for LMC X-1) as frozen parameters [10, 20]

Table 19.4 Best-fitted spectral parameters of LMC X–3

*Model 1: [const*tbabs*(po+diskbb+ga)]*									
ID	const	N_H	T_{in}	N_{DBB}	Γ	N_{PO}	LineE	$\frac{\chi^2}{dof}$	
		$(10^{22}\ cm^{-2})$	(keV)						
11Ni	–	0.045	1.12 ± 0.003	28.45 ± 0.23	1.81 ± 0.01	0.02 ± 0.0005	–	800.12/965	
12Ni	–	0.02 ± 0.005	1.21 ± 0.0047	24.55 ± 0.39	1.98 ± 0.12	0.01 ± 0.004	–	819.12/944	
13X	–	0.02 ± 0.05	1.17 ± 0.006	18.45 ± 0.41	2.67 ± 0.32	0.02 ± 0.01	0.76 ± 0.23	1682.91/1644	
14X	–	0.045	0.95 ± 0.004	19.13 ± 0.44	2.85 ± 0.08	0.02 ± 0.003	0.85 ± 0.16	1784.38/1637	
15Nu	0.003	0.045	1.15 ± 0.002	25.53 ± 0.27	2.68 ± 0.09	0.03 ± 0.009	–	1272.17/1343	
*Model 2: [const*tbabs*(po+kerrbb+ga)]*									
ID	const	N_H	a	$\dot{M}$	N_{kerrbb}	Γ	N_{PO}	LineE	$\frac{\chi^2}{dof}$
		$(10^{22}\ cm^{-2})$		$(10^{18}\ gm\ s^{-1})$					
11Ni	–	0.045	0.29 ± 0.19	7.76 ± 3.74	0.72 ± 0.24	2.21 ± 0.28	0.007 ± 0.002	–	719.05/964
12Ni	–	0.05	0.30 ± 0.17	9.39 ± 3.98	0.66 ± 0.19	2.38 ± 0.12	$0.02 = 0.002$	–	1319.66/963
13X	–	0.12 ± 0.05	0.30 ± 0.13	6.87 ± 2.51	0.64 ± 0.16	5.9 ± 1.9	0.02 ± 0.02	0.76	1722.7/1636
14X	–	0.045	0.24 ± 0.33	3.47 ± 2.7	0.56 ± 0.31	3.20 ± 0.14	0.02 ± 0.002	–	1717.72/1439
15Nu	0.003	0.045	0.29 ± 0.26	5.79 ± 4.15	1.01 ± 0.52	1.89 ± 0.18	0.003 ± 0.002	–	1324.46/1342

Kerr model fitting done taking i=67.00°, $M_{Bh} = 7.00 M_{\odot}$. $D_{Bh} = 48.00$ kpc, hd=1.7 (for LMC X-3) as frozen parameters [10, 21, 22]

Table 19.5 Best-fitted spectral parameters of GX339-4

*Model 1: [const*tbabs*(po+diskbb+ga)]*

ID	N_H	T_{in}	N_{DBB}	Γ	N_{PO}	LineE	$\frac{\chi^2}{dof}$	
	$(10^{22}\ cm^{-2})$	(keV)						
G1Ni	0.52 ± 0.007	0.26 ± 0.007	10000 ± 3016	1.74 ± 0.01	0.81 ± 0.01	6.59 ± 0.05	112.73/154	
G2Ni	0.64 ± 0.03	0.79 ± 0.002	2223.94 ± 75.05	3.07 ± 0.06	4.98 ± 0.74	6.31 ± 0.08	71.23/151	
G3X	0.26 ± 0.01	0.43 ± 0.03	61.64 ± 26.35	1.55 ± 0.01	0.07 ± 0.01	1.04 ± 0.01	1999.05/1836	
G4X	0.25 ± 0.01	0.49 ± 0.05	20.28 ± 10.68	1.52 ± 0.01	0.05 ± 0.001	1.07 ± 0.02	2120.01/1836	
G5X	0.35 ± 0.005	1.03 ± 0.09	0.22 ± 0.13	1.50 ± 0.01	0.02 ± 0.0007	—	2147.85/1859	
G6Nu	0.4	1.39 ± 0.05	0.56 ± 0.08	1.51 ± 0.007	0.04 ± 0.0009	6.48 ± 0.71	2216.92/2090	
G7Nu	0.4	0.72 ± 0.09	5.13 ± 4.02	1.56 ± 0.007	0.02 ± 0.0005	6.21 ± 0.09	1834.15/1740	

*Model 2: [const*tbabs*(po+kerrbb+ga)]*

ID	N_H	a	$\dot{M}$	N_{kerrbb}	Γ	N_{PO}	LineE	$\frac{\chi^2}{dof}$
	$(10^{22}\ cm^{-2})$		$(10^{18}\ gm\ s^{-1})$					
G1Ni	0.54 ± 0.007	0.997 ± 0.004	0.0020	34.81 ± 4.55	1.73 ± 0.01	0.80 ± 0.01	6.59 ± 0.05	110.72/154
G2Ni	0.51 ± 0.01	0.998 ± 0.001	0.1795	4.00 ± 0.06	2.06 ± 0.35	0.48 ± 0.38	6.43 ± 0.08	88.74/151
G3X	0.27 ± 0.01	$0.998 \pm .0.016$	0.0135	0.12 ± 0.03	1.55 ± 0.01	$.0.07 \pm 0.001$	1.04 ± 0.01	1999.25/1836
G4X	0.26 ± 0.01	0.998 ± 0.020	0.0253	0.03 ± 0.01	1.51 ± 0.01	0.05 ± 0.001	1.03 ± 0.03	2052.59/1836
G5X	0.35 ± 0.005	0.998 ± 0.022	1.0114	0.0001 ± 0.0001	1.50 ± 0.02	0.02 ± 0.0009	—	2147.65/1854
G6Nu	0.4	0.998 ± 0.01	0.788	0.002 ± 0.0003	1.52 ± 0.006	0.04 ± 0.0008	6.5 ± 0.04	2151.98/2090
G7Nu	0.4	0.997 ± 0.02	0.106737	0.009 ± 0.005	1.56 ± 0.007	0.02 ± 0.0005	6.22 ± 0.09	1834.01/1740

The value of constant factor for NuSTAR ID are 0.99 ± 0.002 and 0.98 ± 0.003 for G6Nu and G7Nu respectively. Kerr model fitting done taking i=39.2°, $M_{Bh} = 10.00 M_{\odot}$, $D_{Bh} = 8.00$ kpc, hd=1.7 (for GX339-4) as frozen parameters [17, 23]

Fig. 19.7 L1Ni

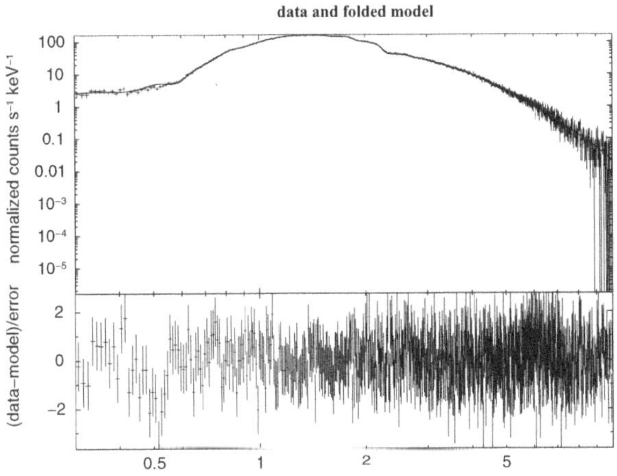

Fig. 19.8 L4X

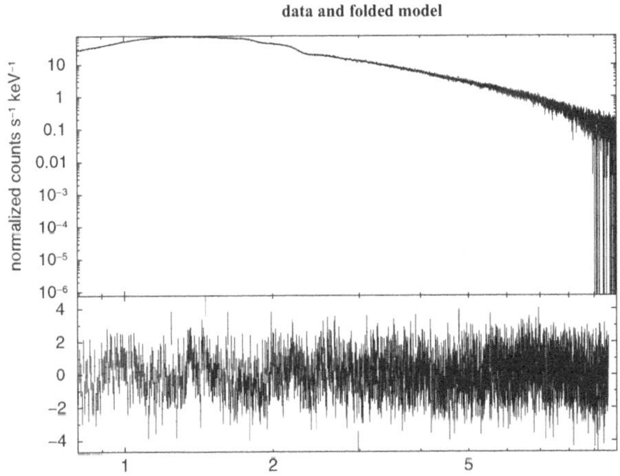

for Obs IDs l2Ni, l3X, l5Nu of LMC X-3 in Figs. 19.10, 19.11, and 19.12 for Model 1 and in Figs. 19.19, 19.20, and 19.21 for Model 2, respectively.

Description Figures 19.7, 19.8, 19.9, 19.10, 19.11, 19.12, 19.13, 19.14, and 19.15: Broadband spectral fitting was performed for the Obs IDs listed below each figure for Model1, covering three sources—LMC X-1, LMC X-3, and GX339-4—across three missions: NICER, XMM-Newton, and NuSTAR. The first, second, and third rows correspond to LMC X-1, LMC X-3, and GX339-4, respectively, while the first, second, and third columns represent the NICER, XMM-Newton, and NuSTAR missions in different energy corresponding to different mission mentioned in Sect. 19.2.3. Here the folded models (which represent the expected counts from the best-fit model convolved with the instrument response) are shown along with the "delchi" plot, which shows the residuals of your spectral fit against energy in keV.

Fig. 19.9 L7Nu

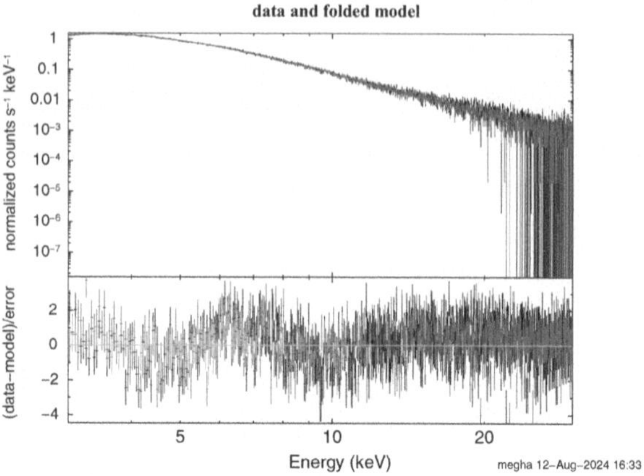

Fig. 19.10 l2Ni

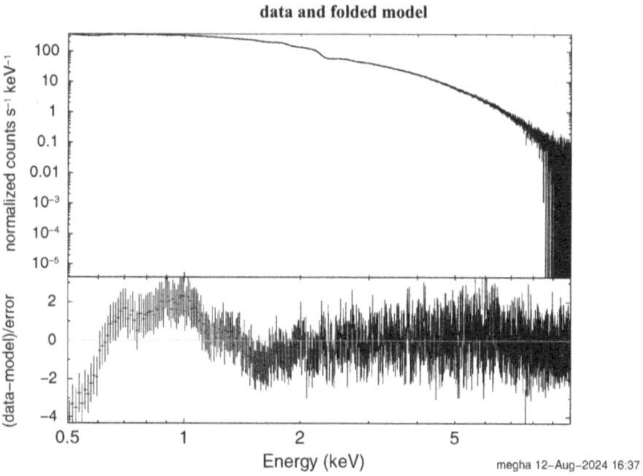

Fig. 19.11 l3X

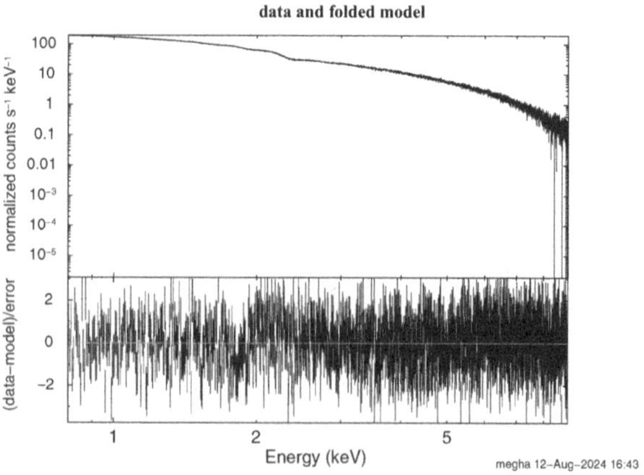

Fig. 19.12 I5Nu

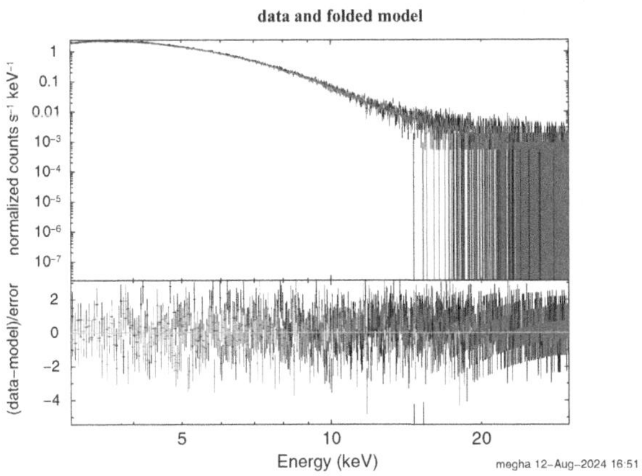

Fig. 19.13 G2Ni

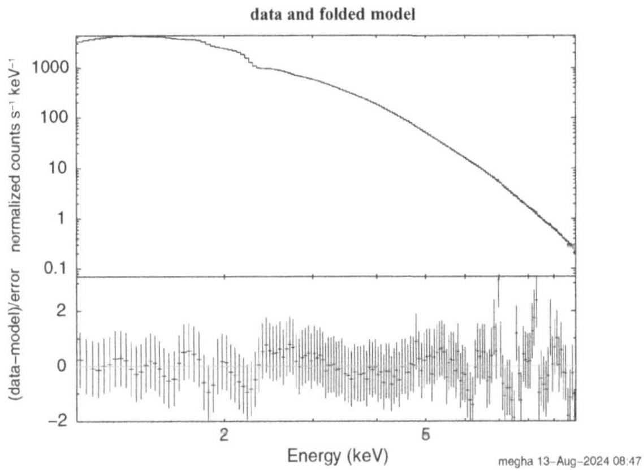

Fig. 19.14 G3X

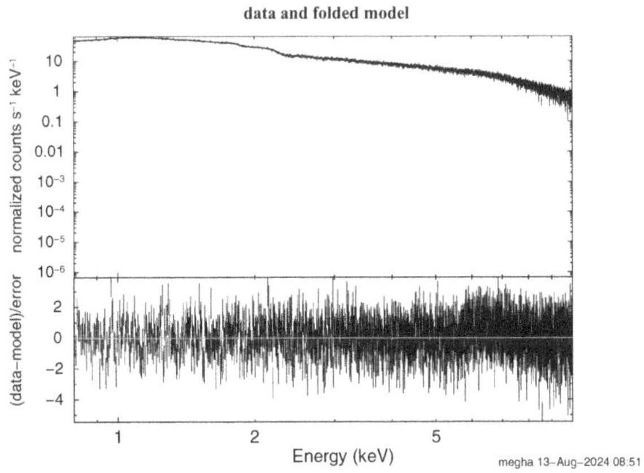

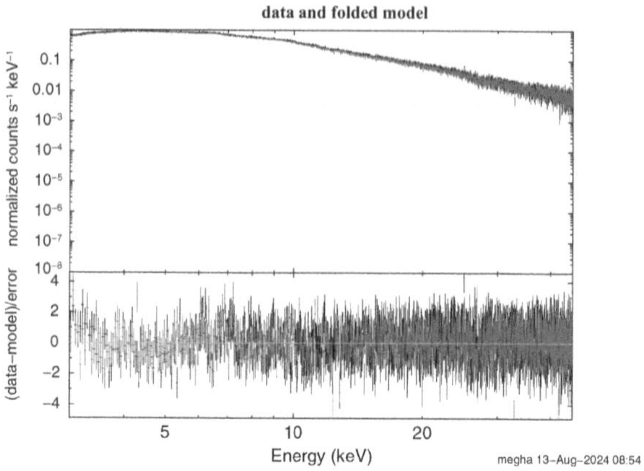

Fig. 19.15 G6Nu

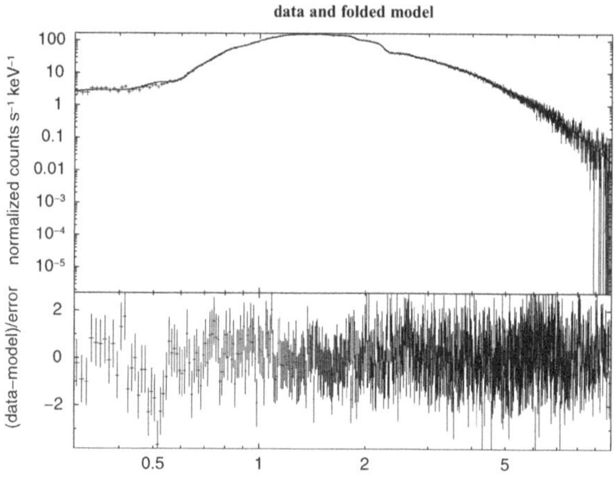

Fig. 19.16 L1Ni

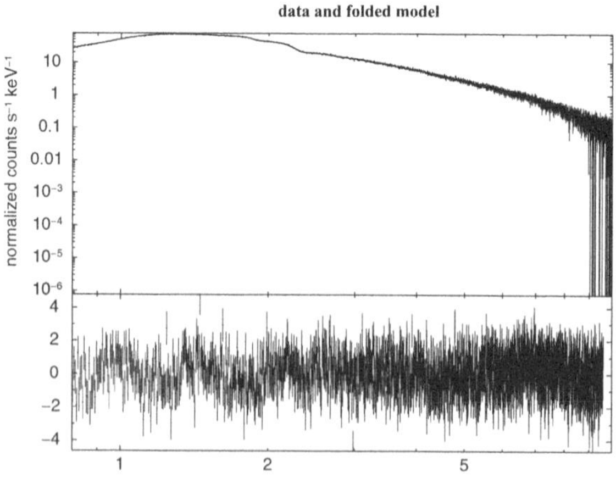

Fig. 19.17 L4X

Fig. 19.18 L7Nu

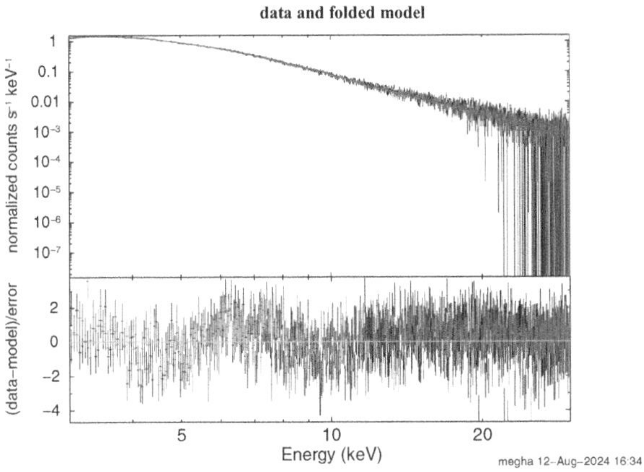

Fig. 19.19 l2Ni

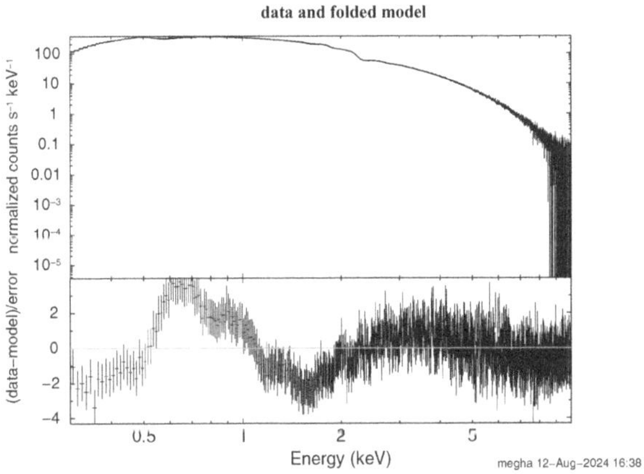

Fig. 19.20 l3X

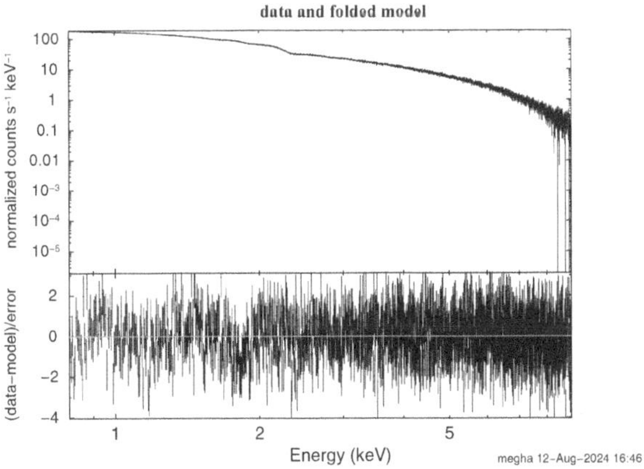

Fig. 19.21 I5Nu

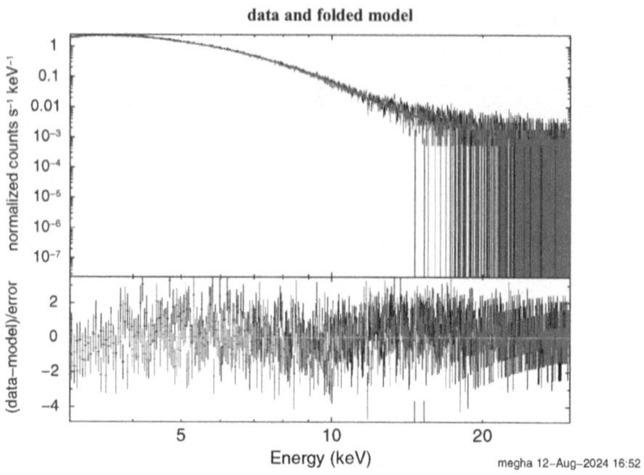

Fig. 19.22 G2Ni

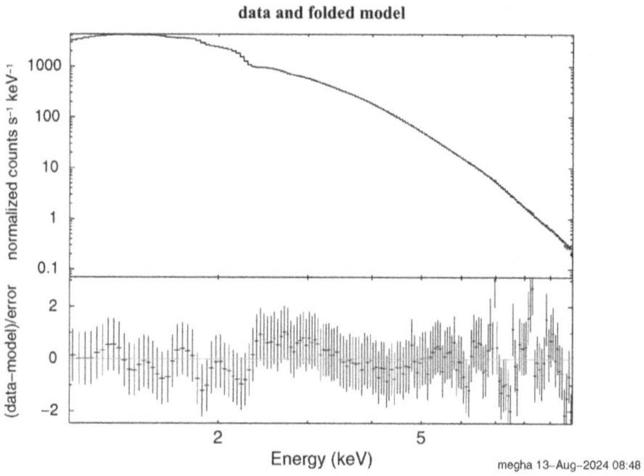

Fig. 19.23 G3X

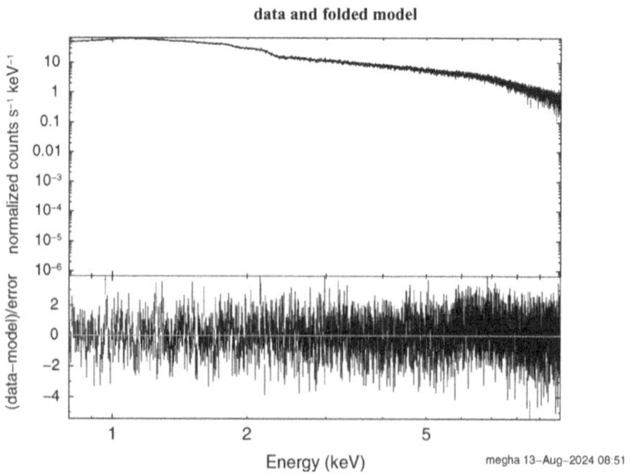

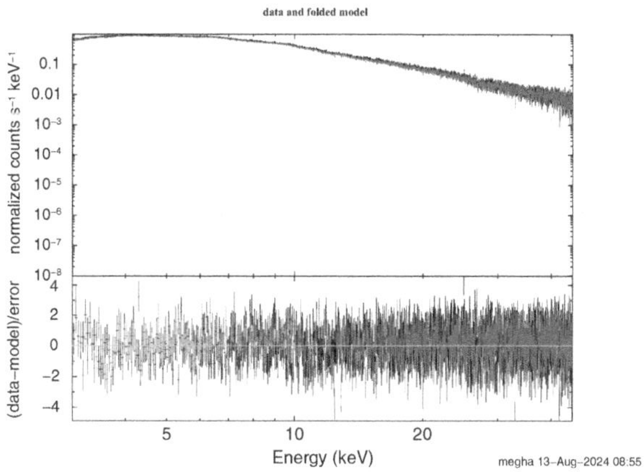

Fig. 19.24 G6Nu

GX339-4 We have also done spectral analysis of GX339-4 using both of the models Model-1 and Model-2. After fitting with Model-1, we got hydrogen column density as $N_H = 0.25–0.64 \times 10^{22}$ cm^{-2}. For Nu STAR data, we fixed it at 0.4×10^{22} cm^{-2}. The spectral analysis revealed that the inner disk temperature (T_{in}) spans from 0.26 to 1.39 keV, while the photon index, Γ varying roughly between 1.50 and 3.07 for model-1. An Iron line is clearly visible in NICER and NuSTAR observation while an Neon line is observed in XMM data. The lines are slightly shifted from its actual value. These effects can include relativistic broadening, Doppler shifts from the motion of material in the accretion disk, and other physical processes related to the geometry and dynamics near the black hole. When Model-2 is applied to fit the data, we got the information about its spin ranging from 0.997–0.998, similar to that used in the chapter by Wang-Ji et al. [17]. During the model fitting, we initially fixed the spin at 0.998 [17] and observed that the mass accretion rate had a satisfactory value, with all other parameters also falling within an acceptable range with minimal error[Table 19.5]. Subsequently, we fixed the value of the mass accretion rate at that level and unfroze the spin parameter, then listed the resulting values. Here We also plotted the folded spectrum and delta chi(delchi) against energy in keV for Obs IDs G2Ni, G3X, G6Nu of GX339-4 in Figs. 19.13, 19.14, and 19.15 for Model 1 and in Figs. 19.22, 19.23, and 19.24 for Model 2 , respectively.

Description Figures 19.16, 19.17, 19.18, 19.19, 19.20, 19.21, 19.22, 19.23, and 19.24: Broadband spectral fitting was performed for the Obs IDs listed below each figure for Model2, covering three sources—LMC X-1, LMC X-3, and GX339-4—across three missions: NICER, XMM-Newton, and NuSTAR. The first, second, and third rows correspond to LMC X-1, LMC X-3, and GX339-4,respectively, while the first, second, and third columns represent the NICER, XMM-Newton, and NuSTAR missions in different energy corresponding to different mission mentioned in Sect. 19.2.3. Here the folded models (which represent the expected counts from the best-fit model convolved with the instrument response) are shown along with the "delchi" plot, which shows the residuals of your spectral fit against energy in keV.

Comparison of Result with Other Group In Sect. 19.2.1, we have given a brief review of the works done by other groups with the X-ray sources: LMC X-1, LMC X-3 and GX 339-4, where most of the groups have reported LMC X-1 (a $\sim$ 0.9–0.97)and GX 339-4 (a > 0.94) as high spinning X-ray black hole binaries and LMC X-3 (a $\sim$ 0.25–0.30) as Low spinning X-ray black hole binaries. We faced many problems constraining the mass accretion rate for GX 339-4 to some physically acceptable value while determining the spin. So, we fixed the mass accretion rate to some physically acceptable value and then fitted the 'Kerrbb' model to obtain the spin. Our estimated spin for GX 339-4 matches well with the works of [17].

19.4 Conclusions

The spin parameter estimations for LMC X-1, LMC X-3, and GX 339-4 provide valuable insights into the formation and evolution of these stellar mass black holes. The high spin parameters measured for LMC X-1 and GX 339-4 (a > 0.95) suggest that these black holes may have been formed through the collapse of a rapidly rotating stellar progenitor or have undergone significant mass and angular momentum accretion during their lifetime [24]. In contrast, the lower spin parameter measured for LMC X-3 (a $\approx$ 0.3) may indicate a different formation history or accretion process. The discrepancies between the spin parameter estimates for GX 339-4 highlight the importance of using robust spectral models and accounting for various physical effects, such as the presence of the companion star and the reflection component, when applying the continuum fitting method. This underscores the need for continued refinement of the continuum fitting technique and a better understanding of the underlying assumptions and uncertainties. In conclusion, the spin parameter estimations of LMC X-1, LMC X-3, and GX 339-4 using the continuum fitting method have provided valuable insights into the astrophysics of stellar mass black holes. The ongoing efforts to improve the accuracy and reliability of these measurements, as well as the incorporation of additional observational constraints, will further our understanding of the formation, evolution, and accretion processes of these compact objects.

19.5 Limitation of Kerrbb and Future Scope of Work

The Kerrbb model, frequently used to estimate black hole spin, assumes a geometrically thin, steady-state disk and does not account for factors like magnetic fields, disk winds, or Comptonization effects. It simplifies relativistic effects, including ray tracing and limb darkening [19], and typically assumes the inner disk radius is at the ISCO. These assumptions can significantly influence the emitted spectrum. Additionally, Kerrbb is highly sensitive to precise measurements of black hole mass, distance, and inclination angle, with errors in these parameters potentially leading to substantial inaccuracies in spin estimation [25].

Recent studies suggest a considerable portion of the disk emission may originate from ISCO region. The Kerrbb model, based on the Novikov-Thorne thin-disk model, generally cuts off at the ISCO, potentially missing crucial contributions to the observed spectra [25].

Another limitation involves the model's treatment of the ISCO stress. The Kerrbb model typically assumes a simplistic approach where the ISCO stress is either zero or minimal. However, simulations and extended models suggest that there could be significant stress at the ISCO, influencing the dynamics and emission properties of the accretion disk. This oversimplification can lead to inaccuracies in the estimation of black hole parameters, such as spin [25].

So far, our analysis covers three sources from three missions using two models. Future work will explore additional Comptonization models such as Simpl, CompTT, Comptb, and Thcomp, as Only Kerrbb does not fully address Comptonization effects. We also plan to incorporate data from missions like AstroSat, Suzaku, and Swift. While this chapter focuses on spectral analysis, our next steps will emphasize timing analysis, which is crucial, and we will investigate spin estimation through iron line emission and X-ray spectropolarimetric analysis.

Acknowledgments TS like to thank IUCAA, Pune, India for the hospitality extended to him under their Visiting Associateship program.

Competing Interests The authors have no conflicts of interest to declare that are relevant to the content of this chapter.

References

1. Ghosh, S., Sarkar, T., & Bhadra, A. (2014). Newtonian analogue of corresponding space–time dynamics of rotating black holes: Implication for black hole accretion. *Monthly Notices of the Royal Astronomical Society, 445*(4), 4460–4476.
2. Shafee, R., McClintock, J. E., Narayan, R., Davis, S. W., Li, L.-X., & Remillard, R. A. (2006). Estimating the spin of stellar-mass black holes by spectral fitting of the x-ray continuum. *The Astrophysical Journal, 636*, L113–L116.
3. Narayan, R., & McClintock, J. E. (2012). Observational evidence for a correlation between jet power and black hole spin. *Monthly Notices of the Royal Astronomical Society: Letters, 419(1)*, L69–L73.
4. Gou, L., McClintock, J. E., Liu, J., Narayan, R., Steiner, J. F., Remillard, R. A., Orosz, J. A., & Davis, S. W. (2009). A determination of the spin of the black hole primary in LMC X-1. *The Astrophysical Journal, 701*, 1076–1090.
5. García, J. A., Steiner, J. F., McClintock, J. E., Remillard, R. A., Grinberg, V., & Dauser, T. (2015). X-ray reflection spectroscopy of the black hole GX 339–4: Exploring the hard state with unprecedented sensitivity. *The Astrophysical Journal, 813*(2), 84.
6. Liu, B. F., & Qiao, E. (2022). Accretion around black holes: The geometry and spectra. *iScience, 25*, 103544.
7. Gúrpide, A., Godet, O., Koliopanos, F., Webb, N., & Olive, J. F. (2021). Long-term X-ray spectral evolution of ultraluminous X-ray sources: Implications on the accretion flow geometry and the nature of the accretor. *Astronomy & Astrophysics, 649*, A104.

8. Orosz, J. A., McClintock, J. E., Aufdenberg, J. P., Remillard, R. A., Reid, M. J., Narayan, R., & Gou, L. (2011). The mass of the black hole in Cygnus X-1. *The Astrophysical Journal, 742*(2), 84.
9. Steiner, J. F., Reis, R. C., Fabian, A. C., Remillard, R. A., McClintock, J. E., Gou, L., Cooke, R., Brenneman, L. W., & Sanders, J. S. (2012). A broad iron line in LMC X-1. *Monthly Notices of the Royal Astronomical Society, 427*(3), 2552–2561.
10. Jana, A., Naik, S., Chatterjee, D., & Jaisawal, G. K. (2021). NuSTAR and swift observations of the extragalactic black hole X-ray binaries. *Monthly Notices of the Royal Astronomical Society, 507*(4), 4779–4787.
11. Bhuvana, G. R., Radhika, D., & Nandi, A. (2022). Multi-mission view of extragalactic black hole X-ray binaries LMC X-1 and LMC X-3: Evolution of broadband spectral features. *Advances in Space Research, 69*(1), 483–498.
12. Paczynski, B. (1983). Mass of large magellanic cloud X-3. *Astrophysical Journal, Part 2-Letters to the Editor (ISSN 0004-637X), 273*, L81–L84.
13. Davis, S. W., Blaes, O. M., Hubeny, I., & Turner, N. J. (2005). Relativistic accretion disk models of high-state black hole X-ray binary spectra. *The Astrophysical Journal, 621*(1), 372.
14. Steiner, J. F., McClintock, J. E., Orosz, J. A., Remillard, R. A., Bailyn, C. D., Kolehmainen, M., & Straub, O. (2014). The low-spin black hole in LMC X-3. *The Astrophysical Journal Letters, 793*(2), L29.
15. Hynes, R. I., Steeghs, D., Casares, J., Charles, P. A., & O'Brien, K. (2003). Dynamical evidence for a black hole in GX 339–4. *The Astrophysical Journal, 583*(2), L95.
16. Miller, J. M., Fabian, A. C., Reynolds, C. S., Nowak, M. A., Homan, J., Freyberg, M. J., Ehle, M., Belloni, T., Wijnands, van der Klis, M., Charles, P. A., & Lewin, W. H. G. (2004). Evidence of black hole spin in GX 339–4: XMM-Newton/EPIC-pn and RXTE spectroscopy of the very high state. *The Astrophysical Journal, 606*(2), L131.
17. Wang-Ji, J., García, J. A., Steiner, J. F., Tomsick, J. A., Harrison, F. A., Bambi, C., Petrucci, P.-O., Ferreira, J., Chakravorty, S., & Clavel, M. (2018). The Evolution of GX 339-4 in the low-hard state as seen by NuSTAR and swift. *The Astrophysical Journal, 855*(1), 61.
18. Mitsuda, K., Inoue, H., Koyama, K., Makishima, K., Matsuoka, M., Ogawara, Y., Shibazaki, N., Suzuki, K., Tanaka, Y., & Hirano, T. (1984). Energy spectra of low-mass binary X-ray sources observed from TENMA. *Astronomical Society of Japan, 36*(4), 741–759.
19. Li, L. X., Zimmerman, E. R., Narayan, R., & McClintock, J. E. (2005). Multitemperature blackbody spectrum of a thin accretion disk around a Kerr black hole: Model computations and comparison with observations. *The Astrophysical Journal Supplement Series, 157*(2), 335.
20. Orosz, J. A., Steeghs, D., McClintock, J. E., Torres, M. A., Bochkov, I., Gou, L., Narayan, R., Blaschak, M., Levine, A. M., Remillard, R. A., Bailyn, C. D., Dwyer, M. M., Buxton, M., & Buxton, M. (2009). A new dynamical model for the black hole binary LMC X-1. *The Astrophysical Journal, 697*(1), 573.
21. Orosz, J. A., Steiner, J. F., McClintock, J. E., Buxton, M. M., Bailyn, C. D., Steeghs, D., Guberman, A., & Torres, M. A. (2014). The mass of the black hole in LMC X-3. *The Astrophysical Journal, 794(2)*, 154.
22. Davis, S. W., Done, C., & Blaes, O. M. (2006). Testing accretion disk theory in black hole X-ray binaries. *The Astrophysical Journal, 647*(1), 525.
23. Munoz-Darias, T., Casares, J., & Martínez-Pais, I. G. (2008). On the masses and evolutionary status of the black hole binary GX 339-4: A twin system of XTE J1550–564? *Monthly Notices of the Royal Astronomical Society, 385*(4), 2205–2209.
24. Blandford, R. D., & Znajek, R. L. (1977). Electromagnetic extraction of energy from Kerr black holes. *Monthly Notices of the Royal Astronomical Society, 179*(3), 433–456.
25. Mummery, A., Ingram, A., Davis, S., & Fabian, A. (2024). Continuum emission from within the plunging region of black hole discs. *Monthly Notices of the Royal Astronomical Society, 531*(1), 366–386.

Chapter 20
Decoding the Deterministic Nature of Black Hole IGR J17091–3624

Anindya Guria and Banibrata Mukhopadhyay

Abstract The differentiation between chaotic and stochastic systems has long been scrutinized, particularly in observations where data is often noise-contaminated and finite. Our research examines the dual nature of the black hole X-ray binary IGR J17091–3624, an object whose behavior has been closely studied in parallel to GRS 1915+105. Remarkable similarities in the temporal classes of these two objects are explored in literature. However, this was not the case with their non-linear dynamics: GRS 1915+105 shows signs of determinism, while IGR J17091-3624 was found to be stochastic. In this study, we confront the inherent challenge of noise contamination, as in IGR J17091–3624, faced by previous studies, particularly Poisson noise, which adversely impacts the reliability of non-linear results. We employ several denoising techniques to mitigate noise effects and employ methods like Autoencoder, Principal Component Analysis (PCA), Singular Value Decomposition (SVD), and Correlation Integral (CI) to isolate the deterministic signatures. We have found signs of determinism in IGR J17091–3624 after denoising, thus supporting the hypothesis of it being similar to GRS 1915+105, even as a dynamical system.

20.1 Introduction

GRS 1915+105, a well-studied black hole X-ray binary, exhibits a high spin and quasi-periodic oscillations (QPOs) across a wide frequency range [1, 2]. It is categorized into 12 temporal classes [3] and identified in three spectral states. Certain classes demonstrate deterministic, chaotic, or fractal characteristics, contrasting with others that appear stochastic (S) [4].

IGR J17091-3624, often considered a twin to GRS1915+105 due to shared features like "heartbeat" oscillations and high-frequency QPOs [5, 6], is however fainter and exhibits lower mass and spin [5]. Nevertheless, its temporal classes were

A. Guria (✉) · B. Mukhopadhyay
Indian Institute of Science, Bengaluru, India
e-mail: anindyaguria@iisc.ac.in; bm@iisc.ac.in

S. Ghosh, A. R. Rao (eds.), *The Relativistic Universe: From Classical to Quantum*, Astrophysics and Space Science Proceedings 61,
https://doi.org/10.1007/978-3-031-90186-7_20

previously found to be S or noise-dominated [7]. This calls into question the true nature of its dynamics, especially since X-ray lightcurves are heavily contaminated by Poisson noise. Previous studies suggested increasing bin size might increase fractal dimensions, though conclusive results are lacking due to insufficient data [7].

We plan here to revisit the problem and rigorously analyze the timeseries of the temporal classes of IGR J17091-3624, with special considerations to address the noise. Our approach is two-pronged: first, we apply a variety of filtering techniques to reduce the effect of Poisson noise in the lightcurves. These methods include moving averages, non-local means [8, 9], and an adaptive algorithm optimized for detecting determinism in heavy-noise environments [10]. Second, we explore beyond the traditional correlation integral (CI) method [4, 7, 11–13] to falsify the hypothesis that all IGR J17091-3624 temporal classes are S. Some of these techniques include principal component analysis (PCA) [14], singular value decomposition (SVD) [12, 14], and a novel machine learning-based autoencoder [15] that has been optimized to search for scale-invariant features in lightcurves (both time and frequency domains). Interestingly, we find that several temporal classes of IGR J17091-3624 show promising signs of being non-stochastic (NS), hence refuting the earlier hypothesis. Through this chapter, we attempt to highlight the importance of utilizing filtering techniques and designing novel tests (e.g., the autoencoder, used in this chapter) while performing non-linear analysis of noisy astrophysical timeseries data.

20.2 Difficulties Encountered Earlier with IGR J17091-3624

Determining whether a dynamical system is S or NS presents several challenges in experiments or observations where data are finite and typically noise-contaminated. In earlier studies of IGR J17091-3624, no NS behaviors were evident, contrasting with the well-established deterministic nature of certain temporal classes of GRS 1915+105. However, the latter exhibits about 20 times more average photon counts than IGR J17091-3624, as recorded by RXTE (PCA). This argues for a significant influence of Poisson noise due to the lower photon counts in IGR J17091-3624. Poisson noise, inherent in photon counting devices and increasing with fewer photons, complicates the interpretation of nonlinear dynamics from such data.

Previous study [7] noted potential deviations from stochasticity with increased binning times from 0.125 to 0.5 s, although the reduction in data points hampered definitive conclusions using the CI method. Nevertheless, increasing the bin size can reduce some effects of Poisson noise, suggesting that advanced noise-reduction techniques might allow a more accurate analysis of the nonlinear characteristics of these sources.

20.3 Proposed Filtering Methods

Based on the experience from previous efforts, we can list some key properties that any filtering method should ideally satisfy. The rationale behind many of these points was discussed succinctly earlier [16], given below as:

1. There should be no implicit or explicit dependence on the signal's power spectrum. This is because chaotic timeseries are expected to have a broad spectral spread. Also, the power spectrum of Poisson noise of an ideal Poisson process is uniform: traditional low pass or band pass filtering would not be effective in our application.
2. Our filtering technique should avoid introducing additional local correlation in our data. It would ensure that our conclusions are not affected by the filtering process.
3. The number of data points should not be reduced drastically by the filtering algorithm. As already mentioned, this enables the use of the CI method.

We shall begin by describing some trivial filters that satisfy our requirements and gradually move on to some of the more sophisticated ones.

20.3.1 Convolution-Based Filters (BOX and GAU)

Convolution filters are foundational in Digital Signal Processing (DSP) for signal manipulation and enhancement. A detailed exploration of convolution kernels can be found in standard texts such as [17]. Consider a normalized kernel g_i, where $\sum_{i=-k}^{k} g_i = 1$, applied to an input series x_n to produce an output y_n defined by $y_n = \sum_{i=-k}^{k} g_i x_{n+i}$. Popular kernels include boxcar (BOX) and Gaussian (GAU), each chosen for their specific advantages in various applications. In this study, we have made use of both. BOX is nothing but a moving average; although it is extremely simple ($g_i = 1/(2k+1)$, $y_n = \sum_{i=-k}^{k} x_{n+i}/(2k+1)$), due to its discontinuous edges it is not ideal for preserving spectral properties. To remedy this, the GAU kernel is used ($g_i = Ce^{-\frac{i^2}{2\sigma^2}}$, $y_n = C\sum_{i=-k}^{k} e^{-\frac{i^2}{2\sigma^2}} x_{n+i}$). Due to the central limit theorem, GAU is mathematically identical to an infinite number of passes by a BOX filter.

Initially devised for image denoising, the non-local means (NLM) algorithm mitigates noise by leveraging similar samples or pixels, regardless of their spatial proximity. A detailed discussion of the algorithm can be found in [8, 9]. To adapt NLM for one-dimensional signals, we begin with a boxcar convolution to generate an auxiliary series $B_n = \sum_{i=-k}^{k} x_{n+i}/(2k+1)$, using a kernel of size 9. The similarity between data points at indices m and n is quantified using a Gaussian weight function $f_{mn} = \exp\left(-(B_m - B_n)^2/h^2\right)$, where h is set based on the noise standard deviation. We choose h based on the standard deviation within each boxcar smoothed "patch." The denoised timeseries is then computed as $y_n = \frac{1}{C_n} \sum_{m=1}^{N} f_{mn} x_m$ where $C_n = \sum_{m=1}^{N} f_{mn}$ is a normalization factor. Although NLM may leave behind low-amplitude white noise, its impact on nonlinear analysis is minimal [8].

20.3.2 Adaptive Denoising Algorithm (ADA)

This technique is suitable for detecting chaos/determinism within heavy noise [10]. Initially, the algorithm divides a timeseries into segments or windows of length $w = 2n + 1$ points, with $n + 1$ points overlapping among them. A kth order polynomial is fitted for each segment. For the ith and $(i + 1)$th segments let them be $p_i^k(l_1)$ and $p_{i+1}^k(l_2)$, where $l_1, l_2 \in \{1, 2, \cdots 2n + 1\}$. The denoised data within this ith overlapped region are found using the weights $w_1 = \left[1 - \frac{(l-1)}{n}\right]$ and $w_2 = \left[\frac{(l-1)}{n}\right]$ as

$$y_{ni+l} = w_1 p_i^k(l + n) + w_2 p_{i+1}^k(l) \; l \in \{1, 2, \cdots n + 1\}.$$

This choice of weighting eliminates discontinuities at the boundaries of neighboring segments; this fitting ensures continuity all across the data. Two free parameters for fine-tuning are the segment size (w) and polynomial order (k). We have followed the recommended protocol [10] to determine their optimum values.

20.3.3 Correlation Integral Method and Surrogate Analysis

In the delay-embedding technique of CI originally proposed by Grassberger-Procaccia (GP), first vectors are constructed in an M-dimensional reconstructed phase space with delay τ, where each vector $\boldsymbol{\xi}_\mathbf{i} = [x_i, x_{i+\tau}, \ldots, x_{i+(M-1)\tau}]$ represents the system's state at different times [18]. The embedding parameters are chosen based on the first minimum of the mutual-information function [7].

The correlation integral $C_M(r)$ estimates the probability that two randomly selected vectors are within a distance r with the scaling law of $C_M(r) \sim r^{D_2}$ such that D_2 is the correlation dimension [11]. The scaling of D_2 linear with M suggests S behavior; an eventual saturation of D_2 suggests deterministic dynamics.

Surrogate analysis tests the null hypothesis that the timeseries is generated by a linear stochastic process [12]. We generate a set of surrogate timeseries that matches the original data in probability distribution and power spectrum using the IAAFT method [19]. The Normalised Mean Sigma Deviation (NMSD) measures differences in D_2 determined from the original and surrogate data:

$$nmsd^2 = \frac{1}{M_{\max} - M_{\min}} \sum_{M=M_{\min}}^{M_{\max}} \left[\frac{D_2(M) - \langle D_2^{\mathrm{surr}}(M) \rangle}{\sigma_{\mathrm{SD}}^{\mathrm{sur}}(M)} \right]^2 .$$

An $nmsd > 3$ for 19 surrogates suggests NS processes [13] (with a confidence of 95%).

20.3.4 Singular Value Decomposition (SVD)

SVD is employed to analyze the temporal dynamics of timeseries data using matrix techniques. The first two components, $E1$ and $E2$, of the right singular vector describe the dominant temporal dynamics, visualized by plotting $E1$ versus $E2$, and analyzed using Betti numbers as topological descriptors [20]. For a two-dimensional manifold, we consider Betti numbers, β_0 (the number of connected components) and β_1 (the number of 1-dimensional holes). S timeseries typically show a simple topology $(\beta_0, \beta_1) = (1, 0)$, whereas NS signals exhibit more complex topologies with multiple components and holes.

The S and NS timeseries are classified based on the sum of Betti numbers. If $\beta_0 + \beta_1 = 1$, then we call it S, else if $\beta_0 + \beta_1 > 1$ then it is NS. The number of connected components is identified using DBSCAN [21] from the scikit-learn package [22], with more complex structures assessed manually. Detailed results are presented in Sect. 20.4.2.

20.3.5 Autoencoder Based Approach for Measuring Deviation from Stochasticity (DS)

Utilizing a method proposed recently [15], we employ autoencoder-based techniques to detect deviations from stochasticity in timeseries. This method leverages time-invariant features across multiple timescales in frequency and time domains through multi-scale reconstruction and peak analysis in dissimilarity curves.

The process involves segmenting the original timeseries, applying Discrete Fourier Transform (DFT) on each segment, and training two autoencoders—one for the time domain and another for the frequency domain data. These autoencoders, trained on a mixture of stochastic, deterministic, and noise-contaminated signals (including GRS 1915+105 lightcurves), reconstruct features to generate dissimilarity (D_t) and prominence-of-peaks (P_t) curves.

For multi-scale analysis, these curves are segmented into windows (W_k), and both bias-corrected dissimilarity ($\tilde{d}_t^k$) and the prominence (p_t^k), are used to compute the Kullback-Leibler divergence and the coefficients of variance ($CV1$ and $CV2$), respectively. $CV1$ reflects the variance across reconstructions at different scales, while $CV2$ assesses changes in peak prominence, with lower values expected for S signals.

The deviation from stochasticity DS is calculated as:

$$DS = \frac{CV1 \times CV2}{100},$$

with a critical threshold at $DS = 1.5$ to distinguish between S and NS timeseries, as determined using an SVM classifier [15]. A value of $DS \geq 1.5$ indicates NS behavior, while $DS < 1.5$ suggests S characteristics. Detailed results of this approach are discussed in Sect. 20.4.3.

20.3.6 Principle Component Analysis and DBSCAN

PCA is employed for dimensionality reduction, focusing on eigenvalue ratios from the covariance matrix of timeseries data to detect dominant directions indicative of NS behavior. If the respective largest eigenvalues of two segments, λ_1 and λ_2, of the covariance matrix [14] follow $\lambda_1 \geq \lambda_2$, the ratio $r = \lambda_1/\lambda_2$ signifies the presence of a dominant direction, a characteristic of NS timeseries.

The key features derived from the eigenvalue ratio (ER) include:

1. **MER**: Maximum ER, lower in S.
2. **VAR**: Variance of ER, also lower in S.
3. **Area**: Total area under the ER curve, higher in NS.

Each timeseries is represented as a point in 3D (MER, VAR, Area) space using these features. DBSCAN, an unsupervised clustering algorithm, is then applied to segregate S and NS signals by identifying dense clusters and outliers. The formalized use of PCA and DBSCAN resolves prior ambiguities in classifying timeseries data, with detailed results discussed in Sect. 20.4.4.

20.4 Results

20.4.1 Correlation Integral Results

Tables 20.1 and 20.2 present *nmsd* and D_2 values for unfiltered and denoised signals. We follow the criterion that D_2 values are reported only when $nmsd \geq 3$ [13]. A class is labeled as S or NS based on a consensus of at least three filtering techniques. Classes with conflicting results from the methods are labeled "S/NS" (Figs. 20.1 and 20.2).

Table 20.1 Correlation integral results of IGR J17091-3624 (Part 1: original, GAU, and BOX data)

ObsID	Class	*nmsd* (Original)	D_2 (Original)	*nmsd* (GAU)	D_2 (GAU)	*nmsd* (BOX)	D_2 (BOX)
96420-01-01-00	I	1.072	NA	0.675	NA	0.658	NA
96420-01-11-00	II	1.366	NA	0.839	NA	1.212	NA
96420-01-04-01	III	0.557	NA	0.699	NA	0.711	NA
96420-01-05-00	IV	2.234	NA	3.623	6.485	1.705	NA
96420-01-06-03	V	1.805	NA	3.315	4.492	2.165	NA
96420-01-09-00	VI	0.928	NA	2.440	NA	3.587	6.330
96420-01-18-05	VII	2.126	NA	8.388	3.015	1.388	NA
96420-01-19-03	VIII	2.977	NA	4.950	4.731	5.217	4.731
96420-01-35-02	IX	0.438	NA	2.823	NA	3.399	7.925

Table 20.2 Correlation Integral results of IGR J17091-3624 (Part 2: ADA and NLM Data)

ObsID	Class	*nmsd* (ADA)	D_2 (ADA)	*nmsd* (NLM)	D_2 (NLM)	Behavior
96420-01-01-00	I	0.547	NA	1.160	NA	S
96420-01-11-00	II	0.514	NA	0.807	NA	S
96420-01-04-01	III	1.650	NA	1.011	NA	S
96420-01-05-00	IV	2.410	NA	3.386	4.568	S/NS
96420-01-06-03	V	5.204	3.247	0.596	NA	S/NS
96420-01-09-00	VI	1.583	NA	4.347	5.242	S/NS
96420-01-18-05	VII	5.486	3.190	4.149	3.243	NS
96420-01-19-03	VIII	11.492	4.552	7.112	4.440	NS
96420-01-35-02	IX	2.595	NA	6.444	6.043	S/NS*

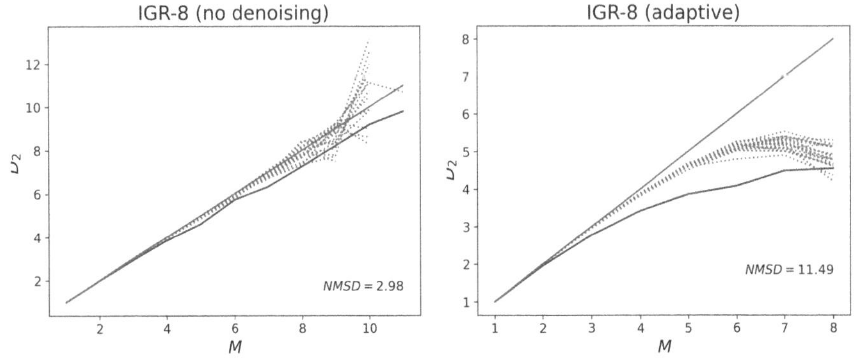

Fig. 20.1 The variation of D_2 with M for the temporal class VIII (determined to be S unanimously by all filtering techniques). The red dotted curves are the surrogates, and the dark blue solid curve is the true data. The light blue straight line shows $D_2 = M$, the expected result for an ideal S signal

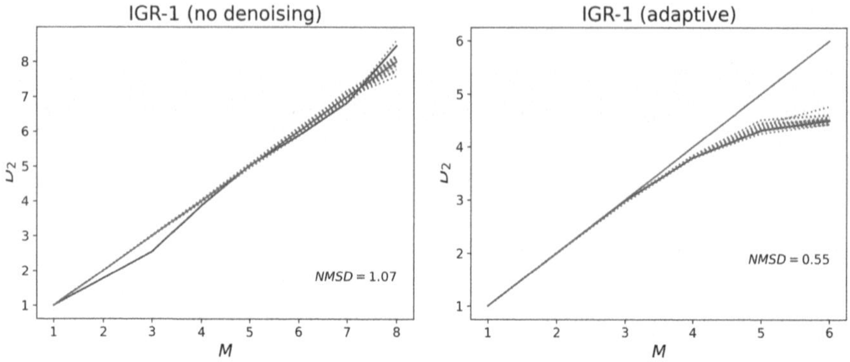

Fig. 20.2 The variation of D_2 with M for the temporal class I (determined to be S unanimously by all filtering techniques)

Table 20.3 Singular Value Decomposition results of IGR J17091-3624

ObsID	Class	Original	ADA	NLM	GAU	BOX	Overall
96420-01-01-00	I	S	S	S	S	S	S
96420-01-11-00	II	S	S	S	S	S	S
96420-01-04-01	III	S	S	S	S	S	S
96420-01-05-00	IV	S	NS	S	S	S	S
96420-01-06-03	V	S	NS	NS	NS	NS	NS
96420-01-09-00	VI	S	NS	NS	NS	NS	NS
96420-01-18-05	VII	S	NS	NS	NS	NS	NS
96420-01-19-03	VIII	S	NS	NS	NS	NS	NS
96420-01-35-02	IX	S	NS	S	S	S	S

20.4.2 Results from SVD

SVD analysis reveals that classes V, VI, VII, and VIII exhibit complex trajectories with multiple loops and dense regions indicative of attractor behavior, reflecting high values of both β_1 and β_0. In contrast, classes I, II, III, IV, and IX display simple topologies expected of S signals, specifically $(\beta_0, \beta_1) = (1, 0)$. We have listed our results in Table 20.3. Figures 20.3 and 20.4 contrast the complex phase portraits of class VIII with the simpler blob-like structure of the S-type IGR J17091-362-I, highlighting the effectiveness of our denoising techniques.

20.4.3 Autoencoder Results

The used autoencoders are the same as in [15] (code obtained from GitHub[1]). As mentioned in Sect. 20.3.5, the cutoff for classifying a signal as NS is 1.5, which

[1] https://github.com/csai-arc/blackhole_stochasticity_measure.

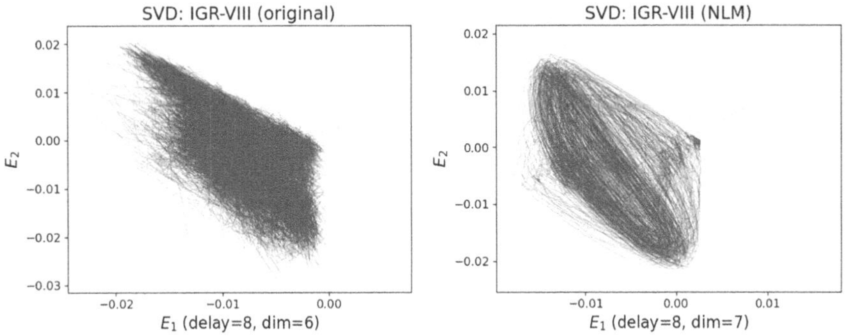

Fig. 20.3 SVD of IGR J17091-3624-VIII without filtering and with NLM. The optimal embedding dimensions and delay are mentioned for each of them

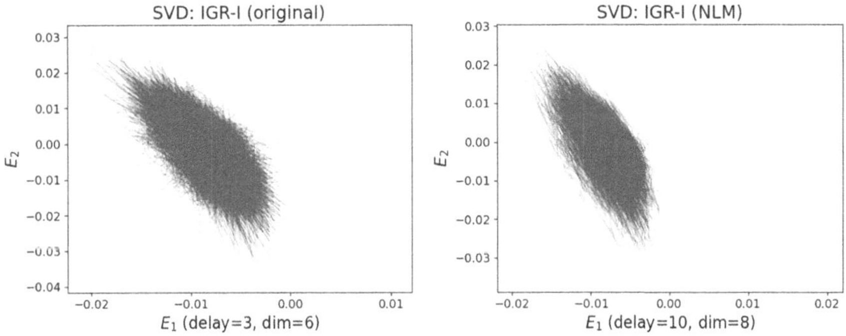

Fig. 20.4 SVD of IGR J17091-3624-I without filtering and with NLM. It shows no complex structure post-filtering

Table 20.4 Deviation from stochasticity (DS) metric for IGR J17091-3624

ObsID	Class	DS_{org}	DS_{ADA}	DS_{NLM}	DS_{GAU}	DS_{BOX}	Behavior
96420-01-01-00	I	0.13	0.79	0.78	0.53	0.52	S
96420-01-11-00	II	0.11	0.66	0.63	0.46	0.46	S
96420-01-04-01	III	1.04	**4.80**	**5.55**	**3.79**	**3.81**	NS
96420-01-05-00	IV	1.27	**2.54**	**3.06**	**2.28**	**2.30**	NS
96420-01-06-03	V	1.16	**1.93**	**2.06**	**1.77**	**1.79**	NS
96420-01-09-00	VI	1.12	**2.11**	**3.21**	**2.15**	**2.17**	NS
96420-01-18-05	VII	0.98	1.23	1.40	1.21	1.22	S
96420-01-19-03	VIII	0.71	0.85	1.31	0.98	0.99	S
96420-01-35-02	IX	0.69	0.84	**2.55**	1.26	1.27	S

Bold font highlights $DS > 1.5$, indicating NS behavior

the authors have determined to provide optimum classification with test data. Our results are listed in Table 20.4.

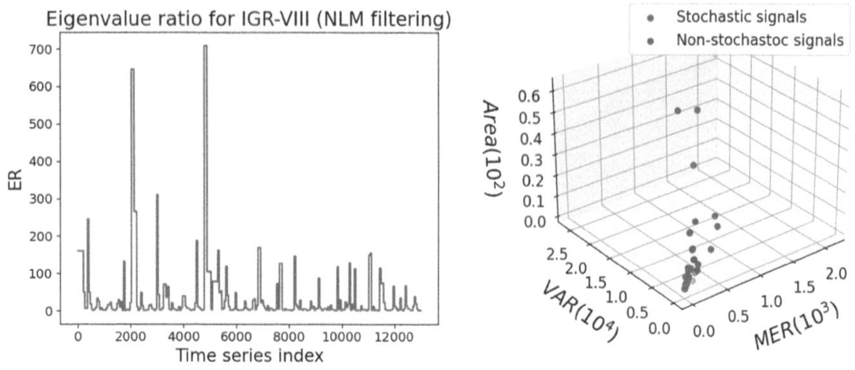

Fig. 20.5 ***Left*****:** A typical eigenvalue ratio (ER) plot. ***Right*****:** Scatter plot of PCA parameters from all (un)filtered signals

Table 20.5 Summary of PCA classification from different denoising methods

ObsID	Class	Original	ADA	NLM	GAU	BOX
96420-01-01-00	I	S	S	S	S	S
96420-01-11-00	II	S	NS	S	S	S
96420-01-04-01	III	S	S	S	S	S
96420-01-05-00	IV	S	NS	NS	NS	S
96420-01-06-03	V	S	NS	NS	NS	NS
96420-01-09-00	VI	S	NS	S	S	S
96420-01-18-05	VII	S	NS	NS	NS	NS
96420-01-19-03	VIII	S	S	NS	NS	NS
96420-01-35-02	IX	S	S	S	S	S

20.4.4 PCA and DBSCAN Results

As outlined in Sect. 20.3.6, the PCA algorithm has been applied to the original and filtered timeseries set, and the following results are obtained. The optimal eigenvalue ratio cutoff and ϵ for DBSCAN are chosen by maximizing the silhouette score [23]. Figure 20.5 shows the actual scatter of the cluster and the outliers. The red points are the S timeseries, while the green points are the outliers, which we identify here as the NS timeseries. Table 20.5 summarizes the results.

20.5 Summary

We have addressed Poisson noise contamination in X-ray lightcurves of IGR J17091-3624 using four filtering techniques: ADA, NLM, GAU, and BOX. These methods have facilitated the application of PCA, autoencoders, CI, and SVD to

Table 20.6 Summary of S/NS behavior across different methods

ObsID	Class	CI	SVD	Autoencoder	PCA	Overall
96420-01-01-00	I	S	S	S	S	S
96420-01-11-00	II	S	S	S	S	S
96420-01-04-01	III	S	S	NS	S	S
96420-01-05-00	IV	S/NS	S	NS	NS	S/NS
96420-01-06-03	V	S/NS	NS	NS	NS	NS
96420-01-09-00	VI	S/NS	NS	NS	S	S/NS
96420-01-18-05	VII	NS	NS	S	NS	NS
96420-01-19-03	VIII	NS	NS	S	NS	NS
96420-01-35-02	IX	S/NS*	S	S	S	S

detect NS behavior in the filtered signals. We have summarized our findings in Table 20.6 (we conclude upon S/NS division for a temporal class when three or more of our applied methods agree). IGR J17091-3624 classes I, II, and IX are consistently found to be S across all methods. Class III is also predominantly S. In contrast, classes V, VII, and VIII are primarily NS. The classes IV and VI remain uncertain in their nonlinear features. Thus, we have uncovered potential complex dynamics in IGR J17091–3624 using multiple denoising methods.

Therefore, like GRS 1915+105, which sometimes exhibits NS or fractal (it could be deterministic and even chaotic) and sometimes S, IGR J17091-3624 also shows transient behavior switching between NS and S in various temporal classes. Such nature was unclear in the previous analysis [7], which could not denoise the signal appropriately. Nevertheless, earlier, both sources were classified into several temporal classes based on lightcurves and related combined temporal and spectral analyses, arguing their transient nature. The present study confirms a similar transient nature for them from the nonlinear dynamics point of view. This supports the argument of their twin nature. Many temporal classes of IGR J17091-3624 show the same dynamical behavior, e.g., S or NS, as their corresponding equivalent GRS 1915+105 classes, e.g., I & χ and V & μ. However, one of the pairs of classes, i.e., III & ν, exhibits different dynamical behavior. The future plan is to combine the spectral behavior with the timeseries properties to understand the true nature of the underlying accretion flows, as was done earlier for GRS 1915+105 [4]. The details will be reported elsewhere (Guria & Mukhopadhyay, in preparation).

References

1. Mirabel, I. F., & Rodríguez, L. F. (1994). *Nature, 371*(6492), 46–48. http://dx.doi.org/10.1038/371046a0
2. Belloni, T., Méndez, M., & Sánchez-Fernández, C. (2001). *Astronomy and Astrophysics, 372*(2), 551. https://doi.org/10.1051/0004-6361:20010480

3. Belloni, T., Klein-Wolt, M., Méndez, M., van der Klis, M., & van Paradijs, J. (2000). *Astronomy and Astrophysics, 355*, 271.
4. Adegoke, O., Dhang, P., Mukhopadhyay, B., Ramadevi, M. C., & Bhattacharya, D. (2018). *Monthly Notices of the Royal Astronomical Society, 476*(2), 1581. https://doi.org/10.1093/mnras/sty263
5. Rao, A., & Vadawale, S. V. (2012). *The Astrophysical Journal Letters, 757*(1), L12. https://dx.doi.org/10.1088/2041-8205/757/1/L12
6. Altamirano, D., Belloni, T., Linares, M., van der Klis, M., Wijnands, R., Curran, P. A., Kalamkar, M., Stiele, H., Motta, S., Muñoz-Darias, T., Casella, P., & Krimm, H. (2011). *The Astrophysical Journal Letters, 742*(2), L17. https://dx.doi.org/10.1088/2041-8205/742/2/L17
7. Adegoke, O., Mukhopadhyay, B., & Misra, R. (2020). *Monthly Notices of the Royal Astronomical Society, 492*(3), 4033. https://doi.org/10.1093/mnras/staa071
8. Buades, A., Coll, B., & Morel, J .M. (2004, preprint). *International Conference on Machine Learning and Applications (CMLA)*, 19. https://api.semanticscholar.org/CorpusID:17780447
9. Buades, A., Coll, B., & Morel, J. M. (2005). *2005 IEEE Computer Society Conference on Computer Vision and Pattern Recognition (CVPR'05)* (vol. 2, pp. 60–65). https://doi.org/10.1109/CVPR.2005.38
10. Tung, W., Gao, J., Hu, J., & Yang, L. (2011). *Physical Review E, 83*, 046210. https://link.aps.org/doi/10.1103/PhysRevE.83.046210
11. Grassberger, P., & Procaccia, I. (1983). *Physica D: Nonlinear Phenomena, 9*(1), 189. https://doi.org/10.1016/0167-2789(83)90298-1. https://www.sciencedirect.com/science/article/pii/0167278983902981
12. Misra, R., Harikrishnan, K. P., Ambika, G., & Kembhavi, A. K. (2006). *The Astrophysical Journal, 643*, 1114 (2006). https://doi.org/10.1086/503094
13. Harikrishnan, K., Misra, R., Ambika, G., & Kembhavi, A. (2006). *Physica D: Nonlinear Phenomena, 215*(2), 137.
14. Chakka, S. P., & Sinha, N. (2024). *Proceedings of 15th International Conference on Signal Processing and Communications (SPCOM)*. Indian Institute of Science. https://doi.org/10.48550/arXiv.2307.07703
15. Chakka, S. P., Sinha, N., & Mukhopadhyay, B. (2023). *ICASSP 2023–2023 IEEE International Conference on Acoustics, Speech and Signal Processing (ICASSP)* (pp. 1–5). https://doi.org/10.1109/ICASSP49357.2023.10095755
16. Kostelich, E. J., & Schreiber, T. (1993). *Physical Review E, 48*, 1752 (1993). https://link.aps.org/doi/10.1103/PhysRevE.48.1752
17. Smith, S. W. (1997). *The scientist & engineer's guide to digital signal processing*, Hardcover edn. California Technical Pub.
18. Fraser, A. M., & Swinney, H. L. (1986). *Physical Review A, 33*, 1134 (1986). https://link.aps.org/doi/10.1103/PhysRevA.33.1134
19. Venema, V., Ament, F., & Simmer, C. (2006). *Nonlinear Processes in Geophysics, 13*(3), 321. https://npg.copernicus.org/articles/13/321/2006/
20. Betti, E. (1870). *Annali di Matematica Pura ed Applicata, 4*(1), 140–158. http://dx.doi.org/10.1007/BF02420029
21. Ester, M., Kriegel, H. P., Sander, J., Xu, X. (1996). *Proceedings of the Second International Conference on Knowledge Discovery and Data Mining (KDD'96)* (pp. 226–231). AAAI Press.
22. Pedregosa, F., Varoquaux, G., Gramfort, A., Michel, V., Thirion, B., Grisel, O., Blondel, M., Prettenhofer, P., Weiss, R., Dubourg, V., Vanderplas, J., Passos, A., Cournapeau, D., Brucher, M., Perrot, M., & Duchesnay, E. (2011). *Journal of Machine Learning Research, 12*, 2825.
23. Rousseeuw, P. J. (1987). *Journal of Computational and Applied Mathematics, 20*, 53–65. http://dx.doi.org/10.1016/0377-0427(87)90125-7

Part V
Gamma-Ray Bursts

Chapter 21
Recent Observational and Theoretical Progress in the Study of Gamma-Ray Bursts

Asaf Pe'er

Abstract I highlight here some of the recent and most exciting observational and theoretical progress in understanding gamma-ray bursts. Observationally: (1) GRB221009A—the 'brightest of all times' (BOAT) GRB with TeV emission and the debate about the origin of this emission—leptonic (inverse Compton) or hadronic (proton-synchrotron); (2) GRB230307A: a long GRB originating from merger, which challenges GRB progenitor models; and (3) Polarization angle change observed, which does not have a convincing theoretical explanation. Theoretically, I discuss (4) The realization that GRB jets are structured, and its observational consequences; (5) Novel ways of explaining a power law spectra by photon energy gain, without the need for particle acceleration; and (6) New ideas about the origin of the "plateau" phase, and the possibility of GRB jet Lorentz factor of a few tens.

21.1 Introduction

Gamma-ray bursts, the most energetic explosions known, have been the subject of an extensive research in the past 30 years and more. These monsters release typically $10^{50} - 10^{53}$ erg in the form of gamma-rays over a few second duration. They show an extremely rich phenomenology, with highly variable lightcurve (see, e.g., [1–4] for recent reviews). While the prompt phase lasts a typical duration of a few seconds, variability is seen, in some bursts, down to $\sim$ 10 ms scale. Their spectra is further observed to extend, in some bursts (when data is available) over many orders of magnitude in energy, from the keV up to the TeV range in some extreme cases. Phenomenologically, they are classified into two categories: the "short" (or short/hard) GRBs, with an average duration of a few tenths of a second; and the "long" (or long/soft) GRBs, with an average duration of 20 s. The dividing line is typically found at $\sim$ 2 s [5–9].

A. Pe'er (✉)
Department of Physics, Bar Ilan University, Ramat-Gan, Israel
e-mail: asaf.peer@biu.ac.il

S. Ghosh, A. R. Rao (eds.), *The Relativistic Universe: From Classical to Quantum*, Astrophysics and Space Science Proceedings 61,
https://doi.org/10.1007/978-3-031-90186-7_21

The extreme high energy photons detected, at energies far exceeding the 0.511 MeV threshold for pair production, combined with the rapid variability that necessitates a compact emission region (using the light crossing time argument, $\delta r \lesssim c\delta t = 3 \times 10^8 \ (\delta t/10 \text{ ms})$ cm) puzzled the community in the early days. The solution came with the realization that the emitting region must be propagating at relativistic speeds towards the observer, implying that the observed high energy photons are Doppler boosted [10–12]. Therefore, in the comoving frame, they are not energetic enough to produce $e^{\pm}$ pairs. This realization led to the construction of the GRB "fireball" model, that, despite being incomplete, its general framework is well established [13–15].

According to the "fireball" model, a vast explosion, associated with the formation of a black hole, leads to the ejection of material that forms a relativistic jet. At a second stage, some of the jet kinetic energy is dissipated, e.g., by internal shock waves [15], magnetic reconnection (e.g., [16–18] and many more) or other, unspecified method, producing the observed prompt emission signal. Following the dissipation, the material in the jet continues to expand relativistically into the interstellar medium (ISM), driving a relativistic shock wave. This shock wave, in turn, heats the ISM material, which then radiate the afterglow emission. This afterglow is routinely detected since 1997 [19–21].

The nature of the explosion that triggers this chain of events is somewhat more uncertain. The two leading models are the collapse of a massive star (the so-called "collapsar" model; [22–24]), and merger of binary stars [25, 26], e.g., a neutron star (NS) and a black hole (BH). Evidence for the connection of massive star collapse with GRBs exist for 20 years, since the discovery of absorption lines in the afterglow of GRB030329 [27, 28]. Such absorption lines are typical for core collapse supernova (known as SN type Ib/c), and their existence is a clear indication that the generation of the long GRBs is associated with a core collapse.

As for the origin of short GRBs, for many years only indirect evidence connected them with the merger of binaries. Such evidence included their observed occurrence location in the outskirts of their host galaxies [29–31]. This changed in 2017, with the discovery of gravitational waves associated with GRB170817 [32, 33]. Such gravitational wave signal, which preceded this GRB by 1.7 s, could only originate from the merger of binary NS system. This discovery, thus, served as a smoking gun proving the merger origin of this GRB. However, this GRB was peculiar in several ways, including being $10^2 - 10^4$ times less luminous than typical short GRB, which may challenge the claim of universality.

A unique observational consequence of this scenario is the rapid production of heavy nuclei. The huge temperatures during the merger event, $\gtrsim 10^{13}$ k, implies a release of a huge number of neutrons and protons. These recombine into α-particles, which merge with free neutrons to assemble heavy seed nuclei (with nuclei larger than the Iron nucleus) These heavy nuclei then radioactively decay very rapidly, producing bright emission on a time scale of $\sim$ day, which became known as "kilonova". The origin of the name is that it is approximately 1000 times brighter than regular novae [34, 35].

Although the ‘skeleton’ of the fireball model is well accepted by the community, many, possibly most of the details remain open questions, even after three decades of extensive research. GRBs represent a very complicated environment, in which several independent processes of energy exchange exist: from gravitational energy release that leads to an explosion, through conversion of the explosion energy to kinetic energy in the form of relativistic jet, and finally kinetic energy dissipation leading to the observed radiative signal. Many of the details of each of these processes is still actively debated in the literature. The main reason for that is that we only see the final outcome of this complicated chain of processes—the radiative signal (spectra and lightcurve), which vary from burst to burst, from which we try to deduce the full physics of the entire chain,

We are fortunate that a plethora of data continuously streams, due to a large number of telescopes that are active, as well as planned, at all wavelengths. As a result, in recent years we detected several unique bursts, which challenge what is already considered as “common knowledge” in this field. Furthermore, the continuous flow of data serves as a great platform, on which new theoretical ideas flourish. Here, I highlight some recent observational and theoretical results which seem to contradict some of the already accepted paradigms.

21.2 GRB230307A: A Long GRB from Merger

With luminosity of 3×10^{-3} erg/s and redshift $z = 0.065$, GRB230307A is the second brightest GRB ever detected [36–38]. It was detected by both Fermi and GECAM [39]. Its observed duration, 41.5 s puts it very deep inside the “long” GRB category. This is further supported by the soft to hard spectral evolution observed, which is typical for long GRBs. As such, evidence for supernova was expected.

However, starting 2.4 days after the burst and more than two months later, clear evidence for kilonova emission was observed [40]. Such a signal is a clear indication for a binary neutron star (BNS) merger origin, and is not expected when collapse of a massive star generates the GRBs. Thus, this GRB challenges the accepted division between the long and short GRB populations, as well as the common belief that long GRBs originate from collapse of a single (massive) star.

It is not clear whether this GRB forms a unique category or not. Somewhat earlier, GRB211211A was also identified as showing evidence for a kilonova during its afterglow phase [41, 42], which led to the suggestion that it also originates from binary merger (e.g., [43]). The evidence for neutron star merger origin in GRB211211A was excess of optical and near-infrared emission consistent with the kilonova observed after the gravitational wave detected GW170817. However, [44] argue that an unusual collapsar could explain both the duration of GRB 211211A and the r-process powered excess in its afterglow. The nature of the progenitor of this GRB, and possibly also of GRB230307A is therefore not yet fully understood.

21.3 Origin of Very High Energy Emission: Leptonic or Hadronic?

One of the most important development in the past decade, is the mature of high energy (GeV- TeV range) detectors. Detectors such as MAGIC [45,46], H.E.S.S [47, 48] and recently LHAASO [49, 50] provide unparallel data in the GeV-TeV band, which opens a new window into the study of GRBs [51]. As a concrete example, the MAGIC collaboration published recently lightcurve and spectra of GRB190114C [46, 52, 53], starting about a minute after the onset of this burst, and lasting for about 40 minutes. MAGIC data shows a comparable flux at the TeV band to that of longer wavelength, in particular the GeV (Fermi-LAT band) and X-rays (XRT and Fermi-GBM bands).

These new data naturally called for theoretical interpretation. A basic model that was suggested as a way of explaining the TeV data is that of inverse-Compton (IC) scattering (see, e.g., [54]). Indeed, this process is naturally expected: as energetic electrons are needed to explain the lower energy (x- and gamma-rays) signal observed by synchrotron emission, they must be accelerated to high energies inside the plasma. These energetic electrons will therefore up-scatter the synchrotron photons radiated to higher energies, which will include the TeV band. Additional advantages of this model is that the energy budget needed is relatively not very high, and the required magnetic field energy is relatively low.

However, a close look reveals that this model requires some additional assumptions in order to provide good fits to the TeV data. Both the flux and the observed spectral and temporal slopes predicted do not match very well those observed. This led [54] to suggest addition of various 'freedom factors', which enable some fine tuning of the model parameter. Such freedom parameters include a certain freedom factor in connecting the emission radius and the observed time, the emitted and observed frequency or the dependence of the observed luminosity on the jet kinetic energy; it turned out that the use of the classical, basic theoretical relations do not provide sufficient fits.

An alternative model that was proposed is that of synchrotron emission from accelerated protons [55, 56]. While somewhat similar ideas were already proposed in the 1990s (e.g., [57–59]), these were less appealing due to the fact that protons are much less efficient radiators than electrons. As a result, total high energy budget is required, needed to be provided to the energetic protons to reproduce the observed flux.

As pointed out recently [55], in fact there is no real problem: it is sufficient to assume that only a small fraction, of $\approx 10\%$ of the protons are accelerated. This is both aligned with the results of particle-in-cell simulations, that show that only a few % of the particles are accelerated in shock waves [60], and lead to a dramatic decrease in the required overall energy budget, which is $\sim 10^{54.5}$ erg—high, but not unreasonable.

According to this model, both electrons and protons are accelerated by the propagating shock. The electrons though have a lower energy, and therefore are in the slow cooling regime, while the more energetic protons are in the fast cooling regime (see, e.g., [61]). Radiation in the X-ray and gamma-ray band is explained by synchrotron emission from the electrons, while the TeV emission is due to synchrotron emission from the accelerated protons. The condition for proton-synchrotron to dominate over IC scattering is that the fraction of post-shock thermal energy converted to the magnetic field is much higher than the fraction of energy given to the electrons, namely, $\epsilon_B \gg \epsilon_e$. Generation of strong magnetic field in shock waves are indeed predicted as due to the two stream or Weibel instabilities [62–64], or, alternatively, by turbulent amplification [65], which may be associated with acceleration of particles [66, 67]. Indeed values that are found to fit the data are $\epsilon_B = 0.13$ and $\epsilon_e = 0.003$. Both values are consistent with the results of PIC simulations, as well as with fits of late time afterglow data of various GRBs. In the opposite regime, $\epsilon_e \gg \epsilon_B$, IC emission from the electrons is found to dominate proton-synchrotron contribution. These results seem to be universal: similar fitting holds also for the spectra and lightcurve of GRB221009A.

21.4 GRB221009A: The Brightest of All Times

A stunning GRB recently detected is GRB221009A. With energy as high as 1.2×10^{55} erg, this GRB became known as the "brightest of all times" (BOAT) GRB [37]. It was so bright, that the NaI detector was saturated for a duration of 58 seconds, between 219–277 s from the burst trigger. It was observed throughout the entire electromagnetic spectrum, from 10 TeV's to radio [68]. This GRB was flagged as having the highest prompt fluence and peak flux ever identified by both the Fermi gamma-ray burst (GBM) monitor and the Konus-Wind instrument: its fluence was found to be 0.21 ± 0.02 erg cm^{-2} [69], which, for redshift $z = 0.15$ imply an isotropic peak luminosity of $L = 3.4 \times 10^{54}$ erg s^{-1}.

For comparison, this observed fluence is about 50 times higher than that of the second brightest GRB (GRB 230307A), and its intrinsic isotropic energy is more than twice of the second most energetic GRB identified (GRB00323; [37]). These facts make it extremely rare: population analysis implies that such a bright GRB should be seen at earth at a rate of $\sim$ once in 10,000 years.

In addition to being so extremely bright, this GRB showed clear evidence of an emission line, starting approximately 80 seconds after the onset of the afterglow (at 226 s after the explosion; [70]). This line was detected at $\sim$ 10 MeV, and its peak energy showed a clear decay in time, as $\epsilon_{peak}(t) \propto t^{-1}$. Such a discriminated line can result from the annihilation of pairs. However, direct calculations would result in Lorentz factor of $\sim$ 20 (the Doppler boost needed to reach this energy). A more sophisticated calculation, that takes into account the high latitude emission (i.e., that the emission originates off the line of sight, therefore the Doppler boost needs to consider the changing angle) reveals a more realistic Lorentz factor, of $\Gamma \approx 600$ [71].

21.5 Evidence for Jet Structure

While many models of GRB jets are limited to 1-d flow, in reality, jets are expected to have an angular profile, of the form $\Gamma = \Gamma(\theta)$. In fact, numerical simulations of jets propagating through the envelope of a collapsing star [72, 73], suggest a profile of the form

$$\Gamma(\theta) = \frac{\Gamma_0}{\sqrt{1 + (\theta/\theta_j)^{2p}}}.$$

Such a profile has 3 parameters: the Lorentz factor at the center, Γ_0; a typical jet angle, θ_j; and the decay profile of the Lorentz factor at angles $\theta > \theta_j$, which assume a power law of index p, $\Gamma(\theta) \propto \theta^{-p}$. In addition to these three free parameters, when calculating the observed signal, one has to add the observer viewing angle, θ_v as a fourth parameter, as the signal will depend on the observer's location relative to the jet axis [74].

In recent years, it turned out that these more realistic jet profiles provide much better fits to late time data; e.g., of the afterglow of the boat GRB221009A [75, 76].

This profile provides a unique environment that enables to introduce an alternative way of explaining the non-thermal spectra observed. In the classical, synchrotron emission explanation, particles are accelerated to a non-thermal distribution (typically, a power law) and then emit synchrotron radiation. The power law distribution of particles therefore produces the power law spectra often observed.

However, an alternative is that thermal photons gain energy directly by scattering deep inside the jet [74, 77, 78]. As the photons scatter back and fourth between regions of different velocities (i.e., in the shear layer of the jet), on the average a photon gains energy. This energy comes at the expense of the differential kinetic energy of the different propagating jet regions. If the jet velocity profile is steep enough, namely $p > 2$, this energy gain overcomes the adiabatic losses. It was recently shown that the result is a power law distribution of the emerging photon spectral index [78]. The repeated scattering that leads to that is an analogue to the Fermi acceleration process of charged particles reflected back and fourth between the upstream and downstream region of collisionless shocks.

The outcome of the Monte-Carlo simulation shows the formation of a high energy spectral slope, with power low index $\beta = -2.54$ (for a realistic jet profile), which is consistent with existing observations [78]. A prediction of this model is the power law index of the high energy spectral slope, which is expected to be $\beta \leq -1.5$, obtained for various values of the jet velocity profile $p \geq 2$. For $p = 2$, adiabatic losses overcome the energy gain, and so no high energy slope is expected.

21.6 Polarization Angle Change

Another observational field which matured in recent years is that of x- and γ ray polarization. Following the launch of AstroSat, a plethora of polarization information became available. A unique example is GRB160821A. This was a very energetic burst ($E \gtrsim 10^{53}$ erg), which show high degree of polarization—more than 30% in the gamma-ray lightcurve. However, the most interesting feature was a polarization angle change, which was detected twice: once during the rise phase and once during the decay phase of a bright pulse that was seen after ~ 120 s. Each of these polarization angle changes is consistent with 90° [79].

Such a polarization angle change challenges existing models. While current models can explain a 90° change for an observer located close to the jet edge (e.g., [80]), this is accompanied by a sharp decrease in the flux. The reason for this flux decay is that in order to obtain such a 90° change, the observer needs to be located close to the jet edge. Initially, it observes a small, magnetized region, which produces a polarized signal. As time elapses, the observed region (i.e., the region from which photons reach the observer) grows, and part of it is outside of the jet. Thus, while the observed parts that remain within the jet can produce a signal polarized in 90° to the initial polarized signal, most of the viewing region is outside of the jet opening angle, and therefore the flux is expected to sharply drop.

As of today, a convincing explanation to this observational result is still lacking, and it requires some 'out of the box' ideas, such as unconventional jet geometry.

21.7 X-Ray Plateau

Following the launch of the Swift satellite, it was realized that GRB lightcurves are more complicated than expected. It was found that in a significant fraction, of $\sim$ 60% of GRBs [81], immediately after the prompt emission phase the signal does not evolve directly to the self-similar phase characterizing the afterglow phase. Rather, at the end of the prompt emission, a flat x-ray lightcurve is seen. This "plateau" phase is observed to last between a few minutes to a few hours [82, 83]. During this phase, the x-ray lightcurve does not decay (or decays very slowly), hence the name "plateau". Only at later times, after few 100's – 10,000 seconds from the trigger, the lightcurve begins to decay in accordance to the expectation from a self-similar expansion that characterizes the afterglow phase [84].

Over the years, several explanations to this phenomenon were proposed. The leading ones are (1) late time energy injection [82, 83]; (2) interaction with the environment—either directly [85] or via a reverse shock [86]; and (3) viewing angle effect [87–89].

An alternative idea was that of emission during the coasting phase (i.e., before sufficient material is collected from the interstellar medium (ISM) to cause shock deceleration). This can fit the data provided the explosion occurs into a "wind" medium, characterized by decaying density profile, $n(r) \propto r^{-2}$ [90], as is expected to be emitted by a massive star prior to its explosion.

Interestingly enough, in a recent analysis we carried [91] it was found that excellent fits to the plateau data are obtained, provided that the explosion occurs into a wind medium, and that the bulk GRB Lorentz factor is relatively low: "only" a few tens, rather than a few hundreds as frequently assumed.

Indeed, the idea that the Lorentz factor of GRB jets is of few hundreds results from the opacity argument (e.g., [92–94]). The fact that energetic photons (above the threshold for pair production—e.g., photons observed in the GeV range) are detected, implies that the optical depth for pair production by these photons must be low. The GRB "fireball" model suggests that in order to avoid attenuation due to pair production, the source of these energetic photons is Doppler boosted with respect to us by a large factor (up to several hundreds). Therefore, in the outflow (comoving) frame, these photons are not sufficiently energetic to produce pairs, and they therefore escape to infinity. This can only occur if the bulk motion of GRB jets is $\Gamma \geq 100$. An independent clue for a high Lorentz factor in GRB jets is a detection of a significant thermal component in the prompt emission spectra. Theory shows that such a component is, by itself, a good indication for a high Lorentz factor [95].

A close analysis of GRBs which show an x-ray plateau as part of their lightcurve reveals that none of these constraints exist for these GRBs. Indeed, only $\sim 15\%$ of GRBs show any evidence for emission above the MeV threshold, and they anti-correlate with those who show x-ray plateau. Only 3/186 GRBs detected by the Fermi-LAT show any evidence of a plateau; and in addition, non of the plateau bursts show evidence for thermal emission (spectral index above the "non-dissipative photosphere" (NDP) line; [96]). Thus, "plateau" GRBs seem to anti-correlate with the requirement of a high value of Lorentz factor, $\Gamma \geq 100$ [91]. Furthermore, when comparing the properties of GRBs with and without plateaus, one finds that the peak energy of the plateau GRBs is, on the average, considerably lower than that of non-plateau GRBs.

As explained above, the existence of plateau in the x-ray lightcurve is a natural outcome of the synchrotron theory, provided that the Lorentz factor of the flow is constant, rather than decays in time. A constant Lorentz factor is expected following the initial outflow acceleration, and before sufficient material from the ISM is collected, and slows down the flow. In this phase, known as the "coasting" phase of the GRB jets, in the fast cooling regime (observed frequency above the cooling frequency, $\nu > \nu_c$) the observed flux is expected to be constant in time, as is indeed seen by the data.

Matching the data with this model, one can deduce the properties of the flow. Using a sample of 13 plateau-GRBs with high quality data, we found an average GRB jet Lorentz factor of $\langle \Gamma_j \rangle \simeq 50$, with a range between a few and 220. Furthermore, in order for this model to fit the data, an explosion into a decaying, "wind" density profile is needed. This wind is ejected from the star prior to its explosion. The values found are approximately 1% of the typical values quoted for Wolf-Rayet stars, which are believed to be the GRB progenitors [91]. This result is further strengthen by an analysis of the flares seen during this phase (Dereli-Begue, Pe'er et al., 2024, submitted [97]).

This result thus bridges an observational gap: while the fastest jets from active galactic nuclei (AGN) can reach $\approx$10–20 [98], and that of bright GRBs values of $\approx 100 \ldots 1000$ [99, 100], there is an observational gap for Γ of fews tens. The results obtain bridges exactly this gap.

21.8 Summary and Conclusions

The field of gamma-ray bursts (GRBs) though is not so new anymore, still continues to surprise. After more than three decades of an extensive research, there is still plenty of new results, which cannot be explained within the framework of existing theoretical models. This is a sign of a very healthy field, in which new observational data continues to stream, and stimulate novel theoretical ideas.

In this short review, I highlighted some of the latest observational breakthroughs, as well as some of the novel ideas at the forefront of current theoretical research in this field. On the observational side: (1) the brightest GRBs challenge theoretical models of progenitor stars; similar challenge to the nature of progenitor stars is given by (2) the identification of long GRBs with clear indication for merger origin. (3) Matureness of detectors in the TeV range challenge models aimed at explaining emission at these bands, from which theoretical models of particle acceleration needs to be revised; (4) New detectors, such as the AstroSat that provides polarization data, challenge theoretical models from a new angle; as of today, there is no theoretical idea that can explain the polarization angle change detected.

Theoretical ideas which were introduced in recent years include (1) the realization that jets have lateral structure, $\Gamma = \Gamma(\theta)$, and that such a structure has observational consequences; specifically, (2) the ability to explain a power law spectra by multiple scattering of photons below the photosphere provides a valid alternative to the need for charged particle acceleration—which may not even be needed to explain the broad, non-thermal spectra seen; and (3) the idea that the existence of plateau may be interpreted as a proof that many GRBs have a Lorentz factor of "only" a few tens, rather than a few hundreds. If proven correct, this idea will lead to a paradigm shift in our understanding of the field.

Due to the limited time and space, there are several key recent observations that I didn't even discuss, such as the connection between GRBs and gravitational waves (GW), that was proven by GRB/GW170817, or the continuous search for non-electromagnetic signals (mainly neutrinos) from GRBs.

New detectors planned in the foreseen future, such as the SVOM [101], ULTRA-SAT [102] or THESEUS [103] guarantees a continuous flow of new data in the foreseen future. This provides the ideal ground for the growth of novel theoretical ideas, which are expected to further deepen our understanding, and possibly result in new paradigm shifts in the coming decade.

References

1. Pe'er, A. (2015). Physics of gamma-ray bursts prompt emission. *Advances in Astronomy, 2015*, 907321.
2. Kumar, P., & Zhang, B. (2015). The physics of gamma-ray bursts & relativistic jets. *Physics Reports, 561*, 1–109.
3. Zhang, B. (2018). *The physics of gamma-ray bursts*. Cambridge University Press.
4. Bošnjak, Ž., Barniol Duran, R., & Pe'er, A. (2022). The GRB prompt emission: An unsolved puzzle. *Galaxies, 10*(2), 38.
5. Kouveliotou, C., Meegan, C. A., Fishman, G. J., Bhat, N. P., Briggs, M. S., Koshut, T. M., Paciesas, W. S., & Pendleton, G. N. (1993). Identification of two classes of gamma-ray bursts. *The Astrophysical Journal Letters, 413*, L101–L104.
6. Paciesas, W. S., et al. (1999). The fourth BATSE gamma-ray burst catalog (revised). *The Astrophysical Journal Supplement, 122*, 465.
7. Gruber, D., et al. (2014). The fermi GBM gamma-ray burst spectral catalog: Four years of data. *The Astrophysical Journal Supplement, 211*, 12.
8. von Kienlin, A., et al. (2014). The second fermi GBM gamma-ray burst catalog: The first four years. *The Astrophysical Journal Supplement, 211*, 13.
9. Lien, A., et al. (2016). The third swift burst alert telescope gamma-ray burst catalog. *The Astrophysical Journal, 829*(1), 7.
10. Cavallo, G., & Rees, M. J. (1978). A qualitative study of cosmic fireballs and gamma-ray bursts. *Monthly Notices of the Royal Astronomical Society, 183*, 359–365.
11. Paczynski, B. (1990). Super-Eddington winds from neutron stars. *The Astrophysical Journal, 363*, 218–226.
12. Shemi, A., & Piran, T. (1990). The appearance of cosmic fireballs. *The Astrophysical Journal Letters, 365*, L55–L58.
13. Rees, M. J., & Meszaros, P. (1992). Relativistic fireballs - Energy conversion and time-scales. *Monthly Notices of the Royal Astronomical Society, 258*, 41P–43P.
14. Meszaros, P., Laguna, P., & Rees, M. J. (1993). Gasdynamics of relativistically expanding gamma-ray burst sources - Kinematics, energetics, magnetic fields, and efficiency. *The Astrophysical Journal, 415*, 181–190.
15. Rees, M. J., & Meszaros, P. (1994). Unsteady outflow models for cosmological gamma-ray bursts. *The Astrophysical Journal Letters, 430*, L93–L96.
16. Thompson, C. (1994). A model of gamma-ray bursts. *Monthly Notices of the Royal Astronomical Society, 270*, 480–+.
17. Katz, J. I. (1997). Yet another model of gamma-ray bursts. *The Astrophysical Journal, 490*, 633–641.
18. Mészáros, P., & Rees, M. J. (1997). Poynting jets from black holes and cosmological gamma-ray bursts. *The Astrophysical Journal Letters, 482*, L29–L32.
19. van Paradijs, J., et al. (1997). Transient optical emission from the error box of the γ-ray burst of 28 February 1997. *Nature, 386*, 686–689.
20. Wijers, R. A. M. J., Rees, M. J., & Meszaros, P. (1997). Shocked by GRB 970228: The afterglow of a cosmological fireball. *Monthly Notices of the Royal Astronomical Society, 288*, L51–L56.
21. Galama, T. J., et al. (1998). An unusual supernova in the error box of the γ-ray burst of 25 April 1998. *Nature, 395*, 670–672.
22. Woosley, S. E. (1993). Gamma-ray bursts from stellar mass accretion disks around black holes. *The Astrophysical Journal, 405*, 273–277.
23. MacFadyen, A. I., & Woosley, S. E. (1999). Collapsars: Gamma-ray bursts and explosions in "failed supernovae". *The Astrophysical Journal, 524*, 262–289.
24. MacFadyen, A. I., Woosley, S. E., & Heger, A. (2001). Supernovae, jets, and collapsars. *The Astrophysical Journal, 550*, 410–425.

25. Eichler, D., Livio, M., Piran, T., & Schramm, D. N. (1989). Nucleosynthesis, neutrino bursts and gamma-rays from coalescing neutron stars. *Nature, 340*, 126–128.
26. Narayan, R., Paczynski, B., & Piran, T. (1992). Gamma-ray bursts as the death throes of massive binary stars. *The Astrophysical Journal Letters, 395*, L83–L86.
27. Hjorth, J., et al. (2003). A very energetic supernova associated with the γ-ray burst of 29 March 2003. *Nature, 423*, 847–850.
28. Stanek, K. Z., et al. (2003). Spectroscopic discovery of the supernova 2003dh associated with GRB 030329. *The Astrophysical Journal Letters, 591*, L17–L20.
29. Gehrels, N., et al. (2005). A short γ-ray burst apparently associated with an elliptical galaxy at redshift z = 0.225. *Nature, 437*, 851–854.
30. Fong, W., & Berger, E. (2013). The locations of short gamma-ray bursts as evidence for compact object binary progenitors. *The Astrophysical Journal, 776*, 18.
31. Berger, E. (2014). Short-duration gamma-ray bursts. *Annual Review of Astronomy and Astrophysics, 52*, 43–105.
32. Kasliwal, M. M., et al. (2017). Illuminating gravitational waves: A concordant picture of photons from a neutron star merger. *Science, 358*(6370), 1559–1565.
33. Abbott, B. P., et al. (2017). GW170817: Observation of gravitational waves from a binary neutron star inspiral. *Physical Review Letters, 119*(16), 161101.
34. Metzger, B. D., Martínez-Pinedo, G., Darbha, S., Quataert, E., Arcones, A., Kasen, D., Thomas, R., Nugent, P., Panov, I. V., & Zinner, N. T. (2010). Electromagnetic counterparts of compact object mergers powered by the radioactive decay of r-process nuclei. *Monthly Notices of the Royal Astronomical Society, 406*(4), 2650–2662.
35. Metzger, B. D., & Berger, E. (2012). What is the most promising electromagnetic counterpart of a neutron star binary merger? *The Astrophysical Journal, 746*(1), 48.
36. Dalessi, S., Roberts, O. J., Meegan, C., & Fermi GBM Team. (2023). GRB 230307A: Fermi GBM observation of a very bright burst. *GRB Coordinates Network, 33411*, 1.
37. Burns, E., et al. (2023). GRB 221009A: The boat. *The Astrophysical Journal Letters, 946*(1), L31.
38. Dichiara, S., Tsang, D., Troja, E., Neill, D., Norris, J. P., & Yang, Y. H. (2023). A luminous precursor in the extremely bright GRB 230307A. *The Astrophysical Journal Letters, 954*(1), L29.
39. Li, Y., et al. (2020). The GECAM and its payload. *Scientia Sinica Physica, Mechanica & Astronomica, 50*(12), 129508.
40. Yang, Y.-H., et al. (2024). A lanthanide-rich kilonova in the aftermath of a long gamma-ray burst. *Nature, 626*(8000), 742–745.
41. Rastinejad, J. C., et al. (2022). A kilonova following a long-duration gamma-ray burst at 350 Mpc. *Nature, 612*(7939), 223–227.
42. Troja, E., et al. (2022). A nearby long gamma-ray burst from a merger of compact objects. *Nature, 612*(7939), 228–231.
43. Zhong, S.-Q., Li, L., & Dai, Z.-G. (2023). GRB 211211A: A neutron star-white dwarf merger? *The Astrophysical Journal Letters, 947*(2), L21.
44. Barnes, J., & Metzger, B. D. (2023). A collapsar origin for GRB 211211A is (just barely) possible. *The Astrophysical Journal, 947*(2), 55.
45. Lorenz, E., & Martinez, M. (2005). High energy astrophysics: The MAGIC telescope. *Astronomy and Geophysics, 46*(6), 6.21–6.25.
46. Melandri, A., et al. (2022). The supernova of the MAGIC gamma-ray burst GRB 190114C. *Astronomy & Astrophysics, 659*, A39.
47. H.E.S.S. Collaboration, Abramowski, A., et al. (2014). Search for TeV gamma-ray emission from GRB 100621A, an extremely bright GRB in X-rays, with H.E.S.S. *Astronomy & Astrophysics, 565*, A16.
48. H. E. S. S. Collaboration, Abdalla, H., et al. (2021). Revealing x-ray and gamma ray temporal and spectral similarities in the GRB 190829A afterglow. *Science, 372*(6546), 1081–1085.
49. Cao, Z., et al. (2019). The large high altitude air shower observatory (LHAASO) science book (2021 edition). arXiv:1905.02773.

50. Cao, Z., et. al. (2023). Very high-energy gamma-ray emission beyond 10 TeV from GRB 221009A. *Science Advances, 9*(46), eadj2778.
51. LHAASO Collaboration, Cao, Z., et al. (2023). A tera-electron volt afterglow from a narrow jet in an extremely bright gamma-ray burst. *Science, 380*(6652), 1390–1396.
52. MAGIC Collaboration, Acciari, V. A., et al. (2019). Teraelectronvolt emission from the γ-ray burst GRB 190114C. *Nature, 575*(7783), 455–458.
53. Ravasio, M. E., Oganesyan, G., Salafia, O. S., Ghirlanda, G., Ghisellini, G., Branchesi, M., Campana, S., Covino, S., & Salvaterra, R. (2019). GRB 190114C: From prompt to afterglow? *Astronomy & Astrophysics, 626*, A12.
54. Derishev, E., & Piran, T. (2021). GRB afterglow parameters in the era of TeV observations: The case of GRB 190114C. *The Astrophysical Journal, 923*(2), 135.
55. Isravel, H., Pe'er, A., & Bégué, D. (2023b). Proton synchrotron origin of the very-high-energy emission of GRB 190114C. *The Astrophysical Journal, 955b*(1), 70.
56. Isravel, H., Bégué, D., & Pe'er, A. (2023b). Hybrid emission modeling of GRB 221009A: Shedding light on TeV emission origins in long GRBs. *The Astrophysical Journal, 956a*(1), 12.
57. Böttcher, M., & Dermer, C. D. (1998). High-energy gamma rays from ultra-high-energy cosmic-ray protons in gamma-ray bursts. *The Astrophysical Journal Letters, 499*, L131–L134.
58. Totani, T. (1998). TEV burst of gamma-ray bursts and ultra-high-energy cosmic rays. *The Astrophysical Journal Letters, 509*, L81–L84.
59. Razzaque, S., Dermer, C. D., & Finke, J. D. (2010). Synchrotron radiation from ultra-high energy protons and the fermi observations of GRB 080916C. *The Open Astronomy Journal, 3*, 150–155.
60. Sironi, L., & Spitkovsky, A. (2011). Particle acceleration in relativistic magnetized collisionless electron-ion shocks. *The Astrophysical Journal, 726*, 75–+.
61. Sari, R., Piran, T., & Narayan, R. (1998). Spectra and light curves of gamma-ray burst afterglows. *The Astrophysical Journal Letters, 497*, L17+.
62. Medvedev, M. V., & Loeb, A. (1999). Generation of magnetic fields in the relativistic shock of gamma-ray burst sources. *The Astrophysical Journal, 526*, 697–706.
63. Silva, L. O., Fonseca, R. A., Tonge, J. W., Dawson, J. M., Mori, W. B., & Medvedev, M. V. (2003). Interpenetrating plasma shells: Near-equipartition magnetic field generation and nonthermal particle acceleration. *The Astrophysical Journal Letters, 596*, L121–L124.
64. Nishikawa, K.-I., Hardee, P., Richardson, G., Preece, R., Sol, H., & Fishman, G. J. (2005). Particle acceleration and magnetic field generation in electron-positron relativistic shocks. *The Astrophysical Journal, 622*, 927–937.
65. Bell, A. R. (2004). Turbulent amplification of magnetic field and diffusive shock acceleration of cosmic rays. *Monthly Notices of the Royal Astronomical Society, 353*, 550–558.
66. Niemiec, J., Pohl, M., Stroman, T., & Nishikawa, K.-I. (2008). Production of magnetic turbulence by cosmic rays drifting upstream of supernova remnant shocks. *The Astrophysical Journal, 684*(2), 1174–1189.
67. Caprioli, D., & Spitkovsky, A. (2014). Simulations of ion acceleration at non-relativistic shocks. II. Magnetic field amplification. *The Astrophysical Journal, 794*(1), 46.
68. Williams, M. A., et al. (2023). GRB 221009A: Discovery of an exceptionally rare nearby and energetic gamma-ray burst. *The Astrophysical Journal Letters, 946*(1), L24.
69. Frederiks, D., et al. (2023). Properties of the extremely energetic GRB 221009A from Konus-WIND and SRG/ART-XC observations. *The Astrophysical Journal Letters, 949*(1), L7.
70. Edvige Ravasio, M., et al. (2023). A bright megaelectronvolt emission line in γ-ray burst GRB 221009A. arXiv:2303.16223.
71. Pe'er, A., & Zhang, B. (2024). Physical conditions that lead to the detection of the pair annihilation line in the BOAT GRB221009A. arXiv:2407.16241.
72. Zhang, W., Woosley, S. E., & MacFadyen, A. I. (2003). Relativistic jets in collapsars. *The Astrophysical Journal, 586*, 356–371.

73. Parsotan, T. M., & Lazzati, D. (2023). Monte Carlo simulations of photospheric emission in gamma ray bursts. In R. Ruffino & G. Vereshchagin (Eds.), *The Sixteenth Marcel Grossmann Meeting. On Recent Developments in Theoretical and Experimental General Relativity, Astrophysics, and Relativistic Field Theories* (pp. 2972–2981).
74. Lundman, C., Pe'er, A., & Ryde, F. (2013). A theory of photospheric emission from relativistic, collimated outflows. *Monthly Notices of the Royal Astronomical Society, 428*, 2430–2442.
75. O'Connor, B., et al. (2023). A structured jet explains the extreme GRB 221009A. *Science Advances, 9*(23), eadi1405.
76. Gill, R., & Granot, J. (2023). GRB 221009A afterglow from a shallow angular structured jet. *Monthly Notices of the Royal Astronomical Society, 524*(1), L78–L83.
77. Ito, H., Nagataki, S., Ono, M., Lee, S.-H., Mao, J., Yamada, S., Pe'er, A., Mizuta, A., & Harikae, S. (2013). Photospheric emission from stratified jets. *The Astrophysical Journal, 777*, 62.
78. Vyas, M. K., & Pe'er, A. (2023). Photons' scattering in relativistic plasma with velocity shear: Generation of high energy power-law spectra. *The Astrophysical Journal Letters, 943*(1), L3.
79. Sharma, V., et al. (2019). Time-varying polarized gamma-rays from GRB 160821A: Evidence for ordered magnetic fields. *The Astrophysical Journal Letters, 882*(1), L10.
80. Gill, R., Kole, M., & Granot, J. (2021).GRB polarization: A unique probe of GRB physics. *Galaxies, 9*(4), 82.
81. Srinivasaragavan, G. P., Dainotti, M. G., Fraija, N., Hernandez, X., Nagataki, S., Lenart, A., Bowden, L., & Wagner, R. (2020). On the investigation of the closure relations for gamma-ray bursts observed by swift in the post-plateau phase and the GRB fundamental plane. *The Astrophysical Journal, 903*(1), 18.
82. Zhang, B., Fan, Y. Z., Dyks, J., Kobayashi, S., Mészáros, P., Burrows, D. N., Nousek, J. A., & Gehrels, N. (2006). Physical processes shaping gamma-ray burst X-ray afterglow light curves: Theoretical implications from the swift x-ray telescope observations. *The Astrophysical Journal, 642*, 354.
83. Nousek, J. A., et al. (2006). Evidence for a canonical gamma-ray burst afterglow light curve in the swift XRT data. *The Astrophysical Journal, 642*, 389–400.
84. Blandford, R. D., & McKee, C. F. (1976). Fluid dynamics of relativistic blast waves. *Physics of Fluids, 19*, 1130–1138.
85. Toma, K., Ioka, K., Yamazaki, R., & Nakamura, T. (2006). Shallow decay of early x-ray afterglows from inhomogeneous gamma-ray burst jets. *The Astrophysical Journal Letters, 640*(2), L139–L142.
86. Genet, F., Daigne, F., & Mochkovitch, R. (2007). Can the early x-ray afterglow of gamma-ray bursts be explained by a contribution from the reverse shock? *Monthly Notices of the Royal Astronomical Society, 381*(2), 732–740.
87. Eichler, D., & Granot, J. (2006). The case for anisotropic afterglow efficiency within gamma-ray burst jets. *The Astrophysical Journal Letters, 641*(1), L5–L8.
88. Oganesyan, G., Ascenzi, S., Branchesi, M., Salafia, O. S., Dall'Osso, S., & Ghirlanda, G. (2020). Structured jets and x-ray plateaus in gamma-ray burst phenomena. *The Astrophysical Journal, 893*(2), 88.
89. Beniamini, P., Granot, J., & Gill, R. (2020). Afterglow light curves from misaligned structured jets. *Monthly Notices of the Royal Astronomical Society, 493*(3), 3521–3534.
90. Shen, R., & Matzner, C. D. (2012). Coasting external shock in wind medium: An origin for the x-ray plateau decay component in swift gamma-ray burst afterglows. *The Astrophysical Journal, 744*, 36.
91. Dereli-Bégué, H., Pe'er, A., Ryde, F., Oates, S. R., Zhang, B., & Dainotti, M. G. (2022). A wind environment and Lorentz factors of tens explain gamma-ray bursts X-ray plateau. *Nature Communications, 13*, 5611.
92. Krolik, J. H., & Pier, E. A. (1991). Relativistic motion in gamma-ray bursts. *The Astrophysical Journal, 373*, 277–284.
93. Fenimore, E. E., Epstein, R. I., & Ho, C. (1993). The escape of 100 MeV photons from cosmological gamma-ray bursts. *Astronomy and Astrophysics Supplement Series, 97*, 59–62.

94. Lithwick, Y., & Sari, R. (2001). Lower limits on lorentz factors in gamma-ray bursts. *The Astrophysical Journal, 555*, 540–545.
95. Pe'er, A., Ryde, F., Wijers, R. A. M. J., Mészáros, P., & Rees, M. J. (2007). A new method of determining the initial size and lorentz factor of gamma-ray burst fireballs using a thermal emission component. *The Astrophysical Journal Letters, 664*, L1–L4.
96. Acuner, Z., Ryde, F., Pe'er, A., Mortlock, D., & Ahlgren, B. (2020). The fraction of gamma-ray bursts with an observed photospheric emission episode. *The Astrophysical Journal, 893*(2), 128.
97. Dereli-Bégué, H., Pe'er, A., Bégué, D., & Ryde, F. (2014). Unraveling the origins of GRB X-ray plateaus through a study of X-ray flares. arXiv:2412.11533.
98. Ghisellini, G., Padovani, P., Celotti, A., & Maraschi, L. (1993). Relativistic bulk motion in active galactic nuclei. *The Astrophysical Journal, 407*, 65.
99. Racusin, J. L., et al. (2011). Fermi and swift gamma-ray burst afterglow population studies. *The Astrophysical Journal, 738*, 138.
100. Ghirlanda, G., Nappo, F., Ghisellini, G., Melandri, A., Marcarini, G., Nava, L., Salafia, O. S., Campana, S., & Salvaterra, R. (2018). Bulk lorentz factors of gamma-ray bursts. *Astronomy & Astrophysics, 609*, A112.
101. Atteia, J. L., Cordier, B., & Wei, J. (2022). The SVOM mission. *International Journal of Modern Physics D, 31*(5), 2230008.
102. Shvartzvald, Y., et al. (2024). ULTRASAT: A wide-field time-domain UV space telescope. *The Astrophysical Journal, 964*(1), 74.
103. Amati, L., et al. (2018). The THESEUS space mission concept: Science case, design and expected performances. *Advances in Space Research, 62*(1), 191–244.

Chapter 22
Multi-Temperature Blackbody Emission in GRB 170409A

Soumya Gupta and Sunder Sahayanathan

Abstract The prompt phase emission process of the Gamma-Ray Bursts (GRB) is still an open question. The GRB spectrum is often well fitted by an empirical smooth broken power-law function term as the Band model. However, observations from the Gamma-ray Burst Monitor (GBM) onboard the Fermi Gamma-Ray space telescope suggest the presence of a thermal component along with the non-thermal, whose origin is unintelligible. We perform a detailed study of the relativistically expanding fireball model with its temperature evolving with radius. The effect of the light travel time combined with the radially averaged spectrum results in a much broader spectrum compared to a Planck function. For this scenario, a numerical code was developed and coupled with the XSpec a local model. The time-averaged spectrum of the bright GRB 170409A was studied using the *Fermi*-GBM data. We find the spectrum is comprised of a thermal (multi-temperature black body model) and a non-thermal (cutoff power law) component. Our fit results suggest the emission is dominated by the thermal process with a photospheric radius of $R_{ph} = 12.82^{-0.05}_{0.07}$ log cm, photospheric temperature $T_{ph} = 1242.9^{-110.0}_{141.5}$ keV and bulk flow of the plasma $\Gamma = 457.6^{-124.7}_{142.5}$.

22.1 Introduction

Gamma-ray bursts (GRBs) represent the most energetic and luminous electromagnetic explosions in the Universe. The early detected prompt phase of the Gamma-ray bursts (GRB) in γ-ray, and hard X-ray lasts from a few ms to a hundred seconds. The prompt phase of GRB though widely studied over a few decades the radiation mechanism of the prompt phase is still a debatable topic [1, 2].

S. Gupta (✉) · S. Sahayanathan
Homi Bhabha National Institute, Mumbai, India

Bhabha Atomic Research Center, Mumbai, India

S. Ghosh, A. R. Rao (eds.), *The Relativistic Universe: From Classical to Quantum*, Astrophysics and Space Science Proceedings 61,
https://doi.org/10.1007/978-3-031-90186-7_22

The prompt emission of the GRB was believed to be thermal/quasi-thermal emission due to high optical depth ([3] and energy $\approx 10^{53}$ ergs released during the time scale of ms to few hundred of sec, in very small space r $\approx$7–8 log centimeter (cm)). However, most GRB's observed spectrum detected by BATSE was well-fitted with the Band function [4]. This spectrum obtained has a wider peak and is softer at low energies, which was extensively believed to be associated with the optically thin synchrotron processes. Our understanding of the prompt phase of GRBs is revolutionized after the advent of *Fermi*, where some observed low-energy spectra had harder low-energy spectra compared to the synchrotron model [5]. Further, in the recent detections, there were small samples of GRB having a spectrum as narrow as the Planck function while some observations were unable to be explained either by a thermal or non-thermal component alone.

It was found that the diverse spectrum of the GRB can be explained by the emission from the photosphere surface of a relativistic expanding fireball. The high optical depth due to the e^+e^- pair makes the fireball opaque to radiation. The photon starts to decouple from the plasma beyond the photospheric radius and escapes from the flow such that the observed optical depth reduces to unity. Additionally, the effect of equal-arrival-time volume will significantly modify the instantaneous spectrum of the GRB prompt phase. In this work, we developed the model under this scenario which results in a spectrum which is the superposition of multiple black bodies (mBB) over the temperature, evolving with the viewing angle. The mBB model is governed by four physical parameters in the observer's frame that are the on-axis photospheric radius R_{ph0}, the on-axis photospheric temperature, T_{pho}; the bulk flow of the plasma Γ and temperature variation index α [12].

The bright GRB 170409A detected by *Fermi* having fluence of the order of $\sim 10^{-4}$ ergs cm^{-2} was studied under the expanding fireball scenario. The spectrum of GRB170409A was well explained by a hybrid model of a multi-temperature blackbody model (thermal component) along with a cutoff power law function (non-thermal component). The spectrum suggests its thermal-dominated nature similar to the bursts reported earlier such as GRB 081221 [6] and GRB 220426A [7].

22.2 Relativistically Expanding Fireball and Observed Spectrum

The initial high optical density of the Gamma-Ray Burst results in the optical depth to the baryon-related electron Thompson scattering more than unity. The emerging radiation thus escaping near or at the base of the flow, termed as photospheric radius is thermalized [8]. Considering the scenario of the relativistically expanding spherical symmetric wind, the observed optical depth of the wind between the source and the observer will be $\tau = \tau(R, \theta)$ [9], where θ is the angle with respect to the line of sight of the observer at a distance D_L from the source and, r is the projected radius. Thus the photospheric radius R, can be expressed as

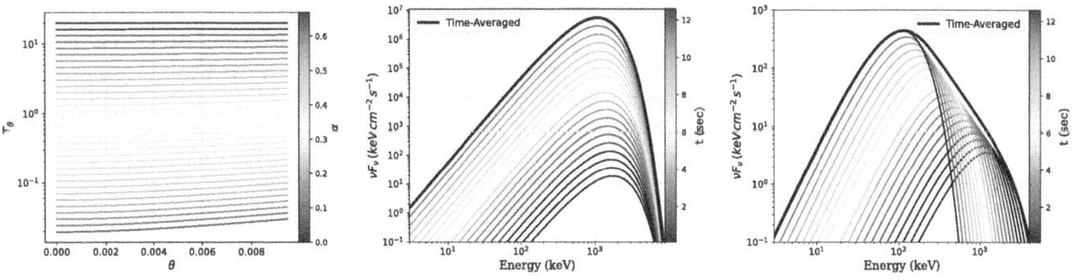

Fig. 22.1 The left plot depicts the variation of temperature with the viewing angle with the temperature variation index, α color coded on right. The middle and right plot shows a time-integrated spectrum (t = 15 s) in the black solid line and the evolution of the instantaneous spectrum, color-coded over time in the right; for the $\alpha = 0.1$ and $\alpha = 0.5$ respectively

$$R_{ph}(\theta) = \frac{\dot{M}\sigma_T}{4m_p\beta c\pi}\left[\frac{\theta}{\sin\theta} - \beta\right], \tag{22.1}$$

where $\dot{M}$ is a constant mass loss rate of the progenitor and β is the wind velocity. This angle dependence of the photospheric radius results in a non-trivial shape of the observed photospheric radius [10]. This influences the observed opening angle of the emission cone θ which can attain the maximum value of $1/\Gamma$. The angle turns to be considerably smaller during the initial evolutionary phase of the fireball and is computed from the condition $R(\theta) \geq R_{ph}(\theta)$; where $R(\theta)$ is the radial distance at an angle θ (Fig. 22.1).

As the fireball expands beyond $R_{ph}(\theta)$, the photons decouple from the inner regions and have a finite probability to be scattered along the line of sight [10, 11]. Thus the modified intensity observed will be proportional to

$$\zeta(\theta) = \int_{R_{in}}^{R(\theta)} [-\tau(z, \theta)]dz\,, \tag{22.2}$$

where, the direction of z is along the line of sight and $R_{in} = \text{MIN}\{[R(\theta) - R_{ph}(\theta)], R_{ph0}\cos(\theta)\}$, where R_{ph0} is on-axis photospheric radius.

The relativistic spherical expansion of the fireball will cause the photons emitted on-axis to arrive earlier than those emitted at higher latitudes. Due to this effect, the observer receives the photons from the *equal time surface* which are the combination of the higher latitude emissions. The evolutionary *equal time surface* is expressed as $[1 - \beta\cos(\theta)]/[1 - \beta]$, which is the ratio of the on-axis radius $R(0)$ to $R(\theta)$ [12]. Hence the observer will see the gradient in temperature over the angle θ such that the off-axis (T_θ) radiation will be hotter compared to the on-axis radiation (T_0). The variation in the temperature can be expressed as

$$T_\theta = T_0\left[\frac{R}{R_\theta}\right]^\alpha, \tag{22.3}$$

where α is the temperature variation index. The dynamics of the fireball as discussed by Piran et al. [13] have a physically acceptable range from 0.0 to 2/3. We can obtain the ratio of the on-axis temperature (T_0) to the on-axis photospheric temperature T_{ph0}, which is equal to $[R_{ph0}/R(0)]^{\alpha}$. Thus we obtain the observed instantaneous flux from the expanding fireball as

$$F_{\nu} = 4\pi h\nu^3 \left[\frac{(1-\beta)R(0)}{cD_L}\right]^2 \times \int_{\cos\phi}^{1} \frac{\mu}{(1-\beta\mu)^2} \frac{\zeta(\theta)\, d\mu}{\left\{\exp\left[\frac{h\nu}{kT_{ph0}}\left(\frac{R(0)}{R_{ph0}}\frac{1-\beta}{1-\beta\mu}\right)^{\alpha}\right]-1\right\}}. \tag{22.4}$$

The evolution of the instantaneous spectrum over the burst duration t over which fireball expands from the radius R_1 to R_2, where $R = 2\beta c\Gamma^2 t + R_{ph0}$. This will result in an observed radially averaged spectrum as a superposition of a series of different temperature black bodies integrated over an opening angle of θ (termed as multi-temperature black body (mBB) spectrum) expressed as

$$\mathscr{F}(\nu) = \frac{\int_{R_1}^{R_2} F_{\nu}(R)dr}{R_2 - R_1}. \tag{22.5}$$

22.3 Observation and Data Reduction

A high-energy emission was detected by Large Area Telescope (LAT) onboard *Fermi* from GRB 170904A [14], triggered by *Fermi* Gamma-ray Burst Monitor (GBM) at 02:41:59.49 on 9 April 2017 (T0) [15]. The best-on-ground location calculated by LAT was 347.53, −7.16 (RA, Dec) degrees (J200) with uncertainty in the radius of 0.15°. GRB 170409A light curve has multiple peaks with a total duration of 120s. The T90 of the GRB was found to be 68s (~T0-29 s to ~T0+95 s) in the energy range of 50–300 keV with prominent partially overlapped peaks. The event fluence in t90 duration was reported to be $(2.956 \pm 0.009)\ 10^{-4}$ erg cm^{-2} in the 10–1000 keV energy band. GRB 170904A was also detected by Konus-Wind [16] with observed fluence of $(5.7 \pm 0.2)\ 10^{-4}$ erg cm^{-2} in the 20 keV–10 MeV energy range. The high fluence of an order, 10^{-4} ergscm^{-2} for the burst (Fig. 22.2) warrants a detailed spectral analysis.

The GRB 170409A spectral study is performed using the *Fermi* -GBM data. The time-tagged event (TTE) mode Fermi GBM data is obtained using gtburst software[1] from the Fermi GBM trigger catalog.[2] The spectral and temporal behavior of the

[1] https://fermi.gsfc.nasa.gov/ssc/data/analysis/scitools/gtburst.html.

[2] https://heasarc.gsfc.nasa.gov/W3Browse/fermi/fermigtrig.html.

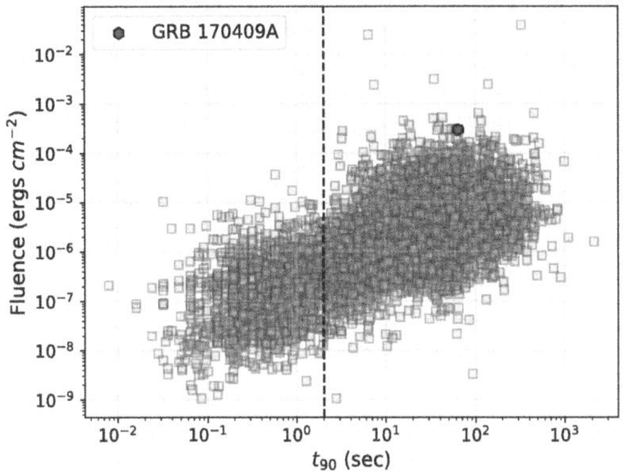

Fig. 22.2 Correlation plot between t90 duration and fluence for Gamma-Ray Bursts from Fermi GBM catalog to compare with the GRB 170409A in blue point

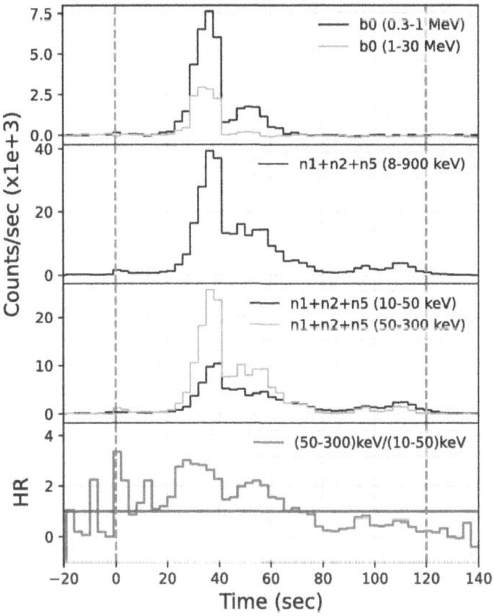

Fig. 22.3 The background-subtracted *Fermi* GBM light curves binned at 4 s in upper three panels in different energy ranges (the top most plot is BGO 0 detector and followed two panels are NaI 2 detector). The 0 on the x-axis is marked as *Fermi* trigger time (T_0) and the vertical grey lines are the start and stop times for the time-averaged analysis based on Bayesian block criteria in the 50–300 keV range. The bottom depicts the HR trend in hard (50–300 keV) to soft (10–50 keV) energy range for NaI 2 detector

prompt phase of the burst is studied using the brightest sodium iodide detectors NaI (1, 2 and 5) with source observing angles, NaI 1: 42°, NaI 2: 25° and NaI 5: 31°, respectively. We also selected the brightest bismuth germanate detector (BGO 0) as this BGO detector is closer to the direction of the burst (an observing angle of 34°). The RMFIT version 4.3.2 software7 was employed to create the energy-resolved prompt emission light curve using Fermi GBM observations. We recalculated the T_{90} of the GRB 170409A, NaI 2 detector in the energy range 50–300 keV using the Bayesian block technique [17].

The Fermi GBM energy-resolved (background-subtracted)light curves along with the evolution of hardness ratio (HR) are presented in Fig. 22.3. The light curve indicates the two well-separated peaks in the BGO 0 detector while the NaI detector hints toward multiple overlapped peaks with decreasing intensity. The hardness ratio (HR) evolution shows the harder to softer trend with the decreasing HR.

22.3.1 Spectral Analysis

The time-integrated analysis was carried out from the NaI and BGO detectors used for the temporal analysis. We extracted Fermi GBM the source spectra, background spectra and the instrument response file (for the time duration of T0-T0+120 s) using Make spectra for the XSPEC tool of gtburst software from Fermi Science Tools. The background spectrum was obtained by fitting the two temporal intervals, one before and the other after the GRB emission. To probe into the possible emission mechanism for the GRB 170407A multiple models were applied to explain the burst spectrum, using the X-ray Spectral fitting software XSPEC. The best model selection was based on the pgstat statistics information criteria for optimization, testing, and finding the best-fit statistics.

The energy range of 8–900 keV was considered for the NaI detector and 350 keV–15 MeV for the BGO detector for the GBM spectrum. The energy band of 33–40 keV was ignored for all the spectra due to the presence of the iodine K-edge at 33.17 keV while analyzing the NaI data. The GRB spectrum is often well-fitted using the smoothly joined broken power-law function known as Band or GRB function [4]. The fit statistics from the Band Function modeling was Bayesian information criterion (BIC_{band}) = 5167.72 and the model parameters obtained were low energy spectral index, α_{band} as $-0.776^{-0.01}_{0.01}$, high energy spectral index, β as $-2.62^{-0.08}_{0.09}$ and peak energy, E_{peak} as $694.18^{-30.3}_{32.0}$ keV (left plot, Fig. 22.4).

The empirical nature of the Band function does not give a proper insight into the physics of the prompt phase emission mechanism. Thus, we fitted the GRB 170409A spectrum using the expanding fireball model (Sect. 22.2) that accounts for the thermal radiation from the burst. The expanding fireball model alone did not explain the spectrum well, hints towards the presence of a non-thermal component. Hence, the cutoff power-law function (CPL) was used to address the non-thermal radiation from the burst (right plot, Fig. 22.4). The fit results achieved from the mBB+CPL model was BIC = 2738.05 ($< BIC_{band}$) which indicates it is a better

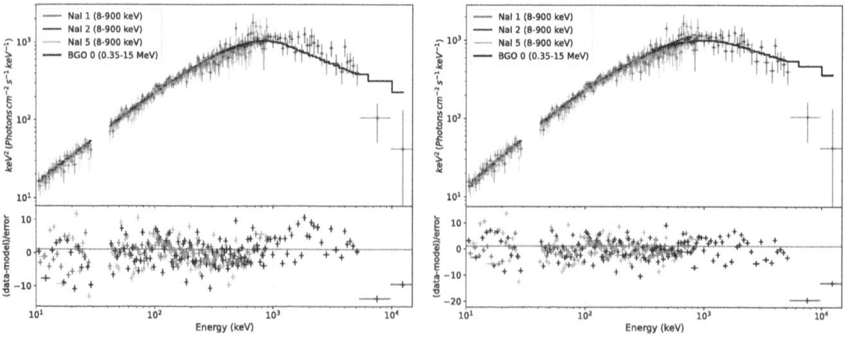

Fig. 22.4 The spectral fit results for the time-integrated region (T0-T0+120 s) fitted with the the Band function (left) and the mBB+CPL model (right)

model compared to the Band Function. We considered the burst in matter-dominated phase [13] and kept $\alpha = 2/3$. During this phase, the fireball coasts with constant $\Gamma = 457.6^{-124.7}_{142.5}$ and the other mBB model parameters attained from the fitting were on-axis photospheric radius (R_{ph}) and photospheric temperature (T_{ph}) of $12.82^{-0.05}_{0.07}$ log cm and $1242.9^{-110.0}_{141.5}$ keV respectively. The non-thermal component parameter for this burst was the photon index of $0.61^{-0.03}_{0.03}$ and the cutoff energy $E_{cut} = 246.47^{-31.75}_{44.7}$ keV.

22.4 Summary and Discussion

Gamma-Ray Bursts vary significantly in temporal as well as spectral behavior. The diverse spectrum of GRBs supported by observations like GRB 130606B [18] expressed by the fast cooling synchrotron emission mechanism, GRB 081221 [6] explained by the thermal radiation while the GRB 090902B [11] have the signature of both thermal and non-thermal components in the burst spectrum. This motivated us to explore the prompt phase emission mechanism from a different perspective. In this work, we performed detailed modeling of a relativistically expanding fireball, to explain the thermal component in the GRB spectrum. A multi-temperature black body spectrum under this scenario was constructed (Sect. 22.2). We developed numerical code and coupled it with XSpec to study the GRB 170409A prompt-phase emission mechanism.

The temporal study shows the harder-to-softer trend of the GRB 170409A during the hardness ratio evolution. The spectral study using the Band function hints at the need for an additional component along a non-thermal component. While pursuing the detailed spectral analysis, we found the mBB model accompanied by a cutoff power law function can explain the GRB 170409A spectrum well. Our results suggest the burst has a Γ of $457.6^{-124.7}_{142.5}$, where the thermal radiation escaped at radius (R_{ph}) of $12.82^{-0.05}_{0.07}$ log cm and photospheric temperature being $1242.9^{-110.0}_{141.5}$ keV.

A large sample of the GRB population can be explained by the hybrid models. The diverse observations of GRBs suggest that there is more than one explanation for the emission process. This also hints that radiation escaping from the fireball occurs at different regions of the burst, depending on the dynamics of the fireball (radiation-dominated or matter-dominated). This opens up more questions about the emission process of the prompt phase and the jet composition of the GRB. Hence, forthcoming sensitive and broad-band observations of the GRB prompt emission spectrum as well as a polarization study will provide a better understanding of the GRB emission mechanism.

Acknowledgments This research has used the Fermi data obtained through the HEASARC Online Service, provided by the NASA-GSFC High Energy Astrophysics Science Archive Research Center Online Service, provided by the NASA/Goddard Space Flight Center.

References

1. Zhang, B., & Yan, H. (2011). *The Astrophysical Journal, 726*, 90. https://doi.org/10.1088/0004-637X/726/2/90
2. Pe'er, A., Barlow, H., O'Mahony, S., et al. (2015). *The Astrophysical Journal, 813*, 127. https://doi.org/10.1088/0004-637X/813/2/127
3. Goodman, J. (1986). *The Astrophysical Journal Letters, 308*, L47. https://doi.org/10.1086/184741
4. Band, D., Matteson, J., Ford, L., et al. (1993). *The Astrophysical Journal, 413*, 281. https://doi.org/10.1086/172995
5. Preece, R. D., Briggs, M. S., Mallozzi, R. S., et al. (1998). *The Astrophysical Journal Letters, 506*, L23. https://doi.org/10.1086/311644
6. Hou, S.-J., Zhang, B.-B., Meng, Y.-Z., et al. (2018). *The Astrophysical Journal, 866*, 13. https://doi.org/10.3847/1538-4357/aadc07
7. Wang, Y., Zheng, T.-C., & Jin, Z.-P. (2022). *The Astrophysical Journal, 940*, 142. https://doi.org/10.3847/1538-4357/aca017
8. Paczynski, B. (1990). *The Astrophysical Journal, 363*, 218. https://doi.org/10.1086/169332
9. Abramowicz, M. A., Novikov, I. D., & Paczynski, B. (1991). *The Astrophysical Journal, 369*, 175. https://doi.org/10.1086/169748
10. Pe'er, A. (2008). *The Astrophysical Journal, 682*, 463. https://doi.org/10.1086/588136
11. Pe'Er, A., Zhang, B.-B., Ryde, F., et al. (2011). *AIP Conference Proceedings, 1358*, 91. https://doi.org/10.1063/1.3621745
12. Gupta, S., & Sahayanathan, S. (2024). *The Astrophysical Journal Letters, 970*, L12. https://doi.org/10.3847/2041-8213/ad5e1e
13. Piran, T., Shemi, A., & Narayan, R. (1993). *Monthly Notices of the Royal Astronomical Society, 263*, 861. https://doi.org/10.1093/mnras/263.4.861
14. Longo, F., Racusin, J. L., & Kocesvki, D. (2017). GRB coordinates network. Circular Service, No. 21004, #1.
15. Burns, E., Bissaldi, E., Poolakkil, S., et al. (2017). GRB coordinates network. Circular Service, No. 21005, #1.
16. Frederiks, D., Golenetskii, S., Aptekar, R., et al. (2017). GRB coordinates network. Circular Service, No. 21007, #1.
17. Scargle, J. D., Norris, J. P., Jackson, B., et al. (2013). arXiv:1304.2818. https://doi.org/10.48550/arXiv.1304.2818
18. Zhang, B.-B., Uhm, Z. L., Connaughton, V., et al. (2016). *The Astrophysical Journal, 816*, 72. https://doi.org/10.3847/0004-637X/816/2/72

Part VI
Active Galactic Nucleus Observations

Chapter 23
Ten Years of Searching for Relics of AGN Jet Feedback Through RAD@home Citizen Science

Ananda Hota, Pratik Dabhade, Prasun Machado, Avinash Kumar, Ck. Avinash, Ninisha Manaswini, Joydeep Das, Sagar Sethi, Sumanta Sahoo, Shilpa Dubal, Sai Arun Dharmik Bhoga, P. K. Navaneeth, Chiranjib Konar, Sabyasachi Pal, Sravani Vaddi, Prakash Apoorva, Megha Rajoria, and Arundhati Purohit

Abstract Understanding the evolution of galaxies cannot exclude the important role played by the central supermassive black hole and the circumgalactic medium (CGM). Simulations have strongly suggested the negative feedback of AGN Jet/wind/outflows on the ISM/CGM of a galaxy leading to the eventual decline of star formation. However, no "smoking gun" evidence exists so far where relics of feedback, observed in any band, are consistent with the time scale of a major decline in star formation, in any sample of galaxies. Relics of any AGN-driven outflows will be observed as a faint and fuzzy structure which may be difficult

A. Hota (✉)
UM-DAE Centre for Excellence in Basic Sciences, University of Mumbai, Mumbai, India

RAD@home Astronomy Collaboratory, Navi Mumbai, India

P. Dabhade
RAD@home Astronomy Collaboratory, Navi Mumbai, India

Instituto de Astrofísica de Canarias, Tenerife, Spain

Universidad de La Laguna (ULL), Departamento de Astrofisica, Tenerife, Spain

Astrophysics Division, National Centre for Nuclear Research, Warsaw, Poland

P. Machado · A. Kumar · C. Avinash · N. Manaswin · J. Das · S. Sethi · S. Sahoo · S. Dubal · S. A. D. Bhoga · P. K. Navaneeth · S. Vaddi · P. Apoorva · M. Rajoria · A. Purohit
RAD@home Astronomy Collaboratory, Navi Mumbai, India

C. Konar
RAD@home Astronomy Collaboratory, Navi Mumbai, India

Amity Institute of Applied Sciences, Amity University Uttar Pradesh, Noida, India

S. Pal
RAD@home Astronomy Collaboratory, Navi Mumbai, India
Midnapore City College, Paschim Medinipur, West Bengal, India

S. Ghosh, A. R. Rao (eds.), *The Relativistic Universe: From Classical to Quantum*, Astrophysics and Space Science Proceedings 61,
https://doi.org/10.1007/978-3-031-90186-7_23

to characterize by automated algorithms but trained citizen scientists can possibly perform better through their intuitive vision with additional heterogeneous data available anywhere on the Internet. RAD@home, launched on 15th April 2013, is not only the first Indian citizen Science Research (CSR) platform in astronomy but also the only CSR publishing discoveries using any Indian telescope. We briefly report 11 CSR discoveries collected over the last eleven years. While searching for such relics we have spotted cases of offset relic lobes from elliptical and spiral, episodic radio galaxies with overlapping lobes as the host galaxy is in motion, large diffuse spiral-shaped emission, cases of jet-galaxy interaction, kinks and burls on the jets, a collimated synchrotron thread etc. Such exotic sources push the boundaries of our understanding of classical Seyferts and radio galaxies with jets and the process of discovery prepares the next generation for science with the upgraded GMRT and Square Kilometre Array Observatory(SKAO).

23.1 Introduction

The current understanding of hierarchical structure formation, galaxy merger and feedback from the active galactic nuclei (AGN) has been widely accepted as prime drivers of galaxy evolution [1, 2]. Incorporation of this AGN-feedback could explain various statistical correlations observed in large populations of galaxies like M-σ relation, color-magnitude relation, luminosity function etc. [3–5]. During the galaxy merger, the nuclear star formation rate increases due to the infall of gas leading to high galactic superwind outflow as well as radiation, and radio-jet outflows from the central AGN [6–8]. These lead to the loss of cold gas and dust content in the merger remnant and eventually transform blue-star-forming late-type galaxies quickly to a red-looking early-type galaxy population. This quick loss within the merger timescale of nearly a billion years shows up as the gap between the "blue cloud" and "red sequence" of galaxies in the color-magnitude diagram [9]

However, this widely accepted standard model of AGN-feedback or outflows causing the decline of star formation as a process has not been observed which can be called "smoking gun" evidence. Massive outflows have been observed, power in it has been estimated to be sufficient, post-merger and post-starburst galaxies have been known, but in no sample so far the relics of the outflow match with the time-scale of decline of star formation [10]. For example, in the post-starburst galaxies, an abrupt decline has been observed in the last one billion years. Massive molecular gas outflows have been observed in various types of galaxies like mergers, post-merger, post-starburst etc. In many cases these outflow rates were found to be significantly above the current star formation rates demonstrating the power of the feedback to cause the eventual decline of star formation following galaxy mergers. However, these outflow time scales are only a few tens of million years [6]. Thus, due to the difference in time scales involved, the truncation of star formation (since a billion years) could not have been caused by the observed outflows (since a few tens of million years of outflows). Hence, there is a need to observe relics of AGN-driven outflow that could be older than the time scale since the decline of star formation in some transitional galaxies. Unlike in any other bands, evidence of multiple episodes of AGN activity is quite common in radio bands, seen clearly as 'Double Double Radio Galaxies' (DDRG; see review by Saikia and Jamrozy [11]) with well-separated synchrotron age of the episodes of jets [12]. In optical bands, the optical relics of quasar ionisation, popularly known as Hanny's Voorwerps, have been found to be a few to ten thousand years old. In the first such objects with low radio frequency follow-up, it has been found that the radio outflow is 100 Million years old that is much older than the optical relic (10 thousand years) and the quasar is no longer active [13]. This demonstrates the large scope in radio bands for the discovery of relic radio lobes and investigates the possible break in the star formation history of the AGN-host to support AGN-feedback models. As an example, we briefly explain here the case of the Cosmic Leaf Blower galaxy, a post-merger galaxy with sub-galactic scale s-shaped radio lobes, NGC3801 [14] with a speculative evolutionary scenario (Fig. 23.1, Left Panel). Over a billion years back an unequal mass merger between two spiral galaxies (Stage-1) led to the formation of a kinematically decoupled core (KDC) galaxy. The outer gaseous/stellar (H I and Hα) disk (r=15 kpc) rotates (v=280 km s^{-1}) along the major axis roughly aligned east-west (Stage-2). This outer gas disk shows some residual young star formation in UV, having an age of nearly 100–500 Million years while most of the galaxy looks yellowish, suggesting a typical early-type galaxy. The molecular gas (and dust) in the inner 4 kpc region rotate orthogonal to the larger scale gas disk. The radio jet emanates orthogonal to the central dust disk but after the hotspot regions, gradually bends to form an S-shape. The synchrotron age of these radio jets suggests it to be very young around 2.4 million years (Stage-4). Surrounding the hotspot regions, nearly 2 kpc away from the nucleus, X-ray emitting shock shells have been found which expand at nearly 850 km s^{-1} speed (Stage-5). Since the jet is expanding in the plane of the sky the shock shell would reach the outer gas disk with some young star formation in the next 10 million years which is like a blink compared to the 330 Million year time-scale of rotation of the large-scale gas disk. So in the next ten million years into the future, the outer young-star-forming gas disk may be disrupted but the overall red color of the galaxy can not be due to the ongoing AGN-driven radio/ionized-gas outflow/feedback and must be from an earlier episode (Stage-3). Thus, if deep low-frequency observation can find a relic radio emission which is around 100–500 million years old, only then, it can be a convincing case of "smoking gun" evidence (see [14] for details).

23.2 Discovery Methods and Results

Since the launch of RAD@home citizen science research collaboratory on 15th April 2013, citizen scientists have been trained to identify host galaxies and active/relic lobes of radio galaxies by making radio-optical overlays [10]. First, they were using NASA Skyview (https://skyview.gsfc.nasa.gov/) for such overlays

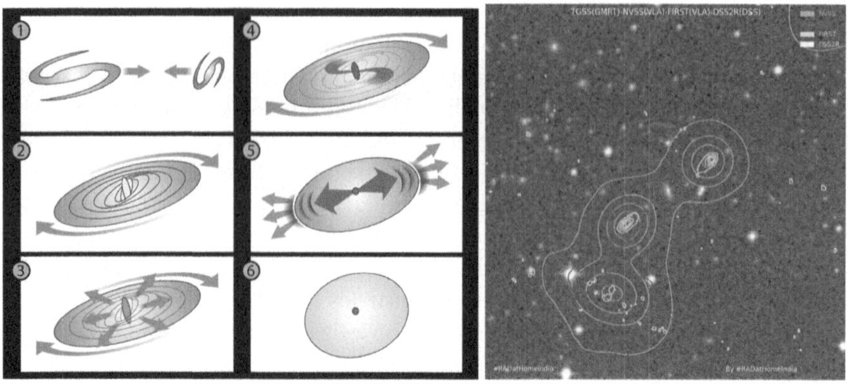

Fig. 23.1 Left panel (NGC3801): Evolutionary scenario of NGC3801 adapted from [14] and associated Press Release https://www.jpl.nasa.gov/news/cosmic-leaf-blower-robs-galaxy-ofstar-making-fuel. Right Panel (Speca): A typical RAD@home RGB-maker Composite Contour image where TGSS (magenta), NVSS(cyan) and FIRST(yellow) are plotted over DSS2R optical image. Contour levels (Jy beam^{-1}) are TGSS:[0.015, 0.031 0.046, 0.062] FIRST:[0.0005, 0.0026, 0.0048, 0.0069] NVSS:[0.0015, 0.01, 0.0185, 0.0271]

and later a Python-based RGB-maker web tool (https://www.radathomeindia.org/rgbmaker), launched by RAD@home on 26th January 2021 [15]. RAD@home training programs [known as RAD@home Discovery Camps (week-long) and RAD@home Astronomy Workshops (one-day event)] were conducted in collaboration with research and educational institutes. After learning to characterize standard radio galaxies (FR I, FR II, wide angle tailed. head-tailed, DDRG, X-shaped radio galaxy etc.) they were asked to identify radio sources from Giant Metrewave Radio Telescope (GMRT) 150 MHz data (TGSS DR5 & TGSS ADR1; [16]) which does not fit into standard classes and/or are "faint and fuzzy" which could be potential cases of relic lobes of past AGN feedbacks. An example of how a comparison of bright NVSS radio emission blobs would be seen faintly in the FIRST data for the relic radio lobe of the episodic radio galaxy Speca [17] is shown in Fig. 23.1 (right panel). In comparison to NVSS the TGSS data being nearly ten times longer in wavelength would also be ten times brighter for a steep spectral relic radio lobe, assuming a spectral index of -1. Thus such a relic lobe detected as an NVSS blob seen at 3σ (1.5 mJy) has a good chance to be seen in the TGSS at 3σ at 15 mJy. Such thumb rules were helpful for citizen scientists without much understanding of angular scale-sensitive radio interferometric images. Such scientifically interesting sources were then followed up with the GMRT telescope through a multi-cycle proposal titled "GMRT Observation of Objects Discovered by RAD@home Astronomy Collaboratory (GOOD-RAC)". One of the potential demonstrators was the discovery of RAD-12, a unique AGN with a one-sided radio jet hitting a companion galaxy and bouncing back to form a radio bubble [18]. Interestingly, unlike Minkowski's object, the companion galaxy shows no sign of enhanced star formation (positive feedback). Here we briefly report some exotic cases where relic radio lobe emissions have been identified which are currently being studied by the collaboratory.

RAD-Q-Mark Galaxy The first practical skill acquired by participants of the RAD@home Discovery Camp, hosted at various institutes, is the ability to identify the host galaxy of a radio source. For a FR I radio galaxy the radio peak matches with an optical peak but for FR II the radio peaks will be on either side of an optical galaxy. RAD12 was a peculiar case where the radio peak was in between two optical galaxies. RAD12, on follow-up observations with the GMRT, turned out to be a case where the one-sided jet hits the companion galaxy and returns to form a bubble (see [18] for details). Similarly, a radio source was found by a trained citizen scientist where the NVSS radio peak coincides roughly with a spiral galaxy, which was suspected as a Speca—spiral-host large episodic radio galaxy [17]. However, the high-resolution FIRST data identifies the core jet with the neighboring elliptical galaxy (RA: 09:18:59.38 Dec: +31:51:40.78), confirming that it is the host (Fig. 23.2 left panel). Interestingly, the FIRST image looks like a question mark sign "?". So this Q-mark galaxy clearly shows compact emission, typically seen in younger radio jets/ lobes but also shows fainter diffuse emission in NVSS, extending beyond the bright/compact features. While the structure seen in FIRST on the northern side suggests a 'kink' [19] the brightened blob on the southern jet, coinciding spiral galaxy, suggests a possible jet-galaxy interaction, without any optical signature, as seen in RAD12. Recent LOFAR-Two-Metre-Sky-Survey data release 2 (LoTSS DR-2 data [20]) of the Q-mark galaxy confirms the structures with higher resolution and higher significance level, strongly supporting our interpretations. A detailed study of the target is currently being drafted.

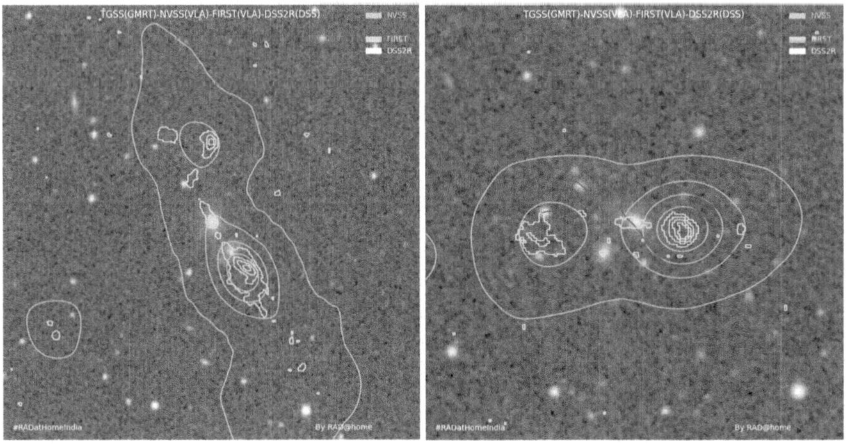

Fig. 23.2 Left Panel (RAD-Q-mark galaxy): Contour levels (Jy beam^{-1}) are FIRST:[0.0005, 0.0023, 0.004, 0.0058], NVSS:[0.0015, 0.0262, 0.051, 0.0757], TGSS (not available). Right Panel (RAD-jet-burl): Contour levels (Jy beam^{-1}) are TGSS:[0.015, 0.04, 0.064, 0.089], FIRST: [0.0005, 0.0021, 0.0037, 0.0054], NVSS: [0.0015, 0.0119, 0.0223, 0.0328]

RAD-Jet-Burl An unusual burl-like growth on the radio jets, other than the lobe and hotspot, was found during a RAD@home e-class conducted online during weekends (Date: 2–7 Oct. 2023). This is unlike the C- or U-shaped kink seen in the previous cases [19]. For the burl seen on the western side of the host galaxy, there is no apparent optical galaxy in the location to suggest possible jet-galaxy interaction and the corresponding structure on the eastern lobe is also missing (Fig. 23.2 right panel). This burl region is even brighter than the entire lobe on the other side. Given the redshift (z =0.140) of the host galaxy (RA: 14 33 40.754 Dec.: +63 58 30.65), clearly showing twin FRI jets in the FIRST data, the burl size can be as big as 25$''$ (60 kpc). Recent LoTSS data clearly shows the burl seen here but it lacks any internal C-/U-shaped structure. A resolved spectral index and polarization image of the jet-burl would be helpful in understanding its nature and origin.

RAD-Rabbit This unusual radio source, which resembles the face of a rabbit, was identified during the RAD@home Discovery Camp at the Institute of Physics (Bhubaneswar, Date: 23–29 June 2014). As shown in Fig. 23.3 this "Rabbit" (RA: 12:19:34.95 Dec: −13:26:11.7, redshift z =0.06950) shows the host galaxy to be offset to the north from most of the radio emission seen in the TGSS data. This is neither consistent with an FR I nor an FR II structure and is likely a relic lobe (Fig. 23.3 left panel). This source was further followed up with the GMRT (GOOD-RAC project) and found to show a diffuse/relic structure indeed. The east-west linear size is nearly 400 kpc with the radio core well-identified in the new GMRT data. The TGSS-NVSS spectral index ranges from −1.4 to −1.8, confirming its relic nature. The host galaxy also shows a merging companion in the Pan-STARRS images

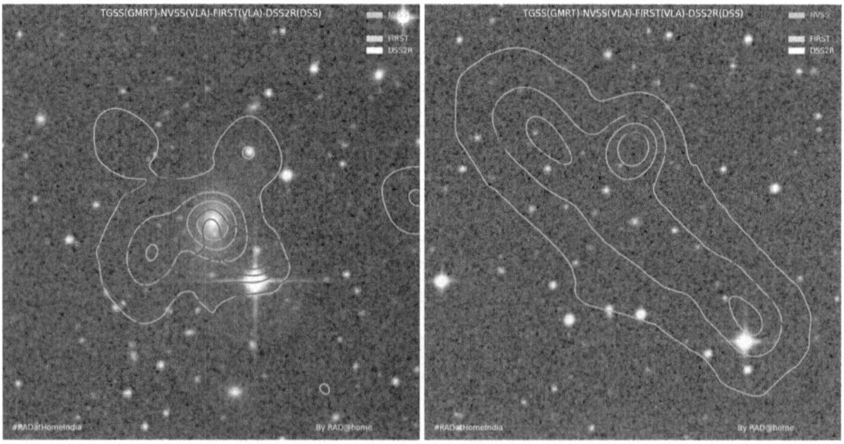

Fig. 23.3 Left panel (RAD-Rabbit): Contour levels (Jy beam^{-1}) are TGSS:[0.015, 0.038, 0.061, 0.083] NVSS: [0.0015, 0.0066, 0.0117, 0.0169], FIRST (not available). Right Panel (RAD-giant DDRG): Contour levels (Jy beam^{-1}) are TGSS:[0.015 0.058 0.1 0.143], NVSS:[0.0015 0.0116 0.0217 0.0319], FIRST not available

and that may have contributed to the highly disturbed circumgalactic environment distorting the radio lobe formation and/or evolution into a relic lobe.

RAD-Giant DDRG This unusual radio source was identified during the Discovery Camp at UM-DAE Centre for Excellence in Basic Sciences, University of Mumbai (Mumbai, Date: 1–8 June 2014). This source showed a compact emission beside a large diffuse emission, linear in NVSS and curved in TGSS DR5 (but straight in ADR1). The compact radio source, suspected to be the core, did not show any optical counterpart in DSS optical data suggesting a high-z galaxy (Fig. 23.3 right panel). If that is associated with the large diffuse source, it would be giant. This was followed up through the GOOD-RAC project and found to be an episodic or a DDRG. The compact source remained compact but in the middle of the linear diffuse emission, a mini double lobe radio source was identified. The mini-double was well aligned with the larger radio source and on either side of the central optical galaxy (RA: 11:11:42.66, Dec: −13:24:11.43, Photo-z $= 0.113 \pm 0.018$). Thus the relic lobe of this DDRG turned out to be a giant 800 kpc size.

RAD-Thumbs Up This unique radio source was identified during the Discovery Camp at the Nehru Planetarium (Delhi; date: 22–26 December 2018). As can be seen in the RAD-composite contour image (Fig. 23.4) where the NVSS (cyan) and TGSS (magenta) contours resemble the "Thumbs up" sign whose host galaxy was identified with a disturbed spiral galaxy (RA: 22:16:56.59 Dec: −04:24:33.4, z =0.095963). The host galaxy was confirmed with core and mini radio lobe detection in FIRST (yellow contours) confirming the nature of the diffuse structure to the west of the galaxy. Although the mini-lobe is seen to be comparable to stellar disk size (15 kpc) the size of the diffuse/relic-lobe emission is nearly 100 kpc, comparable to typical radio galaxies hosted in elliptical galaxies. However, this host galaxy is clearly a disturbed, young star forming (UV bright) spiral and

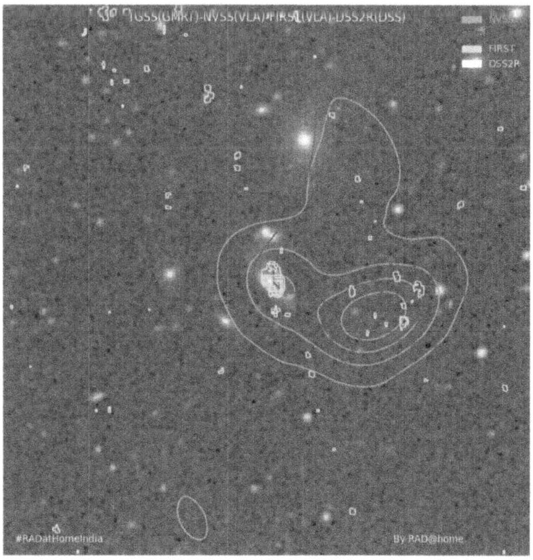

Fig. 23.4 Left panel (RAD-Thumbs Up): Contour levels (Jy beam^{-1}) are TGSS:[0.015, 0.027, 0.038, 0.05], FIRST: [0.0005, 0.0007, 0.0009, 0.0011], NVSS:[0.0015, 0.004, 0.0065, 0.009]

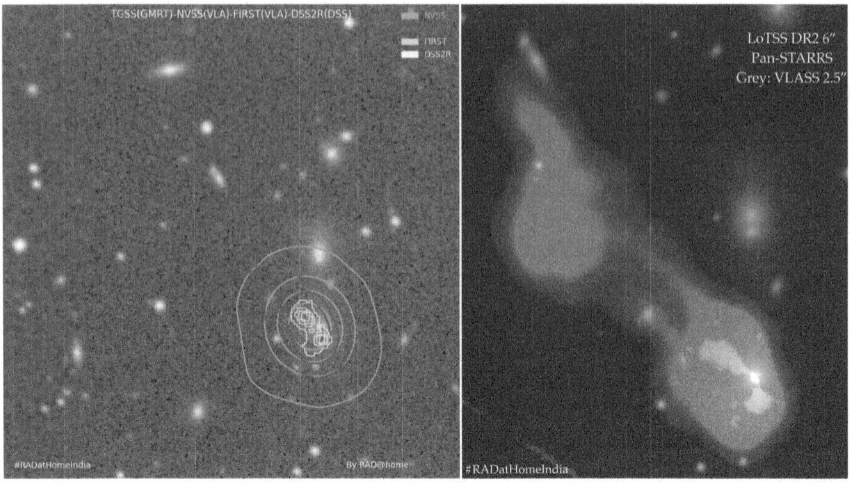

Fig. 23.5 Left panel (RAD-Episodic WAT): Contour levels (Jy beam^{-1}) are TGSS:[0.015, 0.047, 0.079, 0.11], FIRST:[0.0005, 0.0018, 0.0032, 0.0045], NVSS:[0.0015, 0.0113, 0.0211, 0.0308]. Right panel: LoTSS DR2 144 MHz with 6″ angular resolution in inferno color, in grey VLASS 3 GHz with 2.5″ angular resolution overlaid on Pan-STARRS image. This image has been made using the Aladin web tool. The VLASS image clearly resolves the inner structure of the radio galaxy and the LoTSS image shows the overall large extent of the source

even more strangely both the large-scale relic lobes are located on one side of the host in resemblance to wide-angle-tailed radio galaxies. Such large radio sources associated with spirals and leaving both the relic lobes on one side is extremely rare, if not unique. Although the linear size is small, an analogy can be drawn to the Virgo cluster spiral NGC4438 where a mini-double along with diffuse relic emission on one side can be seen due to ram pressure stripping [21].

RAD-Episodic WAT This interesting source was identified (Date: 28th October 2017) during weekend online e-classes discussion sessions conducted by the collaboratory. Comparing all three radio data and optical data (as shown in the composite contour image Fig. 23.5 left panel) it can be seen that the compact radio lobes ($\sim$50 kpc size) are located on either side of the host galaxy (RA: 15:33:20.24 Dec: +31:08:50.2, z =0.066694) which also coincide with the NVSS and TGSS peaks. However, a faint emission tail, 100–200 kpc size, is seen in the TGSS to the north-east of the galaxy which is not detected in NVSS/FIRST suggesting the diffuse and steep spectral nature. Recent LoTSS data [rms noise 80 μJy per beam (6″)] confirms this diffuse tail with a great significance level (Fig. 23.5 right panel). The natural explanation would be an episodic nature along with ram pressure stripping by the group/cluster medium which is strong enough to push back the relic but not enough to disrupt mini radio lobe formation. Such an environment gives rise to wide-angle-tailed (WAT) radio galaxies. Such episodic feedback along with

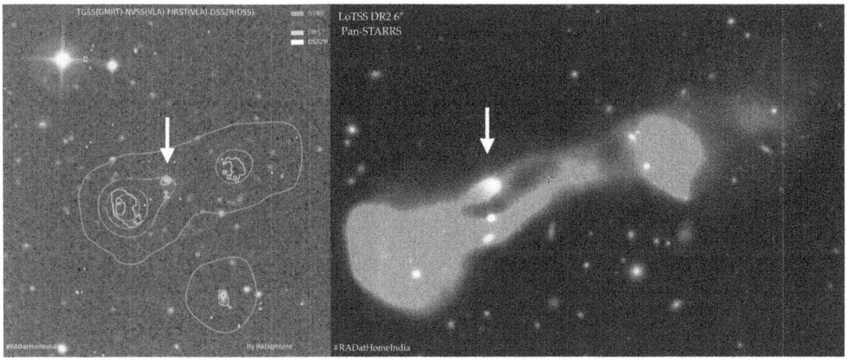

Fig. 23.6 Shows RAD-Collimated Synchrotron Thread source. Left Panel: Contour levels (Jy beam^{-1}) are TGSS:[0.015, 0.036, 0.057, 0.078], FIRST: [0.0005, 0.0028, 0.0051, 0.0074], NVSS:[0.0015, 0.0181, 0.0346, 0.0512]. Right panel: LoTSS DR2 144 MHz with 6″ angular resolution in inferno color overlaid or multi-color Pan-STARRS image. The host galaxy has been marked with an arrow sign on both panels. This image has been made using the Aladin web tool

ram pressure stripping during infall of galaxies into clusters may be depriving the galaxies of fuel for future star formation.

RAD-Collimated Synchrotron Thread Diffuse NVSS emission extending beyond compact emission structures seen in FIRST is a suggestion for relic lobes if the radio source is similar to FR II radio galaxy. Such a case was found during e-class (Date: 19–21 October 2023). The radio core (RA: 08:31:09.55 Dec: +41:47:38.41), eastern lobe and western lobe were identified in FIRST data but only on the western side did the NVSS emission extend farther west (left panel of Fig. 23.6). Similar to previously described cases, if the host galaxy is in motion towards the east, it is possible to explain the structure with two minor episodes of activity where the old and young lobes overlap on the east but are seen separated on the west. The LoTSS DR2 data (right panel of Fig. 23.6) additionally reveals an entirely new linear radio structure connecting the eastern and western lobe, seen south to the radio core. This makes it a rare case of so-called "collimated synchrotron threads" seen remarkably with the MeerKAT telescope in the galaxy ESO 137-006 [22]. We suggest satellite galaxies moving around the host may be creating magnetic channels and these are being energized by the plasma from the radio lobes. A detailed study of the target is in progress.

RAD-Giant-Radio Spiral It is normal to resolve a point source of NVSS to a double-lobe radio source in FIRST data due to better angular resolution (45″ v/s 5″). However, in the following unusual case, spotted on 18th October 2023 the higher resolution data revealed a spiral-galaxy-like structure (Fig. 23.7 left panel) which is neither related to the star-forming arms of the host galaxy (SDSS J133116.48+441851.4, Photometric z =0.250±0.0515) nor the radio lobes are sub-galactic scale where the rotating ISM-could bend the lobes (e.g. NGC3801; [14]).

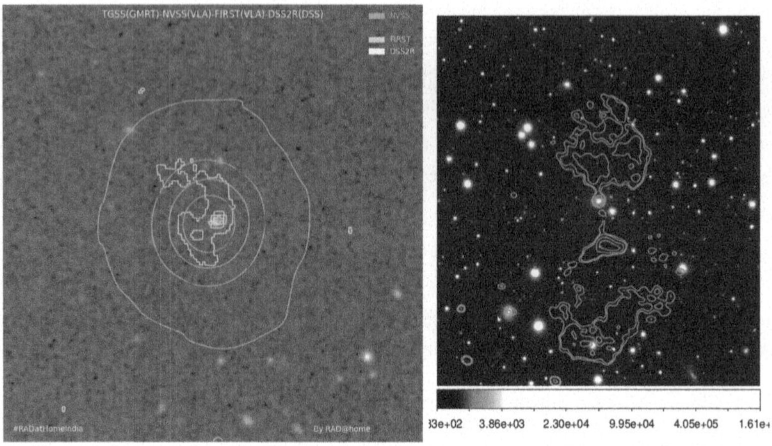

Fig. 23.7 Left panel (RAD-Giant-radio spiral):Contour levels (Jy beam^{-1}) are TGSS:[0.015, 0.1, 0.186, 0.271], FIRST:[0.0005, 0.0093, 0.0181, 0.0269], NVSS:[0.0015, 0.0291, 0.0567, 0.0843]. Right Panel (RAD-double-bubble): LoTSS contour levels in square root scale with values (starting with 3 x rms) are 0.00024, 0.00031875, 0.000555, 0.00094875, 0.0015 Jy beam^{-1}. The background false-color image is from Pan-STARRS in i-band

In Pan-STARRS images the host galaxy is clearly seen as an edge-on disk/lenticular galaxy, roughly aligned east-west. The source structure is confirmed from the LoTSS images where the spiral structure extends nearly 450 kpc, larger than typical FR I and FR II radio galaxies. Analogous to NGC 3801 and CenA, it is possible that the host galaxy is either a merger remnant or a kinematically decoupled core galaxy, with the nuclear gas/dust disk orthogonal to the large-scale gas disk. In future, sensitive SKAO images are likely to reveal more such radio-spirals and help understand the interaction between radio jets and circum-galactic medium which is not possible with other tracers of atomic, molecular and ionized gas.

RAD-Double-Bubble Since TGSS (25″) has a higher resolution than NVSS (45″) some relic lobes can be missed in interferometric TGSS images. So citizen scientists started checking the availability of some of their potential relic lobe cases in recent LoTSS data. One such case resembles two bubble-like lobes, found during e-class (Date: 20 September 2023), which is presented here. Identifying the host galaxy for such sources also becomes challenging, as radio cores may not be visible. However, due to the availability of sensitive surveys like Pan-STARRS host galaxies can now be easily identified from color images. The RGB-maker tool is yet to be updated to add such higher sensitivity images and hence, the SAO ds9 image analysis tool as used by the citizen scientists along with NED and Aladin. The relic lobe presented here neither corresponds to an edge-darkened FR I nor an edge-brightened FR II, rather diffuse bubbles (Fig. 23.7 right panel). The host galaxy is associated with a point source (RA:02:26:31.68, Dec: +42:09:03.28, z =0.144). Hence, the total north-south extent of the relic lobes, 250″, corresponds to 635 kpc, a giant radio

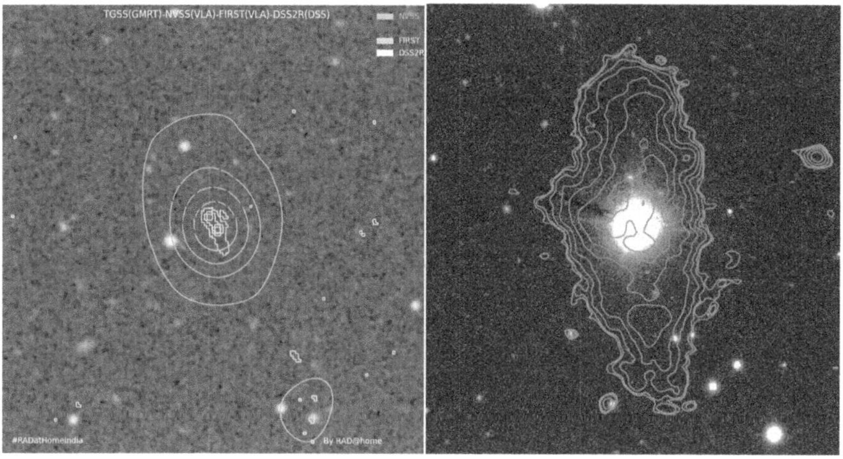

Fig. 23.8 Left panel (RAD-one sided relic lobe): Contour levels (Jy beam^{-1}) are TGSS:[0.015, 0.029, 0.043, 0.057], FIRST:[0.0005, 0.0025, 0.0044, 0.0064], NVSS:[0.0015, 0.0078, 0.0141, 0.0203]. Right panel: LoTSS contour levels in square root scale with values (starting with 3 x rms) are 0.00024, 0.000275918, 0.000383673, 0.000563265, 0.000814694, 0.00113796, 0.00153306, 0.002 Jy beam^{-1}. The background false-color image is from Pan-STARRS in i-band

galaxy. In cases when the core is undetected or lobe orientation is well aligned, it may also become classified as the puzzling class of objects known as Odd Radio Circles [23].

RAD-One-Sided Relic Lobe Yet another similar case was found during an online e-class (Date: 23–28 October 2023). The point source of the NVSS image was resolved to a triple radio source in the FIRST data with a radio core coinciding with the optical host galaxy (RA: 12:19: 11.51 Dec: +61:08:16.7, z –0.552). However, only TGSS shows a clear northern extension suggesting a relic lobe (Fig. 23.8 left panel). Recent LoTSS data resolves the TGSS tail to be a clear relic lobe of linear size near 425 kpc. This relic blob is bigger than the compact lobes of 230 kpc. A resolved imaging study of the relic lobe in any frequency other than around 150 MHz is needed to better understand it. The overall structure suggests a similarity with the previous episodic WAT. Thus it is likely a cluster-infalling galaxy, and episodes of feedback during infall may play an important role in the transformation of galaxies.

RAD-Radio-Fog-Beyond-Star Identifying the host galaxy for any diffuse radio sources has always been challenging. The host galaxy may have moved away after active jet creation. A confusing case, found during e-class (Date: 13–15 October 2022), presented here has a bright point source in the optical data (BD+39 2856, High Proper Motion Star). Furthermore, unlike the above case where the relic lobes are well separated in space, suggesting the host galaxy may be in the geometrical mid-point where the midpoint itself has the star. Bright stars are poor emitters of low-frequency radio emission. With careful analysis of the radio structure, although it has a north-south elongation the middle region has a significant east-west extension, we can see only two sources as possible host galaxies (Fig. 23.8 right panel). The source to the north (RA:15:16:19.91, Dec: +38:34:52.88, photometric z =0.192±0.064) and a fainter one on the north-west (RA:15:16:18.73, Dec: +38:34:46.52, photometric z =0.281± 0.286) are two possible host galaxies. Thus the radio source could be nearly 300 kpc long which is typical of FR II radio galaxies. Our understanding of radio galaxies is prejudiced by bright nearby sources hosted in elliptical galaxies. Such radio sources, not fitting to any class, may reveal new modes (diffuse bubbles) of radio emission in the near future.

23.3 Discussion and Summary

We have briefly presented 11 interesting objects from our decade-old citizen science research program named RAD@home. The participants were initially motivated to discover relic radio lobes as "faint-and-fuzzy" radio blobs in the TGSS 150 MHz radio survey images of the sky obtained using GMRT. In the process, the objects discovered were so remarkable that radio structures far beyond the typical features of jetted Seyferts and classical radio galaxies were identified. Burl-like features have been found on jets that may or may not be due to jet-galaxy interactions. Collimated Synchrotron Threads (CST) have been found that may be re-energized magnetic channels of satellite galaxies around the host radio galaxies. Radio spirals have been found which are larger than the typical FR II radio galaxies and they may have originated in kinematically decoupled core galaxies formed after a merger. Tailed infalling radio galaxies have been observed which show multiple episodes of activity. Some relic radio lobes form in the shape of bubbles, which may explain odd radio circles and some relic lobes can be purely diffuse almost without any structure.

Spiral galaxies exhibit a variety of continuum radio properties, ranging from diffuse radio halos without active galactic nuclei (AGN), such as M82 and NGC 253, to galaxy-scale radio bubbles associated with Seyfert nuclei (e.g., Circinus galaxy, NGC3079, & NGC6764; [24] and references therein). Additionally, some spiral galaxies exhibit relic or FR II radio lobes extending up to Mpc-scales which used to be associated only with giant elliptical galaxies, prior to the discovery of 'Speca' in 2011 [17]. Close to 3 dozen Speca-like galaxies have been discovered since then. This reflects a spectrum of low to high-power AGNs with varying manifestations of radio jets. Some underlying features or intrinsic properties of these galaxies may remain undetected or unobserved due to the limited sensitivity of current observational instruments. However, this limitation will be addressed with the advent of the Square Kilometre Array Observatory (SKAO), which will offer unprecedented sensitivity and superior observational capabilities. This advancement may also uncover details of radio-mode feedback acting on neutral atomic hydrogen gas (HI) located far beyond the stellar light distribution. This will provide new insights into previously unexplored modes of AGN feedback affecting the circumgalactic medium (CGM) of a galaxy. While Artificial Intelligence (AI) and Machine Learning (ML) may lead source classification in large datasets, citizen science research—particularly the Collaboratory model like RAD@home—can provide novel directions and handle complex cases more effectively. Furthermore, it prepares the next generation of astronomers, discovers new or peculiar types of sources (which serve as starting points for training AI/ML models), and contributes towards sustainable development goals (https://sdgs.un.org/goals).

Acknowledgments AH thanks the ISRA2023 conference organizers for the wonderful hospitality and environment conducive to academic discussions. AH acknowledges the University Grants Commission (UGC, Ministry of Education, Government of India) for monthly salary for the last 11 years. The long list of national and international organizations which have helped the establishment and growth of the first Indian astronomical citizen science research platform, RAD@home, are acknowledged in detail at https://radathomeindia.org/brochure.

References

1. Di Matteo, T., Springel, V., & Hernquist, L. (2005). *Nature, 433*, 604. https://doi.org/10.1038/nature03335
2. Hopkins, P. F., Hernquist, L., Cox, T. J., & Kereš, D. (2008). *Astronomy and Astrophysics Supplement Series, 175*, 356. https://doi.org/10.1086/524362
3. McNamara, B. R., & Nulsen, P. E. J. (2012). *New Journal of Physics, 14*, 055023. https://doi.org/10.1088/1367-2630/14/5/055023
4. Fabian, A. C. (2012). *Annual Review of Astronomy and Astrophysics, 50*, 455. https://doi.org/10.1146/annurev-astro-081811-125521
5. Kormendy, J., & Ho, L. C. (2013). *Annual Review of Astronomy and Astrophysics, 51*, 511. https://doi.org/10.1146/annurev-astro-082708-101811
6. Veilleux, S., Cecil, G., & Bland-Hawthorn, J. (2005). *Annual Review of Astronomy and Astrophysics, 43*, 769. https://doi.org/10.1146/annurev.astro.43.072103.150610
7. Croton, D. J., Springel, V., White, S. D. M., et al. (2006). *Monthly Notices of the Royal Astronomical Society, 365*, 11. https://doi.org/10.1111/j.1365-2966.2005.09675.x
8. Hardcastle, M. J., & Croston, J. H. (2020). *New Astronomy Reviews, 88*, 101539. https://doi.org/10.1016/j.newar.2020.101539
9. Heckman, T. M., & Best, P. N. (2014). *Annual Review of Astronomy and Astrophysics, 52*, 589. https://doi.org/10.1146/annurev-astro-081913-035722
10. Hota, A., Konar, C., Stalin, C. S., et al. (2016). *Journal of Astrophysics and Astronomy, 37*, 41. https://doi.org/10.1007/s12036-016-9415-8
11. Saikia, D. J., & Jamrozy, M. (2009). *Bulletin of the Astronomical Society of India, 37*, 63. https://doi.org/10.48550/arXiv.1002.1841
12. Konar, C., Saikia, D. J., Jamrozy, M., & Machalski, J. (2006). *Monthly Notices of the Royal Astronomical Society, 372*, 693. https://doi.org/10.1111/j.1365-2966.2006.10874.x
13. Smith, D. J. B., Krause, M. G., Hardcastle, M. J., & Drake, A. B. (2022). *Monthly Notices of the Royal Astronomical Society, 514*, 3879. https://doi.org/10.1093/mnras/stac1568
14. Hota, A., Rey, S.-C., Kang, Y., et al. (2012). *Monthly Notices of the Royal Astronomical Society, 422*, L38. https://doi.org/10.1111/j.1745-3933.2012.01231.x
15. Kumar, A., Avinash, C., Purohit, A., & Hota, A. (2023). *The multimessenger chakra of blazar jets*, I. Liodakis et al. (Eds.) (p. 375, vol. 40). https://doi.org/10.1017/S1743921323000674
16. Intema, H. T., Jagannathan, P., Mooley, K. P., & Frail, D. A. (2017). *Astronomy & Astrophysics, 598*, A78. https://doi.org/10.1051/0004-6361/201628536

17. Hota, A., Sirothia, S. K., Ohyama, Y., et al. (2011). *Monthly Notices of the Royal Astronomical Society, 417*, L36. https://doi.org/10.1111/j.1745-3933.2011.01115.x
18. Hota, A., Dabhade, P., Vaddi, S., et al. (2022). *Monthly Notices of the Royal Astronomical Society, 517*, L86. https://doi.org/10.1093/mnrasl/slac116
19. Dabhade, P., Shimwell, T. W., Bagchi, J., et al. (2022). *Astronomy & Astrophysics, 668*, A64. https://doi.org/10.1051/0004-6361/202243182
20. Shimwell, T. W., Hardcastle, M. J., Tasse, C., Best, P. N., Röttgering, H. J. A., Williams, W. L., et al. (2022). *Astronomy & Astrophysics, 659*, A1. https://doi.org/10.1051/0004-6361/202142484
21. Hota, A., Saikia, D. J., & Irwin, J. A. (2007). *Monthly Notices of the Royal Astronomical Society, 380*, 1009. https://doi.org/10.1111/j.1365-2966.2007.12114.x
22. Ramatsoku, M., Murgia, M., Vacca, V., et al. (2020). *Astronomy & Astrophysics, 636*, L1. https://doi.org/10.1051/0004-6361/202037800
23. Norris, R. P., Intema, H. T., Kapińska, A. D., et al. (2021). *Publications of the Astronomical Society of Australia, 38*, e003. https://doi.org/10.1017/pasa.2020.52
24. Hota, A., & Saikia, D. J. (2006). *Monthly Notices of the Royal Astronomical Society, 371*, 945. https://doi.org/10.1111/j.1365-2966.2006.10738.x

Chapter 24
UV Star Formation Properties of NGC 1097

Bannanje Ananthamoorthy and Debbijoy Bhattacharya

Abstract We carried out the study of star formation (SF) properties in one of the nearby spiral galaxies NGC 1097 using *AstroSat*-UVIT and *GALEX* observations. We analyzed SF knots in the galaxy, which are identified with UVIT observations within a central 275 arcsec ($\sim$ de Vaucouleurs radius; R_{25}). The 'star formation rate' (SFR) in the nuclear 'starburst ring' is $1.99 \pm 0.03\ M_{\odot}yr^{-1}$, consistent with earlier studies in other wavebands. A good correlation is observed between FUV extinction and the SFR density of the UV knots, and the slope is consistent with a similar relation for global SF properties in AGN host galaxies. Such a correlation could be due to the coincidence of high SF and high dust in the regions of higher gas density, which would then be the case for both active and normal galaxies. An alternative scenario could be a large-scale AGN impact on the host galaxy via radiation pressure on dust, triggering SF.

24.1 Introduction

Galaxies can be broadly classified into early-type and late-type galaxies. The former are associated with a redder color, suggesting that they harbor old stellar populations. In contrast, late-type galaxies harbor blue SF regions [1, 2]. The transition from blue to the red stellar population can be due to secular evolution or due to environmental effects [3]. The central AGN also plays a role in the quenching/triggering of SF in the galaxies [4, 5].

UV radiation is the direct tracer of SF in galaxies. Therefore, the UV study of galaxies plays a crucial role in understanding the SF properties of the galaxies [6–8]. *'Galaxy Evolution Explorer'* [9], a UV mission launched by NASA, played an essential role in revealing crucial insights into SF in galaxies. UV observations have made important discoveries like extended UV disks in galaxies [10], triggered

B. Ananthamoorthy · D. Bhattacharya (✉)
Manipal Centre for Natural Sciences, Manipal Academy of Higher Education, Manipal, India
e-mail: anantha.moorthy@learner.manipal.edu; debbijoy.b@manipal.edu

S. Ghosh, A. R. Rao (eds.), *The Relativistic Universe: From Classical to Quantum*,
Astrophysics and Space Science Proceedings 61,
https://doi.org/10.1007/978-3-031-90186-7_24

SF in merger system [11, 12]. However, the angular resolution was limited to ~ 4–5 arcsec. 'UltraViolet Imaging Telescope' onboard *AstroSat* [13, 14], has not only improved the angular resolution to ~ 1.2 to 1.5 arcsec [15–18] but also can observe in multiple narrow band filters within FUV and NUV. With already available observations, it is providing crucial information [19, 20].

24.1.1 NGC 1097

NGC 1097 is an SBb type of galaxy with a position angle of 141°. The galaxy hosts a strong bar with a primary bar length of 7.9 ± 0.6 kpc [21]. The galaxy hosts a 'Low Ionization Nuclear Emission region (LINER)' [22, 23], with luminosity, $L_{bol} = 4 - 8 \times 10^{41}$ erg s^{-1} [24, 25]. The galaxy is also observed to be interacting with another nearby elliptical galaxy, NGC 1097A. The galaxy is located at a distance of 14.8 Mpc [26], which corresponds to a spatial scale of $1'' \sim 0.72$ pc.

The galaxy also hosts a nuclear SF ring with a radius of 800 pc [27, 28]. Prieto et al. [28] identified 250 clusters associated with the SF ring using *'Hubble Space Telescope' (HST)* observations. They observed dust lanes via a bar connecting the nuclear SF ring. They suggested that the SF ring and the black hole are fed by the kpc scale dust lanes. With the available gas mass and the current SFR, they suggested that the accumulated stellar mass is less than expected at the observed SFR.

Kolcu T. et al. [29] studied the spatially resolved gas properties in the galaxy's central region ($60'' \times 60''$) using optical integrated field spectroscopic observations. They observed a secondary component in gas velocity, which they attributed to the outflows from the AGN. NGC 1097 also hosts a nuclear bar structure with a length of 20 arcsecs [29, 30]. Kolcu T. et al. [29] also observed a shocked gas in the spiral arm in the inner regions, which is aligned with the bar in the nucleus. They suggested that the nuclear bar plays a role in the inflow of matter into the central AGN.

Due to the relatively low inclination angle (54°) and larger angular size (R_{25} = 280 arcsec), a detailed study of star formation properties of the galaxy can be carried out in NGC 1097. NGC 1097 also provides a platform to study the interplay between merger, AGN, and bar influence on the star formation properties of the galaxy. In this chapter, we discuss the star formation properties of the UV knots detected from UVIT observations.

24.2 Data

24.2.1 AstroSat-UVIT

Level-1 data of our observations (PI: B Ananthamoorthy) of NGC 1097 with *AstroSat*-UVIT is obtained from the 'Indian Space Science Data Center' (ISSDC).[1]

[1] https://astrobrowse.issdc.gov.in/astro_archive/archive/Home.jsp.

The source is observed in the FUV BaF2 (F154W) filter. The data is processed through the CCDLAB [31, 32] to convert from Level-1 to Level-2. The final image has a plate scale of $\sim 0.416''$, angular resolution of $\sim 1.3''$, and exposure time of 30.9 ks.

24.2.2 *GALEX*

The *GALEX* has a field of view of $\sim$ 1.25 degrees with an angular resolution of 4.5–5 arcsec [9]. The coadded, calibrated, multiple visit images in FUV and NUV bands from *GALEX* observations are obtained from the 'Mikulski Archive for Space Telescopes' (MAST) portal.[2] The data for the current work is obtained from the 'Nearby Galaxies Survey' (NGS) with exposure of $\sim$ 2.9 ks in both FUV and NUV.

24.3 Analysis and Results

24.3.1 *Detection of SF Knots*

UVIT observations could distinctly resolve the AGN at the center and the SF ring in the galaxy, as provided in Fig. 24.1, due to its better angular resolution than *GALEX*.

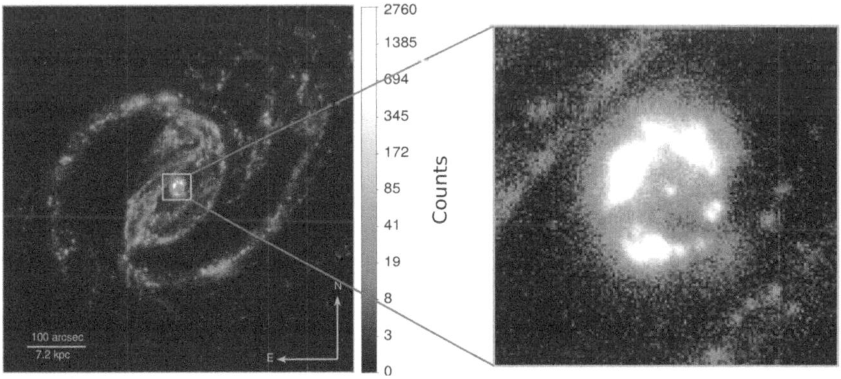

Fig. 24.1 Left panel: *AstroSat-UVIT observations of central* $10' \times 10'$ *region of NGC 1097. Right panel: zoomed in central* $30'' \times 30''$ *region of NGC 1097*

[2] https://galex.stsci.edu/GR6/.

We utilized UVIT-FUV images to identify the SF knots. Identification of the SF knots involves the detection of clumps of SF with intensities above the local fluctuations in the background. 'SExtractor' [33] tool is utilized to detect the SF knots in the galaxy. However, SExtractor assumes the background to be Gaussian in nature. Moreover, Ananthamoorthy et al. [34] showed that it is essential to consider the Poisson distribution of the UVIT background as it has low background counts. Therefore, we utilize the method provided in Ananthamoorthy et al. [34] for SF knots identification from the UVIT observations. The local background is calculated in bins of 128 × 128 pixels by iteratively removing pixels above 3σ equivalent threshold considering Poisson distribution. The 4σ equivalent threshold was set for source detection considering the Poisson distribution at each background bin. We use this calculated background and thresholds to detect sources using SExtractor. As the region is crowded, we require a lower deblending threshold to deblend the sources that are closely lying in the UVIT observation. Therefore, we utilized the deblending parameter of 5e-7 [35] to identify the SF knots. The other values of the SExtractor parameters used in detecting the SF knots are provided in Ananthamoorthy et al. [34]. Due to possible spurious detection with the lower deblending parameters, only sources with a ratio of flux to flux error greater than ten are chosen for further analysis.

A total of 1645 knots are identified with the UVIT observations used for the analysis. In Fig. 24.2, we have provided the number of star-forming knots and the total area under star-forming knots as a function of distance from the galaxy center. It is observed that beyond a radius of 275 arcsec, the number distribution and corresponding area decreases rapidly. Therefore, we considered the knots detected within the radius of 275 arcsec for the analysis. This radius also corresponds to the optical radius ['de Vaucouleurs radius'; R_{25}; $\sim 4.7'$] [36] for NGC 1097. A total of 1482 knots are considered for further analysis.

The FLUX_AUTO calculated using SExtractor in the units of counts per second (cps) is taken as the flux for each of the knots identified. The cps converted to the unitof erg cm^{-2} s^{-1} Å using the conversion factor provided in Tandon et al. [16]. The knots of SF identified with the UVIT observation are provided in Fig. 24.3. The

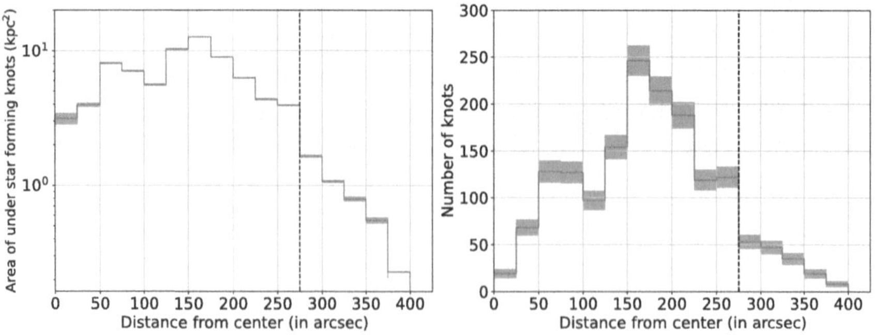

Fig. 24.2 Number of knots and area under the knots as a function of distance from the center

Fig. 24.3 SF knots detected with UVIT observations. Each source is represented by the elliptical aperture with an ellipse parameter derived from the SExtractor

area of the SF knot is calculated from the elliptical parameter derived for each knot using SExtractor. The area of the knots ranges from 0.0013–0.4 kpc^2.

24.3.2 UV Color Map from GALEX

We utilized *GALEX* FUV and NUV images to derive the UV color map. Sine FUV has a resolution ('full width at half maxima'; FWHM) of $\sim$ 4.2 arcsec, and NUV image has a resolution of $\sim$ 5.3 arcsec, both the images are brought into similar resolution by convolving the FUV images with the Gaussian convolution filter of FWHM $\sim$ 3.2 arcsec. The background contribution is subtracted by calculating the count rate in a few source-free regions calculated away from the galaxy. The pixels above 3σ of the local fluctuation of the background in both FUV and NUV are considered to derive the color. The observed count rate is converted to magnitudes using the calibration provided by Morrissey et al. [9]. FUV and NUV magnitudes are corrected for the 'Galactic extinction' [37]. The derived pixel-wise NUV-FUV

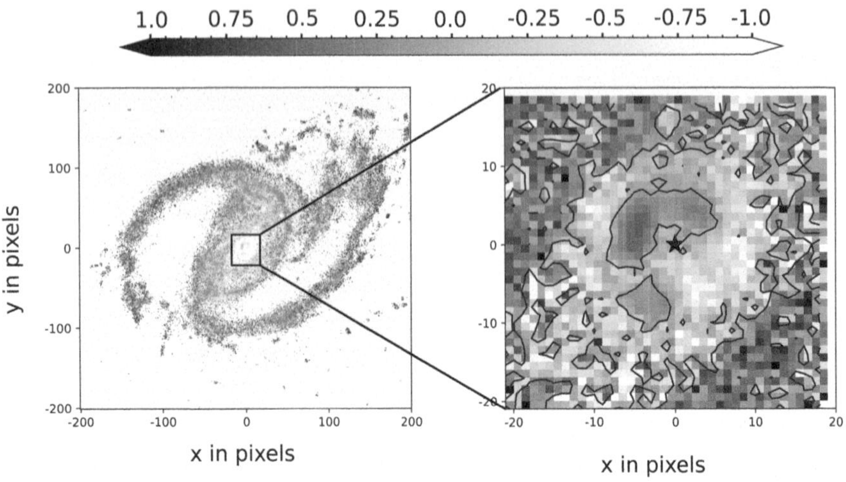

Fig. 24.4 Left panel: NUV-FUV color map obtained from the *GALEX* observation. A median value of −0.35 is subtracted. Right panel: Color in the central 40× 40 pixels of NGC 1097 revealing the FUV excess in the ring of SF. A median value of −1.08 is subtracted. In both the panels the pixels that are excluded from the color estimation have been assigned to −1 for better visibility

magnitude with median color (−0.35) subtracted from the image is provided in the left panel of Fig. 24.4. The Figure shows that the larger FUV emission in the outer region compared to the inner region could be due to higher dust or older stellar population in the central region. However, near the central region, corresponding to the SF ring, is observed to have higher FUV emissions relative to their neighboring region, which is zoomed-in in the right panel of Fig. 24.4.

24.3.3 Properties of the SF Regions

We derived the SFR for each UV knot detected from UVIT observation. The UV flux for the SF knots is corrected for the 'Milkyway extinction' using Cardelli et al. [37] using the visual extinction value of 0.073 [38].

We utilized the UV spectral slope to correct for the internal extinction. The spectral slope is calculated using the method by Nordon et al. [39] as follows:

$$\beta = \frac{m_1 - m_2}{-2.5log\frac{\lambda_1}{\lambda_2}} - 2\,, \qquad (24.1)$$

where m_1 and m_2 are the average *GALEX* FUV and NUV magnitude derived from the *GALEX* image in the box of 5 × 5 pixels centered at UV knot position. λ_1 and λ_2 are *GALEX* FUV and NUV effective wavelengths, respectively [9]. We further calculated the optical color excess $E(B-V)$ from the derived β values using the relation by Reddy et al. [40] as follows,

$$\beta = -2.616 + 4.684E(B - V)\,. \tag{24.2}$$

We utilized Calzetti et al. [41] extinction law to derive the FUV extinction (A_{FUV}) as follows,

$$E_s(B - V) = (0.44 \pm 0.03)E(B - V)\,, \tag{24.3}$$

$$K_\lambda = 2.659(-2.156 + \frac{1.509}{\lambda} - \frac{0.198}{\lambda^2} + \frac{0.011}{\lambda^3}) + R'_V\,, \tag{24.4}$$

$$A_\lambda = E_s(B - V)K_\lambda\,. \tag{24.5}$$

The extinction-corrected UV flux is used to derive SFR using the following relation [42, 43].

$$\mathrm{SFR_{FUV}}(\mathrm{M_\odot yr^{-1}}) = \frac{\mathrm{L_{FUV}}(ergs^{-1})}{3.83 \times 10^{33}} \times 10^{-9.51}\,. \tag{24.6}$$

We derived the SFR for the 'starburst ring' from the sum of SF knots detected between 4 arcsec (280 pc) and 14 arcsec (1 kpc). The choice of the parameters for the ring is obtained from Song et al. [44]. The observed SFR is $1.99 \pm 0.03\ \mathrm{M_\odot yr^{-1}}$ which is in agreement with the derived SFR for the 'starburst ring' by the previous studies using other wavelengths [28, 44].

We plotted the radial dependence of SFR density (Σ_{SFR}) from the center of the galaxy. It is observed that the total SFR density falls off with distance from the center, as shown in the 'left' panel of Fig. 24.5. The FUV extinction also shows a similar trend. We also plotted the position of each knot in the SFR density vs. distance from the galaxy center plane. The knots closer to the center have relatively higher SFR density than those away from the center. The median SFR density in the central 50 arcsecs ($\sim$ 3.6 kpc) region is 0.215 $\mathrm{M_\odot yr^{-1}\ kpc^{-2}}$, whereas in the outskirts in 200–250 arcsec is 0.034 $\mathrm{M_\odot yr^{-1}\ kpc^{-2}}$. The SFR and UV extinction of the SF knots also show a slightly decreasing trend with distance from the center, as shown in the 'top-right' and 'bottom-left' panels of Fig. 24.6, respectively. The area of the SF knots do not show such trends ('bottom-right' panel of Fig. 24.6).

A good correlation with the Pearson correlation coefficient of 0.68 ± 0.01 is also observed between FUV extinction and log (SFR density), as shown in Fig. 24.7. The slope is consistent with Nandi et al. [45], which is derived for the global properties of the AGN host galaxy. The result suggests that the relation between FUV extinction and the SFR density is not only applicable to global properties but also to the local SF regions in AGN host galaxies. It is expected to have such a correlation in both active and normal galaxies due to the coincidence of high dust and SFR in the regions of higher gas density. However, a detailed study of AGN and normal galaxies is essential to understand the effect of AGN feedback. The difference in correlation

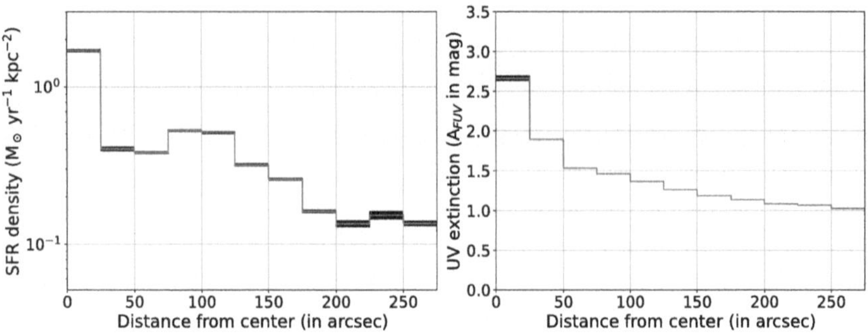

Fig. 24.5 Radial dependence of various properties in the NGC 1097. Left panel: SFR density. Right panel: UV extinction

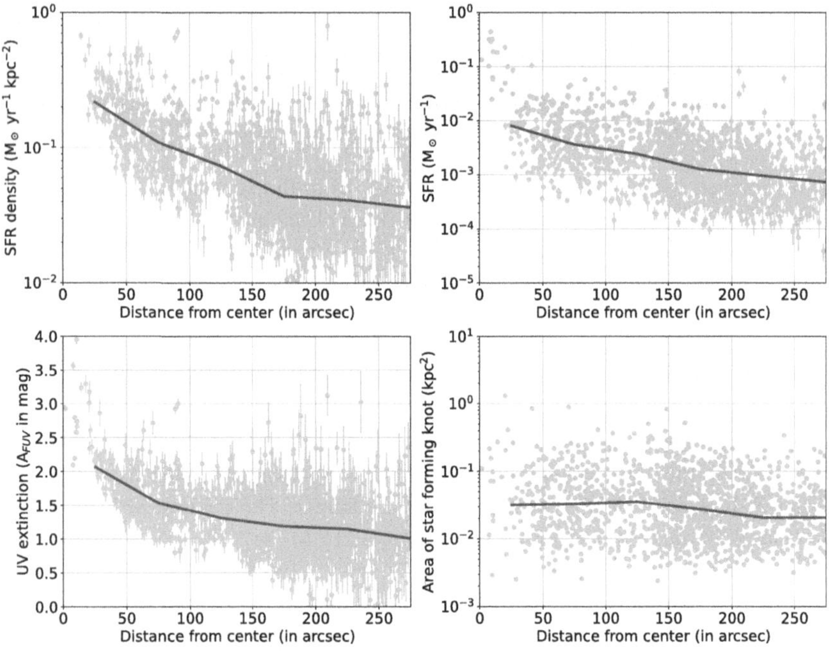

Fig. 24.6 Properties of individual knots in the NGC 1097 as a function of distance from the center. Top-left panel: SFR density. Top-right panel: SFR. Bottom-left panel: UV extinction. Bottom-right: area of SF knots

properties, if any, may indicate the radiation pressure-driven feedback in the AGN host galaxies, as suggested by radiation pressure-driven feedback theories [46, 47].

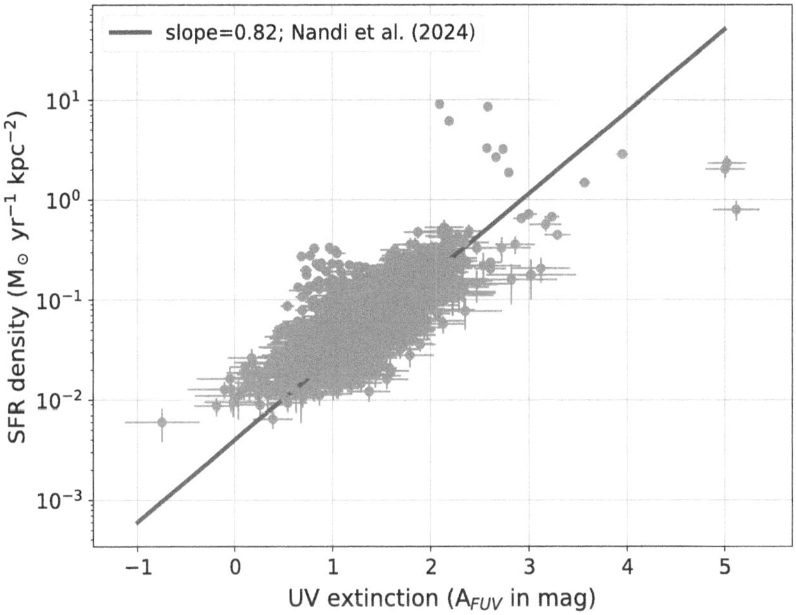

Fig. 24.7 Correlation between A_{FUV} vs. SFR density observed for SF knots of NGC 1097

24.4 Discussion and Summary

In this chapter, we studied the SF properties in one of the barred spiral galaxies, NGC 1097, hosting AGN. We utilized UV observations from *AstroSat*-UVIT and *GALEX*. The summary of the work is as follows.

- We detected the SF knots in NGC 1097 and derived the SF properties for these identified knots using UV observation from *AstroSat*-UVIT and *GALEX* observation.
- We analyzed a total of 1481 knots detected within 275 arcsec of the galaxy center using UVIT observation. The SFR for UV knots is derived using UVIT observations with correction to intrinsic attenuation derived from the *GALEX* FUV-NUV color.
- The UV SF derived from the nuclear star-burst ring is 1.99 ± 0.03 $M_\odot$ yr^{-1} is consistent with earlier observation using other wavebands.
- The area of the SF knots ranged from 0.024 to 1.3 kpc^2 with a median value of 0.027 kpc^2.
- The SFR density decreases as a function of distance from the center to outskirt nearly by a factor of 10, with a median density of 0.215 $M_\odot yr^{-1}$ kpc^{-2} in the central 50 arcsecs ($\sim$ 3.6 kpc) to 0.034 $M_\odot yr^{-1}$ in 200–250 arcsec region.
- A good correlation between FUV extinction and SFR density is observed in the galaxy. The slope of the correlation is consistent with the slope observed

by Nandi et al. [45] used for global galactic correlation observed in AGN host galaxies, indicating the relationship may also be used for local SF regions.

- It is expected to have a correlation between SFR density and UV extinction in both active and normal galaxies due to the coincidence of high dust and SFR in the regions of higher gas density. However, it is interesting to observe the degree of such correlation dependence in the normal and active galaxies to look for the large-scale impact of AGN feedback via radiation pressure as suggested by simulations [46, 47].
- The observed enhanced star formation activity in the nuclear region of this galaxy could be due to the positive AGN feedback. Alternately, the bar could also trigger such enhanced star formation activity. To disentangle these two possible scenarios, one needs to study a sample of the AGN host and a control sample of the non-AGN host galaxy. The results of such studies will be presented elsewhere.

Acknowledgments We acknowledge the anonymous referee for his/her valuable comments. BA acknowledges the financial support by DST, Government of India, under the DST-INSPIRE Fellowship (Application Reference Number: DST/INSPIRE/03/2018/000689; INSPIRE Code: IF190146) program. DB thanks the Department of Space, Govt of India, for the financial support under the ISRO Announcement of Opportunity (AO-3) project (grant number No.DS_2B-13013(2)/2/2021-Sec.2). The research is based to a significant extent on the results obtained from the AstroSat mission of the Indian Space Research Organization (ISRO), archived at the Indian Space Science Data Centre (ISSDC). The Payload Operations Center at IIA processed the UVIT data used here. The UVIT is built in collaboration between IIA, IUCAA, TIFR, ISRO and CSA. We made use of the *GALEX* database in the MAST archive, which is funded by the NASA Office of Space Science. Manipal Centre for Natural Sciences, Centre of Excellence, Manipal Academy of Higher Education (MAHE) is acknowledged for facilities and support.

Competing Interests The authors have no conflicts of interest to declare that are relevant to the content of this chapter.

References

1. Buta, R. J. (2013). *Planets, stars and stellar systems. Volume 6: Extragalactic astronomy and cosmology*, T. D. Oswalt & W. C. Keel (Eds.) (vol. 6, p. 1). https://doi.org/10.1007/978-94-007-5609-0
2. Blanton, M. R., & Moustakas, J. (2009). *Annual Review of Astronomy and Astrophysics, 47*(1), 159. https://doi.org/10.1146/annurev-astro-082708-101734
3. Kormendy, J., & Kennicutt, R. C. Jr. (2004). *Annual Review of Astronomy and Astrophysics, 42*(1), 603. https://doi.org/10.1146/annurev.astro.42.053102.134024
4. Heckman, T. M., & Kauffmann, G. (2011). *Science, 333*(6039), 182. https://doi.org/10.1126/science.1200504
5. Fabian, A. C. (2012). *Annual Review of Astronomy and Astrophysics, 50*, 455. https://doi.org/10.1146/annurev-astro-081811-125521
6. Calzetti, D. (2013). *Star formation rate indicators* (pp. 419–458). Cambridge University Press. Canary Islands Winter School of Astrophysics.

7. Salim, S., et al. (2007). *Astronomy and Astrophysics Supplement Series, 173*(2), 267. https://doi.org/10.1086/519218
8. Kennicutt, R. C. Jr. (1998). *Annual Review of Astronomy and Astrophysics, 36*, 189. https://doi.org/10.1146/annurev.astro.36.1.189
9. Morrissey, P., et al. (2007). *Astronomy and Astrophysics Supplement Series, 173*(2), 682. https://doi.org/10.1086/520512
10. Thilker, D. A., et al. (2007). *Astronomy and Astrophysics Supplement Series, 173*(2), 538. https://doi.org/10.1086/523853
11. George, K., et al. (2018). *Astronomy & Astrophysics, 614*, A130. https://doi.org/10.1051/0004-6361/201832705
12. Hibbard, J. E., et al. (2005). *The Astrophysical Journal Letters, 619*(1), L87. https://doi.org/10.1086/423244
13. Singh, K. P., et al. (2014). *Space telescopes and instrumentation 2014: Ultraviolet to gamma ray. Society of photo-optical instrumentation engineers (SPIE) conference series*, T. Takahashi, J. W. A. den Herder, & M. Bautz (Eds.), (vol. 9144, p. 91441S). https://doi.org/10.1117/12.2062667
14. Agrawal, P. C. (2006). *Advances in Space Research, 38*(12), 2989. https://doi.org/10.1016/j.asr.2006.03.038
15. Tandon, S. N., et al. (2020). *The Astronomical Journal, 159*(4), 158. https://doi.org/10.3847/1538-3881/ab72a3
16. Tandon, S. N., et al. *The Astronomical Journal, 154*, 128. https://doi.org/10.3847/1538-3881/aa8451
17. Subramaniam, A., et al. (2016). *Space Telescopes and Instrumentation 2016: Ultraviolet to Gamma Ray. Society of photo-optical instrumentation engineers (spie) conference series*. J. W. A. den Herder, T. Takahashi, & M. Bautz (Eds.) (vol. 9905, p. 99051F). https://doi.org/10.1117/12.2235271
18. Kumar, A., et al. (2012). *Space telescopes and instrumentation 2012: Ultraviolet to gamma ray*. Society of photo-optical instrumentation engineers (SPIE) conference series. T. Takahashi, S. S. Murray, & J. W. A. den Herder (Eds.) (vol. 8443, p. 84434R). https://doi.org/10.1117/12.924147
19. Saha, K. (2020). *Nature Astronomy, 4*, 1185. https://doi.org/10.1038/s41550-020-1173-5
20. Borgohain, A., Saha, K., Elmegreen, B., Gogoi, R., Combes, F., Tandon, S. N. (2022). *Nature, 607*(7919), 459. https://doi.org/10.1038/s41586-022-04905-9
21. Piñol-Ferrer, N., Fathi, K., Carignan, C., Font, J., Hernandez, O., Karlsson, R., & van de Ven, G. (2014). *Monthly Notices of the Royal Astronomical Society, 438*(2), 971. https://doi.org/10.1093/mnras/stt2162
22. Ho, L. C. (2008). *Annual Review of Astronomy and Astrophysics, 46*, 475. https://doi.org/10.1146/annurev.astro.45.051806.110546
23. Storchi-Bergmann, T., Eracleous, M., Teresa Ruiz, M., Livio, M., Wilson, A. S., & Filippenko, A. V. (1997). *The Astrophysical Journal, 489*(1), 87. https://doi.org/10.1086/304783
24. Nemmen, R. S., Storchi-Bergmann, T., Yuan, F., Eracleous, M., Terashima, Y., & Wilson, A. S. (2006). *The Astrophysical Journal, 643*(2), 652. https://doi.org/10.1086/500571
25. Prieto, M. A., Reunanen, J., Tristram, K. R. W., Neumayer, N., Fernandez-Ontiveros, J. A., Orienti, M., & Meisenheimer, K. (2010). *Monthly Notices of the Royal Astronomical Society, 402*(2), 724. https://doi.org/10.1111/j.1365-2966.2009.15897.x
26. Freedman, W. L., et al. (2001). *The Astrophysical Journal, 553*(1), 47. https://doi.org/10.1086/320638
27. Prieto, M. A., Maciejewski, W., & Reunanen, J. (2005). *The Astronomical Journal, 130*(4), 1472. https://doi.org/10.1086/444591
28. Prieto, M. A., Fernandez-Ontiveros, J.A., Bruzual, G., Burkert, A., Schartmann, M., & Charlot, S. (2019). *Monthly Notices of the Royal Astronomical Society, 485*(3), 3264. https://doi.org/10.1093/mnras/stz579
29. Kolcu, T., et al. (2023). *Monthly Notices of the Royal Astronomical Society, 524*(1), 207 (2023). https://doi.org/10.1093/mnras/stad1896

30. Quillen, A. C., Frogel, J. A., Kuchinski, L. E., & Terndrup, D. M. (1995). *Astronomical Journal, 110*, 156 (1995). https://doi.org/10.1086/117503
31. Postma, J. E., & Leahy, D. (2017). *Publications of the Astronomical Society of the Pacific, 129*(981), 115002. https://doi.org/10.1088/1538-3873/aa8800
32. Postma, J. E., & Leahy, D. (2021). *Journal of Astrophysics and Astronomy, 42*(2), 30 (2021). https://doi.org/10.1007/s12036-020-09689-w
33. Bertin, E., & Arnouts, S. (1996). *Astronomy and Astrophysics Supplement Series, 117*, 393. https://doi.org/10.1051/aas:1996164
34. Ananthamoorthy, B., Bhattacharya, D., P. Sreekumar, Swathi, B. (2024). *The Astrophysical Journal, 168*(1), 22. https://doi.org/10.3847/1538-3881/ad4991
35. Bordoloi, O. P., Ananthamoorthy, B., Shalima, P., Safonova, M., Bhattacharya, D., Shchekinov, Y. A., & Gogoi, R. (2024). *Publications of the Astronomical Society of the Pacific, 136*(7), 074101. https://doi.org/10.1088/1538-3873/ad5d14
36. Higdon, J. L., & Wallin, J. F. (2003). *The Astrophysical Journal, 585*(1), 281. https://doi.org/10.1086/345978
37. Cardelli, J. A., Clayton, G. C., & Mathis, J. S. (1989). *The Astrophysical Journal, 345*, 245. https://doi.org/10.1086/167900
38. Schlafly, E. F., & Finkbeiner, D. P. (2011). *The Astrophysical Journal, 737*(2), 103. https://doi.org/10.1088/0004-637X/737/2/103
39. Nordon, R. (2013). *The Astrophysical Journal, 762*(2), 125. https://doi.org/10.1088/0004-637X/762/2/125
40. Reddy, N. A. (2018). *The Astrophysical Journal, 853*(1), 56. https://doi.org/10.3847/1538-4357/aaa3e7
41. Calzetti, D., Armus, L., Bohlin, R. C., Kinney, A. L., Koornneef, J., & Storchi-Bergmann, T. (2000). *The Astrophysical Journal, 533*(2), 682 (2000). https://doi.org/10.1086/308692
42. Iglesias-Páramo, J., et al. (2006). *The Astrophysical Journal Supplement Series, 164*(1), 38. https://doi.org/10.1086/502628
43. Cortese, L., Gavazzi, G., & Boselli, A. (2008). *Monthly Notices of the Royal Astronomical Society, 390*(3), 1282. https://doi.org/10.1111/j.1365-2966.2008.13838.x
44. Song, Y., et al. (2021). *The Astrophysical Journal, 916*(2), 73 (2021). https://doi.org/10.3847/1538-4357/ac05c2
45. Nandi, P., Stalin, C. S., Dam, P., & Saikia, D. J. (2024). arXiv:2406.18103. https://doi.org/10.48550/arXiv.2406.18103
46. Ishibashi, W., & Fabian, A. C. (2012). *Monthly Notices of the Royal Astronomical Society, 427*(4), 2998. https://doi.org/10.1111/j.1365-2966.2012.22074.x
47. Ishibashi, W., Fabian, A. C., & Canning, R. E. A. (2013). *Monthly Notices of the Royal Astronomical Society, 431*(3), 2350. https://doi.org/10.1093/mnras/stt333

Chapter 25
Simultaneous Optical to X-Ray Spectra of OJ 287 and X-Ray Spectral Changes

Pankaj Kushwaha

Abstract OJ 287 is one of the most dynamic BL Lacertae objects that has exhibited behavior representative of the entire blazar class and is also one of the best sources with simultaneous multi-wavelength coordinated data. Motivated by the strong X-ray variability exhibited by the source, we systematically investigated the simultaneous optical to X-ray emission with a focus on the spectral state of the lowest recorded X-ray flux state to understand the X-ray spectral changes. The optical-UV synchrotron emission directly traces the high-energy end of the underlying particle spectrum and its power-law continuation to X-rays can drastically affect the X-ray spectrum without much change in optical flux. Thus, the combined optical to X-ray provides a potential tool to explore particle spectrum and insights into the highest particle energies. We report the finding of a power-law optical-UV spectrum with a photon spectral index of 2.71 ± 0.03 continuing to X-ray energies during the lowest recorded X-ray state—also a simultaneous lowest of optical and X-rays. Accounting for this contribution at X-ray results in a photon spectral index of 1.15–1.3. We discuss the possible implications of this on X-ray spectral variations and the underlying particle spectrum

25.1 Introduction

Accretion-powered sources are the most dynamic astronomical sources in the sky exhibiting frequent and often drastic changes in their observational behavior, especially the compact systems. The observed behavior and properties of the latter imply a significant contribution from relativistic species, generally in the form of non-thermal components. Studies exploiting different approaches to investigate these sources indicate quite broad similarities in some of the observational properties

P. Kushwaha (✉)
Department of Physical Sciences, Indian Institute of Science Education and Research Mohali, Knowledge City, India
e-mail: pankaj.kushwaha@iisermohali.ac.in

S. Ghosh, A. R. Rao (eds.), *The Relativistic Universe: From Classical to Quantum*, Astrophysics and Space Science Proceedings 61,
https://doi.org/10.1007/978-3-031-90186-7_25

(e.g. temporal: [1–5], and references therein) while completely contrary in other properties (e.g. spectral: [6] and references therein), implying a very diverse range of extreme intrinsic conditions.

Blazar is a sub-category within the active galactic nuclei (AGNs) referring to sources hosting a large-scale powerful relativistic jet almost pointing in the direction of the Earth. It comprised traditionally labeled flat spectrum radio quasars (FSRQs) and BL Lacertae objects (BLLs). Their observed continuum radiation properties are drastically different from the typical galaxies and even the non-jetted AGNs, with an almost exclusively non-thermal component dominated emission spread across the entire accessible electromagnetic (EM) windows from radio up to TeV gamma-rays (e.g. [7, 8]). The emission is characterized by a rapid and strong continuum variability as well as strong and variable polarization at radio and optical in general, and even at X-ray emission in many sources (e.g. [9–11]).

In terms of radiative output in different parts of EM windows, the broadband continuum exhibits a characteristic broad double-hump profile with a peak in near-infrared (NIR) to X-ray energies and the other at sub-MeV to GeV energies (e.g. [12]). The low-energy component spreads from radio up to X-rays and is widely considered to be the synchrotron emission from the relativistic electrons within the jet. The high-energy component, on the contrary, is still not well understood and is argued to be either due to relativistic leptons (electrons/positrons) or hadrons (protons primarily). In the lepton-based scenario, the high-energy component results from inverse Compton scattering of the surrounding photon fields while in the hadron-based explanation, it can result from proton synchrotron or an outcome of cascade triggered due to proton-proton and/or proton-photon interactions. The claim of association of neutrinos in a few blazars (e.g. [13–15]) points towards hadronic contribution but current data support minimal contribution via hadronic channels at MeV-GeV (e.g. [16]), indicating radio to GeV emission is primarily via leptonic channels.

OJ 287 is one of the most dynamic BL Lac objects at the redshift of z=0.306. It is known for strong and frequent variability not only in flux/brightness but spectral and polarization as well (e.g. [17–21]; and references therein). The claim of a $\sim$ 12-year quasi-periodic modulation in the optical band and its interpretation as a binary black hole system with flare resulting from secondary affecting the accretion-disk of primary and prediction of future flares (e.g. [22] and references therein) made it one of the favored source for coordinated monitoring. Later studies using radio data also reported periodicity in knot structure, attributing it to the jet precision of the primary [23] (but see [24]). The outcome so far is mixed with some features consistent with the former while some with the latter (e.g. [12, 25] and references therein). Nonetheless, the appearance of new components [18, 26, 27] indicates a dynamic phenomenon behind it.

OJ287 exhibited one of the most drastic spectral variations during 2016–2017 with spectral changes in all the observational windows of the electromagnetic spectrum, most notable especially at X-rays as reported in many previous works [12, 18, 26, 28–32]. These spectral variations are attributed either to a new emission component (e.g. [18, 29, 32]) or the synchrotron component extending into the X-ray band (e.g. [31] and references therein). In this chapter, we report results from the investigation of simultaneous optical to X-ray spectral variations with a focus on the lowest optical-UV and X-ray state of the source and X-ray spectral changes.

25.2 Simultaneous Optical to X-Ray Data

For transient and highly dynamic high-energy astronomical sources, the *Neils Gehrels Swift* observatory is the best facility for broadband observations with autonomous modes of operation capable of sampling a wide range of brightness states [33]. It hosts three astronomical payloads: the Ultra-Violet Optical Telescope (UVOT) with six astronomical filters covering the optical to UV region, the X-ray Telescope (XRT) sensitive to X-ray photons of energies 0.2–10 keV, and the Burst Alert Telescope (BAT) for photons of energies between 15 to 150 keV, thereby offering simultaneous coverage from optical to hard X-ray bands of the electromagnetic spectrum. For our work here, we have used the public archival data from two payloads—UVOT and XRT from the start of the facility operation in 2005 to 2022 to study the spectral evolution of the BL Lac object OJ 287.

Except for a few pointing observations, most XRT observations have a typical exposure of around a kilo-second with a few hundred seconds in a UVOT filter. Such a typical exposure is sufficient for the spectral and temporal study of OJ 287 most of the time by these instruments. The BAT is not so sensitive for temporal or spectral studies with this kind of exposure unless the source is too bright, like a GRB and thus BAT is not considered (e.g. [34]).

The details of data UVOT and XRT reduction employed for this study are described in [35]. In short, for flux density extraction from the UVOT bands, a standard source region of $5''$ and an annular background region free of any source of inner and outer radii $15''$ and $20''$ was employed. For X-ray, a circular source region of $47''$ was used while the background was extracted using a source-free annular region with inner and other radii respectively of $65''$ and $90''$. The UVOT flux densities were then extracted using the task *uvotsource* while XRT flux points were extracted by fitting the spectra within the XSPEC [36] with an absorption-modified power-law (PL: $dN(E)/dE \equiv K(E/E_0)^{-\Gamma}$) and log-parabola model (LP: $dN(E)/dE \equiv K(E/E_0)^{-\alpha-\beta log(E/E0)}$). The best model was chosen based on the F-test value. For simultaneous optical-UV to X-ray spectral study, the UVOT data files were converted to the respective PHA files using the HEASOFT[1] task *uvot2pha* and the simultaneous fitting was performed in the XSPEC.

[1] https://heasarc.gsfc.nasa.gov/lheasoft/.

25.3 Analysis and Results

Figure 25.1 shows the simultaneous optical to X-ray light curve extracted using the pointed mode data from the XRT and UVOT between 2005 and 2022 (MJD: 53510–59970). In terms of variability, the XRT flux spans > 2 orders of magnitude between the recorded minimum and maximum flux in the 0.3–10 keV while UVOT bands during this period exhibit > 1 order of magnitude in the flux density. The huge flux changes are associated with strong spectral changes as reported in various studies (e.g. [18, 26, 29, 31]) and the 2016–2017 activity was also coincident with the first ever reported very high energy (VHE; E > 100 GeV) activity of the source by the VERITAS observatory [37].

In terms of simultaneous flux/brightness variability trend at optical-UV and X-rays, one can observe a peculiar feature—a simultaneous low during MJD: 54160–54180. Later though optical-UV has gone even lower but not the X-rays. Since many previous studies have reported synchrotron driving X-ray spectral changes (e.g. [31, 38, 39] and references therein), we explored the optical-UV to X-ray spectral state of the lowest optical-UV and X-ray state during this simultaneous low phase. For

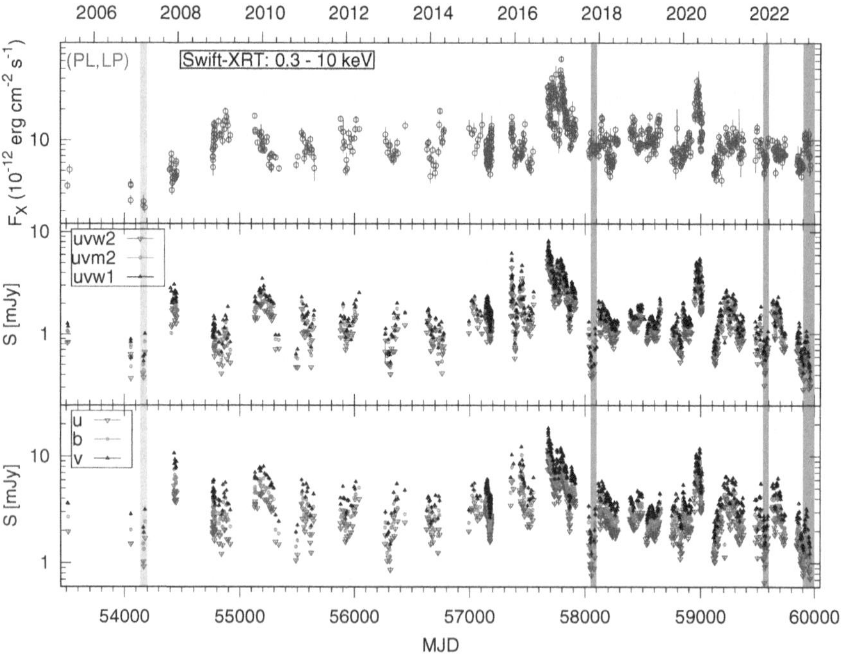

Fig. 25.1 Simultaneous optical to X-ray light curve of OJ 287 from the *Swift* facility observation between 2005 to 2022 (ref Sect. 25.2). The cyan shaded region is the focus of this work while the grey-shaded regions mark the periods with significant spectral changes in the broadband SED e.g. [18, 26, 27, 29, 31]

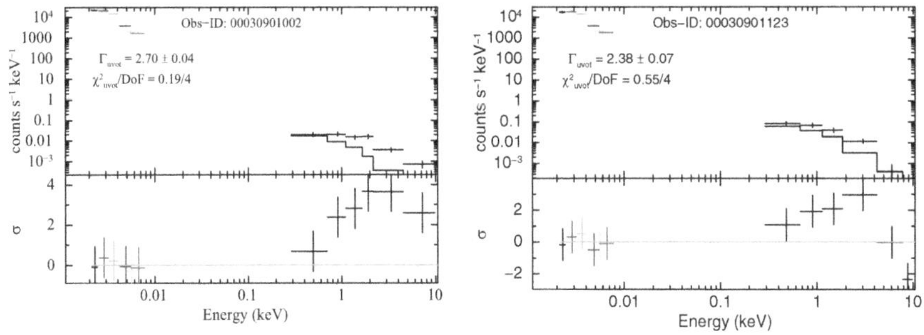

Fig. 25.2 The best-fit power-law optical-UV spectrum and its comparison with the X-ray data for the lowest simultaneous optical-UV and X-ray phase: Obs-ID: 00030901002, and another similar optical-UV state with a different spectral index and also a higher X-ray flux: Obs-ID: 00030901123 (see Sect. 25.3)

this, we first searched the lowest X-ray flux and then the lowest optical-UV within it, resulting in one observation (Obs-ID:00030901002). The details of criteria and fitting are given in [35] and the joint fitting was performed in the XPSEC using the respective source, background, response, and auxiliary files.

First, we independently verified the optical-UV spectrum and found it consistent with a power-law spectrum with a photon spectral index of $\Gamma_{\rm UVOT} = 2.70 \pm 0.04$ (χ^2/DoF $= 0.19/4$; DoF: Degree of Freedom). This best fit and its comparison (assuming power-law extends to X-ray energies) with the X-ray data is shown in Fig. 25.2 (Obs-ID:00030901002). Similarly, we independently verified the corresponding X-ray spectrum and found it consistent with a power-law with a photon spectral index of $\Gamma_{\rm X} = 1.41 \pm 0.15 (C_{\rm stat} = 76.9/84)$. Since the optical-UV best-fit contribution at X-ray is lower/similar to X-ray data, assuming that optical-UV continues to extend to X-ray, we performed a joint-fit and got an acceptable fit with a photon spectral index of $\Gamma_{UVOT} = 2.71 \pm 0.03$ and $\Gamma_X = 1.22 \pm 0.20$. This is the hardest reported X-ray spectrum of OJ 287 to the best of our knowledge and consistent with the inferred X-ray spectrum at hard X-ray energies explored by the Swift-BAT [34]. The uncertainty is high due to low exposure leading to lower counts. However, joint modeling involving similar X-ray spectral states leads to tighter constraints consistent with the current one (e.g. see [35] for more details).

The above joint modeling and inferences were under the assumption that the power-law optical-UV synchrotron spectrum extends unhindered to X-ray energies during the simultaneous low flux state at optical-UV and X-rays. To firmly establish this, we searched other optical-UV states having similar optical-UV flux and examined the X-ray spectra. Interestingly, we found an observation (ID:00030901123) having almost similar optical-UV flux in the U-band but with a different optical-UV and the associated X-ray spectrum compared to the above-reported simultaneous low state spectrum. We found that this different X-ray spectrum can be reproduced with the corresponding synchrotron and the inferred low-hard X-ray spectrum as shown in Fig. 25.3, supporting our assumption of an extended optical-UV spectrum during very-low optical-UV flux states.

25.4 Discussion

X-ray emission from OJ 287 is generally attributed to the synchrotron self-Compton (SSC) i.e. inverse Compton scattering of synchrotron photon (e.g. [39, 40]). However, many studies have also reported a significant contribution of the optical-UV synchrotron high-energy tail to X-rays (e.g. [31, 39]). Without any other contamination, the SSC spectral shape is related to the spectral shape of the low-energy part of the particle distribution and the low-energy cutoff (e.g. [41, 42]).

As stated earlier, OJ 287 has shown significant X-ray spectral variability. With our focus on understanding X-ray spectral changes, especially the role of the optical-UV synchrotron component in driving these changes, we investigated the simultaneous optical-UV to X-ray variation of OJ 287 during the low optical-UV and X-ray state. Variability analysis of the light curve presented in Fig. 25.1 shows that between 2005 and 2022 OJ 287 has exhibited drastic flux/brightness variability, by a factor of > 10 at optical-UV and even more drastic at X-ray—by a factor > 100 between the minimum and the maximum. The drastic flux variability is accompanied by an equally drastic spectral variability across the EM spectrum as reported in previous multi-wavelength studies [12, 18, 26, 28–32, 43]. Focussing on simultaneous[2] optical-UV and X-ray variation, we found a duration: MJD 54160–54180 (cyan-colored band in Fig. 25.1) having the lowest simultaneous state in both and jointly investigated the optical-UV to X-ray spectrum.

Jointly modeling the simultaneous optical-UV to X-ray spectrum of the lowest recorded X-ray flux state (see Sect. 25.3), we found a power-law optical-UV spectrum with $\Gamma_{UVOT} = 2.71 \pm 0.03$ extending to X-ray energies (see Fig. 25.2-left). Accounting for this contribution at X-ray resulted in a very hard X-ray spectrum of $\Gamma_X = 1.22 \pm 0.20$. This inference was factually verified and firmly established with our finding of another similar low-flux optical-UV state with an altogether different optical-UV and a much brighter X-ray flux as shown in Fig. 25.2-right. This different X-ray spectrum is naturally reproduced as the sum of its optical-UV power-law synchrotron component and the lowest inferred hard X-ray spectrum as shown in Fig. 25.3.

Since the optical-UV continuum is synchrotron emission (e.g. [39, 40]) and directly traces the high-energy part of the underlying particle distribution, the finding has two direct implications as far as the X-ray spectral changes are concerned in OJ 287: (a) the optical-UV spectrum extends to X-ray energies and play a significant role in driving X-ray spectral changes during low-brightness phases and (b) the

[2] Within the data and cadence.

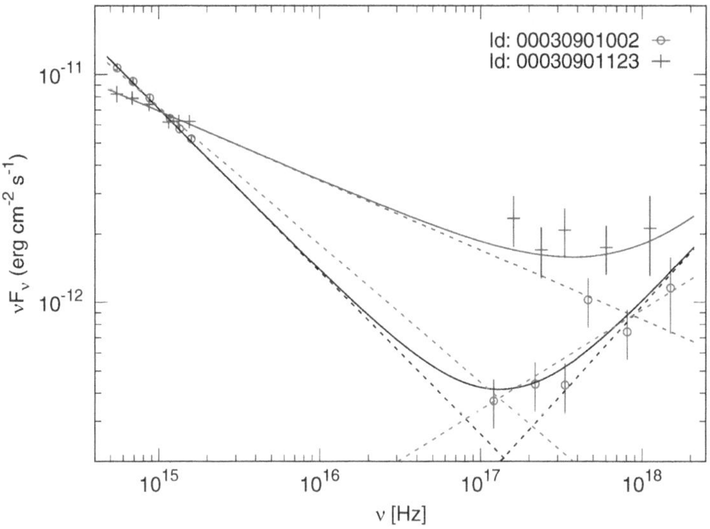

Fig. 25.3 The simultaneous optical-UV to X-ray SED extracted from for the lowest phase (Id: 00030901002) showing best-fit power-law shape from independent fitting (red dashed curves) and from the joint modeling (black dot-dashed curves). The purple data is the other optical-UV spectrum having almost the same UVW1-band flux along with the corresponding X-ray spectrum with curves showing independent optical-UV fit (purple dashed) and the reproduced of X-ray spectrum: purple-dashed optical-UV synchrotron + the inferred hardest X-ray spectrum during the lowest X-ray flux state (see Sect. 25.3).

underlying particle spectrum governing the optical-UV synchrotron extends to much higher energies during low-brightness phases.

Since the jet plasma is very rare (collision-less), the particle acceleration and resulting spectrum are in general related to the size of confinement of the particle within the emission/acceleration region [44] and losses e.g. radiative. An extended particle spectrum indicates a much larger emission region and a relatively much lesser radiative loss. Thus the inferred low and hard X-ray spectrum could be a persistent low-lying component which may be the large-scale jet emission. The inferred optical-UV spectrum implies a corresponding relativistic particle distribution with a power-law spectrum of 4.4 ± 0.6. Similarly using the same relation, the inferred hard X-ray spectrum implies a particle spectrum of power-law index 1.15–1.3. However, it should be noted that the inferred X-ray particle spectrum may not be a true representative of the underlying particle as such a hard state can also be reproduced via a particle distribution with a power-law index closer to the one predicted in standard shock acceleration e.g., [35, 41, 42] with a relatively higher lower energy cutoff—also consistent with a larger acceleration/emission region.

25.5 Summary and Conclusion

OJ 287 has exhibited drastic spectral changes at X-rays which, in general, have been attributed to SSC but optical-UV synchrotron has also been reported to affect the X-ray spectrum in many previous studies. In the quest to understand the role of the optical-UV synchrotron component in causing X-ray spectral changes, we systematically investigated the simultaneous optical-UV and X-ray variability of the source with a focus on the simultaneous low-state in both.

We found a simultaneous low optical-UV and X-ray phase and examined the lowest optical-UV to X-ray state. First, we explored the optical-UV spectrum and found it consistent with a PL spectrum. We then accessed this best fit with the X-ray data by extrapolating the PL to X-ray and found that the lowest energy X-ray data is consistent with the best-fit optical-UV spectrum. We then performed the joint fit of optical-UV to X-ray and inferred a very hard X-ray spectrum with a power-law photon index $\Gamma_X = 1.22 \pm 0.20$—the hardest reported X-ray spectrum for OJ 287 to the best of our knowledge. We also found another observation with an almost similar UVW1 band flux but significantly different optical-UV and X-ray spectra with a much higher X-ray flux. We found that this X-ray spectrum can be naturally explained as the sum of its optics-UV PL synchrotron component and the discovered hard X-ray spectrum, establishing our assumption that optical-UV synchrotron spectrum during very low-optical-UV brightness extends to much higher energies.

The inferred lowest flux yet hard X-ray state and the corresponding optical-UV synchrotron with a high-energy tail imply a much larger acceleration region and could be a persistent emission component associated with the large-scale jet. The corresponding optical-UV spectrum implies a particle distribution (high-energy part) with a power-law index of ~ 4.4 while X-ray implies the low-energy part of a power-law index of $\sim 1.15 - 1.3$. It should be noted that such a hard X-ray spectrum can be reproduced via the standard shock spectrum with a much higher low-energy cutoff in the particle spectrum and may not be reflective of an actual hard underlying particle spectrum.

Acknowledgments The author acknowledge support from the INSPIRE Faculty grant (DST/INSPIRE/04/2020/002586) from the Department of Science and Technology (DST), Government of India.

References

1. Scaringi, S., Maccarone, T. J., Kording, E., et al. (2015). *Science Advances, 1*, e1500686. https://doi.org/10.1126/sciadv.1500686
2. Kushwaha, P., Chandra, S., Misra, R., et al. (2016). *The Astrophysical Journal Letters, 822*, L13. https://doi.org/10.3847/2041-8205/822/1/L13
3. Kushwaha, P., Sinha, A., Misra, R., et al. (2017). *The Astrophysical Journal, 849*, 138. https://doi.org/10.3847/1538-4357/aa8ef5

4. Burke, C. J., Shen, Y., Blaes, O., et al. (2021). *Science, 373*, 789. https://doi.org/10.1126/science.abg9933
5. Zhang, H., Yang, S., & Dai, B. (2024). *The Astrophysical Journal Letters, 967*, L18. https://doi.org/10.3847/2041-8213/ad488d
6. Padovani, P. (2017). *Nature Astronomy, 1*, 0194. https://doi.org/10.1038/s41550-017-0194
7. Abdo, A. A., Ackermann, M., Agudo, I., et al. (2010). *The Astrophysical Journal, 716*, 30. https://doi.org/10.1088/0004-637X/716/1/30
8. Gupta, A. C., Mangalam, A., Wiita, P. J., et al. (2017). *Monthly Notices of the Royal Astronomical Society, 472*, 788. https://doi.org/10.1093/mnras/stx2072
9. Liodakis, I., Marscher, A. P., Agudo, I., et al. (2022). *Nature, 611*, 677. https://doi.org/10.1038/s41586-022-05338-0
10. Hazama, N., Sasada, M., Imazawa, R., et al. (2022). *Publications of the Astronomical Society of Japan Publications of the Astronomical Society of Japan, 74*, 1041. https://doi.org/10.1093/pasj/psac054
11. Gupta, A. C., Kushwaha, P., Valtonen, M. J., et al. (2023). *The Astrophysical Journal Letters, 957*, L11. https://doi.org/10.3847/2041-8213/acfd2e
12. Kushwaha, P. (2022). *Journal of Astrophysics and Astronomy, 43*, 79. https://doi.org/10.1007/s12036-022-09872-1
13. IceCube Collaboration, Aartsen, M. G., Ackermann, M., et al. (2018). *Science, 361*, eaat1378. https://doi.org/10.1126/science.aat1378
14. IceCube Collaboration, Aartsen, M. G., Ackermann, M., et al. (2018). *Science, 361*, 147. https://doi.org/10.1126/science.aat2890
15. IceCube Collaboration, Abbasi, R., Ackermann, M., et al. (2022). *Science, 378*, 538. https://doi.org/10.1126/science.abg3395
16. Gao, S., Fedynitch, A., Winter, W., et al. (2019). *Nature Astronomy, 3*, 88. https://doi.org/10.1038/s41550-018-0610-1
17. Carrasco, L., Dultzin-Hacyan, D., & Cruz-Gonzalez, I. (1985). *Nature, 314*, 146. https://doi.org/10.1038/314146a0
18. Kushwaha, P., Gupta, A. C., Wiita, P. J., et al. (2018). *Monthly Notices of the Royal Astronomical Society, 479*, 1672. https://doi.org/10.1093/mnras/sty1499
19. Gupta, A. C., Gaur, H., Wiita, P. J., et al. (2019). *The Astronomical Journal, 157*, 95. https://doi.org/10.3847/1538-3881/aafe7d
20. Komossa, S., Grupe, D., Kraus, A., et al. (2022). *Monthly Notices of the Royal Astronomical Society, 513*, 3165. https://doi.org/10.1093/mnras/stac792
21. Gupta, A. C., Kushwaha, P., Carrasco, L., et al. (2022). *The Astrophysical Journal Supplement, 260*, 39. https://doi.org/10.3847/1538-4365/ac6c2c
22. Valtonen, M. J., Zola, S., Gopakumar, A., et al. (2023). *Monthly Notices of the Royal Astronomical Society, 521*, 6143. https://doi.org/10.1093/mnras/stad922
23. Britzen, S., Fendt, C., Witzel, G., et al. (2018). *Monthly Notices of the Royal Astronomical Society, 478*, 3199. https://doi.org/10.1093/mnras/sty1026
24. Agudo, I., Marscher, A. P., Jorstad, S. G., et al. (2012). *The Astrophysical Journal, 747*, 63. https://doi.org/10.1088/0004-637X/747/1/63
25. Gopal-Krishna. (2024). *Astronomy & Astrophysics, 688*, L16. https://doi.org/10.1051/0004-6361/202449409
26. Komossa, S., Grupe, D., Schartel, N., et al. (2017). *New Frontiers in Black Hole Astrophysics, 324*, 168. https://doi.org/10.1017/S1743921317001648
27. Kushwaha, P., Gupta, A. C., Wiita, P. J., et al. (2018). *Monthly Notices of the Royal Astronomical Society, 473*, 1145. https://doi.org/10.1093/mnras/stx2394
28. Komossa, S., Grupe, D., Parker, M. L., et al. (2021). *Monthly Notices of the Royal Astronomical Society, 504*, 5575. https://doi.org/10.1093/mnras/stab1223
29. Kushwaha, P., Pal, M., Kalita, N., et al. (2021). *The Astrophysical Journal, 921*, 18. https://doi.org/10.3847/1538-4357/ac19b8
30. Kapanadze, B., Vercellone, S., Romano, P., et al. (2018). *Monthly Notices of the Royal Astronomical Society, 480*, 407. https://doi.org/10.1093/mnras/sty1803

31. Singh, K. P., Kushwaha, P., Sinha, A., et al. (2022). *Monthly Notices of the Royal Astronomical Society, 509*, 2696. https://doi.org/10.1093/mnras/stab3161
32. Huang, S., Hu, S., Yin, H., et al. (2021). *The Astrophysical Journal, 920*, 12. https://doi.org/10.3847/1538-4357/ac0eff
33. Gehrels, N., Chincarini, G., Giommi, P., et al. (2004). *The Astrophysical Journal, 611*, 1005. https://doi.org/10.1086/422091
34. Langejahn, M., Kadler, M., Wilms, J., et al. (2020). *Astronomy & Astrophysics, 637*, A55. https://doi.org/10.1051/0004-6361/202037469
35. Kushwaha P., (2025), Universe, 11, 84. https://doi.10.3390/universe11030084
36. Arnaud, K. A. (1996). *Astronomical Data Analysis Software and Systems V, 101*, 17.
37. Brien, S. O., & VERITAS Collaboration. (2017). *35th International Cosmic Ray Conference (ICRC2017)* (vol. 301, p. 650). https://doi.org/10.22323/1.301.0650
38. Pal, M., Kushwaha, P., Dewangan, G. C., et al. (2020). *The Astrophysical Journal, 890*, 47. https://doi.org/10.3847/1538-4357/ab65ee
39. Isobe, N., Tashiro, M., Sugiho, M., et al. (2001). *Publications of the Astronomical Society of Japan, 53*, 79. https://doi.org/10.1093/pasj/53.1.79
40. Kushwaha, P., Sahayanathan, S., & Singh, K. P. (2013). *Monthly Notices of the Royal Astronomical Society, 433*, 2380. https://doi.org/10.1093/mnras/stt904
41. Inoue, S., & Takahara, F. (1996). *The Astrophysical Journal, 463*, 555. https://doi.org/10.1086/177270
42. Katarzyński, K., Ghisellini, G., Tavecchio, F., et al. (2006). *Monthly Notices of the Royal Astronomical Society, 368*, L52. https://doi.org/10.1111/j.1745-3933.2006.00156.x
43. Prince, R., Agarwal, A., Gupta, N., et al. (2021). *Astronomy & Astrophysics, 654*, A38. https://doi.org/10.1051/0004-6361/202140708
44. Hillas, A. M. (1984). *Annual Review of Astronomy and Astrophysics, 22*, 425. https://doi.org/10.1146/annurev.aa.22.090184.002233

Part VII
Gravitational Wave Astrophysics

Chapter 26
On the Origin of Merging Compact Object Binaries

Tomasz Bulik

Abstract At the time of writing this chapter we know nearly one hundred mergers of compact objects observed in gravitational waves. In this chapter I summarize the observations and their statistical properties. I also review the potential scenarios of formation of merging compact object binaries. Finally I will compare the models with observations. No model can explain the properties of the observed population, suggesting that we are dealing with more than one formation path. I will discuss the possibility to limit these scenarios with further observations.

26.1 Introduction

The discovery of gravitational waves from compact object binary seven years ago marked the beginning of gravitational wave astronomy [1]. It was followed by several other breakthroughs. A merger of binary neutron stars was detected with electromagnetic counterpart [2]. Merger in the mass extremes have been found. One of the merger products clearly exceeded the limit of $100\,M_\odot$ and was the first direct detection of an intermediate mass black hole. Several mergers had the masses in the lower mass gap, either indicating very heavy neutron star or very light black holes. These discoveries led to a number of interesting inferences in fundamental physics confirming validity of General Relativity in the strong field regime. The population of detected binaries includes now nearly 100 objects [3]. This allows some population studies that may help answer the astronomical aspect of these discoveries, namely wat is the origin of the merging binary black holes: how do they form? in what environments do they evolve?

In this chapter I will review the most important observational properties of the merging black holes. Then I will describe the most popular formation paths that lead to the merging black hole binaries. Finally I will confront the two and

T. Bulik (✉)
Astronomical Observatory, University of Warsaw, Warsaw, Poland
e-mail: tb@astrouw.edu.pl

S. Ghosh, A. R. Rao (eds.), *The Relativistic Universe: From Classical to Quantum*, Astrophysics and Space Science Proceedings 61,
https://doi.org/10.1007/978-3-031-90186-7_26

draw conclusions about the population origin. I will also elaborate on how future detections can solve the question of origin of these fascinating objects.

26.2 Observational Properties

For each detected binary a parameter estimation has been done. The main parameter that is estimated is the chirp mass

$$\mathcal{M} = (m_1 m_2)^{3/5} (m_1 + m_2)^{-1/5}. \tag{26.1}$$

Then a more elaborate analysis leads to the estimate of mass ratio and individual masses. As the amplitude of the wave is related to the chirp mass, and the black hole mergers are standard sirens, one can estimate the distance, however this estimate depends on the polarization or in other words inclination of the binary.

Two combinations of spins of the merging black holes can be found. The first is the effective spin which is the mass weighed projection of spins on to the orbital angular momentum

$$\chi_{eff} = \frac{m_1 \mathbf{s}_1 + m_2 \mathbf{s}_2}{m_1 + m_2} \frac{\mathbf{J}}{|J|}, \tag{26.2}$$

and the second is the effective precession. However the latter is known with much worse precision.

The statistical properties of the detected quantities are summarized by the LVK Collaboration in [3]. The distribution of masses of the primary black holes—the primary is the more massive component of the system—shows a peak at about $10 M_\odot$ followed by an exponential fall. The distribution can roughly fitted with a formula

$$\frac{dN}{dm_1} \approx 100 \times \exp(-m_1/8 M_\odot).$$

This exponential fall of is modulated by peaks at approximately $10 M_\odot$, $18 M_\odot$, and $35 M_\odot$. Above $80 M_\odot$ there is o sufficient statistics to asses the shape of the distribution reliably.

As mentioned above the effective spin is the best determined combination of spins. The distribution of the effective spins is centered at a slightly positive value of about $\chi_{eff} \approx 0.1$ and has a full width at half maximum of about 0.3. The precession spin is distribution is centered on the value of about 0.2 and has a width of about 0.2. This indicates that the individual sins of black holes are likely small $|s| < 0.5$ and that there is s light tendency for alignment of the black hole spins with the orbital angular momentum.

The available statistics allows not only to estimate the rate density of the binary black hole merger but also to investigate its changes with redshift. The current estimate of the merger rate density of binary black holes is

$$\mathcal{R}_{BBH} = 17 - 45\text{Gpc}^{-3}\text{yr}^{-1}.$$

The rate density increases with redshift. The rate density at the redshift of $z = 1$ seems to be almost a fact of 10 larger than at locally at $z = 0$ although this estimate is still burdened by a large error bar.

26.3 Modeling the Formation of BBH

26.3.1 Principal Challenges

There are two principal challenges in modeling the population of merging binary black holes.

The first comes from the fact that we need to form a close system ob binary black holes if the system is to merge within the Hubble time. A system of equal mass $20M_{\odot}$ black holes will merge in Hubble time only if its initial separation is about $7R_{\odot}$. If the system is very eccentric with $e \approx 0.99$ this can be increased to about $30R_{\odot}$. This separations are much smaller that the radii of stars from which these black holes might have originated. These stars should also have gone through giant stages when their radii could reach up to $1000R\odot$. Thus the first challenge is to squeeze the binary, extract angular momentum so that the initial separation of black holes is small enough to allow it to merge within the Hubble time.

The second challenge comes from the rates. The rate density estimate should be compared with the formation rate of stellar black holes in our local Universe. The Galactic supernova rate is about one per century i.e. 0.01yr^{-1}, and the density of galaxies is about 0.01Mpc^{-3}. This leads to the supernova rate density

$$\mathcal{R}_{SN} \approx 10^5\text{Gpc}^{-3}\text{yr}^{-1}.$$

The collapse of a star to a black holes happens less frequently. Assuming that the stellar mass function can be described by Salpeter exponent and that the neutron star are formed in the mass range from 8 to $20M_{\odot}$, while black holes are formed for stars with initial mass above $20M_{\odot}$ we obtain that black hole formation rate should be about one tenth of the neutron star formation rate. Hence we can estimate that the black hole formation rate density is

$$\mathcal{R}_{BH} \approx 10^4\text{Gpc}^{-3}\text{yr}^{-1}.$$

Comparing this with the binary black hole merger rate density we find that one out of every 100 to a 1000 black hole must end up in merging binary. This has very profound consequences for the models of the merger formation. First, one must realize that this is quite a large number. It is not easy to make binary black holes with the efficiency that is quite large. There are single stars, there are binaries that may be too large to merge at the start and there may be many systems that will be disrupted instead of forming a merging one. Second we must look for pretty generic scenario that is prevalent in the Universe. If we take an exotic scenario that involves very extreme conditions, than the more exotic conditions these are, the smaller fraction of stellar matter is involved and the larger the required efficiency of binary black hole formation must be in order to explain the observed rates. In particular stellar environments that involve less than 0.1% of stellar mass in the Universe would have to have a 100% efficiency in producing merging binary black holes. Thus these environments can be responsible for forming only a small fraction of the mergers but not the entire population.

26.3.2 Models

I will describe three classes of models that are currently entertained in the community: the binary evolution model, the dense cluster evolution model and the primordial black hole formation model.

However before I do that I will make two remarks regarding the masses and spins of stellar origin black holes. The masses of back holes in X-ray binaries are concentrated around $8 - 10 M_{\odot}$ and there are only a few uncertain outliers with masses close to $20 M_{\odot}$. An interesting case is a recent detection of a $30 M_{\odot}$ by GAIA [4]. On the other hand there is a lot of black holes in merging binaries that have the masses in $30 M_{\odot}$ and above up to even $80 M_{\odot}$. These high mass black holes have been considered by some as a surprise however it must be noted that population of such high mass black holes has been predicted by analyzing the evolution of low mass stars [5]. In the extreme case the Population III stars are predicted to form black holes with the masses reaching even $200 M_{\odot}$.

Additionally there is a fundamental limit on the largest mass of a back hole formed in stellar evolution [6]. It originates in the fact that for stellar cores above $\approx 65 M_{\odot}$ the process of pair creation starts to play a role. Pair creation lead to sudden pressure loss and instability of a star. Simulations show that his instability lead to Pair Creation Supernovae which completely disrupt the star and leave no remnant. The problem is in the fact that the exact limit is not precisely known, it can go as high as $80 - 90 M_{\odot}$ however in my opinion this requires some fine tuning [7, 8].

Spin of black holes as measured indirectly in accretion powered X-ray binaries are typically large and close to their maximal value of unity. However there are arguments that this is an exception not a rule. Stellar black holes form in collapse of massive stars, while neutron stars originate in collapse followed by supernovae explosion. The initial spins of neutron stars observed as pulsar can be estimated in different ways, for example using the energetics of pulsar wind nebulae. The result is that the initial periods of pulsars is of the order of 10–50 ms, which is slow rotation for a pulsar. Therefore by analogy we expect that typical initial black hole sons should not be large. A second argument follows from the analysis of population of long GRBS. GRBS are about 1000 times less frequent than supernovae. In the standard collapsar scenario a quickly rotating black hole must form inside a star to form jest that break through and power a GRB. If all stellar collapses lead to formation of quickly spinning black holes than the population of long GRBs would be much more numerous and their number density would be comparable to that of the supernovae. Thus there are strong arguments that the typical spins of black holes in the unbiased population should be small. The inferences drawn from the population of X-ray binaries are probably flawed by evolutionary history and selection effects.

26.3.2.1 Binary Evolution Model

In the binary evolution model we consider formation binary lack holes in the evolution of solitary binaries in galactic environments. The advantage of this model is that this is a generic as one can imagine. Most of the stellar mass in the Universe resides in galaxics. The typical binary fraction are about 0.5, which means that there is a one binary per each solitary star. The problem with this model is that we need some fine tuning with the initial separation. If the stars are too close they will merge as stars before forming two compact objects. If they are too far away they will either form a non merging binary or even worse, they binary may be disrupted. In the intermediate range one invokes the common envelope process to tighten the binary. The common envelope is a phase of unstable mass transfer from a more massive giant companion to a less massive compact object. It is believed that at this stage the envelope of the giant is expelled at the expense of the orbital energy of the binary. Thus as a result of this process the binary becomes tight and the companions envelope is lost. Following the common envelope phase the stripped giant undergoes collapse and forms a second compact object on a tight orbit.

The common envelope phase is the bottleneck of the process. The binary may merge during this phase and form a Thorne -Żytkow object instead of a merging binary. The common envelope phase is difficult to model and there are very few direct observational constraints on this phase.

This scenario provides a mechanism to align the black hole angular momentum with the orbital one because of the mass transfer phases in the binary. There is a dispute in the literature regarding the level of this alignment as there can be mechanisms that work against it, like separate rotation of the inner and outer cores in stars, or asymmetric neutrino emission during collapse of the star.

26.3.2.2 Dense Cluster Evolution Model

The dense stellar cluster model invoke formation of binary back holes as a result of multiple interaction between stars and binaries in stellar clusters [9–11]. The fraction of stellar mass in globular cluster in our Galaxy is about 1.5% however originally it must have been larger as some cluster have dissolved already. This can be compensated by the potentially larger efficiency of forming binary black holes. Black holes tend to sink towards the center of the cluster, and may form binaries there. The escape velocities from clusters are small and do not exceed $50\mathrm{km\,s}^{-1}$. The newly formed binary black holes may achieve velocities large enough to leave clusters. Additionally merger products may do the same because of the kick velocities due to asymmetric gravitational wave emission at the last stage of the merger.

The expected distribution of effective spins in this models should be centered on zero. There is no reason why there should be any corelation between the back hole spins and orbital angular momentum if they are formed as a results of multiple collisions and exchange of partners.

The dense stellar regions can also be open clusters, although they are short lived, and dense bulges of galaxies. Additionally a model is sometimes put forward where the evolution takes part in AGN accretion disks. The last model has the problem that the fraction of stellar matter that resides in AGN disks is less than 10^{-5} of all stellar matter and therefore it cannot explain all the mergers observed. However, it can be responsible for a few rare exceptional events, that happen once in a hundred or thousand mergers.

26.3.2.3 Primordial Black Hole Formation

The primordial black holes have been postulated as a consequence of phase transition in the early Universe as well as density fluctuations in this epoch. The typical predictions are that their masses should not be constrained by the stellar evolution so that they can have masses stretching below the neutron mass. Additionally the mass spectrum should exhibit features that correspond to mass scales of the transitions in the early Universe. The expected spins are low as we do not expect significant rotation for any horizon sized cell in the early Universe.

The number density of the primordial black holes (PBH) is constrained by the gravitational lensing experiments [12] to be less than one percent of the dark matter. The typical proposed scenario is that the black holes initially evolve as if they were in dense clusters and form binaries via interactions so their merger density scales a density squared. This makes the observed rates hard to explain. Additionally, forming a binary that would live for a Hubble via gravitational wave angular momentum loss in a flyby of two black holes is very difficult, so one probably needs to invoke triple interactions. These are of course much more rare. The formation of binaries should take place immediately after their formation as the Universe is expanding and the density of these black holes decreases and so does interaction probability. For any population the typical merger delay time distribution follows a power law, something like t^{-1} or t^{-2}. so the merger rate that we see today should be just an end of the tail of this distribution.

In summary the predictions for the model are that: spins are low, spins are uncorrelated with orbital angular momentum, mass spectrum may have features and can extend below the stellar masses region, rate should increase with redshift and be uncorrelated with the star formation.

26.4 Models Facing Observations

We are arriving at the point when we can confront the models with observations.

In the case of the binary evolution model the mass distribution is the main problem. There are objects that touch or exceed the pair instability supernova limit which becomes either marginally consistent or plain inconsistent with the model. The distribution of effective spins of merging binaries is shifted to positive values just as expected in this model. Additionally the merger rate increases with redshift in a way similar to the star formation rate history. The value of merger rate density can be well explained by the model.

The dense cluster model has one main problem when confronted with the data—the effective spin distribution. However, the mass distribution can be explained by this model as there is possibility of hierarchical mergers: two black holes merge, make a more massive one that takes a new partner in an interaction with another binary in the cluster. Then we can see the merger of such binary and the mass is no longer limited by the pair instability supernova limit. These event are rare but the observed mass distribution is exponentially decreasing so there is a possibility of consistency. The rates require quite a large efficiency of production of binary black holes given the low amount of stellar mass in the globular clusters. However simulations show that the efficiency of formation of black holes per unit stellar mass in clusters can be larger than that in the isolated evolution even by a factor of ten. This can easily compensate for the fact that there is the fraction of stellar matter in clusters is not higher than a few percent.

The primordial black hole model faces several problem. First the observed effective spin distribution is centered on positive values which can not be explained in this model. The mass distribution has some features, however it falls off quickly for masses below $10M_{\odot}$ and there is no clear reason why there would be such conspiracy to avoid production of low mass black holes. The values of the rates are really unconstrained in this model. The upper limits on the density of black holes from microlensing are very constraining with respect to the expected merger rates in this model. Finally the merger rate density seems to follow the star formation rate. This is not inconsistent with the PBH model, however we would expect that the merger rate density would very weakly depend on redshift especially 13.7Gyrs after formation of all the primordial binaries.

None of the models satisfies all the observational constraints. Therefore it seems that currently more than scenario, or evolutionary path must be responsible for the observed population of compact object binaries. Now the question is how to identify and verify these models and how to identify what was the origin of individual objects that are detected.

26.5 Conclusions

I have reviewed the observations and statistical properties of the population of merging compact object binaries, concentrating on black hole binaries. Then I described briefly the models and the properties of binaries that they imply. The comparison of the model predictions with the data shows that no singe model can explain the observations. Therefore the population must be a mix of different formation paths, or we should look for better models. I would rather reject the latter.

What is the path forward? At the time of writing the next observational round O4 is near the start. This will enhance the statistics of detection by a factor of three. We can expect to see a detailed study of effective spins and perhaps the two components of the distribution will be see: the isotropic and slightly aligned. The large number of detection shall help us in identifying the rare objects at the high mass and investigate the shape of the mass distribution there. This will help identifying how strong a problem with large masses is and to what degree the distribution violates the pair insatiability supernova limit. Additionally at low mass end we may see subsolar mass black mergers—this will be a silver bullet indicating primordial population.

Further in the future we should look for third generation observatories like ET [13] or Cosmic Explorer [14]. They will allow to probe properties of population at very high redshifts. The measured dependence of the merger rate density as function of redshift will be of very high importance for determining the models.

In addition to that detailed models of formation of binary black holes are being developed using the population synthesis and cluster simulations. This is certainly the golden time for this kind of work.

Acknowledgments He author acknowledges the support of the IDUB programme at UW.

References

1. Abbott, B. P., Abbott, R., Abbott, T. D., et al. (2016). *Physical Review Letters, 116*, 061102.
2. Abbott, B. P., Abbott, R., Abbott, T. D., et al. (2017). *Physical Review Letters, 119*, 161101.
3. The LIGO Scientific Collaboration, The Virgo Collaboration, the KAGRA Collaboration, et al. (2021). arXiv:2111.03606.
4. Balbinot, E., Dodd, E., Matsuno, T., et al. (2024). arXiv:2404.11604.
5. Belczynski, K., Bulik, T., Fryer, C. L., et al. (2010). *The Astrophysical Journal, 714*, 1217.
6. Woosley, S. E. (2017). *The Astrophysical Journal, 836*, 244.

7. Belczynski, K., & Banerjee, S. (2020). *Astronomy & Astrophysics, 640*, L20.
8. Belczynski, K., Hirschi, R., Kaiser, E. A., et al. (2020). *The Astrophysical Journal, 890*, 113.
9. Rodriguez, C. L., Chatterjee, S., & Rasio, F. A. (2016). *Physical Review D, 93*, 084029.
10. Askar, A., Szkudlarek, M., Gondek-Rosińska, D., et al. (2017). *Monthly Notices of the Royal Astronomical Society, 464*, L36.
11. Hong, J., Vesperini, E., Askar, A., et al. (2018). *Monthly Notices of the Royal Astronomical Society, 480*, 5645.
12. Mroz, P., Udalski, A., Szymanski, M. K., et al. (2024). arXiv:2403.02386.
13. Punturo, M., Abernathy, M., Acernese, F., et al. (2010). *Classical and Quantum Gravity, 27*, 194002.
14. Hall, E. D. (2022). *Galaxies, 10*, 90.

Chapter 27
Observational Imprints of Accretion Disk on Observed Gravitational Wave Data from LISA

Prasad Basu, Soumen Mondal, and Sangita Chatterjee

Abstract Extreme mass ratio inspirals (EMRIs) are expected to host large accretion disks around the central supermassive black hole within the system. Thus the orbiting smaller mass compact objects (NS, WD, or stellar mass black holes) interact and experience hydrodynamic drag as it swim through the accretion disk of the central black hole. The interaction between the disk and the companion alters the orbital behavior, thereby affecting the profiles of the emitted gravitational waves (GWs). Therefore observing such GW signals one could in principle identify the existence of an accretion disk in such systems. In this chapter we studied such EMRIs with accretion disks. We computed the dynamics of the companion and the resulting GW incorporating the effect of disk interaction, i.e., accretion drag and dynamical friction. We have chosen the transonic disks as the fluid dynamical model to describe accretion flow. We found that the disk's presence is detectable with the present sensitivity of LISA. Detectability increases with the central black hole's accretion rate. For our disk model the disk is detectable for accretion rate as low as $0.1\dot{M}_E$ for 1 year observation by LISA.

27.1 Introduction

With the several successful runs of ground based GW detectors [1] the era of gravitational wave astronomy has finally begun. The science with kilo-hartz GW is in full swing. The planned launch of the future space-based gravitational wave detector LISA presents an exciting opportunity to explore low-frequency gravitational wave observations in greater detail. The primary category of sources for LISA involves extreme mass ratio inspirals (EMRIs), where stellar-mass compact objects (around

P. Basu (✉)
Cotton University, Guwahati, Assam, India

S. Mondal · S. Chatterjee
Jadavpur University, Jadavpur, Kolkata, India

S. Ghosh, A. R. Rao (eds.), *The Relativistic Universe: From Classical to Quantum*, Astrophysics and Space Science Proceedings 61,
https://doi.org/10.1007/978-3-031-90186-7_27

($\sim 10M_{\odot}$)) orbit a central supermassive black hole. In these systems, the masses of the compact objects are insignificant compared to the central supermassive black hole, allowing the emitted gravitational waves to effectively map the surrounding spacetime, reflecting the central SMBH's metric. As the companion approaches the region of the last stable orbit, it completes approximately $\sim 10^5$ orbits before the final merger. This allows one to use EMRIs as the test-bed of various scientific theories (GR, exotic stars, deformed black holes) in the very strong gravitational field [2–6]. It is extremely important to produce extremely accurate GW templates for EMRIs in order to do precision science with GW.

However, GW emitted from real EMRIs are not free from the influence of the astrophysical environment. Presence of other matter sources or dark matter could influence the system dynamics and hence can modify the emitted GW. One of the important environmental matter sources that can be present and can significantly influence the EMRI dynamics is a vast accretion disk surrounding the central supermassive black hole. The accretion disk applies drag forces on the compact objects moving through it by the process of dynamical friction and through the accretion by the orbiting compact object. This results in a modification on the orbits of the companion resulting in a change in the emitted GW profile. Though the change is small for a single orbit, as the compact companion completes a very large number of orbits before plunging, an accumulated effect becomes detectable. Moreover, in order to do precision science with GW one requires to "clean the dirty wavefor" by properly accounting and subtracting the effects of astrophysical environments [7, 8].

The emergence of EMRIs is driven by various dynamical relaxation mechanisms (such as compact object deflection on low angular momentum orbits around a supermassive black hole, binary tidal break-up, or chaotic dynamics around a massive black hole binary), which occur over different timescales. This formation process, known as "dry EMRIs," includes a variety of potential channels for EMRI formation [9–12].

The presence of accretion disk can also influence process of EMRI formation by introducing a new channel called wet channel. This supplementary generation mechanism might lead to a tenfold increase in event rates [13, 14] than previously estimated due to alternative formation channels involving accretion disks surrounding active galactic nuclei (AGN), also known as the "wet channel." The presence of these disks can significantly boost EMRI production rates, potentially increasing them by orders of magnitude. LISA could identify between 10 and 10,000 EMRIs annually through this "wet" channel (accretion-embedded), in contrast to just 1 to 100 per year via the "dry" channel [8].

The companion BH experiences two main types of drag forces within the disk: self-accretion [15] and dynamical friction [16–18]. Although gravitational wave emission is the primary factor in decreasing the energy and angular momentum of the companion black hole, the effect of the accretion disk, though less pronounced individually, becomes notable when assessed collectively [19–23].

The drag effects depend on flow parameters such as density, radial velocity, and accretion rate. As a result, the choice of disk model can lead to variations in the expected behavior of drag forces. Previous analytical, semi-analytical, and simulation studies [4, 7, 8, 19, 21–25] have shown that the effects of the disk can vary considerably depending on the chosen hydrodynamical model [4, 19, 21, 25].

In this study, the hydro-dynamic model of the disc is described by transonic accretion flow, which accurately characterizes the accumulation of matter around a black hole with elevated accretion rates and minimal angular momentum. This approach ensures that matter transitions from subsonic flow to supersonic flow as it nears the core of the supermassive black hole (SMBH), ultimately approaching light speed near the event horizon. Observational evidence for such accretion disks is discussed in [7]. Additionally, theoretical models exploring the subtle distinction of these flows can be found in [26].

In this chapter, we expand upon the study by Barausse and Rezzolla [4], by conducting a complete general relativistic analysis of transonic accretion disks. This model aligns closely with the thin disk model and possesses unique features that differentiate it from other flow models. Expanding the investigation to incorporate transonic accretion flows enhances our comprehension of the disk's hydrodynamics and enables us to evaluate if these non-Keplerian accreting flows exhibit more significant drag effects. Furthermore, the existence of one or more sonic points within the flow can induce sudden variations in the drag force, especially in shock zones, potentially causing noticeable impacts on the gravitational waves (GWs) that LISA could detect.

The rest of the chapter is planned accordingly: in Sect. 27.2, we discuss the geodesic equations, loss rates due to GW and Disk drag of the companion and its detectability. In Sect. 27.3, we examine the outcomes of our research for various elliptical trajectories. We conclude with a discussion in Sect. 27.4.

27.2 Model Equations

27.2.1 EMRI in Vacuum Premises

As noted, we examine an E/IMRI system in which the central supermassive black hole (mass M) shapes the curvature of the surrounding spacetime, while the companion black hole with mass m (where $m \ll M$) behaves like a test particle. The unaffected motion of the companion object is the geodesic of the background spacetime whose metric is completely determined by the central SMBH. This is a reasonable approximation because the mass of the companion $m \ll M$. The companion, treated as a test particle, traces the geodesic of the central black hole, which gradually changes over time due to radiation reaction. When the binding energy of the orbit is much greater than the energy emission rate per cycle due to gravitational wave the evolution of orbit can be determined using adiabatic approximation [27, 28].

We select the rest frame of the central supermassive black hole, and in the Boyer-Lindquist (B-L) coordinates (r, θ, ϕ, t), the metric of the Kerr spacetime assumes the following form:

$$
\begin{aligned}
dS^2 &= g_{\mu\nu}dx^\mu dx^\nu \\
&= (1-\frac{2r}{\Sigma})dt^2 + (\frac{4ar\sin^2\theta}{\Sigma})dtd\phi - \frac{\Sigma}{\Delta}dr^2 - \Sigma d\theta^2 \\
&- (r^2 + a^2 + \frac{2ra^2\sin^2\theta}{\Sigma})\sin^2\theta d\phi^2.
\end{aligned}
\tag{27.1}
$$

In this context, $\Delta = r^2 - 2r + a^2$ and $\Sigma = r^2 + a^2\cos^2\theta$, where a represents the Kerr parameter. We employ units where $M = G = c = 1$. The reference geodesic of the background spacetime can be characterized by two scalar quantities, namely the specific energy $E = u_a\left(\frac{\partial}{\partial t}\right)^a$, and the specific angular momentum $L = u_a\left(\frac{\partial}{\partial \phi}\right)^a$, where $\left(\frac{\partial}{\partial t}\right)^a$ and $\left(\frac{\partial}{\partial \phi}\right)^a$ are the Killing vectors associated with the stationarity and axisymmetry of the background spacetime, respectively. Equation of geodesic then takes the form [28, 29],

$$r^4(\frac{dr}{d\tau})^2 = T^2 - \Delta(r^2 + (L - aE)^2) = V_r, \tag{27.2}$$

$$r^2\frac{d\phi}{d\tau} = -(aE - L) + \frac{aT}{\Delta} = V_\phi, \tag{27.3}$$

$$r^2\frac{dt}{d\tau} = -a(aE - L) + \frac{(r^2 + a^2)T}{\Delta} = V_t. \tag{27.4}$$

In our scenario, we are examining equatorial geodesics; thus,

$$\theta(\tau) = \pi/2, \tag{27.5}$$

where $T = E(r^2 + a^2) - La$, $\Delta = r^2 - 2r + a^2$.

27.2.2 Loss Rates Resulting from Gravitational Waves

The radiated gravitational waves transmit energy (E) and angular momentum(L) from the system, causing the companion to lose four-momentum P^μ. Consequently, the companion's orbit progressively contracts. The corresponding alteration in the companion's four-momentum can be approximated as follows:

$$\left.\frac{dP^t}{d\tau}\right|_{GW} = \left.\frac{dE}{d\tau}\right|_{GW} = \left.\frac{dE}{dt}\right|_{GW}.u^t_{sat}\,, \text{ and}$$

$$\left.\frac{dP^{\phi}}{d\tau}\right|_{GW} = \left.\frac{dL}{d\tau}\right|_{GW} = \left.\frac{dL}{dt}\right|_{GW} .u^{t}_{sat} . \tag{27.6}$$

Under weak-field approximations, the rates of energy and angular momentum loss due to gravitational wave emission are provided in [30].

27.2.3 EMRI Embedded in Transonic Disk

The hydrodynamics of a non-self-gravitating disk in the low viscosity regime can be approximated by the energy momentum tensor of a perfect fluid. We neglected the disk's own gravitational effects to ensure it does not alter the curvature of the surrounding spacetime. The stress-energy tensor is given by,

$$T^{\mu\nu} = (\rho + p)u^{\mu}_{fluid}u^{\nu}_{fluid} + pg^{\mu\nu} , \tag{27.7}$$

where $u^{\mu}_{fluid} = (u^t, u^r, u^{\phi}, 0)$ denotes the four-velocity, p represents the pressure, and $\rho = \rho_0(1 + \mathcal{E})$ signifies the energy density of the fluid. Here, ρ_0 and $\mathcal{E}$ are the rest-mass density and the internal energy density per unit rest mass, respectively.

The hydrodynamics of the fluid is described by the Euler equation $\nabla_{\mu}T^{\mu\nu} = 0$ and Continuity equation $(\rho_0 u^{\mu}_{fluid})_{;\mu} = 0$. In the steady accretion flow, a perfect fluid obeys the two conserved equations: $hu_t^{fluid} = E_{fluid}$ and $hu_{\phi}^{fluid} = -l_{fluid}$ where $h = \frac{p+\rho}{\rho_0}$ is the relativistic enthalpy. Using these equations and integrating $\frac{du^r}{dr}$ and $\frac{da_s}{dr}$ where adiabatic sound speed $a_s = \left(\frac{dP}{d\rho}\right)_s$, we obtain the transonic flow solutions as described in [26] and references therein. One example of this solution is illustrated in Fig. 27.1, where the accretion disk contains a shock. The solid line with arrowheads denotes the profile of the solution (Mach number versus radial distance). The corresponding flow parameters and shock location are given in the Figure caption. The companion is moving within this accretion disk whose velocity profile is obtained from the transonic solutions. The existence of an accretion disk in E/IMRIs has a direct impact on the loss rates. To understand this effect, consider two scenarios: (a) the E/IMRI system without an accretion disk and (b) the E/IMRI system with an accretion disk. Comparing these scenarios will help us assess how the disk affects the companion's motion in E/IMRIs. In scenario (a), the companion loses energy and angular momentum through gravitational wave radiation, leading to a reduction in the orbital radius and extending the time until coalescence. In scenario (b), the presence of the disk introduces additional forces on the companion, resulting in greater energy and angular momentum loss than in scenario (a). This increased loss rate causes the orbital radius to shrink more quickly, leading to faster coalescence. The additional forces acting on the companion are due to (i) self-accretion and (ii) dynamical friction, which alter the companion's 4-momentum.

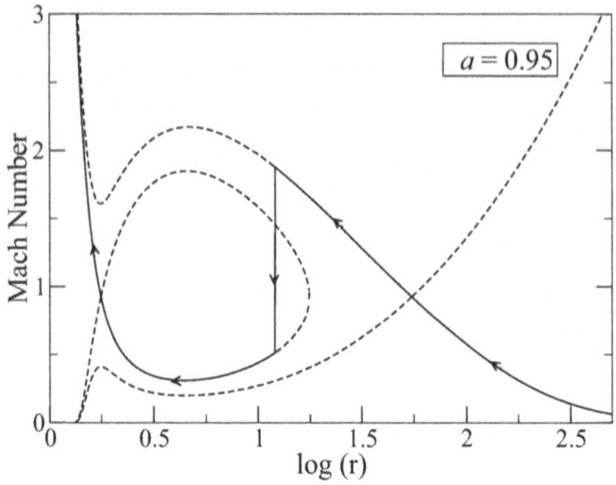

Fig. 27.1 This solution topology (Mach no. i.e., the ratio of velocity and sound speed is plotted with the radial distance for parameters: $a = 0.95$, $E_{disk} = 1.01$, $l_{disk} = 2.25$. The solid arrow-headed line illustrates that at a great distance, the flow is subsonic, transitions to supersonic upon crossing the critical point, and then experiences a shock transition indicated by the vertical arrow. Following the shock, the flow reverts to subsonic, crosses the inner sonic point, and becomes supersonic once more near the black hole. The flow is sub-Keplerian and exhibits significant radial velocity

The overall alteration in momentum caused by the disk and gravitational wave emission is

$$\left.\frac{dP^{\mu}}{d\tau}\right|_{Total} = \left.\frac{dP^{\mu}}{d\tau}\right|_{GW} + \left.\frac{dP^{\mu}}{d\tau}\right|_{Accr} + \left.\frac{dP^{\mu}}{d\tau}\right|_{Defl} . \tag{27.8}$$

The initial term has been covered in Sect. 27.2.2. The remaining elements can be sourced from [4, 18]. Considering all influences, the new 4-momentum of the companion can thus be determined as follows:

$$P(\tau + d\tau) = P(\tau) + \frac{dP}{d\tau}d\tau , \tag{27.9}$$

which can be written as,

$$P^{\mu} = P^{\mu} + \left.\frac{dP^{\mu}}{d\tau}\right|_{total} d\tau , \tag{27.10}$$

where $d\tau$ represents the proper time interval between consecutive orbits of the companion.

27.2.4 Phase Variation Experienced by the Orbiting Companion

The surrounding medium creates a signature on the companion's trajectory due to hydrodynamic resistance. Consequently, the infall time and orbital frequency are altered. Though these changes are minimal, they accumulate, affecting the overall phase of the gravitational wave signal. Hence, the phase differs when the companion is within the disk compared to when it is not. By measuring this phase difference in the presence of the disk, one can infer the existence of a significant accretion disk around the central supermassive black hole. At any given moment, the orbital frequency of the moving companion, denoted as ω , is given by:

$$\omega = \dot{\phi} = \frac{d\phi}{dt} . \tag{27.11}$$

The frequency of the emitted gravitational wave is given by $\omega_{GW} = 2\omega = 2\dot{\phi}$. Consequently, the total phase accrued during the orbital radius reduction from r_0 to r_1 can be calculated by

$$\phi = \int_{\tau_0(r_0)}^{\tau(r_1)} \omega dt . \tag{27.12}$$

Hence, the phase difference between the presence and absence of the disk is:

$$\delta\phi = \phi_{\text{with-disk}} - \phi_{\text{no-disk}} . \tag{27.13}$$

Thus, the resultant phase can be determined using this formula, which also allows for the assessment of the companion's phase shift.

27.2.5 Detectability of Disk Induced Drag

The sensitivity curve for LISA is determined using the formula,

$$S_n(f) = \frac{10}{3L^2}\left(P_{OMS}(f) + \frac{4P_{acc}(f)}{(2\pi f)^4}\right)\left(1 + \frac{6}{10}\left(\frac{f}{f_*}\right)^2\right) + S_c(f) , \tag{27.14}$$

where $L = 2.5Gm$, $f_* = 19.09mHz$, and the formulas for $P_{OMS}(f)$, $P_{acc}(f)$ and $S_c(f)$ can be found in [31].

The signal-to-noise ratio (SNR) is subsequently determined for the gravitational wave signal produced by an EMRI within the frequency range of f_{min} to f_{max} using the methodology outlined in [23],

$$SNR = \sqrt{2.4 \int_{f_{min}}^{f_{max}} df \frac{h_c^2(f)}{S_n(f) f^2}}, \tag{27.15}$$

where h_c is the GW characteristic strain amplitude given by Garg et al. [23],

$$\begin{aligned} h_c &= h\sqrt{f\tau}, && f \leq f_{knee} \\ &= h\sqrt{n}, && f_{knee} < f \leq f_{ISCO} \\ &= 0, && f_{ISCO} < f, \end{aligned} \tag{27.16}$$

where f_{ISCO} defined the observed frequency at the primary BH's innermost stable circular orbit(ISCO) and the other terms are discussed in [7, 23].

With the disk present, a phase shift is introduced, and to detect this shift, we assess the observability of the dephasing using [21, 23],

$$SNR = \sqrt{2.4 \int_{f_{min}}^{f_{max}} df \frac{2h_c^2(f)(1 - \cos(\delta\phi))}{S_n(f) f^2}}, \tag{27.17}$$

where $\delta\phi$ is specified in Eq. (27.13). We assume that the detectability criterion requires a signal-to-noise ratio (SNR) exceeding 8.

27.3 Results

In our research, we investigate the coalescence of a spinning black hole binary system, where the mass ratio between the central supermassive black hole (SMBH) and the smaller companion black hole (CBH) varies from $q = \frac{M_{SMBH}}{m_{CBH}} = 10^{-2}$ to 10^{-4}. The SMBH has a mass of $M_{SMBH} = 10^5$ to $10^8 M_{\odot}$, and the companion's motion occurs within the SMBH's spacetime as a test particle. The companion is presumed to be completely embedded within the plane of the equator of the SMBH's thin, transonic accretion disk. The accretion disk's self-gravity is considered negligible, leaving the metric of the SMBH unchanged. The investigation seeks to examine the inspiral trajectory of the companion, considering the effects of gravitational wave (GW) radiation and the drag induced by the accretion disk. The companion's orbital progression is mainly influenced by its interaction with the central supermassive black hole and the rates of energy and momentum dissipation. These dissipation rates are contingent upon the accretion rate, flow parameters, and orbital characteristics. For our analysis, we fix the accretion rate at $\dot{M} = 1\dot{M}_E$ while keeping other flow parameters constant. We evaluate elliptical orbits with varying initial conditions to assess the maximum impact of disk drag on GW radiation. The results of this analysis are detailed in the following sections.

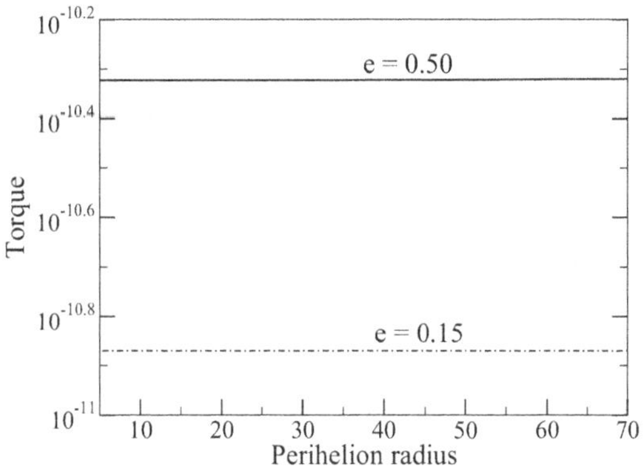

Fig. 27.2 The Figure displays the torque resulting from disk drag. It is observed that the loss rate, or torque, is greater for highly eccentric orbits. The Figure illustrates variations in torque with different eccentricities; however, changes in torque concerning radial separations or time are minimal compared to the variations due to orbital eccentricity.Initial parameters are: $\dot{M} = 1\dot{M}_E$, $a = 0.50$, and semi-major axis $X = 500$

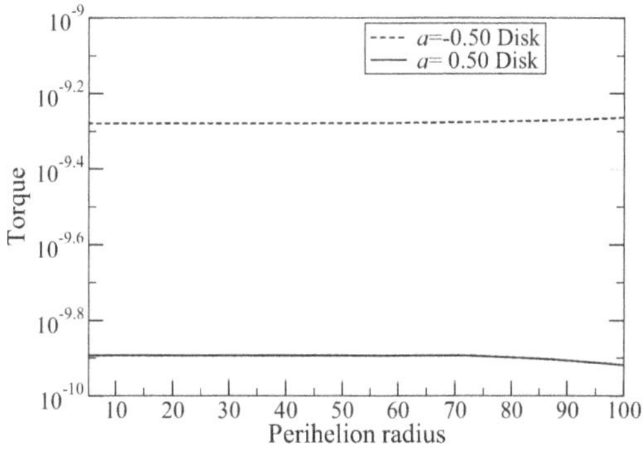

Fig. 27.3 This diagram contrasts the spin dependence of the torque resulting from hydrodynamic drag for two diamctrically opposed spins: $a = 0.5$ versus $a = -0.5$. It illustrates that retrograde orbits encounter a stronger disk-induced torque and have shorter coalescence durations compared to prograde orbits. The accretion rate is $\dot{M} = 1\dot{M}_E$

To examine the effects of the hydrodynamic drag from the disk, we present Fig. 27.2. This Figure illustrates the total torque exerted on the companion, which represents the rate of angular momentum loss $\left(\frac{dL}{dt}\right)_{\text{disk}} = \left(\frac{dL}{dt}\right)_{\text{acc}} + \left(\frac{dL}{dt}\right)_{\text{defl}}$, for eccentricities $e = 0.15$(*dotted line*) and 0.50(*solid line*) respectively.As the orbit evolves and gravitational waves are emitted, the eccentricity of the elliptical orbit diminishes over time, moving toward circularization. The Figure reveals that the torque magnitude $\left(\frac{dL}{dt}\right)_{\text{disk}}$ is greater for orbits with higher eccentricity.

The spin dependence of the angular momentum dissipation rate is clearly illustrated in the plots shown in Fig. 27.3. In this comparison, we examine the torques resulting from hydrodynamic drag for spins of $a = 0.5$ versus $a = -0.5$. It is evident that the magnitude is nearly an order of magnitude greater for a retrograde orbit compared to a prograde orbit (though this magnitude also depends on the disk parameters).

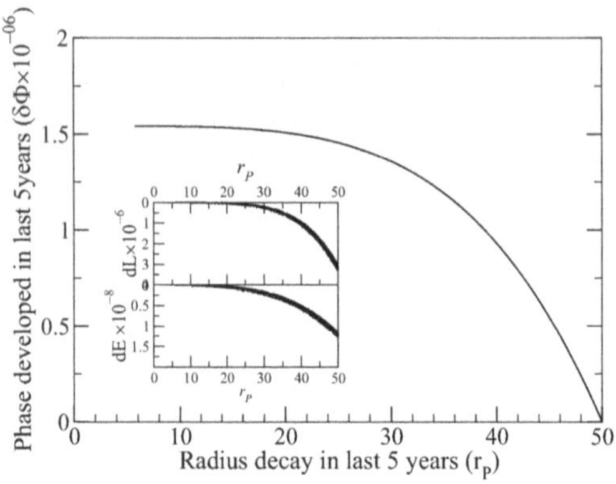

Fig. 27.4 The diagram provides an estimate of the phase accumulated by the companion during the final 5 years before the merger, when the separation between them was sufficiently close, typically around $50r_g$, where r_g represents the Schwarzschild radius. At this distance, we assume the initial phase difference is zero. Thus, the plot illustrates that over the last few years, the phase difference increases steadily as the orbit decays from $50r_g$ onwards. For this plot, we assume the companion masses are $M = 10^8 M_\odot, m = 10^6 M_\odot$, with an eccentricity $e = 0.50$,and a spin parameter $a = 0.50$. The phase difference is approximately 10^{-6}. The effect of the accretion disk on the phase shift is attributed to its contribution to energy and angular momentum loss, calculated as $\Delta E = E_{GW+disk} - E_{GW}$ and $\Delta L = L_{GW+disk} - L_{GW}$ respectively, at each radius. In the subpanel, we display the corresponding variations in energy and angular momentum of the companion black hole during this period, both with and without the disk, providing an estimate of the magnitude of differences. The cumulative effect of losses due to the disk-EMR interaction amplifies the phase shift. Here, the accretion rate is $\dot{M} = 1\dot{M}_E$

We compute the phase gained by the companion both with and without the disk individually and determine the phase difference ($\delta\phi$) in the presence of the accretion disk using Eq. (27.13). In Fig. 27.4, we illustrate the total phase accumulated by the companion black hole as it traverses the disk during the final 5 years before merger, incorporating the combined effects of gravitational waves and disk drag. In the subplot, we display the corresponding variations in energy and angular momentum of the companion black hole during this period, both with and without the disk, to estimate the magnitude of the differences. By considering these variations and setting the initial phase difference at $50r_g$ to zero, we calculate the phase shift. We observe that the phase difference $\delta\phi$ consistently increases throughout the orbital decay.

In Fig. 27.5, the signal-to-noise ratio (SNR) of phase shift is computed for various observational intervals preceding the merger of an EMRI with $M = 10^5 M_\odot$ and $m = 10M_\odot$, situated at a distance of 1 Gpc and surrounded by an accretion disk with an accretion rate of $\dot{M} = 1\dot{M}_E$. The Figure illustrates that the SNR of phase shift consistently declines with the length of the observation period, becoming undetectable in the final 2 years before the merger. Prior to this, the SNR remains ≥ 8, making the system observable (Fig. 27.6).

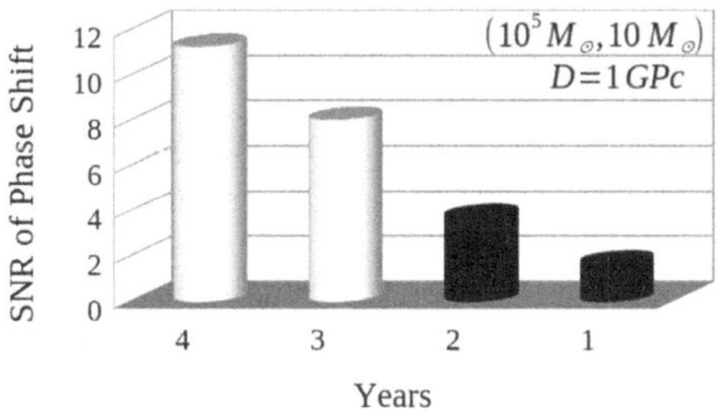

Fig. 27.5 The signal-to-noise ratio (SNR) of the accumulated phase shift is charted against different observation durations. The graph shows that as the observation time lengthens, the SNR rises correspondingly. The accretion rate is $\dot{M} = 1\dot{M}_E$. The gray and black cylindrical bars represent detectable and non-detectable regions, respectively

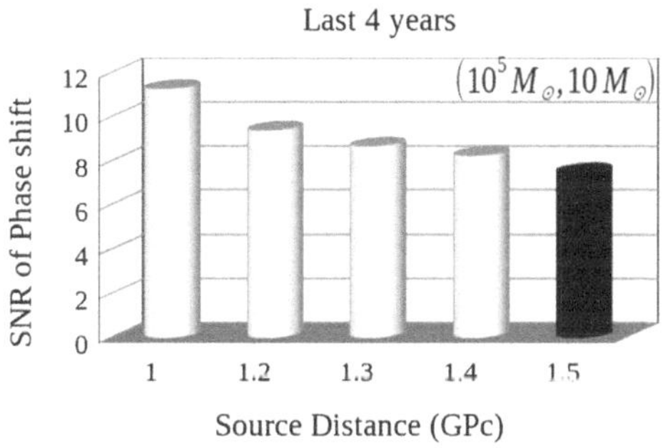

Fig. 27.6 The signal-to-noise ratio (SNR) of the accumulated phase shift is graphed against the source distance DD (in gigaparsecs) for the final four years of observation. With an accretion rate of $\dot{M} = 1\dot{M}_E$, the graph indicates that as the distance between the EMRI and the detector grows, the SNR decreases inversely with D

For the same EMRI with an accretion rate of $\dot{M} = 1\dot{M}_E$, we adjust the source distance and compute the signal-to-noise ratio (SNR) of the phase shift, as shown in Fig. 27.7. It is evident that up to a source distance of 1.4 Gpc, the signal remains within LISA's observational range; however, beyond this distance, it becomes undetectable.

27.4 Discussions

Our findings indicate that there are observable effects of accretion disks on the emitted gravitational waves (GWs), even though they are small relative to the primary amplitude of the GWs. These effects are still detectable by the proposed LISA detector, particularly within the innermost region of the accretion disk, where they become more pronounced during the late stages of inspiral. The impact varies

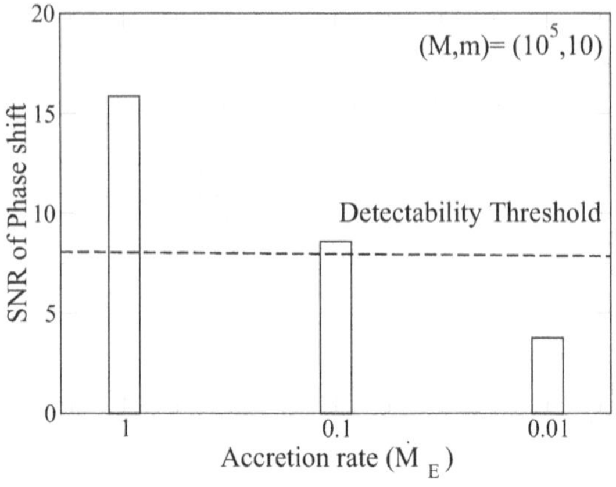

Fig. 27.7 The signal-to-noise ratio (SNR) of the aggregated phase shift is plotted against the accretion rate for the concluding year of observation. The graph illustrates that as the accretion rate diminishes, the SNR correspondingly drops. This image is sourced from P. Basu, S. Chatterjee and S. Mondal, MNRAS, Volume 531, Issue 1, June 2024, Pages 1506–1519, May 2024, 'Eccentric orbits in disc-embedded EMRIs:orbital evolution and observability trend in LISA'

significantly depending on the hydrodynamic configuration of the accretion flow, and the transonic accretion flow exhibits more pronounced drag effects than other disk models.

In the past, the disk's influence was not seen as crucial for creating templates for matched filtering techniques. However, our study utilizes a relativistic framework (e.g., [4]) to measure the interaction between the disk and the companion more accurately than previous Newtonian predictions. This allows us to precisely capture the phase shifts experienced by the companion in real or ideal E/IMRI systems, as supported by earlier findings [21, 22].

The strength of the hydrodynamic drag and the resulting GW profile vary depending on the hydrodynamic model of the accretion flow. Observing GWs from EMRIs with accretion disks can provide insights into the characteristics of the accretion disk. As a result, GWs can be an effective means of exploring the properties of accretion disks.

The presence of super-Eddington accretion flows can shed light on how supermassive black holes could form from seed stellar-mass black holes in the early universe. Numerous numerical and semi-analytical models forecast super-Eddington accretion flows, which exhibit reduced radiative efficiencies of 5–7%. As a result, these flows might be observed as sub-Eddington or low-luminosity objects in optical wavelengths. Despite high accretion rates (over 100 times the Eddington rate), the luminosity increase is not proportional and follows a logarithmic relationship [32]. GW observations, which directly measure accretion rates, could validate this relationship. While our accretion model allows for super-Eddington accretion rates, exploring this in detail is beyond the scope of our current study and can be addressed in future research.

So far, accretion processes have primarily been understood through electromagnetic signals, but the broad range of GW profiles in relation to the disk's hydrodynamic parameters facilitates independent testing of several accretion theories. Thus, concurrent surveillance of both GWs and electromagnetic waves will allow for a more dense knowledge of accretion characteristics and the evaluation of different paradigms in accretion physics.

As future GW detectors become more sensitive, previously neglected effects on GWs may become detectable. Consequently, idealized E/IMRI studies may need to be replaced by more realistic models. GWs are the most efficient probes for understanding near-horizon physics, as they can escape high-density regions without significant loss of information. This level of precision in GW measurements will be crucial for future scientific research, including testing modified gravity paradigms and exploring exotic stars. Such precise measurement (e.g., $h \sim 10^{24-26}$) [33] will make the disk effects more apparent and will need to be accounted for to achieve accurate results.

References

1. Abbott, B. P., et al. (2016). Observation of gravitational waves from a binary black hole merger. *Physical Review Letters, 116*, 061102. https://doi.org/10.1103/PhysRevLett.116.061102
2. Amaro-Seoane P., et al. (2017). Laser interferometer space antenna. arXiv:1702.00786.
3. Gair, J. R., Vallisneri, M., Larson, S. L., et al. (2013). Testing general relativity with low-frequency, space-based gravitational-wave detectors. *Living Reviews in Relativity, 16*, 7.
4. Barausse, E., & Rezzolla, L. (2008). Influence of the hydrodynamic drag from an accretion torus on extreme mass-ratio inspirals. *Physical Review D, 77*, 104027.
5. Armano M., et al. (2009). LISA pathfinder: The experiment and the route to LISA. *Classical and Quantum Gravity, 26*, 094001.
6. Barausse E., et al. (2020). Prospects for fundamental physics with LISA. *General Relativity and Gravitation, 52*, 81.
7. Chatterjee S., Mondal S., & Basu P. (2023). Detectability of gas-rich E/IMRI's in LISA band: Observable signature of transonic accretion flow. *Monthly Notices of the Royal Astronomical Society, 526*, 5612–5627.
8. Speri, L., Antonelli, A., Sberna, L., Babak, S., Barausse, E., Gair, J. R., & Katz M. L. (2023). Probing accretion physics with gravitational waves. *Physical Review X, 13*, 021035.
9. Amaro-Seoane, P., Gair, J. R., Freitag, M., Miller, M. C., Mandel, I., Cutler, C. J., Babak, S. (2007). Topical review: Intermediate and extreme mass-ratio inspirals—astrophysics, science applications and detection using LISA. *Classical and Quantum Gravity, 24*, R113.
10. Babak S., et al. (2017). Science with the space-based interferometer LISA. V. Extreme mass-ratio inspirals. *Physical Review D, 95*, 103012.
11. Gair, J. R., Babak S., Sesana, A., Amaro-Seoane, P., Barausse, E., Berry, C. P. L., Berti, E., & Sopuerta, C. (2017). Prospects for observing extreme-mass-ratio inspirals with LISA. In *Journal of Physics Conference Series* (p. 012021). arXiv:1704.00009. https://doi.org/10.1088/1742-6596/840/1/012021
12. Wang, Y. Y., Wang, F. Y., Zou, Y. C., & Dai Z. G. (2019). A bright electromagnetic counterpart to extreme mass ratio inspirals. *The Astrophysical Journal, 886*, L22.
13. Pan, Z., & Yang, H. (2021). Formation rate of extreme mass ratio inspirals in active galactic nuclei. *Physical Review D, 103*, 103018.

14. Pan, Z., Lyu, Z., & Yang H. (2021). Wet extreme mass ratio inspirals may be more common for spaceborne gravitational wave detection. *Physical Review D, 104*, 063007.
15. Bondi, H. (1952). On spherically symmetrical accretion. *Monthly Notices of the Royal Astronomical Society, 112*, 195.
16. Chandrasekhar, S. (1943). Dynamical friction. I. General considerations: The coefficient of dynamical friction. *The Astrophysical Journal, 97*, 255.
17. Ostriker, E. C. (1999). Dynamical friction in a gaseous medium. *The Astrophysical Journal, 513*, 252.
18. Kim, H., & Kim W.-T. (2007). Dynamical friction of a circular-orbit perturber in a gaseous medium. The *Astrophysical Journal, 665*, 432.
19. Chakrabarti, S. K. (1993). Binary black holes in stationary orbits and a test of the active galactic nucleus paradigm. *Astrophysical Journal, 411*, 610.
20. Basu, P., Mondal, S., & Chakrabarti, S. K. (2008). Gravitational wave emission from a companion black hole in the presence of an accretion disc around a super-massive Kerr black hole. *Monthly Notices of the Royal Astronomical Society, 388*, 219.
21. Kocsis, B., Yunes, N., & Loeb, A. (2011). Observable signatures of extreme mass-ratio inspiral black hole binaries embedded in thin accretion disks. *Physical Review D, 84*, 024032.
22. Derdzinski, A. M., D'Orazio, D., Duffell, P., Haiman, Z., & MacFadyen, A. (2019). Probing gas disc physics with LISA: Simulations of an intermediate mass ratio inspiral in an accretion disc. *Monthly Notices of the Royal Astronomical Society, 486*, 2754.
23. Garg, M., Derdzinski, A., Zwick, L., Capelo, P. R., & Mayer, L. (2022). The imprint of gas on gravitational waves from LISA intermediate-mass black hole binaries. *Monthly Notices of the Royal Astronomical Society, 517*, 1339.
24. Basu, P., Chatterjee, S., & Mondal, S. (2024). Eccentric orbits in disk-embedded EMRIs: Orbital evolution and observability trend in LISA. *Monthly Notices of the Royal Astronomical Society, 531*, 1506.
25. Narayan, R. (2000). Hydrodynamic drag on a compact star orbiting a supermassive black hole. *The Astrophysical Journal, 536*, 663.
26. Chakrabarti, S. K. (1990). *Theory of Transonic Astrophysical Flows*. https://doi.org/10.1142/1091
27. Glampedakis, K., & Kennefick, D. (2002). Zoom and whirl: Eccentric equatorial orbits around spinning black holes and their evolution under gravitational radiation reaction. *Physical Review D, 66*, 044002.
28. Glampedakis, K., Hughes, S. A., & Kennefick, D. (2002). Approximating the inspiral of test bodies into Kerr black holes. *Physical Review D, 66*, 064005.
29. Bardeen, J. M., Press, W. H., & Teukolsky, S. A. (1972). Rotating black holes: Locally nonrotating frames, energy extraction, and scalar synchrotron radiation. *The Astrophysical Journal, 178*, 347–370. https://doi.org/10.1086/151796
30. Ryan, F. D. (1996). Effect of gravitational radiation reaction on nonequatorial orbits around a Kerr black hole. *Physical Review D, 53*, 3064.
31. Robson, T., Cornish, N. J., & Liu C. (2019). *Classical and Quantum Gravity, 36*, 105011.
32. Jian-Min, W., et al. (2013). Super-eddington accreting massive black holes as long-lived cosmological standards. *Physical Review Letters, 110*, 8.
33. Charalampos, B., et al. (2010). Massive, massless and ghost modes of gravitational waves from higher-order gravity. *Astroparticle Physics, 34*, 4.

Chapter 28
Can We Detect Super-Chandrasekhar White Dwarfs via Continuous Gravitational Waves?

Mayusree Das and Banibrata Mukhopadhyay

Abstract The existence of massive white dwarfs (WDs) containing more than Chandrasekhar's maximum mass has been suggested via the detection of peculiar type Ia Supernovae. It had been crucial to directly detect those 'super' (more massive)-Chandrasekhar WDs to confirm their existence. The WD's small size and cold internal environment have been a great disadvantage for detecting them via electromagnetic surveys like GAIA and SDSS. Although many WDs are detected in optical observations, none of the super-Chandrasekhar WDs was observed directly. Mukhopadhyay and his group proposed a decade ago that one of the possibilities to explain the high mass of super-Chandrasekhar WD is to exercise the existence of a strong magnetic field in the WD interior. The group also recently proposed that a high magnetic field, in turn, can nontrivially deform the star from a spherical shape. In the presence of rotation and obliquity angle, such WDs can radiate continuous gravitational waves (CGW). This opens up the exciting possibility of detecting the super-Chandrasekhar WDs with the upcoming detector LISA. However, the gravitational wave (GW) amplitude will decay with time due to electromagnetic radiation and quadrupolar radiation. Altogether, those decay mechanisms will set a timescale for detecting WDs via CGW. Even with a timescale bound, we can expect to detect a few super-Chandrasekhar WDs with a couple of years of cumulative detection, leading to the direct detection of super-Chandrasekhar WDs.

28.1 Introduction

White dwarfs (WDs) are the end state of low to moderately mass main sequence stars. The contraction of the main sequence star comes to a halt after the electrons become degenerate under huge density and support against gravitational force. It is

M. Das (✉) · B. Mukhopadhyay
Joint Astronomy Programme, Department of Physics, Indian Institute of Science, Bangalore, India
e-mail: mayusreedas@iisc.ac.in; bm@iisc.ac.in

S. Ghosh, A. R. Rao (eds.), *The Relativistic Universe: From Classical to Quantum*, Astrophysics and Space Science Proceedings 61,
https://doi.org/10.1007/978-3-031-90186-7_28

famously known that the WDs have a mass limit of $1.4M_{\odot}$- Chandrasekhar's mass limit. If the WD accretes beyond Chandrasekhar's mass limit, then it will result in a type Ia supernova. However, there are overluminous Type Ia supernovae, e.g. SN 2003fg, had been observed with inferred progenitor's mass is as high as $2.8M_{\odot}$ [1], violating Chandrasekhar's mass limit significantly. Mukhopadhyay and his group [2] proposed that the excess mass of the 'super'-Chandrasekhar is held by excess pressure generated due to a high magnetic field. Incidentally, the WDs may retain a high magnetic field in their core from their main sequence progenitors, as suggested by Wickramasinghe and Ferrario [3], or may be generated during the end of the main sequence phase and carried on to WD by flux freezing. A dynamo operating during binary evolution can also generate the WD's magnetic field if they are in a double WD system [4]. It is also widely known that a magnetic field up to 10^9G is observed on WD's surface (see for review [5, 6]). However, the magnetic field can be four orders of magnitude higher near the core. The presence of a high magnetic field, in turn, can deform the star from its spherical shape. Mukhopadhyay's group proposed that in the presence of rotation along a misalignment between the magnetic and rotation axes, the deformed WD can radiate gravitational waves (GW) [7]. Like binary merger systems, this kind of isolated star also poses a non-zero time-varying quadrupole moment, which is a source of GW, very like the binary. This GW is to be called 'continuous' GW (CGW), as it is emitted from the isolated object as long as it is magnetized and rotating along a misaligned axes. WDs, being very lowly luminous, pose a challenge to detect them via optical surveys like GAIA and SDSS. The super-Chandrasekhar WDs are supposed to be smaller, causing even lower luminosity. Possibly, for that reason, we did not detect any super-Chandrasekhar WDs directly. The detection of super-Chandrasekhar WD will become wide open via CGW once LISA is active. However, the GW amplitude will decay with time as the system radiates electromagnetic and GW energy by angular momentum extraction and slows down with time. The magnetic field also decays in different timescales due to the Hall and Ohmic dissipations. Together, those decay mechanisms provide a decay timescale in which the young WDs must be caught up by any detectors before the GW amplitude decays below sensitivity. Here, we aim to discuss the plausibility of detecting super-Chandrasekhar WDs via CGW. The plan of the chapter is as follows. In Sect. 28.2, we discuss the deformed and misaligned rotating magnetized WD model, which is a viable candidate for GW radiation. We discuss the decay mechanisms responsible for GW amplitude decay in Sect. 28.3. We show the overall detection plausibility in Sect. 28.4 depending on the decay timescale. We finally conclude in Sect. 28.5.

28.2 White Dwarf Model

It is well known that any system is to be associated with non-zero time-varying quadrupole moment to generate CGW. The non-zero quadrupole moment can be generated from any fluid body with triaxial symmetry arising from deformation

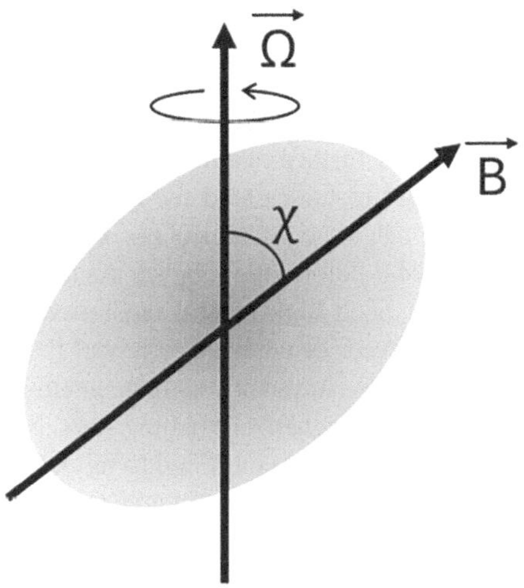

Fig. 28.1 A cartoon diagram of magnetized rotating WD with misalignment between the magnetic field axis and the rotation axis

due to the magnetic field. The poloidal and toroidal magnetic fields provide a directional pressure, deforming the star into respectively an oblate and prolate shape. A rotating body with a misaligned axis and deformation will produce a non-zero time-varying quadrupole moment; the 'time-varying' part is because of rotation. A cartoon diagram of such rotating magnetized WD with misalignment between respective axes is shown in Fig. 28.1. If the rotation rate of the WD is Ω, it is well known that the body will radiate GW strain with two frequencies Ω and 2Ω in two polarizations as follows [7, 8]:

$$h_{+} = h_0 \sin\chi \left[\frac{1}{2}\cos i \sin i \cos\chi \cos\Omega t - \frac{1+\cos^2 i}{2}\sin\chi \cos 2\Omega t\right],$$
$$h_{\times} = h_0 \sin\chi \left[\frac{1}{2}\sin i \cos\chi \sin\Omega t - \cos i \sin\chi \sin 2\Omega t\right]. \tag{28.1}$$

Here, $h_0 = \frac{2G}{c^4}\frac{\Omega^2 \epsilon I_{xx}}{d}\left(2\cos^2\chi - \sin^2\chi\right)$. For $\chi \to 0$ (but $\neq 0$),

$$h_0 = \frac{4G}{c^4}\frac{\Omega^2 \epsilon I_{xx}}{d}, \tag{28.2}$$

where c is the speed of light, G is Newton's gravitational constant, Ω is the angular frequency of the object, d is the distance between the detector and the source object, i is the angle between the rotation axis of the object and our line of sight, and the ellipticity, which is the degree of deformation, is defined as $\epsilon = |I_{zz} - I_{xx}|/I_{xx}$, where I_{xx} and I_{zz} are the principal moments of inertia of the star about x- and z-axes (any two orthogonal axes), respectively. Further, the amplitudes of h_{+} and $h_{\times}$ in Eq. (28.1) will be suppressed by an optimistic factor of $\left(\sin\chi\left[\frac{1}{2}\cos i\sin i\cos\chi\cos\Omega t - \frac{1+\cos^2 i}{2}\sin\chi\cos 2\Omega t\right]\right)$, which can be maximized with respect to i at $\chi = 3°$ (small obliquity angle or angle of misalignment) just during the birth of the WD, i.e., $t = 0$, providing $h = 0.0110297h_0$, which we consider for further calculations.

In order to understand the plausibility of the detection of WDs via CGW, we have to calculate h, hence h_0, which depends on the precise measurement of quadrupolar deformation, hence the moment of inertia. To extract such information, we either need high optical resolution and observational data, which has not yet been achieved, or we need to model the WD with numerical simulation. We use a general relativistic steady state magneto-hydrostatic solver numerical code XNS 4.0 [9] to model magnetized rotating stars. XNS 4.0 was originally designed for neutron stars, which we also modified to model WDs. We will design the WD interior with the relativistic equation of state for highly degenerate electron gas in the polytropic form $P = K\rho^{4/3}$, where $K = 3.1 \times 10^{12}$ in CGS. We are showing some 2D cross-sections and magnetic field topologies (toroidal and poloidal) here for the realization of the deformation (Fig. 28.2). It can be visualized that the WD with toroidal field is slightly prolate in shape, and the WD with poloidal field is oblate in shape. However, the angle between rotation and magnetic axes is zero here, and XNS cannot handle misalignment. Therefore, when we extract the ellipticity from XNS and use it as an input to calculate h_0, it is essential to consider a small misalignment so that the angle does not affect the physics. While extracting the ellipticity, we have to be cautious about only using the effect due to magnetic field deformation and rotation alone cannot provide nonzero quadrupole moment.

28.3 Gravitational Wave Decay

In Sect. 28.2, we have noticed that the GW amplitude is not stationary; rather, it is a function of Ω, χ, ϵ, which in turn are functions of time during the evolution of WD. The ellipticity depends on the magnitude of the central magnetic field. A WD with a non-zero time-varying quadrupole moment will exhibit electromagnetic emission due to dipole radiation and gravitational radiation due to quadrupolar energy loss. Both of these energy losses lead to decay of Ω and χ, meaning it slows down the star and tends to align the rotation and magnetic axes each other. The evolution equations are [10, 11] given by

$$\frac{d(\Omega I_{z'z'})}{dt} = -\frac{2G}{5c^5}(I_{zz} - I_{xx})^2\Omega^5\sin^2\chi(1 + 15\sin^2\chi) - \frac{B_p^2 R_p^6\Omega^3}{2c^3}\sin^2\chi F(x_0) \tag{28.3}$$

and

$$I_{z'z'}\frac{d\chi}{dt} = -\frac{12G}{5c^5}(I_{zz} - I_{xx})^2\Omega^4\sin^3\chi\cos\chi - \frac{B_p^2 R_p^6\Omega^2}{2c^3}\sin\chi\cos\chi F(x_0), \tag{28.4}$$

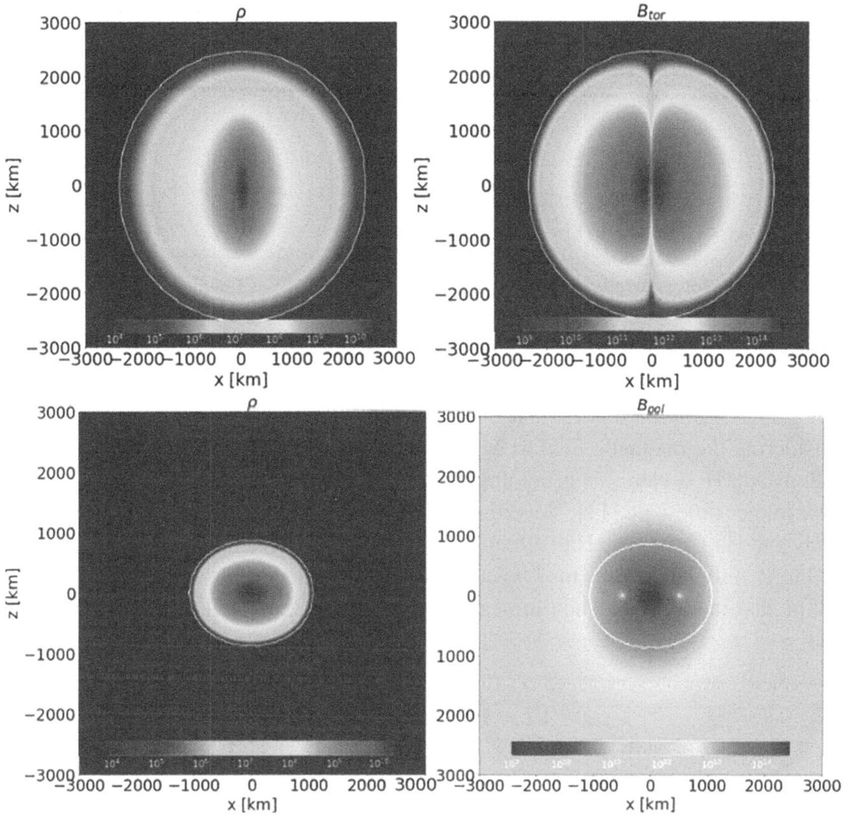

Fig. 28.2 Density isocontours (left) and magnetic field isocontour (right) for toroidally (top) and poloidally (bottom) dominated magnetized WDs of mass $M = 2M_{\odot}$ and $1.6M_{\odot}$ respectively for toroidal and poloidal topologies with $B_{max} = 3 \times 10^{14}$ G. Magnetic to gravitational energy ratio ~ 0.1

where $x_0 = R_0\Omega/c$, B_p is the strength of the magnetic field at the pole, R_p is the polar radius, R_0 is the average radius of WD, and the function $F(x_0)$ is defined as $F(x_0) = \frac{x_0^4}{5(x_0^6-3x_0^4+36)} + \frac{1}{3(x_0^2+1)}$. The set of Eqs. (28.3) and (28.4) is to be solved simultaneously to obtain Ω and χ as functions of time. We need to supply I_{xx}, I_{zz}, B_P, and R_P at $t = 0$, which are the output of particular WD model from the XNS code. In the Fig. 28.3, the evolutions of Ω and χ can be visualized for toroidal field-dominated configuration, where the GW radiation, i.e., the quadrupolar term caused by deformation, plays a role in decay, and B_P is negligible. The corresponding timescale of considerable decay before saturation occurs is around 10^4 years. For poloidally dominated field configurations, the electromagnetic radiation plays an important role, and the decay happens much faster, in a couple of years! However, a stable star, here WD, should sustain a toroidally dominated field [12]. Hence, the former case is more realistic. The spin and obliquity angle decay has been studied

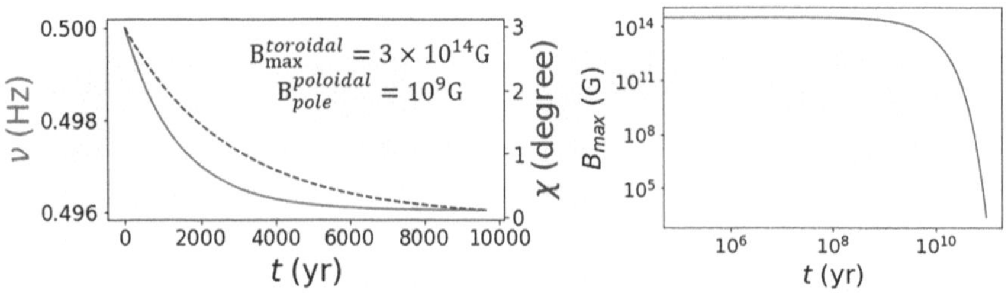

Fig. 28.3 Left: for toroidally dominated WDs, variations of ν, χ as functions of time for initial $\nu = 0.5$ Hz, $\chi = 3°$. Red (solid) and blue (dashed) lines show the variations of ν and χ, respectively. Right: magnetic field decay of the same WD model for the maximum magnetic field 3×10^{14}G

considering the magnetic field to be constant, which enforces B_P and $I_{xx} - I_{zz}$ to be constant. However, this is not unrealistic, as the magnetic field decays in a much more longer timescale [11]. Nevertheless, eventually the field decays via the Ohmic decay and Hall drift. The Hall effect generally conserves magnetic energy. However, the larger turbulent eddies may be cascaded into smaller eddies when Ohmic decay will be dominated [13]. The timescales for the Ohmic and Hall effects are given as [14]

$$t_{\rm Ohm} = \left(7 \times 10^{10}\,{\rm yr}\right) \rho_{c,6}^{1/3} R_4^{1/2} \left(\frac{\rho_{\rm avg}}{\rho_c}\right) \tag{28.5}$$

and

$$t_{\rm Hall} = \left(5 \times 10^{10}\,{\rm yr}\right) l_8^2 B_{0,14}^{-1} T_{c,7}^2 \rho_{c,10} \tag{28.6}$$

respectively, where $\rho_{c,n}$ is the central density in units of 10^n g/cm^3, R_4 is the radius of the star in units of 10^4 cm, $\rho_{\rm avg}$ is the average density, ρ_c is the central density, l_8 is a characteristic length scale in units of 10^8 cm, $B_{0,14}$ is the magnetic field strength in units of 10^{14} G, $T_{c,7}$ is the core temperature in units of 10^7 K. The magnetic field (B) decay in WDs can be studied by solving the decay equation [15]

$$\frac{dB}{dt} = -B\left(\frac{1}{t_{\rm Ohm}} + \frac{1}{t_{\rm Hall}}\right). \tag{28.7}$$

The initial magnetic field profile is taken from the XNS output data. We are showing the evolution of a magnetic field in Fig. 28.3, and the timescale of decay is more than 10 billion years, similar to the universe's age. This confirms that the calculation of spin down, where the magnetic field is considered constant, is correct in the sense that the magnetic field will remain constant indeed during spin down. In further studies of the complete decay mechanisms of GW, we will only take the effect of spin down into account as it is of the smallest timescale and, thus, more relevant.

28.4 Detection Plausibility

In this section, we will show the result of the decay mechanisms discussed in Sect. 28.3 upon detecting GW by detectors. In Fig. 28.4, a toroidally dominated WD with a magnetic field just after its birth and after the decay of ν/χ due to GW emission over time is shown, assuming the distance of the object from the detector to be 5 kpc. It is seen that the WD becomes undetectable by detectors after some time where it was detectable in the beginning. This sets an upper detection time limit, around 10^4 years, which is still well out of the observational timescale (which can be a maximum of 5 years) and is good news. The question remains, however, in this 'active timescale' when the WDs can be detectable in their first few thousands of years of lifetime, what is the probability of detection? To answer this question, we have to rely upon the birthrate of WD, which is 2×10^{-12} pc^{-3} yr^{-1} [16]. The birthrate in 5 kpc volume is $0.25/yr$. Thus, around 2500 WDs will be born in 10^4 years in the nearest 5 kpc volume from the Earth. All these WDs should be detected if they cross the minimum sensitivity of the upcoming detector LISA. However, the WD should be essentially highly magnetized to radiate CGW in the first place, and we know that $\sim$ 10% of the isolated WDs are highly magnetized (e.g. [6]), making the possible number of WD candidates to decrease to 250. Now, it is well known that magnetized WDs are, on average, more massive than their nonmagnetic counterpart [5]. However, no observation guarantees that all the highly magnetized WDs need to be more massive than the nonmagnetized ones. Hence, to target super-Changrasekhar WDs via CGW, we have to target the ratio of several overluminous Type Ia supernovae observed per year to standard Type Ia supernovae per year, which is about 2%. Moreover, there is no guarantee that the birthrate of WDs will be strictly proportional to the supernovae rate. However, one may expect

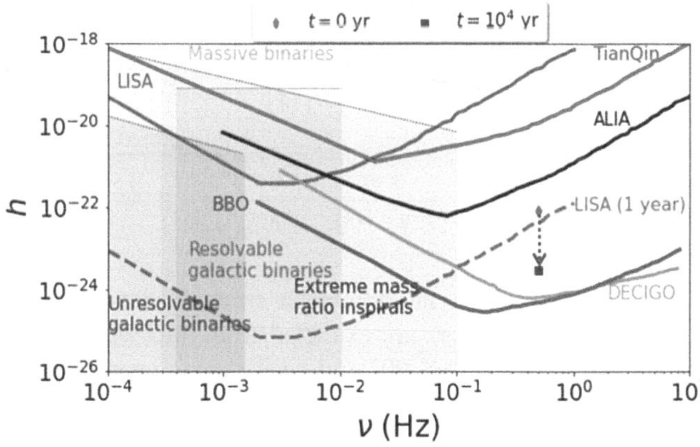

Fig. 28.4 Dimensionless gravitational wave amplitude before ($t = 0$) and after ($t = 10^4$ yrs) gravitational wave decay for a WD with $B_{max}^{Toroidal} = 3 \times 10^{14}$ G, $B_{pole}^{Poloidal} = 10^9$ G, initial $\nu = 0.5$ Hz, initial $\chi = 3°$ and $d = 5$ kpc

the ratio of the standard WD to super-Chandrasekhar WD to be related to the correct percentage to consider when calculating how many WDs should turn out to be super-Chandrasekhar. Therefore, the number of super-Chandrasekhar WDs we can catch via CGW is around 5 as soon as LISA starts operating.

28.4.1 Method to Enhance the Detection Possibility

The challenges of detecting WDs with LISA are to be appreciated. However, the probability of detection can be increased effectively by calculating the signal-to-noise ratio (SNR) cumulatively over a long observational time. The SNR in GW detectors measures how accurately a gravitational wave signal can be distinguished from the background noise in the detector. The calculation of SNR (S/N) is essential for determining the detectability of a GW event. If the GW strain remains constant over time, the SNR can be calculated as the ratio of the GW signal to GW noise as

$$S/N = \frac{h_{signal}}{h_{noise}}, \tag{28.8}$$

where h_{signal} is basically the GW strain of the object and h_{noise} is a function of the detector's power spectral density (S_n). The SNR is to be considered constant for a GW instantaneously (for some fleeting seconds) emitted from merger events. However, we are dealing with CGWs, which require a 'continuous' search over T observational time. The SNR will change with time as the rotation rate and obliquity angle change. In this case, we have to discretize the total observational time into fine $\mathcal{N}$ time bins where the change of rotation rate and obliquity angle stays minimal. Then, we can carry out the time integration by simply summing over all the bins cumulatively. However, this addition is incoherent as the phase of GW for each time stack is different because the Ω changes with time (see [17], for details of the technique). Assuming ν, χ and h_0 remain nearly constant over each time-stack, adding $\mathcal{N}$ such stacks, the cumulative SNR is [11]

$$\langle S/N\rangle = \sqrt{\langle S/N^2_{\Omega}\rangle + \langle S/N^2_{2\Omega}\rangle} = \sqrt{\frac{\sin^2\zeta}{100}\frac{h_0^2 T\sin^2 2\chi}{\sqrt{\mathcal{N}}S_n(\Omega)} + \frac{4\sin^2\zeta}{25}\frac{h_0^2 T\sin^4\chi}{\sqrt{\mathcal{N}}S_n(2\Omega)}}, \tag{28.9}$$

where ζ is the angle between the interferometer arms, the data for power spectral density (PSD) of various detectors are extracted from https://gwplotter.com/. For ground-based interferometers $\zeta = 90°$ and space-based interferometers such as LISA $\zeta = 60°$. The SNR increases if we increase the observational time T; however, it decreases if we discretize the total observational time into more bins. Depending on how fast the spin down happens, we can minimize the number of bins so that the rotation rate and obliquity angle remain almost constant; here, we use $\mathcal{N} \sim 300$. If we would have use $\mathcal{N} \sim 30$, the SNR would be $10^{1/4} \sim 2$. Taking care of

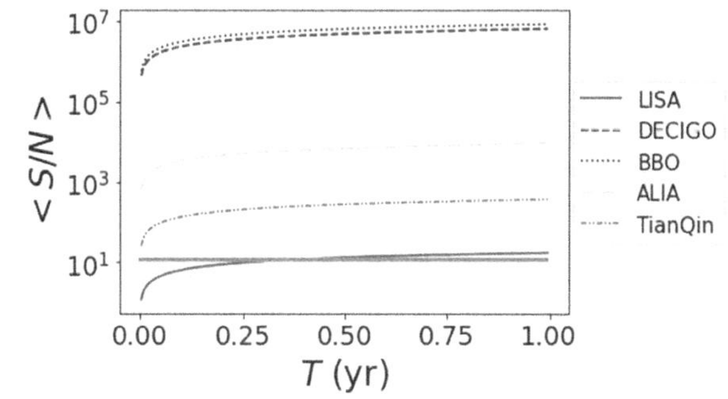

Fig. 28.5 SNR for various detectors as a function of integration time for the toroidally dominated WD (same as Fig. 28.4) with initial $\nu = 0.5$ Hz initial $\chi = 3°$ and $d = 5$ kpc. The orange line corresponds to $< S/N >= 12$

all the hurdles during the long observational time for continuous search, we have considered the SNR threshold to be 12, greater than the instantaneous search's threshold [18]. In the Fig. 28.5, we have shown a case where the WD would not be detectable immediately by LISA, but it can be detectable after couple months of integrated time. The SNR increases with time initially while the GW amplitude is added up in the first few time domains; however, with time, the decreased Ω, χ, and hence the decayed h saturates the SNR with time. The slower the decay is, the more saturation is achieved later, opening the possibility of crossing the SNR threshold even in the later stage of observation. The slowly rotating WDs decay slower, providing the above-mentioned opportunity, though with lower GW emission in the first place. The WDs with higher magnetic fields are most likely to be detected as their h_0, hence SNR, is always higher. The equivalent sensitivity curve, along with the sensitivity of many detectors for a T time-long continuous search, is shown in the Fig. 28.5 considering the sensitivity of the detectors is increased by $\sqrt{T}$ times as the continuous search is considered [19], i.e.

$$h \sim 11.4\sqrt{\frac{S_n}{T}}. \tag{28.10}$$

Using this technique, we have shown the instantaneous LISA sensitivity as well as the LISA sensitivity with 1 year of detection in Fig. 28.4. In fact, the physics meaning of both the Figs. 28.4 and 28.5 is the same.

28.5 Conclusion

We have aimed to discuss the detection plausibility of super-Chandrasekhar WDs via CGW with the upcoming GW detector in WD frequency LISA. By modeling the WD in general relativistic magneto-hydrostatics, we can understand what GW strain range is expected from the WDs, which are 5 Kpc away from us. We also have pointed out that the GW amplitude is not constant with time; rather, it decays due to angular momentum extraction and magnetic energy loss. The effect of all the decays can be visualized in Fig. 28.4. We have shown the result for one single WD model, which is toroidally dominated ($B_{max}^{Toroidal} = 3 \times 10^{14}$G) with observationally relevant surface poloidal field ($B_{pole}^{Poloidal} = 10^{9}$G). The poloidal field, however, is not strong enough to affect the structure of the WD. Hence, the GW amplitude is mostly from the quadrupolar deformation generated by the toroidal field. It is the most realistic model, as any stable WD should be associated with a toroidally dominated magnetic field with negligible poloidal field. In our model, the central toroidal field is at least 10^4 times larger than that of central poloidal field strength. We also have discussed that a few super-Chandrasekhar WDs will be detected with upcoming detectors, keeping in mind the active timescale limit of the WDs. Although the hope for instantaneous detection can be almost ruled out by LISA, the exciting news is that LISA would be able to detect those super-Chandrasekhar WDs with 1 year of continuous detection, leading to direct detection of super-Chandrasekhar WDs for the first time, confirming their existence.

Acknowledgments MD takes this opportunity to thank BM for his support throughout her research so far. MD acknowledges the Prime Minister's Research Fellows (PMRF) scheme, with Ref. No TF/PMRF-22-5442.03.

References

1. Howell, D. A., Sullivan, M., Nugent, P. E., Ellis, R. S., Conley, A. J., Le Borgne, D., Carlberg, R. G., Guy, J., Balam, D., Basa, S., Fouchez, D., Hook, I. M., Hsiao, E. Y., Neill, J. D., Pain, R., Perrett, K. M., & Pritchet, C. J. (2006). The type Ia supernova SNLS-03D3bb from a super-Chandrasekhar-mass white dwarf star. *Nature, 443*(7109), 308–311.
2. Das, U., & Mukhopadhyay, B. (2014). On the nature of highly magnetized white dwarfs: Super-chandrasekhar mass limits. *Journal of Cosmology and Astroparticle Physics, 2014*(06), 050.
3. Wickramasinghe, D. T., & Ferrario, L. (2005). The origin of the magnetic fields in white dwarfs. *Monthly Notices of the Royal Astronomical Society, 356*(4), 1576–1582.
4. Tout, C. A., Wickramasinghe, D. T., Liebert, J., Ferrario, L., & Pringle, J. E. (2008). Binary star origin of high field magnetic white dwarfs. *Monthly Notices of the Royal Astronomical Society, 387*(2), 897–901.
5. Ferrario, L., de Martino, D., & Gänsicke, B. T. (2015). Magnetic white dwarfs. *Space Science Reviews, 191*(1–4), 111–169.
6. Ferrario, L., Wickramasinghe, D., & Kawka, A. (2020). Magnetic fields in isolated and interacting white dwarfs. *Advances in Space Research, 66*(5), 1025–1056.
7. Kalita, S., & Mukhopadhyay, B. (2019). Continuous gravitational wave from magnetized white dwarfs and neutron stars: Possible missions for LISA, DECIGO, BBO, ET detectors. *Monthly Notices of the Royal Astronomical Society, 490*(2), 2692–2705.
8. Bonazzola, S., & Gourgoulhon, E. (1996). Gravitational waves from pulsars: Emission by the magnetic-field-induced distortion. *Astronomy & Astrophysics, 312*, 675–690.
9. Soldateschi, J., Bucciantini, N., & Del Zanna, L. (2021). Quasi-universality of the magnetic deformation of neutron stars in general relativity and beyond. *Astronomy & Astrophysics, 654*, A162.
10. Chau, W. Y., & Henriksen, R. N. (1970). Alignment of an oblique spheroidal rotator through electromagnetic and gravitational radiation. *The Astrophysical Journall, 161*, L137.

11. Das, M., & Mukhopadhyay, B. (2023). Detection possibility of continuous gravitational waves from rotating magnetized neutron stars. *The Astrophysical Journal, 955*(1), 19.
12. Braithwaite, J. (2009). Axisymmetric magnetic fields in stars: Relative strengths of poloidal and toroidal components. *Monthly Notices of the Royal Astronomical Society, 397*(2), 763–774.
13. Goldreich, P., & Reisenegger, A. (1992). Magnetic field decay in isolated neutron stars. *The Astrophysical Journal, 395*, 250.
14. Bhattacharya, M., Hackett, A. J., Gupta, A., Tout, C. A., & Mukhopadhyay, B. (2022). Evolution of highly magnetic white dwarfs by field decay and cooling: Theory and simulations. *The Astrophysical Journal, 925*(2), 133.
15. Heyl, J. S., & Kulkarni, S. R. (1998). How common are magnetars? The consequences of magnetic field decay. *The Astrophysical Journal, 506*(1), L61–L64.
16. Holberg, J. B., Oswalt, T. D., Sion, E. M., & McCook, G. P. (2016). The 25 parsec local white dwarf population. *Monthly Notices of the Royal Astronomical Society, 462*(3), 2295–2318.
17. Maggiore, M. (2007). *Gravitational waves. Vol. 1: Theory and experiments. Oxford master series in physics*. Oxford University Press.
18. Watts, A. L., Krishnan, B., Bildsten, L., & Schutz, B. F. (2008). Detecting gravitational wave emission from the known accreting neutron stars. *Monthly Notices of the Royal Astronomical Society, 389*(2), 839–868.
19. Soldateschi, J., & Bucciantini, N. (2021). Detectability of continuous gravitational waves from magnetically deformed neutron stars. *Galaxies, 9*(4), 101.

Part VIII
Compact Objects

Chapter 29
Was it Really a Kerr Black Hole That EHT Imaged in 2019?

Chandrachur Chakraborty

Abstract By using the working hypothesis of a Kerr black hole (BH), the Event Horizon Telescope was able to map M87* to the Kerr BH, and it was shown that the observed image of M87* is generally consistent with what would be expected for a Kerr black hole's shadow. Here, it is demonstrated that M87* is not certainly a Kerr BH, but rather that it might be a Kerr naked singularity, or a BH/naked singularity with a gravitomagnetic monopole, since the alternatives to the Kerr BH cannot be ruled out.

29.1 Introduction

The gravitational counterpart of Dirac's magnetic monopole is called the gravitomagnetic monopole (GMM) [1, 2]. A spherically symmetric vacuum solution of Einstein's equation with an additional parameter other than mass was found in 1963 by Newmann et al. The new parameter that makes the so-called *static* Schwarzschild metric *stationary*, is known as the NUT (Newmann-Unti-Tamburino) parameter/charge or GMM, and the new solution is nowadays known as the Taub-NUT solution [3, 4]. Note that the Einstein-Hilbert action does not require any modification to include the NUT parameter in the vacuum solution of Einstein equation [1, 2]. The Taub-NUT metric reduces to the Schwarzschild metric, if GMM vanishes. GMM is also called as the magnetic mass, gravitomagnetic mass, dual mass etc. Bonnor [5] physically interpreted it as the 'linear source of pure angular momentum' or 'a massless rotating rod [6]. According to the classical field equation, matter can be divided into two groups: gravitomagnetic matter (also known as magnetic mass or a gravititomagnetic monopole) and gravitoelectric matter (also known as regular matter or ordinary mass) [7, 8]. The idea of ordinary matter does not have any significance for the gravitomagnetic matter because of their

C. Chakraborty (✉)
Manipal Centre for Natural Sciences, Manipal Academy of Higher Education, Manipal, India
e-mail: chandrachur.c@manipal.edu

S. Ghosh, A. R. Rao (eds.), *The Relativistic Universe: From Classical to Quantum*, Astrophysics and Space Science Proceedings 61,
https://doi.org/10.1007/978-3-031-90186-7_29

dissimilar gravitational characteristics [7, 8]. For example, the Komar formula (in the geometrised $G = c = 1$ unit) defines the ordinary mass M and gravitomagnetic monopole (GMM) n as [1, 9]

$$M \equiv \frac{1}{8\pi} \int_{S^2_\infty} *dk \tag{29.1}$$

and

$$n \equiv \frac{1}{8\pi} \int_{S^2_\infty} dk \, , \tag{29.2}$$

respectively, where, the one-form of the timelike Killing vector (∂_t) is $k = (\partial_t)^b = g_{0\mu} dx^\mu$, and the integral is taken over a spacelike 2-surface (S) at spacelike infinity. $*$ signifies the Hodge dual.

While the prospect of GMM's presence is intriguing, no significant attempt has been made to locate it in any astronomical objects. The spectra of supernovae, quasars, or active galactic nuclei are the finest places to search for the GMM, according to Lynden-Bell and Nouri-Zonoz, who were the first to advocate for such an inquiry [10]. The no hair conjecture and cosmic censorship conjecture lead one to believe that any astrophysical collapsed object is a Kerr black hole. Now, if a Kerr spacetime contains GMM, it is called as a Kerr-Taub-NUT (KTN) spacetime. It is noteworthy to mention that, using three main X-ray observational techniques, it has recently been demonstrated that the GMM [11] may be included in the collapsed astronomical object GRO J1655-40 [12], and that the KTN spacetime [13] might better characterize the aforementioned entity. This was first observational sign of GMM in an astronomical object. Now, the question is whether a GMM could be present in any other astrophysical object. Recently, it has been shown that the presence of GMM in nature helps to predict the existence of micro primordial black holes, and those could contribute to the dark matter in the present era [9]. By examining the most recent image released by the Event Horizon Telescope (EHT) collaboration, it is demonstrated here that another astrophysical collapsed object, M87*, could also contain the GMM [14].

29.2 Brief Discussion on Kerr-Taub-NUT Spacetime

Let us first write down the KTN metric [14]

$$ds^2 = -\frac{\Delta}{p^2}(dt - Ad\phi)^2 + \frac{p^2}{\Delta}dr^2 + p^2 d\theta^2 + \frac{1}{p^2}\sin^2\theta\,(adt - Bd\phi)^2 \, , \tag{29.3}$$

with

$$\Delta = r^2 - 2Mr + a^2 - n^2, \qquad p^2 = r^2 + (n + a\cos\theta)^2,$$
$$A = a\sin^2\theta - 2n\cos\theta, \qquad B = r^2 + a^2 + n^2, \tag{29.4}$$

where $n_* = n/M$ is the GMM or NUT parameter of the collapsed object, $a_* = a/M$ is the spin parameter, and M is the mass. The location of the outer horizon's boundary (r_h) is at [14]

$$r_h/M = 1 + \sqrt{1 + n_*^2 - a_*^2}\,, \tag{29.5}$$

and, the singularity is located at [14]

$$r = 0 \quad , \quad \theta_s = \cos^{-1}(-n_*/a_*). \tag{29.6}$$

It is seen from Eq. (29.5) that r_h is real for $0 \leq a_* \leq \sqrt{1 + n_*^2}$. This indicates that the horizon exists, and the KTN spacetime describes a black hole (BH) in such a case. On the other hand, the KTN spacetime describes a naked singularity (NS) for $a_* > \sqrt{1 + n_*^2}$, as the horizon does not exists due to the imaginary value of r_h. Eq. (29.6) indicates that for $|n_*| > |a_*|$, the singularity does not occur. This corresponds a singularity-free KTN black hole. In order to discuss the axisymmetric and stationary KTN spacetime, one can consider four possibilities by dividing the $a_* - n_*$ parameter space as, Quadrant I ($n_* > 0$ and $a_* > 0$), Quadrant II ($a_* < 0$ and $n_* > 0$), Quadrant III ($a_* < 0$ and $n_* < 0$) and Quadrant IV ($a_* > 0$ and $n_* < 0$) [14].

29.3 Shadow Size of a Collapsed Object Containing GMM

Three types of photon trajectories can be found around a collapsed object: scatter to infinity, capture and critical curve. The last one divides the first and second types. A light ray orbiting the compressed object multiple times is unstable if it is almost tangential to the circular photon orbit [14]. This ray is either caught for a tiny disturbance by the central collapsed object, or it escapes to infinity. They produce a brighter area surrounding a centrally dark region in the distant observer's sky plane as they repeatedly orbits the collapsed object. This can be projected as a two-dimensional picture, and the area that is dark is known as the collapsed object's shadow. To examine the collapsed object's shadow, the geodesic equation must be solved. Using the ray tracing code, the geodesic equations are numerically solved for the KTN metric [14].

An object that has collapsed may not always have a perfectly round shadow. It can deviate from the circularity. Therefore, it is vital to identify two key parameters in this context. The average radius ($\bar{R}$) of the shadow is defined as [14]

$$\bar{R}^2 = \frac{1}{2\pi} \int_0^{2\pi} R^2(\phi)\, d\phi\,. \tag{29.7}$$

The definition of asymmetry (Δd) of the shadow is how much the average shadow radius ($\bar{R}$) differs from the RMS distance:

$$\Delta d = \frac{1}{\bar{R}} \sqrt{\frac{1}{2\pi} \int_0^{2\pi} (R(\phi) - \bar{R})^2\, d\phi}\,. \tag{29.8}$$

The asymmetry parameter measures how far the shadow deviates from being circular. Now, using the geodesic equation in a specific spacetime, and the ray tracing code, one can obtain the shadow size in the said spacetime.

Now, let us calculate the shadow size in KTN spacetime. For different (a_*, n_*), one can see that for a Kerr BH, the size of the shadow (radius) shrinks from $10.4M$ to $9.6M$ for the range of Kerr parameter ($a_* \equiv a/M$) of $0 < a_* < 1$. If a_* increases from 1, the Kerr spacetime corresponds to the Kerr near-extremal naked singularity (NS). In such a case, the shadow size decreases to an extremely small (see Figure 2 of [14]) value. For further increment of a_*, the shadow size increases, and can be even greater than $10.4M$ depending on the value of a_*. For the KTN BHs, the size of the shadow grows, if n_* increases. This is true also for the KTN NSs. However, if one increase the value of a_* from 0 to a higher value for a fixed n_* and inclination angle (i), the shadow size first increases for the KTN BHs, decreases for the near-extremal KTN NSs, and increases again, as seen from Fig. 29.1.

29.4 On the Presence of GMM in M87*

The EHT collaboration has recently reported [15] the observational constraints for the size of the shadow in M87*, which are given as [14, 16]

$$D\delta/M \simeq 11.0 \pm 1.5 \tag{29.9}$$

for δ (average diameter of the crescent) $= (42\pm3)$ μarcsec [15], M (mass of M87*) $= (6.5 \pm 0.2|_{\rm stat} \pm 0.7|_{\rm sys}) \times 10^9 M_\odot$ and D (distance of M87*) $= 16.8^{+0.8}_{-0.7}$ Mpc. However, it is considered as $D = 16.8 \pm 0.75$ Mpc in Eq. (29.9) to simplify the calculation. Now, using Eq. (29.9), one may constrain the value of GMM in M87* by taking into account the KTN metric and observed parameter values, including the inclination angle ($i = 17°$). It is noteworthy that the KTN metric is used here rather than the Kerr metric which was used in [16] with a limit of $\Delta d \lesssim 10\%$ [16]. The outcome is spoken below.

It is seen from Fig. 29.2 that although the whole $a_* - n_*$ space indicates the asymmetry (indicated by colors) below $\sim 5\%$, the observed shadow size (9.5 to

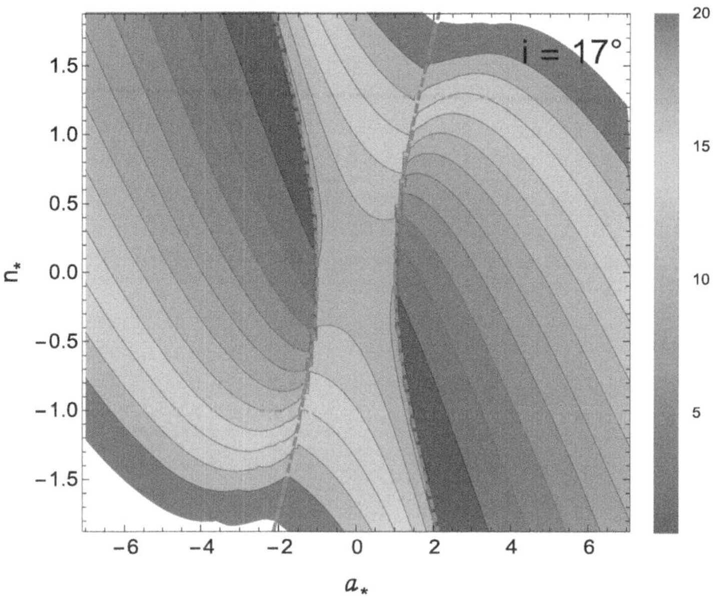

Fig. 29.1 Dependence of a KTN collapsed object's $\bar{R}$ (in 'M') on (a_*, n_*), for $i = 17°$. The shadow size as seen in Quadrants I-IV is indicated by the colors. The pair of slender green dashed lines split the entire plane into the KTN NS and KTN BH regions; that is, the KTN NS region is implied by the remaining portion of the plane, while the KTN BH area is implied by the pair of slender green dashed lines. See Sect. 29.3 for details. The Figure is taken from [14]

12.5) is narrowed down in a very small space which is indicated by the hatched region. Under these circumstances, n_* cannot have an upper limit of more than 1.1, that is, $n_* \lesssim 1.1$ for the prograde rotation ($a_* > 0$), whereas n_* cannot have a lower limit of less than -1.1, that is, $n_* \gtrsim -1.1$ for the retrograde rotation ($a_* < 0$) [14]. This plot also implies that, generally speaking, breaking the degeneracy between a BH and an NS can be aided by measuring the shadow asymmetry with an accuracy of 1%. M87* might be characterized as a KTN BH despite the falsifiability of the KTN NS, with the lower limit of n_* being $n_* > -0.9$ in the case of the retrograde rotation and the upper limit being $n_* < 0.9$ in the case of the prograde rotation. In this case, the plot is created for a small parameter space. The black hatched region of NS can overrun the thin Yellow-Cyan regions, which are next to the green dashed lines of Quadrants II & IV, for a general wide parameter space, provided that the value of $|n_*|$ is increased [14]. If M87* is Kerr and KTN BH, the shadow may be nearly round, according to Fig. 29.2, but the asymmetry in the shadow grows in the Kerr NS and KTN NS region. M87* cannot be a BH at all if the next observation shows that the shadow asymmetry is more than 1%.

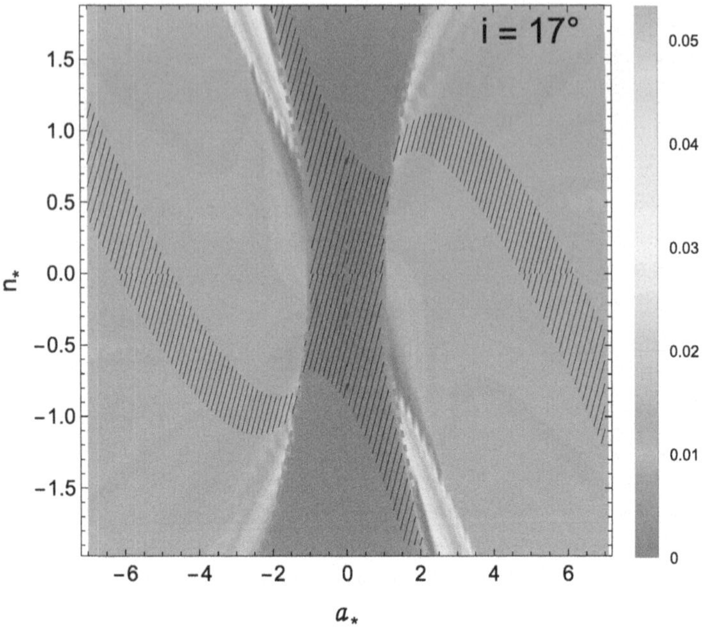

Fig. 29.2 Constraints on (n_*, a_*) is obtained for considering $i = 17°$ and utilizing the observational results for the M87* shadow. As previously mentioned, the plane is divided into four quadrants: I-IV. The pair of slender green dashed lines split the whole plane of the plot into KTN BH and KTN NS sectors. The KTN NS region is implied by the remaining portion of the plane, whereas the KTN BH region is implied by the midway section of the two green lines. $\Delta d \lesssim 10\%$ is represented by the whole colored region of this plot, according to the EHT collaboration's report. The hatched region is compatible with the permitted region for M87*. This plot indicates that M87* might be an NS and that it might include GMM. The Figure is taken from [14]

29.5 Conclusion and Discussion

The explanation above makes clear that M87* is not necessarily a Kerr BH. It could be a Kerr NS. Furthermore, it is impossible to completely rule out the chance that it contains a GMM. As a result, M87* may also be a KTN NS or BH [14]. Resolving degeneracies between BHs and NSs, as well as between different metrics and quadrants, can be facilitated by precise measurements of both shadow size and asymmetry [14]. The NSs would be falsifiable if the asymmetry of the shadow is determined to be smaller than 1% by the next EHT-like observations. Nevertheless, M87* [14] might still be characterised as a KTN BH.

Here we have constrained the (a_*, n_*) parameter space for M87* based on the observational data/parameters released/obtained by the EHT collaboration. This leads to our conclusion based on the kinematical quantities like the size and shape of the shadow of M87*. The issue is dynamical, and could not be settled by a measurement of just the kinematical quantities like size and shape of the observed shadow. Secondly, the formation mechanism of a Kerr BH is well-understood from the stellar collapse point of view, but the alternatives of Kerr BH proposed here still do not. However, in a very recent paper [17], a novel and realistic channel to form a sub-solar-mass Kerr NS is reported (see also [18]) but this is not based on the so-called 'stellar collapse'. A possible formation mechanism of a Taub-NUT BH has also been proposed in [9], but it is specifically related to the primordial BHs. Hopefully, some of these issues will be resolved in the future.

The Taub-NUT spacetime is not asymptotically flat despite the vanishing Riemann tensor ($R_{\mu\nu\alpha\beta} = O\left(1/r^3\right)$) for $r \to \infty$ [4]. Because of n_*, the anisotropy at $r \to \infty$ is caused by the non-vanishing component $g_{t\phi}$. For this reason, a homogeneous and anisotropic model of the universe is described by the Taub-NUT metric [4]. Since the universe is thought to be homogeneous and isotropic, the GMM may also be present in other astrophysical objects in order to lessen or completely negate the influence of the GMM [14]. Interestingly, the string singularity is included in the Taub-NUT spacetime [2]. In order to prevent this string singularity, Misner [4] enforced the time periodicity, which raises the causality violation [2] because of the existence of the closed-timelike curve (CTC). Nevertheless, it has recently been demonstrated [19, 20] that there are no causal pathologies in the Taub-NUT spacetime. More specifically, since CTCs are shown [19, 20] not to be the closed causal geodesics, an observer falling freely should not experience a causality violation (see Refs. [19, 20] for more details). As a result, some long-standing barriers to accepting the Taub-NUT solution as physically relevant are eliminated [9, 19]. It is also possible to formulate a consistent thermodynamics for the Taub-NUT spacetime with Misner string present [21, 22]. In relation to CTC, the causality violation problem [14] in the Taub-NUT BHs can be avoided by applying the Novikov self-consistency conjecture/principle [23, 24].

Acknowledgments I acknowledge the support and kind hospitality of the organizers of International Symposium Relativistic Astrophysics 2023. I take this opportunity to give a special thanks to Shubhrangshu Ghosh and Banibrata Mukhopadhyay for the wonderful meeting. I also thank my home institute (MCNS, MAHE) for supporting my travel expenses to attend this symposium.

References

1. Ramaswamy, S., & Sen, A. (1981). *Journal of Mathematical Physics, 22*, 2612.
2. Ramaswamy, S., & Sen, A. (1986). *Physical Review Letters 57*, 1088.
3. Newman, E., Tamburino, L., & Unti, T. (1963). *Journal of Mathematical Physics, 4*, 915.
4. Misner, C. W. (1963). *Journal of Mathematical Physics, 4*, 924.
5. Bonnor, W. B. (1969). *Proceedings of the Cambridge Philosophical Society, 66*, 145.
6. Dowker, J. S. (1974). *General Relativity and Gravitation, 5*, 603.
7. Shen, J. Q. (2002). *General Relativity and Gravitation, 34*, 1423.
8. Shen, J. Q. (2004). *Annals of Physics (Leipzig), 13*, 532.
9. Chakraborty, C., & Bhattacharyya, S. (2022). *Physical Review, D 106*, 103028.
10. Lynden-Bell, D., & Nouri-Zonoz, M. (1998). *Reviews of Modern Physics, 70*, 427.

11. Sen, G., Chakraborty, C., et al. (2024). *Journal of Cosmology and Astroparticle Physics, 12*, 030.
12. Chakraborty, C., & Bhattacharyya, S. (2018). *Physical Review, D 98*, 043021.
13. Chakraborty, C., & Bhattacharyya, S. (2019). *Journal of Cosmology and Astroparticle Physics, 05*, 034.
14. Ghasemi-Nodehi, M., Chakraborty, C., Yu, Q., & Lu, Y. (2021). *European Physical Journal, C 81*, 939.
15. The Event Horizon Telescope Collaboration. (2019). *Astrophysical Journal, 875*, L6.
16. The Event Horizon Telescope Collaboration. (2019). *Astrophysical Journal, 875*, L1.
17. Chakraborty, C., & Bhattacharyya, S. (2024). *Journal of Cosmology and Astroparticle Physics, 06*, 007.
18. Chakraborty, C., Bhattacharyya, S., & Joshi, P. S. (2024). *Journal of Cosmology and Astroparticle Physics, 07*, 053.
19. Clément, G., Gal'tsov, D., & Guenouche, M. (2015). *Physics Letters, B 750*, 591.
20. Clément, G., Gal'tsov, D., & Guenouche, M. (2016). *Physical Review, D 93*, 024048.
21. Hennigar, R. A., Kubizňák, D., & Mann, R. B. (2019). *Physical Review, D 100*, 064055.
22. Bordo, A. B., Gray, F., & Kubizňák, D. (2019). *Journal of High Energy Physics, 2019*, 119.
23. Novikov, I. D. (1983). *Evolution of the universe*. Cambridge University Press.
24. Friedman, J., et al. (1990). *Physical Review, D 42*, 1915.

Chapter 30
Exploring Magnetic Field Limits in Anisotropic Neutron Stars through GW190814 Tidal Deformability

Debabrata Deb, Banibrata Mukhopadhyay, and Fridolin Weber

Abstract The LIGO/Virgo collaboration detected a compact object in the GW190814 event on August 14, 2019, with an estimated mass of $\sim 2.59^{+0.08}_{-0.09}$ $M_{\odot}$, alongside a 23 $M_{\odot}$ black hole. This observation ignited a scientific discussion due to the object's unique position in the mass gap between neutron stars and black holes. In our study, we investigate the upper limit of the central magnetic field of anisotropic magnetized neutron stars, contextualized by the GW190814 event. We employ the density-dependent relativistic mean-field theory equation of state and assume a specific density dependence for the magnetic field. Our findings indicate that the magnetic field strength, anisotropy, and the magnetic field's orientation significantly affect the physical properties of neutron stars. We examine two orientations: radial (where local magnetic fields point radially) and transverse (where they are perpendicular to the radial direction). Notably, stars with a transverse magnetic field orientation exhibit increased mass with higher anisotropy and magnetic field strength. Additionally, we show that the magnetic field, its orientation, and anisotropy substantially influence the tidal deformability of neutron stars. We aim to constrain the central magnetic field of neutron stars based on the tidal deformability measurement for a canonical 1.4 $M_{\odot}$ neutron star, $\Lambda_{1.4} = 616^{+273}_{-158}$, obtained from GW190814, which suggests a preference for a stiffer equation of state.

D. Deb (✉)
The Institute of Mathematical Sciences, Chennai, India

B. Mukhopadhyay
Department of Physics, Indian Institute of Science, Bangalore, India
e-mail: bm@iisc.ac.in

Fridolin Weber
Department of Physics, San Diego State University, San Diego, CA, USA
e-mail: fweber@sdsu.edu

Department of Physics, University of California at San Diego, La Jolla, CA, USA

S. Ghosh, A. R. Rao (eds.), *The Relativistic Universe: From Classical to Quantum*, Astrophysics and Space Science Proceedings 61,
https://doi.org/10.1007/978-3-031-90186-7_30

30.1 Introduction

Compact stars serve as unique astrophysical laboratories to explore the nature of matter under extreme conditions [1, 2]. Neutron stars (NSs) are examples of such ultra-dense objects. Magnetic flux conservation during stellar collapse results in ultra-strong magnetic fields within these stars. It is hypothesized that the magnetic field at the centers of these inhomogeneous, ultra-dense stars could reach up to $\sim 10^{19}$ G [3–5]. However, observational evidence is insufficient to confirm the magnetic field strength inside compact stars, necessitating theoretical models to investigate the impact of high magnetic fields on their physical parameters. Studying these effects requires a general relativistic (GR) framework. Following the seminal works of Tolman [6] and Oppenheimer and Volkoff [7] (TOV), numerous studies have explored the properties of NSs using the GR hydrostatic equilibrium equation widely known as the TOV equation [8–10].

Since the groundbreaking discovery of radio pulsars by [11], NSs have been intensely studied as potential sources of high-energy emissions. Surface magnetic fields inferred from magnetic dipole models and spin-down rates typically range from 10^8 to 10^{13} G [12, 13]. Notably, PSR J1847−0130 has a strong magnetic field of $B = 9.4 \times 10^{13}$ G [14]. Anomalous X-ray pulsars (AXPs) and soft-γ repeaters (SGRs) exhibit even stronger fields, $10^{14} - 10^{15}$ G [15–17], and are termed magnetars. Magnetars are predicted to be sources of γ-ray bursts, requiring fields of $10^{16} - 10^{17}$ G to produce Poynting flux-dominated jets [18–22]. Though only about 30 magnetars have been detected, they are estimated to constitute roughly 10% of the NS population [23]. The impact of strong magnetic fields on the ultra-dense electron gases in NS interiors has been explored extensively [24–28]. Studies on dense, strongly magnetized nuclear matter include work by [29–35]. Investigations of NSs with different magnetic field configurations−toroidal, poloidal, or mixed— have also been conducted by [36–38]. The magnetic fields of NSs and their effects on mass-radius (M-R) relationship are significant. Magnetic fields influence stars in two main ways: classically, by introducing magnetic pressure that contributes to hydrostatic balance and causes pressure anisotropy, and quantum mechanically, by modifying the equation of state (EoS) through Landau quantization. However, the quantum effect is significant only for fields above 3×10^{18} G [39]. In this study, we show that in the presence of matter anisotropy, the tidal deformability measurement for a canonical 1.4 $M_\odot$ NS, $\Lambda_{1.4} = 616^{+273}_{-158}$, obtained from GW190814 [40], actually constrains the upper limit of the central magnetic field for NSs to well below the critical limit that required for Landau quantization effects on the EoS.

Recent gravitational wave (GW) observations have advanced our understanding of the mass gap between the heaviest NSs and the lightest black holes (BHs) [41, 42]. The GW190814 event [40] detected the merger of two compact objects, one with a mass in the range of $2.50 - 2.67$ $M_\odot$, situated within this mass gap. Some studies [43] have suggested that this event could involve two black holes. However, the exact nature of this object—whether it is a black hole or a NS—remains unresolved, as no electromagnetic counterpart was observed. Other GW events, like GW200210$_0$92254 [44], further support the existence of objects within the mass gap. The possibility of a NS as the secondary component in GW190814 has been explored in several studies [43, 45–54]. Alternatively, the secondary object being a quark or hybrid star has been considered in other studies [55–61]. Huang et al. [46] indicated that the secondary object could be a hadronic NS using the density-dependent relativistic mean field model. Zevin et al. [47] concluded that forming GW190814-like systems requires a supernova engine model with extended timescales for substantial accretion before explosion, suggesting a narrower or nonexistent mass gap between NSs and black holes. Some researchers have also applied modified gravity theories to address this issue [62–64]. Tsokaros et al. [45] demonstrated that a stable 2.6 $M_\odot$ NS could be explained by rapid uniform rotation with moderately stiff EoSs, though not with softer ones. The existing literature motivated us to use the tidal deformability measurement for a canonical 1.4 $M_\odot$ NS in GW190814 to constrain the possible upper limit of the central magnetic field for anisotropic magnetized NSs.

In this study, we explore the properties of anisotropic magnetized NSs by considering the magnetic field profile consistent with the Einstein-Maxwell equations. A common approach is the density-dependent exponential profile proposed by Bandyopadhyay et al. [65], which aligns with these equations. On the other hand, there is ongoing debate regarding whether a strong magnetic field increases or decreases the maximum mass of a compact star [31, 37, 65–72]. Chu et al. [73] addressed this issue by examining the effects of magnetic field orientation. They classified local magnetic fields as radially oriented (RO) when directed towards the radial direction, and transversely oriented (TO) when perpendicular to the radial direction (e.g., along the θ direction). Their study indicated that both the strength and orientation of the magnetic field significantly influence the maximum mass of a compact stellar object. In this chapter, we investigate anisotropic compact stars by incorporating magnetic field orientations and spatial distributions into the TOV equation. We consider anisotropy arising from (i) the local anisotropy of the fluid, and (ii) the presence of a strong magnetic field. The magnetic field strength in magnetars is relatively weak at the surface but increases significantly towards the center [17, 74, 75]. Although pressure anisotropy inside magnetars may not be substantial, it is already demonstrated that accounting for anisotropy provides a more generalized TOV equation, which is crucial for calculating the properties and ensuring the stability of magnetized anisotropic compact stars [76].

The structure of this chapter is as follows: Sect. 30.2 outlines the basic formalism and mathematical framework necessary to describe highly magnetized compact stellar objects. Section 30.3 presents the results of our study. Finally, Sect. 30.4 provides a brief discussion and concludes the work.

30.2 Stellar Structure Equations and Basic Mathematical Formalism

To describe static, anisotropic, spherically symmetric compact objects under a strong magnetic field, the fundamental stellar structure equations are [76]

$$\frac{\mathrm{d}m}{\mathrm{d}r} = 4\pi\left(\rho + \frac{B^2}{8\pi}\right) r^2, \tag{30.1}$$

$$\begin{cases} \dfrac{\mathrm{d}p_r}{\mathrm{d}r} = \dfrac{-(\rho+p_r)\dfrac{4\pi r^3\left(p_r-\frac{B^2}{8\pi}\right)+m}{r(r-2m)}+\frac{2}{r}\Delta}{\left[1-\frac{\mathrm{d}}{\mathrm{d}\rho}\left(\frac{B^2}{8\pi}\right)\frac{\mathrm{d}\rho}{\mathrm{d}p_r}\right]}, & \text{for RO} \\ \dfrac{\mathrm{d}p_r}{\mathrm{d}r} = \dfrac{-\left(\rho+p_r+\frac{B^2}{4\pi}\right)\dfrac{4\pi r^3\left(p_r+\frac{B^2}{8\pi}\right)+m}{r(r-2m)}+\frac{2}{r}\Delta}{\left[1+\frac{\mathrm{d}}{\mathrm{d}\rho}\left(\frac{B^2}{8\pi}\right)\frac{\mathrm{d}\rho}{\mathrm{d}p_r}\right]}. & \text{for TO} \end{cases} \tag{30.2}$$

Here, ρ is the density, B is the magnetic field strength at radius r, and m is the enclosed mass. The radial pressure is denoted by p_r, differing from the transverse pressure p_t. The term Δ represents the effective anisotropy due to both local anisotropy of matter and the magnetic field, specified as $\left(p_t - p_r + \frac{B^2}{4\pi}\right)$ for RO and $\left(p_t - p_r - \frac{B^2}{8\pi}\right)$ for TO.

To address the combined anisotropic effects due to both local fluid anisotropy and the presence of a strong magnetic field, we require a functional form for Δ. Unfortunately, the existing literature does not provide an explicit form derived directly from microscopic theory that captures these combined effects. To address this issue, we modify the phenomenological form introduced by Horvat et al. [77] to include magnetic field effects and their orientation. The modified anisotropy is given by:

$$\Delta = \begin{cases} \kappa\left(p_r - \frac{B^2}{8\pi}\right)\frac{2m}{r}, & \text{for RO} \\ \kappa\left(p_r + \frac{B^2}{8\pi}\right)\frac{2m}{r}, & \text{for TO} \end{cases} \tag{30.3}$$

where κ is a dimensionless parameter that controls the degree of anisotropy within the star. Based on existing works [77–79], we consider κ values in the range [2, −2]. This modified anisotropy form allows us to incorporate both the effects of effective anisotropy and magnetic field orientation into our equations.

This study utilizes the density-dependent relativistic mean-field theory (DD2) EoS model [80, 81] to describe the nuclear matter in NSs, comprising baryons like neutrons (n) and protons (p), and leptons such as electrons (e) and muons ($\bar{\mu}$). The density-dependent Lagrangian incorporates baryon fields ψ_B and lepton fields ψ_l, modeled as isospinors. Baryons interact via the exchange of isoscalar-scalar (σ), isoscalar-vector (ω), and isovector-vector (ρ) mesons. In this model, leptons are considered non-interacting. For simplicity, and because the attractive hyperon-hyperon interaction mediated by the isoscalar scalar σ_s meson has been reported to be very weak [82], we have chosen not to consider the contribution of the σ_s meson in this study. Additionally, the isoscalar vector ϕ meson does not couple with nucleons, and since we are only considering neutrons and protons in this study, hyperons have been excluded. Considering β-equilibrium and local charge neutrality, the energy density (ρ) and radial pressure (p_r) of NS matter, described using a quantum hydrodynamic framework that includes nucleons, mesons, and leptons, can be expressed as follows [83, 84]:

$$\begin{aligned}\rho = {} & \tfrac{1}{2}m_\sigma^2\sigma^2 + \tfrac{1}{2}m_\omega^2\omega_0^2 + \tfrac{1}{2}m_\rho^2\rho_{03}^2 + \tfrac{1}{2}m_\phi^2\phi_0^2 \\ & + \textstyle\sum_B \frac{2J_B+1}{2\pi^2}\int_0^{k_{F_B}}(k^2+m_B^{*2})^{1/2}k^2\,dk \\ & + \textstyle\sum_l \frac{1}{\pi^2}\int_0^{K_{F_l}}(k^2+m_l^2)^{1/2}k^2\,dk, \end{aligned} \tag{30.4}$$

and

$$\begin{aligned}p_r = {} & -\tfrac{1}{2}m_\sigma^2\sigma^2 + \tfrac{1}{2}m_\omega^2\omega_0^2 + \tfrac{1}{2}m_\phi^2\phi_0^2 \\ & + \tfrac{1}{2}m_\rho^2\rho_{03}^2 + \Sigma_B^{(r)}\textstyle\sum_B n_B \\ & + \tfrac{1}{3}\textstyle\sum_B \frac{2J_B+1}{2\pi^2}\int_0^{k_{F_B}}\frac{k^4\,dk}{(k^2+m_B^{*2})^{1/2}} + \tfrac{1}{3}\sum_l \frac{1}{\pi^2}\int_0^{K_{F_l}}\frac{k^4\,dk}{(k^2+m_l^2)^{1/2}}, \end{aligned} \tag{30.5}$$

where m_σ, m_ω, m_ρ, and m_ϕ are the meson masses, while σ, ω_0, ρ_{03}, and ϕ_0 are the meson fields. Baryons B have effective masses m_B^* and degeneracy factors $2J_B+1$. The Fermi momenta for baryons and leptons are k_{F_B} and K_{F_l}, respectively. The term n_B denotes the baryon number density, and $\Sigma_B^{(r)}$ is the rearrangement term for baryons arising from the density dependence of meson-baryon coupling parameters. For further details, see [80, 81, 83, 84] and references therein.

The nucleon-meson density-dependent couplings ($g_{\alpha B}(n)$) are determined using the method outlined by Typel et al. [80, 81]. This formulation, initially introduced by Typel and Wolter [85], is expressed as:

$$g_{\alpha B}(n) = g_{\alpha B}(n_0) f_\alpha(x). \tag{30.6}$$

Here, n denotes the total baryon density. The functional forms of $f_\alpha(x)$ depend on the type of meson. The masses of the neutron and proton are approximately 939.56536 MeV and 938.27203 MeV, respectively. The ω and ρ mesons have masses of 783 MeV and 763 MeV, respectively, as detailed in Table II of Typel et al. [81].

In accordance with previous studies [65, 66], the profile for the density-dependent magnetic field is chosen such that the magnetic field at the stellar core, B_c, satisfies the virial theorem, while the surface magnetic field, B_s, aligns with observed values. This profile is described by:

$$B(\rho) = B_s + B_0 \left[1 - \exp\left\{-\eta \left(\frac{\rho}{\rho_0}\right)^{\gamma}\right\}\right], \tag{30.7}$$

where B_0 is a parameter with the same units as magnetic field strength, and the dimensionless parameters η and γ govern the decay of the magnetic field from its peak at the core to its minimum at the surface. Specifically, η controls the rate of decay at the saturation density, while γ determines the transition width. Here, ρ_0 represents the normal nuclear matter density.

For this study, we adopt $B_s = 10^{15}$ G for NSs. However, it is noted that the specific choice of B_s does not significantly affect our results.

30.3 Results and Discussion

The combined effects of anisotropy and magnetic field orientation significantly influence the structure and physical properties of NSs. This aspect remains largely unexplored. Chu et al. [73] introduced the idea that both the strength and orientation of the magnetic field affect the maximum mass of a compact stellar object. However, their study did not account for the impact of the magnetic field or its orientation on the TOV equation, observing changes only in the total stellar mass.

To investigate this further, we incorporate the effects of magnetic field orientations and their spatial distribution into the TOV equation, as shown in Eq. (30.2). We examine the physical properties of NSs within the framework of the relativistic field theoretical model with density-dependent couplings, determined using the DD2 model by Typel et al. [80, 81]. Following Bandyopadhyay et al. [65, 66], we use density-dependent magnetic field strength profiles for NSs as described in Eq. (30.7).

Any functional forms of pressure anisotropy discussed in the literature [8, 77, 86, 87] are predominantly phenomenological. In our earlier study [76], we extensively analyzed the anisotropic phenomenological form proposed by Bowers and Liang [8]. In this work, however, to solve the system of stellar structure equations, which includes the effects of magnetic field strength and orientation, we modify the phenomenological functional form of anisotropy proposed by Horvat et al. [77], as shown in Eq. (30.3). To demonstrate the impact of anisotropy, magnetic field strength, and its orientation on the microscopic properties of neutron stars—such as matter density, pressure, and enclosed mass—we consider the NS candidate PSR J0740+6620, with a mass of $2.08 \pm 0.07\ M_{\odot}$ [88].

To study NSs, we have considered $B_0 = 1.1 \times 10^{18}$ G and $B_0 = 0.8 \times 10^{18}$ G for TO, and $B_0 = 0.3 \times 10^{18}$ G and $B_0 = 0.5 \times 10^{18}$ G for RO. We have also considered a surface magnetic field of $B_s = 10^{15}$ G. In the upper left panel of Figure 7 in our previous work [76], we demonstrated that while the masses of NSs either increase or decrease depending on the orientation of the magnetic field, these changes are asymmetric when compared to the anisotropic and non-magnetized cases. A similar effect is observed in the present study. Since the impact of the magnetic field on

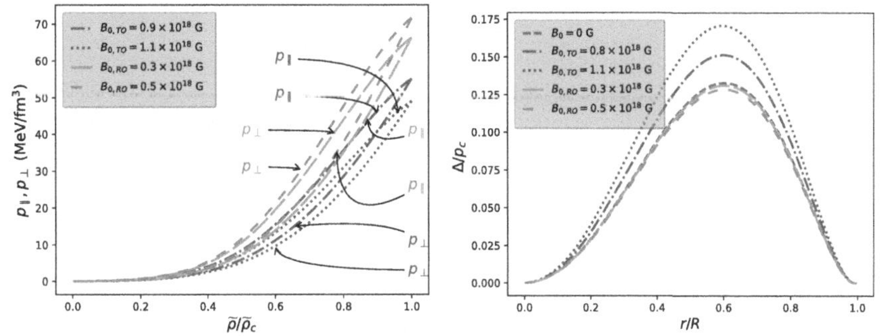

Fig. 30.1 (i) Variation of parallel pressure ($p_{\|}$) and transverse pressure ($p_{\perp}$) with system density ($\tilde{\rho}$), normalized to central density ($\tilde{\rho}_c$) (left panel) and (ii) anisotropy profiles Δ, normalized to central pressure (p_c) (right panel), for the 2.08 ± 0.07 $M_{\odot}$ NS candidate PSR J0740+6620 [88]. The dotted, dash-dotted, long-dashed, and short-dashed curves correspond to $B_0 = 1.1 \times 10^{18}$ G (TO), $B_0 = 0.8 \times 10^{18}$ G (TO), $B_0 = 0.3 \times 10^{18}$ G (RO), and $B_0 = 0.5 \times 10^{18}$ G (RO), respectively. The dashed curve in the right panel represents $B = 0$ G. Here, we have used $\eta = 0.2$, $\gamma = 2$, and $\kappa = 1$

stellar mass differs between TO and RO configurations, as in our previous work, we have opted to use different values of B_0 for the RO and TO cases to ensure a clearer graphical representation.

In the left panel of Fig. 30.1, we present the variation of $p_{\|}$ and $p_{\perp}$ with the normalized system density $\tilde{\rho}/\tilde{\rho}_c$ for both RO and TO of the magnetic field. The Figure indicates that as B_0, i.e., the magnetic field strength increases in the TO configuration, the matter becomes progressively softer. In contrast, for the RO configuration, the increase in B_0 leads to stiffer neutron matter under anisotropic and magnetized conditions. The right panel of Fig. 30.1 illustrates the variation of the anisotropy profiles normalized to the central pressure p_c. For the TO case, normalized anisotropy increases with B_0, whereas for the RO case, normalized anisotropy decreases as B_0 increases.

The left panel of Fig. 30.2 shows the $M - R$ relationship for NSs with varying B_0, $\eta = 0.2$, $\gamma = 2$, and $\kappa = 2$. As B_0 increases, the maximum mass of NSs in the TO case gradually increases compared to the anisotropic but non-magnetized NSs. Conversely, in the RO case, the maximum mass of NSs decreases with increasing B_0. For the TO case with $B_0 = 1.1 \times 10^{18}$ G, the maximum mass and corresponding radius are 2.942 $M_{\odot}$ and 13.040 km, respectively. These values are 4.11 and 5.35% higher compared to the maximum mass for anisotropic but non-magnetized NSs, and 19.52 and 9.36% higher compared to the maximum mass for isotropic and non-magnetized NSs. In contrast, for the RO case with $B_0 = 5 \times 10^{18}$ G, the maximum mass and corresponding radius are 2.756 $M_{\odot}$ and 12.009 km, respectively, which are lower than those of anisotropic but non-magnetized NSs. It is noteworthy that in all cases, based on the chosen model parameters $\eta = 0.2$, $\gamma = 2$, and $\kappa = 2$, the achieved maximum mass of NSs is well above the mass of the secondary stellar object in the GW190814 observation, which is approximately $2.59^{+0.08}_{-0.09}$ $M_{\odot}$.

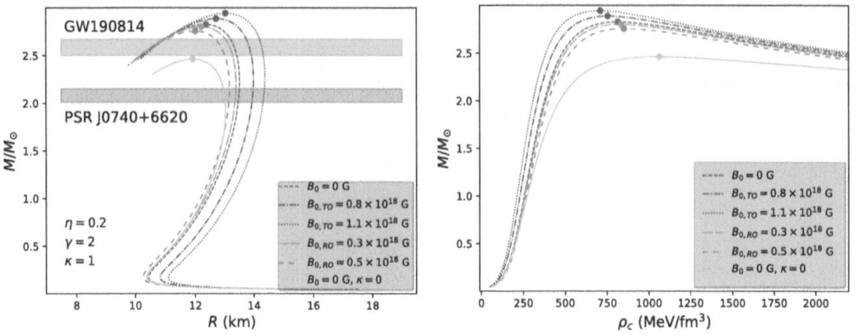

Fig. 30.2 (i) The relationship between the total mass (M/M$_\odot$) of neutron stars and the stellar radius R (left panel), and (ii) the relationship between the total mass (M/M$_\odot$) and the central density ρ_c (right panel), for various B_0 values, with $\eta = 0.2$, $\gamma = 2$, and $\kappa = 1$. Solid dots indicate the points of maximum mass

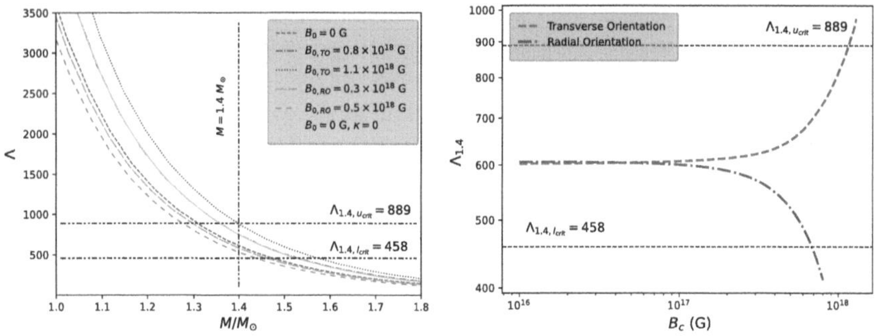

Fig. 30.3 Variation of (i) tidal deformability Λ with respect to total mass M/M$_\odot$ for NSs (left panel), and (ii) tidal deformability for a canonical 1.4 M$_\odot$ NS with varying central magnetic field strength B_c (right panel). Here, we have used $\eta = 0.2$, $\gamma = 2$, and $\kappa = 1$

The right panel of Fig. 30.2 presents the variation of total mass with central densities (ρ_c). The stability criterion $dM/d\rho_c > 0$ is met up to the maximum mass for all cases. Our findings indicate that in the RO cases, the maximum mass is achieved at progressively higher ρ_c with increasing $B_{0,RO}$ compared to anisotropic but non-magnetized NSs. For the TO cases, the maximum mass is achieved at progressively lower ρ_c with increasing $B_{0,TO}$. Therefore, it can be concluded from both panels of Fig. 30.2 that as B_0 increases, NSs in the TO case become more massive and larger, while in the RO case, NSs become less massive but more compact. The TO case provides a more interesting scenario, potentially explaining highly massive NSs.

In the left panel of Fig. 30.3 we present profile of tidal deformability of NSs for varying B_0. We also found for all the cases tidal diformability $\Lambda_{1.4}$ for a canonical 1.4 M$_\odot$ NS lies well within the updated tidal deformability constraint

$458 < \Lambda_{1.4} < 889$ imposed through the recent GW190814 observation [40]. Interestingly, the updated tidal deformability constraints derived from the recent GW190814 observation result from a reanalysis of the GW170817 data, assuming the secondary object in GW190814 is a massive NS. Specifically, the spectral EoS distribution from GW170817 was reweighted by the probability that its maximum mass aligns with the mass of the secondary object observed in GW190814 [40]. This adjustment favors stiffer EOS, resulting in an updated tidal deformability constraint of $458 < \Lambda_{1.4} < 889$. This is particularly notable because, unlike the previously reported constraints of $\Lambda_{1.4} = 190^{+390}_{-120}$, which favored softer EOS, the updated constraint aligns with the stiffer EoS predicted by NICER [89] and PREX-2 [90] data. In fact through the present study (see left panel of Fig. 30.3) we found that presence of matter anisotropy lowers the tidal deformability, whereas for the fixed κ the tidal deformability increases with the increasing magnetic field for TO.

In the right panel of Fig. 30.3, we illustrate the variation of tidal deformability $\Lambda_{1.4}$ for a canonical 1.4 $M_{\odot}$ NS as a function of the central magnetic field strength (B_c) for both TO and RO magnetic field orientations. For the TO case, $\Lambda_{1.4}$ increases with increasing B_c, while for the RO case, $\Lambda_{1.4}$ decreases with increasing B_c. Using model parameters $\eta = 0.2$, $\gamma = 2$, and $\kappa = 1$, we find that the upper critical limit of tidal deformability, $\Lambda_{1.4,u_{crit}} = 889$, corresponds to a maximum possible B_c of approximately 1.1×10^{18} G for TO. For the RO case, the lower critical limit, $\Lambda_{1.4,l_{crit}} = 458$, indicates a maximum possible B_c of around 0.65×10^{18} G.

While we can constrain B_c using the critical values of $\Lambda_{1.4}$ from the GW190814 observation, different choices of the anisotropy parameter κ could lead to other sets of B_c values. If the secondary object observed in GW190814 is indeed a neutron star, the combination of κ and B_0 must be such that the maximum mass of the NS exceeds $\sim 2.59^{+0.08}_{-0.09}$ $M_{\odot}$. Despite these considerations, our study provides a robust method to constrain the upper limit of B_c for both RO and TO configurations based on the predicted range of $\Lambda_{1.4}$ obtained from the GW190814 observation. Additionally, it can also be noted that the result also may exhibits dependence on the model parameters η and γ. Therefore, the applicability of this constraint may vary based on the chosen model parameters.

To better understand the effects of anisotropy and magnetic field orientations on NSs, Table 30.1 presents numerical values of various stellar properties. These include maximum mass, corresponding radius, central magnetic field (B_c), $\widetilde{p}_c$, ($\widetilde{p}_c$, Buchdahl ratio, surface redshift (Z_s), the ratio of magnetic energy (E_{mag}) to gravitational energy (E_{grav}), and $\Lambda_{1.4}$. It is observed from Table 30.1 that for TO magnetic fields, both the maximum mass and the corresponding radius increase as B_0 increases. Conversely, for RO magnetic fields, these quantities decrease gradually with increasing B_0. In each scenario, the ratio $2M/R$ remains well below the critical value of 8/9, ensuring stellar stability [91]. Furthermore, the gravitational energy $\mid E_{grav} \mid$ significantly exceeds E_{mag} in all cases, reinforcing the stability of these NSs. Table 30.1 also demonstrates that the surface redshifts of maximum-mass stars fall within expected ranges for different magnetic field configurations. Additionally, the tidal deformability $\Lambda_{1.4}$ for a 1.4 $M_{\odot}$ star consistently lies within the permissible range $458 < \Lambda_{1.4} < 889$ across all scenarios.

Table 30.1 Numerical values of physical parameters for the NSs with $\kappa = 0.5$ for different values of B_0

Orientation of magnetic field	Value of B_0 (Gauss)	Value of maximum mass ($M_\odot$)	Corresponding predicted radius (km)	Central magnetic field B_c (Gauss)	Central density $\widetilde{\rho}_c$ (g/cm^3)	Central pressure $\widetilde{p}_c$ (dyne/cm^2)	$\frac{2M}{R}$	Surface redshift (Z_s)	$\frac{E_{mag}}{E_{grav}}$	$\Lambda_{1.4}$
TO	1.1×10^{18}	2.942	13.040	1.096×10^{18}	1.319×10^{15}	4.456×10^{35}	0.67	0.73	0.085	879.89
	0.8×10^{18}	2.886	12.725	0.798×10^{18}	1.371×10^{15}	4.722×10^{35}	0.67	0.74	0.043	750.54
$B = 0$	–	2.826	12.378	–	1.452×10^{15}	5.011×10^{35}	0.67	0.75	–	614.36
RO	0.3×10^{18}	2.800	12.218	0.3×10^{18}	1.495×10^{15}	5.277×10^{35}	0.68	0.76	0.006	581.02
	0.5×10^{18}	2.756	12.009	0.5×10^{18}	1.529×10^{15}	6.017×10^{35}	0.68	0.76	0.015	529.23

30.4 Conclusion

The present study investigates the complex interplay between anisotropy, magnetic field orientations, and their effects on NS properties using a relativistic mean field theory EoS. By modifying the TOV equations to include magnetic field contributions and employing the DD2 model for density-dependent couplings, the study explores how varying magnetic field strengths and orientations impact NS characteristics. Key findings include the observation that increasing magnetic field strength in the TO configuration leads to softer neutron matter, while in the RO configuration, it results in stiffer matter. In this chapter, we have considered different functional forms for anisotropy and EoS compared to our previous work [76] but achieved results completely aligned with our previously reported work [76] and the recent study done by Zuraiq et al. [92].

Numerical simulations illustrate that TO configurations enhance the maximum mass and size of NSs, contrasting with RO configurations where NSs become more compact but less massive as B_0 increases. Notably, all computed maximum masses surpass the observed mass of the secondary object in GW190814, suggesting present work is suitable in explaining highly massive NSs.

Tidal deformability profiles show distinct trends: $\Lambda_{1.4}$ increases with B_c in TO cases and decreases in RO cases, aligning with theoretical predictions and observational limits set by GW190814. The updated constraint on $\Lambda_{1.4}$ favors stiffer EoS, corroborated by NICER and PREX-2 data, underpinning the robustness of the present model in predicting NS behavior under extreme magnetic conditions.

Overall, the study provides insights into how anisotropy and magnetic field strength and its orientations influence NS properties, offering a methodological framework to constrain upper limits of B_c in different magnetic configurations, crucial for understanding highly massive NSs observed in astrophysical scenarios. The results of this study have significant implications for future gravitational wave observations. By providing stringent constraints on the internal magnetic fields of neutron stars, our findings can help guide the interpretation of tidal deformability measurements in upcoming gravitational wave events. This will ultimately contribute to a deeper understanding of the fundamental properties of neutron stars and the extreme conditions within their interiors.

Acknowledgments DD acknowledges the support from the Department of Atomic Energy, Government of India through 'Apex Project—Advance Research and Education in Mathematical Sciences' at The Institute of Mathematical Sciences. FW is supported through the U.S. National Science Foundation under Grant No. PHY-2012152. We thank the anonymous referee for their valuable comments, which significantly improved our work.

References

1. Glendenning, N. K. (1996). *Nuclear physics, particle physics and general relativity*. Astronomy and Astrophysics Library. Springer-Verlag.
2. Weber, F. (2017). *Pulsars as astrophysical laboratories for nuclear and particle physics*. Series in high energy physics, cosmology and gravitation. Institute of Physics Publishing.
3. Yuan, Y. F., & Zhang, J. L. (1998). *Astronomy and Astrophysics, 335*, 969.
4. Tatsumi, T. (2000). *Physics Letters B, 489*, 280.
5. Ferrer, E. J., de La Incera, V., Keith, J. P., Portillo, I., & Springsteen, P. L. (2010). *Physics Review C, 82*, 065802.
6. Tolman, R. C. (1939). *Physics Review, 55*, 364.
7. Oppenheimer, J. R., & Volkoff, G. M. (1939). *Physics Review, 55*, 374.
8. Bowers, R. L., & Liang, E. P. T. (1974). *The Astrophysical Journal, 188*, 657.
9. Hillebrandt, W., & Steinmetz, K. O. (1976). *Astronomy and Astrophysics, 53*, 283 (1976).
10. Mak, M. K., & Harko, T. (2002). *Chinese Journal of Astronomy and Astrophysics, 2*, 248.
11. Hewish, A., Bell, S. J., Pilkington, J. D. H., Scott, P. F., & Collins, R. A. (1968). *Nature, 217*, 709.
12. Alpar, M. A., Cheng, A. F., Ruderman, M. A., & Shaham, J. (1982). *Nature, 300*, 728.
13. Taylor, J. H., Manchester, R. N., & Lyne, A. G. (1993). *The Astrophysical Journal Supplement Series, 88*, 529.
14. McLaughlin, M. A., Stairs, I. H., Kaspi, V. M., Nice, D. J., Lommen, A. N., Lorimer, D. R., Manchester, R. N., & Camilo, F. (2003). *The Astrophysical Journal Letter, 591*, L135.
15. Paczynski, B. (1992). *Acta Astronomica, 42*, 145.
16. Thompson, C., & Duncan, R. C. (1996). *The Astrophysical Journal, 473*, 322.
17. Melatos, A. (1999). *The Astrophysical Journal Letter, 519*, L77.
18. Usov, V. V. (1992). *Nature, 357*, 472.
19. Kluźniak, W., & Ruderman, M. (1998). *The Astrophysical Journal Letter, 508*, L113.
20. Wheeler, J. C., Yi, I., Höflich, P., & Wang, L. (2000). *The Astrophysical Journal, 537*, 810.
21. Starling, R. L. C., Rol, E., van der Horst, A. J., Wijers, R. A. M. J., Wiersema, K., & Kouveliotou, C. (2009). *Monthly Notices of the Royal Astronomical Society, 400*, 90.
22. Cenko, S. B., Frail, D. A., Harrison, F. A., Nakar, E., Ofek, E. O., Fox, D. B., Soderberg, A. M., Gal-Yam, A., Price, P. A., Kulkarni, S. R., Horesh, A., Cobb, J. B., Chandra, R., & Foley, R. J. (2010). *The Astrophysical Journal, 711*, 641.
23. Kouveliotou, C., Dieters, S., Strohmayer, T., van Paradijs, J., Fishman, G. J., Meegan, C. A., Hurley, K., Kommers, J., Smith, I., Frail, D., & Murakami, T. (1998). *Nature, 393*, 235.
24. Canuto, V. (1977). *Annals of the New York Academy of Sciences, 302*, 514.
25. Fushiki, I., Gudmundsson, E. H., & Pethick, C. J. (1989). *The Astrophysical Journal, 342*, 958.
26. Abrahams, A. M., & Shapiro, S. L. (1991). *The Astrophysical Journal, 374*, 652.
27. Fushiki, I., Gudmundsson, E. H., Pethick, C. J., & Yngvason, J. (1992). *Annals of Physics (N.Y.), 216*, 29.
28. Roegnvaldsson, O. E., Fushiki, I., Gudmundsson, E. H., Pethick, C. J., & Yngvason, J. (1993). *The Astrophysical Journal, 416*, 276.
29. Chakrabarty, S., Bandyopadhyay, D., & Pal, S. (1997). *Physical Review Letters, 78*, 2898.
30. Bandyopadhyay, D., Chakrabarty, S., Dey, P., & Pal, S. (1998). *Physical Review D, 58*, 121301.
31. Broderick, A., Prakash, M., & Lattimer, J. M. (2000). *The Astrophysical Journal, 537*, 351.
32. Suh, I.-S., & Mathews, G. J. (2001). *The Astrophysical Journal, 546*, 1126.
33. Harding, A. K., & Lai, D. (2006). *Reports on Progress in Physics, 69*, 2631.
34. Chen, W., Zhang, P.-Q., & Liu, L.-G. (2007). *Modern Physics Letters A, 22*, 623.
35. Rabhi, A., Providência, C., & Da Providência, J. (2008). *Journal of Physics G: Nuclear and Particle Physics, 35*, 125201.
36. Bocquet, M., Bonazzola, S., Gourgoulhon, E., & Novak, J. (1995). *Astronomy and Astrophysics, 301*, 757.
37. Cardall, C. Y., Prakash, M., & Lattimer, J. M. (2001). *The Astrophysical Journal, 554*, 322.

38. Pili, A. G., Bucciantini, N., & Del Zanna, L. (2014). *Monthly Notices of the Royal Astronomical Society*, *439*, 3541.
39. Sinha, M., Mukhopadhyay, B., & Sedrakian, A. (2013). *Nuclear Physics A*, *898*, 43.
40. Abbott, R., et al., LIGO Scientific Collaboration and Virgo Collaboration. (2020). *The Astrophysical Journal Letter*, *896*, L44.
41. Abbott, B. P., et al., The LIGO Scientific Collaboration and the Virgo Collaboration. (2017). *Physical Review Letters*, *119*, 161101.
42. Abbott, R., et al., The LIGO Scientific Collaboration and the Virgo Collaboration. (2020). *The Astrophysical Journal Letter*,*896*, L44.
43. Sedrakian, A., Weber, F., & Li, J. J. (2020). *Physical Review D*, *102*, 041301.
44. Abbott, R., Abbott, T. D., Acernese, F., et al. (2023). *Physical Review X*, *13*, 041039.
45. Tsokaros, A., Ruiz, M., & Shapiro, S. L. (2020). *The Astrophysical Journal*, *905*, 48.
46. Huang, K., Hu, J., Zhang, Y., & Shen, H. (2020). *The Astrophysical Journal*, *904*, 39.
47. Zevin, M., Spera, M., Berry, C., & Kalogera, V. (2020). *The Astrophysical Journal Letter*, *899*, L1.
48. Fattoyev, F. J., Horowitz, C. J., Piekarewicz, J., & Reed, B. (2020). *Physical Review C*, *102*, 065805.
49. Essick, R., & Landry, P. (2020). *The Astrophysical Journal*, *904*, 80.
50. Safarzadeh, M., & Loeb, A. (2020). *The Astrophysical Journal Letter*, *899*, L15.
51. Biswas, B., Nandi, R., Char, P., et al. (2021). *Monthly Notices of the Royal Astronomical Society*, *505*, 1600.
52. Zhang, N. B., & Li, B. A. (2020). *The Astrophysical Journal*, *902*, 38.
53. Most, E. R., Papenfort, L. J., Weih, L. R., & Rezzolla, L. (2020). *Monthly Notices of the Royal Astronomical Society*, *499*, L82.
54. Tan, H., Hostler-Noronha, J., & Yunes, N. (2020). *Physical Review Letters*, *125*, 261104.
55. Zhang, C., & Mann, R. B. (2021). *Physical Review D*, *103*, 063018.
56. Bombaci, I., Drago, A., Logoteta, D., et al. (2021). *Physical Review Letters*, *126*, 162702.
57. Demircik, T., Ecker, C., & Järvinen, M. (2021). *The Astrophysical Journal Letter*, *907*, L37.
58. Cao, Z., Chen, L.-W., Chu, P.-C., et al. (2022). *Physical Review D*, *106*, 083007.
59. Dexheimer, V., Gomes, R. O., Klähn, T. et al. (2021). *Physical Review C*, *103*, 025808.
60. Roupas, Z., Panotopoulos, G., & Lopes, I. (2021). *Physical Review D*, *103*, 083015.
61. Rather, I. A., Usmani, A. A., & Patra, S. K. (2021). *Journal of Physics G*, *48*, 085201.
62. Moffat, J. W. (2020). arXiv:gr-qc/2008.04404.
63. Astashenok, A. V., Capozziello, S., Odintsov, S. D., & Oikonomou, V. K. (2020). *Physical Letters B*, *811*, 135910.
64. Nunes, R. C., Coelho, G. G., & Araujo, J. C. N. (2020). *European Physical Journal C*, *80*, 1115.
65. Bandyopadhyay, D., Chakrabarty, S., & Pal, S. (1997). *Physical Review Letters*, *79*, 2176.
66. Bandyopadhyay, D., Pal, S., & Chakrabarty, S. (1998). *Journal of Physics G*, *24*, 1647.
67. Menezes, D. P., Pinto, M. B., Avancini, S. S., et al. (2009). *Physical Review C*, *79*, 035807.
68. Paulucci, L., Ferrer, E. J., de La Incera, V., et al. (2011). *Physical Review D*, *83*, 043009.
69. Ryu, C.-Y., Cheoun, M.-K., Kajino, T., et al. (2012). *Astroparticle Physics*, *38*, 25.
70. Dexheimer, V., Menezes, D. P., & Strickland, M. (2014). *Journal of Physics G*, *41*, 015203.
71. Hou, J.-X., Peng, G.-X., Xia, C.-J., et al. (2015). *Chinese Physics C*, *39*, 015101.
72. Kayanikhoo, F., Naficy, K., & Bordbar, G. H. (2020). *The European Physical Journal A*, *56*, 2.
73. Chu, P.-C., Chen, L.-W., & Wang, X. (2014). *Physical Review D*, *90*, 063013.
74. Makishima, K., Enoto, T., Hiraga, J. S., et al. (2014). *Physical Review Letters*, *112*, 171102.
75. Dexheimer, V., Franzon, B., Gomes, R. O., et al. (2017). *Physical Letters B*, *773*, 487.
76. Deb, D., Mukhopadhyay, B., & Weber, F. (2021). *The Astrophysical Journal*, *922*, 149.
77. Horvat, D., Ilijić, S., & Marunović, A. (2011). *Classical and Quantum Gravity*, *28*, 025009.
78. Doneva, D. D., & Yazadjiev, S. S. (2012). *Physical Review D*, *85*, 124023.
79. Folomeev, V. (2018). *Physical Review D*, *97*, 124009.
80. Typel, S. (2005). *Physical Review C*, *71*, 064301.

81. Typel, S., Röpke, G., Klähn, T., Blaschke, D., & Wolter, H. H. (2010). *Physical Review C*, *81*, 015803.
82. Schaffner, J., & Mishustin, I. N. (1996). *Physical Review C*, *53*, 1416.
83. Banik, S., & Bandyopadhyay, D. (2002) *Physical Review C*, *66*, 065801.
84. Banik, S., Hempel, M., & Bandyopadhyay, D. (2014). *The Astrophysical Journal Supplement Series*, *214*, 22.
85. Typel, S., & Wolter, H. H. (1999). *Nuclear Physics A*, *656*, 331.
86. Herrera, L., & Barreto, W. (2013). *Physical Review D*, *88*, 084022.
87. Raposo, G., et al. (2019). *Physical Review D*, *99*, 104072.
88. Miller, M. C., Lamb, F. K., Dittmann, A. J., et al. (2021). *The Astrophysical Journal Letter*, *918*, L28.
89. Raaijmakers, G., et al. (2020). *The Astrophysical Journal Letter*, 893, L21.
90. Reed, B. T., Fattoyev, F. J., Horowitz, C. J., et al. (2021). *Physical Review Letters*, 126, 172503.
91. Chandrasekhar, S., & Fermi, E. (1953). *The Astrophysical Journal*, *118*, 116.
92. Zuraiq, Z., Mukhopadhyay, B., & Weber, F. (2024). *Physical Review D*, *109*, 023027.

Chapter 31
Observational Mass-Radius Relation for Magnetized White Dwarfs

Drisya Karinkuzhi and Banibrata Mukhopadhyay

Abstract The Mass-Radius (M–R) relation for a sample of 47 moderately magnetized white dwarfs (WD) is presented. Instead of traditional methods for deriving the mass and radius, we have used their location in the Hertzsprung-Russel diagram to estimate their masses. The atmospheric parameters, effective temperature (T_{eff}), surface gravity (log g), and luminosity, are first derived by the Virtual Observatory Spectral Energy Distribution Analyzer (VOSA). The observed M–R relation is then compared with theoretical predictions from finite temperature models considering different magnetic field strengths. The agreement between the observations and model prediction at lower field strength may suggest the existence of Super-Chandrasekhar White dwarfs at stronger fields.

31.1 Introduction

The effect of magnetic fields on the physical properties of white dwarfs (WDs), such as on their mass-radius (M–R) relation as well as on their overall formation and evolution is still an open question. Especially in the case of highly magnetized WDs, all the physical properties are expected to be modified by the strong magnetic field strength. A robust understanding of the origin and evolution of these WDs can even provide information about their progenitors as a whole. Even though a handful of magnetized WDs are identified now, thanks to large field photometric surveys, detailed studies on their physical and chemical characteristics are scarce

D. Karinkuzhi (✉)
Department of Physics, University of Calicut, Malappuram, India
e-mail: drdrisyak@uoc.ac.in

B. Mukhopadyay
Department of Physics, Indian Institute of Science, Bangalore, India
e-mail: bm@iisc.ac.in

S. Ghosh, A. R. Rao (eds.), *The Relativistic Universe: From Classical to Quantum*, Astrophysics and Space Science Proceedings 61,
https://doi.org/10.1007/978-3-031-90186-7_31

in the literature. One difficulty is the lack of theoretical models which consider the effect of fields on their evolution. A further complicating factor is the composition of their photospheric layers and the core. Some of their photospheres are hydrogen-rich, the so-called DA WDs and some have helium-rich atmospheres and are called DB-WDs. In addition, not-so-common DQ-WDs with the enhancement of weak metal lines also exist in nature. We have recently attempted to understand the M–R relation for 47 magnetized WDS with field strength (B) ranging between 1 to 773 MG, selected from the Sloan Digital Sky Survey (SDSS) [1]. The physical properties of WDs are very difficult to measure observationally. Instead of the traditional method of deriving masses using surface gravities, we depend on their location in the Hertzsprung–Russel (HR) diagram to obtain the masses. Other basic parameters such as temperature and luminosities are derived by fitting spectral energy distribution (SED) with model atmospheres using the virtual observatory SED fitting tool, VOSA. Finally, we compare our results with the theoretical M–R relation predicted considering a finite temperature gradient below the interface. The methodology is detailed in the next section.

31.2 Data and Methodology

The selected sample in this analysis belongs to the WD catalog of [2] constructed based on their spectra from SDSS data release 7. The magnetic field strength in these WDs is measured by [3]. Our selection is purely random but we ensured that our sample's measured B values were well separated. The details about the sample are presented in [1], but here we provide only the necessary information for its clarity.

The atmospheric parameters T_{eff} and log g are derived using VOSA, [4] using the model atmospheres provided by [5]. Due to the unavailability of magnetic models we used non-magnetic models in VOSA for deriving the atmospheric parameters. But to avoid the uncertainties caused by this selection, the flux points corresponding to the spectral regions such as Balmer regions are avoided for finding the best fit with the models as these regions are known to be affected by the existence of higher fields (see Fig. 31.1). The caveats of using non-magnetic models for the analyses of magnetized WDs are detailed in [1] especially in the case of highly magnetized WDs with $B \approx 100$ MG, as they are known to be most affected by the presence of high magnetic fields [6]. Along with the atmospheric parameters, VOSA SED fitting also provides the luminosities and surface gravities of the sample WDs. We do not adopt the mass values provided by VOSA since they are derived from surface gravities. Instead, the mass of the WDs is derived using their location in the HR diagram. The cooling tracks constructed for different masses and core composition by [7] are overplotted to estimate the masses.

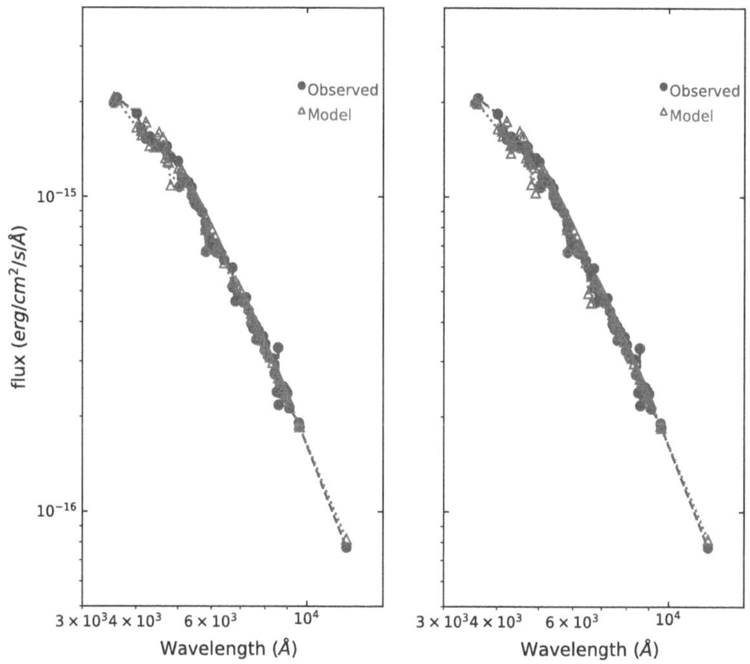

Fig. 31.1 The SEDs are shown for the WD with the highest field strength (773 MG). The left and right panels show the SED fitting including and excluding the flux points corresponding to the Balmer regions respectively. Though the temperature remains the same in both cases, the surface gravity (log g) differs by 0.25 dex indicating very low uncertainties at these moderate levels of field strength

31.3 Results and Discussion

The derived M–R values are then compared with the theoretically derived M–R curve of [8]. Instead of assuming a constant temperature from the core to the core-envelop interface, [8] had considered the possibility of the finite temperature gradient while deriving the M–R relation. The effect of luminosities on the M–R relation is also explored in detail. The observational derived M–R values mostly agree with the theoretical predictions, assuming $dT/dr = 0$ below the interface, except for the region corresponding to higher masses and smaller radii. Further, the values are also compared with theoretical values calculated assuming $dT/dr = 0$ and different luminosity values. To have a clear picture, the comparison is done after grouping our sample WDs into different groups based on their luminosity values. The most important comparison is after including the magnetic fields in the calculation. The luminosity value is kept constant at 10^{-4} $L_{\odot}$and B values at the center and surface are varied between (10^{12}, 10^{7}) to (10^{14}, 10^{9}) respectively. Our results match the theoretical predictions corresponding to the lowest magnetic field as expected (see Fig. 31.2).

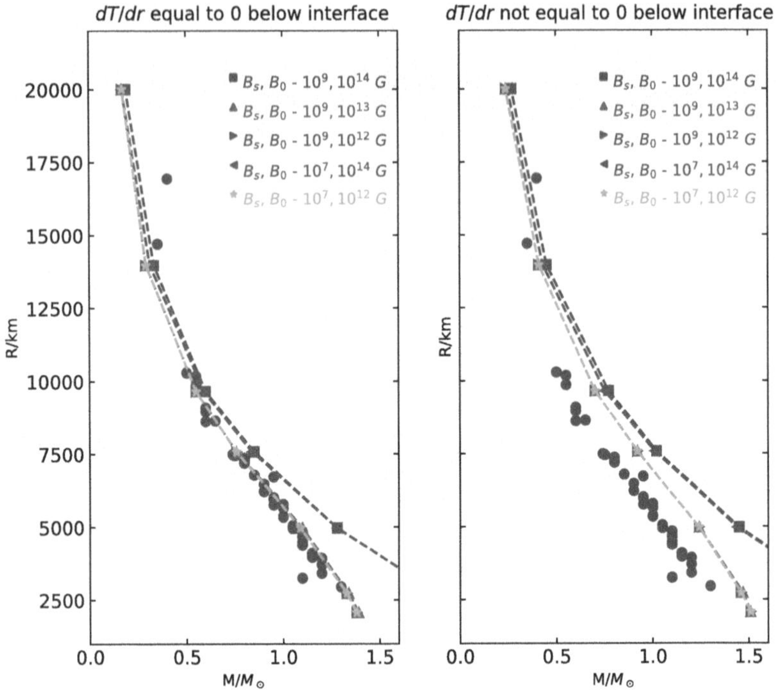

Fig. 31.2 The comparison of observed M–R values with finite temperature $dT/dr = 0$ model is shown in the left panel for different field strengths. In the right panel, observed (M–R) values are compared with model predictions considering $dT/dr, \neq 0$

Another important point to consider is the effect of chemical composition on the M–R relation. Unfortunately, we do not have any clear indication of the core composition of the WDs. A few studies such as [9] earlier predicted that the deviation of the M–R relation from standard predictions could also be due to the chemical composition. This needs to be explored in detail after the proper identification of the core composition.

We have also explored the effect of magnetic fields on the physical parameters of the WDs such as mass, luminosity, and temperature. We have carried out multiple comparisons of these parameters with B and could not draw a robust conclusion on their dependence as we could not see a direct relation between them. Moreover, we do not have the information about the central B for our sample to understand their role in the modification of the stellar structure.

31.4 Conclusions

We present the M–R relation for a sample of 47 moderately magnetized WDs. The physical parameters except the mass of these WDs are derived using the VOSA SED fitting tool. The mass is derived using their location in the HR diagram with the help of theoretical cooling tracks constructed assuming the non-magnetic effects. Though there exist many uncertainties, when we use non-magnetic models for constraining the atmospheric parameters of magnetized WDs, we try to keep them minimal with necessary modifications to the estimating methods. The main aim of this analysis is to show the feasibility of deriving M–R relation observationally for moderately magnetized WDs. Further, we compared our results with theoretical predictions considering a finite temperature model. For the moderate field strength considered for the sample in this chapter, we found a satisfactory fit of the observation with the theory. This probably indicates the existence of super Chandrasekhar WDs at higher magnetic field strengths.

Acknowledgments DK acknowledges the financial support from CSIR-India through file No.13(9086-A)2019-Pool. BM acknowledges partial support from a project of the Department of Science and Technology (DST-SERB) with research Grant No. DSTO/PPH/BMP/1946 (EMR/2017/001226).

Competing Interests The authors have no conflicts of interest to declare that are relevant to the content of this chapter.

References

1. Karinkuzhi, D., Mukhopadhyay, B., Wickramasinghe, D., & Tout, C. A. (2024). Mass-radius relation for magnetized white dwarfs from SDSS. *Monthly Notices of the Royal Astronomical Society, 529*(4), 4577–4584.
2. Kleinman, S. J., et al. (2013). SDSS DR7 White Dwarf Catalog. *The Astrophysical Journal Supplement, 204*(1), 5.
3. Kepler, S. O., et al. (2013). Magnetic white dwarf stars in the Sloan Digital Sky Survey. *Monthly Notices of the Royal Astronomical Society, 429*(4), 2934–2944.
4. Bayo, A., Rodrigo, C., Barrado Y Navascués, D., Solano, E., Gutiérrez, R., Morales-Calderón, M., & Allard, F. (2008). VOSA: virtual observatory SED analyzer. An application to the Collinder 69 open cluster. *Astronomy & Astrophysics, 492*(1), 277–287.
5. Koester, D. (2010). White dwarf spectra and atmosphere models. *Memorie della Societa Astronomica Italiana, 81*, 921–931.
6. Hardy, F., Dufour, P., & Jordan, S. (2023). Spectrophotometric analysis of magnetic white dwarf - I. Hydrogen-rich compositions. *Monthly Notices of the Royal Astronomical Society, 520*(4), 6111–6134.
7. Renedo, I., Althaus, L. G., Miller Bertolami, M. M., Romero, A. D., Córsico, A. H., Rohrmann, R. D., & García-Berro, E. (2010). New cooling sequences for old white dwarfs. *The Astrophysical Journal, 717*(1), 183–195.
8. Gupta, A., Mukhopadhyay, B., Tout, C. A. (2020). Suppression of luminosity and mass-radius relation of highly magnetized white dwarfs. *Monthly Notices of the Royal Astronomical Society, 496*(1), 894–902.
9. Tremblay, P. E., et al. (2017). The Gaia DR1 mass-radius relation for white dwarfs. *Monthly Notices of the Royal Astronomical Society, 465*(3), 2849–2861.

Chapter 32
Simulating Super-Chandrasekhar White Dwarfs

Zenia Zuraiq, Achal Kumar, Alexander J. Hackett, Surendra Bhattarai, Christopher A. Tout, and Banibrata Mukhopadhyay

Abstract Over the last few decades, there has been considerable interest in the violation of the *sacred* "Chandrasekhar" mass limit of white dwarfs (WDs). Peculiar over-luminous type Ia supernovae (such as SNLS-03D3bb) lend observational support to the idea that these super-Chandrasekhar WDs exist. Our group, for more than a decade, has been actively working on the theoretical possibility of these objects through the presence of the star's magnetic field. The magnetic field greatly contributes to the existence of these massive WDs, both through classical and quantum effects. In this chapter, we explore super-Chandrasekhar WDs, formed via evolution from a main sequence star, as a result of the classical effects of the star's magnetic field. We obtain super-Chandrasekhar WDs and new mass limit(s), depending on the magnetic field geometry. We explore the full evolution and stability of these objects from the main sequence stage through the one-dimensional stellar evolution code STARS. In order to do so, we have appropriately modified the

Z. Zuraiq (✉) · B. Mukhopadhyay
Department of Physics, Indian Institute of Science, Bengaluru, India
e-mail: zeniazuraiq@iisc.ac.in; bm@iisc.ac.in

A. Kumar
Department of Physics, University of Florida, Gainesville, FL, USA
e-mail: achal.kumar@ufl.edu

A. J. Hackett
Institute of Astronomy, University of Cambridge, Cambridge, UK
e-mail: ajh291@cam.ac.uk,hackett@fzu.cz

CEICO, Institute of Physics of the Czech Academy of Sciences, Prague, Czechia

S. Bhattarai
Department of Physical Sciences, Indian Institute of Science Education and Research, Kolkata, India
e-mail: sb19ms136@iiserkol.ac.in

C. A. Tout
Institute of Astronomy, University of Cambridge, Cambridge, UK
e-mail: cat@ast.cam.ac.uk

S. Ghosh, A. R. Rao (eds.), *The Relativistic Universe: From Classical to Quantum*, Astrophysics and Space Science Proceedings 61,
https://doi.org/10.1007/978-3-031-90186-7_32

given codes by introducing magnetic effect and cooling. Our simulation confirms that massive WDs are possible in the presence of a magnetic field satisfying underlying stability.

32.1 Introduction

Type Ia supernovae (SNe Ia) have long been an integral part of the cosmic distance ladder, as standard candles used for measuring distances to far-away galaxies. It was the measurements using a population of SNe Ia that led to the Nobel prize winning discovery of the Universe's accelerated expansion.

However, over the last few decades or so, there have been observations that indicate that these standard candles are not so standard after all. Specifically, there have been multiple observations of over-luminous SNe Ia, which have higher luminosities than the expected standard for these objects, along with a varied shape of the lightcurve and low ejecta velocity. These observations indicate that the progenitors of these objects are massive white dwarfs (WDs), that exceed the conventional Chandrasekhar limit, with some even as massive as $2.8M_\odot$ [1]. These are the proposed "super-Chandrasekhar" WDs.

Mukhopadhyay and group have investigated the possible existence of these objects through various theoretical studies over the last decade or so (starting with the paper [2]). In one of the series of avenues, they have found that the magnetic fields of WDs can lead to them having super-Chandrasekhar masses.

Magnetic fields can lead to quantum effects on the WD structure through Landau quantization effects, as explored in previous work [3]. On the other hand, classical effects on the WDs through magnetic pressure and tension effects [4] can also lead to effects on the WD's mass. The consequence of these magnetic effects on the WDs is that instead of having a single Chandrasekhar mass limit, we have a series of mass limits, depending on the exact magnetic field included in the star. Nevertheless, they may have an ultimate (upper) mass limit, when we consider the star to be free from all instabilities.

In this chapter we continue exploring magnetized WDs (BWDs), and more specifically, their formation and time-dependent evolution. We explore in detail the time-dependent simulation for the evolution of the magnetized main sequence (MS) stars to BWDs using the Cambridge Stellar evolution code: STARS [5]. In the original STARS, there were no magnetic fields. We have introduced fields in it by means of a profile of magnetic fields, which is a valid assumption as long as the star is approximately spherical.

There are two ways in which this magnetic effect could affect the star's evolution. BWDs may be massive enough through their formation/evolution from the MS star itself. This may be possible when we consider the evolution of higher mass stars ($\approx 10M_\odot$), which are sufficiently magnetized during the MS phase itself. An alternate formation scenario is possible when we consider binary systems. A WD with a low or dormant magnetic field may accrete matter (perhaps from a binary companion), leading to strengthening of its field and a higher mass limit for the WD. In the current work, we will be focusing on the second formation scenario, i.e., the "accretion cases".

This chapter is structured as follows. In Sect. 32.2, we introduce the STARS code. We introduce the magnetic field profile and the modifications to the set of stellar structure equations to include magnetic fields in the simulation. In Sect. 32.3, we highlight the major results, including the evolutionary track from the MS star to BWD and the properties of the BWDs obtained. We also investigate the possibility of how the evolution is affected by the decay of the magnetic field. In Sect. 32.4, we lay out the future directions of this work. We end by summarizing our results and with our conclusions in Sect. 32.5.

32.2 STARS Code: Formalism

The goal of this work is to explore the formation and evolution of BWDs and, by extension, super-Chandrasekhar WDs. For this, we have made use of the one-dimensional stellar evolution code, first developed at Cambridge, known as STARS. [5]. We evolve MS stars and study this evolution through various phases like the asymptotic giant branch (AGB) up to its formation as a BWD.

Throughout our simulations, we have the following assumptions and considerations:

1. The star is spherical and one-dimensional.
2. The star is nonrotating.
3. We do not solve the hydrostatic balance and Maxwell's equations simultaneously, rather solve the hydrostatic balance equation with a modification to the pressure arising due to the model magnetic field profile.

The stellar structure equations used in our simulation (in STARS), keeping the same symbols as used in the original STARS code, are:

$$\frac{d\log P}{dm} = -\frac{Gm}{4\pi r^4 P}, \tag{32.1}$$

$$\frac{d\log T}{dm} = \frac{d\log P}{dm}\nabla, \tag{32.2}$$

$$\frac{d\log r}{dm} = \frac{1}{4\pi r^3 m}, \tag{32.3}$$

and

$$\frac{dL}{dm} = \epsilon - \epsilon_\nu - \frac{Du}{Dt} + \frac{P}{\rho^2}\frac{D\rho}{Dt}. \tag{32.4}$$

Here, ϵ is the energy generation rate due to hydrogen burning, ϵ_ν is the energy loss rate from neutrino emission, and ∇ represents the logarithmic gradient of temperature against pressure, which varies for convective and radiative regions.

32.2.1 Inclusion of Magnetic Field

In our simulations, the magnetic field is calculated at every point in the star using the density-dependent profile [6]

$$B(\rho) = B_s + B_0 \left[1 - \exp\left\{ -\eta \left(\frac{\rho}{\rho_0} \right)^\gamma \right\} \right], \tag{32.5}$$

where B_s corresponds to the surface field of the star, B_0 and ρ_0 control the field at the center, and η and γ are model parameters that control how the field decays from the center to surface. In this work, we take $B_s = 10^7\ G$, $B_0 = 10^{14}\ G$ and $\rho_0 = 10^9$ g/cm^3.

We incorporate magnetic fields by modifying the pressure to have contributions from both the matter (P_m) and the magnetic field ($P_B = B^2/8\pi$). We consider the star to have a toroidal field, leading to $P = P_m + P_B$. Note here that the magnetic density ρ_B is throughout significantly less than the mass density for the fields under consideration. This form of magnetized pressure modification was followed in earlier work, where it was also shown that this form of the profile is consistent with Maxwell's equations under the assumption of approximate spherical symmetry. [4]. Further, toroidal fields lead to minimal deviation as shown in previous (time-independent) two-dimensional results from XNS code, e.g. [7].

32.3 Results

32.3.1 Formation of BWD Models

As mentioned previously, our goal is to simulate the complete formation and evolution of BWDs from the MS stars. For this, we have modified the STARS code to include magnetic field. We follow a similar implementation to form the WD as described in [8]. The general scheme for the formation of a carbon-oxygen (CO) WD through the code is described below.

We use the STARS code to generate a zero-age main sequence (ZAMS) model. In the case of a CO WD, we would need to start with a MS model of mass $\approx 3 - 8 M_\odot$.We evolve this star from the MS up to the AGB stage until its CO core has grown to a sufficiently large amount. At this point, all nuclear reactions in the star (chemical evolution) are halted, and a mass-loss mechanism in terms of the code's stellar wind parameter is enabled. This strips the outer layers of the star, leaving us with a mostly degenerate star. We thus obtain a CO WD. To obtain WDs of other compositions, one must start with appropriate ZAMS/MS models. An oxygen-neon WD would require a higher mass ZAMS model, such that conditions are favorable for carbon burning in the core. On the other hand, to form the lighter helium WDs, one must start with a sufficiently low mass ZAMS model, such that the core mass is not sufficient enough for the formation of a CO core in the first place.

In this chapter, we assume the magnetic field effects are minimal in the MS phase. It is only on the star's collapse and field's growth (through either a flux freezing mechanism or perhaps some kind of dynamo effect) that the magnetic effects become important.

Including a large magnetic field (either by B_0 or B_s) at this late stage of the star's evolution can lead to destabilizing effects. Keeping that in mind, the field is often introduced in stages so that the model has sufficient time to relax to the final BWD stage. The scale of this extra relaxation time is much smaller than the typical timescale of cooling or other subsequent processes (such as accretion/mass addition) introduced to the WD.

Here, we focus on a particular evolution scenario: the formation of a $0.4-0.5M_{\odot}$ BWD formed from a $3M_{\odot}$ MS star. We can visualize the various stages of the evolution process towards the formation of the BWD on the Hertzsprung-Russell (HR) diagram, as shown in Fig. 32.1. The solid red line indicates the formation of $0.53M_{\odot}$ BWD. Following this, we consider accretion onto the WD (described in detail in next subsection). Accretion onto a WD having magnetic field is shown

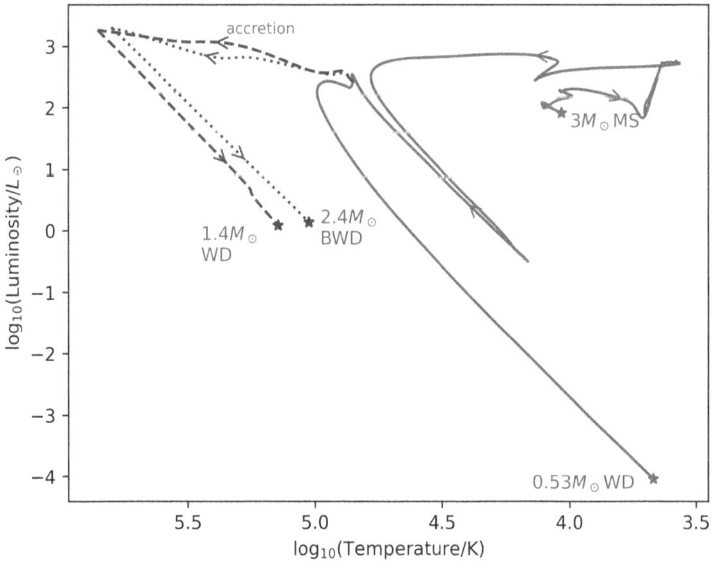

Fig. 32.1 Evolution of MS star to BWD as shown by changing position on the HR diagram. The subsequent cooling track of BWD is not shown here

with the dotted line, while the dashed line shows accretion onto a non-magnetic WD. Depending on whether the field is there or not, the evolution ends up with different limiting masses for the BWDs.

32.3.2 Super-Chandrasekhar WDs: Series of Mass Limits for a BWD

As described in the previous subsection, through the generation and subsequent evolution of an appropriate mass ZAMS model, one can evolve a CO WD using STARS. Further, on introducing the pressure modification by the magnetic field, one obtains a BWD model.

The magnetic field of a BWD can lead to interesting effects. As mentioned previously, the magnetic field leads to a modification of the conventional Chandrasekhar limit through both classical and quantum effects. In our modification, we have focused on the modification to the hydrostatic equilibrium (pressure balance) through the inclusion of a toroidal magnetic field in the star.

To examine this change of mass and mass limit through our simulation, we need to examine the maximum limit of how much mass our BWD can support against gravity. We introduce accretion to our star, through the code's (spherical) mass addition control. By adding mass to our BWD model, we obtain higher and higher mass BWDs. Higher mass BWDs have denser cores, and thus higher fields at the center as well. Thus, along with the growth of mass in subsequent models, one can also see the growth of the central field of the BWD.

One can examine the change in mass limit through the various mass-radius (M-R) curves for the WDs obtained by mass addition. By comparing the M-R curves obtained for magnetized and non-magnetized cases, we see that there is a clear enhancement of saturation mass in the former, as shown in Fig. 32.2. The mass limit is now much higher than the Chandrasekhar limit—in one case, it is as high as $2.8M_{\odot}$. These are the theorized super-Chandrasekhar WDs, finally realized in a time-dependent simulation, via evolution from the MS star!

Further, as shown in the Figure, different magnetic field profiles (represented here through the change of η) lead to different saturated masses/mass limits. It is not just the strength of the magnetic field, but the exact profile/geometry that it follows within the star, that finally determines the exact mass limit obtained.

Figure 32.3 shows another interesting consequence of the change of field profile. While both the field profiles considered here give the same saturated mass/mass limit, due to the difference in their gradients, their M-R curves are different. The blue dashed curve ($\eta = 0.8, \gamma = 2$) and the green dotted curve ($\eta = 0.7, \gamma = 0.9$) have the same limiting mass, but the green dotted case gives higher radii at lower masses. This shows how the effect of magnetic field could be seen in sub-Chandrasekhar mass WDs in specific cases, and this could ultimately have important implications for observations.

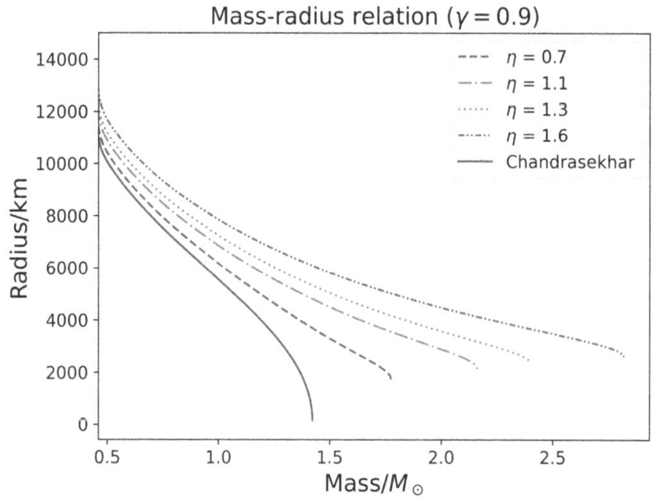

Fig. 32.2 M-R curves showing different mass limits from different magnetic field profiles (different values of η). The mass accretion rate is $10^{-9} M_\odot$/Yr

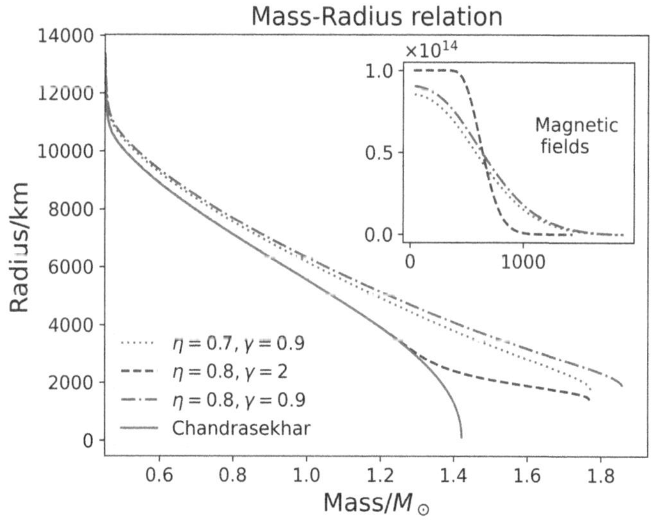

Fig. 32.3 M-R curves showing different radii in the low mass (sub-Chandrasekhar) regime based on different profiles

Thus, in the presence of magnetic field, one obtains a series of mass limits for a WD, depending on the exact field strength and geometry introduced. This was earlier shown analytically by [4] and we are now able to reproduce the result through time-dependent simulations as well. Further, the magnetic field can affect sub-Chandrasekhar WDs, and this could be an important observational discriminant.

We are able to establish the accretion formation scenario for super-Chandrasekhar WDs, where an initially lowly magnetized WD accretes matter, leading to the field of the star being strengthened, e.g. by flux conservation. Eventually, accretion can lead to a maximal mass beyond the Chandrasekhar limit. As we show here, the exact maximum mass is very much dependent on the details of the magnetic field present within the star.

32.3.3 *Magnetic Field Decay*

Given that we are examining how the magnetic field can lead to a WD supporting super-Chandrasekhar masses, it is interesting to examine what happens when this field changes. Particularly, we know that the magnetic field of stars decays over time through a combination of different processes. A natural question to ask then is: *What happens to the extra mass, which was held due to the stronger field, when the field decays?* Two possible scenarios could arise here—the star may shed the extra mass through some mechanism and end up as a Chandrasekhar WD; or the star may collapse, unable to support the extra mass against gravity in absence of the field. The latter scenario points towards a SNe type evolution.

There are three main processes that contribute to field decay in compact stars: ambipolar diffusion, Ohmic decay, and Hall effect [9]. In WDs, generally ambipolar diffusion does not feature, and it is the other two processes which are dominant. Field decay by a combination of Ohmic and Hall effects was explored analytically in previous work by our group [8].

Following previous work [10], we assume the magnetic field decay to take the following form:

$$\frac{dB}{dt} = -B\left(\frac{1}{t_{Ohm}} + \frac{1}{t_{Hall}}\right), \tag{32.6}$$

where B is the magnitude of magnetic field, as calculated by Eq. (32.5). We use the expressions for the characteristic time-scales of these equations from previous analytical work [8], given by

$$t_{Ohm} = (7 \times 10^{10})\rho_{c,6}^{1/3} R_4^{1/2} \frac{\rho_{av}}{\rho} \text{ years}, \tag{32.7}$$

and

$$t_{Hall} = (5 \times 10^{10}) l_8^2 B_{0,14}^{-1} T_{c,7}^2 \rho_{c,10} \text{ years}, \tag{32.8}$$

where $\rho_{c,n} = \rho_c/10^n$ g/cm^3, ρ_c is the central density of the WD, $R_4 = R/10^4$ km, $T_{c,7} = T_c/10^7\ K$, $B_{0,14} = B_0/10^{14}\ G$ and $l = l_8 \times 10^8$ cm is characteristic length scale of the flux loops through the outer core of the WD.

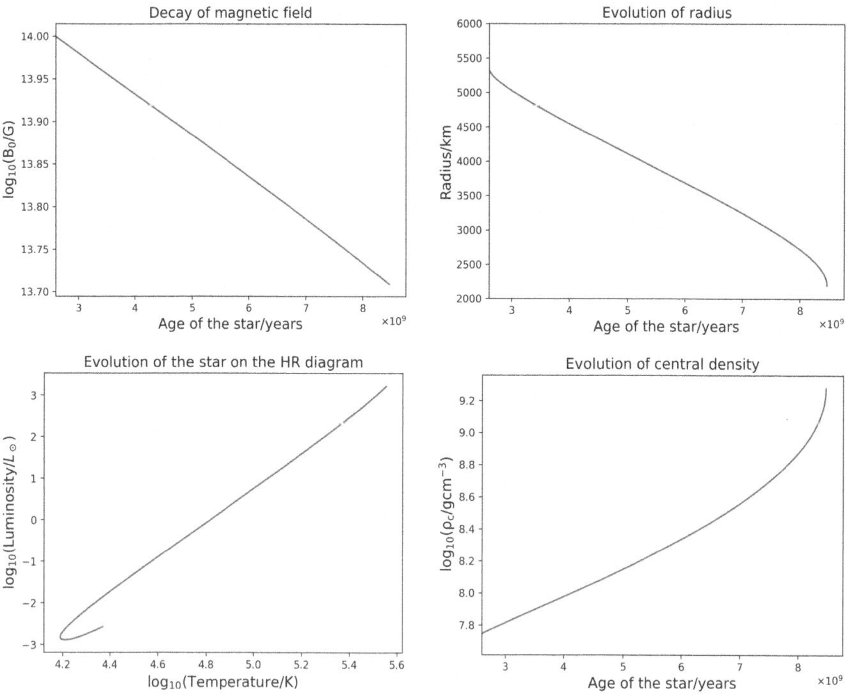

Fig. 32.4 Change in various stellar properties as the magnetic field decays

Given the constraints of a one-dimensional hydrostatic stellar evolution code, we cannot hope to capture the entirety of either of these extremely dynamic processes. However, as a preliminary analysis, one can introduce field decay as one of the set of equations to be solved. We can then examine what scenario the code points us to.

We find some interesting results before the code crashes (due to triggering of instability). As seen from the top panels of Fig. 32.4, the decay of the magnetic field is accompanied by a contraction of the star, seen by its decrease of radius. The WD becomes nearly half its initial radius before the instability triggers. Further, as shown in the lower panels of Fig. 32.4, this is also accompanied by a simultaneous increase in the density and luminosity (in the last phase) of the star. Both the density and luminosity increase sharply just before the run stops. These are all indications that the WD is moving towards a denser phase of evolution, i.e., these all indicate the WD collapsing to form a SNe Ia. This result is not altered qualitatively on varying the decay rate of the magnetic field.

Thus, our preliminary analysis seems to argue the fact that SNe Ia do not necessarily originate only from the thermonuclear explosions of Chandrasekhar mass WDs, and may be formed from the subsequent evolution of super-Chandrasekhar BWDs.

32.4 Future (and on-going) Work

We have described here a set of preliminary results showing the time-dependent simulation of the formation and evolution of BWDs, and their further evolution to super-Chandrasekhar BWDs. These initial results are extremely promising, as they already point us towards the stability and realizability of these objects.

We have looked at parallel results from another stellar evolution code, MESA [11]. We find consistency with formation mechanisms in the non-magnetic cases. Work is underway to incorporate magnetic fields in a similar fashion in MESA. MESA also allows for more realistic accretion, with controls present to specify the composition of accreted matter.

We are currently working towards making results more robust in a number of ways. For instance, we are currently including the chemical evolution through nuclear reactions throughout the WD phase as well. Although there is minimal burning in the WD once it is fully degenerate, there are cases where sufficient burning of the envelope can lead to some interesting results. We are also examining the effects of having a more realistic mixed-field geometry in our star, as it is long known that purely toroidal and purely poloidal fields are magnetohydrodynamically unstable [12]. These changes to the implementation do not majorly change the qualitative results. Lowly magnetized/dormant field based BWDs can accrete to form super-Chandrasekhar BWDs, even with all these caveats considered.

Finally, we also examine the alternate formation scenario for super-Chandrasekhar WDs mentioned earlier. BWDs may be massive enough through their formation/evolution from a sufficiently massive and magnetized MS star itself.

32.5 Summary and Conclusions

The Chandrasekhar limit is one of the fundamental results of theoretical, stellar and high energy astrophysics. It brings together astrophysics (cosmology), quantum statistics and gravity in a single, beautiful formula. However, observations of over-luminous SNe Ia tell us that it is not the complete picture. Particularly, effects due to finite temperature, rotation, and magnetic fields were not considered in Chandrasekhar's treatment. All these effects, separately and together lead to clear theoretical deviations from the ideal Chandrasekhar M-R relation and mass limit.

Through the current work, we capture the finite temperature effects in the WDs by solving for full stellar structure. Further, through modification of the code, we introduce magnetic effects as well. Through these effects, we are able to reproduce super-Chandrasekhar WDs through time-dependent simulations. We obtain the full evolutionary track from the MS star up to the formation of BWD. By accreting onto both magnetized and non-magnetized WDs, one clearly obtains different M-R relations, and thus different mass limits. Thus, we have now established a stable formation scenario for a super-Chandrasekhar WD, consisting of the initial formation of lowly magnetized BWD, whose field is further enhanced through accretion, leading to it supporting masses beyond the Chandrasekhar limit. This is consistent with the observational fact that a higher percentage (20–25%) of WDs in binary systems are magnetic [13]. On doing a preliminary analysis of magnetic field decay in the star, we see that these super-Chandrasekhar WDs tend towards SNe type evolution, further cementing their link to observations of overluminous SNe Ia.

Our work is definitive proof that the super-Chandrasekhar BWDs, proposed a decade ago by Mukhopadhyay and group, are stable and realizable. Through our simulation, we are able to establish a time-dependent evolutionary track showing the formation of these objects, with some tracks ending in WDs with masses as high as $2.8M_{\odot}$. This is probably the first work of its kind, and the first successful STARS simulation which shows such highly super-Chandrasekhar WDs. Our results have far–reaching implications which may even caution against SNe Ia as standard candles.

Acknowledgments Z. Z. thanks the organizers of the International Symposium on Recent Developments in Relativistic Astrophysics (ISRA) for the opportunity to present this work. Z. Z. also acknowledges the Prime Minister's Research Fellows (PMRF) scheme, with Ref. No. TF/PMRF-22-7307, for providing fellowship. A. J. H. acknowledges the Czech Grant Agency (GAČR) under the grant number 21-16583M.

Competing Interests The authors have no conflicts of interest to declare that are relevant to the content of this chapter.

References

1. Howell, D. A., et. al. (2006). *Nature, 443*, 308–311.
2. Kundu, A., & Mukhopadhyay, B. (2012). *Modern Physics Letters A, 27*(15), 1250084.
3. Das, U., & Mukhopadhyay, B. (2013). *Physical Review Letters, 110*(7), 071102.
4. Deb, D., Mukhopadhyay, B., & Weber, F. (2022). *The Astrophysical Journal, 926*(1), 66.
5. Eggleton, P. P. (1971). *Monthly Notices of the Royal Astronomical Society, 151*, 351.
6. Chakrabarty, S., Bandyopadhyay, D., & Pal, S. (1997). *Physical Review Letters, 78*(15), 2898.
7. Kalita, S., & Mukhopadhyay, B. (2019). *Monthly Notices of the Royal Astronomical Society, 490*(2), 2692–2705.
8. Bhattacharya, M., Hackett, A. J., Gupta, A., Tout, C. A., & Mukhopadhyay, B. (2022). *The Astrophysical Journal, 925*(2), 133. .
9. Goldreich, P., & Reisenegger, A. (1992). *The Astrophysical Journal, 395*, 250–258.
10. Heyl, J. S., & Kulkarni, S. R. (1998). *The Astrophysical Journal Letter, 506*, L61.
11. Paxton, B., et. al. (2011). *The Astrophysical Journal Supplement Series, 192*, 3.
12. Braithwaite, J. (2009). *Monthly Notices of the Royal Astronomical Society, 397*, 763.
13. Ferrario, L., de Martino, D., & Gaensicke, B. (2015). *Space Science Reviews, 191*, 1–4, 111–169.

Part IX
Astroparticle Physics and Cosmological Issues

Chapter 33
Ultralight Dark Matter: A Novel proposal

T. R. Govindarajan

Abstract A novel proposal is made to account for the dark matter component of the Universe. Ultralight dark matter with mass $\leq \mathcal{O}(10^{-22})$ eV is one of the strong candidates for the missing mass which aids the formation of galaxies as well as holding them together. They are also known as fuzzy dark matter(FDM) which will come under Cold Dark matter. The question is what is this particle and its implications. How do we experimentally see it is an outstanding question. We propose to answer some of these questions with some evidences and the estimates.

33.1 Introduction

For the past five decades the origin and the question of dark matter [1] which constitutes 25% of matter in the universe have puzzled both astrophysicists, particle physicists together. The observation that the radial speeds of luminous stars and other objects do not go down sufficiently fast over a large distance in galaxies as expected through Newtonian/Einsteinian gravity ignited physicists to conjecture that there could be matter within the galaxies with only gravitational interactions. In addition galaxy clusters, gravitational lensing, bullet clusters and CMB spectrum [1] offer further support to this conjecture. There were several candidate proposals inspired by Beyond Standard Model (BSM) expectations. Supersymmetry provides several potential massive particles. Sterile neutrinos and primordial blackholes too were considered as potential candidates. There is also the suggestion of an extra dark photon may be lurking without any interaction with matter in the Universe except through gravity. Unfortunately none of them satisfactorily answer all the queries regarding this.

T. R. Govindarajan (✉)
The Institute of Mathematical Sciences, Chennai, India
e-mail: trg@imsc.res.in; govindarajan.thupil@krea.edu.in
Krea University, Sricity, India

S. Ghosh, A. R. Rao (eds.), *The Relativistic Universe: From Classical to Quantum*, Astrophysics and Space Science Proceedings 61,
https://doi.org/10.1007/978-3-031-90186-7_33

The desperation even led to the conjecture that gravity itself will need changes and Newtonian law of Universal gravitation might fail when the acceleration is below a particular threshold [2]. This theory known Modified theory of Newtonian dynamics is also contender which explains rotational curves. But it fails in explaining bullet clusters or gravitational lensing in addition to being arbitrary.

Extremely light bosons from QCD axions [3], and axion like scalar bosons have also been considered. While QCD axions have not succeeded in providing formulation as well as discovery such cold dark matter (CDM) can be potential candidate with ultra light mass.

Having given the background to the dark matter question we will recollect one of the profound questions raised by Schrodinger in 50's namely 'Must the photon mass be zero?'[4]. We will deal with his answer and further developments on this question. Finally we will link the dark matter question with this and provide a novel proposal for this burning question.

33.2 Schrodinger and Photon Mass

Erwin Schrodinger in 1955 at Dublin Institute of Advanced Studies posed the question about the photon mass. Since the massive photon will have three degrees of freedom whereas massless one will have one less, will it affect for example the Stefan's constant value.

! 3 vs 2

While calculating in blackbody radiation, energy density as a function of frequencies we multiply by a factor of 2 to account for the transverse degrees of freedom. Density of modes is given by:

$$\rho(\nu) = \frac{2h\nu^3}{c^2}\frac{1}{e^{\frac{h\nu}{kT}} - 1}. \tag{33.1}$$

We can obtain Stefan's constant by integrating the density

$$E = \int \rho d\nu = \sigma T^4.$$

Schrodinger himself answered if the vector potential for the photon is coupled only to a conserved current it will have very little effect if the mass of the photon is very tiny. That is if the interaction is

$$H_{int} = \int d^4x \, j^\mu \, A_\mu \, , \quad and \;\; \partial_\mu \, j^\mu = 0 \, , \tag{33.2}$$

then corrections to the cross sections due to the longitudinal photons will be extremely small if the mass is very small.

Schrodinger: *Even if we find in Nature the limiting case is realized, we should still feel the urge to adumbrate a theory which agrees with experience on approaching to the limit, not by a sudden.*

He went ahead and estimated the mass of the photon from the data available at that time about geomagnetic fields on the surface of earth. The limit he obtained was:

$$m_\gamma \;\leq\; 10^{-16} eV \, . \tag{33.3}$$

The scale of this mass is easily understood as the only length scale available was the radius of the earth and we can expect the mass in natural units to be the inverse of radius. More modern data about geomagnetic fields satellite based experiments lead to an improvement to the value as $m_\gamma \leq 10^{-18}$ eV. Particle data book provides this as the current upper bound [5].

Proca and Stueckelberg Theory

Massive vector boson theory known as Proca theory has two problems (1) It has additional degree of freedom and massless limit has a discontinuity in the number of degrees of freedom. (2) Mass term explicitly breaks local gauge invariance. But the massive QED is renormalizable and the extra contributions are small if the mass is small [6]. But gauge invariance is our guiding principle and we expect it to be preserved. Stueckelberg theory avoids simultaneously the discontinuity in the degrees of freedom and the lack of gauge invariance nicely [7].

Stueckelberg introduced another scalar field ϕ and obtained a gauge invariant massive QED whose massless limit in addition avoids the discontinuity. The Lagrangian for Stueckelberg theory is:

$$\mathcal{L} = -\frac{1}{4}\left(F_{\mu\nu}\right)^2 + \frac{1}{2}m^2\left(A_\mu - \frac{1}{m}\partial_\mu\phi\right)^2 + \bar{\psi}\Big[\gamma^\mu(i\partial_\mu + eA_\mu) - M\Big]\psi \, , \tag{33.4}$$

where ϕ is the Stueckelberg field and ψ is the electron field. The gauge transformations are:

$$\psi \rightarrow e^{i\lambda(x)}\psi, \ A_\mu \rightarrow A_\mu - \partial_\mu\lambda(x), \ \phi \rightarrow \phi + m\lambda(x)\,. \tag{33.5}$$

We can fix The gauge using: $-\frac{1}{2}(\partial_\mu A^\mu + m\,\phi)^2$.

We can also provide mass to the photon by the well known Higgs mechanism by coupling the abelian gauge field to a complex scalar field Φ. This will have nonzero vacuum expectation value giving mass to the photon. We can write in the symmetry broken phase $\Phi = R\,e^{i\phi}$. And the Lagrangian becomes:

$$\mathcal{L} = -\frac{1}{4}F \wedge F + |D_\mu\Phi|^2 - V(\Phi) + \cdots \tag{33.6}$$

Phase of this field will be the Stueckelberg field and this mechanism in a specific limit of freezing the fluctuations of R (i.e.,make R very massive) gives the Stueckelberg theory. There are other mechanisms like topological mass term by coupling to Kalb Ramond field $B_{\mu\nu}$ which will not be discussed here.

33.3 Stueckelberg Field as the Dark Matter Candidate

We learn the lesson that, Stueckelberg field does not interact with normal matter in a Compton length scale of $\frac{h}{mc}$. But it has energy and contributes extra terms to the energy momentum tensor. This can take part in gravitational interaction. The question we would like to ask is: can it contribute as dark matter?

If the mass of the candidate is too small, they will travel very close to the velocity of 'light' and decouple very early after the bigbang during radiation dominated era. That will not help.That is where Bose and Einstein come to our rescue. Bose [8] wrote his famous paper in 1924 providing the basis for his statistics which was followed by Einstein for a new phase of matter known as the 'Bose Einstein Condensate (BEC)'[9].

In our proposal we may treat the constituent particles as the Stueckelberg particles which do not interact with matter. This can also be taken as the longitudinal photons through a gauge choice. These particles will be such a candidate only if they form a Bose-Einstein condensate. The formation of a Bose-Einstein condensate needs (1) a conservation law for particle number (2) the system should be at a temperature below the critical temperature T_c [10]. For massless scalar fields, the conserved quantity is by shift invariance of the field, which is broken by a mass term and self interactions. It is an approximate symmetry. The T_c is given by (where ρ is the number density of a gas of particles),

$$T_c = \frac{\hbar c}{k_B}\left(\frac{\rho\pi^2}{m_\gamma \zeta(3)}\right)^{1/3}. \tag{33.7}$$

Ideal Bose gas is a quantum state of matter similar to ideal gas in classical statistical mechanics. They obey Bose statistics and have integer spin. This was originally proposed by Bose for photon gas and extended to massive particles by Einstein.

While at large temperature they behave similarly, but at very low temperatures they form a condensate under certain conditions due to Bose Statistics. (After seeing Bose's letter, Einstein realized immediately such a state can exist!) BEC was obtained for Rubidium 87 gas whose mass is 86 amu ($\approx O(86)$ Gev). Critical temperature is $\approx$ nano kelvins. We have very low value, because of high mass of the atoms, making up the gas.

Stueckelberg particles on the other hand have mass $\leq 10^{-20}$ eV. This makes the critical temperature very high (close to the temperature of post Big bang after inflation).

33.3.1 Dark Matter and BEC

We need to provide the relic density and the change in density effected as a consequence of the Friedman expansion. If dark matter is given by an initial density of ρ_0 at the time of decoupling, before the radiation dominated era begins we will know the present density. The epochs are given in Table 33.1:

Using these, we can relate the current density ρ_{final} of dark matter to the initial density as:

$$\rho_{final}^{\frac{1}{3}} = \rho_0^{\frac{1}{3}} \frac{1}{(1.2 \times 10^{12})^{\frac{1}{2}}} \left(\frac{47000}{9.8 \times 10^9} \right)^{\frac{2}{3}} \frac{1}{1.377}. \tag{33.8}$$

Employing this we get the relation between the observed dark matter density ρ and the critical temperature required to achieve Bose-Einstein condensation (m_γ in eV),

$$\rho \sim 10^{-22} m_\gamma \, T_c^3 \, . \tag{33.9}$$

In SI units, the observed dark matter density in our galaxy [5] which is approximately $\sim 10^{-22}\,\mathrm{kg/m^3}$ or 1 proton/cc is recovered if we take $m_\gamma \sim 10^{-19}$ eV and $T_c \sim 10^{17} K$. The corresponding estimate for $m_\gamma = 10^{-22}$ eV would be $T_c \sim 10^{19} K$.

Table 33.1 Epochs

Epoch and time	Scale factor	Temperature
Radiation era $1s$ to $1.2 \times 10^{12} s$	$\propto t^{1/2}$	$10^{12} K$–$10^4 K$
Matter era $4.7 \times 10^4 y$ to $9.8 \times 10^9 y$	$\propto t^{2/3}$	$10^4 K$–$4K$
Dark energy 9.8–13.8 billion y	$\propto e^{Ht}$	$<4K$

Further the condensate once it is formed at the earlier times will remain so for all epochs since the temperature has been cooling all the time.

For our galaxy, Milky way luminous matter is $\approx$ 10^{10} solar masses. But dark matter is estimated to be $\approx$ 10^{12} solar masses. Dark matter (DM) halo is a theoretical model of galaxy that bounds the galactic disk and extends beyond luminous (visible) part. The mass of the halo dominates the galactic mass (nearly 95% !) and only postulated through observation of rotation curves. DM halos are crucial for galaxy formation and evolution.

During galaxy formation temperature of matter is too high to form gravitationally bound objects. Prior formation DM is needed to add additional structure. After the galaxy formation it extends far beyond observable part and required for understanding velocities and lensing.

33.3.2 DM Halo

We have assumed that the current dark matter density can be entirely explained in terms of Stueckelberg particles. Several density profiles are in the literature. There are two crucial parameters for modeling dark matter profile. They are half radius and central density. We can compute half radii of condensates and their masses. With extremely ultralight mass for the particles, we can consider the dark matter as a fluid.

A BEC condensate is there if the temperature is less than T_c. This description is by the ansatz $\phi \sim e^{-imc^2t}\psi$ in a perturbed FRLW universe:

$$i\left(\partial_t + \frac{3}{2}\frac{\dot{a}}{a}\right)\psi = \left(\frac{-\nabla^2}{2m} + mV\right)\psi\,, \tag{33.10}$$

where V is a gravitational potential, in the linear approximation.

Following Hui et al. [11] the expression (equations (29) and (30) in that paper) for the half radius and mass are given by:

$$r_{\frac{1}{2}} = 3.925\frac{\hbar^2}{GMm_\gamma^2}, \tag{33.11}$$

$$\rho_c = 4.4 \times 10^{-3}\left(\frac{Gm_\gamma^2}{\hbar^2}\right)^3 M^4, \tag{33.12}$$

where ρ_c is the central density of the halo and M is the mass of the soliton.

We can parameterize mass as 10^{-17-x} eV and $T_c = 10^{\frac{50}{3}-y}K$ and obtain the sample space for these parameters.

There are couple of points we should remember about the program. First if we consider the abelian Higgs model and obtain the Stueckelberg theory only in the limit, there will be further interactions of the Higgs field R and the longitudinal component ϕ. Secondly due to the fluid/wavy behavior the central density will avoid the well known core-cusp problem which plagues in general all DM models particularly for the dwarf galaxy.

33.4 Arguments In Favor

Now we will see some arguments in favor of the proposal. First there are several dwarf galaxies orbiting Milky way of various sizes from 200 light years to 10^5 light years. They hold stars from 1000 to 10^6. They provide arguments in support of FDM. Recently nano gravitational waves have been observed through radio telescopes arranged to analyze pulsar emission timings. This is known as Pulsar timing array (PTA). They surprisingly provide an argument in support of FDMs. Lastly one can check the Stress energy tensor due to massive vector theory. Due to the mass term there is extra contribution due to the longitudinal photons. These are prominent when we consider long wavelength limit. This provide another evidence in support of FDM. We will now elaborate each of these arguments.

Dwarf Galaxies

Stars are also at the initial stages of formation in some of the galaxies. N body Simulation with dark matter indicate the density grows exponentially at the core. But for FDM models there is no such problem. This is due to the fluid like behavior of FDM. There are also anticipated around 200 such galaxies orbiting Milky way. But only about 40 have been seen so far. The rest are probably too faint or DM content is too high and hence less luminous. There are several with diameters less than 1000 light years. The Compton wavelength of the Stueckelberg particles will range from few light days to 100–150 light years depending on the mass from 10^{-20}–10^{-24} eV. The size of the BEC condensate can be expected to be that of Compton wavelength i.e., $O(\frac{h}{m_\gamma c})$. The size of smallest galaxy will set a limit on the Compton wavelength and hence of the mass of the dark matter candidate.

Segue-2 Galaxy

The smallest galaxy known so far is Segue-2. It is in the Aries constellation. It is 1.1×10^5 light years away. It has a half radius of 115 light years. It has a mass of $O(10^6)M_\odot$. It holds around 1000 stars. We can expect 100 light years as the limit set by dwarf galaxy sizes which will correspond to $m_\gamma \geq 10^{-24}$ eV. This estimate also coincides with expectations from nano gravitational waves from pulsar timing array.

Pulsar Timing Array

Pulsar timing array is essentially radio telescopes looking for Nano hertz gravitational waves due to merger of galaxies in the very early universe. This study is done by analyzing the timing of the Pulsars within our galaxy. They have detected recently with confidence level 98% a hum of the early universe. This background noise is similar to CMB for Electromagnetic spectrum. This discovery has an interesting connection to FDM. The gravitational waves of the galaxy mergers of the early universe are affected by the dark matter between pulsars in our galaxy and affect the noise. For this to be measured, the average distances of Pulsars should be approximately the Compton wavelength of FDM candidates. If this is correct, we expect the mass of FDM candidate should be $\geq 10^{-24}$ eV [12]. Recently with Parkes Pulsar Timing array which has been monitoring 20 millisecond pulsars data for the period 2004–2016 has been analyzed. it is claimed to produce a limit FDM mass should be $\geq 10^{-24}$. Hence the window is narrowing for the FDM candidate proposals.

Massive Photon and Stress Energy Tensor

Ryutov et al. showed the Maxwell Proca electrodynamics with finite photon mass changes stress tensor as expected. This is obviously due to the extra longitudinal photons. This stress under certain conditions develop a 'negative pressure' [13]. This negative pressure imitates gravitational pull and can become dominant. The effect is associated with random magnetic fields with correlation lengths exceeding the photon Compton wavelength. The stresses act predominantly on the interstellar gas and cause an additional force pulling the gas towards the center and towards the galactic plane. This is precisely the scale at which the longitudinal photons make the difference. But the force is not sufficient to explain the contribution to a star like sun in the rotational curves. This might be lacuna of the Proca theory, but it requires to study this theory in the context of spontaneous symmetry breaking through Higgs mechanism and associate the extra pressure directly to the contribution of the Stueckelberg photons.

33.5 CMB and Stueckelberg Field

We started with photon mass query raised by Schrodinger and developed further to link up with fuzzy dark matter. Now we can go back and ask what can we learn from CMB radiation with photon mass. CMB provides evidence for hot big bang ends when radiation decouples with black body radiation at a temperature 2.7 K. While it agrees with black body curve to a great accuracy but have shown discrepancies away from homogeneity. We can ask if we look for massive photon correction to the blackbody. Such a correction was worked out by Julien Heeck [14] and the energy density equation gets modified.

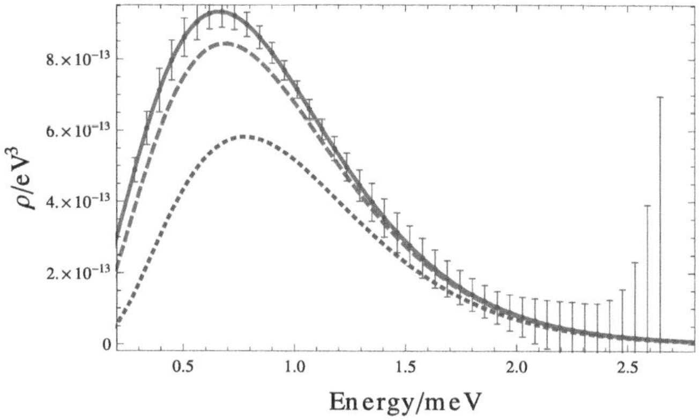

Fig. 33.1 Black body radiation with mass: Source: Phy Rev. Letts 111, 2013

First we go back to the issue of the effect of photon mass on black body radiation. Cosmic microwave background being one of most perfectly known 'black body' we can check its effect on the CMB distribution. One can put limits on the mass of the photon from the deviation from the black body spectrum for CMB. The density of photons with frequency is given by [14]:

$$\rho(E, T) = \frac{E^3}{e^{\frac{E}{kT}} - 1}\sqrt{1 - \frac{m^2}{E^2}}, \tag{33.13}$$

due to the modified dispersion relation $p^2 = E^2 - m^2$. See Fig. 33.1. Such an anlaysis was done and it gives poor estimate! In the Fig. 33.1 the gray line is for $m_\gamma = 0$ and the red line is for $m_\gamma = 10^{-5}$ eV. We expect $m_\gamma < 10^{-6}$ eV. Geo and Solar Magnetic field analysis give much tighter bound as of now.

Obviously the limit is far removed from the limits we obtain from Earth and Solar magnetic field and the current expectation as candidate from FDM. This gives the realization that the part of anisotropies can be at least characterized by the mass parameter. But we have to consider Abelian Higgs model and the role of the interaction between Higgs and the Stueckelberg field also in the description. This is ongoing and will be reported.

33.6 Detection of Longitudinal Photons

Several attempts to estimate the mass of the photon have already been done. See [5]. We can ask whether any particle physics signals can be provided for the longitudinal component of the photon. One such could be the implications for theorem due to Landau and Yang [15]. The theorem predicts a spin -1 massive particle cannot decay to two massless spin -1 particles. This implies the Z boson of Weinberg-Salam model cannot decay to two photons. Proof of the theorem does not require action or Hamiltonian. It simply follows from representation theory of Poincare group and Bose statistics of photons. It follows from the fact that symmetric product of two massless representations of Poincare group does not contain the massive spin -1 representation [16]. This is easy to see in the rest frame of massive spin -1 particle we have two polarizations of photons $\epsilon_i, i = 1, 2$ and $k, -k$ the momenta of outgoing photons along with spin vector **S** of the massive particle. We cannot write a non zero amplitude which is under $k \leftrightarrow -k, \epsilon_1 \leftrightarrow \epsilon_2$. (we need $k \cdot \epsilon_i = 0$ also). If the photon has mass then there will also be longitudinal component, making it possible to violate the theorem. In actual process involving one of the photons with longitudinal component, we will have non zero amplitude, but detection of the that photon will be very difficult. We have to look for missing energy and momentum processes.

33.6.1 Extension to WS Model with Massive Photons

We can easily extend the above scheme to the standard model of Weinberg Salam $SU(2)_L \otimes U(1)$ theory. The action can be written as

$$\mathcal{L} = \mathcal{L}_g + \mathcal{L}_f + \mathcal{L}_S, \tag{33.14}$$

$$\mathcal{L}_g = -\frac{1}{4} B_{\mu\nu} B^{\mu\nu} - \frac{1}{4} Tr(f_{\mu\nu} f^{\mu\nu}) + \frac{1}{2} m_\gamma^2 \left(B_\mu - \frac{1}{m_\gamma} \partial_\mu \phi \right)^2, \tag{33.15}$$

$$\mathcal{L}_S = |D_\mu \Phi|^2 - \lambda \left(|\Phi|^2 - \frac{f^2}{2} \right)^2, \tag{33.16}$$

where B_μ is used to denote the weak hypercharge field and D_μ is the covariant derivative acting on the Higgs. $B_{\mu\nu}$ is the hypercharge field strength and $f_{\mu\nu}$ is the $SU(2)$ field strength.

33.7 Summary and Conclusions

Stueckelberg theory is a powerful precursor to Higgs model and introduces mass to the photon without breaking gauge invariance. The theory is renormalizable in the ultraviolet and does not have infrared problem due to finite mass. Interestingly it achieves freedom from infrared question even in the limit of $m_\gamma \longrightarrow 0$. This is because it naturally maps to Faddeev Kulish theory of asymptotic coherent states in the massless limit. This can be understood from the difference between little group of massless and massive representations of Poincare group. These are E(2) and SO(3). There is a natural Inonu Wigner contraction through a parameter which is in our case the mass itself. The review by Ruegg and Ruiz-Altaba [17] "The Stueckelberg Field", provides all the necessary background. The details of Stueckelbrg particle as dark matter candidate are provided in our contribution [18]. Dvali et al. [19] propose holography can be formulated in terms of the information capacity of Stueckelberg degrees of freedom. These degrees of freedom act as qubits to encode quantum information. The capacity is controlled by the inverse Stueckelberg energy gap to the size of the system. They relate the scaling of the gap of the boundary Stueckelberg edge modes to Bogoliubov modes. Unfortunately our program cannot be easily extended to QCD as there is no Stueckelberg program for non abelian gauge theory which is unitary and renormalizable. Can we get massive gravity by similar analysis? Answer seems to be 'no'. Again unitarity queries seems to be in conflict.

We want to close this summary with profound remarks due to Bernhard Riemann on the ultimate structure of space time in his Habilitation talk [20].

AXIOMS UNDERLYING THE BASIS OF GEOMETRY

MAY NEED CHANGES

AT INFINETESIMAL AND INFINITE LEVEL.

Bernhard Riemann

Acknowledgments This chapter [21] has some implications for the current work which will be followed later. I thank the organizers of the International Symposium on Recent Developments in Relativistic Astrophysics 2023. Special thanks are due to Shubhrangshu Ghosh for a wonderful meeting. I also thank the participants for discussions. Special mention about Banibrata Mukhopadhyay for enlightening me on issues on Astrophysics and Cosmology. SRM University at Gangtok was wonderful place for discussions.

References

1. Zwicky, F. (1937). *Astrophysical Journal, 86*, 217; de Swart, J., Bertone, G., van Dongen, J. (2017). *Nature Astronomy, 1*, 0059; Liu, J., Chen, X., Ji, X. (2017). *Nature Physics, 13*, 212.
2. Milgrom, M. (1983). *Astrophysical Journal, 270*, 365.
3. Peccei, R. D., & Quinn, H. R. (1977). *Physical Review Letters, 38*, 1440.
4. Bass, L., & Schrodinger, E. (1955). *Proceedings of the Royal Society of London. Series A, Mathematical and Physical Sciences, 232*, 1188.
5. Particle Data Group. https://pdg.lbl.gov/listings.
6. Banks, T . (2008). *Modern quantum field theory*. Cambridge University Press.
7. Stueckelberg, E. C. G. (1938). *Helvetica Physica Acta* (in German), *11*, 225.

8. Bose, S. N. (1924). *Zeitschrift für Physik, 26*, 178.
9. Einstein, A. (1925). *Konigliche Preubische Akademie der Wissenschaften, Sitzungsberichte* (in German), 261–267.
10. Grether, M., de Llano, M., & Baker, Jr., G.A. (2007). *Physical Review Letters, 99*, 200406.
11. Hui, L., Ostriker, J. P., Tremaine, S., & Witten, E. (2017). *Physical Review D, 95*, 043541.
12. Khmelnitsky, A., & Rubakov, V. (2014). *Journal of Cosmology and Astroparticle Physics, 02*, 019.
13. Ryutov, D. D., Budker, D., & Flambaum, V. V. (2019). *The Astrophysical Journal, 871*, 218.
14. Heeck, J. (2013). *Physical Review Letters, 111*, 021801.
15. Landau, L. D. (1948). *Doklady Akademii Nauk SSSR, 60*, 207; Yang, C. N. (1950). *Physical Review, 77*, 242.
16. Asorey, M., Balachandran, A. P., Arshad Momen, M., & Qureshi, B. (2023). *Journal of High Energy Physics, 10*, 028.
17. Ruegg, H., & Ruiz-Altaba, M. (2004). *International Journal of Modern Physics A, 19*, 3265.
18. Govindarajan, T. R., & Kalyanapuram, N. (2019). *Modern Physics Letters A, 33*, 1950330; *Modern Physics Letters A, 34*, 1950009.
19. Dvali, G., Gomez, C., & Wintergerst, N. (2016). *Physical Review D, 94*, 084051.
20. Riemann, B. (1998). *Nature, VIII*, 183, 184, 14. Translated by W K Clifford.
21. Khlopov, M. Y., Malomed, B. A., & Zeldovich, Y. B. (1985). *Monthly Notices of the Royal Astronomical Society, 215*, 575.

Chapter 34
The Origin of Cosmic Rays: An Enduring Mystery in Physics

Arunava Bhadra

Abstract Cosmic Rays (CRs) were discovered over a century ago, but their origin is still unclear. CRs with energy up to a few PeV are generally believed to be produced in the galaxy, whereas CRs above a few EeV are considered extra-galactic. Supernova remnants are widely considered the potential sources of galactic cosmic rays, while Active Galactic Nuclei and gamma-ray bursts are potential sources of CRs at the highest energies. However, no clear evidence has emerged from recent astronomical observations in favor of the standard scenarios of cosmic ray sources against the expectations. Here, we shall briefly review the standard model of the origin of CRs and critically analyze the present status of the model in light of recent observations from very high-energy gamma-ray and neutrino astronomy.

34.1 Introduction

Cosmic Rays (CRs) are a highly isotropic flux of relativistic particles (dominantly charged particles) that reach Earth's vicinity from outer space. Where do CRs come from? As charged particles, CRs are deflected by the galactic magnetic field except at the highest energies, preventing their direction from pointing toward their source. The CRs above EeV (10^{18} eV) energies show a slightly dipolar distribution, only indicating they likely originate from other galaxies rather than from sources within the Milky Way [1]. Consequently, the origin of CRs is theoretically modeled to explain their observed features consistently. Despite extensive research over the 112 years since their discovery, the origin of CRs remains unclear (see for instance [2, 3]).

Primary CRs are thought to be accelerated in various astrophysical sources. When these accelerated CRs interact with radiation and matter in their stellar environments, they produce high-energy gamma rays and neutrinos. Consequently, an

A. Bhadra (✉)
High Energy & Cosmic Ray Research Centre, University of North Bengal, Siliguri, India
e-mail: abhadra@nbu.ac.in

S. Ghosh, A. R. Rao (eds.), *The Relativistic Universe: From Classical to Quantum*, Astrophysics and Space Science Proceedings 61,
https://doi.org/10.1007/978-3-031-90186-7_34

alternative approach to studying the origins of CRs involves detecting gamma rays and neutrinos from the potential astrophysical sources. Photons emitted abundantly, can be detected easily, and traceable to their origins. However, photons with energies surpassing 100 TeV (1 TeV = 10^{12} eV) are diminished by interstellar dust, gas, and cosmic background radiation, limiting their survival over extragalactic distances. Additionally, gamma ray observations often fit well with both electromagnetic and hadronic acceleration models, complicating the correlation between gamma rays and CRs. In contrast, neutrinos, which are also electrically neutral, interact minimally with matter and consequently, escape unaffected from the inner regions of the most energetic celestial objects, offering crucial insights into energy release processes. High-energy neutrinos, which are only produced in hadronic interaction (however, see [4]), directly point back to their production sites, establishing a clear connection between the source and CRs. Detecting cosmic neutrinos is, however, challenging due to their weak interaction with matter, necessitating large-scale particle detectors to accumulate sufficient numbers for statistical analysis. Both gamma ray and neutrino observations are pivotal as they collectively provide evidence to discern the origins of CRs.

In the present chapter, we will briefly review the current status of the problem concerning the origin of CRs in light of recent results from very high-energy gamma-ray and neutrino astronomy. We specifically focus on the origin of CR nuclei. In the next section, we will outline the standard (orthodox) paradigm for the origin of galactic cosmic rays. In Sect. 34.3, we will discuss the status of the model in light of recent results from high-energy gamma-ray and neutrino astronomy. Finally, we will conclude in Sect. 34.4.

34.2 Cosmic Rays and Its Origin

34.2.1 Basic Features of Cosmic Rays

CRs cover a vast energy range, from sub-GeV to at least 300 EeV, the highest energy CR event observed to date. The flux of CRs decreases sharply with increasing energy. Direct measurement of CRs using balloon flights or satellite experiments in Low Earth Orbit is limited to about 100 TeV due to the small size of detectors in space missions. For higher energy ranges, indirect studies of primary CRs are conducted through ground-based experiments using extensive air shower (EAS) arrays, which observe the cascade of particles induced by the primary CR interaction at the top of the atmosphere.

CRs are primarily composed of atomic nuclei. Direct measurements indicate that up to 100 TeV, the dominant component (around 90%) consists of protons (hydrogen nuclei), with helium nuclei making up about 10%, and smaller quantities of heavier elements. The abundance of protons and helium in CRs is slightly lower than their universal abundance, whereas elements like boron, beryllium, and lithium are found in significantly higher concentrations. Various EAS experiments have attempted to estimate the primary mass composition of CRs at higher energies, but the results are still ambiguous. The local energy density of CRs is approximately $1\,\mathrm{eV/cm^3}$, comparable to the energy densities of the galactic magnetic field and starlight, and close to that ($\sim 0.26\,\mathrm{eV/cm^3}$) of the Cosmic Microwave Background Radiation [5, 6]. Electrons and positrons contribute about one-hundredth of the hadronic CR energy density [6].

The energy spectrum of CRs follows a steeply declining power law with spectral index ~ -2.7 for energies beyond those influenced by the sun. However, a significant deviation occurs around 3 PeV (1 PeV $\equiv 10^{15}$ eV), where the spectrum sharply steepens, a feature known as the knee. Additionally, around 3 EeV, the spectrum flattens, a feature called the ankle, before reverting to its pre-knee slope. Recent direct observations suggest that even at GeV-TeV energies, the CR spectrum is not entirely smooth but contains a few small structures.

For a detail description of the features of CRs, one may see the monograph/review articles [2, 5–8].

34.2.2 Cosmic Ray Origin: The Supernova Paradigm

The observed isotropy of CRs is likely due to their confinement in the galaxy by the interstellar magnetic field. Consequently, their energy density in the galaxy is expected to be nearly the same as that measured in the vicinity of the Earth. Given the power required to maintain the observed CR energy density, it is widely accepted that CRs, at least up to the knee energy, are of galactic origin. If CRs with the same energy density were also spread over extra-galactic space, the power requirement for sources would be too high to meet. Moreover, in such a case, the gamma-ray background produced by the interaction of CRs with intergalactic gas would exceed the observed levels.

The origin of observed CRs above the ankle energy is likely to be extra-galactic. This is because the Larmor radius of these CRs, in the typical magnetic field of a few microgauss found in the interstellar medium, is larger than the radius of our galaxy, implying they cannot be confined and subsequently isotropized by the galactic magnetic field. Observations of astronomical objects with high energetics (active galactic nuclei (AGN), gamma-ray bursts, tidal disruption events, etc.), which are required for acceleration to ultra-high energies, also suggest that these CRs originate in other galaxies.

The conventional theory of CRs posits that SNRs are the major contributor of non-solar galactic CRs for energies up to around 100 PeV, what is generally referred to as "Supernova remnant paradigm for the origin of galactic Cosmic Rays". The main reasons in favor of SNR paradigm are the followings (for more details see [6, 9]):

1. Power requirement: The total energy (energy density × volume) of CRs in the galaxy is about 10^{55} erg. The residence time (t_R) of CRs in the galaxy can be estimated from the abundances of species of CRs that are entirely secondary in nature (boron, Be, Li for example), and that are mostly produced in spallation reactions involving a heavier primary CR species. In particular, the boron to carbon ration is quite useful to obtain t_R, which is estimated as about 6 million years. So to maintain a steady state the CR power to be injected by the galactic sources is $P_{CR} \sim 10^{41}$ erg/s.

   The typical mechanical energy released during a supernova explosion is $\sim 10^{51}$ erg. Assuming that the supernova explosion rate in the galaxy is three per century, the total energy injected in the interstellar medium per second is $\sim 10^{42}$ erg/s. For an energy conversion efficiency of about 10% (from supernova explosion energy (kinetic energy) to CR energy), the supernova can meet the necessary power for CRs.
2. Power law energy spectrum: The particle acceleration in SNRs is believed to occur through diffusive shock acceleration (DSA), which leads to power law spectra of accelerated particles, with spectral index close to −2.0.
3. The maximum energy of accelerated particles:Within the non-linear DSA framework with magnetic field amplification, the maximum energy of particles accelerated in SNRs may reach around Z × the knee energy.
4. The spectral features: The simple interpretation of the knee as a superposition of the cutoffs in the spectra of individual elements. Thus the knee is the proton knee in this scenario. An iron nucleus will be accelerated to around 10^{17} eV. So the spectrum of Galactic CRs may end at around 10^{17} eV. However, there may be some other class of source in the galaxy that accelerate CRs to the ankle energy.
5. The supports from gamma-ray astronomy: The gamma rays of most likely hadronic origin have been detected from molecular clouds close to several SNRs. The gamma-ray emission detected from a few SNRs also appears to be most likely of hadronic origin (see for instance [10]). However, till now all the evidences are circumstantial, no direct proof has been found yet that SNRs can accelerate CRs up to the knee energy.

Here it is worthwhile to mention that a normal star like the Sun also produces CRs mainly during flaring activities, but the maximum energy of the solar accelerated CRs is restricted to a few GeV. Moreover, the CR generating power of the Sun is about 10^{25} erg/s. There are roughly 10^{11} stars in our galaxy. So the total CR generating power of all normal stars in our galaxy is four to five order less that the powers of galactic CRs. The exact transition energy from galactic to extra-galactic is not yet settled. It may be at around 100 PeV or at the ankle energy. In the former case, a new class of galactic sources have to be present where CRs are accelerated at least up to the ankle energy.

The residence time of CRs in the galaxy suggests that on the average CRs travel about 1 Mpc distance, which exceeds the radius and the thickness of the galactic disk by far. This clearly implies that CRs propagate diffusively in the galaxy. The study of unstable secondaries in CRs, such as the radioactive isotope ^{10}Be indicate that the residence time is two to three times larger from what is estimated from analyzing the abundance of stable secondaries. Moreover, the radio observations suggest the presence of energetic electrons outside the disc. These findings are described by assuming that CRs are diffusively confined within a magnetized Galactic halo, whose volume is much larger than that of the disk.

Various analyses of CR and gamma-ray data suggest that the CR diffusion coefficient is a function of the particle rigidity R (and hence on particle energy) and can be expressed as $D = D_o \beta \, (R/R_o)^{\delta}$ where β is the particle velocity in units of the speed of light, $D_o \sim 10^{28}$–10^{29} cm^2/s, $\delta \sim 0.3$–0.6 for R_o equal to 1 GV.

34.3 Critical Evaluation of SNR Paradigm in the Light of the Recent Observations

The SNR paradigm of CR origin rests on a few sensitive issues which include the spectral index of accelerated particles, the efficiency of conversion of supernova explosion energy to cosmic rays, and the maximum energy that can be attained by a cosmic ray particle in SNR. A few other relevant issues are whether all SNRs accelerate and inject CRs to PeV energy range and whether the maximum attainable energy is (almost) the same for all the SNRs.

Gamma rays are produced in astrophysical objects through interactions involving relativistic particles, primarily protons and electrons. Relativistic electrons generate gamma rays through inverse Compton scattering with low-energy photons. Energetic protons, on the other hand, produce gamma rays by interacting with the surrounding medium, leading to the decay of secondary π^o mesons. A characteristic of gamma-ray emission produced by CRs is that its spectral index closely aligns with that of the CRs themselves [11]. Therefore, taking into account the diffusive propagation of CRs, one would expect the power-law spectral index of gamma-ray emissions induced by CRs at the source to range from -2.1 to -2.4 for primary energies below the knee and from -2.5 to -2.8 otherwise. We shall analyze the aforesaid issues on the basis of the latest gamma-ray observations.

Our understanding of the high-energy gamma-ray universe has significantly advanced over the past two decades due to remarkable improvements in both space-based and ground-based gamma-ray observatories. At high energies ($E > 0.1$ GeV), the Fermi Large Area Telescope has been surveying the entire sky since 2008, detecting 6658 Galactic and extragalactic gamma-ray sources in its first 12 years of observations [12]. For very high energies (VHE, $E > 0.1$ TeV), ground-based gamma-ray detectors are essential due to their large area requirements. The second generation of imaging atmospheric Cherenkov telescopes (IACTs) has so far detected more than 200 VHE gamma-ray sources. However, IACTs are generally not suitable for long-term comprehensive sky surveys because of their limited duty cycle.

Ground-based extensive air shower arrays, capable of distinguishing gamma ray-initiated cascades from those induced by hadrons, are ideal for gamma-ray sky surveys. One such new generation gamma-ray observatory is the Large High Altitude Air Shower Observatory (LHAASO), designed with a sensitivity of 12 milli-Crab units (CU), equivalent to the flux from the Crab Nebula. LHAASO is a dual-purpose facility designed for cosmic ray (CR) and gamma-ray studies at TeV and PeV energies [13]. It comprises three interconnected detectors: the Water Cherenkov Detector Array (WCDA), the Kilometer Square Array (KM2A), and the Wide Field-of-view Cherenkov Telescope Array (WFCTA), all situated at an altitude of 4410 meters above sea level in China. From 508 days of WCDA data and 933 days of KM2A data, LHAASO has detected 90 sources above 1 TeV and 43 sources above 100 TeV [14].

The estimated fluxes of TeV gamma rays from supernova remnants located within a few kiloparsecs are well above the detection sensitivities of modern gamma-ray telescopes [11]. A comprehensive classification of SNRs in gamma rays is challenging. However, certain groups of objects have been identified as SNRs based on their spectral morphology and multiwavelength behavior. To date, about forty SNRs have been firmly identified in GeV gamma-ray energies. In very high energy (VHE) and ultra-high energy (UHE, above 100 TeV) gamma-ray observations, the numbers of such sources are 20 and 3, respectively [15]. The three SNRs detected by LHAASO are W51C, γ-Cygni and G106.3+02.7 (1LHAASO J2228+6100u). SNRs that interact with surrounding molecular clouds provide direct evidence of SNR acceleration of cosmic ray (CR) nuclei. However, before LHAASO the gamma-ray spectrum measurements of these interacting SNRs have been limited to energies below approximately 10 TeV. W51C is one such interacting SNR, notable for its significant π^o-decay bump. W51C is estimated to be 5.5 kpc away and approximately 30,000 years old. It is widely accepted that W51C interacts with the molecular cloud (MC) linked to the nearby star-forming region, W51B. LHAASO has recently detected gamma rays of the order of 100 TeV from the molecular clouds of W51C.

In the Table 34.1 the spectral index of gamma ray emissions from the three LHAASO observed SNRs of the Milky Way are displayed. For the SNR W51, the gamma-ray emission from the nearby region J1922 + 1403 and J1924 + 1609 are given.

Table 34.1 Estimated spectral index of observed PeV gamma-ray emission from SNRs

SNRs	WCDA	KM2A
γ-Cygni J2019 + 407 G78.2 + 02.1	2.91 ± 0.06	3.56 ± 0.23
G106.3 + 02.7–	2.26 ± 0.04	2.95 ± 0.04
W51 J1922 + 1403	2.62 ± 0.07	3.79 ± 0.20
W51 J1924 + 1609	2.54 ± 0.08	3.61 ± 0.22

It has been observed that the spectral indices of different SNRs are neither identical nor similar, even at lower energies. The gamma-ray observations of several SNRs by Fermi-LAT and LHAASO, which reveal steep spectra, do not align with the DSA theory of CRs in SNRs. This is particularly important given that no viable alternative to DSA is available so far.

The maximum energy that can be attained by a cosmic ray particle in an ordinary SNR when the remnant is passing through a medium of density N_H cm^{-3} is [16, 17]

$$E_{max} \simeq 4 \times 10^5 Z \left(\frac{E_{SN}}{10^{51}\ erg}\right)^{1/2} \left(\frac{M_{ej}}{10 M_\odot}\right)^{-1/6} \left(\frac{N_H}{3 \times 10^{-3}\ cm^{-3}}\right)^{-1/3} \left(\frac{B_o}{3\mu G}\right) GeV\ , \tag{34.1}$$

where E_{SN} and M_{ej} are the kinetic energy and mass of the ejecta respectively, and B_o is the magnetic field of the SNR. The maximum energy achievable for proton primaries falls short of the knee of the cosmic ray energy spectrum by more than an order of magnitude. However, this issue is somewhat mitigated by the potential amplification of the effective magnetic field strength at the shock, which can be enhanced through the growth of magnetic waves induced by accelerated cosmic rays [16, 18, 19]. With this amplified field, SNR might be able to reach the knee of the spectrum for protons, and potentially reach the second knee for iron nuclei.

The analysis of GeV to TeV gamma-ray emission from several SNRs, as observed by the Fermi-LAT, VERITAS, and MAGIC observatories, suggests that the maximum attainable energy in these SNRs is significantly lower than the knee energy [20]. The LHAASO observatory reported the detection of very high-energy gamma-ray emission from two SNRs, Cassiopeia A and IC 443, but only through the WCDA detector. The non-detection of gamma-ray flux from Cassiopeia A and IC 443 by the KM2A detector places a stringent upper limit on the flux above 25 TeV. Recently, Banik [21] found that the observed gamma-ray spectra from these two SNRs can be consistently explained when the maximum energy of CR protons in these SNRs is considered to be around 100 TeV. This indicates that the maximum attainable energy in all SNRs is not a few PeV, as assumed in the SNR paradigm model of the origin of CRs.

Recently, by analyzing the Fermi Large Area Telescope's multi-wavelength data of SN 2023ixf, a Type II supernova (SN) in the nearby galaxy M 101, Martí–Devesa et al found [22] that under reasonable assumptions the maximum efficiency on the CR acceleration may be as low as 1%, which is inconsistent with the common estimate of 10% in typical supernovae. Another important point is that if SNRs are indeed the sources of cosmic rays, they should accelerate not only protons but also heavier nuclei in the correct proportions. Moreover, the maximum energy of heavier nuclei should be roughly Z times that of protons, where Z is the atomic number. To account for recent gamma-ray observations, the energy conversion efficiency for a mixed cosmic ray composition must be nearly twice as high as that for a composition composed solely of protons [20]. Although there is no definitive upper limit for conversion efficiency, achieving a high level of it is challenging.

34.4 Other Potential Galactic Sources

Recently, the HAWC observatory detected, for the first time, TeV gamma-rays from the solar disk during the Sun's quiescent phase [23]. This discovery extends the range of intense, hard GeV emission detected by Fermi-LAT from $0.1 - 200$ GeV. The TeV gamma-rays observed by HAWC from the solar disk have a softer spectrum (slope ≈ -3.62) compared to the Fermi-detected gamma-ray spectrum up to 200 GeV (slope ≈ -2.2), suggesting a break in the energy spectrum around 400 GeV. The gamma-ray flux observed by both Fermi-LAT and HAWC is significantly higher than what would be expected from hadronic interactions between galactic cosmic rays and the solar atmosphere. Additionally, the spectral slopes of the gamma-ray energy spectra observed by Fermi and HAWC differ markedly from those of galactic cosmic rays, raising doubts about the current galactic cosmic ray model for solar disk gamma rays. The observed gamma rays from the solar disk may be explained by hadronic interactions between solar cosmic rays and solar matter above the photosphere, provided the Sun accelerates particles to a few TeV energies with an appropriate flux level during its quiet state [24].

If a normal star like the Sun can accelerate CRs up to TeV energies, the maximum attainable energy of CRs in some brighter class of stars or in star cluster may reach the knee energy for protons. LHAASO detected total 90 sources above 1 TeV from 508 days data collected by the WCDA and 933 days data collected by the KM2A. Out of these sources, 54 sources were detected by both the WCDA and the KM2A. In the Fig. 34.1 we plot the spectral indices of these 54 sources as estimated from the KM2A observations.

The minimum energy of primary CR that produces a 100 TeV gamma-ray will be within the knee energy and hence the source power law spectral index of gamma-ray emission produced from primary CRs should be in the range -2.2 to -2.4 considering the diffusive propagation or at least about -2.7, the spectral index of the primary cosmic rays if diffusion is energy independent. It is found that only for a couple of detected source such a condition is fulfilled. Moreover, the spectral index obtained from the WCDA observations does not consistently match that obtained from the KM2A observations for the sources those meet the above-stated condition. Besides, such sources are not observed by other gamma-ray observatories at lower energies.

In 2021, LHAASO reported the detection of 12 PeV sources [25]. Pulsar wind nebulae (PWNe) have been found in the proximity most these detected PeV sources. The PWNe have long been considered as potential sources of CRs [26]. From an energy perspective, pulsars are an even more promising source, as their rotational energy at birth can exceed the energy released in a typical supernova explosion by more than tenfold. However, the spectral index, mass composition, and other

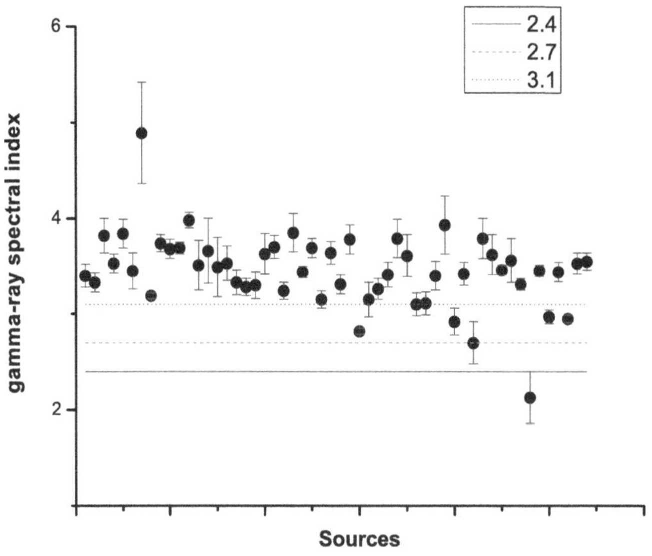

Fig. 34.1 The spectral index of the KM2A observed sources

relevant characteristics are still underdeveloped to accurately describe cosmic ray observations [27, 28].

The IceCube neutrino observatory, located at the South Pole, is a particle detector embedded in Antarctic ice, spanning a cubic kilometer. Positioned about 2500 meters below the surface, it is the first-ever gigaton-scale neutrino detector. Its primary purpose is to detect neutrinos originating from the most extreme astrophysical phenomena in the universe. The IceCube Collaboration has investigated neutrino emissions within the Milky Way and recently discovered evidence of additional neutrinos being emitted along the plane of the Galaxy. However, IceCube has not yet detected any point sources within the Milky Way [29].

34.5 CRs in the EeV Energy Range

The CR energy spectrum exhibits a hardening at a particle energy equal to ~ 3 EeV, which is the 'ankle' of the spectrum. This hardening is thought to arise from the appearance of the extragalactic component of CRs which was sub-dominant at lower energies. However, in the SNR paradigm, the contribution to the total CR spectrum from SNRs should extend up to about 100 PeV, which is roughly 30 times lower than the energy of the ankle. The proposals that the transition from Galactic to extragalactic CRs occurs at 100 PeV or the existence of a third CR population that would contribute between 100 PeV to 3 EeV require an unnatural fine tuning of CR flux at 100 PeV (see, for instance [30]).

The galactic magnetic field cannot confine or isotropize cosmic rays (CRs) at EeV energies due to their larger rigidity. Therefore, CRs at such energies are expected to reveal distinct sources in the sky and provide more direct information about their nature. The Pierre Auger Observatory, a hybrid detector array consisting of water Cherenkov and fluorescence detectors that covers a 3000 km^2 area in western Argentina, initially reported an excess correlation between the arrival directions of CRs above 55 EeV and the positions of nearby AGNs on a small angular scale of 3 degrees [31]. However, as more data became available, this correlation faded. Observations from both the Auger Observatory and the Telescope Array have not revealed any clear anisotropy, particularly on small angular scales, contrary to expectations that would provide direct information about the sources of CRs at the highest energies or at least isolate the main sources across the sky [32].

IceCube has conducted multiple searches for point-like sources of neutrinos. From 10 years of data collected between 2008 and 2018, the IceCube observatory detected four PeV neutrino sources—NGC 1068, TXS 0506+056, PKS 1424+240, and GB6 J1542+6129—in a comprehensive search of the Northern sky [29]. However, none of these are steady sources. Various theoretical models have been proposed in the literature to explain the IceCube point source observations in conjunction with electromagnetic observations (see, for instance, [33–37] and references therein), but the situation remains unclear due to lack of universality.

34.6 Conclusion

The SNR paradigm is considered the standard model for the origin of galactic cosmic rays (CRs). The model is theoretically well-developed and understood, with several appealing features. However, recent observations by modern gamma-ray and neutrino experiments do not align with the model. Below, we summarize the major findings from recent gamma-ray observations on this issue:

1. It is now almost certain that not all SNRs are PeVatrons; the maximum energy of CRs in several SNRs is much lower than the knee energy.
2. The spectral indices of gamma-ray emissions from different SNRs are not consistent (or even close), contrary to expectations.
3. Fermi-LAT and LHAASO gamma-ray observations of several SNRs with steep spectra do not support the DSA of CRs in SNRs.
4. LHAASO has not detected most of the SNRs that were observed at lower gamma-ray energies by other gamma-ray missions.

At the highest energies, the situation regarding the origin of cosmic rays (CRs) is even more complex. CRs at EeV energies are expected to allow direct observation of individual sources in the sky, but the observations do not reveal any clear anisotropy on small angular scales. Additionally, there are challenges in determining the transition energy from galactic to extragalactic dominance.

In conclusion, despite significant efforts on both theoretical and experimental fronts, the origin of CRs remains unclear. The results from the Cherenkov Telescope Array (CTA), the deep-core low-energy extension of the IceCube experiment, and results based on larger statistics from LHAASO may help us better understand the cosmic ray origin problem.

Acknowledgments The author thanks an anonymous reviewer for useful comments.

References

1. The Pierre Auger Collaboration. (2017). *Science, 357* 1266.
2. Gabici, S., Evoli, C., Gaggero, D., et al. (2019). *International Journal of Modern Physics D, 28*, 1930022.
3. Hillas, A. M. (2005). *Journal of Physics G Nuclear Physics, 31* R95.
4. Bhadra, A., & Banik, P. (2018). arXiv:1801.07677v1.
5. Wolfendale, A. W. (1979). *Pramana, 12*, 631.
6. Berezinsky, V. S., Bulanov, S. V., Dogiel, V. A., & Ptuskin, V. S. (1990). V. L. Ginzburg (Ed.), *Astrophysics of cosmic rays*. North-Holland.
7. Ginzburg, V. L., & Ptuskin, V. S. (1976). *Reviews of Modern Physics, 48* 161.
8. Hillas, A. M. (1984). *Annual Review of Astronomy and Astrophysics, 22* 425.
9. Blasi, P. (2013). *The Astronomy and Astrophysics Review, 21*, 70.
10. Aharonian, F. A. (2013). *Astroparticle Physics, 43* 71.
11. O'C Drury, L., Aharonian, F. A., & Völk, H. J. (1994). *Astronomy & Astrophysics, 287*, 959.
12. Abdollahi, S., et al. (2022). *The Astrophysical Journal Supplement Series, 260* 53.
13. Ma, X.-H., Bi, Y.-J., Cao, Z., et al. (2022). *Congenital Heart Public Health Consortium, 46*, 030001.
14. Cao, Z., Aharonian, F., An, Q., et al. (2024). *The Astrophysical Journal Supplement Series, 271*, 25.
15. Giuliani, A., & Cardillo, M. (2024). *Universe, 10* 203.
16. Lagage, P. O., & Cesarsky, C. J. (1983). *Astronomy & Astrophysics, 118*, 223.
17. Berezhko, E. G., & Ksenofontov, L. T. (1999). *Journal of Experimental and Theoretical Physics, 89*, 391.
18. Bell, A. R. (1978). *Monthly Notices of the Royal Astronomical Society, 182*, 147; ibid *182*, 443 (1978).
19. Vink, J. (2012). *The Astronomy and Astrophysics Review, 20*, 49.
20. Banik, P., & Bhadra, A. (2017). *Physical Review D, 95*, 123014.
21. Banik, P. (2024). *Journal of High Energy Astrophysics, 43*, 61.
22. Martí-Devesa, G., et al. (2024). *Astronomy & Astrophysics, 686*, A254.
23. Albert, A., et al. (2023). *Physical Review Letters, 131*, 051201.
24. Banik, P., Bhadra, A., & Ghosh, S. (2023). arXiv:2305.17086v2.
25. Cao, Z., Aharonian, F., An, Q., et al. (2021). *Nature, 594*, 34.
26. Goldreich, P., & Julian, W.J. (1969). *The Astrophysical Journal, 157*, 869.
27. Bhadra, A. (2006). *Astroparticle Physics, 25*, 226.
28. Bhadra, A., & Dey, R. K. (2009). *Monthly Notices of the Royal Astronomical Society, 395*, 1371.
29. Aartsen, M. G., Ackermann, M., Adams, J., et al. (2020). *Physical Review Letters, 124*, 051103.
30. Parizot, E. (2014). *Nuclear Physics B - Proceedings Supplements, 256–257*, 197.
31. Pierre Auger Collaboration. (2007). *Science, 318*, 938.
32. di Matteo, A., Bister, T., Biteau, J., et al. (2019). *PoS*, 439.

33. Ansoldi, S., et al. (2018). *The Astrophysical Journal*, *863*, L10.
34. Liu, R., Wang, K., Xue, R., et al. (2019). *Physical Review D*, *99*, 063008.
35. Banik, P., & Bhadra, A. (2019). *Physical Review D*, *99*, 103006.
36. Banik, P., Bhadra, A., Pandey, M., & Majumdar, D. (2020). *Physical Review D*, *101*, 063024.
37. Banik, P., & Bhadra, A. (2022). *European Physical Journal C*, *82*, 517.

Chapter 35
A Glitch in Gravity: Cosmic Lorentz-Violation from Fiery Big Bang to Glacial Heat Death

Robin Y. Wen, Lukas T. Hergt, Niayesh Afshordi, and Douglas Scott

Abstract One regime where we might see departures from general relativity is at the largest accessible scales, with a natural choice in cosmology being the cosmological horizon (or Hubble) scale. We investigate a single-parameter extension to the standard cosmological model with a different strength of gravity above and below this scale—a "cosmic glitch" in gravity. Cosmic microwave background observations favor weaker superhorizon gravity, easing both the Hubble and clustering tensions with other cosmological data. This compounds evidence for an even stronger glitch during Big Bang nucleosynthesis (from helium abundance observations), suggesting that symmetries of general relativity are maximally violated at the Big Bang, but gradually recovered as we approach the present-day cosmological de Sitter scale, associated with the observed dark energy.

35.1 Introduction

During the last century, the general theory of relativity (GR) has become a funda mental pillar of modern physics, successfully passing every empirical examination

R. Y. Wen
California Institute of Technology, Pasadena, CA, USA
e-mail: ywen@caltech.edu

L. T. Hergt
Department of Physics & Astronomy, University of British Columbia, Vancouver, BC, Canada
e-mail: lthergt@phas.ubc.ca

N. Afshordi (✉)
Waterloo Centre for Astrophysics, University of Waterloo, Waterloo, ON, Canada
e-mail: nafshordi@pitp.ca

Department of Physics & Astronomy, University of Waterloo, Waterloo, ON, Canada
Perimeter Institute For Theoretical Physics, Waterloo, ON, Canada

D. Scott
Department of Physics & Astronomy, University of British Columbia, Vancouver, BC, Canada

S. Ghosh, A. R. Rao (eds.), *The Relativistic Universe: From Classical to Quantum*, Astrophysics and Space Science Proceedings 61,
https://doi.org/10.1007/978-3-031-90186-7_35

across a wide range of scales and regimes [1]. However, several reasons, most importantly the inconsistency of the principles of general relativity with those of quantum theory, have motivated extensive studies on deviations or modifications beyond GR. Additionally, the presence of a preferred cosmological reference frame, in which the cosmic microwave background (CMB) anisotropies possess no dipole, could point towards alternative gravity theories that can allow a genuinely unique cosmological frame. The existence of such a special frame could imply that minimal Lorentz-violating deviations from GR would only appear on cosmological scales [2, 3].

Instead of requiring a new fundamental length scale, an alternative implementation of a minimal Lorentz violation at cosmological distances is to introduce a "glitch" between Newton's constant of gravitation G_{N} that determines the sub-horizon dynamics and the gravitational constant G_{cosmo} that affects cosmology on super-horizon scales. More specifically, we can write the Friedmann equation as

$$H^2 = \frac{8\pi G_{\mathrm{cosmo}}}{3}\rho_{\mathrm{tot}} = \frac{8\pi G_{\mathrm{N}}}{3}\left[\rho_{\mathrm{tot}} + \left(\frac{G_{\mathrm{cosmo}}}{G_{\mathrm{N}}} - 1\right)\rho_{\mathrm{tot}}\right], \tag{35.1}$$

where G_{cosmo} connects the total energy density ρ_{tot} in the Universe to the Hubble expansion rate H. Such a cosmic glitch can be realized through alternative gravity theories such as cuscuton and Hořava–Lifshitz proposals for Lorentz-violating gravity [4, 5] or the Einstein-Aether theory [6], with the ratio $G_{\mathrm{N}}/G_{\mathrm{cosmo}}$ depending on the theories' respective parameters ($G_{\mathrm{N}}/G_{\mathrm{cosmo}} = 1$ in GR without a "glitch"). It can be rigorously shown that all of these theories in appropriate limits are indistinguishable from general relativity in asymptotically flat spacetimes [7], so they can only be examined on scales that are close to or larger than the Hubble scale [8].

35.2 The Glitch Model

Phenomenologically, this cosmic glitch in gravity (hereafter referred to as CGG) can be reformulated as an additional dark energy component on top of the usual cosmological constant Λ and cold-dark-matter fluids within the concordance model of cosmology (commonly known as ΛCDM). Analogously to how the cosmological constant Λ is typically reinterpreted as a vacuum energy density $\rho_\Lambda = \Lambda/8\pi G_{\mathrm{N}}$ in the ΛCDM model under GR, we assume that the second term in the brackets of Eq. (35.1) corresponds to some glitch energy density ρ_{g} that contributes to the overall critical density $\rho_{\mathrm{crit}} \equiv 3H^2/8\pi G_{\mathrm{N}}$ of GR. From Eq. (35.1), the total density can be expressed in terms of the critical density,

$$\rho_{\mathrm{tot}} = \frac{G_{\mathrm{N}}}{G_{\mathrm{cosmo}}}\rho_{\mathrm{crit}}\,. \tag{35.2}$$

Substituting the above Eq. (35.2) into the second term of Eq. (35.1), we obtain

$$H^2 = \frac{8\pi G_N}{3}\left[\rho_{tot} + \left(1 - \frac{G_N}{G_{cosmo}}\right)\rho_{crit}\right]. \tag{35.3}$$

We can reinterpret the CGG (the second term of the above Eq. (35.3)) as an additional dark energy component with a constant parameter proportional to the critical density:

$$\Omega_g \equiv \frac{\rho_g}{\rho_{crit}} = \frac{\rho_{DE} - \rho_\Lambda}{\rho_{crit}} = 1 - \frac{G_N}{G_{cosmo}}. \tag{35.4}$$

The total *effective* dark energy component now has the energy density constituted by the combination of the constant dark energy density (the cosmological constant) and the energy density of the glitch:

$$\rho_{DE} = \rho_\Lambda + \Omega_g \rho_{crit}. \tag{35.5}$$

Given that ρ_{crit} also contains ρ_{DE}, we can rearrange Eq. (35.5) to isolate ρ_{DE}:

$$\rho_{DE} = \frac{\Omega_g \rho_{nonDE} + \rho_\Lambda}{1 - \Omega_g}, \tag{35.6}$$

where ρ_{nonDE} includes the densities of all the components other than the dark energy. As ρ_Λ becomes dominant over ρ_{nonDE} in the late Universe, the behavior of this modified dark energy model closely resembles that of ΛCDM during late times. However, the dynamics of ρ_{DE} undergo considerable changes in the early Universe due to its proportionality with the critical density; the modified dark energy component tracks the behavior of radiation and matter densities, respectively, when the radiation and matter dominate over the total energy budget of the Universe.

Using Eq. (35.6), we can compute the equation of state (EoS) parameter w_{DE} (defined as the ratio between pressure and energy density $w \equiv p/\rho$) for the effective dark energy component and confirm that the CGG model tracks the dominant component contributing to the Universe's density. As plotted in the left panel of Fig. 35.1, w_{DE} evolves from $\frac{1}{3}$ (EoS for radiation) to 0 (EoS for matter) and then to around -1 (EoS for the constant dark energy Λ). The EoS parameter w_{DE} changes smoothly when Ω_g is positive. However, for negative Ω_g values, w_{DE} becomes divergent at the positive-to-negative transition of ρ_{DE}. A negative energy density ρ_{DE} typically leads to ghost instabilities in dynamical dark energy models [9, 10], but here we have formulated the model by taking a negative glitch value Ω_g to be a phenomenological approximation to modified gravity theories. This means that such Lorentz-violating theories do not have new dynamical degrees of freedom that could suffer from ghost instabilities [2, 3]. We also note that the divergence of ω_{DE} in the case of negative Ω_g does not lead to any divergent behavior in the deceleration parameter ($q \equiv -\ddot{a}a/\dot{a}^2 = \frac{1}{2}\sum_i \Omega_i(1 + 3\omega_i)$, where the sum

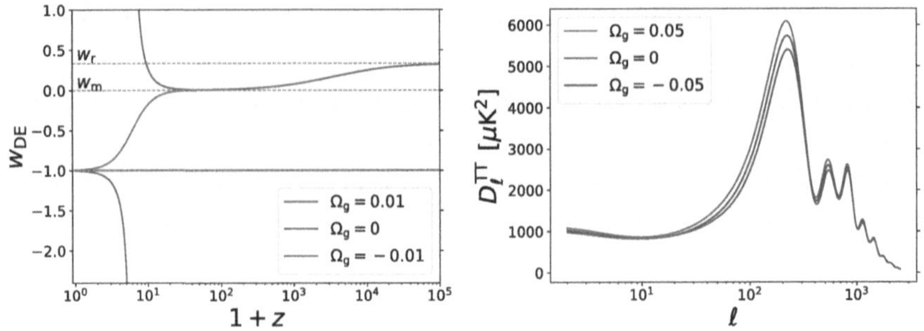

Fig. 35.1 The cosmic glitch in gravity (CGG) model. The left panel shows the EoS parameter for the dark energy component $w_{\mathrm{DE}}(z)$ for different values of the glitch parameter Ω_{g} (with all other cosmological parameters fixed at the best-fit values for the Planck18 data under the ΛCDM model). The blue dashed lines plot the EoS parameters for radiation ($w_{\mathrm{r}} = \frac{1}{3}$) and matter ($w_{\mathrm{m}} = 0$). The right panel shows the CMB temperature power spectrum $D_\ell^{TT} \equiv (\ell(\ell+1)/2\pi)C_\ell^{TT}$ for different glitch parameters

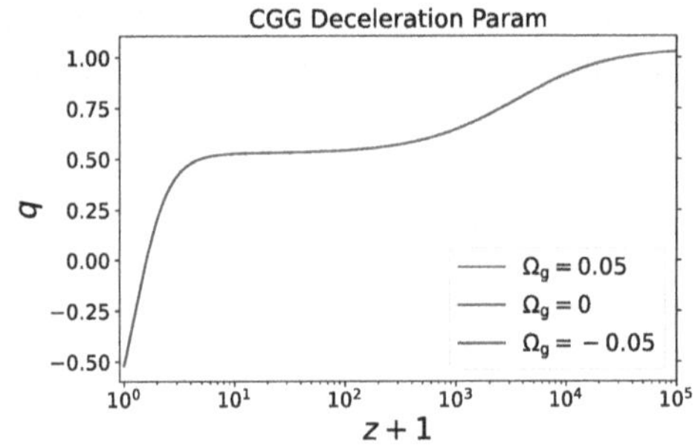

Fig. 35.2 The deceleration parameter q of the CGG model for different Ω_{g} values

occurs over all the cosmological components), as shown in Fig. 35.2. In fact, the deceleration parameter remains the same across time, despite varying Ω_{g} values, indicating that the CGG model does not change the background dynamics at or beyond second order.

We can confront our CGG model with cosmological data to see how well it fits. To compute the CMB anisotropy power spectra under the CGG model, we modify the cosmological Boltzmann code CAMB [11]. We treat the CGG component as a perfect fluid at the linear perturbation level. For negative Ω_{g}, the model crosses the so-called "phantom divide", with $w < -1$ at late times. To ensure gravitational stability at such a crossing, a double-field description,[1] known as the parameterized post-Friedmann (PPF) framework, has been used to solve the DE perturbation equations [12, 13], allowing for a generic dark energy sector with an arbitrary EoS parameter $w(z)$ as a function of redshift. We use the PPF framework for numerical implementation when Ω_{g} takes negative values, and we assume the sound speed

[1] A single scalar field with minimal coupling will be insufficient for stability.

of the effective dark energy fluid in the rest frame to be approximately the speed of light, that is $c_s^2 = 1$ [13]. The right panel of Fig. 35.1 illustrates the effects of Ω_g on the CMB temperature power spectrum D_ℓ^{TT}: a positive Ω_g enhances the integrated Sachs-Wolfe (ISW) effect at large scales (corresponding to low angular multipoles ℓ) and slightly suppresses the small-scale power in D_ℓ^{TT}, while a negative Ω_g exhibits the exactly opposite behavior.

35.3 Observational Constraints

In the absence of a theoretically favored value for Ω_g, it is best to constrain its value using the existing CMB and large-scale structure (LSS) data. Our CGG model is a single-parameter extension to the 6-parameter ΛCDM model, and we can constrain Ω_g using the *Planck* 2018 TT, TE, EE+low$E$+lensing likelihoods (hereafter abbreviated as Planck18) [14, 15].[2] In Table 35.1, we summarize the constraints on the cosmological parameters for the ΛCDM and CGG models.[3] We find that the *Planck* 2018 data prefer Ω_g to stay in the negative region, with $\Omega_g = -0.0087 \pm 0.0046$. The ΛCDM model is almost $2\,\sigma$ away from the mean Ω_g value measured under CGG. Using Eq. (35.4), we can recast the constraint on Ω_g as a constraint on the ratio between the strengths of super-horizon and sub-horizon gravity, $G_{\rm cosmo}/G_{\rm N} = 0.9914 \pm 0.0045$, indicating that the *Planck* CMB data favor superhorizon gravity that is about a percent weaker than in GR, at the roughly $2\,\sigma$ significance level.

Table 35.1 Mean and $1\,\sigma$ uncertainties of the parameters in the ΛCDM and CGG models using the *Planck* 2018 data. The first seven parameters are the main parameters of the models used in the nested sampling, while the last three are derived according to the seven main parameters

Parameter	Planck18 [ΛCDM]	Planck18 [CGG]
$\Omega_b h^2$	0.02237 ± 0.00014	0.02248 ± 0.00016
$\Omega_c h^2$	0.1200 ± 0.0012	0.1168 ± 0.0020
H_0	67.36 ± 0.54	68.58 ± 0.86
$\tau_{\rm reio}$	$0.0542^{+0.0071}_{-0.0082}$	$0.0499^{+0.0079}_{-0.0071}$
$\ln(10^{10} A_s)$	3.044 ± 0.015	3.033 ± 0.016
n_s	0.9649 ± 0.0042	0.9690 ± 0.0046
Ω_g	0	-0.0087 ± 0.0046
Ω_m	0.3155 ± 0.0074	0.298 ± 0.012
σ_8	0.8112 ± 0.0061	0.831 ± 0.012
$S_8 \equiv \sigma_8 \sqrt{\Omega_m/0.3}$	0.832 ± 0.013	0.828 ± 0.012

[2] Our results were computed using the nested sampler `PolyChord` [16, 17] interfaced with `Cobaya` [18] and a modified version of `CAMB` [11].

[3] All parameter uncertainties given in this chapter are expressed as $\pm 1\,\sigma$.

This preference for a negative glitch parameter remains at a similar significance level when we consider different constant values for the sound speed (instead of the default $c_s^2 = 1$) of the effective dark energy components in the PPF framework. When $c_s^2 = 0.1$, a sound speed substantially smaller the speed of light, we obtain $\Omega_g = -0.0098^{+0.0055}_{-0.0050}$ under the Planck18 data, while increasing the sound speed to $c_s^2 = 10$ yields $\Omega_g = -0.0063 \pm 0.0034$. We see that increasing (decreasing) the DE sound speed increases (decreases) the mean constraint on the glitch parameter. Since the $1\,\sigma$ error of the constraint decreases (increases) with a similar percentage to the mean value, the preference for the negative glitch is kept stable at around the $2\,\sigma$ level, regardless of the c_s^2 values.

35.3.1 A Logarithmic Running of the Glitch

The preference of the *Planck* CMB data for a negative Ω_g is interestingly echoed by recent measurements on Big Bang nucleosynthesis (BBN). The EMPRESS collaboration finds an apparent discrepancy between the standard model of BBN and the measurement of ^{4}He abundance in 10 extremely metal-poor galaxies [19]. The BBN predictions can be reconciled with the EMPRESS data if $\Omega_g = -0.085 \pm 0.027$ during nucleosynthesis [20]. However, this value needed to explain the abundance measurements of ^{4}He is much lower than the constraints obtained by fitting the Planck18 CMB data. Since these different Ω_g values influence cosmic dynamics at vastly different epochs, it raises the possibility of a logarithmic running of the cosmic glitch with scale, analogous to how other dimensionless constants behave in renormalizable theories. In Fig. 35.3, we illustrate the constraints on the glitch parameter Ω_g from BBN and CMB data, along with the potential linear extrapolations based on renormalization group flow. The BBN and CMB constraints suggest a possible scenario where the glitch almost vanishes and we recover near-exact de Sitter symmetry on the scale of the observed cosmological constant (or de Sitter radius) today. In stark contrast, at the Big Bang, where the curvature approaches the Planck scale, the glitch parameter becomes $\mathcal{O}(1)$, signaling a substantial violation of Lorentz symmetry potentially due to the quantum nature of gravity. Such a scenario may further hint at a genuine quantum gravity solution to the cosmological horizon problem, which is traditionally tackled under an inflationary paradigm [21].

35.3.2 Cosmic Tensions

In addition to the discrepancy in the abundance of ^{4}He, the increased precision of cosmological measurements has uncovered several other discrepancies among different cosmological probes [22], with the most notable being the so-called

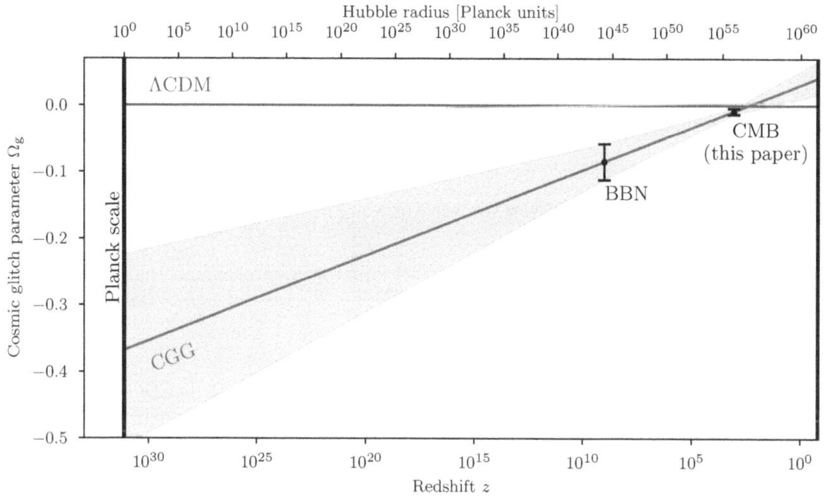

Fig. 35.3 Current measurements of a time-varying cosmic glitch parameter Ω_g, as a function of Hubble radius at the measurement time. The lines and gray region show the linear extrapolation ($\pm 1\,\sigma$) out to Planck and de Sitter scales

"Hubble tension". This tension refers to the difference between the measurements of the present-day cosmic expansion rate H_0 based on the distance ladder in the late Universe and those obtained from CMB anisotropies analyzed through the standard cosmological model. Specifically, the discrepancy between the value inferred from the *Planck* data [14, 15, 23] under the ΛCDM model and the value from the SH0ES collaboration using Cepheid-calibrated Type Ia supernovae [24–26] has now surpassed the significance level of $4\,\sigma$. By exploiting the degeneracy between Ω_g and H_0, where a negative Ω_g allows for a lower matter density and a higher H_0, our CGG model can somewhat alleviate the Hubble tension. We determine H_0 to be 68.58 ± 0.86 km/s/Mpc for our CGG model under Planck18, which is higher than the ΛCDM value and has a larger uncertainty, thus mildly decreasing the Hubble tension from $4.1\,\sigma$ to $3.0\,\sigma$.

Adding baryonic acoustic oscillation (BAO) measurements [27–29] from the previous Stage-III galaxy surveys to the Planck18 data leads to $H_0 = 68.11 \pm 0.46$ km/s/Mpc and $\Omega_g = -0.0063 \pm 0.0031$. The Hubble parameter measured in this case is closer to the ΛCDM result than the CGG fit under the Planck18 data only, despite the addition of the BAO data slightly strengthening the evidence for the glitch. Substituting the default Planck18 likelihoods with a newer analysis of the Planck Public Release 4 (PR4) data [23, 30] and replacing the Stage-III BAO measurements with the newer BAO results coming from the recent

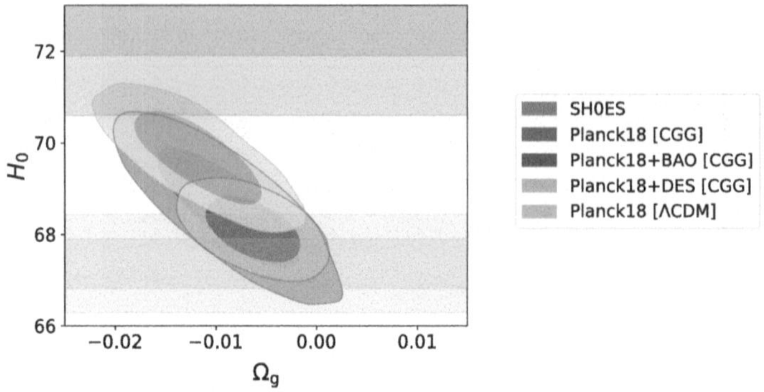

Fig. 35.4 Constraints on the Hubble parameter H_0 and glitch parameter Ω_g in CGG (ΛCDM+Ω_g) under different scenarios. The magenta band gives the distance-ladder measurement of H_0 from SH0ES [24], while the blue band shows the measurement of H_0 from Planck18 data under the ΛCDM model (i.e. setting $\Omega_g = 0$)

DESI Y1 analysis [31, 32],[4] the glitch and Hubble parameters slightly tighten to $\Omega_g = -0.0067 \pm 0.0029$ and $H_0 = 68.49 \pm 0.41$ km/s/Mpc, respectively. The combined CMB and BAO constraints are comparable across these two different combinations. Including the 3×2-point analysis (combined galaxy clustering and weak gravitational lensing measurements based on photometric surveys) from the Dark Energy Survey (DES) [33] along with the Planck18 data yields $H_0 = 69.69 \pm 0.66$ km/s/Mpc and $\Omega_g = -0.0118 \pm 0.0042$. This further raises the fitted H_0 values to within $2.4\,\sigma$ of the local SH0ES measurement. These constraints on the Hubble and glitch parameters are summarized in Fig. 35.4 and Table 35.2, respectively.

Another tension related to the matter clustering in the Universe is found between galaxy clustering or weak lensing surveys and CMB measurements [22, 34]. This clustering tension is typically characterized through the matter density parameter Ω_m and the matter power spectrum amplitude parameter σ_8, or their combination $S_8 \equiv \sigma_8(\Omega_m/0.3)^{0.5}$. We plot the constraints on Ω_m and S_8 under Planck18 and DES data in Fig. 35.5. Allowing Ω_g to be negative in the CGG model lowers the constraints for Ω_m under the Planck18 data. Consequently, the $S_8 - \Omega_m$ confidence regions of DES and Planck18 exhibit a greater area of overlap under the CGG model in comparison to ΛCDM, which is illustrated in Fig. 35.5. The clustering tension is therefore reduced by the CGG model when we view the tension in the 2D plane of S_8 and Ω_m parameters. The clustering tension is similarly reduced in the σ_8–Ω_m plane Fig. 35.5.

[4] Note that the DESI Y1 BAO measurements have comparable (if not slightly better) constraining power to the previous stage-III galaxy surveys, and they will significant exceed the statistical power of previous results in the upcoming data releases.

Table 35.2 Mean and 1 σ uncertainties of the cosmic glitch parameter Ω_g under different combinations of data and likelihoods. For the last row, the "All" case includes the combination Planck PR4, DESI Y1 BAO, Pantheon+ SNe (supernovae), DES Y1, and SH0ES likelihoods

Likelihoods	Ω_g Constraints
Planck18 [PR3]	-0.0087 ± 0.0046
Planck18+Pantheon SNe	-0.0083 ± 0.0041
Planck18+DES Y1	-0.0118 ± 0.0042
Planck18+Stage-III BAO	-0.0063 ± 0.0031
Planck PR4	-0.0054 ± 0.0042
Planck PR4+DESI Y1 BAO	-0.0067 ± 0.0029
All	-0.0059 ± 0.0027

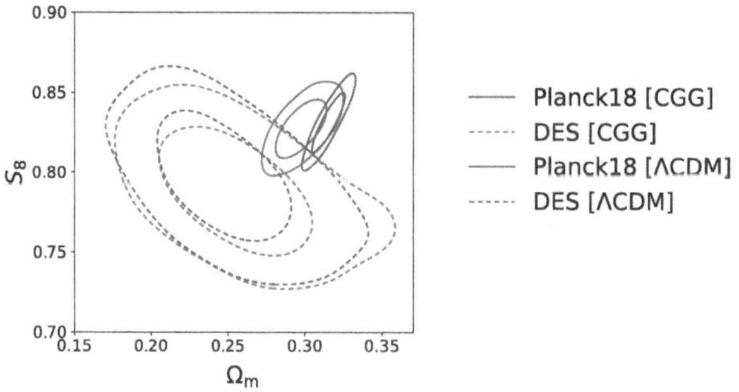

Fig. 35.5 Constraints on S_8 and Ω_m in the ΛCDM and CGG models, using Planck18 only and DES data only

35.4 Conclusions

In summary, we have studied a *cosmic glitch in gravity*, i.e. a model characterized by gravity behaving differently on super-horizon and sub-horizon scales, from both theoretical and observational perspectives. We find that the *Planck* data and a combination of various other data sets prefer a negative cosmic glitch parameter Ω_g. The significance of the evidence for this glitch varies between $1.3\,\sigma$ and $2.8\,\sigma$, depending on the additional LSS data included in the Bayesian analysis. The currently tightest constraint on the glitch is $\Omega_g = -0.0059 \pm 0.0027$, which is obtained from combining all of the CMB (Planck PR4 [23]), BAO (DESI Y1 [31]), Supernovae (Pantheon+ [35]), galaxy clustering and weak lensing (DES Y1 [33]) data, along with the distance ladder (SH0ES [24]) measurements. These constraints indicate that the current cosmological data mildly favor superhorizon gravity to be slightly weaker than the case in GR. The glitch model alleviates both the Hubble and the clustering tensions when considering the *Planck* 2018 data, while the H_0 measurement obtained under Planck18 combined with DES Y1 data becomes more consistent with the SH0ES measurement. The preference for a negative glitch parameter and the reduction of cosmic tensions motivate further study of the CGG model in the future, when increasingly larger volumes of cosmological data become available from multiple new surveys.

To assess how upcoming cosmological measurements can improve the constraints on the glitch parameter Ω_g, we follow [36] and use the combined Fisher-information forecasts from a cosmic-variance limited CMB experiment and a *Euclid*-like BAO survey [37]. We forecast the 1σ error of the parameter to be below 10^{-3}, which is roughly a fourfold reduction in uncertainty over the current results. The precision of the upcoming Stage IV CMB [38] and LSS surveys [37, 39–41] will allow us to determine whether Ω_g is genuinely negative (or just due to a statistical fluctuation or unknown systematics in the current cosmological data sets) and shed light on whether the CGG model resolves some of the cosmic tensions witnessed by different probes. At a more fundamental level, future constraints on the cosmic glitch will provide a new window into the nature of the symmetries that govern gravitational dynamics across the Cosmos, from its fiery Big Bang to its glacial heat death.

Acknowledgments This research was supported by the Natural Sciences and Engineering Research Council of Canada. NA is further supported by the Perimeter Institute for Theoretical Physics. Research at Perimeter Institute is supported in part by the Government of Canada through the Department of Innovation, Science and Economic Development Canada and by the Province of Ontario through the Ministry of Colleges and Universities. LTH was supported by a Killam Postdoctoral Fellowship and a CITA National Fellowship. Computing resources were provided by the Digital Research Alliance of Canada/Calcul Canada (alliancecan.ca). Parts of this work are based on observations obtained with *Planck* (www.esa.int/Planck), an ESA science mission with instruments and contributions directly funded by ESA Member States, NASA and Canada, and results from Dark Energy Spectroscopic Instrument (DESI, www.desi.lbl.gov). This work used the codes `CAMB` (camb.readthedocs.io/en/latest/) and `Cobaya` (cobaya.readthedocs.io/en/latest/).

References

1. Will, C. M., & Yunes, N. (2020). *Is Einstein still right?: Black holes, gravitational waves, and the quest to verify einstein's greatest creation*. Oxford University Press.
2. Afshordi, N., Chung, D. J. H., Doran, M., & Geshnizjani, G. (2007). *Physical Review D, 75*(12), 123509. https://doi.org/10.1103/PhysRevD.75.123509
3. Mukohyama, S., & Noui, K. (2019). *Journal of Cosmology and Astroparticle Physics, 2019*(07), 049.
4. Hořava, P. (2009). *Physical Review D, 79*(8), 084008.
5. Afshordi, N. (2009). *Physical Review D, 80*(8), 081502. https://doi.org/10.1103/PhysRevD.80.081502.
6. Jacobson, T. (2010). *Physical Review D, 81*(10), 101502.
7. Loll, R., & Pires, L. (2014). *Physical Review D, 90*(12), 124050. https://doi.org/10.1103/PhysRevD.90.124050.
8. Robbers, G., Afshordi, N., & Doran, M. (2008). *Physical Review Letters, 100*, 111101. https://doi.org/10.1103/PhysRevLett.100.111101.
9. Ebrahimi, E., & Sheykhi, A. (2011). *International Journal of Modern Physics D, 20*(12), 2369. https://doi.org/10.1142/S021827181102041X.
10. Sbisà, F. (2015). *European Journal of Physics, 36*(1), 015009. https://doi.org/10.1088/0143-0807/36/1/015009.
11. Lewis, A., Challinor, A., & Lasenby, A. (2000). *The Astrophysical Journal, 538*, 473. https://doi.org/10.1086/309179.

12. Hu, W. (2008). *Physical Review D, 77*(10), 103524. https://doi.org/10.1103/PhysRevD.77.103524.
13. Fang, W., Hu, W., & Lewis, A. (2008).*Physical Review D, 78*(8), 087303. https://doi.org/10.1103/PhysRevD.78.087303.
14. Planck Collaboration V. (2020). *Astronomy & Astrophysics, 641*, A5. https://doi.org/10.1051/0004-6361/201936386.
15. Planck Collaboration VI. (2020). *Astronomy & Astrophysics, 641*, A6. https://doi.org/10.1051/0004-6361/201833910.
16. Handley, W. J., Hobson, M. P., & Lasenby, A. N. (2015). *Monthly Notices of the Royal Astronomical Society, 453*(4), 4384. https://doi.org/10.1093/mnras/stv1911.
17. Handley, W. J., Hobson, M. P., & Lasenby, A. N. (2015). *Monthly Notices of the Royal Astronomical Society, 450*, L61. https://doi.org/10.1093/mnrasl/slv047.
18. Torrado, J., & Lewis, A. (2020) arXiv e-prints.
19. Matsumoto, A., Ouchi, M., Nakajima, K., Kawasaki, M., Murai, K., Motohara, K., Harikane, Y., Ono, Y., Kushibiki, K., Koyama, S., et al. (2022). *The Astrophysical Journal, 941*(2), 167. https://doi.org/10.3847/1538-4357/ac9ea1.
20. Kohri, K., & Maeda, K.I. (2022). *Progress of Theoretical and Experimental Physics, 2022*(9), 091E01. https://doi.org/10.1093/ptep/ptac114.
21. Afshordi, N., & Magueijo, J. (2016). *Physical Review D, 94*(10), 101301. https://doi.org/10.1103/PhysRevD.94.101301.
22. Abdalla, E., Abellán, G. F., Aboubrahim, A., et al. (2022). *Journal of High Energy Astrophysics, 34*, 49. https://doi.org/10.1016/j.jheap.2022.04.002.
23. Tristram, M., Banday, A. J., Douspis, M., Garrido, X., Górski, K. M., Henrot-Versillé, S., Ilić, S., Keskitalo, R., Lagache, G., Lawrence, C. R., Partridge, B., & Scott, D. (2024). *Astronomy & Astrophysics, 682*, A37. https://doi.org/10.1051/0004-6361/202348015.
24. Riess, A.G., Casertano, S., Yuancob, W., Bowers, J. B., Macri, L., Zinn, J. C., & Scolnic, D. (2021). *The Astrophysical Journal Letter, 908*(1), L6. https://doi.org/10.3847/2041-8213/abdbaf.
25. Riess, A. G., Casertano, S., Yuan, W., Macri, L. M., & Scolnic, D. (2019). *The Astrophysical Journal, 876*(1), 85. https://doi.org/10.3847/1538-4357/ab1422.
26. Riess, A. G., Yuan, W., Macri, L. M., Scolnic, D., Brout, D., Casertano, S., Jones, D. O., Murakami, Y., Anand, G. S., Breuval, L., Brink, T. G., Filippenko, A. V., Hoffmann, S., Jha, S. W., D'arcy Kenworthy, W., Mackenty, J., Stahl, B. E., & Zheng, W. (2022). *The Astrophysical Journal Letter, 934*(1), L7. https://doi.org/10.3847/2041-8213/ac5c5b.
27. Beutler, F., Blake, C., Colless, M., Jones, D. H., Staveley-Smith, L., Campbell, L., Parker, Q., Saunders, W., & Watson, F. (2011). *Monthly Notices of the Royal Astronomical Society, 416*(4), 3017. https://doi.org/10.1111/j.1365-2966.2011.19250.x.
28. Ross, A. J., Samushia, L., Howlett, C., Percival, W. J., Burden, A., & Manera, M. (2015). *Monthly Notices of the Royal Astronomical Society, 449*(1), 835. https://doi.org/10.1093/mnras/stv154.
29. Alam, S., Ata, M., Bailey, S., Beutler, F., Bizyaev, D., et al. (2017). *Monthly Notices of the Royal Astronomical Society, 470*(3), 2617. https://doi.org/10.1093/mnras/stx721.
30. Planck Collaboration, & Akrami, Y., et al. (2020). *Astronomy & Astrophysics, 643*, A42. https://doi.org/10.1051/0004-6361/202038073.
31. Adame, A. G., Aguilar, J., Ahlen, S., Alam, S., Alexander, D. M., Alvarez, M., Alves, O., Anand, A., Andrade, U., Armengaud, E., et al. (2025). DESI 2024 VI: Cosmological constraints from the measurements of baryon acoustic oscillations. *Journal of Cosmology and Astroparticle Physics, 2025*(02), 021.
32. Adame, A. G., Aguilar, J., Ahlen, S., Alam, S., Alexander, D. M., Alvarez, M., Alves, O., Anand, A., Andrade, U., Armengaud, E., et al. (2025). DESI 2024 III: Baryon acoustic oscillations from galaxies and quasars. *Journal of Cosmology and Astroparticle Physics, 2025*(04), 012.
33. Abbott, T. M. C., et al. (2018). *Physical Review D, 98*(4), 043526. https://doi.org/10.1103/PhysRevD.98.043526.

34. Handley, W., & Lemos, P. (2019). *Physical Review D, 100*(4), 043504. https://doi.org/10.1103/PhysRevD.100.043504.
35. Scolnic, D., et al. (2022). *The Astrophysical Journal, 938*(2), 113. https://doi.org/10.3847/1538-4357/ac8b7a.
36. Wen, Y., Scott, D., Sullivan, R., & Zibin, J. P. (2021). *Physical Review D, 104*(4), 043516. https://doi.org/10.1103/PhysRevD.104.043516.
37. Laureijs, R., Amiaux, J., Arduini, S., Augueres, J. L., & Brinchmann, J., et al. (2011). arXiv e-prints.
38. Abazajian, K. N., Adshead, P., Ahmed, Z., Allen, S. W., Alonso, D., et al. (2016). arXiv e-prints.
39. Ivezić, Ž., et al. (2019). *The Astrophysical Journal, 873*(2), 111. https://doi.org/10.3847/1538-4357/ab042c.
40. Eifler, T., et al. (2021). *Monthly Notices of the Royal Astronomical Society, 507*(1), 1514. https://doi.org/10.1093/mnras/stab533.
41. Doré, O., et al. (2014). arXiv e-prints arXiv:1412.4872. https://doi.org/10.48550/arXiv.1412.4872.

Chapter 36
Inference of Hubble Constant using Standard Sirens and Reconstructed Matter Density Field

Supranta S. Boruah, Ghazal Geshnizjani, and Guilhem Lavaux

Abstract We summarize a new approach for measuring the Hubble constant using standard sirens and the reconstructed matter density field obtained from observed galaxy surveys. Specifically, we describe and test this method using Bayesian forward-modeled software BORG. This software evolves the initial density field to the present, sampling plausible density fields from the posterior, accounting for peculiar velocities, and automatically incorporating higher-order correlation functions. The advantage of adding additional information from correlations is expected to make this method more effective in low-signal-to-noise regimes, such as those with modest galaxy number density or incomplete surveys. Our results show that developing a cross-correlation framework between gravitational waves and galaxy surveys, based on the forward-modeled reconstructed density field, to measure H_0 is promising.

36.1 Introduction

This chapter is based on the research conducted as part of Boruah's Ph.D. thesis [1] The groundbreaking observation of gravitational waves (GWs) [2] resulting from the mergers of stellar binary compact objects ranks among the most extraordinary milestones in contemporary physics. These detections introduce a novel method for exploring space that does not rely on electromagnetic waves or particle emissions.

S. S. Boruah
University of Pennsylvania, Philadelphia, PA, USA
e-mail: supranta@sas.upenn.edu

G. Geshnizjani (✉)
Perimeter Institute for Theoretical Physics, University of Waterloo, Waterloo, ON, Canada
e-mail: ggeshnizjani@pitp.ca

G. Lavaux
Institut d'Astrophysique de Paris (IAP)/CNRS, Paris, France
e-mail: guilhem.lavaux@iap.fr

S. Ghosh, A. R. Rao (eds.), *The Relativistic Universe: From Classical to Quantum*, Astrophysics and Space Science Proceedings 61,
https://doi.org/10.1007/978-3-031-90186-7_36

This new avenue offers many opportunities for studying various areas of physics. Of special interest to cosmologists, it presents a novel tool for determining the luminosity distance at cosmological scales, termed *standard sirens* [3, 4]. This method is especially intriguing because of the increasing conflict between Hubble constant values derived from early and late universe observations.

Unlike traditional distance measurements from standard candles [5, 6], standard sirens do not require intermediate calibration with a distance ladder. Instead, distances are directly estimated using general relativity to calculate a direct distance measurement. The concept of using standard siren to measure the Hubble constant was first proposed by Schutz [7], and has since been refined.

The first practical application of GW-based cosmological inference came with the detection of a binary neutron star merger event by LIGO, GW170817 [8], which included an electromagnetic counterpart that helped measure $H_0 = 70^{+12}_{-8}$ [9]. Although this is not yet competitive with values obtained from CMB and SNe observations, it is in agreement with them. This precision could improve significantly if more events with electromagnetic counterparts, such as gamma-ray bursts or kilonovae providing redshift measurements, were observed. Yet, most of the events observed so far match the profile of binary black holes (BBHs) lacking electromagnetic counterparts, *dark sirens*, making it challenging to determine the redshifts of their host galaxies. In fact, as data collection from the LIGO-Virgo-KAGRA network continues to expand and with the forthcoming addition of LIGO-India, a substantial increase in BBH events is anticipated.

However, deficiencies in electromagnetic data could potentially be compensated for by robust statistical methods, provided that effective statistical frameworks are established. For example, one method to start with is to assume that binary events are associated with galaxies in an observational catalog, and then adjust their redshifts statistically to those of the host galaxies to estimate the Hubble constant or other cosmological parameters. This approach was first applied to the GW170817 data without assuming an electromagnetic counterpart [10] and later used for dark sirens [11], leading to the estimate $H_0 = 68^{+8}_{-6}$. This statistical method while promising is limited by the ability to identify potential non-visible host galaxies in the siren's sky localization region. Incorrect modeling of the sky localization of the GW data can lead to biased cosmological parameter measurements. Interestingly, we know that galaxies do not have a random distribution. Therefore, alternative methods have been explored to take advantage of large-scale clustering with dark sirens [12–15]. These methods use the fact that both galaxies and sirens trace the underlying matter distribution, employing correlation functions to statistically infer cosmological parameters.

In what follows, we discuss a new approach that also relies on the assumption that dark sirens trace the underlying matter distribution. The central feature of this approach is to reconstruct an ensemble of density fields consistent with observed galaxy surveys and then apply a bias model to fit the dark siren population, thereby obtaining posteriors for cosmological parameters, specifically H_0. That is, statistically measuring the Hubble constant using standard sirens, based on a reconstructed dark-matter density field from a galaxy catalog. Here, we present a summary of the method skipping some technical details, which can be found in [1]. This reconstruction is performed with the Bayesian forward-modeled software BORG [16, 17], which uses a physical model of structure formation to evolve the initial density field to the present day, sampling an ensemble of plausible density fields from the posterior. This method accounts for peculiar velocities [18], which can introduce systematic uncertainties in redshift estimation at low redshifts, and incorporates higher-order correlation functions induced by structure formation. The forward-modeled reconstruction framework is particularly effective in low signal-to-noise regimes, such as when the number density of the galaxy is modest or surveys are incomplete. Joint reconstruction of the density and true positions of galaxies from photometric surveys has also proven powerful for improving photometric redshift estimates. Therefore, a framework relying solely on the reconstructed density field to measure H_0 is promising.

36.2 H_0 Inference from Gravitational Wave Event and Galaxy Catalogs Within BORG Framework

As mentioned before the goal of this method is to build a probabilistic framework for estimating the Hubble constant (H_0) by cross-correlating gravitational wave data with galaxy surveys through a reconstructed density field. To be more specific, use a Bayesian forward modeling software named BORG [16, 17] to reconstruct density fields that optimize the likelihood fit for H_0 with given galaxy catalog data $\{x_{\rm gal}\}$ and a set of gravitational wave events $\{x_{\rm GW}\}$. In the following sections, we describe testing this approach using mock catalogs and actual observed data. However, the first step is to derive a posterior probability $P(H_0|\{x_{\rm GW}\}, \{x_{\rm gal}\})$ or alternatively a likelihood $P(\{x_{\rm GW}\}|H_0, \{x_{\rm gal}\})$ using the Bayes rule, which would be compatible with the reconstructed matter density field, δ, from BORG.

We start by assuming that the likelihoods of individual standard siren events are independent of each other, and the final joint likelihood can be expressed as a product of individual event likelihoods. The data in the galaxy catalogs from observation provide their redshifts and angular positions of the galaxies, whereas the GW events provide luminosity distances from merger events, and the relation between these coordinates depends on H_0. Moreover, the likelihood of dark sirens is influenced by the clustering details (density reconstruction) derived from the galaxy catalog data. Therefore, the posterior can be written as

$$\begin{aligned} P(H_0|\{x_{\rm GW}\}, \{x_{\rm gal}\}) &\propto P(H_0) P(\{x_{\rm GW}\}|\{x_{\rm gal}\}, H_0) \\ &= P(H_0) \prod_{i=1}^{N_{\rm GW}} P(x^i_{\rm GW}|\{x_{\rm gal}\}, H_0)\,. \end{aligned} \tag{36.1}$$

Here, $P(H_0)$ is the prior for H_0. Now for a specific $x_{\rm GW}$ event given a specific δ realization, the probability that the host galaxy is located at comoving coordinate r and angular position Ω can be written as

$$P(x_{\rm GW}|\{x_{\rm gal}\}, H_0) \propto P(x_{\rm GW}|\hat{d}_L(r, H_0), \Omega)P(r, \Omega|\delta)P(\delta|\{x_{\rm gal}\}). \tag{36.2}$$

The probability distribution $P(\delta|\{x_{\rm gal}\})$, for the observed galaxy catalog $\{x_{\rm gal}\}$ is derived from the reconstruction framework within BORG. BORG produces a set of density fields $\{\delta\}$ that statistically agree with the galaxy catalog.

The probability $P(r, \Omega|\delta)$ in Eq. (36.2) represents the probability that the host galaxy of the GW event is at the coordinate position (r, Ω), given a certain overdensity profile. This is modeled as

$$p(r, \Omega|n_{\rm GW}, b_{\rm GW}, \delta) \propto n_{\rm GW}(1 + \delta_{\rm GW}(r, \Omega)), \tag{36.3}$$

where the the new quantity $n_{\rm GW}$ gives the average rate of mergers per unit volume in the Universe, $\delta_{\rm GW}$ is the overdensity of merger rate events and $b_{\rm GW}$, denotes the 'bias' parameter quantifying dependence of the merger events overdensity, $\delta_{\rm GW}$, with respect to the matter overdensity, δ. For our mock simulations, assuming that the binary events are uniformly placed on the galaxies we take $b_{\rm GW}$ to be the same as galaxy bias and marginalize over that. However, the 'bias' of binary events could be a more complicated function of astrophysical properties of host galaxies.

The density field is calculated in comoving coordinates r and is expressed in units of h^{-1} pc. However, as pointed out, gravitational wave signals from standard sirens can be used to estimate the luminosity distance to the GW event. The conversion from r to the luminosity distance, $\hat{d}_L$, depends on H_0 through $\hat{d}_L(r, H_0) = r(1 + z)/h$ and $z(r, H_0) = H_0 r/c$.

To evaluate $P(x_{\rm GW}|\hat{d}_L(r, H_0), \Omega)$, we implemented the gravitational wave data analysis tool BAYESTAR [19]. This application provides a 3D reconstruction of the posterior $P(d_L, \Omega|x_{\rm GW})$ with respect to d_L and Ω for given gravitational wave event data. The likelihood can then be obtained using Bayes' theorem as $p(x_{\rm GW}|d_L, \Omega) \propto \frac{p(d_L, \Omega|x_{\rm GW})}{p(d_L, \Omega)}$. The prior $p(d_L, \Omega)$ is considered to be a volumetric prior on the luminosity distance, $P(d_L, \Omega) \propto d_L^2$.

Integrating Eq. (36.2) over the full ensemble of realizations $\{\delta\}$ the likelihood of a single gravitational wave event can be written as follows:

$$P(x_{\rm GW}|\{x_{\rm gal}\}, H_0) \propto \int d\{\delta\}\, d\Omega\, dr\; P(x_{\rm GW}|\hat{d}_L(r, H_0), \Omega)P(r, \Omega|\{\delta\}) \\ P(\{\delta\}|\{x_{\rm gal}\}). \tag{36.4}$$

Lastly, substituting these relations and performing the marginalization over δ using the Monte Carlo samples supplied by BORG, Eq. (36.4), would lead to

$$P(x_{\rm GW}|\{x_{\rm gal}\}, H_0) = \frac{H_0^2}{\alpha(H_0)} \int db_{\rm GW} dr d\Omega \Big[\frac{1}{(1+z(r, H_0))^2}(1 + \langle \delta_{\rm GW}^i \rangle)$$
$$P(\hat{d}_L(r, H_0), \Omega | x_{\rm GW}) P(b_{\rm GW}) \Big], \tag{36.5}$$

where $\alpha(H_0)$ is the normalization term and $\langle \delta_{GW}^i \rangle$ indicates the mean density field from BORG sampling. We assume that the galaxy catalog has a high degree of completeness and normalization is only due to the volume effects and $\alpha(H_0) \propto H_0^3$. However, the normalization factor can generally be different corrections to be considered [20, 21]. This is the likelihood that was used in the inference calculations.

36.3 BORG Methodology

We now present an overview of the BORG framework and highlight key information on how the BORG runs were utilized in our study. For a more comprehensive understanding of BORG, we refer the readers to the original publications [16, 17] and for details of the implementations in this study to [1]. BORG is an advanced software designed to simultaneously achieve multiple objectives with respect to an observational galaxy catalog. In particular, for our purposes, it can generate a posterior distribution for the large-scale density field of the universe $\{\delta_i\}$, estimate the parameters of the bias model $\Theta_{\rm bias}$, and apply the corrections due to peculiar velocities along the way.

The approach implemented in BORG is forward modeling, which, as the name indicates, infers the late-time nonlinear density field $\{\delta_i\}$ starting from a sample for the initial density field $\{\delta_i^{\rm IC}\}$ and based on a structure formation model. That is, the initial densities, $\delta_i^{\rm IC}$ are forwarded through the structure formation process in a simulation box to the late-time densities, δ_i. Therefore, $\{\delta_i\}$ is some deterministic function of $\{\delta_i^{\rm IC}\}$. The density field profile is constructed in the simulation box so that the density in each voxel is one parameter. To efficiently sample such a high-dimensional parameter space, an efficient sampling scheme known as Hamiltonian Monte Carlo (HMC) [22, 23] is incorporated.

Associating the number of galaxies/halos to the density field requires a bias model. The bias model implemented in this study is based on the nonlinear model proposed in [24], such that the number of observed galaxies are Poisson samples from the density field δ with Poisson rate $\lambda_g(\delta)$, which has a power law term times an exponential cutoff in order to capture the low number density in the void regions. To ensure consistent sampling of both bias parameters and densities, a block sampling algorithm is used. That is, given the observed galaxy numbers $\{N_i^{\rm obs}\}$ in each voxel, initially the bias parameters $\Theta_{\rm bias}$ are kept constant while the density field is sampled from the conditional posterior $\mathcal{P}(\{\delta_i^{\rm IC}\}|\{N_i^{\rm obs}\}, \Theta_{\rm bias})$. Subsequently, using the updated density field $\{\delta_{i+1}^{\rm IC}\}$, the bias parameters are conditionally sampled from the posterior bias $\mathcal{P}(\Theta_{\rm bias}|\{N_i^{\rm obs}\}, \{\delta_{i+1}\})$.

Also, as we pointed out earlier in the process of simulation from the initial position of the particles to the final positions, their velocity is also computed, and given that the observation of galaxies provides their redshift, BORG estimates the density in redshift space and systematically addresses the redshift space distortions in the reconstruction.

36.4 Testing the H_0 Inference Model Using Mock Data Sets

To test our methodology, we applied it to a simulated data set consisting of a mock galaxy catalog and mocked gravitational wave data corresponding to a sample of 500 sirens hosted in the galaxy catalog. In this section, we will start by providing the details of the simulated galaxy catalog and the BORG run executed on this catalog to reconstruct the density field. Following that, we will explain our approach for simulating the GW data employed in the inference, and then present the results of using our inference framework on these simulated catalogs.

Mock Galaxy Catalog The mock galaxy catalog was generated using the `VELMASS` simulation suite [25]. It consists of 10 cosmological simulations probing variations of cosmological parameters and different initial phases. In this work, we have just used the central simulation results, which assume Planck 2015 [26] values for the cosmological parameters: $\Omega_m = 0.315$, $\Omega_b = 0.049$, $H_0 = 68$ km s^{-1} Mpc^{-1}, $\sigma_8 = 0.81$, $n_s = 0.97$, and $Y_{\rm He} = 0.248$. The simulated volume is a cubic box with a side length of 2000 h^{-1} Mpc and contains 2048^3 dark matter particles. It was initialized at redshift of $z = 50$ and evolved to the present epoch using `GADGET2` [27]. Then, the `|ROCKSTAR|` halo finder algorithm [28] was used to find the halos of the final density profile, resulting in a halo catalog. Afterwards, a smaller box of size $(500\ h^{-1}\ {\rm Mpc})^3$ was extracted from the full simulation volume. Within this smaller box, individual galaxies were assigned to halos using an abundance matching technique [29, 30] which involves ranking halos and galaxies by halo mass and galaxy brightness, respectively, in descending order and pairing them accordingly. The galaxy luminosities were derived from a Schechter function, based on the best-fit Schechter parameters reported by [31]. Lastly, galaxies fainter and with apparent magnitude above the threshold of 12.5 were excluded from the dataset, resulting in the mock data for the "observed" galaxy catalog, $\{x_{\rm gal}\}$, with $\sim 8 \times 10^4$ galaxies.

Mock GW Catalog To produce the GW catalog, $\{x_{\rm GW}\}$, we first randomly select 500 galaxies within 100 h^{-1} Mpc from our mock catalog as host galaxies for GW events. Assuming the Hubble constant of $H_0 = 68$ km s^{-1} Mpc^{-1}, the luminosity distances to these sources were calculated. Given these distances, we then simulated GW data events $\{x_{\rm GW}\}$ corresponding to $1.4M_\odot$–$1.4M_\odot$ binary systems one in each host galaxy using BAYESTAR software [19]. The inclinations for the binaries were randomly selected, and only events with a signal-to-noise ratio greater than 12 were kept in the "observed" GW catalog. Note that BAYESTAR software has dual application, allowing it to be used in this step as part of the Mock GW data production, but, as pointed out earlier in section Eq. (36.2) and explained below, it can also be used as part of the inference process.

H_0 Inference Having generated the mock data for $\{x_{\text{gal}}\}$ and $\{x_{\text{GW}}\}$ we performed the likelihood analysis outlined in section Eq. (36.2) on these data set to infer the value of H_0 and compared it to injected value of $H_0 = 68$ km s^{-1} in the simulations.

The steps from Eqs. (36.4) to (36.5) involved performing BORG to reconstruct the density field samples corresponding to the mock catalog $\{x_{\text{gal}}\}$ and marginalizing over them. Given the potential luminosity dependence of the galaxy bias, galaxies with absolute magnitude between -23.25 and -25.25 were divided into eight equal width bins of absolute magnitude, and the galaxy bias parameters Θ_{bias} for each bin were separately modeled while consistently accounting for red-shift space distortions introduced by peculiar velocity. Figure 36.1 illustrates how the resulting power spectrum from BORG samples compares to the prior ΛCDM power spectrum used in the N-body simulation (scaled to $z = 0$ using linear perturbation theory).

Next, evaluating $P(x_{\text{GW}}|\hat{d}_L(r, H_0), \Omega)$ in Eq. (36.5) as eluded before involved using BAYESTAR again but this time as part of the inference process. That is BAYESTAR generates a 3 dimensional volume probability for the distance and the sky localization of each of the GW events in the mock $\{x_{\text{GW}}\}$. This probability distribution was constructed considering two scenarios: *i)* Active GW detectors

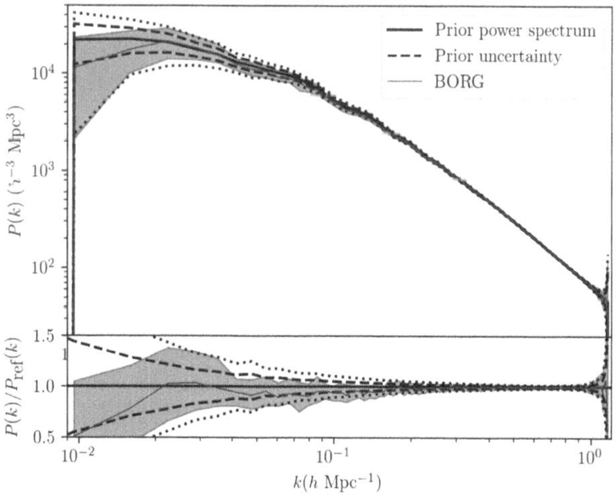

Fig. 36.1 The power spectrum of the initial density fields, adjusted to $z = 0$ using linear perturbation theory, is illustrated. The black solid line depicts the ΛCDM prediction, and the dashed and dotted lines represent the expected 1σ and 2σ uncertainties, assuming Poissonian uncertainty while the blue solid line represents the mean of all samples reconstructed by BORG and the blue shaded area indicates the 2σ area for BORG samples [1]

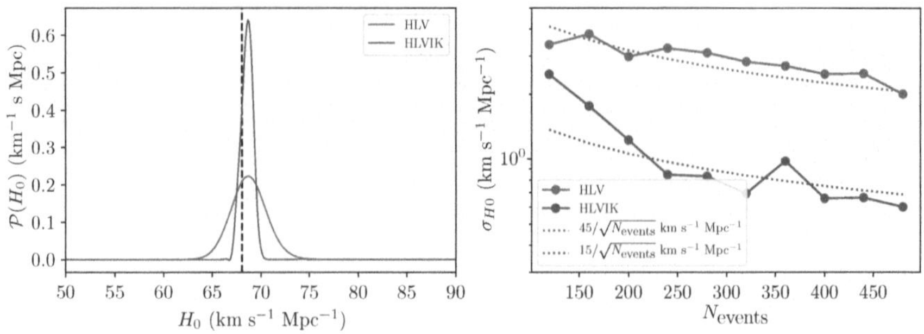

Fig. 36.2 Inference on H_0 using the mock $\{x_{\rm GW}\}$ and $\{x_{\rm GW}\}$ data set. *(Left):* Posterior distribution of H_0 for both HLV and HLVIK data sets. The true injected H_0 value is marked by the black vertical line. *(Right):* The variation in uncertainty of H_0 as a function of the number of events for either cases

include only the LIGO and VIRGO detectors (HLV), and *ii)* in addition to HLV, the LIGO India and KAGRA detectors are also active (HLVIK).

The posteriors of these data sets for H_0 are illustrated in the left panel of Fig. 36.2. In both scenarios, an unbiased value of the Hubble constant is recovered. For 500 HLV events, we obtain $H_0 = 68.75 \pm 1.83$ km s^{-1} Mpc^{-1}. Similarly, for 500 HLVIK events, the recovered value is $H_0 = 68.53 \pm 0.64$ km s^{-1} Mpc^{-1}. Moreover, as observed in the right panel of Fig. 36.2, with an increase in the number of events, the error appears to decrease approximately in proportion to $1/\sqrt{N_{\rm events}}$ similar to a Poisson distribution.

36.5 H_0 Inference for GW170817 and 2M++

Following initial validation with simulated data, we extended our framework's evaluation to real observational data for gravitational waves (GWs) and galaxies. Presently, the computational restrictions of the Bayesian reconstruction from galaxies (BORG) framework impose limitations on the box size, thereby constraining the redshift range. Consequently, our current assessment is restricted to the low-redshift domain. Nevertheless, we anticipate that this method will demonstrate its full potential in the future, given the fast pace of advances in computational efficiency and the availability of extensive high-redshift galaxy surveys and GW data. With that in mind, we applied our inference method on galaxy catalog 2M++ [31] and the GW data from the binary neutron star event GW170817 using BORG reconstruction for the density. Although an optical counterpart was identified for GW170817, our analysis does not incorporate this information. An alternative measurement of the Hubble constant, H_0, which does not rely on the optical counterpart of GW170817 and correlations to the density field, is described in [32].

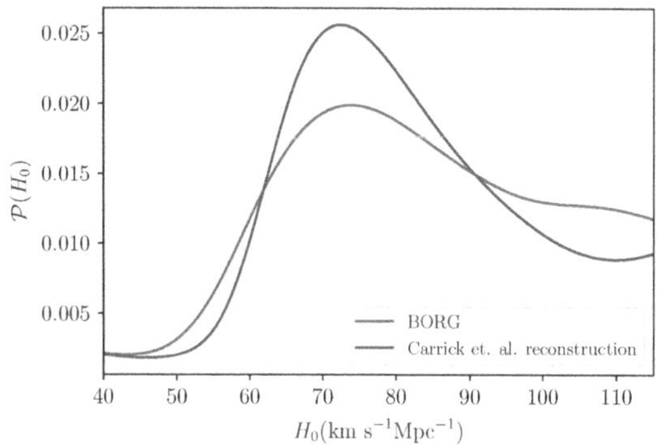

Fig. 36.3 Posterior for H_0 given the data corresponding to GW170817 event and 2M++ using two distinct reconstruction techniques. The red curve represents the posterior obtained via the BORG reconstruction applied to the 2M++ catalog while the blue curve reconstruction method described in [33]

The detailed account of BORG reconstruction of the 2M++ catalog is presented in [17]. The galaxy bias is again modeled assuming luminosity dependence and by splitting galaxies into eight bins of equal width based on their absolute magnitude in this case between -21 and -25. Additionally, the galaxies in the 2M++ catalog with $K_{2M++} < 11.5$ and those in $11.5 < K_{2M++} < 12.5$ were categorized separately due to different selections.

The inference for H_0 resulted in $H_0 = 84^{+23}_{-18}$ km s^{-1} Mpc^{-1} (see Fig. 36.3), consistent with other cosmological measurement methods for H_0, but not offering competitive precision. Furthermore, we also derived a similar result using our approach, but substituting the reconstructed galaxy over-density field based on the luminosity weighting method presented in [32] in place of the BORG reconstructed matter density. The result was $H_0 = 82^{+20}_{-14}$ km s^{-1} Mpc^{-1}, which is comparable to BORG (see Fig. 36.3) and also aligns with the findings in [33] . This is expected, as the matter density correlation function should be more effective at higher redshifts, where survey completeness is more limited.

36.6 Conclusion

We presented a new approach for inferring cosmological parameters through the cross-correlation of gravitational wave data with galaxy surveys. More specifically provided an analytical likelihood that can employ a forward-modeled reconstructed density field obtained from observed galaxy surveys via BORG software and using standard sirens to estimate the Hubble constant (H_0). This framework automatically incorporates higher-order correlation functions and we predict to be effective in low-signal-to-noise regimes, such as modest galaxy number density or incomplete surveys, where the correlation could provide additional information. We tested this framework on simulated datasets for galaxies and gravitational wave (GW) events, showing that our method provides an unbiased estimate of H_0. We also applied it to data from the binary neutron star event GW170817 and density reconstructions from the 2M++ galaxy redshift catalog using BORG. Our method consistently predicted

$H_0 = 84^{+23}_{-18}$ kms^{-1} Mpc^{-1}, in agreement with other cosmological measurements, but not very competitive. We also derived a similar result using our approach, but instead of BORG substituting the reconstructed galaxy over-density field based on the luminosity weighting method. This strongly suggests that reconstruction methods can provide unbiased inferences. In fact, reconstructing the density field jointly while inferring photometric redshifts is known to improve the photometric redshift estimate [34]. This would integrate well with our method for statistically inferring H_0 from high redshift GW events. A similar analysis can be applied to Type Ia supernovae when the host galaxy cannot be uniquely identified. This will be particularly relevant for future large-scale surveys like the Vera-Rubin Observatory (VRO), which will detect a substantial number of supernovae. Future work will need to compare our method with data that will become available in the coming decades for both GWs and galaxy surveys, as well as numerical advancements in high-redshift density reconstruction simulations.

Acknowledgments The research carried out during this study by SB and GG was supported by a Discovery Grant from the Natural Science and Engineering Research Council of Canada (NSERC). GG is also partly supported by the Perimeter Institute for Theoretical Physics (PI). Research at PI is supported by the Government of Canada through the Department of Innovation, Science and Economic Development Canada and led by the Province of Ontario through the Ministry of Research, Innovation and Science. GG is also grateful for invitation to the International Symposium on Relativistic Astrophysics held in Gangtok, Sikkim, India to present this work.

References

1. Boruah, S. S. (2020). Topics in early and late Universe cosmology. *UWSpace*. University of Waterloo Institutional Repository.
2. Abbott, B. P., et al. (2016). Observation of Gravitational Waves from a Binary Black Hole Merger. *Physical Review Letters, 116*(6), 061102.
3. Holz, D. E., & Hughes, S. A. (2005). Using gravitational-wave standard sirens. *The Astrophysical Journal, 629*, 15–22.
4. Dalal, N., Holz, D. E., Hughes, S. A., & Jain, B. (2006). Short GRB and binary black hole standard sirens as a probe of dark energy. *Physical Review D, 74*, 063006.
5. Perlmutter, S., et al. (1999). Measurements of Ω and Λ from 42 high redshift supernovae. *The Astrophysical Journal, 517*, 565–586.
6. Riess, A. G., Casertano, S., Yuan, W., Macri, L. M., & Scolnic, D. (2019). Large magellanic cloud cepheid standards provide a 1% foundation for the determination of the hubble constant and stronger evidence for physics beyond ΛCDM. *The Astrophysical Journal, 876*(1), 85.
7. Schutz, B. F. (1986). Determining the hubble constant from gravitational wave observations. *Nature, 323*, 310–311.
8. Abbott, B. P., et al. (2017). A gravitational-wave standard siren measurement of the Hubble constant. *Nature, 551*(7678), 85–88.
9. Abbott, B. P., et al. (2017). Gravitational waves and gamma-rays from a binary neutron star merger: GW170817 and GRB 170817A. *The Astrophysical Journal Letter, 848*(2), L13.
10. Fishbach, M., et al. (2018). A standard siren measurement of the hubble constant from GW170817 without the electromagnetic counterpart. *The Astrophysical Journal Letter, 871*(1), L13, 2019.
11. Abbott, R., et al. (2021) Constraints on the cosmic expansion history from GWTC-3. *The Astrophysical Journal*.

12. Oguri, M. (2016). Measuring the distance-redshift relation with the cross-correlation of gravitational wave standard sirens and galaxies. *Physical Review D, 93*(8), 083511.
13. Mukherjee, S., & Wandelt, B. D. (2018). *Beyond the classical distance-redshift test: Cross-correlating redshift-free standard candles and sirens with redshift surveys*. https://arxiv.org/abs/1808.06615
14. Mukherjee, S., Wandelt, B. D., Nissanke, S. M., & Silvestri, A. (2021). Accurate precision Cosmology with redshift unknown gravitational wave sources. *Physical Review D, 103*(4), 043520.
15. Bera, S., Rana, D., More, S., & Bose, S. (2020). Incompleteness matters not: Inference of H_0 from binary black hole–Galaxy cross-correlations. *The Astrophysical Journal, 902*(1), 79.
16. Jasche, J., & Wandelt, B. D. (2013). Bayesian physical reconstruction of initial conditions from large-scale structure surveys. *Monthly Notices of the Royal Astronomical Society, 432*(2), 894–913.
17. Jasche, J., & Lavaux, G. (2019). Physical Bayesian modelling of the non-linear matter distribution: New insights into the nearby universe. *Astronomy and Astrophysics, 625*, A64.
18. Hui, L., & Greene, P. B. (2006). Correlated fluctuations in luminosity distance and the importance of peculiar motion in supernova surveys. *Physical Review D, 73*(12), 123526.
19. Singer, L. P., & Price, L. R. (2016). Rapid Bayesian position reconstruction for gravitational-wave transients. *Physical Review D, 93*(2), 024013.
20. Mandel, I., Farr, W. M., & Gair, J. R. (2019). Extracting distribution parameters from multiple uncertain observations with selection biases. *Monthly Notices of the Royal Astronomical Society, 486*(1), 1086–1093.
21. Abbott, B. P., et al. (2017). A gravitational-wave standard siren measurement of the Hubble constant. *Nature, 551*(7678), 85–88.
22. Neal, R. (1993). Probabilistic inference using Markov chain Monte Carlo methods. *Technical Report CRG-TR-93-1. Dept. of Computer Science, University of Toronto*, 1993.
23. Neal, R. (1996). *Bayesian learning for neural networks*. Lecture Notes in Statistics (Vol. 118). Springer-Verlag.
24. Neyrinck, M. C., Aragón-Calvo, M. A., Jeong, D., & Wang, X. (2014). A halo bias function measured deeply into voids without stochasticity. *Monthly Notices of the Royal Astronomical Society, 441*(1), 646–655.
25. Ramanah, D. K., Charnock, T., & Lavaux, G. (2019). Painting halos from cosmic density fields of dark matter with physically motivated neural networks. *Physical Review D, 100*(4), 043515.
26. Ade, P. A. R., et al. (2016). Planck 2015 results. XIII. Cosmological parameters. *Astronomy & Astrophysics, 594*, A13.
27. Springel, V. (2005). The cosmological simulation code GADGET 2. *Monthly Notices of the Royal Astronomical Society, 364*, 1105–1134.
28. Behroozi, P. S., Wechsler, R. H., & Wu, H.-Y. (2013). The ROCKSTAR phase-space temporal halo finder and the velocity offsets of cluster cores. *The Astrophysical Journal, 762*(2), 109.
29. Marinoni, C., & Hudson, M. J. (2002). The mass-to-light function of virialized systems and the relationship between their optical and X-Ray properties. *The Astrophysical Journal, 569*(1), 101–111.
30. Kravtsov, A. V., Berlind, A. A., Wechsler, R. H., Klypin, A. A., Gottlöber, S., Brand on Allgood, & Primack, J. R. (2004). The dark side of the halo occupation distribution. *The Astrophysical Journal, 609*(1), 35–49.
31. Lavaux, G., & Hudson, M. J. (2011). The 2M++ galaxy redshift catalogue. *Monthly Notices of the Royal Astronomical Society, 416*(4), 2840–2856.
32. Fishbach, M., et al. (2019). A standard siren measurement of the hubble constant from gw170817 without the electromagnetic counterpart. *The Astrophysical Journal, 871*(1), L13.
33. Carrick, J., Turnbull, S. J., Lavaux, G., & Hudson, M. J. (2015). Cosmological parameters from the comparison of peculiar velocities with predictions from the 2M++ density field. *Monthly Notices of the Royal Astronomical Society, 450*(1), 317–332.
34. Jasche, J., & Wandelt, B. D. (2012). Bayesian inference from photometric redshift surveys. *Monthly Notices of the Royal Astronomical Society, 425*(2), 1042–1056.

Chapter 37
Novel Understanding of Cosmological Phenomena Using Fast Radio Bursts

Surajit Kalita, Shruti Bhatporia, and Amanda Weltman

Abstract Fast radio bursts (FRBs) offer unique probes of diverse cosmological phenomena due to their characteristic properties, including short duration timescale and high dispersion measure. This study investigates two distinct theoretical frameworks: the Gertsenshtein-Zel'dovich (GZ) mechanism for ultra-high-frequency gravitational waves (GWs) and fraction of dark matter in primordial mass black holes. We explore the hypothesis that ultra-high-frequency GWs could be responsible for FRB generation. Consequently, the detection of continuous GWs signal from the vicinity of an FRB by current or future detectors would disfavor merger-based FRB formation models and lend significant credence to the GZ theory, which postulates the existence of high-frequency GWs. Moreover, we examine the effects of modified gravity on the gravitational lensing of FRBs and thereby put constraints on the fraction of primordial mass black holes made up of dark matter. Our analysis suggests that modified gravity introduces a screening effect on lensing, analogous to the scattering effect by plasma on light rays. We further discuss the expected detection rates of FRBs as well as lensed FRBs with upcoming radio telescopes, primarily HIRAX.

S. Kalita (✉)
High Energy Physics, Cosmology and Astrophysics Theory (HEPCAT) Group, Department of Mathematics and Applied Mathematics, University of Cape Town, Cape Town, South Africa

Astronomical Observatory, University of Warsaw, Warszawa, Poland
e-mail: skalita@astrouw.edu.pl

S. Bhatporia · A. Weltman
High Energy Physics, Cosmology and Astrophysics Theory (HEPCAT) Group, Department of Mathematics and Applied Mathematics, University of Cape Town, Cape Town, South Africa
e-mail: amanda.weltman@uct.ac.za

S. Ghosh, A. R. Rao (eds.), *The Relativistic Universe: From Classical to Quantum*, Astrophysics and Space Science Proceedings 61,
https://doi.org/10.1007/978-3-031-90186-7_37

37.1 Introduction

Fast radio bursts (FRBs), characterized by millisecond radio transients, exhibit high fluxes on the order of hundreds of Jy. Despite their recent discovery, FRBs have garnered significant scientific interest due to their unique properties and potential to unveil new astrophysical processes. Extragalactic distances for most FRBs are suggested by their high dispersion measures (DMs) and isotropic sky distribution, favoring a cosmological origin. However, pinpointing their exact host environments is challenging. Associations with dwarf galaxies have been observed for some FRBs, while others might originate from regions with strong magnetic fields, such as those surrounding magnetars or neutron stars (NSs). The recent discovery of a repeating FRB (FRB 20201124A) [1] within a young stellar cluster suggests a possible connection to massive star formation regions.

The true origin of all FRBs remains largely enigmatic although the Galactic FRB 20200428 confirms an association with the magnetar SGR 1935+2154 [2–4]. Their immense luminosity suggests a cataclysmic event; however, the lack of a confirmed electromagnetic (EM) counterpart hinders definitive progenitor identification. Proposed candidates include giant magnetar flares, collapsars during core-collapse supernovae, or even processes involving NS mergers [5]. Notably, all these mechanisms involve compact objects like white dwarfs (WDs), NSs, or black holes (BHs). Therefore, multi-wavelength observations or detections of gravitational waves (GWs) and neutrinos from FRBs are crucial to understanding their progenitor mechanism. Unveiling the FRB progenitor holds the potential to revolutionize our understanding of stellar evolution, interstellar medium (ISM) properties, and even fundamental physics concepts like magnetar behavior or the existence of exotic new objects.

FRBs hold immense potential as powerful tools for cosmological investigations and testing gravity theories. By studying DM, the delay experienced by different radio frequencies due to plasma along the propagation path, FRBs can map the total electron content along their path. This information, combined with redshift measurements (when possible), allows for mapping the distribution of matter across vast cosmic distances. Additionally, studying FRB polarization and potential interactions with the magneto-ionic medium can provide clues about the local environment around their source. Ultimately, FRBs offer a unique opportunity to map the large-scale structure of the Universe, constrain cosmological parameters, and potentially shed light on fundamental physics at the extreme environments from which they originate.

In this chapter, we aim to explore the applications of FRBs in cosmology, particularly with gravitational lensing and high-frequency gravitational waves. In Sect. 37.2, we discuss how GW astronomy would help in understanding the progenitor mechanism of FRBs. Moreover, in Sect. 37.3, using lensing in FRBs, we obtain the constraints on the fraction of dark matter in primordial mass black holes in modified gravity theory. Furthermore, in Sect. 37.4, we estimate the total FRB and lensed FRB rates detectable with the upcoming Hydrogen Intensity and Real-time Analysis eXperiment (HIRAX) telescope in South Africa. Finally, we put our concluding remarks in Sect. 37.5.

37.2 Gravitational Waves to Distinguish Progenitor Theories of FRBs

As mentioned in the previous section, several astrophysical mechanisms have been proposed to explain the origins of FRBs. These mechanisms can be broadly categorized into two regimes: those involving the coalescence of compact objects and those involving isolated compact objects. In the merger scenario, a powerful magnetohydrodynamic shock wave is generated during the collision. Electrons accelerated within this shock dissipate a significant amount of their energy within the surrounding magnetosphere, potentially leading to the production of an FRB. Notably, the final merger remnant, likely a black hole, would no longer be expected to emit significant gravitational radiation. In contrast, FRBs arising from isolated compact objects would leave the object in its original location. Future generation GW detectors with enhanced sensitivity may be capable of directly detecting these objects, provided they retain rotation-powered emissions similar to those of pulsars.

We now focus on a specific scenario for the generation of FRBs from isolated compact objects: the Gertsenshtein-Zel'dovich (GZ) effect. It proposes a partial conversion of incoming GWs into EM waves upon interaction with the magnetic field of the pulsar magnetosphere. If the resulting EM radiation possesses characteristics consistent with observed FRBs, the GZ mechanism becomes a viable candidate for FRB production. Figure 37.1 illustrates a schematic representation of this effect. Let us assume the polarization modes over time t of infall GWs propagating in z-direction are given by

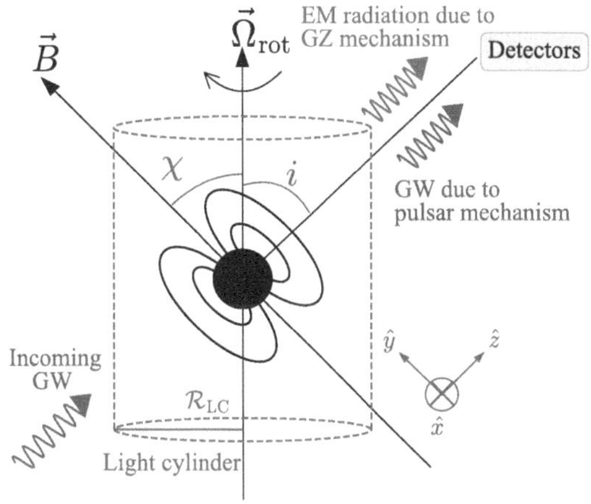

Fig. 37.1 Schematic diagram of GZ mechanism showing green GWs get converted to red EM waves due to its interaction with the pulsar magnetosphere. The blue GWs are emitted due to the pulsar mechanism

$$h_+ = A_+ e^{i(k_g z - \Omega_g t)} \quad \text{and} \quad h_\times = i A_\times e^{i(k_g z - \Omega_g t)}, \tag{37.1}$$

where Ω_g is the GW frequency, k_g is its wave number, A_+ and $A_\times$ are their amplitudes. Assuming $A_+ = A_\times$ and the time-varying magnetic field of the pulsar in y-direction as $B_y(t) = B_y^{(0)} + \delta B_y \sin(\Omega_{\rm rot} t)$ with $\Omega_{\rm rot}$ being the rotation frequency of the pulsar, outgoing EM waves can be represented the following wave equations [6]

$$\frac{1}{c^2}\frac{\partial^2 \tilde{E}_x}{\partial t^2} - \partial_z^2 \tilde{E}_x = f_E(z,t) \quad \text{and} \quad \frac{1}{c^2}\frac{\partial^2 \tilde{B}_y}{\partial t^2} - \partial_z^2 \tilde{B}_y = f_B(z,t), \tag{37.2}$$

where $f_E, f_B \sim e^{i(k_g z - \Omega_+ t)}$ and $\Omega_\pm = \Omega_g \pm \Omega_{\rm rot}$. Note that as $\Omega_\pm \sim$ MHz – GHz (due to observed frequency of FRBs) and $\Omega_{\rm rot} \sim$ Hz – kHz, we must have $\Omega_g \sim$ MHz – GHz. This implies that the infall gravitational radiation must be cosmological in nature. For a detailed discussion on different physical mechanisms that emit GWs at frequencies MHz – GHz, one may refer [7].

Let us now illustrate the concept with FRB 20180817, an FRB observed at a frequency of 501.1 MHz with pulse width $\delta = 0.01769$ s and peak flux = 2.4 Jy. This translates to a light cylinder radius $\mathcal{R}_{\rm LC} = \delta c/2 = 2.65 \times 10^8$ cm and thereby $\Omega_{\rm rot} = c/\mathcal{R}_{\rm LC} = 113.1$ rad s^{-1}. Now, the peak flux is theoretically given by

$$S_z = \frac{A_+^2 \left|B_y^{(0)}\right|^2 c}{128\pi}\left[\sqrt{\frac{24c^2 \nu_g^2 \alpha_{\rm tot}}{\pi G \left|B_y^{(0)}\right|^2} - 51} - \frac{6c^2 \nu_g \nu_{\rm rot} \alpha_{\rm tot}}{\pi G \left|B_y^{(0)}\right|^2} - 1\right], \tag{37.3}$$

where c is the speed of light, G is Newton's gravitational constant, $\nu_{g,\rm rot} = \Omega_{g,\rm rot}/2\pi$, and $\alpha_{\rm tot}$ is the total amount of energy converted from GWs to EM waves. Substituting the known values results in two unknowns: A_+ and $\left|B_y^{(0)}\right|$. Assuming $A_+ = 10^{-24}$, we obtain $\left|B_y^{(0)}\right| \approx 3 \times 10^{10}$ G and if $A_+ = 10^{-29}$, we obtain $\left|B_y^{(0)}\right| \approx 3 \times 10^{15}$ G. These values suggest the compact object could either be a NS or a magnetar. To definitively distinguish between the two, we require GW detection from the FRB's location. A detailed analysis of GW detection for such scenarios can be found in [8]. Figure 37.2 illustrates the possibility of detecting GWs from NS- and magnetar-based scenarios. For magnetars, detectors like the Cosmic Explorer and the Einstein Telescope could potentially detect the signal immediately. However, for NS, none of the currently proposed detectors is sensitive enough. Therefore, GW observations from the FRB's location hold the key to differentiating between the merger and isolated progenitor mechanisms for FRBs.

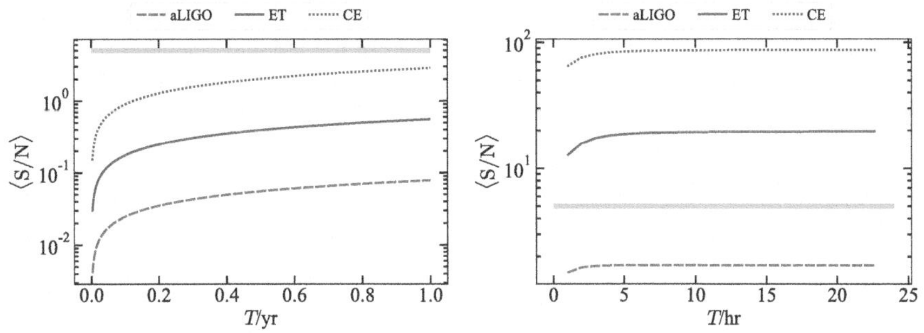

Fig. 37.2 Cumulative signal-to-noise ratio (S/N) over time for different GW detectors. The left Figure shows a typical case of NS whereas the right Figure is for a magnetar. The thick cyan line represents the threshold S/N above which the signal can be detected

37.3 FRBs to Constrain Dark Matter Fraction in Primordial Mass Black Holes

Beyond their origin, FRBs serve as powerful cosmological probes. Their high DM information can shed light on the properties of the intergalactic medium (IGM) and the large-scale structure of the cosmos. Additionally, studying lensed FRBs, where the light from an FRB is bent by the gravity of a massive object along its path, can provide unique insights into dark matter distribution and the nature of gravity itself. In strong lensing events, a single FRB source can be magnified and split into multiple images. Considering spherical polar coordinates (t, r, θ, ϕ), let us now assume the following spherically symmetric background metric

$$ds^2 = \left(1 - \frac{2GM_L}{c^2 r} + \Psi r^2\right) c^2 dt^2 - \frac{1}{1 - \frac{2GM_L}{c^2 r} + \Psi r^2} dr^2 - r^2 d\theta^2 - r^2 \sin^2\theta d\phi^2, \tag{37.4}$$

where M_L is the lensed mass and Ψ is a parameter characterizing modified gravity. We consider this particular metric as it is a solution within the framework of a general scalar-tensor gravity theory. Thereby, employing Shapiro delay on the lensed images, we can obtain the differential time delay between the two images, which is given by [9]

$$\Delta t = (1 + z_L) \left[\frac{1}{c\sqrt{\Psi}} \tan^{-1}\left(\sqrt{\Psi} \frac{4GM_L}{c^2} \frac{y}{2} \sqrt{y^2 + 4}\right) + \frac{4GM_L}{c^3} \ln\left(\frac{\sqrt{y^2 + 4} + y}{\sqrt{y^2 + 4} - y}\right)\right], \tag{37.5}$$

where z_L is the redshift of the lens plane, $y = r/\theta_E (D_L + D_{LS})$ with θ_E being the Einstein radius, and D_L and D_{LS} correspond to the angular diameter distances from the observer to the lensing plane and from the lensing plane to the source, respectively.

This analysis utilizes the Canadian Hydrogen Intensity Mapping Experiment (CHIME) dataset,[1] encompassing 636 FRBs. The total DM can be decomposed into individual contributions from various sources as follows:

$$\mathrm{DM} = \mathrm{DM_{MW}} + \mathrm{DM_{Halo}} + \mathrm{DM_{IGM}}(z_s) + \frac{\mathrm{DM_{Host}}}{1 + z_s}, \tag{37.6}$$

where $\mathrm{DM_{MW}}$, $\mathrm{DM_{Halo}}$, $\mathrm{DM_{IGM}}$, and $\mathrm{DM_{Host}}$ are DM contributions from our Milky way galaxy, its circumgalactic halo, IGM, and host galaxy, respectively with z_s being the source redshift. Assuming standard ΛCDM cosmology, the mean DM contribution from IGM is given by [10]

$$\langle \mathrm{DM_{IGM}} \rangle = \frac{3cH_0\Omega_b}{8\pi G m_p} \int_0^{z_s} \frac{f_{\mathrm{IGM}}(z)\chi(z)(1+z)}{\sqrt{\Omega_m (1+z)^3 + \Omega_\Lambda}}\, dz\,, \tag{37.7}$$

where H_0 is the Hubble constant, Ω_b is the baryonic matter density, m_p is the proton mass, $\chi(z)$ is the ionization fraction along the line of sight, and f_{IGM} is the baryon mass fraction in the IGM. Employing fixed value of $\langle \mathrm{DM_{Host}} \rangle = 117\,\mathrm{pc\,cm^{-3}}$ and calculating $\mathrm{DM_{MW}}$, $\mathrm{DM_{Halo}}$ using the NE2001 model of Galactic distribution of free electrons, we determine z_s for all 636 FRBs in the dataset. Note that minor adjustments to $\langle \mathrm{DM_{Host}} \rangle$ are expected to have a negligible impact on our overall results.

Considering the possibility of the lensed object being a primordial black hole (PBH), we now define the optical depth of an FRB source [11]

$$\begin{aligned}\tau(M_L, z_s) = \frac{3}{2c} f_{\mathrm{PBH}}\Omega_c \int_0^{z_S} dz_L \frac{H_0}{\sqrt{\Omega_m (1+z_L)^3 + \Omega_\Lambda}} \frac{D_L D_{LS}}{D_S} (1+z_L)^2 \\ \times \left[y_{\max}^2(\mu) - y_{\min}^2(M_L, z_L) \right],\end{aligned} \tag{37.8}$$

where f_{PBH} is the fraction of dark matter made up of PBHs, Ω_c is the current cold dark matter density, D_S is the angular diameter distance to the source, $y_{\min}$ and $y_{\max}$ are respectively the minimum and maximum impact parameters. Taking into account all FRBs, we define the following integrated optical depth

$$\bar{\tau} = \frac{1}{N_{\mathrm{FRB}}} \sum_{i=1}^{N_{\mathrm{FRB}}} \tau(M_L, z_{S,i}). \tag{37.9}$$

[1] https://www.chime-frb.ca/catalog.

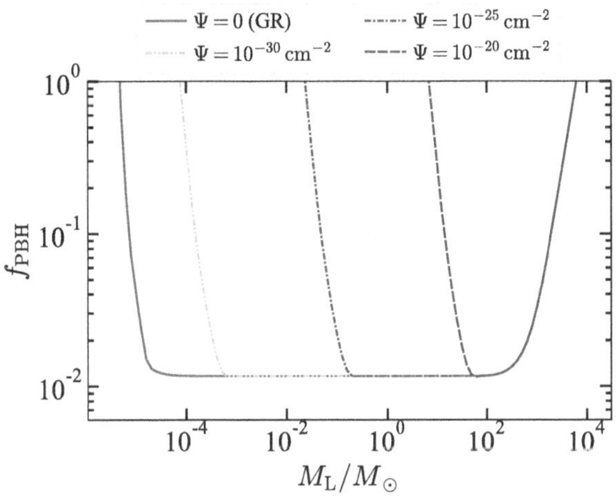

Fig. 37.3 Constraints on fraction of dark matter in primordial mass black holes for different values of modified gravity parameter Ψ

As the number of lensed FRBs is expected to be quite small in comparison to the total number of FRBs, we can safely use the Poisson statistics as follows:

$$N_{\text{lensed, FRB}} = \left(1 - e^{-\bar{\tau}}\right) N_{\text{FRB}}. \tag{37.10}$$

Since no lensed FRB has been confirmed so far, the above relation implies

$$f_{\text{PBH}} < \frac{1}{\tau_1} \ln\left(\frac{N_{\text{FRB}}}{N_{\text{FRB}} - 1}\right), \tag{37.11}$$

where $\bar{\tau} = f_{\text{PBH}}\tau_1$. Figure 37.3 presents constraints on f_{PBH} across a range of modified gravity parameter values within the standard ΛCDM cosmological framework. Notably, the bound on f_{PBH} exhibits a significant dependence on the value of the parameter Ψ. The minimum detectable time delay of the observing telescope, approximately 10^{-9} s for CHIME [12], defines the left-hand cutoff for each curve. Conversely, the right-hand cutoff is determined by the maximum achievable magnification ratio of the lensed image pair, which essentially reflects the detectability of the fainter image alongside the brighter one. Interestingly, our results demonstrate similarity to those observed for plasma lensing [12]. It suggests that modified gravity, in this context, mimics the behavior of a scattering screen positioned along the path of the light ray. It is noteworthy that current experiments can, at best, place upper limits on f_{PBH} parameter, but do not yet provide definitive evidence for PBHs.

37.4 New Generation Telescope: HIRAX

The quest to unravel the mysteries of FRBs is ongoing. In this study, we investigate the detection prospects of FRBs manifested by gravitational lensing with an under-construction HIRAX telescope in South Africa.[2] The effective detection rate of any object can be extrapolated observationally from the telescope's field of view (FoV) and several detected sources for given flux/fluence limits [13]. Assuming similarity between FRB and gamma-ray burst (GRB) detection rates, the number of GRBs detectable by a given instrument above its sensitivity flux limit S is given by [14]

$$N(>S) = \frac{\Omega T}{4\pi} \int_0^{z(L_{\max}, S)} \int_{L_{\min}(z,S)}^{L_{\max}} \Phi(L, z) \frac{\Psi(z)}{1+z} \frac{\mathrm{d}V}{\mathrm{d}z} \, \mathrm{d}L \, \mathrm{d}z \,, \tag{37.12}$$

where Ω and T are respectively system FoV and Temperature, $\Psi(z)$ is luminosity function, $\mathrm{d}V/\mathrm{d}z$ is the differential comoving volume, and $\Phi(L, z)$ is GRB formation rate with L being its luminosity and z its redshift. This equation can be modified in FRB astronomy as [15]

$$R_k = R_l \frac{\Omega_k}{\Omega_l} \left(\frac{\mathrm{SEFD}_l}{\mathrm{SEFD}_k} \sqrt{\frac{B_k n_{p,k}}{B_l n_{p,l}}} \right)^{\alpha} . \tag{37.13}$$

Here the rates R are divided between two frequency bands and Ωs are FoVs for those bands, SEFD is the corresponding system equivalent flux density, B is bandwidth and n is a number of polarizations. For current estimations, the source-count parameter $\alpha = 3R_l/2$, which is a known quantity, and R_k is derived from Eq. (37.13). HIRAX is predicted to detect a few hundred FRBs per unit solid angle per day. In this regard, it has been proposed to develop a pipeline capable of detecting FRBs at various DMs in real-time. The following equation gives an estimation of FRB detection rate above an observed fluence F_{obs} [16]

$$\mathcal{N}(>F_{\mathrm{obs}}) = \frac{1.2 \times 10^4}{4\pi} \left(\frac{F_{\mathrm{obs}}}{1.8\,\mathrm{Jy\,ms}} \right)^{-\alpha} \mathrm{sr}^{-1}\,\mathrm{day}^{-1}. \tag{37.14}$$

Figure 37.4 shows the FRB detection rate with HIRAX when the known rate is taken from the Parkes survey and all the calculations are done with uniform FRB source distribution with Euclidean sky.

The wider beamwidth of HIRAX facilitates a survey of large areas of the southern hemisphere. Current estimates predict HIRAX will detect on average several bursts daily utilizing its 256 elements. We analyze the likelihood of detecting lensed FRBs across a redshift range using the HIRAX/FRB search. We calculate the number of lensed FRBs expected at various redshifts considering different lensing sources

[2] https://hirax.ukzn.ac.za.

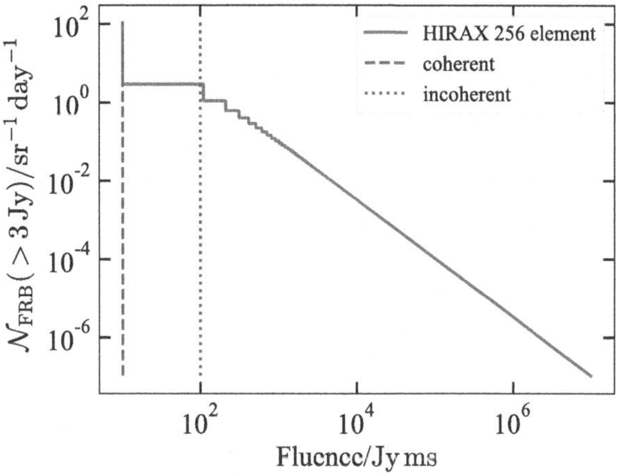

Fig. 37.4 Expected FRB rates with HIRAX 256 elements. We show flux cutoff/telescope sensitivity for incoherent and coherent beam sets of FRBs. Even though computationally expansive, coherent beam configuration provides higher sensitivity of telescope

throughout the year.[3] Our calculations incorporate microlensing [15] from sources like massive compact halo objects (MACHOs), PBHs, stars, free-floating planets, along with lensing by heavier objects like intermediate-mass black holes, dark matter halos, galaxy lensing at extended distances, and galaxy clusters. We assume a uniform distribution of both lensing sources and FRBs across the celestial sphere. Based on this, we predict roughly one lensing event per year detectable by HIRAX. We also incorporate results from CHIME and The Canadian Hydrogen Observatory and Radio-transient Detector (CHORD) in this analysis [17].

All these telescopes have ability to probe distant FRBs at high redshifts as illustrated in Fig. 37.5. This Figure represents the distribution of rate of FRBs detected by CHIME, CHORD, and HIRAX (solid curves). The shaded regions represent the normalized probability distribution function for all lensed FRBs across the redshift range. Consequently, future telescopes like HIRAX, specifically positioned in the southern hemisphere, offer significant potential for precise and sensitive FRB searches, along with predictions of lensed FRB occurrences.

37.5 Conclusions

This study investigates the application of FRBs for understanding various cosmological phenomena. We posit that high-frequency GWs of cosmological origin could be a potential FRB progenitor mechanism. In this regard, the GZ effect, where GWs convert to EM waves upon interaction with the magnetosphere of a pulsar, offers a viable explanation for FRB generation. This mechanism is particularly

[3] Our results are based on the publicly available code https://github.com/liamconnor/frb-grav-lensing.

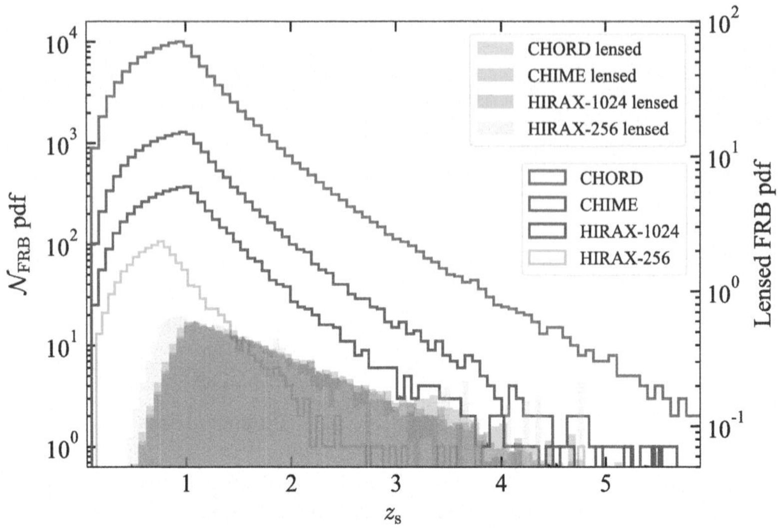

Fig. 37.5 Distribution of FRBs over redshifts. We perform a comparative analysis of three radio telescopes: CHIME, its upgrade CHORD, and HIRAX. The total number of FRBs detected by HIRAX, as well as the number of lensed FRBs, exhibits a peak around a redshift of 1

attractive as it can accommodate both repeating and apparently non-repeating FRBs. Furthermore, the future detection of GWs co-located with FRBs could differentiate between merger-based and isolated progenitor scenarios. Additionally, it could provide insights into the nature of the compact object, which is currently challenging with radio observations alone. The second part of this chapter explores how the lensing properties of FRBs can be utilized to constrain the fraction of PBHs composed of dark matter. We further demonstrate that modified gravity can mimic a scattering screen for light rays, analogous to plasma lensing.

Pinpointing the location of FRBs with high precision can help us understand their environments. Studying the host galaxies of FRBs can reveal information about the conditions favorable for their formation. In this context, several next-generation radio telescopes with enhanced sensitivity and resolution are under construction, including HIRAX, the Square Kilometre Array (SKA), CHORD, Deep Synoptic Array (DSA)-2000, and Bustling Universe Radio Survey Telescope in Taiwan (BURSTT) are being built. These telescopes hold immense promise for acquiring a wealth of new data, potentially leading to significant breakthroughs in FRB research. Our calculations estimate the FRB detection rate for HIRAX, revealing a rate comparable to CHORD, an upgraded version of the existing CHIME telescope. Both HIRAX and CHORD offer the capability to probe higher redshifts compared to CHIME, enabling the detection of fainter and more distant FRBs. Considering the rarity of astronomical objects detected at cosmological distances, the anticipated increase in FRB detections, with a significant number becoming localized, opens exciting avenues for exploring new physics, such as underscoring the possibility of resolving the Hubble tension or placing stronger limits on the photon mass in the near future.

Acknowledgments We gratefully acknowledge support from the University of Cape Town Vice Chancellor's Future Leaders 2030 Awards program which has generously funded this research and support from the South African Research Chairs Initiative of the Department of Science and Technology and the National Research Foundation.

References

1. Wang, F. Y., Zhang, G. Q., Dai, Z. G., & Cheng, K. S. (2022). *Nature Communications*, *13*, 4382. https://doi.org/10.1038/s41467-022-31923-y.
2. Bochenek, C. D., McKenna, D. L., Belov, K. V., Kocz, J., Kulkarni, S. R., Lamb, J., Ravi, V., & Woody, D. (2020). *Publications of the Astronomical Society of the Pacific*, *132*(1009), 034202. https://doi.org/10.1088/1538-3873/ab63b3.
3. Bochenek, C. D., Ravi, V., Belov, K. V., Hallinan, G., Kocz, J., Kulkarni, S. R., & McKenna, D. L. (2020). *Nature*, *587*(7832), 59. https://doi.org/10.1038/s41586-020-2872-x.
4. CHIME/FRB Collaboration, Andersen, B.C., et al. (2020). *Nature*, *587*(7832), 54. https://doi.org/10.1038/s41586-020-2863-y.
5. Platts, E., Weltman, A., Walters, A., Tendulkar, S. P., Gordin, J. E. B., & Kandhai, S. (2019). *Physics Reports*, *821*, 1. https://doi.org/10.1016/j.physrep.2019.06.003.
6. Kushwaha, A., Malik, S., & Shankaranarayanan, S. (2024). *Monthly Notices of the Royal Astronomical Society*, *527*(3), 4378. https://doi.org/10.1093/mnras/stad3545.
7. Aggarwal, N., et al. (2021). *Living Reviews in Relativity*, *24*(1), 4. https://doi.org/10.1007/s41114-021-00032-5.
8. Kalita, S., & Weltman, A. (2023). *Monthly Notices of the Royal Astronomical Society*, *520*(3), 3742. https://doi.org/10.1093/mnras/stad392.
9. Kalita, S., Bhatporia, S., & Weltman, A. (2023). *Journal of Cosmology and Astroparticle Physics*, *2023*(11), 059. https://doi.org/10.1088/1475-7516/2023/11/059.
10. Macquart, J. P., et al. (2020). *Nature*, *581*(7809), 391. https://doi.org/10.1038/s41586-020-2300-2.
11. Muñoz, J. B., Kovetz, E. D., Dai, L., & Kamionkowski, M. (2016). *Physical Review Letters*, *117*(9), 091301. https://doi.org/10.1103/PhysRevLett.117.091301.
12. Leung, C., et al. (2022). *Physical Review D*, *106*(4), 043017. https://doi.org/10.1103/PhysRevD.106.043017.
13. Keane, E. F., & Petroff, E. (2015). *Monthly Notices of the Royal Astronomical Society*, *447*(3), 2852. https://doi.org/10.1093/mnras/stu2650.
14. Pescalli, A., Ghirlanda, G., Salvaterra, R., Ghisellini, G., Vergani, S. D., Nappo, F., Salafia, O. S., Melandri, A., Covino, S., & Götz, D. (2016). *Astronomy & Astrophysics*, *587*, A40. https://doi.org/10.1051/0004-6361/201526760.
15. Connor, L., & Ravi, V. (2023). *Monthly Notices of the Royal Astronomical Society*, *521*(3), 4024. https://doi.org/10.1093/mnras/stad667.
16. Vedantham, H. K., Ravi, V., Hallinan, G., & Shannon, R. M. (2016). *The Astrophysical Journal*, *830*(2), 75. https://doi.org/10.3847/0004-637X/830/2/75.
17. Vanderlinde, K., et al. (2019). *Canadian Long Range Plan for Astronomy and Astrophysics White Papers, 2020*, 28. https://doi.org/10.5281/zenodo.3765414.

Chapter 38
Cosmic Phase Transitions in a Supercomputer?

Soumya Kanti Ganguly

Abstract In the moments immediately following the Big Bang, the universe rapidly expanded and then cooled down. This rapid cooling may have resulted in the creation of topological defects, which can be seen as remnants of the past. By studying the behavior of these defects in a computer model, we can gain insight into the early stages of the universe's formation. In this chapter, we will present the numerical results of a simple model for topological phase transitions mediated by loops, and discuss its relevance to cosmic phase transitions.

38.1 Introduction

The Big Bang theory is widely accepted as the model for the creation of the universe. According to this theory, the universe started out as a highly dense plasma which expanded and then cooled down. Approximately $10^{-12} - 10^{-10}$ seconds after the universe was formed, the electromagnetic and weak forces were unified at energy scales close to $\sim 10^2$ GeV or temperature scales $T_c \sim G_F^{1/2}$ (Fermi coupling). As the universe cooled down, it underwent a series of phase changes, marked by the formation of strings, boundaries, and domain walls that separated these phases. These objects are called topological defects (TD) and are topologically very stable, which means they cannot be removed by any continuous deformation of the field variables. TDs are the main focus of this work because they contain a lot of information about the early universe.

In the following sections, we will briefly discuss Landau's theory of phase transition and the idea of TDs. After that, we will explain the well-known Kibble-Zurek mechanism [1–3] for the formation of cosmic topological defects. Finally, we will present some numerical results obtained by us in the last few sections of the report.

S. K. Ganguly (✉)
ONGIL.AI, Chennai, India
e-mail: skg@ongil.ai

S. Ghosh, A. R. Rao (eds.), *The Relativistic Universe: From Classical to Quantum*, Astrophysics and Space Science Proceedings 61,
https://doi.org/10.1007/978-3-031-90186-7_38

38.2 Landau-Ginzburg Theory of Phase Transitions

Paul Ehrenfest was the first to perform a systematic study of phase transitions by examining the analytic properties of the thermodynamic free energy and its derivatives. However, Lev Landau laid the modern foundations of this field in the early 1930s by arguing that phase transitions can be understood based on the symmetries in the problem. During a phase transition, the system transits from a more symmetric disordered phase to a less symmetric ordered phase by *spontaneously breaking* the symmetry. For instance, in the disordered phase, the free energy of an Ising ferromagnet will have the complete discrete $\mathbb{Z}_2$ symmetry. But below the *Curie* temperature, the ± 1 symmetry is spontaneously broken, and the system is driven to an ordered state where it makes a specific choice of the magnetization state ($+1$ or -1).

Landau's theory is primarily a mean-field description where the field variables are uniform and homogeneous. However, such a theory cannot account for spatial or temporal fluctuations, which result in correlations that are crucial for phase transitions. To address this issue, Landau and Ginzburg [4] wrote down a free energy with gradient terms. This free energy successfully provided the first phenomenological description of superconductivity, which was later extended to describe superfluidity in He^4 and other important condensed matter systems [5] (Fig. 38.1).

38.3 Topological Defects and Cosmic Strings

Previously, we had discussed the example of the Ising ferromagnet with discrete $\mathbb{Z}_2$ symmetry. As the system approaches the critical region from the disordered state, the system develops correlations in the magnetization state which scales as its size. Different parts of the system respond to the thermal changes differently.

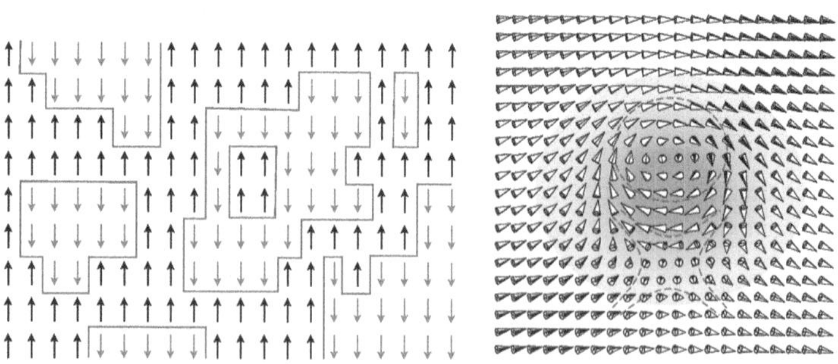

(a) (b)

Fig. 38.1 Topological defects in various physical systems: (**a**) Domain walls in a two dimensional Ising spin system; (**b**) Vector field cartoon of vortices in a superfluid medium

As a result of which the order is not established globally at once. Rather domains of $+1$ and -1 magnetization states form disjointly which are separated by well-defined boundaries called domain walls. Across these domain walls, the field variable experiences a jump discontinuity that can be energetically favored and difficult to remove. For a d-dimensional system where L is the typical length of a domain wall, its energy will scale as $E_L \sim L^{d-1}$. In the physics literature, these domain walls are known as *topological defects* (TD) [6, 7]. For systems with continuous symmetry such as the complex scalar field theory or the $O(2)$ model (XY model) in two spatial dimensions, the TDs are point vortices whose strength is determined by their winding numbers. For the same model in three dimensions we will have vortex strings (see Sects. 38.4.2 and 38.5.1). Other kinds of TDs such as dislocations and disclinations occur in solids where they are responsible for the breakdown of translational and orientational orders, respectively.

Condensed matter systems, such as magnets and superfluids, are not the only systems where phase transitions can occur. The Salam-Weinberg theory of electroweak interactions suggests a phase transition at temperatures close to 100 GeV. Above the transition temperature, there will be electroweak unification, while below this, there will be the W and Z bosons. Similarly, the quantum chromodynamics of strong interactions predicts a quark-confined or quark-gluon plasma phase above 100 MeV. Below this, there are individual hadrons, which are bound states such as protons and neutrons. Moments during and after the Big Bang, when the energy scales were that of the grand unification, or the less likely electroweak unification phase, the thermodynamic state of the universe was driven out of equilibrium as it rapidly expanded and cooled down. It is speculated that this "quenched" state of the universe gave birth to string-like topological defects (TDs) whose mass per unit length was around $\mu = 1.35 \times 10^{21} \mu_o$ Kg/m (μ_o is an adjustable parameter) [1, 2]. These cosmic strings are very similar to the three-dimensional vortex lines in superfluids (3D XY model), or the flux tubes in type-II superconductors. Careful analysis of these string fields can give us a lot of insights into the early phases of the universe. In the following section, we will discuss the dynamics of the formation of these TDs in the light of the well-known Kibble-Zurek mechanism (KZM) [7].

38.4 Quenching of Topological Defects and the Kibble-Zurek Mechanism

All physical systems have a natural timescale at which they adapt to changes in their environment. If changes in the environmental conditions occur at a rate slower than the natural frequency of the system (which is the inverse of the system's natural timescale), then the system will smoothly transition into another equilibrium or a quasi-steady state. The situation is different when the timescales are comparable or coincide. In such a scenario, the system will be driven far from equilibrium and may take a very long time to reach equilibrium again.

Moments after the Big Bang (10^{-30} seconds), the universe started expanding at an exponential rate, which rapidly cooled the system driving it into a *quenched* state. As a result of this, all the cosmic strings that were created during the process got diluted and pushed beyond the event horizon.

38.4.1 Kibble's Model of the Early Universe

In his original work, Kibble [1] considered a simple model for the universe. He wrote down a Lagrangian for an N-component real scalar field Φ^a ($a \in [1, .., N]$), minimally coupled to $N(N-1)/2$ vector gauge fields B_μ, by a coupling constant e, as

$$\mathscr{L}(\Phi^a, B_\mu) = D_\mu \Phi^a D^\mu \Phi^a - \frac{g^2}{8}(\Phi^a \Phi^a - \eta^2)^2 + \frac{1}{8} Tr(B_{\mu\nu} B^{\mu\nu}) \quad (38.1)$$

$$\text{where,} \quad D_\mu \Phi^a \equiv (\partial_\mu - eB_\mu)\Phi^a,$$

$$\text{and} \quad B_{\mu\nu} \equiv \partial_\mu B_\nu - \partial_\nu B_\mu + e[B_\mu, B_\nu].$$

The constant g is comparable to e. This model undergoes spontaneous symmetry breaking below the transition temperature. The resulting model has a broken $O(N-1)$ symmetry, whose vacuum expectation value is

$$\langle 0|\Phi^a|0\rangle = \eta, \quad a \in [1, .., N-1]. \quad (38.2)$$

Above the transition temperature, the system exhibits complete symmetry, where all points on the N-dimensional sphere S^N are equally likely. Once the symmetry is broken, the manifold of the degenerate vacua is reduced to S^{N-1}. For $N = 2$, the two scalar field components Φ^1, Φ^2 will have a broken symmetry, leading to a degenerate vacuum manifold in the shape of a circle S^1, where all points have the same energy (Fig. 38.2).

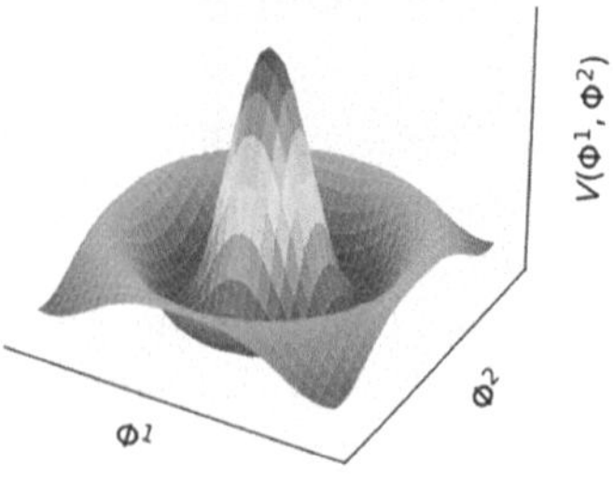

Fig. 38.2 Spontaneous symmetry breaking in a the two component scalar field potential $V(1, 2)$. The axes are labeled as (Φ^1), (Φ^2), and $(V(\Phi^1, \Phi^2))$

38.4.2 Equivalence Between the free $U(1)$ Complex Scalar Field Theory and the $O(2)$ or the XY Model

The $O(2)$ model or the XY model in statistical mechanics, is completely equivalent to the $U(1)$ model for complex scalar fields ψ, which is used to describe superfluidity in He^4. To see this, let us consider the Lagrangian for the free complex scalar field theory, given by

$$\mathscr{L}(\psi, \psi^*) = \frac{1}{2}|\partial_\mu \psi|^2 - \frac{1}{2}m_\psi^2|\psi|^2. \tag{38.3}$$

We can write down the field variable in the following polar form $\psi(\mathbf{x}) = e^{i\theta(\mathbf{x})}\psi_0, \theta \in [0, 2\pi]$, where the amplitude ψ_0 is a constant in $\mathbf{x}$ and t, and the gradient of the phase '$\nabla\theta(\mathbf{x})$' is the superfluid velocity. On a lattice, the discrete version of the gradient of ψ is given by

$$\psi(\mathbf{x}+\hat{\boldsymbol{\mu}}) - \psi(\mathbf{x}) = \psi_0\left(e^{i\theta(\mathbf{x}+\hat{\boldsymbol{\mu}})} - e^{i\theta(\mathbf{x})}\right), \quad \hat{\mu} \in [1, 3], \tag{38.4}$$

$$\text{or,} \quad |\psi(\mathbf{x}+\hat{\boldsymbol{\mu}}) - \psi(\mathbf{x})|^2 = 2\psi_0^2\left[1 - \cos\left(\theta(\mathbf{x}+\hat{\boldsymbol{\mu}}) - \theta(\mathbf{x})\right)\right].$$

The Hamiltonian for the Lagrangian in Eq. (38.3) on a lattice is simply

$$\mathscr{H}(\theta) = \psi_0^2\left[1 - \cos\left(\theta(\mathbf{x}+\hat{\boldsymbol{\mu}}) - \theta(\mathbf{x})\right)\right] + m_\psi^2\psi_0^2. \tag{38.5}$$

Apart from the constant density terms involving ψ_0, Eq. (38.5) represents the Hamiltonian or the Hamiltonian density for planar spins on a lattice, also known as the XY-model. In the ground state of such a system, all the spins point in a given direction. When all the spins are uniformly rotated by a constant angle, there is no change in energy. At low temperatures close to the ground state, non-uniform changes in θ will result in spin wave-like excitations. In the limit when fluctuations among neighboring spins are small, we can make the following approximation, as

$$1 - \cos\left(\theta(\mathbf{x}+\hat{\boldsymbol{\mu}}) - \theta(\mathbf{x})\right) \approx \frac{1}{2}\left(\theta(\mathbf{x}+\hat{\boldsymbol{\mu}}) - \theta(\mathbf{x})\right)^2, \tag{38.6}$$

$$= \frac{1}{2}(\mathbf{a}.\nabla\theta(\mathbf{x}))^2 ,$$

where $\mathbf{a}$ is the lattice constant. Within this approximation, the Hamiltonian density for the spin system will be

$$\mathscr{H}(\theta) = \frac{\rho_s}{2}|\nabla\theta(\mathbf{x})|^2 , \tag{38.7}$$

where $\rho_s = ca^2\psi_0^2/2$, is known as the thermodynamic stiffness constant, and 'c' is the co-ordination number.

38.4.3 Zurek's Extension of Kibble's Work

The superfluid velocity $\nabla\theta$, can have a *regular* part, and a *singular* part, arising due the vortices. This means that the XY model can exhibit topological defects (TDs) in the form of stable vortices [5]. In two dimensions, the TDs are point vortices, while in three dimensions, they are vortex lines. The dynamics of the superfluid order parameter are governed by the Gross-Pitaevskii equation [8, 9]. Zurek considered this as the model system to study the quenching of vortices during the transition from normal to superfluid in He^4 [3]. Even though $N = 2$ is a special case, it exhibits critical behavior characterized by universal exponents. This means that two different physical phenomena, such as the superfluid transition mediated by vortices and the formation of cosmic strings during the Big Bang, are characterized by the same underlying Hamiltonian with the same symmetry properties. Therefore, they exhibit the same critical behavior and have the same critical exponents. In this section, we will explore Zurek's theory to understand how these cosmic vortex strings could have formed during the early stages of the universe's formation.

Let us consider a physical system that exhibits critical behavior and has an intrinsic time scale 'τ_0' determined by the microscopic properties of the system. Let 'ξ' be the equilibrium correlation length of the system near the critical region (T_c) which has the following scaling relation

$$\xi \sim |\epsilon|^{-\nu}, \tag{38.8}$$

where $\epsilon \equiv 1 - T/T_c$ and 'ν' is the correlation length exponent. If 'τ_Q' be the quench time, then for linear quench,

$$\epsilon = \frac{t}{\tau_Q}. \tag{38.9}$$

If 'τ_R' be the characteristic time of the system to adapt to thermal changes, then near the critical region,

$$\tau_R(\epsilon) = \frac{\tau_0}{|\epsilon|^{\nu z}}, \tag{38.10}$$

where 'z' is the *dynamical critical exponent* [10, 11]. When the system is far from criticality, it responds adiabatically to its surroundings, but near the critical region, it fails to do so. As $T \to T_c$, this response time 't' has a lower bound given by

$$t \geq (\tau_0 \tau_Q^{\nu z})^{\frac{1}{1+\nu z}}. \tag{38.11}$$

Close to the critical region, fluctuations in the ordered state occur at length scales determined by ξ. From Eq. (38.10), one finds

$$\epsilon(t) = \left(\frac{\tau_0}{\tau_Q}\right)^{\frac{1}{1+\nu z}}. \tag{38.12}$$

Substituting this relation in Eq. (38.8), one obtains the relation between the correlation length and the quench time τ_Q, as

$$\xi = \xi_0 \left(\frac{\tau_Q}{\tau_0}\right)^{\frac{\nu}{1+\nu z}}. \tag{38.13}$$

As previously discussed, during the onset of the symmetry-breaking phase, the system will experience the formation of topological defects throughout the region. In this scenario, ξ offers a direct estimation of these defect densities. For example, in a system with periodic boundary conditions that are proliferated by defects such as loops (Fig. 38.5) of different sizes, one can use the well-known *depth-first search* (DFS) algorithm [12, 13], which is a graphical algorithm to find all the connected components of different sizes and return their frequencies [7, 14]. If '$N(s)$' be the number of defects of a given size 's', then the average size of the defect is given by

$$M = \frac{1}{L^3}\sum_s s N^{(c)}(s). \tag{38.14}$$

From Eq. (38.13), we will have

$$M \propto \xi \propto \tau_Q^p, \quad \text{where,} \quad p = \frac{\nu}{1+\nu z}. \tag{38.15}$$

The connected components are a direct measure of the average loop size 'M' in Eq. (38.14). The exponent p in Eq. (38.15), which is directly related to M can be calculated from computer experiments. According to the Kibble-Zurek theory, the density of topological defects can be directly calculated using Eq. (38.15).

In the following section, we will see how vortices can be transformed into loops by a technique known as duality. The loops can be easily simulated by a computer and show the same critical behavior as the vortices (Fig. 38.3).

38.5 Vortices, Loops and Duality

Vortices behave similarly to electric charges. They have long-range interactions and their strength is determined by their *winding numbers*. Depending on the nature or chirality of the winding, they can be positive or negative. The long-range interactions among vortices often make it difficult to study their critical behavior.

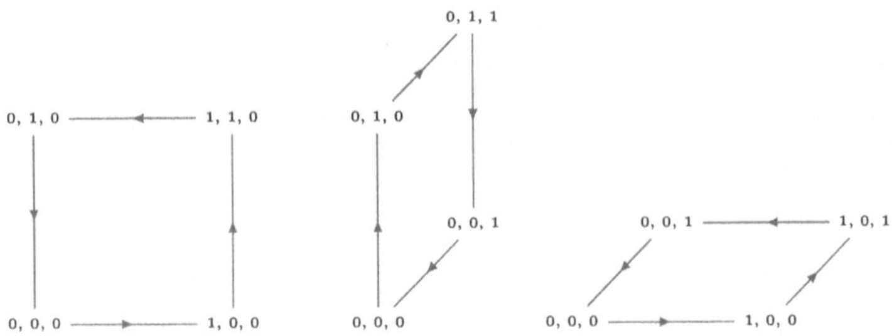

Fig. 38.3 Loops or plaquettes oriented along different planes in three dimensions. (Left) Anti-clockwise plaquette along XY-plane. (Middle) Clock-wise plaquette along YZ-plane, (Right) Anti-clockwise plaquette along XZ-plane

Physicists have used the duality trick to convert the vortices into loops that can be easily simulated in a computer to overcome this problem. In this section, we will outline the method.

38.5.1 Villain's Dual Theory for Vortices

In the previous section, we discussed how the superfluid phase in He^4 can be explained by the XY model, or its equivalent to the $U(1)$ complex scalar field theory. To understand how the loop description can be derived from the XY model, let's examine the canonical partition function for the Hamiltonian (excluding the additive constant factor $m_\psi^2 \psi_0^2$) in Eq. (38.5)

$$Z(\beta) = \int \mathcal{D}\theta \exp\Big[-\beta\psi_0^2 \sum_{\mu,\mathbf{x}} \Big(1 - \cos\big(\mathbf{\Delta}_\mu \theta(\mathbf{x})\big)\Big)\Big], \quad \beta = 1/k_B T,$$

where,

$$\mathbf{\Delta}_\mu \theta(\mathbf{x}) \equiv \theta(\mathbf{x} + \hat{\mu}) - \theta(\mathbf{x}) \quad \text{and,} \quad \int \mathcal{D}\theta \equiv \prod_{\mathbf{x}} \int_0^{2\pi} d\theta(\mathbf{x}). \quad (38.16)$$

Since the Hamiltonian in the exponent of Eq. (38.16) is a periodic function in the θ variables, it is invariant under the transformation

$$\mathbf{\Delta}_\mu \theta(\mathbf{x}) \to \mathbf{\Delta}_\mu \theta(\mathbf{x}) + 2\pi n_{\hat{\mu},\mathbf{x}}, \quad (38.17)$$

where, '$n_{\hat{\mu},\mathbf{x}}$' are integers. Following the work of Villain [5, 15, 16], we can transform the partition function in Eq. (38.16) that involves the θ degrees of freedom into a partition function expressed in terms of the integer variables $n_{\hat{\mu},\mathbf{r}}$, on the *dual lattice* '$\mathbf{r}$' (Fig. 38.4), obeying a continuity condition, given by

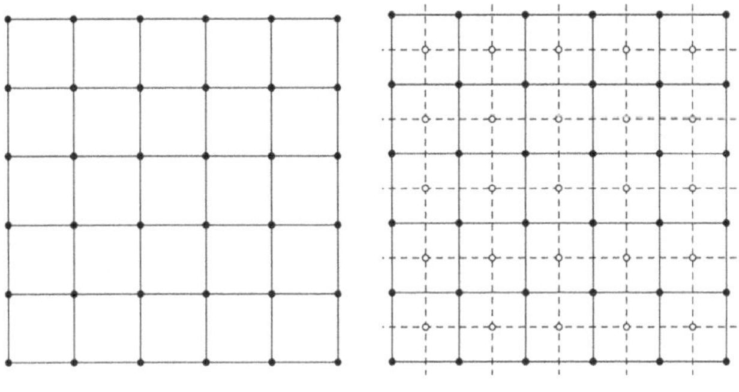

Fig. 38.4 (**a**) original lattice in two dimensions; (**b**) dual lattice with similar grid structure in dashed lines and smaller open circles

$$Z = \sum_{n_{\hat{\mu},\mathbf{r}}} \exp\Big[-\sum_{\mu,\mathbf{r}} \frac{1}{2\beta} n^2_{\hat{\mu},\mathbf{r}}\Big] \delta_{\boldsymbol{\Delta}_{\hat{\mu}} n_{\hat{\mu},\mathbf{r}},0}. \tag{38.18}$$

In this equation, the sum in the partition function is constrained over configurations of $n_{\hat{\mu},\mathbf{r}}$ that satisfy the following divergenceless condition on the dual lattice $\mathbf{r}$

$$\boldsymbol{\Delta}_{\hat{\mu}} n_{\hat{\mu},\mathbf{r}} = 0. \tag{38.19}$$

The resulting model is a *discrete Gaussian* [17] model involving *closed loop* configurations (Fig. 38.3). Moreover, the partition function for the XY-model in Eq. (38.16) and the one for the loops in Eq. (38.18) have the following relation

$$Z_{XY}(\beta) = Z_{loop}(1/\beta). \tag{38.20}$$

This means that if the XY-model exhibits critical behavior at β_c, then the loop model will exhibit the same critical behavior at $1/\beta_c$. This is a property of dual models in statistical mechanics, where the high-temperature properties of one model are related to the low-temperature properties of its dual model.

In the following section, we will analyze the results of computer simulations that we conducted on a variation of these loop models (Fig. 38.5).

38.6 Simulating Loops in a Supercomputer

Computer simulations play a crucial role in studying physical phenomena. In the field of high-energy physics, Monte Carlo simulations have been widely used to investigate lattice gauge theories [18–20]. We conducted computer experiments on a specific loop model using Monte Carlo simulations on a supercomputer (NVIDIA,

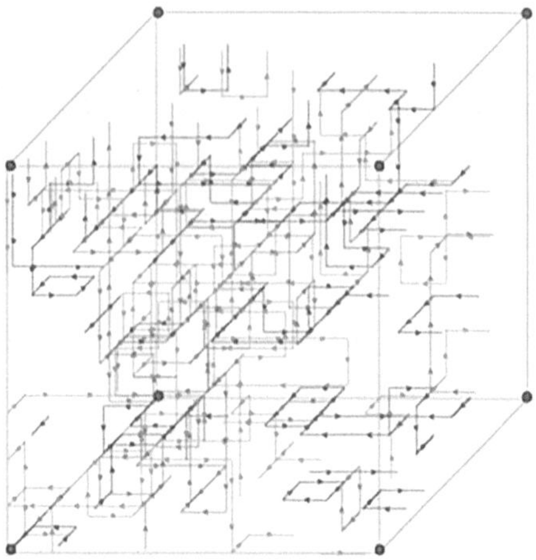

Fig. 38.5 High temperature disordered state of the NS loop system

Tesla GPU clusters) [14, 21]. In this section, we will provide a brief description of this model andi discuss its quenching properties.

38.6.1 The Multi-color Loop Models

Our primary objective was to investigate order-disorder transitions in regular three-dimensional lattice systems. The loop model we examined consists of integer-valued tensors 'η_{ij}', where the indices $i, j \in [1, 3]$, represent the direction and an additional property referred to as "*color*" respectively (distinct from color charges in QCD). Also, η_{ij} may or may not be symmetric in the indices. Interestingly, we have found that the thermodynamic properties of the two models are very different. The model with non-symmetric indices shows a second-order phase transition, while the model with symmetric indices shows a first-order phase transition [21]. In this section we will briefly describe the two models. Hereafter, the non-symmetric model will be referred to as the NS loop model, while its symmetric cousin will be called the SY loop model. In Eq. (38.19), we saw that the loops of each color obey a lattice continuity equation in the dual space $\mathbf{r}$, as

$$\Delta_i \eta_{ij}(\mathbf{r}) = 0, \quad j \in \{1, 2, 3\}. \tag{38.21}$$

The Hamiltonian for both the models is a quadratic function of the η_{ij}s, which is given by

$$H(\eta_{ij}) = \sum_{i,j,\mathbf{r}} [A\eta_{ii}^2(\mathbf{r}) + B\eta_{ij}^2(\mathbf{r}) + D\eta_{ii}(\mathbf{r})\eta_{jj}(\mathbf{r})], \tag{38.22}$$

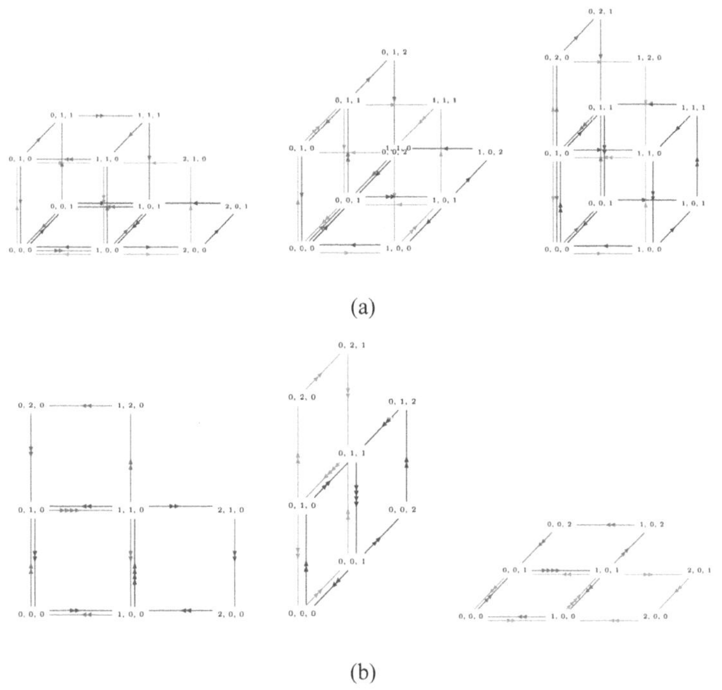

Fig. 38.6 Elementary excitations in SY loops: (**a**) first excited states with 15 units of energy (kBT units); (**b**) second excited states with 32 units of energy

where, $A = 0.5$, $B = 0.25$, and $D = 0.1$, in units of k_BT. Since we are focusing on studying the order-disorder transition in this loop model using computer simulations, understanding the computational mechanism for generating these loops in the system is crucial. In the following sub-section we will discuss the simulation method (Fig. 38.6).

38.6.2 Monte Carlo Update Rules for the Loop Models

In this sub-section, we will describe the simulation method by discussing the update rules for one of the loop models, namely the NS loop model. The update rules for the SY loops are more complex, and we refer to our work [21] for details. As mentioned earlier, the loop variables are integer-valued quantities. The ground state of the lattice system does not contain any loops, as shown below

$$\eta_{ij}(\mathbf{r}) = 0 \quad \forall i, j, \mathbf{r}. \tag{38.23}$$

With the ground state as the initial state, suppose that a red-colored loop ($j = 1$) with anti-clockwise chirality and orientation along the XY plane, occurs spontaneously at the point $\mathbf{r} = (0, 0, 0)$ (Fig. 38.3 leftmost plaquette). This loop will be represented by four bonds between the lattice points $(0, 0, 0)$, $(1, 0, 0)$, $(1, 1, 0)$, and $(0, 1, 0)$, respectively. The bond between the points $(0, 0, 0)$ and $(1, 0, 0)$ will be occupied by the loop variable $\eta_{11}(0, 0, 0) = +1$. It is $+1$ because the flow is in the positive x-direction. Similarly, the link between the lattice points $(1, 0, 0)$ and $(1, 1, 0)$ will be occupied by the loop variable $\eta_{21}(1, 0, 0) = +1$. For the lattice point $(0, 1, 0)$, there is an inward flow from $(1, 1, 0)$, hence $\eta_{11}(0, 1, 0) = -1$. Finally, we complete the circuit by a flow from $(0, 1, 0)$ to $(0, 0, 0)$ by $\eta_{21}(0, 0, 0) = -1$. At an arbitrary lattice point $\mathbf{r}$, such a loop will be described by the following set of update equations

$$
\begin{aligned}
\eta_{11}(\mathbf{r}) &\rightarrow \eta_{11}(\mathbf{r}) + 1, \\
\eta_{21}(\mathbf{r}) &\rightarrow \eta_{21}(\mathbf{r}) - 1, \\
\eta_{11}(\mathbf{r} + \hat{y}) &\rightarrow \eta_{11}(\mathbf{r} + \hat{y}) - 1, \\
\eta_{21}(\mathbf{r} + \hat{x}) &\rightarrow \eta_{21}(\mathbf{r} + \hat{x}) + 1,
\end{aligned}
\tag{38.24}
$$

where $\hat{x}$ and $\hat{y}$ are the unit vectors along the x and the y directions, respectively. The arrow signs indicate that the instantaneous values of the loop variables are updated. These update equations readily obey the continuity condition given by Eq. (38.21). The update equations for the other loops of different colors can be written in a similar manner. The SY loops can be obtained from the NS loops through the process of *symmetrization*. This involves expressing the following *non-local* relationship

$$
\bar{\eta}_{ij}(\mathbf{x}) = \Delta_m[\epsilon_{mjl}\eta_{il}(\mathbf{x}) + \epsilon_{mil}\eta_{jl}(\mathbf{x})]. \tag{38.25}
$$

Here, $\bar{\eta}_{ij}$ represents the SY loops, $\Delta_m f(\mathbf{x}) \equiv f(\mathbf{x} + \hat{m}) - f(\mathbf{x})$, and ϵ is the third order anti-symmetric tensor. It is easy to check that $\bar{\eta}_{ij} = \bar{\eta}_{ji}$, and since $\Delta_i \eta_{ij}(\mathbf{x}) = 0$ we will have $\Delta_i \bar{\eta}_{ij}(\mathbf{x}) = 0$.

38.6.3 The Metropolis Algorithm for Generating the Loops

We utilize the Metropolis algorithm, the update rules specified in Eq. (38.24), and the Hamiltonian presented in Eq. (38.22) to generate a canonical ensemble of loops. According to the algorithm, the evolution of the canonical ensemble can be represented by a *Markov chain* of events [22]. The *transition probability* from state a to b is represented as '$W(a \rightarrow b)$'. If ΔE be the change in the energy of the system while transiting from a to b, then we accept the new state with a probability which is given by the $\min(1, \exp(-\beta \Delta E))$. In other words

$$W(a \to b) = \begin{cases} \exp(-\beta \Delta E) & \text{if} \quad \Delta E > 0 \\ 1 & \text{otherwise} . \end{cases} \tag{38.26}$$

We have explored how to simulate the thermodynamic properties of loops by generating a canonical ensemble of these loops and studying their statistical properties. In the following section, we will examine their quenching properties and calculate the relevant exponents.

38.7 Quenching of NS Loops

For specific parameters $A = 0.5$, $B = 0.25$, and $D = 0.1$, we observed that the NS loop model exhibited critical behavior at $T_c \approx 0.2$ [21]. To investigate the quenching of these loops, we initialized a system of size $L = 8$ to a high-temperature disordered state ($T > T_c$) (Fig. 38.5), and then quenched to a temperature of $T = 0.1$. The loops disintegrated smoothly and the system entered the ordered state (Fig. 38.7a). Due to the continuous nature of the transition seen in NS loops, this kind of behavior is not surprising. For NS loops the correlation length exponent $\nu = 0.73(0053)$ [14]. If the dynamical critical exponent $z = 2$ [10], then from Eq. (38.15), the exponent related to the average loop size $p = 0.59(0052)$. Our computer experimental data indicates that $p = 0.52(003)$ (Fig. 38.7b), which is accurate within 7%. The discrepancy may be attributed to the inter-color coupling $D = 0.1$ which is relevant for the critical behavior in NS loops. Experiments on quenched vortices in hexagonal Manganites, which act as a model system for the three-dimensional XY-model, suggest a value close to $p \approx 0.59$ [11]. Additionally, the value of p obtained from numerical studies of the three-dimensional XY-model is found to be close to $p = 0.57$ [23].

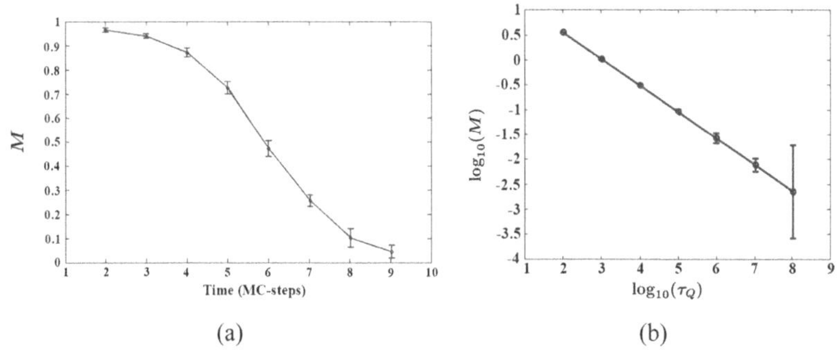

Fig. 38.7 Quenching in NS loops (Color 2): (**a**) M plotted as function of time (Monte Carlo steps); (**b**) Log-Log plot for M vs τ_Q ($p = 0.52\,(003)$)

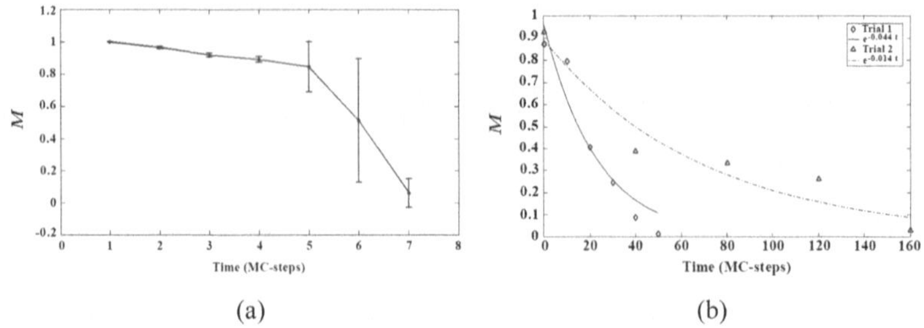

Fig. 38.8 Quenching in SY loops (Color 2): (**a**) M plotted as function of time (Monte Carlo steps); (**b**) M vs time (short time limit) Two sets of data points with fitted exponential decay curves. The legend indicates "Trial 1" with circles and a solid line, and "Trial 2" with triangles and a dashed line. The equations ($e^{-0.044t}$) and ($e^{-0.014t}$) represent the decay rates

38.8 Quenching of SY Loops

The protocol for studying quenching in SY loops is the same as that for the NS loops. Unlike the NS loops, the SY loops undergo a first-order discontinuous transition, where the transition temperature is $T_c \approx 2.14$ [21]. We have conducted many quenching experiments with different initial configurations. We found that during many instances, the loops would not disintegrate as they would be trapped in a *metastable* state. The entrapment of the SY loops in this state is evident from the M vs t plot (Fig. 38.8a), showing a significant variance in the average loop size. On the other hand, when the loops disintegrated into an ordered state, they did so with different time constants, as shown in Fig. 38.8b. We performed an exponential fit for two such curves and found the time constants to be 4.4×10^{-2} and 1.4×10^{-2} respectively.

38.9 Summary and Conclusion

The occurrence of phase transitions in all thermodynamic systems can be influenced by topological defects, which are widespread from the Big Bang to Superfluidity in He^4. This chapter primarily focuses on studying the thermodynamic properties of these transitions through computer-aided simulations. Specifically, we have examined systems where topological defects can be represented by loops using a technique known as duality. These loops undergo order-disorder transitions in a regular three-dimensional lattice. The non-symmetric loops undergo a continuous transition, while the symmetric loops display a first-order phase transition. Quenching studies of these loops reveal that the non-symmetric loops decay smoothly into the ordered state, whereas the symmetric loops can be trapped in a metastable state. The quench exponent for non-symmetric loops has been calculated using the Kibble-Zurek theory, and it is found to be within 7% when compared with experiments and previous studies on the XY model in three dimensions. This confirms the accuracy of both the method and the model.

38.10 Future Directions

Phase transitions caused due to topological defects or equivalently loops, are like expanding networks or domains. Computationally, one counts the number of lattice sites that belong to a cluster (connected components), and then report their sizes (see Sect. 38.4.3). From the geometrical point of view, this is like a percolation transition [24, 25], where disjoint components of different sizes join one another to form giant clusters or loops. On the other hand, one can also treat the boundaries or walls of these expanding or contracting domains as independent degrees of freedom with their dynamics. For instance, in a two-dimensional Ising model, the domain walls can be seen as one-dimensional wiggling polymer chains, resembling extended objects. Similarly, in three dimensions, the loop dual to a vortex string resembles a closed polymer loop [26, 27]. In cosmology, determining the density of these dynamic domain walls during a continuous transition, or calculating the dynamical critical exponents, remains an open question. Additionally, do these domain walls form lattices during a phase transition? These are some of the related questions that need exploration (see [7]).

Acknowledgments I would like to express my gratitude to the organizers of ISRA 2023. I am especially thankful to Prof. Shubhrangshu Ghosh and Prof. Banibrata Mukhopadhyay. I would also like to extend my thanks to Prof. C. Dasgupta (IISc, ICTS) for suggesting the problem and to Dr. S. Mukherjee (IBM Research Labs, Bangalore) for engaging in numerous scientific discussions. Additionally, I acknowledge the financial support from the Council of Scientific & Industrial Research (CSIR), the Indo-U.S. Science & Technology Forum (IUSSTF), India. I am also grateful for the computational support from the Supercomputer Education & Research Centre (SERC) at IISc, Bangalore. I would like to thank the entire ONGIL technical team for their constant technical and moral support. Finally, I would like to express my gratitude to all the anonymous referees for their valuable suggestions.

References

1. Kibble, T. W. B. (1976). *Journal of Physics A, 9*, 1387.
2. Hindmarsh, M. B., & Kibble, T. W. B. (1995). *Reports on Progress in Physics, 58*, 477.
3. Zurek, W. H. (1985). *Nature, 317*, 10.
4. Ginzburg, V. L., & Landau, L. D. (1950). *Journal of Experimental and Theoretical Physics, 20*.
5. Chaikin, P. M., & Lubensky, T. C. (1995). *Principles of condensed matter physics*. Cambridge University Press.
6. Mermin, N. D. (1979). *Reviews of Modern Physics, 51*, 3.
7. Vachaspati, T. (2006). *Kinks and domain walls*. Cambridge University Press.
8. Gross, E. P. (1961). *Nuovo Cimento, 20*, 454.

9. Pitaevskii, L. P. (1961). *Soviet Physics Journal of Experimental and Theoretical Physics, 13*(2), 451.
10. Hohenberg, P., & Halperin, B. I. (1977). *Reviews of Modern Physics, 49*, 435.
11. Lin, S. Z., et al. (2014). *Nature Physics, 10*, 970.
12. Sedgewick, R. (1990). *Algorithms in C*. Addison-Wesley.
13. Hartmann, A., & Rieger, H. (2002). *Optimization algorithms in physics*. Wiley-VCH.
14. Ganguly, S. K., Mukherjee, S., & Dasgupta, C. (2024). *Physical Review E, 110*, 034117.
15. Villain, J. (1975). *Journal de Physique, 36*(6), 581–590.
16. Savit, R. (1980). *Reviews of Modern Physics, 52*, 2.
17. Chui, S. T., & Weeks, J. D. (1976). *Physical Review B, 14*, 4978.
18. Creutz, M., et al. (1979). *Physical Review D, 20*, 1915.
19. Creutz, M. (1980). *Physical Review D, 21*, 2308.
20. Rebbi, C. (1983). *Lattice Gauge theories and Monte Carlo simulations*. World Scientific.
21. Ganguly, S. K., Mukherjee, S., & Dasgupta, C. (2024). *Physical Review E, 110*, 034116.
22. Landau, D., & Binder, K. (2009). *A guide to Monte Carlo simulations in statistical physics*. Cambridge University Press.
23. Campostrini, M., et al. (2001). *Physical Review B, 63*, 214503.
24. Broadbent, S. R., & Hammersley, J. M. (1957). Percolation processes. *Mathematical Proceedings of the Cambridge Philosophical Society, 53*, 629.
25. Stauffer, D., & Aharony, A. (1991). *Introduction to percolation theory*. Taylor and Francis.
26. Kierfeld, J., & Vinokur, V. (2000). *Physical Review B* (Rapid Communications), *61*, R14928(R).
27. Burakovsky, L., Preston, D., & Silbar, R. (2000). *Physical Review B, 61*, 15011.

Part X
Quantum Physics in Space

Chapter 39
Quantum Correlations in Neutrino and Neutral Meson Oscillations

Subhashish Banerjee

Abstract We discuss the impact of ideas of open quantum systems and quantum information to various facets of neutrino and neutral meson oscillations. These oscillations are characterized by a number of quantum correlations, both spatial as well as temporal. For neutrinos, the correlations are shown to be simple functions of the product of neutrino survival and oscillation probabilities. The quantum correlations in the neutral mesons are seen to be non-trivially different from their stable counterparts.

39.1 Introduction

The theory of open quantum systems addresses the issues of damping and dephasing by positing that all practical systems interact with their surrounding environments, thereby classifying them as 'open' systems [1]. Quantum optics was among the first fields to apply and test the concept of open quantum systems [2]. The application of this theory to other fields was significantly advanced by the work of [3], [4], and others. The recent increase in interest in open quantum systems is driven by remarkable progress in manipulating quantum states of matter, and in encoding, transmitting, and processing quantum information [5–7]. Understanding and controlling environmental impacts is crucial for all these advances. This enhances the importance of open system notions for quantum computation and information.

In recent years, there has been an active intersection between ideas of open quantum systems and quantum information with sub-atomic physics. Here, we will focus on this issue. In particular, we will study neutrino and neutral meson oscillations in this framework. Pauli proposed the existence of neutrinos to bring about energy and momentum conservation in beta decay [8]. Neutrinos are now understood to come in three flavors ν_e, ν_μ, ν_τ, that is, electron, muon, and tau

S. Banerjee (✉)
Indian Institute of Technology Jodhpur, Jodhpur, Rajasthan
e-mail: subhashish@iitj.ac.in

S. Ghosh, A. R. Rao (eds.), *The Relativistic Universe: From Classical to Quantum*, Astrophysics and Space Science Proceedings 61,
https://doi.org/10.1007/978-3-031-90186-7_39

neutrinos. They oscillate because they have non-zero mass [8]. Oscillating neutrinos fundamentally involve three flavors but can often be approximated as two-flavor scenarios. Prominent issues related to neutrinos are the neutrino mass hierarchy, CP (charge-parity) violation, among others. The investigation of these issues are the primary focus of the current neutrino experiments including the (long-baseline) accelerator [9].

Neutral mesons K^0, B_d^0, B_s^0 are unstable massive systems undergoing oscillations. The study of single and decaying correlated mesons provide an interesting platform for open system ideas to decaying quantum systems. Further, correlated neutral mesons facilitate the interplay of quantum correlations and CP-symmetry.

The chapter is laid out in the following manner. Section 39.2 introduces open quantum systems and their connection to quantum channels. Section 39.3 provides a brief introduction to several measures of quantum correlations, both spatial and temporal, that will be subsequently used on neutrino and meson oscillations. Quantum correlation in neutrino oscillations are then discussed, followed by neutral meson systems. We then make our conclusions.

39.2 Open Quantum Systems

Here, we will make explicit what we mean by open quantum systems [1]. To illustrate the formalism, we will provide simple examples of open systems, of relevance to quantum information. Following this, we will connect open system ideas to quantum information. Consider the Hamiltonian of the total (closed system):

$$H = H_S + H_R + H_{SR}, \tag{39.1}$$

where S represents the system, R denotes the reservoir (bath), and SR signifies the interaction between them.

The given Hamiltonian of the system-reservoir follows unitary evolution as: $\rho(t) = e^{-\frac{i}{\hbar}Ht}\rho(0)e^{\frac{i}{\hbar}Ht}$. We focus on the reduced dynamics of the system S, while accounting for its environmental influence. This is accomplished by tracing over the reservoir degrees of freedom to obtain the reduced dynamics, which is *non-unitary*: $\rho^S(t) = \mathrm{Tr}_R[\rho(t)]$. Open quantum systems are commonly categorized into two main types: (a) Quantum non-demolition (QND), where $[H_S, H_{SR}] = 0$, leading to decoherence without dissipation [10, 11] and (b) Quantum dissipative systems, where $[H_S, H_{SR}] \neq 0$ resulting in decoherence accompanied by dissipation [3, 12].

In quantum information theory, the noise generated by a QND open system is referred to as a "phase damping channel", whereas the noise from a dissipative (Lindblad) evolution is termed a "(generalized) amplitude damping channel".

39.2.1 System-Reservoir Initial Conditions

The evolution of the open system begins with the system and the reservoir either initially uncorrelated or correlated.

(A) Separable Initial Condition: It is assumed that the system and the environment (reservoir) are initially independent of each other [3, 13]. Under these conditions, the initial density matrix can be expressed as $\rho(0) = \rho^S(0) \otimes \rho^R(0)$, where $\rho^S(0)$ and $\rho^R(0)$ represents the system's and reservoir's initial density matrices, respectively.

(B) Non-Separable Initial Condition: in many scenarios, the system and the reservoir are interconnected as parts of a larger system, and their interaction is beyond our control. This situation prompts the use of a specific class of initial conditions known as 'generalized initial conditions' [12, 14]. A broad category of these initial conditions can be expressed as $\rho_0 = \sum_j O_j \rho_\beta O'_j$, where $\rho_\beta = Z_\beta^{-1} \exp(-\beta H)$ is the canonical density matrix that describes the equilibrium state of the interacting system in the presence of a time-independent potential V, with Z_β^{-1} being the partition function. Here $\beta = (k_B T)^{-1}$, where T is the equilibrium temperature of the interacting system. The operators O_j, O'_j act solely on the system's coordinates, leaving the reservoir's coordinates (environment) unchanged, though they can be selected arbitrarily otherwise.

Time Scales Associated with the Open System Evolution Various time scales characterize the open system evolution, with the most notable being:

(a) Scale related to the system's natural frequency
(b) Relaxation time scale governed by the strength of the $S - R$ coupling
(c) Reservoir correlation time (memory time), linked to the high-frequency cutoff in the reservoir's spectral density and the time scale related to the reservoir temperature, which reflects the relative significance of quantum effects compared to thermal effects

Dissipative Form of Evolution Equation: Lindblad Form A widely used form of the evolution equation for the reduced density matrix of a system, applicable when the system and interaction Hamiltonians do not commute, is the Lindblad master equation. This equation is based on the physical assumptions of the *Born* (weak coupling), *Markov* (memoryless), and *Rotating Wave Approximation* (rapid system dynamics relative to the relaxation time).

A common prototype for the Lindblad evolution would be the decay of a two-level system $H_S = \frac{1}{2}\hbar\omega_0\sigma_z$, interacting with a radiation field (bath) in the weak Born-Markov, rotating wave approximation. In the interaction picture (ignoring the Lamb shift terms), the Lindblad master equation for $\rho^S(t)$ is

$$\frac{d}{dt}\rho^S(t) = \gamma_0 \left[(N_{th} + 1)\chi_1(t) + N_{th}\chi_2(t)\right], \tag{39.2}$$

where

$$\chi_1(t) = \left[\sigma_- \rho^S(t)\sigma_+ - (1/2)\sigma_+\sigma_-\rho^S(t) - (1/2)\rho^S(t)\sigma_+\sigma_-\right],$$

$$\chi_2(t) = \left[\sigma_+ \rho^S(t)\sigma_- - (1/2)\sigma_-\sigma_+\rho^S(t) - (1/2)\rho^S(t)\sigma_-\sigma_+\right].$$

Here, γ_0 is spontaneous emission rate $\gamma_0 = \frac{4\omega^3|\mathbf{d}|^2}{3\hbar c^3}$, and σ_+, σ_-: standard raising and lowering operators, respectively given by $\sigma_+ = |1\rangle\langle 0| = \frac{1}{2}\left(\sigma_x + i\sigma_y\right)$; $\sigma_- = |0\rangle\langle 1| = \frac{1}{2}\left(\sigma_x - i\sigma_y\right)$. In the RHS of Eq. (39.2), the first parenthesis, involving $\gamma_0(N_{th} + 1)$, describes both spontaneous emission (γ_0) and thermal emission ($\gamma_0 N_{th}$). The second parenthesis, which includes $\gamma_0 N_{th}$, represents thermal absorption. Additionally, $N_{th} = \frac{1}{e^{\frac{\hbar\omega}{k_B T}} - 1}$ represents the Planck distribution, which gives the count for thermal photons at the frequency ω.

The two-level density matrix using Pauli operators can be expressed as

$$\rho^S(t) = \left(\frac{I + \langle \boldsymbol{\sigma}(t)\rangle . \boldsymbol{\sigma}}{2}\right) = \begin{pmatrix} \alpha_+(t) & \langle \sigma_-(t)\rangle \\ \langle \sigma_+(t)\rangle & \alpha_-(t) \end{pmatrix}, \tag{39.3}$$

where $\alpha_\pm(t) = (1/2)\left(1 \pm \langle \sigma_z(t)\rangle\right)$. Using Eq. (39.3), we can readily derive Eq. (39.2). This allows us to solve for the Bloch vectors, resulting in

$$\langle \sigma_x(t)\rangle = e^{-\frac{\gamma_0}{2}(2N_{th}+1)t}\langle \sigma_x(0)\rangle, \quad \langle \sigma_y(t)\rangle = e^{-\frac{\gamma_0}{2}(2N_{th}+1)t}\langle \sigma_y(0)\rangle,$$

$$\langle \sigma_z(t)\rangle = e^{-\gamma_0(2N_{th}+1)t}\langle \sigma_z(0)\rangle - \frac{1}{(2N_{th}+1)}\left(1 - e^{-\gamma_0(2N_{th}+1)t}\right). \tag{39.4}$$

39.2.2 Connection to Quantum Noise Processes

Now, we seek to relate the above findings to common noisy channels: In what ways can quantum computing be influenced by environmental factors? In operator-sum representation, the action of a superoperator O due to environmental interaction is

$$\rho \longrightarrow O(\rho) = \sum_k \langle e_k|\mathcal{U}(\rho \otimes |f_0\rangle\langle f_0|)\mathcal{U}^\dagger|e_k\rangle = \sum_j E_j \rho E_j^\dagger.$$

This expression describes the unitary operator $\mathcal{U}$, which accounts for the free evolution of the system with initial state ρ, the environment, and their interaction. Here, $|f_0\rangle$ denotes the initial state of the environment, and $\{|e_k\rangle\}$ represents a basis for the environment. It is assumed that the initial states of the system and environment are separable. $E_j \equiv \langle e_k|\mathcal{U}|f_0\rangle$ implements the dynamical mapping;

and a partition of unity: $\sum_j E_j^{\dagger} E_j = I$. Any transformation representable as operator-sum is a completely positive (CP) map [15]. This method is referred to as the quantum operations formalism [16–18].

39.3 Measures of Quantum Correlations

Quantum correlations are a *many-faceted* entity. Some of the prominent ones are discussed below.

39.3.1 Bell's Inequality

This measures the non-local characteristics of the system. If ρ is the density matrix for a pair of qubits, the correlation matrix T with its matrix elements can be expressed as $T_{mn} = Tr\,[\rho(\sigma_m \otimes \sigma_n)]$. The eigenvalues u_i (where, $i = 1, 2, 3$) of the matrix $T^{\dagger}T$ quantify the Bell-CHSH inequality as $M(\rho) < 1$ [19]. Here, $M(\rho) = \max(u_i + u_j)$ $(i \neq j)$.

39.3.2 Entanglement: Concurrence

This is the most well known measure of quantum correlation. The concurrence (C) is an entanglement measure for a mixed state ρ of two qubits and is defined as $C = \max(\lambda_1 - \lambda_2 - \lambda_3 - \lambda_4, 0)$. To compute concurrence, first, find the eigenvalues of the matrix: $\sqrt{\rho}(\sigma_y \otimes \sigma_y)\rho^*(\sigma_y \otimes \sigma_y)\sqrt{\rho}$ and take their square roots. These square roots arranged in decreasing order are λ_i. Here, ρ^* is calculated on the two-qubit computational basis [20]. The entanglement of formation is equivalent to the concurrence in a two-qubit system and can be expressed as a monotonic function of the concurrence C given by $E_F = -\frac{1+\sqrt{1-C^2}}{2}\log_2(\frac{1+\sqrt{1-C^2}}{2}) - \frac{1-\sqrt{1-C^2}}{2}\log_2(\frac{1-\sqrt{1-C^2}}{2})$.

39.3.3 Teleportation Fidelity

Teleportation gives practical significance to the concept of entanglement. When $F_{\max} > \frac{2}{3}$, teleportation becomes possible. $F_{\max}$ can be determined using the eigenvalues $\{u_i\}$ of $T^{\dagger}T$. $F_{\max} = \frac{1}{2}\left(1 + \frac{1}{3}N(\rho)\right)$ where $N(\rho) = \sqrt{u_1}+\sqrt{u_2}+\sqrt{u_3}$ [19]. An inequality involving $M(\rho)$ and $F_{\max}$ is $F_{max} \geq \frac{1}{2}\left(1 + \frac{1}{3}M(\rho)\right) \geq \frac{2}{3}$ if $M(\rho) > 1$.

39.3.4 Measurement Induced Disturbance (QMID)

QMID measures the extent of quantum correlation between the bipartite states shared by two parties, say Alice and Bob. For the given $\rho'_{A,R}$, if ρ'_A and ρ'_R are the reduced density matrices, then the mutual information that quantifies the correlation between Alice and Rob is $I = S(\rho'_A) + S(\rho'_R) - S(\rho'_{A,R})$, where, $S(.)$ is the von Neumann entropy. If $\rho'_A = \sum_i \lambda'^i_A \Pi'^i_A$ and $\rho'_R = \sum_j \lambda'^j_R \Pi'^j_R$ denotes the spectral decomposition of ρ'_A and ρ'_R, respectively, then the state $\rho'_{A,R}$ after measuring in joint basis $\{\Pi_A, \Pi_R\}$ is $\Pi(\rho'_{A,R}) = \sum_{i,j}(\Pi'^i_A \otimes \Pi'^j_R)\rho'_{A,R}(\Pi'^i_A \otimes \Pi'^j_R)$. QMID [21] is $M(\rho'_{A,R}) = I(\rho'_{A,R}) - I(\Pi(\rho'_{A,R}))$ is a measure of quantumness of the correlation.

39.3.5 Discord

Quantum discord (QD) [22, 23] quantifies the disparity between total and classical correlations. It is derived from two distinct methods of extending classical mutual information into the quantum domain: one adheres to conventional mutual information, and the other represents a quantum interpretation of conditional entropy influenced by the measurement process. It is defined as $\mathcal{J}(\rho_P|\rho_C) = \mathcal{S}(\rho_P) - \mathcal{S}(\rho_P|\Pi_i^C)$ where $\mathcal{S}(\rho_P|\Pi_i^C)$ is the quantum conditional entropy, defined with respect to a set of projective measurements $\{\Pi_i\}$ performed on the state ρ_C. QD is $\mathcal{D}(\rho_P|\rho_C) = \mathcal{I}(\rho_P : \rho_C) - \max_{\Pi_i^C} \mathcal{J}(\rho_P|\rho_C)$. A conceptually similar though computationally easier measure is the geometric discord D_G [24]. An optimal version of QMID can be viewed as QD. QMID characterizes quantum correlations using spectral projections of measurement operators without any optimization and makes it relatively straightforward to compute. In contrast, QD requires optimization over all possible projectors to calculate the correlations between states, resulting in high computational complexity.

39.3.6 Leggett-Garg Inequality

This epitomizes the temporal correlations [25]. It is based on the assumptions: (a) Macroscopic realism (MR): A macroscopic system with two or more distinctly different states will always be in one of these states at any given time; (b) Non-Invasive measurability (NIM): In principle, it allows to perform the measurement without disturbing the future dynamics of the system. A central quantity here is the two-time correlation and involves dichotomic observable $Q = \pm 1$.

The two-time correlation functions have the form $C_{ij} = \frac{1}{N}\sum_{q=1}^{N} Q_i^q Q_j^q$, where $-1 \leq C_{ij} \leq 1$. $C_{ij} = \pm 1$ implies perfectly (anti-)correlated case, while $C_{ij} = 0$ implies no correlation. The simplest form of the Leggett-Garg Inequality is: $K_3 = C_{12} + C_{23} - C_{13} = \langle Q_1 Q_2 \rangle + \langle Q_2 Q_3 \rangle - \langle Q_1 Q_3 \rangle$. It can be easily shown that $-3 \leq K_3 \leq 1$.

39.4 Quantum Correlations in Neutrino Oscillations

Quantum correlations serve a vital role in understanding & harnessing the power of quantum mechanics across various natural systems. Traditionally, studies of quantum mechanics foundations have centered on optical and electronic systems, where different measures of quantum correlations are well-established. However, recent technical advances in high-energy physics, particularly in neutrino oscillation experiments, open new avenues for exploring these correlations in neutrinos. Here, we are interested in examining various aspects of quantum correlations in neutrino oscillation, including non-locality, entanglement, and weaker measures like discord.

The neutrino system is especially intriguing because it experiences minimal decoherence compared to other particles commonly used in quantum information processing. This characteristic makes neutrinos a valuable platform for exploring the foundations of quantum mechanics.

The coherent time evolution of neutrino flavor eigenstates indicates that a flavor state ν_α ($\alpha = e, \mu, \tau$) can be represented as a linear combination of its mass eigenstates ν_j ($j = 1, 2, 3$). Neutrino oscillations can be effectively described using transition probabilities between flavor states. The three flavor states (ν_α) of neutrinos are eigenstates of weak interaction that are detectable in the laboratory and can be mixed via a 3×3 unitary matrix to form the three mass eigenstates ν_j. These mass eigenstates propagate in space over time, resulting in the time-evolved flavor state. Neutrino oscillations occur only if the three corresponding masses, m_1, m_2, and m_3, are non-degenerate. Among the three mass-squared differences $\Delta_{ij} = m_i^2 - m_j^2$ (where $i, j = 1, 2, 3$ with $i > j$), only two are independent. Oscillation measurements show that $\Delta_{21} \approx 0.03 \times \Delta_{32}$, suggesting $\Delta_{31} \approx \Delta_{32}$.

In general, neutrino oscillations are described by the full three-flavor oscillation equations. However, in certain scenarios, these equations can be simplified to an effective two-flavor model by setting one or both of the small parameters, Δ_{21}/Δ_{32} and θ_{13}, to zero. For instance, in long-baseline accelerator experiments, these parameters can often be ignored for leading-order calculations. This simplification reduces the problem to two-flavor mixing, where flavor neutrinos ν_α and ν_β mix to produce two mass eigenstates, ν_j and ν_k. In this simplified model, oscillations are characterized by a single mixing angle θ ($\equiv \theta_{23}$) and a single mass-squared difference Δ ($\equiv \Delta_{32}$). The flavor and mass eigenstates are related through 2×2 rotation matrix, $U(\theta) = \begin{pmatrix} \cos(\theta) & \sin(\theta) \\ -\sin(\theta) & \cos(\theta) \end{pmatrix}$, as

$$\begin{pmatrix} \nu_\alpha \\ \nu_\beta \end{pmatrix} = U(\theta) \begin{pmatrix} \nu_j \\ \nu_k \end{pmatrix} . \tag{39.5}$$

Thus, each flavor state $|\nu_\alpha\rangle$ represents a superposition of mass eigenstates: $|\nu_\alpha\rangle = \sum_j U_{\alpha j} |\nu_j\rangle$, where $\alpha = \mu$ or τ and $j = 2, 3$. The time evolution of the mass eigenstates $|\nu_j\rangle$ is $|\nu_j(t)\rangle = e^{-iE_j t} |\nu_j\rangle$, where $|\nu_j\rangle$ are the mass states at time $t = 0$. Therefore,

$$|\nu_\alpha(t)\rangle = \sum_j e^{-iE_j t} U_{\alpha j} |\nu_j\rangle . \tag{39.6}$$

It is also possible to project the evolving flavor neutrino state $|\nu_\alpha\rangle$ onto the flavor basis as

$$|\nu_\alpha(t)\rangle = \tilde{U}_{\alpha\alpha}(t) |\nu_\alpha\rangle + \tilde{U}_{\alpha\beta}(t) |\nu_\beta\rangle , \quad \text{where } |\tilde{U}_{\alpha\alpha}(t)|^2 + |\tilde{U}_{\alpha\beta}(t)|^2 = 1 , \tag{39.7}$$

and $|\nu_\alpha\rangle$ is the flavor state at time $t = 0$. The flavor eigenstates are clearly explained in the framework of quantum mechanics, where all detectable neutrinos are ultra-relativistic. This allows us to establish a correspondence with two-qubit states by using the occupation number of neutrinos as [26, 27],

$$|\nu_\alpha\rangle \equiv |1\rangle_\alpha \otimes |0\rangle_\beta \equiv |10\rangle , \quad |\nu_\beta\rangle \equiv |0\rangle_\alpha \otimes |1\rangle_\beta \equiv |01\rangle . \tag{39.8}$$

Using Eq. (39.8) in Eq. (39.7), we get

$$|\nu_\alpha(t)\rangle = \tilde{U}_{\alpha\alpha}(t) |1\rangle_\alpha \otimes |0\rangle_\beta + \tilde{U}_{\alpha\beta}(t) |0\rangle_\alpha \otimes |1\rangle_\beta , \tag{39.9}$$

where,

$$\begin{aligned} \tilde{U}_{\alpha\alpha}(t) &= \cos^2(\theta) e^{-iE_2 t} + \sin^2(\theta) e^{-iE_3 t}; \\ \tilde{U}_{\alpha\beta}(t) &= \sin(\theta)\cos(\theta)(e^{-iE_3 t} - e^{-iE_2 t}), \end{aligned} \tag{39.10}$$

which implies that $|\nu_\alpha(t)\rangle$ exhibits an entangled state. The survival and oscillation probabilities of the initial state $|\nu_\alpha\rangle$ are given by $P_{sur} = |\tilde{U}_{\alpha\alpha}(t)|^2 = \frac{1}{4}(3 + \cos(4\theta) + 2\cos(\phi)\sin^2(2\theta))$ and $P_{osc} = |\tilde{U}_{\alpha\beta}(t)|^2 = \sin^2(2\theta)\sin^2(\phi/2)$, respectively, where, $\phi = \frac{\Delta t}{2E}$. Here, $P_{sur} + P_{osc} = 1$. Since the density matrix $\rho_\alpha(t) = |\nu_\alpha(t)\rangle \langle \nu_\alpha(t)|$ incorporates the mixing angle (θ) and the mass-squared difference (Δ), which are known, various quantum correlation measures can now be evaluated using this matrix [28, 29].

1. Non-locality:

$$\begin{aligned} M(\rho) &= 1 + \Big[3 + \cos(4\theta) + 2\cos(\phi)\sin^2(2\theta)\Big] \sin^2(2\theta)\sin^2(\phi/2) \\ &= 1 + 4 P_{sur} P_{osc}. \end{aligned} \tag{39.11}$$

$M(\rho)$ is tied up with neutrino oscillation probabilities. When $\theta = 0$ (no mixing), $M(\rho) = 1$.

2. Concurrence:

$$C = \sqrt{(3 + \cos(4\theta) + 2\cos(\phi)\sin^2(2\theta))}\sin(2\theta)\sin(\phi/2)$$
$$= 2\sqrt{P_{sur} P_{osc}}. \qquad (39.12)$$

In case of no oscillation, there is no entanglement. Also, $M(\rho) = 1 + C^2$.

3. Geometric discord:

$$D_G(\rho) = \frac{2}{3}\left[3 + \cos(4\theta) + 2\cos(\phi)\sin^2(2\theta)\right]\sin^2(2\theta)\sin^2(\phi/2)$$
$$= \frac{8}{3} P_{sur} P_{osc}. \qquad (39.13)$$

Further, $D_G(\rho) \neq 0$ for $P_{osc} \neq 0$.

4. Teleportation fidelity:

$$F_{\max} = \frac{1}{3}\left[2 + (3 + \cos(4\theta) + 2\cos(\phi)\sin^2(2\theta))^{1/2}\sin(2\theta)\sin(\phi/2)\right]$$
$$= (2/3)(1 + \sqrt{P_{sur} P_{osc}}). \qquad (39.14)$$

For nonzero P_{osc}, $F_{\max} > 2/3$.

Thus, we can see that Bell's inequality is consistently violated, demonstrating the non-local nature of neutrino evolution. The teleportation fidelity remains above 2/3, which obeys the typical relationship between Bell's inequality violation and teleportation fidelity observed in electronic and photonic systems. These quantum correlations, which are functions of the product of neutrino survival and oscillation probabilities, can exhibit classically forbidden values for non-zero oscillation probabilities. The strength of these correlations is closely linked to the neutrino mixing angle. This analysis can be extended to the case of three flavor neutrino oscillations and has been used to study both spatial and temporal correlations [30, 31], neutrino mass degeneracy problem [32], CP-violation in neutrino oscillations [33], neutrino oscillations in curved space-time [34], among others.

The various quantum correlations used in this chapter are indeed helpful for studying bipartite entanglement in oscillating neutrinos. A comprehensive analysis is made in [35], where quantum correlations such as three-tangle and three-π have been used to illustrate genuine tripartite entanglement in oscillating neutrinos. Additionally, since qutrits have higher dimensions than qubits, quantum correlations such as generalized concurrence (an extension of concurrence in a higher dimension) have been quantified by treating neutrinos as qutrits [36]. Inspired by these findings, implementing neutrino entanglement on the IBM quantum computer appears promising for future studies of quantum correlations associated with neutrino oscillations in realistic situations which may be difficult to treat otherwise [37].

39.5 Quantum Correlations in Neutral Meson Systems

In this section, we analyze various widely recognized quantum correlation measures, including Bell's inequality violations, teleportation fidelity, concurrence, and geometric discord, for the correlated $B\bar{B}$ and $K\bar{K}$ systems. Additionally, we explore the interplay between these measures [38, 39]. B factories, electron-positron colliders designed specifically for studying the production and decay of B mesons, and ϕ factories, which serve the same purpose for K mesons, offer an ideal testing ground. The collider energy is set to the Υ resonance in B factories, which initiates the process $e^+e^- \rightarrow \Upsilon$. The Υ then decays into $b\bar{b}$, which subsequently form $B_q\bar{B}_q$ $(q = s, d)$ pair through hadronization. This process occurs almost instantaneously, while the B mesons separate and decay for a much longer period. An important aspect of these systems for studying correlations is the oscillation of bottom and strangeness flavors, leading to $B\bar{B}$ and $K\bar{K}$ oscillations. Since these systems are unstable and decaying, it is essential to investigate quantum correlations in the context of these unstable, decaying $B\bar{B}$ and $K\bar{K}$ systems.

Using the state of the two-particle decaying system at time t [39], we have the following quantum correlations in neutral mesons:

1. *Non-locality*: $M(\rho) = (1 + e^{-4\lambda t})$. Here λ characterizes the decoherence, that in turn arises due to the irreversibility caused by the decay.
2. *Concurrence*: $C = e^{-2\lambda t}$. The corresponding entanglement of formation is $E_F = -\frac{1+\sqrt{1-C^2}}{2}\log_2(\frac{1+\sqrt{1-C^2}}{2}) - \frac{1-\sqrt{1-C^2}}{2}\log_2(\frac{1-\sqrt{1-C^2}}{2})$. Also, $M(\rho) = 1 + C^2$.
3. *Geometric discord*: $D_G(\rho) = M(\rho)/3$.
4. *Teleportation fidelity*: $F_{\max} = \frac{1}{12}\left[6 + 2e^{-2\lambda t} + \sqrt{2}\sqrt{\alpha - \sqrt{\beta}} + \sqrt{2}\sqrt{\alpha + \sqrt{\beta}}\right]$, where $\alpha = 1 + \cosh(4\lambda t) - \sinh(4\lambda t)$, and $\beta = 3 - 2\alpha + \cosh(8\lambda t) - \sinh(8\lambda t)$.

To incorporate the effects of decay in the systems under investigation, the various correlations are modified by the probability of survival of the particle pair up to a given time, which is expressed as $e^{-2\Gamma t}$, where Γ represents the meson decay width. For the K meson, the decay width is given by $\Gamma = \frac{1}{2}(\Gamma_S + \Gamma_L)$, with Γ_S and Γ_L being the decay widths of the short and long neutral kaon states, respectively; its value is $5.59 \times 10^9\,\mathrm{s}^{-1}$ [40]. The decay widths for B_d and B_s mesons are $6.58 \times 10^{11}\,\mathrm{s}^{-1}$ and $6.61 \times 10^{11}\,\mathrm{s}^{-1}$, respectively [41]. For the K meson system, the decoherence parameter λ determine by the KLOE collaboration through the interference between initially entangled kaons and their decay products in the

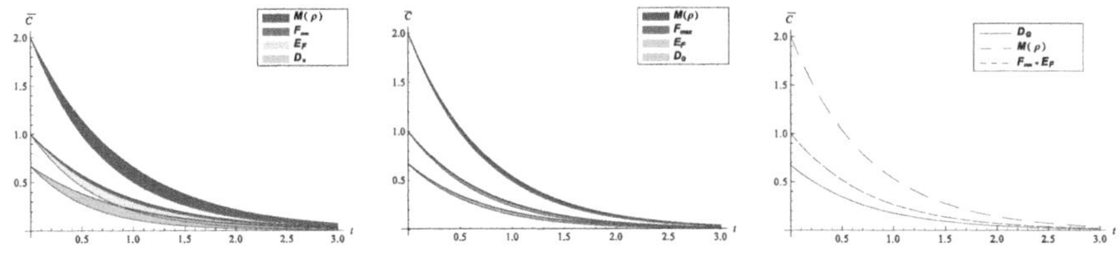

Fig. 39.1 Average correlation measures i.e., the various measures modulated by the exponential factor $e^{-2\Gamma t}$, are plotted as a function of time t. The left, middle, and right plots show the correlations for $K\bar{K}$, $B_d\bar{B}_d$ and $B_s\bar{B}_s$ pair created at $t = 0$, respectively. The four correlation measures displayed from top to bottom are $M(\rho)$ (Bell's inequality; blue band), $F_{\max}$ (teleportation fidelity; red band), E_F (entanglement of formation; grey band) and D_G (geometric discord; green band). For $K\bar{K}$ pairs (left plot), time is measured in units of 10^{-10} seconds whereas for $B_d\bar{B}_d$ and $B_s\bar{B}_s$ pairs (middle and right plot), time is in units of 10^{-12} seconds. These reflect the approximate lifetimes of the particles. The bands in the left and middle plots represent the effects of decoherence, with decoherence parameter (λ) set to a 3σ upper bound. The right plot does not include such bands due to the lack of experimental evidence for λ in B_s mesons; in this case, $F_{\max} = E_F$

channel $\phi \to K_S K_L \to \pi^+\pi^-\pi^+\pi^-$ [42]. The value of λ is constrained to be no more than $1.58 \times 10^9\ \mathrm{s}^{-1}$ at the 3σ level. For B_d meson systems, the decoherence parameter is typically evaluated through time-integrated dilepton events. The value of λ for B_d mesons is calculated by the measurement of R_d, the ratio of the total same-sign to opposite-sign dilepton rates in the decays of coherent $B_d - \bar{B}_d$ coming from the $\Upsilon(4S)$ decays; the upper bound is $2.82 \times 10^{11}\ \mathrm{s}^{-1}$ at 3σ [43]. In [44], several methods were proposed to calculate λ in the B meson systems. For B_s mesons, due to the lack of experimental evidence for λ, we assume it to be zero in this study (Fig. 39.1).

Typically, Bell's inequality is violated in these correlated meson systems for approximately half of the meson lifetime. We observe that the quantum correlations in these systems can be *nontrivially different* from those in stable systems, as illustrated by the relationship between Bell's inequality violations and teleportation fidelity. A *notable observation* is that the teleportation fidelity does not surpass the classical threshold value of 2/3 despite the violation of Bell's inequality. As in the case of neutrinos, the temporal quantum correlations can be studied by the LG inequality. For the initial state $|\bar{B}^o\rangle$, the three measurement LG parameter takes the form [45]:

$$K_3 = 1 - 4P_{\bar{B}^o \to B^o}(\Delta t) + 2P_{\bar{B}^o \to B^o}(2\Delta t). \tag{39.15}$$

As illustrated in Fig. 39.2, the LG inequality is violated for some time intervals. A knowledge of the transition probability at Δt and $2\Delta t$ is needed to compute the function K_3.

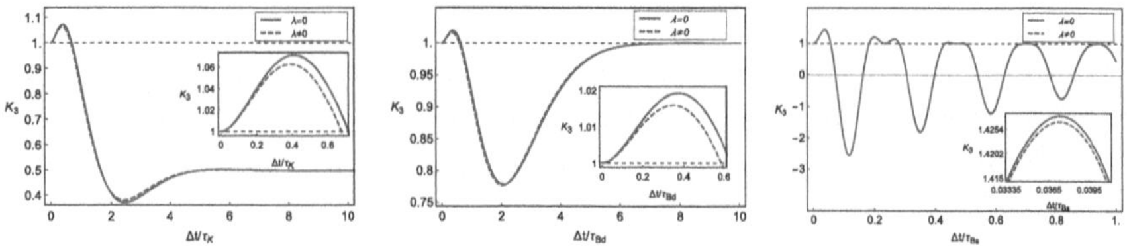

Fig. 39.2 Depicting the LG parameter K_3 as a function of the time interval Δt between the successive measurements. The left, middle, and right plots illustrate the K, B_d and B_s meson systems, respectively. The dashed line indicates the presence of decoherence, whereas the solid line represents the absence of decoherence

39.6 Conclusions

We illustrated the role played by ideas in Open Quantum Systems and Quantum Information to various facets of neutrino and neutral meson oscillations. The ubiquity of these tools is highlighted by the range of their applications, from quantum information to particle physics. The evolution of neutrinos was seen to be highly non-local in nature. The quantum correlations were closely tied to the neutrino mixing energy. The correlations were shown to be simple functions of the product of neutrino survival and oscillation probabilities. The quantum correlations in the neutral mesons were seen to be non-trivially different from their stable counterparts.

Acknowledgments I thank the organizers of ISRA 2023. My special thanks are due to Shubhrangshu Ghosh and Banibrata Mukhopadhyay.

References

1. Banerjee, S. (2018). *Open quantum systems: Dynamics of nonclassical evolution*. Springer.
2. Louisell, W. H. (1973). *Quantum statistical properties of radiation*. John Wiley and Sons.
3. Caldeira, A. O., & Leggett, A. J. (1983). *Physica A*, *121*, 587.
4. Zurek, W. H. (1991). *Physics Today*, *44*, 36; *Progress of Theoretical and Experimental Physics*, *87*, 281 (1993).
5. Turchette, Q. A., Myatt, C. J., King, B. E., Sackett, C. A., et al. (2000). *Physical Review A*, *62*, 053807.
6. Myatt, C. J., King, B. E., Turchette, Q. A., Sackett, C. A., et al. (2000). *Nature*, *403*, 269.
7. Brune, M., et al. (1996). *Physical Review Letters*, *77*, 4887.
8. Giunti, C., & Kim, C. W. (2007). *Fundamentals of neutrino physics and astrophysics*. Oxford Academic. Online edn, 2010.
9. Kudryavtsev, V. A., & For the DUNE Collaboration (2016). *Journal of Physics: Conference Series*, *718*, 062032.
10. Braginsky, V. B., Vorontsov, Y. I., & Thorne, K. S. (1980). *Science*, *209*, 547.
11. Caves, C. M., et al. (1980). *Reviews of Modern Physics*, *52*, 341.
12. Grabert, H., Schramm, P., & Ingold, G. L. (1988). *Physics Reports*, *168*, 115.

13. Feynman, R. P., & Vernon, F. L. (1963). *Annals of Physics (N.Y.)*, *24*, 118.
14. Hakim, V., & Ambegaokar, V. (1985). *Physical Review A*, *32*, 423.
15. Nielsen, M., & Chuang, I. (2000). *Quantum computation and quantum information*. Cambridge University Press.
16. Stinespring, W. F. (1955). Positive functions on C*-algebras. *Proceedings of the American Mathematical Society*, 211–216.
17. Sudarshan, E. C. G., et al. (1961). *Physical Review*, *121*, 920.
18. Kraus, K. (1983) *States, effects and operations: Fundamental notions of quantum theory*. Springer Verlag.
19. Horodecki, R., Horodecki, M., & Horodecki, P. (1996). *Physics Letters A*, *222*, 21–25.
20. Wootters, W. K. (1998). *Physical Review Letters*, *80*, 2245.
21. Luo, S. (2008). *Physical Review A*, *77*, 022301.
22. Ollivier, H., & Zurek, W. H. (2001). *Physical Review Letters*, *88*, 017901.
23. Henderson, L., & Vedral, V. (2001). *Journal of Physics A: Mathematical and General*, *34*, 6899.
24. B. Dakić, Vedral, V., & Brukner, Č. (2010). *Physical Review Letters*, *105*, 190502.
25. Leggett, A. J., & Garg, A. (1985). *Physical Review Letters*, *54*, 857.
26. Blasone, M., Dell'Anno, F., De Siena, S., Di Mauro, M., & Illuminati, F. (2008). *Physical Review D*, *77*, 096002.
27. Blasone, M., Dell'Anno, F., De Siena, S., & Illuminati, F. (2009). *Europhysics Letters*, *85*, 50002.
28. Alok, A. K., Banerjee, S., & Uma Sankar, S. (2016). *Nuclear Physics B*, *909*, 65.
29. Banerjee, S., Alok, A. K., Srikanth, R., & Hiesmayr, B. C. (2015). *European Physical Journal C*, *75*, 487.
30. Naikoo, J., Alok, A. K., Banerjee, S., & Uma Sankar, S. (2019). *Physical Review D*, *99*, 095001.
31. Naikoo, J., Kumari, S., Banerjee, S., & Pan, A. K. (2020). *Journal of Physics G*, *47*, 095004.
32. Dixit, K., Naikoo, J., Banerjee, S., & Alok, A. K. (2018). *European Physical Journal C*, *78*, 914.
33. Bouri, S., Jha, A. K., & Banerjee, S. (2024). arXiv:2405.13114 [hep-ph]
34. Dixit, K., Naikoo, J., Mukhopadhyay, B., & Banerjee, S. (2019). *Physical Review D*, *100*, 055021.
35. Kumar Jha, A., Mukherjee, S., & Bambah, B. A. (2021). *Modern Physics Letters A*, *36*, 2150056.
36. Jha, A. K., Chatla, A., & Bambah, B. A. (2024). *The European Physical Journal Plus*, *139*, 68.
37. Jha, A. K., & Chatla, A. (2022). *The European Physical Journal Special Topics*, *231*, 141.
38. Alok, A. K., & Banerjee, S. (2013). *Physical Review D*, *88*, 094013.
39. Banerjee, S., Alok, A. K., & MacKenzie, R. (2016). *The European Physical Journal Plus*, *131*, 129.
40. Particle Data Group Collaboration, Olive, K. A., et. al. (2014). *Chinese Physics C*, *38*, 090001.
41. Heavy Favor Averaging Group Collaboration, Amhis, Y., et. al. (2012). arXiv:1907.1158.
42. KLOE Collaboration, Ambrosino, F., et. al. (2006). *Physical Letter B*, *642*, 315.
43. Bertlmann, R. A., & Grimus, W. (2001). *Physical Review D*, *64*, 056004.
44. Alok, A. K., Banerjee, S., & Uma Sankar, S. (2015). *Physical Letter B*, *749*, 94.
45. Naikoo, J., Alok, A. K., & Banerjee, S. (2018). *Physical Review D*, *97*, 053008.

Chapter 40
Influence of Gravity on the Quantum Speed Limit in Neutrino Oscillations

Abhishek Kumar Jha, Mriganka Dutta, Subhashish Banerjee, and Banibrata Mukhopadhyay

Abstract The quantum speed limits (QSLs) determine the minimal amount of time required for a quantum system to evolve from an initial to a final state. We investigate QSLs for the unitary evolution of the neutrino-antineutrino system in the presence of a gravitational field. It is known that the transition probabilities between neutrino and antineutrino in the framework of one and two flavors depend on the strength of the gravitational field. The behavior of the QSL time in the two-flavor system indicates fast flavor transitions as the gravitational field strength increases. Subsequently, we observe quick suppression of entanglement by exploring the speed limit for entanglement entropy of two-flavor oscillations in the neutrino-antineutrino system in the proximity of a spinning primordial black hole.

40.1 Introduction

Several astrophysical and cosmological problems are involved with neutrino physics [1]. Examples include neutrino-cooled accretion disks [2], r-process nucleosynthesis in supernova explosions [3], leptogenesis-baryogenesis [4], etc. On the other hand, neutrino oscillation is a quantum phenomenon where a neutrino, starting in one flavor state, can transit into a different flavor state as it progresses over time [5, 6]. This flavor transition occurs because neutrinos have non-zero mass. Any initial neutrino flavor states ν_α (where, $\alpha = e, \mu, \tau$) can be represented as a linear superposition of three non-degenerate mass eigenstates. The study of neutrino oscillations in the combined two-flavor neutrino-antineutrino system has already been explored in the presence of gravity [7].

A. K. Jha (✉) · M. Dutta · B. Mukhopadhyay
Department of Physics, Indian Institute of Science, Bangalore, India
e-mail: kjabhishek@iisc.ac.in; mrigankad@iisc.ac.in; bm@iisc.ac.in

S. Banerjee
Indian Institute of Technology Jodhpur, Jodhpur, India
e-mail: subhashish@iitj.ac.in

S. Ghosh, A. R. Rao (eds.), *The Relativistic Universe: From Classical to Quantum*, Astrophysics and Space Science Proceedings 61,
https://doi.org/10.1007/978-3-031-90186-7_40

In a locally flat coordinate system, gravitational interaction can be treated as an effective interaction [8], when the neutrino Lagrangian density under gravity can be expressed as $\mathcal{L} = \mathcal{L}_f + \mathcal{L}_I$, where $\mathcal{L}_f$ corresponds to the free part, akin to the Lagrangian density in flat space, while the gravitational interaction part $\mathcal{L}_I$ results in an effective extension of the flat-space description leading to Lorentz invariance (LI), given by $\gamma^a \gamma^5 B_a$. When B_a remains constant within a locally flat coordinate system, LI is responsible for the charge-parity-time reversal (CPT) violation in gravity-induced neutrino-antineutrino systems [9–14]. This occurs in, e.g., the vicinity of rotating black holes (BH) [8, 12], the anisotropic early universe phase [15].

In the Weyl representation, the term γ^5 in $\mathcal{L}_I$ leads to gravitational interactions represented as $\bar{\psi}\gamma^a\psi$ and $-\bar{\psi}^c\gamma^a\psi^c$ for neutrino and antineutrino, respectively. The superscript "c" on ψ indicates a charge-conjugate spinor or the spinor for the antiparticle. Consequently, the dispersion relations for neutrinos and antineutrinos differ and can be described as follows [7, 8]:

$$E_\psi = \sqrt{(\mathbf{p} - \mathbf{B})^2 + m^2} + B_0 \, , \qquad E_{\psi^c} = \sqrt{(\mathbf{p} + \mathbf{B})^2 + m^2} - B_0 \, , \tag{40.1}$$

where B_a is 4-vector gravitational potential and $\mathbf{p}$ the momentum of particles. Equation (40.1) effectively denotes gravitational Zeeman effect (GZE) [9–16]. The detailed neutrino-antineutrino mixing and flavor oscillations in a two-flavor scenario were undertaken earlier [7, 8]. The fundamental question arises: how quickly does the transition of the two-flavor neutrino-antineutrino states occur with the change of the gravitational field strength. In this chapter, we consider the unitary dynamics of two-flavor neutrino-antineutrino oscillations. We employ the quantum speed limit (QSL) time technique as a primary analytical tool to estimate the minimum time required for the evolution of two-flavor neutrino-antineutrino system.

The initial discovery of QSL stems from the uncertainty relationship between conjugate variables in quantum mechanics [17, 18]. The QSL time formulation depends on factors such as the shortest distance (or geodesic distance) between the initial and final states and the variance (or fluctuation) of the driving Hamiltonian [19]. Since, one of the most important aspects of quantum mechanics is the existence of superposition and entanglement, the deep understanding of the QSL has also led to the fundamental limitations on the rate of change of various entanglement measures [20, 21].

In recent times, entanglement in neutrino oscillations has been extensively discussed in various references [22–27]. Additionally, quantification of entanglement measures in two-flavor neutrino-antineutrino oscillations within curved spacetime has also been done [28, 29]. Their results emphasize the existence of entanglement during the evolution of neutrino-antineutrino flavor states in the presence of gravity. In this chapter, we treat the two-flavor neutrino-antineutrino system as a bipartite quantum system. We delve into the behavior of entanglement entropy and the capacity of entanglement in two-flavor oscillations in the neutrino-antineutrino system, specifically near a spinning BH. We probe the QSL time for entanglement in the presence of low and high gravitational fields.

40.2 Quantum Speed Limit for Unitary Quantum Evolution and for Entanglement

In the realm of quantum mechanics, the wave function $|\psi\rangle$ describes the state of a quantum system, while the dynamic governs its temporal evolution as

$$i\hbar\frac{d\,|\psi\rangle}{dt} = H\,|\psi\rangle\ , \tag{40.2}$$

such that $|\psi(t)\rangle = e^{-iHt/\hbar}\,|\psi\rangle \equiv U_t\,|\psi\rangle$, where U_t is the unitary time evolution operator, and the driving Hamiltonian H is Hermitian and time independent. The QSL time, $T_{\rm QSL}$, for such a system is defined as [19]

$$T \geq T_{\rm QSL}\ , \qquad T_{\rm QSL} = \frac{\hbar S_0}{\Delta H}\ , \tag{40.3}$$

where $S_0 = \cos^{-1}(|\langle\psi(0)|\psi(T)\rangle|) = \arccos(\sqrt{P_s})$, is the geodesic distance between an initial and final states, and $P_s = |\langle\psi(0)|\psi(T)\rangle|^2$ is the survival probability of the initial state $|\psi\rangle$ at time T. A general representation of energy fluctuation ΔH when quantum system undergoes unitary dynamics is given by

$$\Delta H \equiv \sqrt{[\langle\dot{\psi}(t)|\dot{\psi}(t)\rangle - (i\langle\psi(t)|\dot{\psi}(t)\rangle)^2]}\ , \tag{40.4}$$

where $\left|\dot{\psi}(t)\right\rangle \equiv d\,|\psi(t)\rangle\,/dt$. Equation (40.3) suggests that $T_{\rm QSL}$ is the minimum time required for the quantum system to evolve from $|\psi(0)\rangle$ to $|\psi(T)\rangle$ by unitary evolution.

A bipartite quantum system, in the context of quantum mechanics, is a composite quantum system that consists of two individual subsystems or components [30]. Each of these subsystems can be thought of as a separate quantum system, and they are often described by their own quantum states and associated Hilbert spaces. For unitary quantum evolution, the quantum state of the entire bipartite system is typically described using a mathematical framework called a density matrix: $\rho = |\psi\rangle\,\langle\psi|$ and it is often represented as a tensor product of the quantum states of the individual subsystems: $\rho = \rho_A \otimes \rho_B \in \mathcal{H}_A \otimes \mathcal{H}_B$. It is worth mentioning that due to the unitary evolution, ρ always exists in a pure quantum state with properties such as $\rho = \rho^2$ (idempotent matrix) and $\mathrm{Tr}(\rho^2) = 1$. The reduced density matrix $\rho_A = \mathrm{Tr}_B(\rho)$ and $\rho_B = \mathrm{Tr}_A(\rho)$ follows $\mathrm{Tr}(\rho_A^2) < 1$ and $\mathrm{Tr}(\rho_B^2) < 1$, which are in the mixed states. An example of a bipartite pure quantum system is Bell's state. Furthermore, ρ accounts for the entanglement and correlations that can exist between the two subsystems. Within a bipartite pure quantum system, QSL for generating entanglement is intricately tied not only to fluctuations in the Hamiltonian but also inversely linked to the time-averaged square root of the capacity of entanglement. In scenarios where the driving Hamiltonian remains time-independent, the quantum speed limit for entanglement adheres to the following bound [21]

$$T \geq T^E_{\text{QSL}} = \frac{\hbar|S_{EE}(T) - S_{EE}(0)|}{2\Delta H \frac{1}{T}\int_0^T \sqrt{C_E(t)}dt}\,, \tag{40.5}$$

where the entanglement entropy S_{EE} is defined as

$$S_{EE} = S(\rho_A) = -\text{Tr}(\rho_A log_2 \rho_A)\,, \tag{40.6}$$

for the reduced density matrix ρ_A. The variance (or fluctuation) in the entanglement entropy operator S_{EE} is defined as capacity of entanglement C_E, given as

$$C_E = \sum_i \lambda_i log_2^2 \lambda_i - S_{EE}^2\,, \tag{40.7}$$

where $\lambda_i' s$ are the eigenvalue of ρ_A (or ρ_B). Equation (40.5) establishes a QSL on the production and decay of entanglement in the time-evolved state $|\psi(t)\rangle$.

40.3 Four-Vector Gravitational Potential Around a Rotating Compact Object

We consider a rotating BH described by the Kerr geometry. We derive the analytical expression of the four-vector gravitational potential for a rotating BH, taking into account Kerr geometry in the Kerr-Schild form [31]. Our variables are denoted by $t\ (= x_0)$, $x\ (= x_1)$, $y\ (= x_2)$, and $z\ (= x_3)$. In this Cartesian form, the Kerr metric with signature [+ - - -] can be written as

$$ds^2 = \eta_{ij}dx^i dx^j - \left[\frac{2\alpha}{\rho} s_i v_j + \alpha^2 v_i v_j\right] dx^i dx^j\,, \tag{40.8}$$

where

$$\alpha = \frac{\sqrt{2Mr}}{\rho}, \qquad \rho^2 = r^2 + \frac{a^2 z^2}{r^2}, \tag{40.9}$$

$$s_i = \left(0, \frac{rx}{\sqrt{r^2+a^2}}, \frac{ry}{\sqrt{r^2+a^2}}, \frac{z\sqrt{r^2+a^2}}{r}\right),$$
$$v_i = \left(1, \frac{ay}{a^2+r^2}, -\frac{ax}{a^2+r^2}, 0\right). \tag{40.10}$$

Here a and M are respectively the specific angular momentum and mass of the Kerr BH and r is a positive definite coordinate satisfying the following equation,

$$r^4 - r^2(x^2 + y^2 + z^2 - a^2) - a^2z^2 = 0. \tag{40.11}$$

Here we choose $G = M = c = \hbar = 1$. The four-vector B^d can be calculated by [14]

$$B^d = \epsilon^{abcd} e_{b\lambda} \left(\partial_a e_c^{\lambda} + \Gamma^{\lambda}_{\alpha\mu} e_c^{\alpha} e_a^{\mu} \right), \tag{40.12}$$

which leads to

$$B^0 = -\frac{2\sqrt{2M}az}{\left[\rho^2\sqrt{r(a^2 + r^2)}\right]}, \tag{40.13}$$

$$\begin{aligned}
B^1 &= a\alpha^2 xz \left[\frac{a^2\alpha^2(x^2 + y^2) + (\alpha^2 + 3)(a^2 + r^2)^2}{\rho^2(a^2 + r^2)^3} \right], \\
B^2 &= -a\alpha^2 yz \left[\frac{a^2\alpha^2(x^2 + y^2) + (\alpha^2 + 3)(a^2 + r^2)^2}{\rho^2(a^2 + r^2)^3} \right], \\
B^3 &= -\frac{a\alpha^2 r^2}{\rho^2(a^2 + r^2)^4} \left[(a^2 + r^2)^2(x^2 + y^2)(3 + \alpha^2) + a^2\alpha^2(x^2 + y^2)^2 \right].
\end{aligned} \tag{40.14}$$

40.4 Quantum Speed Limit for the Evolution of Two-Flavor Neutrino-Antineutrino System in the Presence of Gravity

We strictly follow earlier results [7] to obtain QSL for a evolved neutrino-antineutrino flavor state under gravity. The evolving flavor states $|\psi_e^c(t)\rangle$, $|\psi_\mu^c(t)\rangle$, $|\psi_e(t)\rangle$ and $|\psi_\mu(t)\rangle$ can also be projected to flavor basis in the form

$$\begin{pmatrix} |\psi_e^c(t)\rangle \\ |\psi_\mu^c(t)\rangle \\ |\psi_e(t)\rangle \\ |\psi_\mu(t)\rangle \end{pmatrix} = \mathbf{T}\,\mathrm{diag}(e^{-iE_1t}, e^{-iE_2t}, e^{-iE_3t}, e^{-iE_4t})\mathbf{T}^{-1} \begin{pmatrix} |\psi_e^c\rangle \\ |\psi_\mu^c\rangle \\ |\psi_e\rangle \\ |\psi_\mu\rangle \end{pmatrix}, \tag{40.15}$$

where $|\psi_e^c\rangle$, $|\psi_\mu^c\rangle$, $|\psi_e\rangle$ and $|\psi_\mu\rangle$ are flavor states at time $t = 0$. Thus, using Eq. (40.15), the time evolved electron flavor neutrino state in superposition of flavor basis can be obtained as

$$|\psi_e(t)\rangle = T_{ee^c}(t)\,|\psi_e^c\rangle + T_{e\mu^c}(t)\,|\psi_\mu^c\rangle + T_{ee}(t)\,|\psi_e\rangle + T_{e\mu}(t)\,|\psi_\mu\rangle\,. \tag{40.16}$$

The mixing angles for neutrino-antineutrino and electron-muon neutrino mixing are related to the masses and the gravitational scalar potential, respectively, given by [7]

$$\tan\theta_{e,\mu} = \frac{m_{e,\mu}}{B_0 + \sqrt{(B_0)^2 + m_{e,\mu}^2}} , \tag{40.17}$$

$$\tan\phi_{1,2} = \frac{\mp 2m_{e\mu}}{m_{e(1,2)} - m_{\mu(1,2)} + \sqrt{(m_{e(1,2)} - m_{\mu(1,2)})^2 + 4m_{e\mu}^2}} . \tag{40.18}$$

The masses corresponding to the mass eigenstates are given as

$$M_{1,2} = \frac{1}{2}[(m_{e1} + m_{\mu 1}) \pm \sqrt{(m_{e1} - m_{\mu 1})^2 + 4m_{e\mu}^2}] ,$$
$$M_{3,4} = \frac{1}{2}[(m_{e2} + m_{\mu 2}) \pm \sqrt{(m_{e2} - m_{\mu 2})^2 + 4m_{e\mu}^2}] , \tag{40.19}$$

with

$$m_{(e,\mu)1} = -\sqrt{(B_0)^2 + m_{e,\mu}^2}, \qquad m_{(e,\mu)2} = \sqrt{(B_0)^2 + m_{e,\mu}^2} . \tag{40.20}$$

Assuming neutrinos to be ultra-relativistic particles, energies are

$$E_1 = E_2 = |\mathbf{p}| + |\mathbf{B}| - B_0 , \qquad E_3 = E_4 = |\mathbf{p}| - |\mathbf{B}| + B_0 . \tag{40.21}$$

Thus, we find the survival probability of the initial electron flavor neutrino state $|\psi_e\rangle$ using Eq. (40.16) as

$$P_s = |T_{ee}(t)|^2 . \tag{40.22}$$

Furthermore, using Eqs. (40.4) and (40.16), we compute energy fluctuation ΔH. Subsequently, using Eqs. (40.3) and (40.22), we estimate T_{QSL} for the initial electron flavor neutrino state $|\psi_e\rangle$ as

$$T_{\mathrm{QSL}} = \frac{\arccos(\sqrt{|T_{ee}(t)|^2})}{\Delta H} . \tag{40.23}$$

We assume that neutrinos are moving around a rotating BH at fixed radii. In the ultra-relativistic limit, $t \approx L$, when L is the propagation length. Figure 40.1 illustrates T_{QSL} of the initial electron flavor neutrino state $|\psi_e\rangle$ as a function of propagation length L for different values of the gravitational scalar potential, as mentioned in Table 40.1. We observe that the behavior of T_{QSL} vs. L at low B_0 (blue dashed and green dot-dashed lines) is tight, which means T_{QSL}/L equals 1. Therefore, the evolutionary speed of the initial state $|\psi_e\rangle$ is unaltered and the minimum time taken to oscillate is an extremum. However, as the B_0 changes from

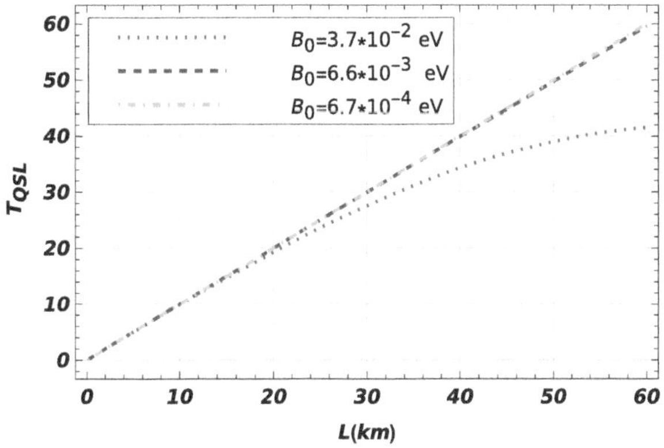

Fig. 40.1 In the two-flavor scenario, T_{QSL} of the initial state $|\psi_e\rangle$ as a function of propagation length L(km) at three distinct values of the gravitational potential, i.e. at three different distance from BH, given in Table 40.1. The other parameters used are: $m_e = 5 \times 10^{-3}$ eV, $m_\mu = 6.5 \times 10^{-3}$ eV, $m_{e\mu} = 3.5 \times 10^{-3}$ eV

Table 40.1 Three different gravitational scalar-potential, B_0(eV), and vector potential magnitude, $|\mathbf{B}|$(eV), corresponding to three different radii, 100, 200 and 500, from the BH

r	B_0 (eV)	$\|\mathbf{B}\|$ (eV)
100	3.7×10^{-2}	1.3×10^{-2}
200	6.6×10^{-3}	1.7×10^{-3}
500	6.7×10^{-4}	1.1×10^{-4}

a smaller to a higher value (red dotted line), the initial neutrino state $|\psi_e\rangle$ begins to oscillate faster because the ratio $T_{\mathrm{QSL}}/L < 1$, causing the dynamical evolution to speed up as a smaller value of T_{QSL}/L implies a faster evolution of the quantum state. Hence, it is evident from Eq. (40.23) that the QSL time of two-flavor neutrino-antineutrino oscillations changes significantly depending on the gravitational field strength.

40.5 Quantum Speed Limit for Entanglement in Two-Flavor Neutrino-Antineutrino Oscillations in the Presence of Gravity

Now, we treat the system of the two-flavor neutrino-antineutrino system as a four-qubit system under the impact of neutrino-antineutrino mixing caused by the gravitational field. The representation of the occupation number can be expressed as [28]

$$\begin{aligned}
\left|\psi_e^c\right\rangle &\equiv |1\rangle_{e^c} \otimes |0\rangle_{\mu^c} \otimes |0\rangle_e \otimes |0\rangle_\mu \equiv |1000\rangle, \\
\left|\psi_\mu^c\right\rangle &\equiv |0\rangle_{e^c} \otimes |1\rangle_{\mu^c} \otimes |0\rangle_e \otimes |0\rangle_\mu \equiv |0100\rangle, \\
|\psi_e\rangle &\equiv |0\rangle_{e^c} \otimes |0\rangle_{\mu^c} \otimes |1\rangle_e \otimes |0\rangle_\mu \equiv |0010\rangle, \\
\left|\psi_\mu\right\rangle &\equiv |0\rangle_{e^c} \otimes |0\rangle_{\mu^c} \otimes |0\rangle_e \otimes |1\rangle_\mu \equiv |0001\rangle.
\end{aligned} \tag{40.24}$$

Thus, using Eq. (40.16), the time evolution of electron flavor neutrino state in four-qubit system can be written as

$$|\psi_e(t)\rangle = T_{ee^c}(t)\,|1000\rangle + T_{e\mu^c}(t)\,|0100\rangle + T_{ee}(t)\,|0010\rangle + T_{e\mu}(t)\,|0001\rangle \ . \tag{40.25}$$

Furthermore, using Eq. (40.25), we construct a 16×16 density matrix, $\rho(t) = |\psi_e(t)\rangle\,\langle\psi_e(t)|$. Here, $\rho^2(t) = \rho(t)$ and $\mathrm{Tr}(\rho^2(t)) = 1$, which means that the $\rho(t)$ is in a pure state. We look for a bipartite scenario where the reduced density matrix is obtained by tracing $\rho(t)$ over the other two qubits. The required 4×4 reduced density matrix can be expressed as

$$\rho_{\mathrm{red}}(t) = \mathrm{diag}(0, 0, \lambda_1, \lambda_2) \ , \tag{40.26}$$

where $\rho^2_{\mathrm{red}}(t) \neq \rho_{\mathrm{red}}(t)$ and $\mathrm{Tr}(\rho^2_{\mathrm{red}}(t)) < 1$, signifying $\rho_{\mathrm{red}}(t)$ is in a mixed state. Consequently, the state $|\psi_e(t)\rangle$ is a bipartite pure state. Here, λ_1 and λ_2 represent the two non-zero eigenvalues of $\rho_{\mathrm{red}}(t)$. Subsequently, putting λ_1 and λ_2 values in Eqs. (40.6) and (40.7), we calculate entanglement entropy

$$S_{EE} = -\lambda_1 log_2\lambda_1 - \lambda_2 log_2\lambda_2 \ , \tag{40.27}$$

and capacity of entanglement

$$C_E = (\lambda_1 log_2^2\lambda_1 + \lambda_2 log_2^2\lambda_2) - (-\lambda_1 log_2\lambda_1 - \lambda_2 log_2\lambda_2)^2 \ , \tag{40.28}$$

of the state $|\psi_e(t)\rangle$.

In the ultra-relativistic limit, $t \approx L$. In Fig. 40.2, at a fixed propagation length of $L = 300$ km, we depict the variation of S_{EE} (red dotted line) and C_E (blue dashed line) for the initial state $|\psi_e\rangle$ as functions of gravitational scalar potential B_0. We observe a decrease in entanglement entropy (S_{EE}) as B_0 increases. In Figs. 40.3 and 40.4, we estimate S_{EE} (red dotted line) and C_E (blue dashed line) as functions of propagation length L for the initial state $|\psi_e\rangle$ at low and high values of B_0. At $L = 0$, λ_1 and λ_2 tend to 0 and 1, respectively. This implies that at $L = 0$, both S_{EE} and C_E vanish. At $L > 0$, $S_{EE} > 0$ and $C_E > 0$. In Fig. 40.3, we observe that at a low value of $B_0 = 6.7 \times 10^{-4}$ eV, S_{EE} (red dotted line) develops monotonically towards the maximum. However, C_E (blue dashed line) goes to zero in both the maximally entangled and separable state limits, while reaching a maximum in a partially entangled state with varying L. This result shows that the bipartite entanglement persists during the time evolution of the initial electron flavor neutrino state $|\psi_e\rangle$. However, the comparison between Figs. 40.3 and 40.4 reveals that as we vary the strength of B_0 from a low ($B_0 = 6.7 \times 10^{-4}$ eV) to a high value ($B_0 = 3.7 \times 10^{-2}$ eV), entanglement in the state $|\psi_e(t)\rangle$ gets suppressed.

Further, using Eqs. (40.27) and (40.28) in Eq. (40.5), QSL for entanglement can be obtained for the state $|\psi_e(t)\rangle$ as

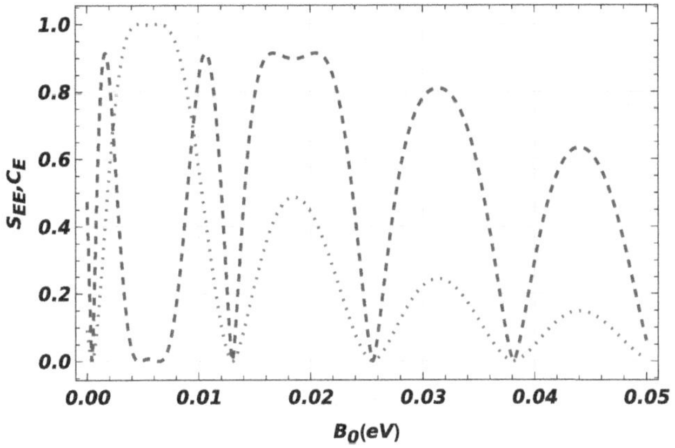

Fig. 40.2 In the two-flavor scenario, the entanglement entropy, S_{EE} (red dotted line), and the capacity of entanglement, C_E (blue dashed line), of the initial state $|\psi_e\rangle$ as functions of gravitational scalar potential B_0. The other parameters are: $m_e = 5 \times 10^{-3}$ eV, $m_\mu = 6.5 \times 10^{-3}$ eV, $m_{e\mu} = 3.5 \times 10^{-3}$ eV and $L = 300$ km (Color figure online)

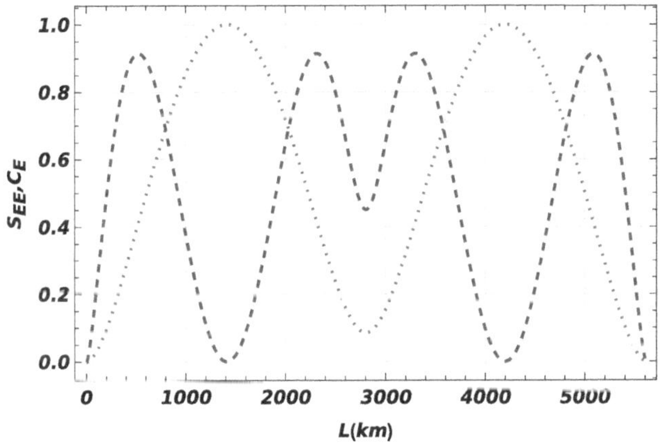

Fig. 40.3 In the two-flavor scenario, the entanglement entropy, S_{EE} (red dotted line), and the capacity of entanglement, C_E (blue dashed line), of the initial state $|\psi_e\rangle$ as functions of propagation length L(km) at low value of gravitational scalar potential $B_0 = 6.7 \times 10^{-4}$ eV. The other parameters are: $m_e = 5 \times 10^{-3}$ eV, $m_\mu = 6.5 \times 10^{-3}$ eV, $m_{e\mu} = 3.5 \times 10^{-3}$ eV (Color figure online)

$$T^E_{\mathrm{QSL}} = \frac{|-\lambda_1 log_2\lambda_1 - \lambda_2 log_2\lambda_2|}{\Delta H \frac{1}{T}\int_0^T \sqrt{C_E(t)}dt}. \tag{40.29}$$

In Fig. 40.5, we depict T^E_{QSL} vs. L at three different values of gravitational scalar potential. Under the time bound condition $T^E_{\mathrm{QSL}}/L < 1$, we observe quick suppression of entanglement with the increase of gravitational field.

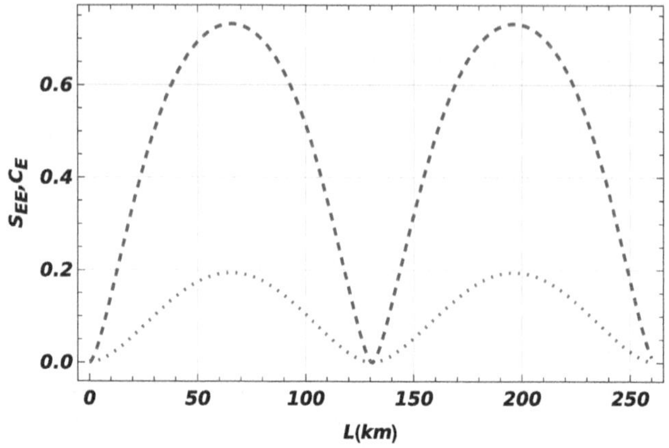

Fig. 40.4 In the two-flavor scenario, the entanglement entropy, S_{EE} (red dotted line), and the capacity of entanglement, C_E (blue dashed line), of the initial state $|\psi_e\rangle$ as functions of propagation length L(km) at high value of gravitational scalar potential $B_0 = 3.7 \times 10^{-2}$ eV. The other parameters are: $m_e = 5 \times 10^{-3}$ eV, $m_\mu = 6.5 \times 10^{-3}$ eV, $m_{e\mu} = 3.5 \times 10^{-3}$ eV (Color figure online)

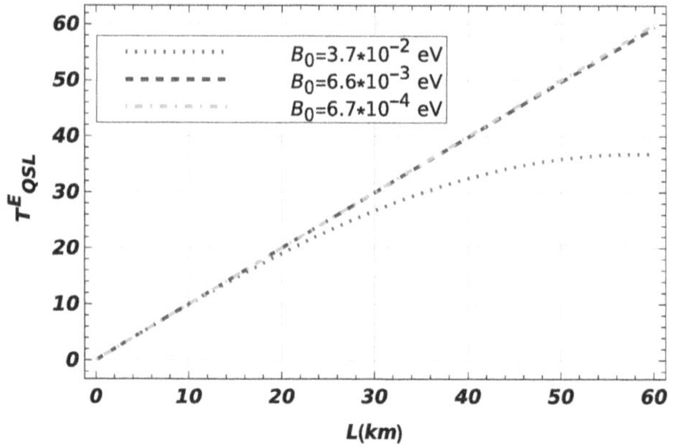

Fig. 40.5 Same as Fig. 40.1, except T^E_{QSL}

40.6 Conclusion

We have estimated the QSL time for two-flavor oscillations in the neutrino-antineutrino system in the presence of a gravitational field. It was observed that the QSL time of the initial neutrino flavor state is influenced by the strength of the gravitational field. The investigation of QSL time for the initial flavor state evolution suggests fast-flavor neutrino-antineutrino transitions can occur when the gravitational field strength increases from low to high values. Additionally, we have examined two bipartite entanglement measures: entanglement entropy and capacity of entanglement. The non-zero values of these entanglement measures have revealed that the two-flavor oscillation of the neutrino-antineutrino system exhibits pure bipartite entanglement. Further, we have explored the QSL time for entanglement entropy for the same system. In the proximity of a spinning BH, with the transition from a low to high gravitational field, there was a faster suppression of entanglement during the evolution of a two-flavor neutrino-antineutrino system. However, throughout the work, we assume that neutrinos are moving around a rotating BH at fixed radii.

Acknowledgments The authors would like to acknowledge the project funded by SERB, India, with Ref.No. CRG/2022/003460, for supporting this research.

References

1. Balantekin, A. B., & Fuller, G. M. (2013). Neutrinos in cosmology and astrophysics. *Progress in Particle and Nuclear Physics, 71*, 162–166. https://doi.org/10.1016/j.ppnp.2013.03.008.
2. Chen, W. X., & Beloborodov, A. M. (2007). Neutrino-cooled accretion disks around spinning black hole. *Astrophysical Journal, 657*, 383–399. https://doi.org/10.1086/508923.
3. Surman, R., et al. (2008). The role of neutrinos in r-process nucleosynthesis in supernovae and gamma-ray bursts. *Journal of Physics G: Nuclear and Particle Physics, 35*, 014059. https://doi.org/10.1088/0954-3899/35/1/014059.
4. Luty, M. A. (1992). Baryogenesis via leptogenesis. *Physical Review D, 45*, 455–465. https://doi.org/10.1103/PhysRevD.45.455.
5. Pontecorvo, B. (1957). Mesonium and anti-mesonium. *Soviet Physics JETP, 6*, 429.
6. Bilenky, S. M. (2005). The history of neutrino oscillations. *Physica Scripta T, 121*, 17–22. https://doi.org/10.1088/0031-8949/2005/T121/001.
7. Sinha, M., & Mukhopadhyay, B. (2008). CPT and lepton number violation in neutrino sector: Modified mass matrix of neutrino coupled to gravity. *Physical Review D, 77*, 025003. https://doi.org/10.1103/PhysRevD.77.025003.
8. Mukhopadhyay, B. (2007). Gravity induced neutrino-antineutrino oscillation: CPT and lepton number non-conservation under gravity. *Classical and Quantum Gravity, 24*, 1433–1442. https://doi.org/10.1088/0264-9381/24/6/004.
9. Barger, V. D., Learned, J. G., Pakvasa, S., & Weiler, T. J. (1999). Neutrino decay as an explanation of atmospheric neutrino observations. *Physical Review Letters, 82*, 2640–2643. https://doi.org/10.1103/PhysRevLett.82.2640.
10. Barenboim, G., Beacom, J. F., Borissov, L., & Kayser, B. (2002). CPT violation and the nature of neutrinos. *Physics Letters B, 537*, 227–232. https://doi.org/10.1016/S0370-2693(02)01947-0.
11. Mohanty, S., Mukhopadhyay, B., & Prasanna, A. R. (2002). Experimental tests of curvature couplings of fermions in general relativity. *Physical Review D, 65*, 122001. https://doi.org/10.1103/PhysRevD.65.122001.
12. Singh, P., & Mukhopadhyay, B. (2003). Gravitationally induced neutrino asymmetry. *Modern Physics Letters A, 18*, 779–785. https://doi.org/10.1142/S0217732303009691.
13. Ahluwalia, D. V., & Grumiller, D. (2005). Dark matter: A Spin one half fermion field with mass dimension one?. *Physical Review D, 72*, 067701. https://doi.org/10.1103/PhysRevD.72.067701.
14. Mukhopadhyay, B. (2005). Neutrino asymmetry around black holes: Neutrinos interact with gravity. *Modern Physics Letters A, 20*, 2145–2156. https://doi.org/10.1142/S0217732305017640.

15. Debnath, U., Mukhopadhyay, B., & Dadhich, N. (2006). Space-time curvature coupling of spinors in early universe: Neutrino asymmetry and a possible source of baryogenesis. *Modern Physics Letters A, 21*, 399–408. https://doi.org/10.1142/S0217732306019542.
16. Mukhopadhyay, B., Ghosh, T., & Ganguly, S. K. (2023). Gravitational geometric phase. In *The Sixteenth Marcel Grossmann Meeting* (pp. 689–698). https://doi.org/10.1142/9789811269776_0052.
17. Mandelstam, L., & Tamm, I. (1991). The uncertainty relation between energy and time in non-relativistic quantum mechanics. In B. M. Bolotovskii, V. Y. Frenkel, & R. Peierls (Eds.), *Selected Papers*. Springer. https://doi.org/10.1007/978-3-642-74626-0_8.
18. Margolus, N., & Levitin, L. B. (1998). The maximum speed of dynamical evolution. *Physica D, 120*, 188–195. https://doi.org/10.1016/S0167-2789%2898%2900054-2.
19. Thakuria, D., Srivastav, A., Mohan, B., Kumari, A., & Pati, A. K. (2024). Generalised quantum speed limit for arbitrary time-continuous evolution. *Journal of Physics A, 57*, 025302. https://doi.org/10.1088/1751-8121/ad15ad.
20. Pandey, V., Shrimali, D., Mohan, B., Das, S., & Pati, A. K. (2023). Speed limits on correlations in bipartite quantum systems. *Physical Review A, 107*, 052419. https://doi.org/10.1103/PhysRevA.107.052419.
21. Shrimali, D., Bhowmick, S., Pandey, V., & Pati, A. K. (2022). Capacity of entanglement for a nonlocal Hamiltonian. *Physical Review A, 106*, 042419. https://doi.org/10.1103/PhysRevA.106.042419.
22. Blasone, M., Dell'Anno, F., De Siena, S., & Illuminati, F. (2009). Entanglement in neutrino oscillations. *Europhysics Letters, 85*, 50002. https://doi.org/10.1209/0295-5075/85/50002.
23. Banerjee, S., Alok, A. K., Srikanth, R., & Hiesmayr, B. C. (2015). A quantum information theoretic analysis of three flavor neutrino oscillations. *European Physical Journal C, 75*, 487. https://doi.org/10.1140/epjc/s10052-015-3717-x.
24. Alok, A. K., Banerjee, S., & Sankar, S. U. (2016). Quantum correlations in terms of neutrino oscillation probabilities. *Nuclear Physics B, 909*, 65–72. https://doi.org/10.1016/j.nuclphysb.2016.05.001.
25. Jha, A. K., Mukherjee, S., & Bambah, B. A. (2021). Tri-Partite entanglement in neutrino oscillations. *Modern Physics Letters A, 36*, 2150056. https://doi.org/10.1142/S0217732321500565.
26. Jha, A. K., & Chatla, A. (2022). Quantum studies of neutrinos on IBMQ processors. *European Physical Journal Special Topics, 231*, 141. https://doi.org/10.1140/epjs/s11734-021-00358-9.
27. Jha, A. K., Chatla, A., & Bambah, B. A. (2024). Neutrinos as qubits and qutrits. *European Physical Journal Plus, 139*, 68. https://doi.org/10.1140/epjp/s13360-024-04861-5.
28. Dixit, K., Naikoo, J., Mukhopadhyay, B., & Banerjee, S. (2019). Quantum correlations in neutrino oscillations in curved spacetime. *Physical Review D, 100*, 055021. https://doi.org/10.1103/PhysRevD.100.055021.
29. Mukhopadhyay, B., & Ganguly, S. K. (2020). Gravity-induced geometric phases and entanglement in spinors and neutrinos: Gravitational Zeeman effect. *Universe, 6*, 160. https://doi.org/10.3390/universe6100160.
30. Nielsen, M. A., & Chuang, I. L. (2012). *Quantum computation and quantum information*. Cambridge University Press. https://doi.org/10.1017/cbo9780511976667.
31. Doran, C. (2000). A new form of the Kerr solution. *Physical Review D, 61*, 067503. https://doi.org/10.1103/PhysRevD.61.067503.

Chapter 41
Increasing Quantum Speed Limit of Relativistic Electron via Non-uniform Magnetic Field

Srishty Aggarwal, Banibrata Mukhopadhyay, Subhashish Banerjee, Arindam Ghosh, and Gianluca Gregori

Abstract Quantum speed limit (QSL) defines the theoretical upper bound on how fast a quantum system can evolve between states. It imposes a fundamental constraint on the rate of quantum information processing. For a relativistic spin-up electron in a uniform magnetic field, QSL increased with the magnetic field strength till around 10^{15} Gauss, before saturating at a saturated QSL (SQSL) of $0.2407c$, where 'c' is the speed of light. We show that by using variable magnetic fields, it is possible to surpass this limit, achieving SQSL up to 0.4–$0.6c$. To attain this quantum phenomenon, we solve the evolution equation of relativistic electron in spatially varying magnetic fields and find that the energies of various electron states become non-degenerate as opposed to the constant magnetic field case. This redistribution of energy is the key ingredient to accomplish higher QSL and, thus, a high information processing speed. We further explore how QSL can serve as a bridge between relativistic and non-relativistic quantum dynamics, providing insights via the Bremermann-Bekenstein bound, a quantity which constrains the maximal rate of information production. We also propose a practical experimental setup to realize these advancements. These results hold immense potential for propelling fields of quantum computation, thermodynamics and metrology.

S. Aggarwal (✉) · B. Mukhopadhyay · A. Ghosh
Department of Physics, Indian Institute of Science, Bengaluru, India
e-mail: srishtya@iisc.ac.in; bm@iisc.ac.in; arindam@iisc.ac.in

S. Banerjee
Department of Physics, Indian Institute of Technology Jodhpur, Jodhpur, India
e-mail: subhashish@iitj.ac.in

G. Gregori
Department of Physics, University of Oxford, Oxford, UK
e-mail: gianluca.gregori@physics.ox.ac.uk

S. Ghosh, A. R. Rao (eds.), *The Relativistic Universe: From Classical to Quantum*, Astrophysics and Space Science Proceedings 61,
https://doi.org/10.1007/978-3-031-90186-7_41

41.1 Introduction

The quantum speed limit (QSL) imposes an elemental constraint on the rate at which a quantum system can evolve. This limit, rooted in the Heisenberg uncertainty principle, is pivotal in quantum algorithms and information processing speed [1].

The QSL concept has illuminated diverse facets of quantum science, including information processing, open system dynamics, quantum control, and quantum thermodynamics, where precise manipulation and measurement of quantum systems are essential. Further, QSL time can be related to the energy cost per bit of information via Bremermann-Bekenstein (BB) bound [2, 3].

Magnetic fields are crucial for controlling quantum systems. Scientists have used uniform magnetic fields to manipulate chains of molecular qubits [4]. The QSL of relativistic electron reached to $0.2407c$, when subjected to uniform magnetic field (c is the speed of light) [5]. Can non-uniform magnetic fields be used to increase QSL, thereby accelerating quantum information processing?

We, therefore, investigate the QSL for a relativistic electron in a superposition of its ground and first excited states under spatially varying magnetic field. Further, we identify the critical magnetic field representing the boundary between non-relativistic and relativistic regimes by using the BB bound. In the end, we propose a feasible experimental setup to generate such magnetic field configurations.

41.2 Energy Levels of Relativistic Electron in Non-uniform Magnetic Field

We utilize a strictly spatially varying magnetic field profile, as detailed in [6], which is consistent with Maxwell's electromagnetic laws and given by

$$\mathbf{B} = B_0 \rho^n \hat{z}, \tag{41.1}$$

in cylindrical coordinates (ρ, ϕ, z). The magnetic field is uniform if $n = 0$, and is strictly spatially increasing or decreasing depending on whether $n > 0$ or $n < 0$ respectively. However, the solution of Eq. (41.1) is converging only till $n > -1$ [6].

We choose the vector potential $\mathbf{A}$ as

$$\mathbf{A} = B_0 \frac{\rho^{n+1}}{n+2} \hat{\phi} = A\hat{\phi}. \tag{41.2}$$

To investigate the dynamics of a relativistic electron within a non-uniform magnetic field, we solve the Dirac equation, given by

$$i\hbar \frac{\partial \Psi}{\partial t} = \left[c\boldsymbol{\alpha} \cdot \left(-i\hbar \nabla - \frac{q\mathbf{A}}{c} \right) + \beta m_e c^2 \right] \Psi, \tag{41.3}$$

where m_e and q $(-e)$ are electron's mass and charge respectively, $\hbar$ is the reduced Planck constant, $\mathbf{A}$ is the vector potential and $\boldsymbol{\alpha}$ and β are Dirac matrices. For stationary states, the wavefunction can be written as

$$\Psi = e^{-i\frac{Et}{\hbar}} \begin{bmatrix} \chi \\ \phi \end{bmatrix}, \tag{41.4}$$

ϕ and χ are 2-component spinors. The antiparticle wavefunction $\phi = -\chi$ [7].

Defining $\boldsymbol{\pi} = -i\hbar\nabla - q\mathbf{A}/c$ and using $(\boldsymbol{\sigma} \cdot \boldsymbol{\pi})(\boldsymbol{\sigma} \cdot \boldsymbol{\pi}) = \pi^2 - q\hbar\boldsymbol{\sigma} \cdot \boldsymbol{B}/c$, the decoupled equation obtained for χ is

$$(E^2 - m_e^2 c^4)\chi = \left[c^2\left(\pi^2 - \frac{q\hbar}{c}\boldsymbol{\sigma} \cdot \mathbf{B}\right)\right]\chi. \tag{41.5}$$

Let

$$\chi = e^{i\left(m\phi + \frac{p_z}{\hbar}z\right)} R(\rho), \tag{41.6}$$

where '$m\hbar$' is the angular momentum, p_z is the eigenvalue of momentum in the z-direction and $R(\rho)$ is a two-component matrix. $p_z = 0$ as the electron is confined to a plane perpendicular to the z-direction. Substituting χ from Eq. (41.6) into Eq. (41.5), and dividing it by $m_e^2 c^2$, Eq. (41.5) becomes

$$\begin{aligned} \alpha \tilde{R}_\pm = &-\lambda_e^2 \left[\frac{\partial^2}{\partial \rho^2} + \frac{1}{\rho}\frac{\partial}{\partial \rho} - \frac{m^2}{\rho^2}\right]\tilde{R}_\pm \\ &+ \left[\left(\frac{kB_0\rho^{n+1}}{n+2}\right)^2 + k\lambda_e\left(-\frac{2m}{n+2} \pm 1\right)B_0\rho^n\right]\tilde{R}_\pm, \end{aligned} \tag{41.7}$$

α is the eigenvalue of the problem and is given by $\epsilon^2 - 1$, $\epsilon = E/m_e c^2$ (dimensionless energy), $\lambda_e = \hbar/m_e c$ (Compton wavelength of electrons), $k = e/m_e c^2$ [6]. Hence, energy of level ν becomes

$$E_\nu = m_e c^2 \sqrt{1 + \alpha_\nu}. \tag{41.8}$$

In absence of Zeeman splitting and with $m = 0$, Eq. (41.7) becomes

$$\alpha' \tilde{R} = -\lambda_e^2 \left[\frac{\partial^2}{\partial \rho^2} + \frac{1}{\rho}\frac{\partial}{\partial \rho} - \right]\tilde{R} + \left(\frac{kB_0\rho^{n+1}}{n+2}\right)^2 \tilde{R}, \tag{41.9}$$

where α' is the energy level excluding Zeeman effect.

Solving Eq. (41.7) using Runge-Kutta method [6], we can obtain the eigenspectrum for different n. The mathematical relation for eigenvalues with n has been shown in [8]. The eigenspectrum, corresponding to α'_ν, and the split states α_ν for

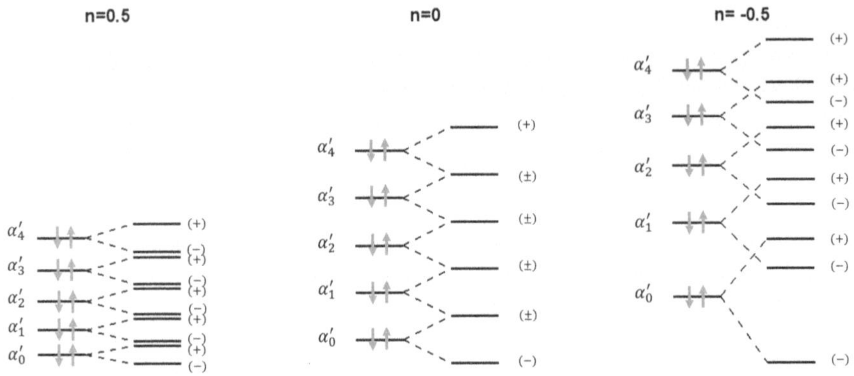

Fig. 41.1 Schematic representation for comparison of first five eigenlevels before splitting (α'_υ) and after splitting into $-\sigma.B$ and $+\sigma.B$ taking $m = 0$ for $n = 0.5$, 0 and $n = -0.5$

first five levels are given in Fig. 41.1. As depicted, with increasing eigenstates, eigenvalue difference for α'_υ between the two consecutive states stays the same for $n = 0$, increases for $n > 0$ and decreases for $n < 0$. The main reason for this difference is that in first case (growing magnetic field), electron observes a weak magnetic field close to the center, thereby, has a lesser gap in lower levels. As the field strength increases, the discretion in the energy levels increases and, therefore, the gap between the energy levels increases in case of positive n. Similar reasoning holds for negative n leading to an initial large gap in the eigenlevels and its subsequent decrease.

Moreover, the spin degeneracy inherent in a uniform magnetic field is broken in the case of a non-uniform field. One of the important differences between the radially growing and decaying magnetic fields is the corresponding alignment of spin-down and spin-up energy levels. In the case of a growing magnetic field (positive n), the lower energy level of the spin-up electron lies below the subsequent energy level of the spin-down electron. On the other hand, for decaying magnetic field (negative n), the former is positioned above the latter, as is illustrated for $n = 0.5$ and -0.5 in Fig. 41.1. These different alignments can be used to generate different closed systems of interests. For example, using negative n, one can generate a system of spin-down electrons confined to the ground and first excited states. Similarly, one can build a two-level system with only spin transitions in case of positive n.

41.3 QSL in the Non-uniform Magnetic Field

In general, for the evolution from one state to the other, the QSL of an electron ($\tilde{v}$) can be evaluated as the ratio of the its radial displacement (ρ_{disp}) to the minimum time of evolution (τ_{QSL}), given by

$$\tilde{v} = \frac{\rho_{disp}}{\tau_{QSL}}. \tag{41.10}$$

The initial ($t = 0$) and final ($t = \tau_{QSL}$) states of electron, assuming it to be in the superposition of the ν_{th} and $(\nu + 1)_{th}$ states at all times, are

$$\Psi(r, 0) = \frac{1}{\sqrt{2}} \left[\psi_\nu(r) + \psi_{\nu+1}(r)\right], \tag{41.11}$$

and

$$\Psi(r, \tau_{QSL}) = \frac{1}{\sqrt{2}} \left[\psi_\nu(r) e^{\frac{iE_\nu \tau_{QSL}}{\hbar}} + \psi_{\nu+1}(r) e^{\frac{iE_{\nu+1}\tau_{QSL}}{\hbar}}\right]. \tag{41.12}$$

τ_{QSL} between two orthogonal states is

$$\tau_{QSL} = max \left\{\frac{\pi\hbar}{2\Delta H}, \frac{\pi\hbar}{2\langle H\rangle}\right\}, \tag{41.13}$$

where the first and second terms are the Mandelstam-Tamm (MT) [9] and Margolous-Levitian (ML) [10] bounds respectively. As all the energy levels are positive, ΔH is smaller than $\langle H\rangle$. Therefore, τ_{QSL} is given by MT bound. Hence,

$$\tau_{QSL} = \frac{\pi\hbar}{2\Delta H}, \tag{41.14}$$

with

$$\Delta H = \frac{E_{\nu+1} - E_\nu}{2}. \tag{41.15}$$

The mean radial position at a time t is represented by

$$\langle\rho\rangle = \frac{1}{2}\left[\langle\nu|\rho|\nu\rangle + \langle\nu + 1|\rho|\nu + 1\rangle + 2\langle\nu|\rho|\nu + 1\rangle cos(\mathcal{E}t)\right], \tag{41.16}$$

where

$$\mathcal{E} = \frac{E_{\nu+1} - E_\nu}{\hbar}.$$

Therefore,

$$\begin{aligned} \rho_{disp} &= |\langle\rho\rangle_{\tau_{QSL}} - \langle\rho\rangle_0| \\ &= 2\left|\int_0^\infty \rho D_S(\rho) d\rho\right|, \end{aligned} \tag{41.17}$$

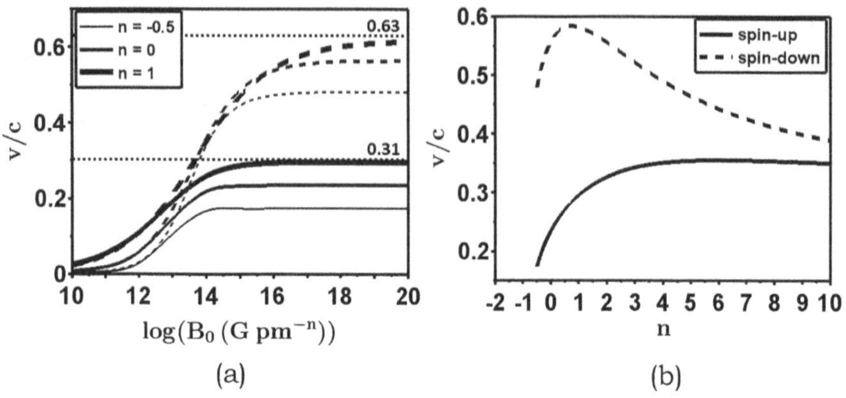

Fig. 41.2 (**a**) Variation of QSL with magnetic field strength (B_0) for increasing ($n = 1$), uniform ($n = 0$) and decreasing ($n = -0.5$) magnetic field profile for spin-up (solid lines) and spin-down (broken lines) electron. The line thickness is in order of increasing n. Here, for $n = 1$, SQSLs are highlighted by horizontal line. (**b**) Variation of QSL with n at $B_0 = 10^{17}\ G\ pm^{-n}$

where

$$D_s(\rho) = \psi_\nu^\dagger\ \rho\ \psi_{\nu+1}\ . \tag{41.18}$$

Thus, QSL can be evaluated as the ratio of Eq. (41.17) to Eq. (41.14) [11].

The variation of QSL with the magnetic field strength is shown in Fig. 41.2a for uniform ($n = 0$), decreasing ($n = -0.5$) and increasing ($n = 1$) fields for spin-up and spin-down electrons. QSL increases with the increase in B_0 and then tends to saturated QSL (SQSL). The SQSL is higher for $n = 1$ as compared to $n = 0$ and -0.5. To determine the change of SQSL with n, the variation of QSL with n is depicted in Fig. 41.2b at very high magnetic field strength of $B_0 = 10^{17}\ G\ pm^{-n}$, such that QSL tends to SQSL. The SQSL tends to increase with n for spin-up electron, while for spin-down electron, it increases up to $n \sim 2$, and then decreases with n. Thus growing magnetic field type variation leads to higher QSL as compared to uniform magnetic field.

41.4 Bremermann-Bekenstein Bound: Finding Critical Magnetic Field

The relativistic treatment becomes important only when the energy associated with magnetic field exceeds the rest mass energy of the particle. The magnetic field threshold beyond which the relativistic dynamics dominates is known as the critical magnetic field. For electrons, the critical magnetic field is given by $m_e^2 c^3/\hbar e$ Gauss in case of a uniform magnetic field and achieved when the gyromagnetic radius approximates its Compton wavelength. However, such an analysis is much more involved for a non-uniform magnetic field. Interestingly, we can find the critical magnetic field for the latter case using a unique tool of quantum information: the BB bound.

BB bound is a fundamental theoretical limit on the fastest possible speed of information processing or storage. It provides a link between the energy cost $\langle H \rangle$ for a bit of information I and the QSL time, and is given by [12]

$$\frac{\langle H \rangle}{I} > \frac{\hbar \ln 2}{\pi \tau_{QSL}}. \tag{41.19}$$

This bound arises from the intersection of quantum mechanics, general relativity, and information theory, offering deep insights into the fundamental constraints of physical systems [2]. It has profound implications in quantum information, cosmology and black hole physics. It also leads to the development of concepts such as black hole entropy and the holographic principle.

Figure 41.3 shows the variation of LHS (taking $I = 1$) and RHS of Eq. (41.19) with B_0 for two cases: (a) $n = 0$ and (b) $n = 2$. The two sub-figures are split into

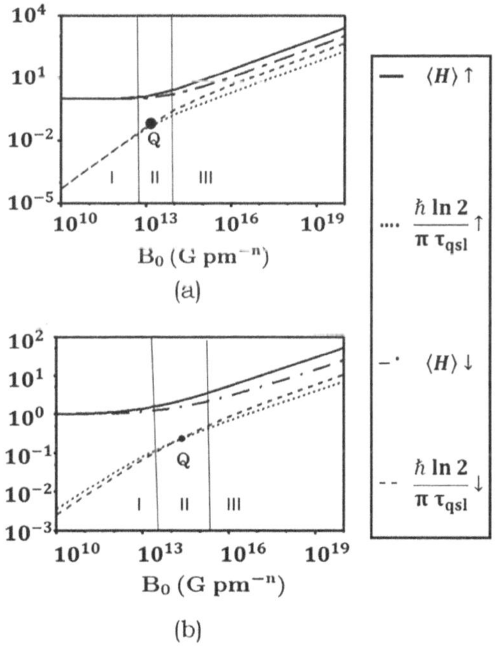

Fig. 41.3 Variation of $\langle H \rangle$ (LHS) and $\hbar\ ln\ 2/\pi\tau_{QSL}$ (RHS) of Eq. (41.19) with B_0 for spin-up ($\uparrow$) and spin-down ($\downarrow$) electrons as an illustration of the difference between non-relativistic and relativistic regimes for (**a**) $n = 0$ and (**b**) $n = 2$. I, II and III regions represent 'non-relativistic', 'transition' and 'relativistic' regimes, respectively. Q represents the critical magnetic field. While in case of uniform magnetic field in (**a**), $(\hbar ln 2)/(\pi\tau_{QSL})$ is same for both the spins, in non-relativistic regime, it is higher for spin-down electron in the relativistic regime. Further, it is smaller for the spin-down electron in non-relativistic regime in the case of non-uniform magnetic field (**b**), while larger in the relativistic one

three regions. In region I, characterized by low B_0 and non-relativistic dynamics, $\langle H \rangle$ and $(\hbar ln2)/(\pi \tau_{QSL})$ exhibit a clear separation. As we transition to region II, with a stronger magnetic field, this separation diminishes, indicating a shift towards relativistic behavior. In region III, dominated by a strong magnetic field, the two quantities become nearly parallel, signifying a fully relativistic regime. Thus, the difference in the regimes of non-relativistic and relativistic is indeed evident.

The point of separation of RHS for spin-down and spin-up electrons, highlighted as **Q** $(4.414 \times 10^{13} G)$ in Fig. 41.3a, depicts the already known critical magnetic field for uniform case ($n = 0$). In Fig. 41.3b, **Q** ($B_0 = 1.35 \times 10^{14}\ G\ pm^{-n}$) is the point of intersection for spin-down and spin-up electrons. Using a similar comparison for non-uniform magnetic field, **Q** may indicate the critical magnetic field for $n = 2$. Hence, BB bound furnishes an easy technique for estimating the critical magnetic field.

41.5 Experimental Design for Non-uniform Magnetic Field

Generating non-uniform magnetic fields requires customised approaches based on the specific quantum system under investigation. To attain varying magnetic field, a solenoid with its ferrite core equipped with curved pole pieces can be taken as shown in Fig. 41.4. Utilizing high-permeability materials like ferrite or soft iron for the core enhances field strength. The concavity (convexity) of the pole pieces will lead to smaller (larger) magnetic field strength at the center, and larger (smaller) at the periphery. Thus, the curvature can be adjusted to attain the desired n.

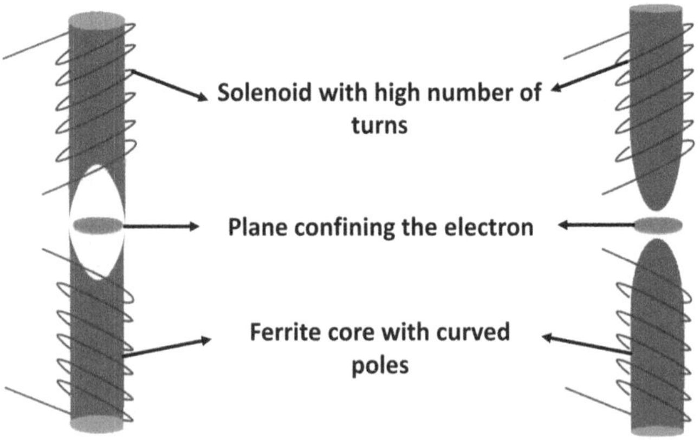

Fig. 41.4 The experimental design to attain non-uniform magnetic field. The solenoids with ferrite core having concave (left) and convex (right) pole pieces could be applied to achieve growing ($n > 0$) and decaying ($n < 0$) magnetic fields respectively

To compare the QSL for electrons in linearly varying magnetic fields against uniform case, we have simulated a laboratory setting. Let us confine the electron at the center within 1 μm in a plane of diameter 1 mm and put it between the solenoids with concave pole pieces. The size of the electron confinement (1 μm) is comparable to the scale of quantum dots [13] or Josephson junction [14]. The solenoid with ferrite core has 100 turns/cm and carries 1 A current. Assuming a linearly increasing magnetic field profile, the magnetic field will be 0 at the center and approximately $10^4\, G$ at the edge of the plane with its strength at the periphery of electron's circular plane being 10 G. Mathematically, $|\mathbf{B}| = 10\, G$ at $\rho = 0.5\, \mu\text{m} = 5 \times 10^5\, pm$ when $n = 1$. So, $B_0 = 10/(5 \times 10^5) = 2 \times 10^{-5}\, G\, pm^{-n}$ (using Eq. (41.1)). QSL for spin-up and spin-down electrons are $3.2 \times 10^{-7}c$ and $3 \times 10^{-7}c$, respectively. For uniform magnetic field, the similar condition with $B_0 = 10\, G$ provides a QSL of $1.9 \times 10^{-7}c$ for both the electrons' spins. Thus, spatially increasing magnetic field leads to higher QSL of electron.

Additionally, spatially varying magnetic fields can be realized through various experimental techniques. For instance, a three-dimensional architecture employing lithographically patterned films with varying thickness and strong perpendicular magnetic anisotropy can generate such fields [15]. Also, local magnetic fields with controlled spatial distribution can be induced in materials like graphene or transition metal dichalcogenides by exploiting grain boundaries or topological defects [16]. Hence, these methods offer promising avenues for experimentally realizing the increased QSL observed in our model with spatially increasing magnetic fields.

41.6 Conclusions

We have shown that a non-uniform magnetic field is crucial in attaining higher QSL till $0.6c$. Further, we demonstrate how, adopting a foundational perspective, we can find the critical magnetic field by using the BB bound. Thus, the BB bound provides a unified framework for examining both relativistic and non-relativistic treatments, enabling a comprehensive understanding of their interrelationship. We further present an experimental blueprint to realize the proposed concepts in a laboratory setting.

Thus, the present endeavor not only accentuates the use of relativistic dynamics to quantify information theory via an increase of QSL in a non-uniform magnetic field, but also probes into the facets of the former using the latter via the use of BB bound to determine the critical magnetic field. Its applications potentially impact a broader spectrum of scientific disciplines, including quantum information and condensed matter physics.

Competing Interests The authors have no conflicts of interest.

Acknowledgments The authors acknowledge ISRA 2023 organizing committee for providing a platform to exchange ideas about relativity, astrophysics and quantum physics.

References

1. Deffner, S., & Campbell, S. (2017). Quantum speed limits: From Heisenberg's uncertainty principle to optimal quantum control. *Journal of Physics A Mathematical General, 50*, 453001.
2. Bremermann, H. J. (1967). *Quantum Noise and Information Proc. of the 5th Berkeley Symposium on Mathematical Statistics, Probability* (Vol. 4). Biology, Problems of Health. University of California Press.
3. Bekenstein, J. D., & Schiffer, M. (1990). Quantum limitations on the storage and transmission of information. *International Journal of Modern Physics C, 1*, 355–422.
4. Santini, P., Carretta, S., Troiani, F., & Amoretti, G. (2011). Molecular nanomagnets as quantum simulators. *Physical Review Letters, 107*, 230502.
5. Villamizar, D. V., & Duzzioni, E. I. (2015). Quantum speed limit for a relativistic electron in a uniform magnetic field. *Physical Review A, 92*, 042106.
6. Aggarwal, S., Mukhopadhyay, B., & Gregori, G. (2021). Relativistic Landau quantization in non-uniform magnetic field and its applications to white dwarfs and quantum information. *SciPost Physics, 11*, 93.
7. Strange, P. (1998). *Relativistic quantum mechanics*. Cambridge University Press.
8. Aggarwal, S., & Mukhopadhyay, B. (2023). *Dynamics of relativistic electrons in non-uniform magnetic fields and its applications in quantum computing and astrophysics* (pp. 4362–4373). World Scientific.
9. Mandelstam, L., & Tamm, I. (1945). The energy–time uncertainty relation in non-relativistic quantum mechanics. *Journal of Physics USSR, 9*, 249–254.
10. Margolus, N., & Levitin, L. B. (1998). The maximum speed of dynamical evolution. *Physica D: Nonlinear Phenomena, 120*(1), 188–195. Proceedings of the Fourth Workshop on Physics and Consumption.
11. Aggarwal, S., Banerjee, S., Ghosh, A., & Mukhopadhyay, B. (2022). Non-uniform magnetic field as a booster for quantum speed limit: Faster quantum information processing. *New Journal of Physics, 24*, 085001.
12. Deffner, S., & Lutz, E. (2013). Quantum speed limit for non-markovian dynamics. *Physical Review Letters, 111*, 010402.
13. Kouwenhoven, L. P., Austing, D. G., & Tarucha, S. (2001). Few-electron quantum dots. *Reports on Progress in Physics, 64*, 701–736.
14. Ren, H., Pientka, F., Hart, S., Pierce, A. T. Kosowsky, M., Lunczer, L., Schlereth, R., Scharf, B., Hankiewicz, E. M., Molenkamp, L. W., Halperin, B. I., & Yacoby, A. (2019). Topological superconductivity in a phase-controlled Josephson junction. *Nature, 569*, 93–98.
15. Rougemaille, N., N'Diaye, A. T., Coraux, J., Vo-Van, C., Fruchart, O., & Schmid, A. K. (2012). Perpendicular magnetic anisotropy of cobalt films intercalated under graphene. *Applied Physics Letters, 101*(14), 142403.
16. Zhang, Z., Zou, X., Crespi, V. H., & Yakobson, B. I. (2013). Intrinsic magnetism of grain boundaries in two-dimensional metal dichalcogenides. *ACS Nano, 7*(12), 10475–10481. PMID: 24206002.

Part XI
Gravity and Quantum Physics Interface

Chapter 42
Affine Connection and Quantum Theory

Kaushik Ghosh

Abstract We have previously demonstrated that an affine connection resulting in non-metricity can bring in two new scalar fields into the Einstein-Palatini action. These scalar fields can play crucial roles in elucidating inflation and dark energy. In this chapter, we have explored potential coupling of these fields with scalar, gauge and Dirac fields. We have demonstrated that the above mentioned scalar fields may be excluded from the actions of gauge theories interacting with Dirac fields through conserved currents. A similar aspect holds true for scalar fields. The connection-scalars can be coupled with ordinary matter and gauge fields by incorporating suitable interaction terms. We have also found that Stokes' theorem necessitates the inclusion of right handed neutrinos in the electroweak theory in curved spacetime, even when we use only the Levi-Civita connection. This inclusion is essential for obtaining consistent equations of motion and a conserved vector current for the neutrinos. Axial vector currents for different Dirac fields may remain anomalous depending on the theory. The right handed neutrinos can be significant in explaining neutrino oscillation and dark matter.

42.1 Introduction

Significant progress has been made over the past few decades regarding two periods of cosmic acceleration: inflation and dark energy [1–6]. Present models to explain inflation require one or more scalar fields [7–9]. Dark energy has yet to be directly detected. Unlike ordinary matter, it is not localized and is presumed to possess negative pressure to account for the current cosmic acceleration. Furthermore, there is dark matter, observed through its gravitational effects [6, 10–12]. CMBR data and supernova surveys indicate that approximately 70% of the energy of universe is dark energy and 20% is dark matter. The remainder is ordinary matter, such as different leptons, hadrons and gauge fields. The primary theory for dark energy

K. Ghosh (✉)
Vivekananda College (University of Calcutta), Kolkata, India

S. Ghosh, A. R. Rao (eds.), *The Relativistic Universe: From Classical to Quantum*, Astrophysics and Space Science Proceedings 61,
https://doi.org/10.1007/978-3-031-90186-7_42

incorporates the cosmological constant Λ in Einstein's equation [6, 13, 14]. Other similar models introduce fields like cosmons that come with negative pressure in Einstein's equation [15–17]. A different approach is to modify the Einstein-Hilbert action. This includes $f(R)$ gravity [12, 18], scalar-tensor theories [19, 20], and braneworld models [21, 22].

Connection is the fundamental field to study the geometry of curved spacetime. In the Einstein's theory of gravity, a classical theory of gravity, we restrict the connection to metric compatible Levi-Civita connection, the later being an affine connection. There is no problem with the principle of general covariance if we use connection more general than $\Gamma^{\alpha}_{\ \mu\nu}$ [23–30]. Such a connection introduces non-metricity [27, 31] that can be an integral part of quantum gravity [23–25]. In the next section, we will use the Einstein-Palatini formalism to show that we can use non-metricity to introduce two scalar fields ω and ψ. Connection is expressed as: $\Theta^{\alpha}_{\ \mu\nu} = \Gamma^{\alpha}_{\ \mu\nu} + C^{\alpha}_{\ \mu\nu}$, and the scalar fields can be associated with the traces: $C^{\alpha}_{\ \alpha\nu}$ and $C^{\alpha\ \mu}_{\ \mu}$. One of these scalar fields, ω, is akin to a dilaton, while the other behaves like a phantom scalar [32–34].

In Sect. 42.3, we will investigate the potential coupling between ω, ψ and ordinary fields associated with various representations of $SL(2, c)$. We will develop a Lagrangian formalism that maintains general covariance, the principle of equivalence, and is consistent with Stokes' theorem. We will discover that the connection scalar fields mentioned earlier can be removed from the actions and stress tensors of scalar fields, Dirac fields, and gauge fields. This aligns with the experimental observations regarding dark energy, which does not interact with ordinary fields. Additionally, we may need to incorporate right handed neutrinos as a non-interacting field within electroweak theory in curved spacetime, regardless of the connection. This inclusion is necessary for deriving consistent equations of motion and expressions for non-anomalous vector currents for the neutrinos. They can play a significant role in dark matter research, and there are further advantages related to neutrino oscillations.

In Sect. 42.4, we will explore potential modifications to Einstein's theory of gravity with finite ω and ψ. We will change ω and ψ to Φ and Ψ with the later being linear combinations of ω and ψ. This is required to decouple the equations for ω and ψ interconnected through derivative terms. Φ contributes a positive stress tensor to Einstein's equation, while Ψ contributes a negative stress tensor and may represent a phantom/ghost scalar [6]. When necessary, we can enhance the Einstein-Palatini action with potential terms dependent on Φ or Ψ. We can couple Φ and Ψ. Additionally, we can couple Φ and Ψ with matter and gauge fields. This results in a general version of scalar-tensor theories and can be utilized to develop various models for inflation and dark energy [6].

Vacuum energies and quantum fluctuations in Φ and Ψ can be utilized to explore inflation and dark energy, akin to the Casimir effect. Another approach to generalizing Einstein's theory involves incorporating Φ and Ψ and Λ. This could lead to a potential theory of inflation alongside dynamic dark energy that may not be homogeneous and isotropic. The presence of Λ introduces deviations from

Newtonian gravity in the weak field, slow-motion limit [29]. Furthermore, Φ and Ψ may provide deviation from the conventional splitting of spacetime into space and time [24, 25]. Consequently, Λ, Φ and Ψ exhibit similar effects on spacetime. Various astrophysical observations suggest that the current non-metricity induced by Φ and Ψ is minimal [25]. This is analogous to Λ in the ΛCDM theory. In the following, we have adopted the geometrized units where, $\hbar = G = c = 1$. The metric is of signature $+2$.

42.2 Scalar Fields from Affine Connection

In the Einstein-Palatini action, the connection is given by [29]: $\Theta^{\alpha}{}_{\mu\nu} = \Gamma^{\alpha}{}_{\mu\nu} + C^{\alpha}{}_{\mu\nu}$, where $\Gamma^{\alpha}{}_{\mu\nu}$ is the Levi-Civita connection and $C^{\alpha}{}_{\mu\nu}$ can introduce non-metricity. In this section we will consider $C^{\alpha}{}_{\mu\nu}$ that is symmetric in the lower indices. The curvature and Ricci tensors are given by the following expressions:

$$
\begin{aligned}
\nabla'_{\mu} A_{\nu} &= \partial_{\mu} A_{\nu} - \Theta^{\alpha}{}_{\mu\nu} A_{\alpha} \\
(\nabla'_{\mu}\nabla'_{\nu} - \nabla'_{\nu}\nabla'_{\mu}) A^{\alpha}{}_{\beta} &= -R'_{\mu\nu\kappa}{}^{\alpha} A^{\kappa}{}_{\beta} + R'_{\mu\nu\beta}{}^{\kappa} A^{\alpha}{}_{\kappa} \\
R'_{\mu\nu\alpha}{}^{\kappa} &= R_{\mu\nu\alpha}{}^{\kappa} + 2\nabla_{[\nu} C^{\kappa}{}_{\mu]\alpha} + 2[C^{\lambda}{}_{[\mu|\alpha|} C^{\kappa}{}_{\nu|\lambda|]}] \\
R'_{\mu\alpha} &= R_{\mu\alpha} + 2\nabla_{[\kappa} C^{\kappa}{}_{\mu]\alpha} + 2[C^{\lambda}{}_{[\mu|\alpha|} C^{\kappa}{}_{\kappa|\lambda|]}] \\
R' &= g^{\mu\alpha}[R_{\mu\alpha} + 2\nabla_{[\kappa} C^{\kappa}{}_{\mu]\alpha} + 2[C^{\lambda}{}_{[\mu|\alpha|} C^{\kappa}{}_{\kappa|\lambda|]}] ,
\end{aligned}
\tag{42.1}
$$

where ∇'_{μ} uses the complete connection $\Theta^{\alpha}{}_{\mu\nu}$, and ∇_{μ} uses $\Gamma^{\alpha}{}_{\mu\nu}$. $R_{\mu\nu\alpha}{}^{\kappa}$ is the curvature tensor for $\Gamma^{\alpha}{}_{\mu\nu}$ and $R_{\mu\alpha}$ is the Ricci tensor associated with the Levi-Civita connection. The Lagrangian density is given by $\sqrt{-g}R'$, where R' is the scalar curvature evaluated using the connection $\Theta^{\alpha}{}_{\mu\nu}$. Classically, it yields null solutions for $C^{\alpha}{}_{\mu\nu}$ [29]. In [23–25], we have expressed $C^{\alpha}{}_{\mu\nu}$ using covariant derivatives of lower rank tensors. This is similar to $\Gamma^{\alpha}{}_{\mu\nu}$ constructed from $g_{\mu\nu}$. It will be found in the following after Eq. (42.6) as point (1) that scalar fields are primary candidates for this purpose. The resulting Einstein's equation is consistent with the contracted Bianchi identity in $\Gamma^{\alpha}{}_{\mu\nu}$. Here, we express $C^{\alpha}{}_{\mu\nu}$ as:

$$
\begin{aligned}
\Theta^{\alpha}{}_{\mu\nu} &= \Gamma^{\alpha}{}_{\mu\nu} + C^{\alpha}{}_{\mu\nu} , \\
C^{\alpha}{}_{\mu\nu} &= \delta^{\alpha}{}_{(\mu}\nabla_{\nu)}\omega - \frac{1}{2} g_{\mu\nu}\nabla^{\alpha}\omega + g_{\mu\nu}\nabla^{\alpha}\psi .
\end{aligned}
\tag{42.2}
$$

The reason for this choice is that the first two terms of $C^{\alpha}{}_{\mu\nu}$ gives the contribution of the conformal scalar field when we construct the connection to be compatible with:

$$
\tilde{g}_{\mu\nu} = (1 + \phi(x)) g_{\mu\nu}, \quad \omega = ln(1 + \phi). \tag{42.3}
$$

We can associate ω/ϕ and ψ with two traces of $C^{\alpha}_{\ \mu\nu}$. We find that: $C^{\alpha}_{\ \alpha\nu} = \nabla_\nu(2\omega - \psi)$ and $C^{\alpha\mu}_{\ \ \mu} = \nabla^\alpha(4\psi - \omega)$. Equation (42.3) gives us a symmetric Ricci tensor. The Einstein-Palatini action takes the following form:

$$S = \int \sqrt{-g}R'\mathbf{e} + \sum \kappa_M S_M(\chi_a, g_{\mu\nu}) = \int \sqrt{-g}\mathcal{R}'\mathbf{e} + \sum \kappa_M S_M(\chi_a, g_{\mu\nu}), \tag{42.4}$$

where $\mathbf{e}$ is $dx^0 \wedge dx^1 \wedge dx^2 \wedge dx^3$ [29], χ_a includes various Dirac and gauge fields with S_M being corresponding action. The second term, $2\nabla_{[\kappa} C^{\kappa}_{\ \mu]\alpha}$, present in the expression of $R'_{\mu\alpha}$ given by Eq. (42.2), gives a four-divergence in S and is discarded. Comments on corresponding boundary conditions can be found in [25]. We note that it is always possible to modify the action by adding a boundary term with the integrand expressed as: $-C^{\alpha\ \mu}_{\ \mu} + C^{\mu\ \alpha}_{\ \mu}$ [25]. A similar approach is required for $g_{\mu\nu}$ [29]. The modified Ricci tensor $\mathcal{R}'_{\mu\nu}$ and modified curvature scalar $\mathcal{R}'$ are given by:

$$\begin{aligned} \mathcal{R}' = g^{\mu\alpha}\mathcal{R}'_{\mu\alpha} &= g^{\mu\alpha}\left\{R_{\mu\alpha} + 2[C^{\lambda}_{\ [\mu|\alpha|}C^{\kappa}_{\ \kappa|\lambda|]}]\right\} \\ &= R - \frac{3}{2}(\nabla\omega)^2 + 3(\nabla\psi)^2 + 3[(\nabla_\kappa\omega)(\nabla^\kappa\psi)], \end{aligned} \tag{42.5}$$

where $R_{\mu\alpha}$ and R are the Ricci tensor and curvature scalar obtained from $\Gamma^{\alpha}_{\ \mu\nu}$. Field equations are given by various functional derivatives. $C^{\alpha}_{\ \mu\nu}(\omega, \psi)$ is assumed not to be present in S_M. This is discussed further in the next section. The modified Einstein's equation is given by [23–25]:

$$G_{\mu\alpha} = 8\pi[P_{\mu\alpha}(\chi_a) + \frac{3}{16\pi}P_{\mu\alpha}(\omega) - \frac{3}{8\pi}P_{\mu\alpha}(\psi) - \frac{3}{8\pi}P_{\mu\alpha}(\omega, \psi)] \tag{42.6}$$

$$\begin{aligned} P_{\mu\alpha}(\omega, \psi) &= (\nabla_{(\mu}\psi)(\nabla_{\alpha)}\omega) - \frac{1}{2}g_{\mu\alpha}(\nabla_\kappa\psi)(\nabla^\kappa\omega) \\ &= \frac{1}{(1+\phi)}[(\nabla_{(\mu}\psi)(\nabla_{\alpha)}\phi) - \frac{1}{2}g_{\mu\alpha}(\nabla_\kappa\psi)(\nabla^\kappa\phi)] \\ \nabla_\kappa\nabla^\kappa\omega &= \frac{1}{1+\phi}\nabla_\kappa\nabla^\kappa\phi - (\frac{1}{1+\phi})^2(\nabla\phi)^2 = 0 \\ \nabla_\kappa\nabla^\kappa\psi &= 0, \end{aligned}$$

where $P_{\mu\alpha}(\chi_a)$ gives the stress tensor of ordinary fields. ω and ψ obey the massless Klien-Gordon equation and may remain finite for most spacetimes. This results in the non-metricity discussed in [23–25]. Equation (42.6) agrees with the contracted Bianchi identity in $\Gamma^{\alpha}_{\ \mu\nu}$: $\nabla^\alpha G_{\mu\alpha} = 0$. This aspect will be explored in detail in the next section. $P_{\mu\alpha}(\omega)$ and $P_{\mu\alpha}(\psi)$ are the stress tensors of massless scalar fields. ψ, with a negative stress tensor to Einstein's equation, behaves like a phantom scalar field. We will elaborate on this in the next section. Discussions on negative stress tensor can be found in [35, 36]. Coupling of ψ with ω in $\mathcal{R}'$ gives a coupled term in Einstein's equation. We will separate the equations in the next section. $C^{\alpha}_{\ \mu\nu}$ in Eq. (42.3) is obtained from non-trivial solutions of Eq. (42.6). ω and ψ lead to non-metricity and could be required in quantum gravity [23–25].

42.3 Matter Fields in Presence of Non-metricity Scalar Fields

We now explore the potential coupling between (ω, ψ) and ordinary fields. We must consider several points when attempting to construct an interacting theory of gravity with a connection different from $\Gamma^{\alpha}_{\ \mu\nu}$ [23–25].

(1) **Einstein-Palatini action, variational derivatives and Bianchi identity**:
We begin with the Euler-Lagrange's equations and Bianchi identities. No problem arises with Bianchi identities for $R'^{\ \ \kappa}_{\mu\nu\alpha}$ given by Eq. (42.2) when the connection is $\Theta^{\alpha}_{\ \mu\nu}$. ω and ψ in Eq. (42.3) can be arbitrary regular fields. Generally, problems will arise when we discard four-divergence term in R' given by $2g^{\mu\alpha}\nabla_{[\kappa}C^{\kappa}_{\ \mu]\alpha}$ and proceed to apply the principle of least action to S given by Eq. (42.4). The modified Einstein's equation given by Eq. (42.6) contains ω and ψ as sources. Contracted Bianchi identity *w.r.t* the Levi-Civita connection necessitates to have: $\nabla^{\alpha}G_{\mu\alpha} = 0$. This remains valid for ψ and ω individually. $(\nabla\omega)^2$ and $(\nabla\psi)^2$ are the only scalar quadratic in the first order derivatives of the corresponding fields. Thus, their contribution in $\sqrt{-g}\mathcal{R}'$ is same as the Lagrangian density of a massless scalar field apart from numerical factors and, we have in Eq. (42.6), stress tensor of two massless scalar fields accompanied by numerical factors. We also have derivative coupling between ψ and ω in Eq. (42.6). The contracted Bianchi identity in the Levi-Civita connection remains valid for Eq. (42.6). We will discuss this further in the next section. The above arguments go in favour to include ψ and ω. We will later introduce self interaction terms in ω and ψ.

(2) **Restrictions on Stokes' theorem and construction of conserved currents**:
Next comes the issue of removing $\int\sqrt{-g}g^{\mu\alpha}\nabla_{[\kappa}C^{\kappa}_{\ \mu]\alpha}d^4x$ by applying the Stokes' theorem. This is possible when we can express this term as: $\int d^4x\nabla_{\mu}(\sqrt{-g}V^{\mu})$, where $\sqrt{-g}V^{\mu}$ is a relative contravariant vector field of weight $+1$, [27, 29]. This is the case with $\Gamma^{\alpha}_{\ \mu\nu}$ as the connection [27, 29]. Thus, we have used $\Gamma^{\alpha}_{\ \mu\nu}$ in ∇_{μ}. Similar restriction remains valid when we try to derive the field equations for scalar, Dirac and gauge fields from respective actions in curved spacetime. The details can be found in [25].

In [25], we have discussed potential coupling between ω, ψ and matter fields that are different representations of $SL(2, c)$ [37]. This includes scalar, Dirac and gauge fields. It can be shown that ω and ψ do not appear in the Lagrangian densities of scalar, electromagnetic and non-abelian gauge fields [25]. Here, we discuss the case of Dirac fields. The Lagrangian density is given by [38–40]:

$$\mathcal{L}' = -\sqrt{-g}\{\frac{1}{2}[\bar{\chi}\gamma^{\mu}\nabla'_{\mu}\chi - (\nabla'_{\mu}\bar{\chi})\gamma^{\mu}\chi] + m\bar{\chi}\chi\},\ \gamma^{\mu} = V^{\mu}_{a}\gamma^{a}, \tag{42.7}$$

where χ are the $SL(2, c)$ spinors, γ^a are the flat spacetime Dirac matrices and V^{μ}_a are the vierbeins [39, 40]. We locally have: $V^{\mu}_a(x)V_{b\mu}(x) = \eta_{ab}$. However, the vierbeins at two distant points may not be related through parallel transport when $C^{\alpha}{}_{\mu\nu}$ is finite. Covariant derivative on χ can be expressed by the following equation [39, 40]:

$$\begin{aligned}\nabla'_{\mu} &= \partial_{\mu} + \frac{1}{2}\sigma^{bc}V^{\nu}_{b}(\nabla'_{\mu}V_{c\nu}) \\ &= \nabla_{\mu} - \frac{1}{2}\sigma^{bc}V^{\nu}_{b}V_{c\alpha}C^{\alpha}{}_{\mu\nu},\quad \sigma^{bc} = -\frac{i}{4}[\gamma^{b}, \gamma^{c}] \\ &= \nabla_{\mu} + C_{\mu}.\end{aligned} \tag{42.8}$$

Equations (42.3) and (42.7) give the following expression for $\mathcal{L}'$:

$$\mathcal{L}' = -\sqrt{-g}\{\frac{1}{2}[\bar{\chi}\gamma^{\mu}\nabla_{\mu}\chi - (\nabla_{\mu}\bar{\chi})\gamma^{\mu}\chi] + m\bar{\chi}\chi - \frac{3i}{4}(\bar{\chi}\gamma^{\mu}\chi)\nabla_{\mu}(\omega - \psi)\}. \tag{42.9}$$

The additional term, dependent on ω and ψ, can be discarded when the current is conserved. Thus, as in the case of scalar and gauge fields, we can replace $\Theta^{\alpha}{}_{\mu\nu}$ with $\Gamma^{\alpha}{}_{\mu\nu}$ in Eq. (42.7). $\mathcal{L}'$ contributes a stress-tensor in Einstein's equation, with the later remaining consistent with the Bianchi identity in $\Gamma^{\alpha}{}_{\mu\nu}$. To discuss the problems associated with weak interactions, we consider the following Lagrangian density:

$$\begin{aligned}\tilde{\mathcal{L}}_D = \sqrt{-g}[&-\bar{\chi}_e\gamma^{\mu}\nabla_{\mu}\chi_e \\ &-\frac{1}{2}\bar{\chi}_{\nu}\gamma^{\mu}\nabla_{\mu}\chi_{\nu} + \frac{1}{2}\bar{\chi}_{\nu}\gamma^{\mu}\gamma_5\nabla_{\mu}\chi_{\nu}],\end{aligned} \tag{42.10}$$

where χ_e and χ_{ν} are the Dirac fields for the electrons and neutrinos respectively. No problem arises with χ_e. However, $\sqrt{-g}[\bar{\chi}_{\nu}\gamma^{\mu}\gamma_5(\delta\chi_{\nu})]$ being a vector, we cannot discard $\int \mathbf{e}\nabla_{\mu}[\sqrt{-g}(\bar{\chi}_{\nu}\gamma^{\mu}\gamma_5\delta\chi_{\nu})]$ completely. This is related with the application of Stoke's theorem mentioned before. $\frac{\delta\tilde{S}_D}{\delta\chi_{\nu}(x)}$ becomes non-covariant [25]. It is easy to show that the equations for χ_{ν} and $\bar{\chi}_{\nu}$ do not remain consistent with each other. To solve the above problems, we add ν_R in Eq. (42.10):

$$\mathcal{L}_D = -\frac{1}{2}\sqrt{-g}[\bar{\chi}_e\gamma^{\mu}\nabla_{\mu}\chi_e - (\nabla_{\mu}\bar{\chi}_e)\gamma^{\mu}\chi_e + \bar{\chi}_{\nu}\gamma^{\mu}\nabla_{\mu}\chi_{\nu} - (\nabla_{\mu}\bar{\chi}_{\nu})\gamma^{\mu}\chi_{\nu}]. \tag{42.11}$$

Equations for χ_{ν} and $\bar{\chi}_{\nu}$ can be used to derive those for ν_L and $\bar{\nu}_L$. There are further notable advantages. $\mathcal{L}_D$ can be employed to derive an expression for the conserved vector current of the neutrino, whereas this does not apply to $\tilde{\mathcal{L}}_D$ [25]. Similar situation continue to remain valid in flat spacetime with curvilinear coordinates.

In flat spacetime, right handed neutrinos can effectively resolve triangle anomalies in the lepton sector [41] and may become crucial for explaining neutrino oscillation [42]. They may also contribute to dark matter research. To conclude the present discussions, we write down the Lagrangian density of electroweak theory in curved spacetime that includes the right handed neutrinos but exclude the complex scalar fields:

$$\mathcal{L} = \sqrt{-g}\{-\frac{1}{4}\mathbf{A}_{\mu\nu}.\mathbf{A}^{\mu\nu} - \frac{1}{4}B_{\mu\nu}B^{\mu\nu} \tag{42.12}$$
$$-\frac{1}{2}[\bar{\chi}_e\gamma^\mu\nabla_\mu\chi_e - (\nabla_\mu\bar{\chi}_e)\gamma^\mu\chi_e + \bar{\chi}_\nu\gamma^\mu\nabla_\mu\chi_\nu - (\nabla_\mu\bar{\chi}_\nu)\gamma^\mu\chi_\nu]$$
$$+ ig'\bar{e}_R\gamma^\mu B_\mu e_R + i\frac{g'}{2}\bar{L}_e\gamma^\mu B_\mu L_e - i\frac{g}{2}\bar{L}_e\sigma_i\gamma^\mu A^i_\mu L_e\},$$

where, similar to the case of Eq. (42.9), ω and ψ can be discarded from the above Lagrangian density even when we begin with $\Theta^\alpha{}_{\mu\nu}$ as the connection. Thus, when $\Theta^\alpha{}_{\mu\nu}$ is given by Eq. (42.3), we can couple the standard model with gravity using only $\Gamma^\alpha{}_{\mu\nu}$. This somewhat validates our attempts to elucidate dark energy using ψ. There is no issue with the contracted Bianchi identity in extending the standard model to curved spacetime with $\Theta^\alpha{}_{\mu\nu}$ given by Eq. (42.3) as the connection, as long as we have modified the action to prevent problems with $\Gamma^\alpha{}_{\mu\nu}$. The right handed neutrino, ν_R, remains a free field. The axial vector currents for various Dirac fields can become anomalous even with the addition of the right handed neutrinos. The action derived from Eq. (42.11) remains invariant under the phase transformation: $\chi \to e^{-i\alpha(x)\gamma_5}\chi$ classically by the equations of motion. However, we can not define a conserved axial vector current in the quantum regime by claiming invariance of the action under such transformations. This is again due the restriction on Stoke's theorem mentioned before. Anomaly in axial vector currents may depend on further details of the theory [25, 37, 43–45]. As an alternate to Eq. (42.12), we can construct a theory that couples ω and ψ with the neutrinos, with the coupling given by: $\sqrt{-g}[\bar{\chi}_\nu\gamma^\mu(1-\gamma_5)\chi_\nu]\nabla_\mu(\omega-\psi)$ [25]. This can introduce novel features.

(3) Principle of covariance and a minimally coupled theory:

In order to construct covariant physical theories in curved spacetime, we replace various partial derivatives of matter and gauge fields by appropriate covariant derivatives. Looking at the present observations on on dark energy, we can assume that ω and ψ do not couple with non-gravitational fields and they do not appear in various covariant derivatives of matter and gauge fields. This is a minimally coupling of gravity with other fields where the effect of the connection scalars ω and ψ in $C^\alpha{}_{\mu\nu}(\omega, \psi)$ is only to introduce $P_{\mu\alpha}(\omega)$, $P_{\mu\alpha}(\psi)$ and $P_{\mu\alpha}(\omega, \psi)$ in Einstein's equation such as that given by Eq. (42.6). A similar approach is adopted to define the gauge invariant field tensors in presence of torsion.

To summarize the present section, we find that we may need to incorporate right handed neutrinos as free fields in the standard model. There are several advantages to this in flat spacetime. With this addition in curved spacetime, we can exclude $C^{\alpha}{}_{\mu\nu}(\omega, \psi)$ from the actions of the scalar fields, Dirac fields, and gauge fields in the standard model while including gravity. This results in a covariant Lagrangian formalism that incorporates the principle of equivalence.

Cosmological observations suggest that non-metricities produced by ω and ψ are currently very small. This aligns with the minimal value of the cosmological constant in the ΛCDM model. ω and ψ manifest themselves in long-range geodesics and global structure of spacetime. However, local inhomogeneities and anisotropies in ω and ψ can cause significant effects on geodesic motions in the corresponding regions. In the last section, we will find it more beneficial to consider linear combinations of ω and ψ when discussing their effects, such as inflation and dark energy.

42.4 Extension of Einstein's Theory of Gravity

We can use the semi-classical theories like quantum fields in curved spacetime to determine the influence of ω and ψ on spacetime. This will be utmost important in cosmology. We can compare Eqs. (42.4), (42.6) with the action of scalar-tensor theories [6, 46]:

$$S = \int \mathbf{e}\sqrt{-g}\left[f(\phi, R) - \zeta(\phi)(\nabla\phi)^2\right] + S_m(\chi_a, \nu_R, g_{\mu\nu}) \,, \tag{42.13}$$

where χ_a represents scalars, Dirac and gauge fields. Additionally, there can be dark matter. Equation (42.6) gives: $\zeta(\omega) = \frac{3}{2}$, $\zeta(\psi) = -3$ and $f(\phi, R) = R$, for both ω and ψ. Thus, inflation and dark energy may be of geometrical origin and are required in quantum gravity [23–25]. Generally, the aforementioned Lagrangian density will yield derivatively coupled equations in ω and ψ. We get a decoupled theory by performing linear transformations on ω and ψ to get the following action, [25, 47]:

$$\mathcal{R}' = R - \frac{1}{2}[(\nabla\Phi)^2 - (\nabla\Psi)^2]. \tag{42.14}$$

The above equation offers a more useful method to elaborate inflation and dark energy using ω and ψ. Ψ is a phantom scalar [6, 34]. We consistently obtain a phantom scalar in Einstein's equation regardless of the combinations of $\nabla^{\mu}\omega$ and $\nabla^{\nu}\psi$ in Eq. (42.3), as long as they are linked to two different traces of $C^{\alpha}{}_{\mu\nu}$. Depending on observations, we can add suitable self-interaction terms for Φ and Ψ. We then obtain:

$$\mathcal{L} = \sqrt{-g}(\mathcal{R}' - \mathcal{V}) \,. \tag{42.15}$$

Ψ contributes a negative stress-tensor. The corresponding issue of stability is discussed in [25, 47]. We have shown that there is no problem in both classical and quantum theory so long as Ψ does not couple with matter fields. Coupling between Φ and Φ requires special attention and will be discussed later. This is also consistent with the non-interacting nature of dark energy with matter fields. We can include interactions like the slow-rolling potential in Φ as $\mathcal{V}$.

It is possible to construct a theory that includes Φ, Ψ and Λ. This could result in a potential theory of inflation alongside dynamic and spatially varying dark energy. Einstein's equation is extended to:

$$G_{\mu\nu} + \Lambda g_{\mu\nu} - \Lambda_{\mu\nu}(\Phi, \Psi) = 8\pi P_{\mu\nu}(\chi_a) \,, \tag{42.16}$$

where $\Lambda_{\mu\nu}(\Phi, \Psi)$ includes $P_{\mu\nu}(\Phi)$, $P_{\mu\nu}(\Psi)$, potential terms for Φ and Ψ as well as possible coupling terms between the two fields. $P_{\mu\nu}(\chi_a)$ is the stress tensor of ordinary matter and radiation including the right handed neutrinos. We can construct new non-trivial solutions of the vacuum Einstein's equation [29, 48, 49], *i.e*, for $P_{\mu\nu}(\chi_a) = 0$, with different choices of $\Lambda_{\mu\nu}(\Phi, \Psi)$. We have found in the last section that Φ and Ψ do not couple with χ_a through $C^{\alpha}_{\ \mu\nu}(\Phi, \Psi)$. We can include possible interaction terms between (Φ, Ψ) and χ_a through $\mathcal{V}(\Phi, \Psi, \chi_a)$. A similar approach has been used in many models to explain inflation and dark energy [6, 32, 33]. We note that we have to be careful in coupling Ψ with matter fields, radiation and Φ to have a stable theory. This is due to the negative stress-tensor associated with Ψ. This aspect partly justifies absence of $C^{\alpha}_{\ \mu\nu}(\Phi, \Psi)$ in various matter and radiation field actions. Φ and Ψ can be useful to construct the k-essence and phantom models of dark energy [6, 50–52]. Ψ can also remain as a free field similar to Λ to give relatively stable cosmological solutions with dark energy. An interesting case will be to consider a self-interacting theory for Ψ including quadratic potential which gives stable solutions [6, 51, 52]. A scalar field with negative energy density is also required to construct the *steady state* model of the expanding universe where this scalar field can create matter and radiation [48, 53–56]. The issue of stability is discussed in [47]. There remains the issue of unstable vacuum [56]. In the present theory, Ψ does not couple with ordinary matter and scattering between the two sectors causing instability of the vacuum is not possible. Processes of the form: $Vacuum \longrightarrow 2\tilde{\gamma} + 2\tilde{\psi}.\ \tilde{\chi}$, where Ψ is a photon and ψ gives a phantom quanta need careful analysis. Here, we illustrate this with the simpler case of the process: $Vacuum \longrightarrow 2\tilde{\phi} + 2\tilde{\Psi}$, where $\tilde{\phi}$ is a Φ field quantum. Firstly, we don't allow negative energy states for the phantom quantum $\tilde{\psi}$. Such states don't exist in a quantum field theory. Also, such states won't contribute negatively to the total energy in the matter field part of Eq.(6) and the above process is not possible. These processes use the kinetic terms of the form: $\nabla(\varphi)^2$ and a graviton propagator [56]. In a perturbative theory of spin-two particle in flat spacetime background, the non-vanishing matrix elements are of the form $\langle\tilde{\phi}\tilde{\phi}\tilde{\psi}\tilde{\psi}|(a_{\tilde{\phi}}a_{\tilde{\phi}}a_{\tilde{\psi}}a_{\tilde{\psi}})^{\dagger}|0\rangle$ and are obtained from: $\langle\tilde{\phi}\tilde{\phi}\tilde{\psi}\tilde{\psi}|\int\int d^4x d^4y\ :\ \nabla^{\alpha}\Psi\nabla^{\nu}\Psi(x) S^F_{\alpha\beta,\mu\nu}(x-y)\nabla^{\mu}\varphi(y)\ :\ |0\rangle$, where $S^F_{\alpha\beta,\mu\nu}(x-y)$ gives the translationally invariant Feynman propagator for the spin -two graviton in flat spacetime. This requires negative energy states for both $\tilde{\psi}$ and $\tilde{\chi}$. Thus, the decay: $Vacuum \longrightarrow 2\tilde{\phi} + 2\tilde{\psi}$ is not possible in a perturbative quantum field theory. Replacing φ by higher spin fields won't change the situation. Similar situation will remain valid in any stationary background. In addition, such processes break conservation of linear momentum in a spacetime translationally invariant in one or more spatial direction(s). This is the case with the flat spacetime.

Φ and Ψ are quantum gravitational in origin [23, 24] and it is expected that they can play important roles in the very early universe. Various free field configurations of Φ and Ψ, their vacuum energies, quantum fluctuations and classical configurations when interactions are added can be useful to explain inflation and dark energy [6, 32, 33]. In this context, we note that unlike conventional matter and radiation, Φ and Ψ do not couple with metric through the gravitational constant. Thus, their stress tensors and corresponding quantum fluctuations are more effective than that of ordinary matter and radiation in determining the dynamics of metric and global structure of spacetime. This is similar to the Jordan-Brans-Dicke theory [32]. The minimal value of Λ in the ΛCDM model indicates that Φ and Ψ are small. The same remains valid for the non-metricity associated with them. Φ and Ψ manifest themselves in long-range geodesics and global structure of spacetime.

References

1. Starobinisky, A. A. (1980). A new type of isotropic cosmological models without singularity. *Physics Letters B, 91*, 99.
2. Guth, A. H. (1981). Inflationary universe: A possible solution to the horizon and flatness problems. *Physical Review D, 23*, 347.
3. Spergel, D. N., et al., [WMAP Collaboration]. (2003). First year Wilkinson microwave anisotropy probe (WMAP) observations: Determination of cosmological parameters. *Astrophysical Journal Supplement, 148*, 175.
4. Spergel, D. N., et al., [WMAP Collaboration]. (2007). Three-Year Wilkinson microwave anisotropy probe (WMAP) observations: Implications for cosmology. *Astrophysical Journal Supplement, 170*, 377.
5. Komatsu, E., et al., [WMAP Collaboration]. (2009). Five-Year Wilkinson microwave anisotropy probe observations: Cosmological interpretation. *Astrophysical Journal Supplement, 180*, 330.
6. Amendola, L., & Tsujikawa, S. (2010). *Dark energy*. Cambridge University Press.
7. Linde, A. D. (1991). Axions in inflationary cosmology. *Physics Letters B, 259*, 38.
8. Kaloper, N. (1991). Lorentz Chern-Simons terms in Bianchi cosmologies and the cosmic no hair conjecture. *Physical Review D, 44*, 2380.
9. Golovnev, A., Mukhanov, M., & Vanchurin, V. (2008). Vector inflation. *Journal of Cosmology and Astroparticle Physics, 0806*, 009.
10. Zwicky, F. (1933). Die Rotverschieb ung von extragalaktischen Nebeln. *Helvetica Physica Acta, 6*, 110.
11. De Felice, A., & Tsujikawa, S. (2010). f(R) theories. *Living Reviews in Relativity, 13*, 3.
12. Vollick, D. N. (2003). 1/R curvature corrections as the source of the cosmological acceleration. *Physical Review D, 68*, 063510.
13. Weinberg, S. (1989). The cosmological constant problem. *Reviews of Modern Physics, 61*, 1.

14. Kobayashi, S., & Nomiju, K. (1991). *Foundation of differential geometry* (Vol. I). John Wiley and Sons, Inc.
15. Fujii, Y. (1982). Origin of the gravitational constant and particle masses in a scale-invariant scalar-tensor theory. *Physical Review D, 26*, 2580.
16. Chiba, T., Okabe, T., & Yamaguchi, M. (2000). Kinetically driven quintessence. *Physical Review D, 62*, 023511.
17. Kamenshchik, A. Y., Moschella, U., & Pasquier, V. (2001). An alternative to quintessence. *Physics Letters B, 511*, 265.
18. Capozzillo, S. (2002). Curvature quintessence. *International Journal of Modern Physics D, 11*, 483.
19. Amendola, L. (1999). Scaling solutions in general nonminimal coupling theories. *Physical Review D, 60*, 043501.
20. Uzan, J. P. (1999). Cosmological scaling solutions of nonminimally coupled scalar fields. *Physical Review D, 59*, 123510.
21. Dvali, G. R., Gabadadze, G., & Porrati, M. (2000). 4-D gravity on a brane in 5-D Minkowski space. *Physics Letters B, 485*, 208.
22. Sahni, V., & Shtanov, Y. (2003). Braneworld models of dark energy. *Journal of Cosmology and Astroparticle Physics, 311*, 014.
23. Ghosh, K. (2018). Non-metric fields from quantum gravity. In E. S. Battistelli, R. T. Jantzen, & R. Ruffini (Eds.), *Proceedings of the MG15 Meeting on General Relativity, Part A, University of Rome "La Sapienza", Italy, 1–7 July 2018*. https://doi.org/10.1142/12843
24. Ghosh, K. (2019). Affine connections in quantum gravity and new scalar fields. *Physics of the Dark Universe, 26*, 100403. https://hal.science/hal-02105422v4
25. Ghosh, K. (2024). Affine connections in quantum gravity and new fields. *Quantum Studies: Mathematics and Foundations, 11*, 625. https://hal.science/hal-04086265
26. Hocking, J. G., & Young, G. S. (1961). *Topology*. Dover Publications, Inc.
27. Lovelock, D., & Rund, H. (1989). *Tensors, differential forms, and variational principals*. Dover Publications, Inc.
28. Landau, L. D., & Lifshitz, E. M. (1998). *The classical theory of fields*. Butterworth-Heinenann.
29. Wald, R. M. (1984). *General relativity*. The University of Chicago Press.
30. Misner, C. W., Thorne, K. S., & Wheeler, J. A. (1970). *Gravitation*. W.H. Freeman and Company.
31. Hehl, F. W., et al. (1976). General relativity with spin and torsion: Foundations and prospects. *Reviews of Modern Physics, 48*, 3641.
32. Liddle, A. R., & Lyth, D. H. (2000). *Cosmological inflation and large scale structures*. Cambridge University Press.
33. Dodelson, S., & Schmidt, F. (2021). *Modern cosmology*. Academic Press.
34. Gannouji, R., Polarski, D., ranquet, A., & Starobinski, A. A. (2006). Scalar-tensor models of normal and phantom dark energy. *Journal of Cosmology and Astroparticle Physics, 0609*, 016.
35. Narlikar, J. V. (1974). Mini-bangs in cosmology and astrophysics. *Pramana, 2*(3), 158–170.
36. Ford, L. H., & Roman, T. A. (2000). Negative energy, wormholes and warp drive. *Scientific American, 282*, 46.
37. Itzykson, C., & Zuber, J. B. (2005). *Quantum field theory*. Dover Publications, Inc.
38. Weinberg, S. (1996). *The quantum theory of fields* (Vol. I). Cambridge University Press.
39. Birrell, N. D., & Davies, P. C. W. (1982). *Quantum fields in curved spaces*. Cambridge University Press.
40. Weinberg, S. (2004). *Gravitation and cosmology*. John Wiley & Sons.
41. Yanagida, T. (1979). Horizontal symmetry and mass of the t quark. *Physical Review D, 20*, 2986.
42. Yanagida, T. (1980). Horizontal symmetry and masses of neutrinos. *Progress of Theoretical Physics, 64*, No. 3, 1103.
43. Preskin, M. E., & Schroeder, D. V. (1995). *An introduction to quantum field theory*. Addison-Wesley Publishing Company.
44. Weinberg, S. (1996). *The quantum theory of fields* (Vol. II). Cambridge University Press.

45. Ramond, P. (1990). *Field theory: A modern primer*. Addison-Wesley Publishing Company.
46. Gasperini, M., & Veneziano, G. (2003). The pre-big bang scenario in string cosmology. *Physics Report, 373*, 1.
47. Ghosh, K. (2023). *Affine connections in quantum gravity and new fields*. Quantum Studies: Foundations and Mathematics (10.1007/s40509-024-00340-9). https://hal.science/hal-04086265
48. Hawking, S. W., & Ellis, G. F. R. (1973). *The large scale structure of space-time*. Cambridge University Press.
49. Stephani, H., et al. (2003). *Exact solutions of Einstein's field equations*. Cambridge University Press.
50. Carroll, S. M., Hoffman, M., & Trodden, M. (2003). Can the dark energy equation-of-state parameter w be less than -1? *Physical Review D, 70*, 023509.
51. Singh, P., Sami, M., & Dadhich, N. (2003). Cosmological dynamics of phantom field. *Physical Review D, 68*, 023522.
52. Sami, M., & Toporensky, A. (2004). Phantom field and the fate of universe. *Modern Physics Letters A, 19*, 1509.
53. Bondi, H., & Gold, T. (1948). The steady-state theory of the expanding universe. *Monthly Notices of the Royal Astronomical Society, 108*, 252–270.
54. Hoyle, F. (1948). A new model for the expanding universe. *Monthly Notices of the Royal Astronomical Society, 108*, 372–382.
55. Pirani, F. A. E. (1955). On the energy-momentum tensor and the creation of matter in relativistic cosmology. *Proceedings of the Royal Society of London A, 228*, 455–462.
56. Hoyle, F., & Narlikar, J. V. (1964). A new theory of gravitation. *Proceedings of the Royal Society of London A, 277*, 1–23.

Chapter 43
Holographic Approach to Anomalous Transport in a Massive $U(1)$ Gauge Theory

Nishal Rai and Eugenio Megías

Abstract In this chapter, we investigate a massive $U(1)$ gauge holographic model that includes both pure gauge and mixed gauge-gravitational Chern-Simons terms. By taking into account the complete backreaction of the gauge field on the metric, we study the vortical and energy transport sectors. Our results for the chiral vortical conductivity, σ_V, and the chiral magnetic/vortical conductivity of the energy current, σ_B^ε, showed that $\sigma_B^\varepsilon = \sigma_V^\varepsilon$. Notably, we identify a contribution to σ_V that originates from the mixed Chern-Simons term in the massive theory, a feature that is absent in the massless case.

43.1 Introduction

The AdS/CFT correspondence, first proposed by Maldacena [1] and further developed by Witten [2], posits a duality between a large-N_c, $\mathcal{N} = 4$ supersymmetric Yang-Mills (SYM) theory in four-dimensional spacetime and type IIB string theory formulated on the $AdS_5 \times S^5$ background in the low-energy limit. This duality has emerged as a theoretical framework for probing strongly coupled quantum systems. The AdS/CFT correspondence has found diverse applications across various domains of physics, including condensed matter systems, hydrodynamics, and quantum chromodynamics (QCD) [3–22] to name a few. In this study, we focus on the fluid/gravity correspondence, an application of AdS/CFT in the field of

N. Rai (✉)
Department of Physics, St. Joseph's College, North Point, Darjeeling, India

Center for Astrophysics Gravitation and Cosmology (CAGC), SRM University Sikkim, Upper Tadong, Sikkim, India

E. Megías
Departamento de Física Atómica, Moleculary Nuclear and Instituto Carlos I de Física Teóricay Computacional, Universidad de Granada, Granada, Spain
e-mail: emegias@ugr.es

S. Ghosh, A. R. Rao (eds.), *The Relativistic Universe: From Classical to Quantum*, Astrophysics and Space Science Proceedings 61,
https://doi.org/10.1007/978-3-031-90186-7_43

hydrodynamics, with an emphasis on investigating anomalous transport phenomena within these systems.

Quantum chiral anomalies, which arises beyond perturbation theory of relativistic field theories of chiral fermions, play a fundamental role in relativistic hydrodynamics [23–26]. Since the 1980s, extensive research has been conducted on anomaly-induced transport mechanisms [27]. Notable effects include the chiral magnetic effect, where a charge current is generated parallel to an external magnetic field [28], and the chiral vortical effect, where a vortex in a charged fluid induces a current [29–32]. These phenomena are of particular interest for their potential observability in heavy ion collisions at RHIC and LHC [33], as well as their contribution to anomalous transport properties in materials such as Weyl semimetals [34, 35]. In this chapter, we investigate the chiral vortical effects within a massive $U(1)$ gauge holographic model. Our approach includes accounting for the full backreaction and incorporating a mixed gauge-gravitational Chern-Simons term. To evaluate the transport coefficients, we use the Kubo formalism, wherein the anomalous conductivities are determined by the following relations:

$$\begin{aligned}\sigma_V &= \lim_{k_c\to 0}\frac{i}{2k_c}\sum_{a,b}\epsilon_{abc}\langle J^a T^{0b}\rangle|_{\omega=0}\,,\\ \sigma_V^{\varepsilon} &= \lim_{k_c\to 0}\frac{i}{2k_c}\sum_{a,b}\epsilon_{abc}\langle T^{0a} T^{0b}\rangle|_{\omega=0}\,,\end{aligned} \tag{43.1}$$

where σ_V and σ_V^{ε} corresponds to the chiral vortical conductivity and chiral vortical conductivity of energy current, respectively. Similarly, the chiral magnetic conductivities for charge, σ_B, and energy, σ_B^{ε}, current are given by

$$\begin{aligned}\sigma_B &= \lim_{k_c\to 0}\frac{i}{2k_c}\sum_{a,b}\epsilon_{abc}\langle J^a J^{b}\rangle|_{\omega=0}\,,\\ \sigma_B^{\varepsilon} &= \lim_{k_c\to 0}\frac{i}{2k_c}\sum_{a,b}\epsilon_{abc}\langle T^{0a} J^{b}\rangle|_{\omega=0}\,.\end{aligned} \tag{43.2}$$

These correlators can be evaluated using the AdS/CFT dictionary [36–38].

43.2 Holographic Model

The holographic action for a massive $U(1)$ gauge boson comprising of a pure gauge and a mixed gauge-gravitational Chern-Simon term is given by [7, 36, 39]

$$S = \frac{1}{16\pi G}\int d^5x\sqrt{-g}\Big[R + 2\Lambda - \frac{1}{4}F_{MN}F^{MN}$$

$$
\begin{aligned}
&-\frac{m^2}{2}(A_M - \partial_M\theta)(A^M - \partial^M\theta) \\
&+\epsilon^{MNPQR}(A_M - \partial_N\theta)\left(\frac{\kappa}{3}F_{NP}F_{QR} + \lambda R^{A}{}_{BNP}R^{B}{}_{AQR}\right)\Big] \\
&+S_{\mathrm{GH}} + S_{\mathrm{CSK}}\,, \qquad (43.3)
\end{aligned}
$$

where S_{GH} is the Gibbons-Hawking boundary term and S_{CSK} corresponds to the boundary term induced by the mixed gauge-gravitational anomaly, respectively. To ensure the gauge invariance we introduce a θ field . As mentioned in [40–42], the Stückelberg term arises as the holographic realization of dynamical anomalies. For convenience we will define a new field $B_M \equiv A_M - \partial_M\theta$, such that θ does not appear explicitly in the equations of motion. From now onwards we will be working with the field B_μ instead of A_μ.

43.2.1 Background Solution

The ansatz for the background metric and fields is chosen as follows in Fefferman-Graham coordinates [43, 44],

$$
\begin{aligned}
ds^2 &= -\frac{\ell^2}{\rho}g_{\tau\tau}(\rho)d\tau^2 + \frac{\ell^2}{\rho}g_{xx}(\rho)d\mathbf{x}^2 + \frac{\ell^2}{4\rho^2}d\rho^2\,, \\
B_\mu dx^\mu &= B_t(\rho)dt\,,
\end{aligned} \qquad (43.4)
$$

where the boundary and the horizon lies at $\rho = 0$ and ρ_h respective. The equations of motion for the background are derived from this ansatz as follows

$$
\begin{aligned}
&g''_{xx}(\rho) - \frac{g'_{xx}(\rho)}{\rho} + \frac{1}{6\ell^2\rho}\frac{g_{xx}(\rho)}{g_{\tau\tau}(\rho)}\left(\frac{m^2\ell^2}{4}B_t(\rho)^2 + \rho^2 B'_t(\rho)^2\right) = 0\,, \\
&g'_{\tau\tau}(\rho)\left(1 - \rho\frac{g'_{xx}(\rho)}{g_{xx}(\rho)}\right) + g_{\tau\tau}(\rho)\frac{g'_{xx}(\rho)}{g_{xx}(\rho)}\left(3 - \rho\frac{g'_{xx}(\rho)}{g_{xx}(\rho)}\right) \\
&\qquad -\frac{1}{3}\frac{\rho^2}{\ell^2}B'_t(\rho)^2 + \frac{1}{12}m^2 B_t(\rho)^2 = 0\,, \\
&B''_t(\rho) + \frac{1}{2}\left(3\frac{g'_{xx}(\rho)}{g_{xx}(\rho)} - \frac{g'_{\tau\tau}(\rho)}{g_{\tau\tau}(\rho)}\right)B'_t(\rho) - \frac{\ell^2 m^2}{4\rho^2}B_t(\rho) = 0\,.
\end{aligned} \qquad (43.5)
$$

We numerically solve the above equations of motion (43.5) subjected to the following boundary conditions

$$
B_t(\rho_h) = 0\,, \qquad \lim_{\rho\to 0}\left(\rho^{\Delta/2}B_t(\rho)\right) = \mu_5\,, \qquad (43.6)
$$

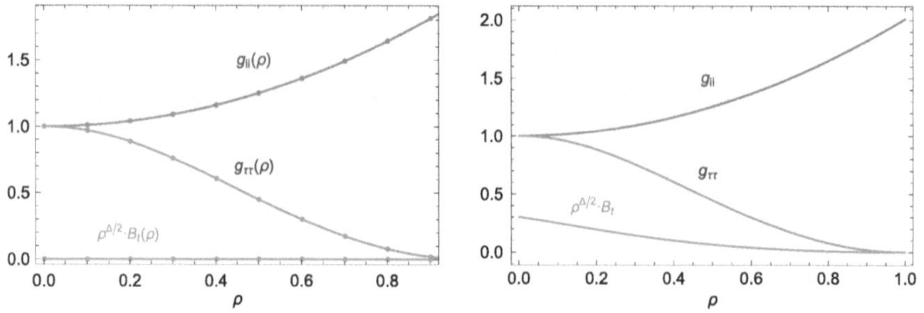

Fig. 43.1 We display the results for $g_{xx}(\rho)$ (blue), $g_{\tau\tau}(\rho)$ (orange), and $\rho^{\Delta/2}B_t(\rho)$ (green). Left panel: Solid line being the analytic solution and the dots the numerical one for $\mu_5 = 0$. Right panel: We have chosen $\Delta = 0.1$ and $\mu_5 = 0.3$ (Color figure online)

where μ_5 corresponds to the source of the gauge field and Δ being the anomalous dimension of the dual current [39] related to the mass of the gauge field through the following relation $m^2\ell^2 = \Delta(\Delta+2)$. In addition to this, we demand the solution to the metric tensor to be regular at the horizon.

With $\mu_5 = 0$ one can obtain an analytic solution of the background which is given by

$$g_{\tau\tau}(\rho) = \frac{1}{\rho_h^2}\frac{(\rho_h^2-\rho^2)^2}{\rho_h^2+\rho^2}\,, \qquad g_{xx}(\rho) = 1+\frac{\rho^2}{\rho_h^2}\,, \qquad B_t(\rho) = 0\,. \tag{43.7}$$

In Fig. 43.1 (left) we have shown the numerical solution of the equations of motion for $\mu_5 = 0$ (dots), and compare it with the corresponding analytical solution (solid lines). We display also in the right panel the numerical solution for $\mu_5 = 0.3$.

43.3 Anomalous Conductivity

To compute these correlators, we will consider perturbation of the fields on top of the background, and subsequently apply the AdS/CFT correspondence to compute the correlators.

43.3.1 Fluctuations

Within this approach, we perturb the fields on top of the numerical solution for the background displayed in Fig. 43.1. We will consider the linear response of the fluctuations by decomposing the metric and gauge field into their background components and linear perturbations, i.e.,

$$g_{MN} = g^{(0)}_{MN} + \epsilon h_{MN}, \qquad B_M = B^{(0)}_M + \epsilon b_M , \tag{43.8}$$

where the terms with $^{(0)}$ correspond to the background metric/field, and the terms with ϵ the fluctuations of this metric/field. Next, we will proceed with the standard method of decomposing into Fourier modes [45]

$$h_{MN}(\rho, x^\mu) = \int \frac{d^d k}{(2\pi)^d} h_{MN}(\rho) e^{-i\omega t + i\mathbf{k}.\mathbf{x}} , \tag{43.9}$$

$$b_M(\rho, x^\mu) = \int \frac{d^d k}{(2\pi)^d} b_M(\rho) e^{-i\omega t + i\mathbf{k}.\mathbf{x}} . \tag{43.10}$$

Without loss of generality, we focus on perturbations characterized by a frequency ω and momentum k along the z-direction. To compute correlators at zero frequency, we set the terms involving ω to zero and solve the equations up to first order in k. In this limit, the fields h^i_z decouple from the system and remain constant. Consequently, the differential equations governing the shear sector can be expressed as follows

$$b_i''(\rho) + \frac{1}{2}\left(\frac{g_{xx}'(\rho)}{g_{xx}(\rho)} + \frac{g_{\tau\tau}'(\rho)}{g_{\tau\tau}(\rho)}\right) b_i'(\rho) - \frac{\Delta(\Delta+2)}{4\rho^2} b_i(\rho) \tag{43.11}$$
$$+ \left(\frac{4i\kappa k \epsilon_{ij} h_j(\rho)}{\sqrt{g_{xx}(\rho) g_{\tau\tau}(\rho)}} + \frac{g_{xx}(\rho) h^{i\prime}_t(\rho)}{g_{\tau\tau}(\rho)}\right) B_t'(\rho) + i\lambda k \epsilon_{ij} h^{j\prime}_t(\rho) \Omega(\rho) = 0,$$

$$h^{i\prime\prime}_t(\rho) - \left(\frac{g_{\tau\tau}'(\rho)}{2g_{\tau\tau}(\rho)} - \frac{5g_{xx}'(\rho)}{2g_{xx}(\rho)} + \frac{1}{\rho}\right) h^{i\prime}_t(\rho) + \frac{\rho B_t'(\rho)}{g_{xx}(\rho)} b_i'(\rho)$$
$$+ \frac{\Delta(\Delta+2) B_t(\rho)}{4\rho g_{xx}(\rho)} b_i(\rho) + i\lambda k \epsilon_{ij} \Phi_j(\rho) = 0, \tag{43.12}$$

where $i, j = x, y$. The detailed expressions for the functions $\Omega(\rho)$ and $\Phi_j(\rho)$ can be found in Appendix A of [7].

The asymptotic behavior of the fields up to the first subleading term are given by

$$b_i(\rho) = b_i^{(0)} \rho^{-\frac{\Delta}{2}} + b_i^{(1)} \rho^{\frac{\Delta}{2}+1} + \cdots , \tag{43.13}$$

$$h^i_t(\rho) = h^{i\,(0)}_t + h^{i\,(1)}_t \rho^2 + \cdots , \tag{43.14}$$

where the leading-order terms $b_i^{(0)}$ and $h^{i\,(0)}_t$ act as sources.

43.3.1.1 Conductivities

From the holographic description of the correlation functions, the one-point functions can be evaluated as

$$\langle J_a \rangle = \frac{\delta S_{\text{ren}}}{\delta b_a^{(0)}} = -\frac{2}{16\pi G}(\Delta + 1) b_a^{(1)}$$

$$\langle T_{0a} \rangle = \frac{\delta S_{\text{ren}}}{\delta h_t^{a\,(0)}} = \frac{1}{16\pi G}\left(2h_t^{a\,(0)} + h_t^{a\,(1)}\right) , \tag{43.15}$$

with $(a = x, y)$. The renormalized action is given by $S_{\text{ren}} = S + S_{\text{ct}}$, where S is the action defined in Eq. (43.3) and S_{ct} represents the counterterm. The two-point functions are derived by varying the one-point function with respect to the corresponding source term, and the conductivities are calculated using the Kubo formulae provided in Eqs. (43.1) and (43.2), i.e.

$$\sigma_V = \lim_{k\to 0}\frac{1}{k}\text{Im}\langle J_x T_{0y}\rangle , \qquad \sigma_V^{\varepsilon} = \lim_{k\to 0}\frac{1}{k}\text{Im}\langle T_{0x} T_{0y}\rangle , \tag{43.16}$$

$$\sigma_B = \lim_{k\to 0}\frac{1}{k}\text{Im}\langle J_x J_y\rangle , \qquad \sigma_B^{\varepsilon} = \lim_{k\to 0}\frac{1}{k}\text{Im}\langle T_{0x} J_y\rangle . \tag{43.17}$$

By comparing the variation of the action under axial gauge transformations with the consistent form of the anomaly for chiral fermions [23], we can determine the value of the anomaly coefficients to be

$$\kappa = -\frac{G}{2\pi} , \qquad \lambda = -\frac{G}{48\pi} . \tag{43.18}$$

In this regard we will be setting $G = 2\pi$, so that $\kappa = -1$ and $\lambda = -1/24$. In Fig. 43.2 we have plotted the anomalous conductivities vs Δ (a parameter related to the mass of the gauge field). From this Figure, we observe that the chiral vortical conductivity and the chiral magnetic conductivity for energy current are the same even at finite mass, i.e., $\sigma_V = \sigma_B^{\varepsilon}$, and both increase with Δ. Additionally, we see that the chiral vortical conductivity of energy current, σ_V^{ε}, decreases with Δ, but the rate decreases rapidly. Conversely, in the case of the chiral magnetic conductivity, σ_B, it increases with Δ as shown in Fig. 43.2. In the limit of vanishing mass, our results lead to $\frac{\sigma_B}{\mu_5|\kappa|} \simeq 16/3$, which exactly coincides with the results in [36, 39] where α has been set to μ_5 in both references.[1]

Acknowledgments N.R. thanks the International Centre for Theoretical Sciences (ICTS) for the support through the program—Field Theory and Turbulence (ICTS/FTT2023/12). The works of N.R. and E.M. are supported by the project PID2020-114767GB-I00 and by the Ramón y Cajal Program under Grant RYC-2016-20678 funded by MCIN/AEI/10.13039/501100011033 and by "FSE Investing in your future", as well as by the FEDER/Junta de Andalucía-Consejería de Economía y Conocimiento 2014–2020 Operational Program under Grant A-FQM-178-UGR18. The work of E.M. is also supported by Junta de Andalucía under Grant FQM-225, and by the "Prórrogas de Contratos Ramón y Cajal" Program at the University of Granada.

[1] α corresponds to the asymptotic value of the gauge field A_t for $\rho \to 0$. In our case, we assume $\alpha = \mu_5$ for $\Delta = 0$.

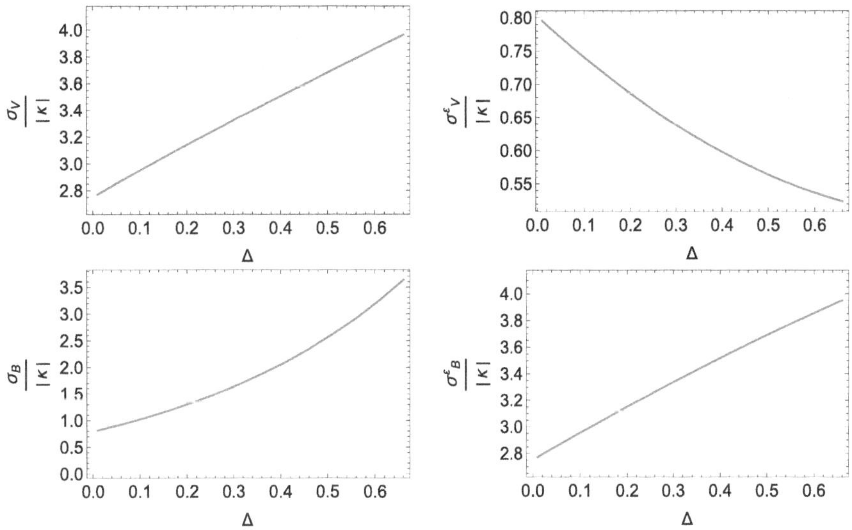

Fig. 43.2 Upper panels: Plot of σ_V (left) and σ_V^ε (right) as a function of Δ. Lower panels: Plot of σ_B (left) and σ_B^ε (right) as a function of Δ. In all panels, we have taken $\mu_5 = 0.15$. We have divided the conductivities by $|\kappa|$ to make them dimensionless

References

1. Maldacena, J. M. (1999). *International Journal of Theoretical Physics, 38*, 1113 [*Advances in Theoretical and Mathematical Physics, 2*, 231]. https://doi.org/10.1023/A:1026654312961
2. Witten, E. (1998). *Advances in Theoretical and Mathematical Physics, 2*, 253.
3. Grieninger, S., & Morales-Tejera, S. (2023). *Physical Review D, 108*, 126010. https://doi.org/10.1103/PhysRevD.108.126010
4. Baggioli, M., Bu, Y., & Ziogas, V. (2023). *Journal of High Energy Physics, 09*, 019. https://doi.org/10.1007/JHEP09(2023)019
5. Morales-Tejera, S., Ambruş, V. E., & Chernodub, M. N. [arXiv:2403.19755 [hep-th]]
6. Landsteiner, K., Morales-Tejera, S., & Saura-Bastida, P. (2023). *Physical Review D, 107*, 125003. https://doi.org/10.1103/PhysRevD.107.125003
7. Rai, N., & Megías, E. (2023). *Journal of High Energy Physics, 06*, 215. https://doi.org/10.1007/JHEP06(2023)215
8. Mukhopadhyay, S., & Rai, N. (2021). *Journal of High Energy Physics, 09*, 160. https://doi.org/10.1007/JHEP09(2021)160
9. Rai, N. (2019). *European Physical Journal C, 79*, 972. https://doi.org/10.1140/epjc/s10052-019-7469-x
10. Mukhopadhyay, S., & Rai, N. (2017). *Physical Review D, 96*, 066001. https://doi.org/10.1103/PhysRevD.96.066001
11. Mukhopadhyay, S., & Rai, N. (2017). *Physical Review D, 96*, 026005. https://doi.org/10.1103/PhysRevD.96.026005
12. Ammon, M., Arean, D., Baggioli, M., Gray, S., & Grieninger, S. (2022). *Journal of High Energy Physics, 03*, 015. https://doi.org/10.1007/JHEP03(2022)015
13. Grieninger, S., & Morales-Tejera, S. (2022). *EPJ Web of Conferences, 258*, 10007. https://doi.org/10.1051/epjconf/202225810007

14. Ahn, Y., Baggioli, M., Liu, Y., & Wu, X. M. (2024). *Journal of High Energy Physics, 03*, 124. https://doi.org/10.1007/JHEP03(2024)124 [arXiv:2401.07772 [hep-th]].
15. Chu, H., Ji, X., & Sun, Y. W. (2024). *Journal of High Energy Physics, 05*, 166. https://doi.org/10.1007/JHEP05(2024)166
16. Landsteiner, K., & Liu, Y. (2016). *Physics Letters B, 753*, 453–457. https://doi.org/10.1016/j.physletb.2015.12.052
17. Landsteiner, K., & Melgar, L. (2012). *Journal of High Energy Physics, 10*, 131. https://doi.org/10.1007/JHEP10(2012)131
18. Megías, E., & Pena-Benitez, F. (2013). *Journal of High Energy Physics, 05*, 115. https://doi.org/10.1007/JHEP05(2013)115
19. Newman, G. M., & Son, D. T. (2006). *Physical Review D, 73*, 045006. https://doi.org/10.1103/PhysRevD.73.045006
20. Newman, G. M. (2006). *Journal of High Energy Physics, 01*, 158. https://doi.org/10.1088/1126-6708/2006/01/158
21. Megías, E. (2017). *EPJ Web of Conferences, 164*, 08001. https://doi.org/10.1051/epjconf/201716408001
22. Landsteiner, K., Megías, E., & Pena-Benitez, F. (2014). *Physical Review D, 90*, 065026. https://doi.org/10.1103/PhysRevD.90.065026
23. Bertlmann, R. A. (1996). *International series of monographs on physics* (Vol. 91, 566 pp.). Clarendon
24. Bastianelli, F., & van Nieuwenhuizen, P. (2006). *Path integrals and anomalies in curved space* (379 pp.). University Press.
25. Fujikawa, K., & Suzuki, H. (2004). *Path integrals and quantum anomalies* (284 pp.). Clarendon.
26. Landsteiner, K., Megías, E., & Pena-Benitez, F. (2013). *Lecture Notes in Physics, 871*, 433–468. https://doi.org/10.1007/978-3-642-37305-3_17 [arXiv:1207.5808 [hep-th]].
27. Vilenkin, A. (1979). *Physical Review D, 20*, 1807. Vilenkin, A. (1980). *Physical Review D, 21*, 2260. Vilenkin, A. (1980). *Physical Review D, 22*, 3067. Vilenkin, A. (1980). *Physical Review D, 22*, 3080.
28. Fukushima, K., Kharzeev, D. E., & Warringa, H. J. (2008). *Physical Review D, 78*, 074033. https://doi.org/10.1103/PhysRevD.78.074033
29. Kharzeev, D., & Zhitnitsky, A. (2007). *Nuclear Physics A, 797*, 67–79. https://doi.org/10.1016/j.nuclphysa.2007.10.001
30. Banerjee, N., Bhattacharya, J., Bhattacharyya, S., Dutta, S., Loganayagam, R., & Surowka, P. (2011). *Journal of High Energy Physics, 01*, 094. https://doi.org/10.1007/JHEP01(2011)094
31. Son, D. T., & Surowka, P. (2009). *Physical Review Letters, 103*, 191601. https://doi.org/10.1103/PhysRevLett.103.191601
32. Landsteiner, K., Megías, E., & Pena-Benitez, F. (2011). *Physical Review Letters, 107*, 021601. https://doi.org/10.1103/PhysRevLett.107.021601
33. Abelev, B. I., et al. [STAR]. (2009). *Physical Review Letters, 103*, 251601. https://doi.org/10.1103/PhysRevLett.103.251601
34. Basar, G., Kharzeev, D. E., & Yee, H. U. (2014). *Physical Review B, 89*, 035142. https://doi.org/10.1103/PhysRevB.89.035142
35. Landsteiner, K. (2014). *Physical Review B, 89*, 075124. https://doi.org/10.1103/PhysRevB.89.075124
36. Landsteiner, K., Megías, E., Melgar, L., & Pena-Benitez, F. (2011). *Journal of High Energy Physics, 09*, 121. https://doi.org/10.1007/JHEP09(2011)121
37. Son, D. T., & Starinets, A. O. (2002). *Journal of High Energy Physics, 09*, 042. https://doi.org/10.1088/1126-6708/2002/09/042
38. Herzog, C. P., & Son, D. T. (2003). *Journal of High Energy Physics, 03*, 046. https://doi.org/10.1088/1126-6708/2003/03/046
39. Jimenez-Alba, A., Landsteiner, K., & Melgar, L. (2014). *Physical Review D, 90*, 126004. https://doi.org/10.1103/PhysRevD.90.126004

40. Klebanov, I. R., Ouyang, P., & Witten, E. (2002). *Physical Review D, 65*, 105007. https://doi.org/10.1103/PhysRevD.65.105007
41. Gürsoy, U., & Jansen, A. (2014). *Journal of High Energy Physics, 10*, 092. https://doi.org/10.1007/JHEP10(2014)092
42. Casero, R., Kiritsis, E., & Paredes, A. (2007). *Nuclear Physics B, 787*, 98–134. https://doi.org/10.1016/j.nuclphysb.2007.07.009
43. Megías, E., & Valle, M. (2018). *Fortschritte der Physik, 66*, 1800035. https://doi.org/10.1002/prop.201800035
44. Megías, E. (2019). *Journal of Physics: Conference Series, 1416*, 012022. https://doi.org/10.1088/1742-6596/1416/1/012022
45. Amado, I., Landsteiner, K., & Pena-Benitez, F. (2011). *Journal of High Energy Physics, 05*, 081. https://doi.org/10.1007/JHEP05(2011)081

Chapter 44
Holographic Fermions in Nonlinear Arcsin Electrodynamics

Ashis Tamang, Shubhrangshu Ghosh, and Nishal Rai

Abstract We consider a Holographic model for non-linear arcsin electrodynamics coupled with gravity and axions which is dual to some strongly coupled field theory in (2+1) dimension. We have studied the fermionic response on the background solution given by this model, particularly through the analysis of fermionic spectral functions for backreacted solutions at finite temperatures. Our results reveal that varying the charge density induces a temperature dependent phase transition from a normal metal to strange metal phase, offering potential insights into the electronic properties of materials and mechanisms underlying non-Fermi liquid behavior in condensed matter physics and quantum field theory.

44.1 Introduction

In condensed matter physics, holographic approaches based on gauge/gravity duality often known as AdS/CFT correspondence [1–3] have greatly increased our knowledge of strongly coupled systems [4–6]. These techniques, which make use of a gravitational framework with black holes, have been very useful in the study of

A. Tamang (✉)
Department of Physics, SRM University Sikkim, Upper Tadong, Sikkim, India

S. Ghosh
Center for Astrophysics, Gravitation and Cosmology (CAGC), SRM University Sikkim, Upper Tadong, Sikkim, India

Department of Physics, SRM University Sikkim, Upper Tadong, Sikkim, India
e-mail: shubhrangshugosh.r@srmus.edu.in

N. Rai
Department of Physics, St. Joseph's College, North Point, Darjeeling, India

Center for Astrophysics, Gravitation and Cosmology (CAGC), SRM University Sikkim, Upper Tadong, Sikkim, India

S. Ghosh, A. R. Rao (eds.), *The Relativistic Universe: From Classical to Quantum*, Astrophysics and Space Science Proceedings 61,
https://doi.org/10.1007/978-3-031-90186-7_44

fermionic excitations [7–10], providing new insights into the dual theory dimensions and scaling exponents as well as non-Fermi liquid scaling behaviors [11].

The behavior of strongly coupled fermionic systems, especially in the strange metal phase, remains an important but challenging area of study. Conventional linear electrodynamic models, which comprised of only the Maxwell term, while successful in capturing certain characteristics of condensed matter physics, are not able to explain some key phenomena, such as the linear temperature dependence of resistivity observed in cuprate high T_c superconductors [12]. Due to this inadequacy of the linear model to explain such phenomena, people started exploring the non-linear electrodynamics model. Among these, the Dirac-Born-Infeld (DBI) model [13] stands out for demonstrating both linear resistivity scaling with temperature and DC conductivity [14]. This success has paved the way for the exploration of other non-linear models, such as arcsin electrodynamics, introduced in [15] which remains underexplored, particularly about fermionic behavior and spectral functions.

Although previous studies have successfully utilized non-linear models such as those in [13] to examine transport properties such as DC conductivity, a detailed survey of fermionic excitations and transitions between the Fermi liquid and non-Fermi liquid phase is yet to be carried out using arcsin electrodynamics, particularly for backreacted solutions at finite temperatures. The objective of this chapter is to address this gap by examining the fermionic response and analyzing fermionic spectral functions within this model. The focus is on the temperature-dependent transitions from a normal metal to a strange metal phase, which are influenced by charge density variations. This will shed light on anomalous scaling behaviors that are observed in condensed matter systems. This transition has significant implications for understanding the anomalous properties of cuprates in the strange metal phase. It indicates a broad spectrum of behaviors, making arcsin electrodynamics a compelling area of study.

44.2 Holographic Model

In this section, we consider a nonlinear arcsin electrodynamics model, which interacts with gravity and axions. The action is given by [14]

$$S = \int d^4x\sqrt{-g}\left[R - \frac{1}{2}(\partial\phi)^2 - V(\phi) - \frac{Y}{2}\sum_{I=1}^{2}(\partial\psi^I)^2\right] + S_0, \tag{44.1}$$

with

$$S_0 = \int d^4x\sqrt{-g}\left[-Z_1(\phi)\sin^{-1}(Z_2(\phi)\mathcal{F})\right], \tag{44.2}$$

and $\mathcal{F} = \frac{1}{4}F_{\mu\nu}F^{\mu\nu}$. The fields ψ^I are axions, ϕ is a scalar field, and $Z_1(\phi)$, $Z_2(\phi)$ are coupling functions dependent on the scalar field. The term involving

$\sin^{-1}(Z_2(\phi)\mathcal{F})$ introduces nonlinearity into the electrodynamics, which modifies the behavior of electric and magnetic fields.

44.2.1 Background Solution

We will consider the following ansatz for the background metric and the fields

$$\begin{aligned} ds^2 &= -D(r)dt^2 + B(r)dr^2 + C(r)(dx^2 + dy^2), \\ A_\mu dx^\nu &= A_t(r)dt + \frac{h}{2}(xdy - ydx), \\ \phi &= \phi(r), \quad \psi^1 = k_p x, \quad \psi^2 = k_p y\,. \end{aligned} \tag{44.3}$$

The background equation of motion are simplified by considering $Z_1(\phi) = z_1$, $Z_2(\phi) = Y = 1$, $V(\phi) = -V_0$, $\phi(r) = 0$ and $B(r)D(r) = 1$. Here z_1 and V_0 are constants where we consider $z_1 = 1$. The solution for the metric coefficients are given in [14], where $C(r) = r^2$. With these choices the background metric and the field turns into

$$\begin{aligned} ds^2 &= -r^2 f(r)dt^2 + \frac{dr^2}{r^2 f(r)} + r^2 d\mathbf{x}^2, \\ A_t(r) &= \int_r^{r_h} \sqrt{-2\mathcal{F}}\,dr \quad \text{with} \quad \mathcal{F} = \frac{r^4 - \sqrt{r^8 + \rho^4}}{\rho^2}, \end{aligned} \tag{44.4}$$

where $f(r) = D(r)/r^2$. The horizon is defined by a point r_h where $D(r_h) = 0$ or $f(r_h)$ vanishes, similarly we will define a point r_* such that the derivative of $f(r)$ at r_*, vanishes ($f'(r_*) = 0$). However, in the limit of zero temperature, these two points coincide with each other i.e. $r_* = r_h$. The temperature of the black hole is given by

$$T = \frac{1}{8\pi r_h^3}\left(\mathcal{F}\Omega r_h^4 - 2\sin^{-1}(\mathcal{F})r_h^2 - k_p^2 r_h^2 - V_0 r_h^4\right), \tag{44.5}$$

where $\Omega = \frac{1}{\sqrt{1-\mathcal{F}^2}}$ and $\mathcal{F}$ being evaluated at the horizon $r = r_h$.

The plot of $f(r)$ for different values of momentum dissipation k_p has been shown below in Fig. 44.1. We have also presented the $3D$ plot of temperature of blackhole as given in Eq. (44.5) over the range of momentum dissipation k_p in Fig. 44.2a. The variation of gauge field A_t as a function of r is also shown alongside in Fig. 44.2b.

In the limit of strong momentum dissipation, the system admits a hyperscaling violating geometry solution given by

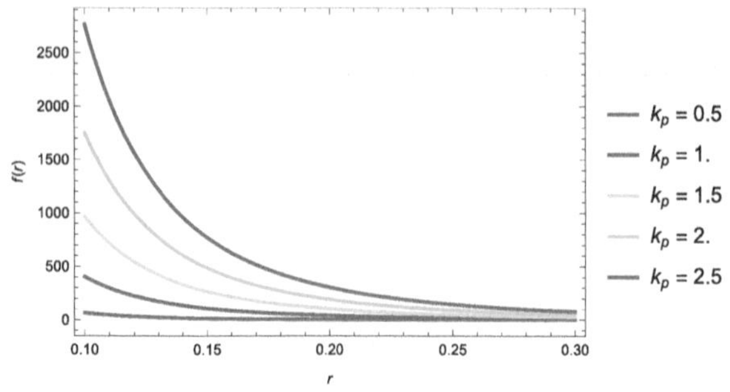

Fig. 44.1 The plot of $f(r)$ vs. r for various k_p values. Color code: Purple ($k_p = 0.5$), Red ($k_p = 1.0$), Yellow ($k_p = 1.5$), Green ($k_p = 2.0$), Blue ($k_p = 2.5$) (Color figure online)

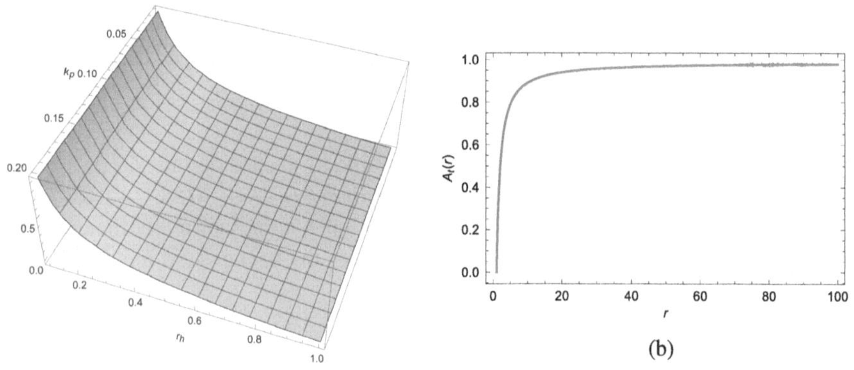

(a) (b)

Fig. 44.2 Left: 3D plot of black hole temperature as a function of r_h and k_p. Right: Plot of gauge field $A_t(r)$ as a function of r. (**a**) T vs. r_h and k_p. (**b**) $A_t(r)$ vs. r

$$
\begin{aligned}
ds^2 &= -u^2 F(u) dt^2 + \frac{\hat{L}^2}{u^2 F(u)} du^2 + u^{\frac{2-\theta}{z-\theta}} d\mathbf{x}^2, \\
F(u) &= u^{\frac{\theta}{z-\theta}} \left(1 - \left(\frac{u_h}{u}\right)^{\frac{2+z-\theta}{z-\theta}}\right),
\end{aligned}
\tag{44.6}
$$

where u_h is defined as $F(u_h) = 0$ and $\hat{L} = L/(z-\theta)$. In the near horizon limit both backreacted and probe limit metric solution turn out to be a blackhole in $AdS_2 \times R^2$.

44.3 Green's Functions

The fermionic Green's function $G(\omega, k)$ is computed by solving the Dirac equation in the given gravitational and gauge field background. The action for fermions in the bulk is introduced, and the corresponding Dirac equation is derived. The action is given by

$$S = i \int d^4x \sqrt{-g}\, \bar{\psi} \left(\Gamma^a D_a - m - \frac{1}{2} i p \Gamma^{\mu\nu} (e_\mu)^a (e_\nu)^b F_{ab} \right) \psi, \tag{44.7}$$

where,

$$D_a = \partial_a + \frac{1}{4} (\omega_{\mu\nu})_a \Gamma^{\mu\nu} - i q A_a, \tag{44.8}$$

$$\Gamma^a = e^a_\mu \Gamma^\mu, \quad \Gamma^{\mu\nu} = \frac{1}{2} [\Gamma^\mu, \Gamma^\nu], \tag{44.9}$$

$(e_\mu)^a$ is the vielbein, $\omega_{\mu\nu}$ is the spin connection, and q is the charge of the spin $1/2$ particle. In addition, we use the spatial rotational symmetry to set $k_i = k$ for $i = x$ and zero for $i = y$. The equation of motion for the above action is given by,

$$\begin{aligned} \left(\partial_r + m\sqrt{g_{rr}}\right) \lambda_{1\pm} &= \left(\sqrt{\frac{g_{rr}}{g_{tt}}} (\omega + q A_t) \pm \sqrt{\frac{g_{rr}}{g_{xx}}} k + p \sqrt{g_{tt}} \partial_r A_t \right) \lambda_{2\pm}, \\ \left(\partial_r - m\sqrt{g_{rr}}\right) \lambda_{2\pm} &= \left(-\sqrt{\frac{g_{rr}}{g_{tt}}} (\omega + q A_t) \pm \sqrt{\frac{g_{rr}}{g_{xx}}} k + p \sqrt{g_{tt}} \partial_r A_t \right) \lambda_{1\pm}, \end{aligned} \tag{44.10}$$

where p is dipole coupling. In the near horizon limit, the equation further reduces to

$$\xi \partial_\xi \begin{pmatrix} \lambda_{1\pm} \\ \lambda_{2\pm} \end{pmatrix} = U \begin{pmatrix} \lambda_{1\pm} \\ \lambda_{2\pm} \end{pmatrix}, \tag{44.11}$$

$$U = \begin{pmatrix} -m L_2 & qe + \frac{k}{r_h} L_2 + \frac{pe}{L_2} \\ -qe + \frac{k}{r_h} L_2 + \frac{pe}{L_2} & m L_2 \end{pmatrix}. \tag{44.12}$$

The solution to this equation is given by $\lambda_\pm = a E_\pm \xi^{-\nu} + b E_\mp \xi^\nu$, where $E_\pm$ are the real eigenvectors of U and for the finite temperature, the eigenvalue ν of U is given by[1]

$$\nu = \sqrt{\left(m^2 + \left(\frac{k}{r_*} + \frac{pe}{L_2} \right)^2 \right) L_2^2 - q^2 e^2} \quad (T > 0). \tag{44.13}$$

The retarded Green's function $G_R(\omega, k)$ is related to the response of the fermionic operator in the dual field theory. The boundary retarded Green's function is given by

[1] A detailed discussion can be found in [16].

$$G_R = \lim_{r\to\infty} r^{2m} \begin{pmatrix} \eta_+ & 0 \\ 0 & \eta_- \end{pmatrix}. \tag{44.14}$$

The phase diagram as presented below for $\nu \geq 1$ features a normal metal phase, where quasiparticles follow Landau Fermi liquid theory. For $1/2 < \nu < 1$, quasiparticles are stable but have a different lifetime scale than in a traditional Fermi liquid. When $\nu < 1/2$, the quasiparticles are short-lived, characteristic of a strange metal phase, known as the non-Fermi liquid region. The transition point at $\nu = 1/2$ represents a marginal Fermi liquid regime, marking the boundary between stable and unstable quasiparticles.

44.4 Results

In this chapter, all calculations and numerical analysis were performed using *Mathematica*. We used the shooting method to numerically solve the set of differential equations, including the Dirac equation for fermions. We shot from the horizon toward the asymptotic boundary and looked for the normalizable solution. The Green's function, which defines the spectral function, was calculated by solving the flow equation for fermionic modes.

In this section we will present the results. To see the change in the nature of fermionic excitations over the range of momentum dissipation with variation of temperature, we present a density plot in the $T/\sqrt{\rho}/k_p$ vs. $\sqrt{\rho}$ plane in Fig. 44.3,

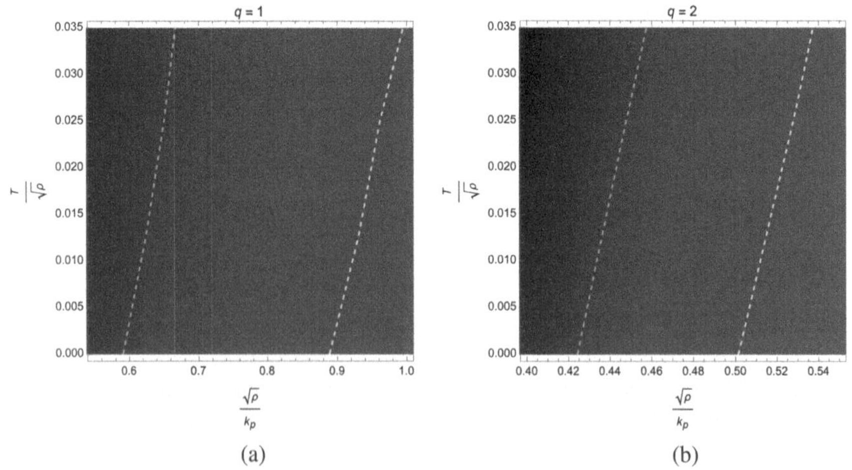

Fig. 44.3 Density plot of ν in $T/\sqrt{\rho}$ vs. $\sqrt{\rho}$ plane. The color gradient corresponds to the value of ν which increases from left to right. The green and yellow dashed lines correspond to $\nu = 1/2$ and $\nu = 1$ respectively. (**a**) $T/\sqrt{\rho}$ vs. $\sqrt{\rho}/k_p$ for $q = 2$. (**b**) $T/\sqrt{\rho}$ vs. $\sqrt{\rho}/k_p$ for $q = 1$ (Color figure online)

illustrating the change in the nature of fermionic excitations over the range of momentum dissipation with temperature variation. The plot shows a color gradient representing value of ν for two different values of charge $q = 1$ and $q = 2$, with darker and lighter shades corresponding to lower and higher values of ν. The plot extends for higher values of $\nu > 1$, indicating a normal metal phase. As temperature increases, the critical value of momentum dissipation for transitioning from strange metal to normal metal phase decreases. The qualitative behavior does not depend on the particle charge, similar to the normal metal to strange metal transition in [17] where the transition occurred due to tuning of doping rather than variation of the magnitude of axionic scalar.

44.5 Discussion

In this chapter we have studied the fermionic behavior of the non linear arcsin electrodynamics model. We presented the profile of $f(r)$ and blackhole temperature T over the range of momentum dissipation k_p. Additionally, we have studied the nature of Fermi surface in relation to the system's charge density ρ and momentum dissipation k_p in the finite temperature limit. We observe that with an increase in temperature the critical value of momentum dissipation for the transition decreases. When the charge density ρ varies with temperature, the critical charge density for the transition increases. For a given value of momentum dissipation and charge density with the increase in temperature, there is a transition from normal metal to strange metal phase, which is similar to the case in cuprates high T_c superconductors where the transition depends on the doping.

Comparing our results with the results of [18], we see that the boundary charge density ρ plays an important role in both the phases of the dual boundary theory. In the case of [18], ρ determines the critical phase transition temperature of the superconductor and also in the present case ρ determines the critical transition temperature from the regime of Fermi to non-Fermi. In both cases, the transition temperature increases with the increase of charge density ρ. In particular, we have observed that in [18] for a given temperature as β increases, the critical value of charge density increases similarly in the present model if we keep the temperature fixed at $T = 0$ and increase the analogous parameter $Z_2(\phi)$, the critical value of charge density also increases. We have also studied the fermionic behavior with the variation of the arcsin parameter $Z_2(\phi)$, where we find that with the increase of the arcsin parameter there is a transition from Fermi to non-Fermi regime. Comparing this result with another model of nonlinear electrodynamics studied in [19] where the nonlinearity was controlled by BI parameter, we see that in both the cases with the increase of nonlinear parameter the Fermi momenta decreases and the system which is already in the non-Fermi regime moves toward "more non-Fermi" regime.

It would be interesting to consider more general configurations with non-zero neutral scalar fields and varied coupling parameters to explore new solutions and fermionic behavior. One can also investigate the effects of magnetic fields on metal-insulator transitions and explore fermionic behavior in the superconducting phase at both finite and zero temperatures, including potential domain wall solutions. Also one can extend the model to include $U(1) \times U(1)$ gauge fields and introduce doping to study the impact on the system.

Acknowledgments The work of A.T is supported by Ministry of Tribal Affairs, Government of India, for the financial support provided through the National Fellowship for Scheduled Tribes (NFST). N.R. thanks the International Centre for Theoretical Sciences (ICTS) for the support through the program—Field Theory and Turbulence (ICTS/FTT2023/12).

References

1. Maldacena, J. M. (1998). The large N limit of superconformal field theories and supergravity. *Advances in Theoretical and Mathematical Physics, 2*, 231–252.
2. Witten, E. (1998). Anti-de Sitter space and holography. *Advances in Theoretical and Mathematical Physics, 2*, 253–291.
3. Gubser, S. S., Klebanov, I. R., & Polyakov, A. M. (1998). Gauge theory correlators from noncritical string theory. *Physics Letters B, 428*, 105–114.
4. Hartnoll, S. A. (2009). Lectures on holographic methods for condensed matter physics. *Classical and Quantum Gravity, 26*, 224002.
5. Herzog, C. P. (2009). Lectures on holographic superfluidity and superconductivity. *Journal of Physics A, 42*, 343001.
6. Horowitz, G. T. (2011). Introduction to holographic superconductors. *Lecture Notes in Physics, 828*, 313–347.
7. Lee, S.-S. (2009). A non-Fermi liquid from a charged black hole: A critical Fermi ball. *Physical Review D, 79*, 086006.
8. Cubrovic, M., Zaanen, J., & Schalm, K. (2009). String theory, quantum phase transitions and the emergent Fermi-liquid. *Science, 325*, 439–444.
9. Liu, H., McGreevy, J., & Vegh, D. (2011). Non-Fermi liquids from holography. *Physical Review D, 83*, 065029.
10. Mukhopadhyay, S., & Rai, N. (2017). Holographic Fermi surfaces in the six-dimensional (2, 0) theory. *Physical Review D, 96*(2), 026005.
11. Faulkner, T., Liu, H., McGreevy, J., & Vegh, D. (2011). Emergent quantum criticality, Fermi surfaces, and AdS(2). *Physical Review D, 83*, 125002.
12. Edalati, M., Leigh, R. G., Lo, K. W., & Phillips, P. W. (2011). Dynamical gap and cuprate-like physics from holography. *Physical Review D, 83*, 046012.
13. Alam, Y. F., & Behne, A. (2021). Review of Born-Infeld electrodynamics. arXiv: 2111.08657.
14. Rai, N., & Mukhopadhyay, S. (2019). DC conductivity from backreacted non-linear arcsin electrodynamics. *Annals Physics, 411*, 167974.
15. Kruglov, S. I. (2015). Nonlinear arcsin-electrodynamics. *Annalen Physics, 527*, 397–401.
16. Rai, N. (2019). Fermionic response in nonlinear arcsin electrodynamics. *European Physical Journal C, 79*(11), 972.
17. Giordano, G., Grandi, N., Lugo, A., & Soto-Garrido, R. (2018). Strange metal crossover in the doped holographic superconductor. *Journal of High Energy Physics, 10*, 068.
18. Kruglov, S. I. (2018). Holographic superconductors with nonlinear arcsin-electrodynamics. *Annalen Physics, 530*(8), 1800070.
19. Wu, J.-P. (2016). Holographic fermionic spectrum from Born–Infeld AdS black hole. *Physics Letters B, 758*, 440–448.

Chapter 45
A Covariant Approach to the Berry Phase Arising due to Einstein Gravity

Achal Kumar and Banibrata Mukhopadhyay

Abstract The Berry phase is a widely recognized phenomenon, extensively studied for Dirac particles interacting with an electromagnetic field. We expect that Dirac particles interacting with gravity should also cause a similar effect. We find a covariant formalism of Berry-like connection in the presence of Einstein gravity, using the WKB approximation. Using this Berry connection we can describe the Berry phase that is acquired by the Dirac particle. We perform the same analysis on massless Dirac particles (Weyl particles). Our findings indicate that, in the massless case, the Berry connection resembles that of electromagnetic waves, differing only by a proportionality factor of 1/2.

45.1 Introduction

The Berry phase is a well known phenomenon which was first introduced in quantum mechanics to describe the evolution of spin in case of a Hamiltonian that varies adiabatically in time. The motion of a Dirac particle in an electromagnetic field will also cause this effect and covariant formalism of this has been adequately studied [1]. We expect that Dirac particles interacting with gravity should also cause a similar effect. Even though there exists some work describing this phase for gravity [2, 3], this subject lacks a completely covariant formalism. We present a covariant formalism for this for Dirac and Weyl particles using WKB approximation. We find a covariant form of the Berry-like connection. Using this we can explain the evolution of the Berry-phase in gravity.

A. Kumar (✉)
Department of Physics, University of Florida, Gainesville, FL, USA
e-mail: achal.kumar@ufl.edu

B. Mukhopadhyay
Department of Physics, Indian Institute of Science, Bengaluru, India
e-mail: bm@iisc.ac.in

S. Ghosh, A. R. Rao (eds.), *The Relativistic Universe: From Classical to Quantum*,
Astrophysics and Space Science Proceedings 61,
https://doi.org/10.1007/978-3-031-90186-7_45

The WKB approximation leads to a semi-classical analysis of the problem, where we are unable to capture the wave nature of the particle itself but we can capture the effect of the motion of the particle on the spin degree of freedom. We are also not able to capture the effect of spin on the motion of the particle but further details of this effect can be seen in [4, 5].

The chapter is structured as follows. We start in Sect. 45.2 by introducing the Dirac equation on the curved spacetime. In Sect. 45.3, we define our WKB approximation. In Sect. 45.4, we consider the case of massive Dirac particles and analyze the semi-classical WKB equation, thus introducing the covariant Berry connection. The same analysis is performed in Sect. 45.5 for massless Dirac particles. In addition, we relate the results found here to already known results for electromagnetic fields. We conclude in Sect. 45.6.

45.2 The Dirac Equation in the Presence of Gravity

We start with the Dirac equation in the presence of gravity. A Dirac field $\Psi(x)$ with a mass m satisfies

$$\big(i\hbar\gamma^{\mu}(\nabla_{\mu}) - m\big)\Psi(x) = 0, \tag{45.1}$$

where γ^{μ}-s are the spacetime gamma matrices. They are related to the Minkowski spacetime gamma matrices by,

$$\gamma^{\alpha} = e^{\alpha}_{a}\gamma^{a},$$

where e^{α}_{a} is the tetrad. Here the Greek indices imply curved coordinates and Latin indices imply flat coordinates, and ∇_{μ} is the covariant derivative for spinors which is defined by a spin connection. The spinor derivative acts on the spinor field as

$$\nabla_{\mu}\Psi(x) = (\partial_{\mu} + \Omega_{\mu})\Psi(x), \tag{45.2}$$

where Ω_{μ} is defined as

$$\Omega_{\mu} = -\frac{i}{4}\omega_{ab\mu}(x)\sigma^{ab} = \frac{1}{8}\omega_{ab\mu}(x)[\gamma^{a}, \gamma^{b}]. \tag{45.3}$$

In this chapter, we will use ∇_{μ} to denote both the covariant derivative for tensors, which includes the Christoffel connection, and the covariant derivative for spinors, which involves the spin connection.

45.3 WKB Approximation

We assume that the Dirac field can be expanded using the WKB ansatz as follows,

$$\Psi(x) = \psi \exp^{-\frac{i\varphi}{\hbar}}, \tag{45.4}$$

where $\psi = \psi_0 + \hbar\psi_1 + \hbar^2\psi_2 + \ldots$, Also $\partial_\mu\varphi = p_\mu = (E, -\mathbf{p})$ and $\psi_i \equiv \psi_i(x, p_\mu)$. We use the Planck's constant $\hbar$ as the small expansion parameter and consider the expansion of the Dirac equation in curved spacetime upto the order $\hbar$. Here we have assumed background spacetime varies slowly.

45.4 Massive Dirac Fields

In this section, we present our semi-classical analysis for the case of a massive Dirac field. Our aim is to first get the solution of the Dirac equation in curved spacetime under the WKB approximation and then identify the term which corresponds to the Berry-like connection, thus giving us a covariant formalism for the same.

45.4.1 *Leading Order WKB Equations*

At order $\hbar^0$ the Dirac equation in curved spacetime reduces to

$$(\gamma^\mu p_\mu - m)\psi_0 = 0. \tag{45.5}$$

We define the zeroth order four-current as $j_0^\mu = \bar{\psi}_0\gamma^\mu\psi_0$. Using Eq. (45.5) and its conjugate, the continuity equation for j_0^μ can be shown as follows,

$$\nabla_\mu j_0^\mu = 0. \tag{45.6}$$

Now using the anti-commutator relation of the gamma matrices, $\gamma^\mu\gamma^\nu + \gamma^\nu\gamma^\mu = -2g^{\mu\nu}$, we obtain

$$j_0{}^\mu = \bar{\psi}_0\gamma^\mu\psi_0 = \frac{1}{m}\mathcal{I}_0 p^\mu, \tag{45.7}$$

where we define $\mathcal{I}_0 = \bar{\psi}_0\psi$. Therefore it can be shown that the most general ψ_0 which satisfies Eqs. (45.5)–(45.7) can be written as,

$$\psi_0(x, p) = \sqrt{\mathcal{I}_0}(u_0 z_0 + u_1 z_1) = \sqrt{\mathcal{I}_0} z_\alpha u_\alpha, \tag{45.8}$$

where u^α-s are the orthonormal and complete set of eigenspinor solutions of the equation,

$$(\gamma^\mu p_\mu - m)u_\alpha = 0. \tag{45.9}$$

Here $\alpha \in \{0, 1\}$ representing the two positive energy solutions of the Dirac equation in the presence of gravity. Here z_α-s are complex coefficients which satisfy

$$\bar{z}_0 z_0 + \bar{z}_1 z_1 = \bar{z}_\alpha z_\alpha = 1. \tag{45.10}$$

45.4.2 Next-to-Leading Order Equation

We have obtained the form of our solution in Eq. (45.8) using the zeroth order approximation. We will use the first order equation to determine the coefficients z_α, which contain the information about the phase of the particle. The zeroth order solution completely determines the eigenspinors u_α-s, which do not have any effect of gravity. The effect of gravity enters only in the first order equation, thus only affecting the coefficients z_α-s. Thus we capture the effect of gravity in the evolution of the spin but we do not capture the effect of spin on the trajectory of the Dirac particle. Hence, the trajectory of these particles, under this approximation, is just the trajectory of a spinless particle in gravity.

At h^1 we obtain,

$$i(\gamma^\mu \nabla_\mu \psi_0) + (\gamma^\mu p_\mu - m)\psi_1 = 0 \implies \bar{u}_\beta(\gamma^\mu \nabla_\mu \psi_0) = 0. \tag{45.11}$$

Using Eq. (45.8) in Eq. (45.11) we solve for the coefficients z_α-s. Utilizing the properties of the eigenspinors and after some algebra it can be shown that z_α-s follow,

$$\left[\delta_{\alpha\beta} V^\mu \nabla_\mu + \frac{1}{2} V^\mu (\bar{u}_\alpha \nabla_\mu u_\beta - \nabla_\mu \bar{u}_\alpha u_\beta)\right] z_\alpha = 0, \tag{45.12}$$

where $V^\mu = p^\mu/m$ is the four velocity. Thus the evolution of the coefficients z_α can be described as,

$$\dot{\mathbf{z}} = i V^\mu B_\mu \mathbf{z} \quad \text{where} \quad z \equiv \begin{pmatrix} z_0 \\ z_1 \end{pmatrix}. \tag{45.13}$$

Here $\mathbf{z}$ is a complex unit vector. Thus we define the covariant Berry connection as,

$$(B_\mu)_{\alpha\beta} = \frac{i}{2}(\bar{u}_\alpha \nabla_\mu u_\beta - \nabla_\mu \bar{u}_\alpha u_\beta). \tag{45.14}$$

We also performed these calculations in the presence of an electromagnetic field which can be found in [5]. This Berry connection can be used to determine the Berry phase itself which can also be seen in [4, 5]. The above expression for the Berry connection reduces to the well known expression for flat spacetime as follows,

$$iV^{\mu}B_{\mu\alpha\beta} = \frac{i}{2}V^{\mu}(\bar{u}_{\alpha}\partial_{\mu}u_{\beta} - \partial_{\mu}\bar{u}_{\alpha}u_{\beta}) = i\bar{u}_{\alpha}V^{\mu}\partial_{\mu}u_{\beta} = i\bar{u}_{\alpha}\frac{d}{d\tau}u_{\beta}. \tag{45.15}$$

In this case, B_{μ} becomes diagonal giving,

$$\gamma_{\alpha} = i\int_{o}^{\tau}\bar{u}_{\alpha}\frac{d}{d\tau'}u_{\alpha}d\tau', \tag{45.16}$$

which is our well-known Berry phase.

45.5 Massless Dirac Fields (Weyl Particles)

In this section we will repeat the same analysis for a massless Dirac field or Weyl particles. In this case the Dirac equation reduces to,

$$\gamma^{\mu}\nabla_{\mu}\Psi(x) = 0. \tag{45.17}$$

We can not just put $m = 0$ in our previous calculations, because for $m = 0$, the four-current defined in Sect. 45.4 diverges. We will see that we get a different Berry connection in this case. We will then relate the results found for this case to the pre-existing results for electromagnetic fields.

45.5.1 *Leading Order WKB Equations*

At the lowest order we would obtain,

$$\gamma^{\mu}p_{\mu}\psi_{0} = 0. \tag{45.18}$$

Therefore, similar to the last case we can write the most general solution in terms of the eigenspinors as follows,

$$\psi_{0}(x, p) = \sqrt{I_{0}}z_{\alpha}\Sigma_{\alpha}(x, k), \tag{45.19}$$

where Σ_{α}-s are the eigenspinors of Eq. (45.18). It is easier to solve this equation in the Weyl basis in which it decouples, the eigenspinors in the Weyl basis is given as follows,

$$\Sigma_0 = \begin{pmatrix} 0 \\ u \end{pmatrix}, \qquad \Sigma_1 = \begin{pmatrix} v \\ 0 \end{pmatrix}, \tag{45.20}$$

where u and v are 2-spinors satisfy,

$$p_\mu \sigma^\mu u = 0, \qquad p_\mu \bar{\sigma}^\mu v = 0, \tag{45.21}$$

and gamma matrices in this basis is given as,

$$\gamma^\mu = \begin{bmatrix} 0 & \sigma^\mu \\ \bar{\sigma}^\mu & 0 \end{bmatrix}. \tag{45.22}$$

In Eqs. (45.20) and (45.22), the '0' represents a null vector and a null matrix respectively. Also, we use the spacetime Pauli 4-vectors $\sigma^\mu = (e_a)^\mu \sigma^a$ and $\bar{\sigma}^\mu = (e_a)^\mu \bar{\sigma}^a$, where $\sigma^a = \left(\mathbb{I}_2, \sigma^i\right)$ and $\bar{\sigma}^a = \left(\mathbb{I}_2, -\sigma^i\right)$. In this basis Σ_0 represents a fermion with a right-hand chirality, and Σ_1 represents a fermion with left-hand chirality, both having positive energies.

Using the properties of the Pauli 4-vectors, we can obtain a relation for the four-current as follows,

$$j_0^\mu = \bar{\psi}_0 \gamma^\mu \psi_0 = \mathcal{I}_0 \frac{p^\mu}{p_0} = \mathcal{I}_0 \frac{p^\mu}{p_\nu e_0^\nu}. \tag{45.23}$$

We observe that the four-current in this scenario is dependent on the reference frame, as it includes the tetrad e_0^ν, which differs from the case of a massive particle. This is because a massless particle lacks a rest frame, making the four-current dependent on the observer's frame. Similar to the massive case we have the continuity equation as follows,

$$\nabla_\mu j_0^\mu = \nabla_\mu \left(\mathcal{I}_0 \frac{p^\mu}{p_\nu e_0^\nu} \right) = 0. \tag{45.24}$$

45.5.2 Next-to-Leading Order Equation

At the next order we get,

$$\gamma^\mu \nabla_\mu \psi_0 - i \gamma^\mu p_\mu \psi_1 = 0 \implies \bar{\Sigma}_\beta (\gamma^\mu \nabla_\mu \psi_0) = 0. \tag{45.25}$$

Now similar to massive case we would use the above relation to obtain a relation between the coefficients z_α. Expanding everything in terms of 2-spinors, and using the properties of the 4-Pauli vectors, it can be shown that the Berry connection turns out to be,

$$B_\mu = \frac{i}{2}\begin{pmatrix} \bar{u}\nabla_\mu u - \nabla_\mu \bar{u} u & 0 \\ 0 & \bar{v}\nabla_\mu v - \nabla_\mu \bar{v} v \end{pmatrix}. \tag{45.26}$$

Since the Weyl equation can be decoupled in the Weyl basis, the Berry connection turns out to be diagonal and the coefficients z_α, evolve as,

$$z_0(\tau) = e^{i\gamma_0(\tau)} z_0(\tau_0), \qquad z_1(\tau) = e^{i\gamma_1(\tau)} z_1(\tau_0), \tag{45.27}$$

where,

$$\gamma_0(\tau) = \int_{\tau_0}^{\tau} d\tau' (p^\mu B_\mu)_{00}, \qquad \gamma_1(\tau) = \int_{\tau_0}^{\tau} d\tau' (p^\mu B_\mu)_{11}. \tag{45.28}$$

Since the equations decouple in the massless case, the Berry phase can be calculated explicitly, as shown in Eq. (45.28). Using the null-tetrad formalism, this Berry connection can be related to the one found for electromagnetic field, these results can be found in [5, 6]. It can be shown that the Berry connection here has the same form as that in the case of electromagnetic field except for a proportionality factor of 1/2.

45.6 Conclusion

We have developed a covariant formalism for the Berry connection of spin-half particles, using the WKB approximation as the foundation of our analysis. We have considered both the leading-order and next-to-leading-order terms in the WKB approximation. We have described the effect of gravity on the dynamics of the spin. More detailed calculations which also include the effects of an electromagnetic field can be found in [4, 5]. For the massless case, we have found that our Berry connection turned out to be an observer-dependent quantity, unlike the massive case. We also show that the Berry connection found for the massless case has the same form as that of the electromagnetic field, except for a proportionality factor of $1/2$.

Acknowledgments A.K. thanks the organizers of the International Symposium on Recent Developments in Relativistic Astrophysics (ISRA) for the opportunity to present this work.

References

1. Stone, M., Dwivedi, V., & Zhou, T. (2015). *Physical Review D, 91*, 025004.
2. Ghosh, T., & Mukhopadhyay, B. (2021). *International Journal of Modern Physics D, 30*, 2150090. arXiv:2010.05945.
3. Mukhopadhyay, B., & Ganguly, S. K. (2020). *Universe, 6*(10), 160. arXiv:1802.10377.

4. Oancea, M. A., Paganini, C. F., Joudioux, J., & Andersson, L. (2019). An overview of the gravitational spin Hall effect. Gravitational spin Hall effect. arXiv:1904.09963.
5. Oancea, M. A., & Kumar, A. (2023). Semiclassical analysis of Dirac fields on curved spacetime. *Physical Review D, 107*, 044029.
6. Oancea, M. A., Joudioux, J., Dodin, I. Y., Ruiz, D. E., Paganini, C. F., & Andersson, L. (2020). Gravitational spin Hall effect of light. *Physical Review D, 102*, 024075.

Chapter 46
Black Hole Thermodynamics in Modified Gravity

Mriganka Dutta and Banibrata Mukhopadhyay

Abstract The theory of general relativity is often considered under the framework of modified Einstein gravity to explain different phenomena under strong curvature. The strong curvature effect plays a main role near black holes, where the gravitational field is strongest. The idea of black hole thermodynamics is to describe the strong field curvature properties of a black hole in the effective thermodynamical framework, e.g. entropy, temperature, heat capacity etc. In this chapter, our aim is to explore how the effect of modified gravity changes the thermodynamic properties of black hole. We show that even a small modification to Einstein gravity affects the thermodynamical properties of a black hole.

46.1 Introduction

The theory of general relativity is one of the finest and most successful theories that we have, for over 100 years, to describe the spacetime structure. The theory explains how mass affects the spacetime curvature and in turn how spacetime curvature affects the path of mass transportation. Einstein's general relativity has been verified by many phenomena such as gravitational lensing [1], perihelion precession of Mercury's orbit [2], gravitational wave generation [3], etc. However, most of these are done based on the weak field approximation of general relativity. It is still uncertain whether the theory holds its validity in the presence of strong curvature in spacetime.

Einstein's gravity theory also predicts that if we have concentrated mass at a very small volume, it can form a one-way membrane in spacetime. This membrane is called the event horizon surface. If classically, whatever crosses the event horizon getting trapped inside it, we call that structure black hole. Karl Schwarzschild gave the spacetime metric structure for non-rotating and later Roy Kerr deduced the

M. Dutta (✉) · B. Mukhopadhyay
Department of Physics, Indian Institute of Science, Bangalore, India
e-mail: mrigankad@iisc.ac.in; bm@iisc.ac.in

S. Ghosh, A. R. Rao (eds.), *The Relativistic Universe: From Classical to Quantum*, Astrophysics and Space Science Proceedings 61,
https://doi.org/10.1007/978-3-031-90186-7_46

spacetime metric structure for rotating black holes. Spacetime near black holes is such a strong field regime that the validity of general relativity remains uncertain. Hence one may need to consider the modification of gravity theory.

In 1971, Roger Penrose showed that theoretically that it should be possible to extract energy from a rotating black hole. Consequently, it was understood that the area of a black hole (i.e. the area of the event horizon surface) never decreases. Based on these ideas, Jacob Bekenstein further formulated a connection between the black hole area and its entropy [4]. He also proposed the idea of the generalized second law of thermodynamics (GSL). Later in 1975, Steven Hawking showed that due to quantum mechanical effects, black holes can radiate particles and have some finite temperature [5]. All of these works are done assuming Einstein's gravity theory holds well in strong spacetime curvature.

One of the frameworks of modified gravity theory is that asymptotically, hence in a weaker spacetime curvature, a modified gravity leads to the results of Einstein's gravity theory. However, the modified gravity has significant variations in thermodynamic properties near black holes. In this chapter, we shall try to explain what happens to the thermodynamic properties of black holes in the presence of modified gravity.

46.2 Metric of Rotating Black Hole in Modified Gravity

The modification to general relativity can be done in various ways. Out of various models of modified gravity, we consider the $f(R)$-gravity. The idea is to have asymptotically flat solutions such that on a large scale, the modified gravity behaves as regular gravity theory. The Einstein-Hilbert action for a $f(R)$-gravity (with metric signature [+,-,-,-]) is

$$S = \int \left(\frac{c^4}{16\pi G} f(R) + \mathcal{L}_M \right) \sqrt{-g}\, d^4x. \tag{46.1}$$

Here '$-g$' is the determinant of the metric tensor $g_{\mu\nu}$, $\mathcal{L}_M$ is the Lagrangian density of matter, $f(R)$ is a function of Ricci scalar R. The expression for $f(R)$ has to be such that $df(R)/dR$ behaves as $1 + B/r$ so that at large r the gravity theory acts as the Einstein's gravity theory.

Following a previous work [6], one can consider the metric to describe the rotating black hole spacetime in a modified gravity. We take the metric tensor, in (t, r, θ, ϕ) coordinates, as

$$g_{\mu\nu} = \begin{pmatrix} 1 - \frac{2Mr}{\Sigma} - \frac{\beta}{\Sigma} & 0 & 0 & \frac{Ma\sin^2\theta}{\Sigma}(2r+\beta) \\ 0 & -\frac{\Sigma}{\Delta} & 0 & 0 \\ 0 & 0 & -\Sigma & 0 \\ \frac{Ma\sin^2\theta}{\Sigma}(2r+\beta) & 0 & 0 & -\sin^2\theta\left(r^2 + a^2 + \frac{a^2\sin^2\theta(2Mr+\beta)}{\Sigma}\right) \end{pmatrix}, \tag{46.2}$$

where $\Sigma = r^2 + a^2 \cos^2\theta$, a is the specific angular momentum of black hole, β is $B(B-6)/2$ and $\Delta = \Sigma\, g_{tt}/X(r,\theta) + a^2 \sin^2\theta$. The expression for $X(r,\theta)$ is given as $(r^2 + a^2\cos^2\theta)^2/[(r+B/2)^2 + a^2\cos^2\theta]^2$. It needs to be stated that the value of B has to be negative and we restrict the modification such that $|B| < 1$. For $B > 0$ we have unphysical properties of gravity. Considering the fact that $|B|$ is small; we can take $X(r,\theta) \approx 1$.

46.3 Thermodynamic Properties

46.3.1 Temperature and Entropy of Black Hole in Modified Gravity

The temperature of a black hole can be calculated from the surface gravity, once the line element or the metric tensor of spacetime is determined. For the metric, mentioned in Eq. (46.2), we have two Killing vectors. One for the time-translation in-variance and another for rotational in-variance along ϕ direction. Thus for a rotating black hole the Killing vector can be taken as

$$K^\mu = [1, 0, 0, \Omega_H], \tag{46.3}$$

where $\Omega_H = -g_{t\phi}/g_{\phi\phi}$ at the horizon. To determine the horizon we make $g_{rr} \to \infty$, i.e. Δ to be 0. Considering $|B| < 1$, Δ can be simplified to $r^2 - 2Mr + a^2 - \beta$. This gives the radii of the outer and inner horizons as

$$r_\pm = M \pm \sqrt{M^2 - a^2 + \beta}, \tag{46.4}$$

where r_+ and r_- denote the outer and inner horizons respectively. The area corresponding to r_+ turns out to be

$$A = 4\pi(r_+^2 + a^2). \tag{46.5}$$

The surface gravity of a black hole (κ) can be calculated from the Killing vector by

$$\kappa^2 = \lim_{r\to r_+} \frac{(\nabla_\nu(K^\mu K_\mu)\nabla^\nu(K^\mu K_\mu))}{4(K^\mu K_\mu)}. \tag{46.6}$$

From the surface gravity one can define temperature of black hole as $T'_{BH} = \kappa/(2\pi)$. It turns out that, one can also define the temperature of the black hole just by differentiating the area A [4]. After differentiation we obtain

$$dM = \frac{r_+ - r_-}{16\pi(r_+^2 + a^2)} dA + \frac{a}{(r_+^2 + a^2)} dL, \tag{46.7}$$

where black hole mass M, area A and total angular momentum L (=aM) are taken to be independent variables. Equation (46.7) gives the essence of the second law of thermodynamics for a black hole. Following the earlier works of Bekenstein [4] and Hawking [5], the relation between the area and entropy of a black hole is $S = A/4$. Hence one obtains the temperature from Eq. (46.7) as

$$T_{BH} = \frac{r_+ - r_-}{4\pi(r_+^2 + a^2)}. \tag{46.8}$$

It can be verified that the temperature expressions we obtain from Eqs. (46.6) and (46.8) are same, i.e. $T'_{BH} = T_{BH}$.

46.3.2 Heat Capacity of Black Hole in Modified Gravity

Following the thermodynamic relations we can say that the 'heat capacity' [7] of black hole is defined as

$$C_L = T\left(\frac{\partial S}{\partial T}\right)_L. \tag{46.9}$$

To obtain an analytical form of heat capacity, one can define M as a function of S, T_{BH} and L. From Eq. (46.5), the mass of black hole can be written as

$$M = \left(\frac{S}{4\pi} + \frac{\pi\beta^2}{4S} + \frac{4\pi L^2}{S} - \frac{\beta}{2}\right)^{1/2}. \tag{46.10}$$

Thus temperature in terms of entropy, mass and angular momentum becomes

$$T_{BH} = \frac{1}{2M}\left(\frac{1}{4\pi} - \frac{\frac{\beta^2}{4} + L^2}{S^2}\right). \tag{46.11}$$

Now using Eqs. (46.11) and (46.10) in Eq. (46.9) we obtain 'heat capacity'

$$C_L = \frac{MT_{BH}S}{\frac{1}{4\pi} - 2MT_{BH} - T_{BH}^2 S}. \tag{46.12}$$

46.3.3 Free Energy of Black Hole in Modified Gravity

The free energy of a thermodynamic system is described as the energy available for work done. Similarly considering the thermodynamic properties of black holes, one can define the free energy. Free energy of a black hole determines how strong the spacetime curvature effect is. We define the free energy of black holes as

$$F = M - T_{BH} S. \tag{46.13}$$

To obtain the entropy of a black hole, as a function of M, L and β $(= B(B-6)/2)$, we write Eq. (46.5) as

$$S = 2\pi(M^2 + \beta/2) + 2\pi\sqrt{M^4 + M^2\beta - L^2}. \tag{46.14}$$

46.4 Implications

We have considered the metric for rotating black holes in a modified gravity. From Fig. 46.1 we see that with decreasing B there is an increase of β and hence the increase of outer horizon radius, at a fixed specific angular momentum a. The entropy, as in Fig. 46.2, of the black hole also increases as the outer horizon radius increases. The surface temperature of black hole decreases with increasing specific angular momentum, as shown in Fig. 46.3. The behavior of heat capacity (Fig. 46.4) shows a sudden change from a negative value to a positive value, at some constant specific angular momentum, i.e. a phase transition sets in. The value of

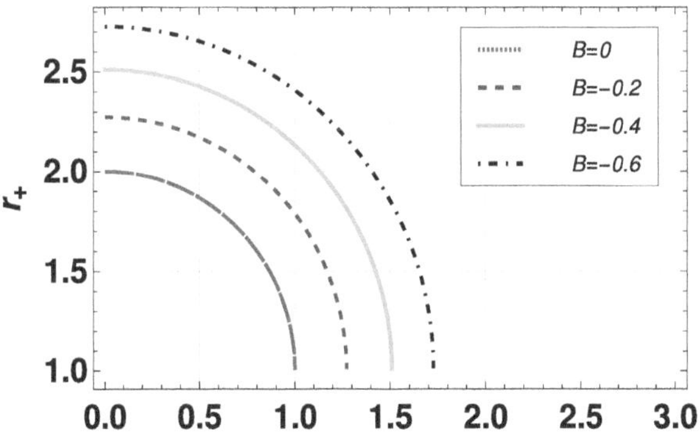

Fig. 46.1 Radius of the outer horizon as a function of spin parameter, for different values of B parameter considering $M = 1$

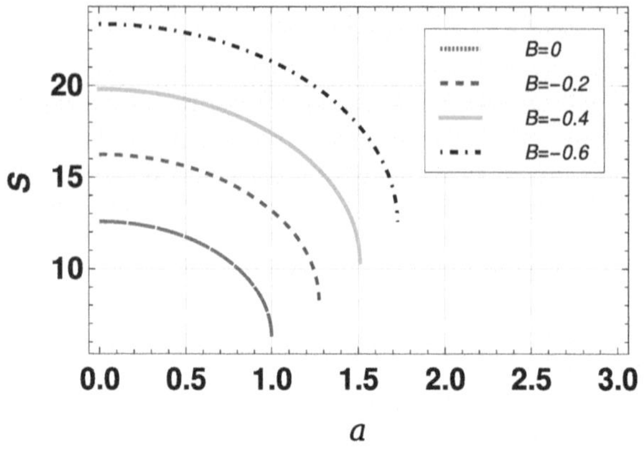

Fig. 46.2 Entropy of the black hole as a function of spin parameter, for different values of B parameter considering $M = 1$

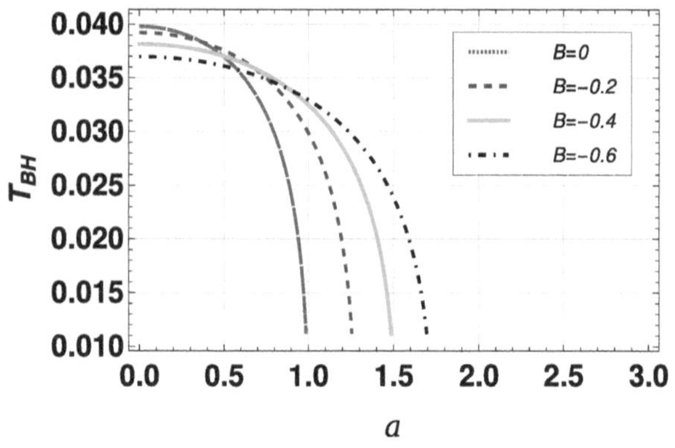

Fig. 46.3 Temperature of the black hole as a function of spin parameter(a) for different values of B parameter considering $M = 1$

specific angular momentum, where this phase transition takes place, shifts towards higher values of specific angular momentum with decreasing B. This effect is seen because the effect of modified gravity opposes the effect of angular momentum, hence supporting a higher value of a. The free energy of a black hole is also a thermodynamic property that is drastically affected by modified gravity. From Fig. 46.5 we see that with decreasing B or increasing β, the free energy decreases. The lower value of the free energy makes the curvature of spacetime less affected. The lesser curvature effect causes the outer horizon radius to increase.

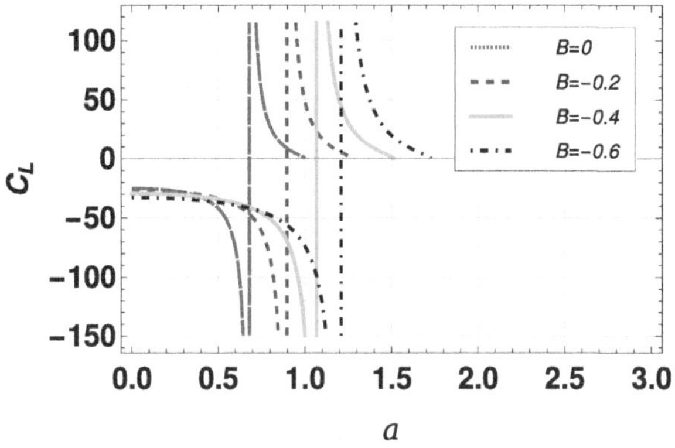

Fig. 46.4 Heat capacity at constant angular momentum of the black hole as a function of the spin parameter, for different values of B parameter considering $M = 1$

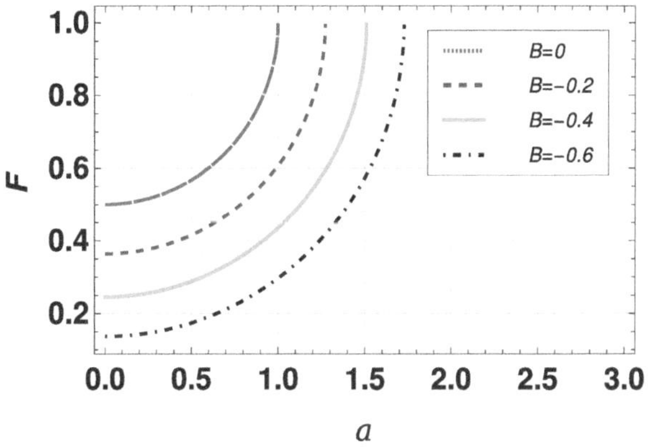

Fig. 46.5 Free energy of the black hole as a function of the spin parameter, for different values of B parameter considering $M = 1$

46.5 Conclusions

We have explored a $f(R)$ gravity in a strong gravitational field regime. Considering the modified metric for a rotating black hole, we obtain analytical expressions for temperature, entropy, heat capacity, and free energy. This work shows significant variation in the thermodynamic properties compared to those in Einstein's gravity. Thus one can not ignore the possibility of a modification of gravity theory, particularly to probe properties of spacetime close to the black hole. As a future work, we plan to look for the modification of black hole evaporation considering modified gravity theory, see [8] in this volume, for a preliminary result.

Acknowledgments MD thanks the organizers of the International Symposium on Recent Developments in Relativistic Astrophysics (ISRA) for giving the opportunity to write this chapter in this special volume.

References

1. Wambsganss, J. (1998). Gravitational lensing in astronomy. *Living Reviews in Relativity, 1*, 12. https://doi.org/10.12942/lrr-1998-12
2. Kraniotis, G. V., & Whitehouse, S. B. (2003). Exact calculation of the perihelion precession of mercury in general relativity, the cosmological constant and Jacobi's inversion problem. *Classical and Quantum Gravity, 20*, 4817–4835. https://doi.org/10.1088/0264-9381/20/22/007
3. Abbott, B. P., et al., [LIGO Scientific and Virgo]. (2016). Observation of gravitational waves from a binary Black Hole merger. *Physical Review Letters, 116*, 061102. https://doi.org/10.1103/PhysRevLett.116.061102
4. Bekenstein, J. D. (1973). Black holes and entropy. *Physical Review D, 7*, 2333–2346. https://doi.org/10.1103/PhysRevD.7.2333
5. Hawking, S. W. (1975). Particle creation by Black Holes. *Communications in Mathematical Physics, 43*, 199–220. [Erratum: *Communications in Mathematical Physics, 46*, 206 (1976)]. https://doi.org/10.1007/BF02345020
6. Das, A. R., & Mukhopadhyay, B. (2022). Asymptotically flat vacuum solution for a rotating black hole in a modified gravity theory. *European Physical Journal C, 82*, 939. https://doi.org/10.1140/epjc/s10052-022-10899-5
7. Davies, P. C. W. (1978). Thermodynamics of black holes. *Reports on Progress in Physics, 41*, 1313–1355. https:://doi.org/10.1088/0034-4885/41/8/004
8. Dey, P., & Mukhopadhyay, B. (2024). Faster rate of Hawking radiation in modified gravity constraining dark matter. In S. Ghosh & A. R. Rao (Eds.), *The relativistic universe: From classical to quantum*. Springer.

Chapter 47
Faster Rate of Hawking Radiation in Modified Gravity Constraining Dark Matter

Panchajanya Dey and Banibrata Mukhopadhyay

Abstract The exact theory of gravity in the strong field regime is still under debate. There are observations implying the need for modification to Einstein's gravity. On the other hand, the exact constituents of dark matter are also a big puzzle, where primordial black holes (PBHs) are argued to be a potential candidate. We explore Hawking radiation in a modified gravity and find that PBHs evaporate faster in a scalar-tensor theory based modified gravity. Subsequently, all the nonrotating BHs of mass $\sim 10^{15}$ g or less should have been evaporated by today, which is an order of magnitude heavier than what the Einstein gravity predicts. This has many consequences including a strict constraint on contributing PBHs to dark matter, widening the debate of dark matter origin.

47.1 Introduction

The most popular theory of cosmology is the Λ-CDM model when the universe contains 5% ordinary matter, 26.8% dark matter, and 68.2% dark energy. The dark matter is assumed to be cold (of low velocity), which exhibits structures of the universe emerging by the gradual accumulation of particles. Nevertheless, there has been no success in the discovery of dark matter particles even over the half a century search. This and more recent observations lying with gravitational wave and James Webb Space Telescope have considerably strengthened the idea of primordial and direct collapse black holes (BHs) to be dark matter [1, 2]. Moreover, intrigued by specific observations not explainable properly by ordinary dark matter (as well as dark energy) hypothesis, various modifications to Einstein's theory of general relativity (GR), including modified Newtonian dynamics (MOND) [3] were proposed. However, no modified gravity (MG) theory seems to describe every piece

P. Dey (✉) · B. Mukhopadhyay
Department of Physics, Indian Institute of Science, Bangalore, India
e-mail: panchajanyad@iisc.ac.in; bm@iisc.ac.in

S. Ghosh, A. R. Rao (eds.), *The Relativistic Universe: From Classical to Quantum*, Astrophysics and Space Science Proceedings 61,
https://doi.org/10.1007/978-3-031-90186-7_47

of observation successfully at the same time. This suggests that even in the presence of MG, the existence of dark matter is inevitable.

GR is the most panoramic theory that explains the theory of gravity. It can easily explain a large number of phenomena where Newtonian gravity falls short, which includes deflection of light in strong gravity, generation of gravitational wave, gravitational redshift of light etc. However, apart from dark matter, a number of other recent observations in cosmology and in the high density/energy regions in the universe, such as the inside/vicinity of compact objects, it seems that the GR may not be the ultimate theory of gravity [4, 5]. Starobinsky first used one of the modified theories of gravity, namely R^2-gravity [6], to explain dark energy era and the very early universe cosmology, thereby did overcome some shortcomings of GR to explain cosmos. Eventually, plenty of different models have been proposed to explain various physical phenomena where GR fails. The question, hence, arises whether globally GR is the ultimate theory of gravitation, or it requires modification in the strong gravity regime. In stellar physics, indeed it was shown that MG theories reveal significant deviations from the GR solutions of neutron stars (NSs) [7–9]. However, MG was also explored in BHs, e.g. asymptotically flat vacuum solutions [10]. The present work centers around MG based BHs.

When a MG is introduced, whatever be the (observational) purposes, it should suffice other experiments/observations, which do not necessarily need a MG. That is, it should pass through the solar system test or at some regime any MG should reduce to GR (and subsequently to special relativity theory). In other words, if MG is true for one physics and astrophysics, it should be true for other gravity involved physics as well. Some of the interesting gravity induced sites are, NS interior, early evolution of universe, Hawking radiation, BH accretion flow, particle creation in curved spacetime, gravitational lensing and waves, to say a few.

One of the very interesting, old and phenomenal physical processes lying with BHs is the Hawking radiation. All the BHs keep radiating continuously and, hence, get evaporated. The rate of evaporation is inversely proportional to the mass of BHs. Hence, the primordial BHs (PBHs) being much lighter than the ones formed by the stellar collapse evaporate faster. According to GR, all the BHs of mass $\lesssim 5 \times 10^{14}$ g forming in early universe have been evaporated by today. On the other hand, a part of PBHs is argued to be the candidate for dark matter [11].

What if the gravity is MG not GR? What is the fate of Hawking radiation then, evaporation of PBHs and its connection to dark matter? Can it resolve the problem of dark matter origin? Can it enlighten the issues of the invisible source of gravity in the universe? We need to explore GR/MG, BH physics, cosmology, and a bit of particle physics in a unified manner to answer this question.

47.2 Brief Description of a Modified Gravity Metric

There are several ways to introduce modification to GR. One of such ideas is based on scalar-tensor theory [12] which further leads to a special class of MG called $f(R)$-gravity. With the metric signature $(+, -, -, -)$, the modified Einstein-Hilbert action is given by

$$S = \int \left[\frac{c^4}{16\pi G} f(R) + \mathcal{L}_M \right] \sqrt{-g} d^4x, \tag{47.1}$$

where $f(R)$ is a function of R, which is Ricci scalar, $-g$ is the determinant of metric tensor $g_{\mu\nu}$, $\mathcal{L}_M$ the matter Lagrangian density, G Newton's gravitation constant, and c the speed of light. This gives MG equation as

$$F(R)G_{\mu\nu} + \frac{1}{2} g_{\mu\nu}[RF(R) - f(R)] - (\nabla_\mu \nabla_\nu - g_{\mu\nu}\Box)F(R) = \frac{8\pi G}{c^4} T_{\mu\nu}, \tag{47.2}$$

where $F(R) = df(R)/dR$, $G_{\mu\nu}$ is the Einstein tensor, and $T_{\mu\nu}$ the energy-momentum tensor. For $f(R) = R$, MG reduces to GR. Assuming a spherically symmetric vacuum solution of MG equation given by

$$g_{\mu\nu} = diag\left(s(r), -p(r), -r^2, -r^2 sin^2\theta \right) \tag{47.3}$$

and also assuming that $F(R) = 1 + B/r$ [10], such that the vacuum solution resembles GR at large distance, i.e. $r \to \infty$, we obtain

$$\begin{aligned} s(r) = 1 - \frac{2}{r} - \frac{B(-6+B)}{2r^2} + \frac{B^2(-66+13B)}{20r^3} \\ - \frac{B^3(-156+31B)}{48r^4} + \frac{3B^4(-57+11B)}{56r^5} \\ - \frac{B^5(-360+67B)}{128r^6} + O(r^{-7}) \end{aligned} \tag{47.4}$$

and $p(r) = X(r)/s(r)$, where $X(r)$ is given by

$$X(r) = \frac{16r^4}{(B+2r)^4}, \tag{47.5}$$

for $B < 0$. Note that $B > 0$ corresponds to $R < 0$ which implies unphysical repulsive gravitational field, hence is ruled out. Equations (47.4) and (47.5) reduce to those of GR (Schwarzschild metric) either for $B = 0$ or at $r \to \infty$.

47.3 Size of Black Hole in Modified Gravity

GR suggests the radius of a nonrotating BH to be $2GM/c^2$, which is called Schwarzschild radius, also is event horizon (r_H). To obtain the event horizon radius in MG, we equate the denominator of g_{rr}, i.e. $-p(r)$, to zero. For the present

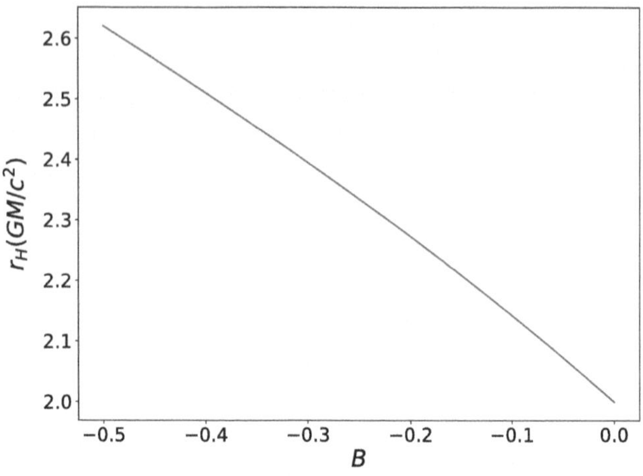

Fig. 47.1 Radius of event horizon of a nonrotating BH as a function of B

purpose, the event horizon corresponds to $s(r) = 0$, leading to r_H in MG up to the first order in B as

$$r_H = 2 - 1.5B + O(B^2). \tag{47.6}$$

In general, the dependence of r_H on the parameter B is shown in Fig. 47.1. It is seen therein that as B decreases, r_H increases. This implies that MG increases the effective mass of BH, if compared with a Schwarzschild BH of the same r_H.

47.4 Hawking Temperature in a Modified Gravity

The Hawking temperature of a BH in GR can be defined as

$$T_H = \frac{\hbar\kappa}{2\pi c k_B}, \tag{47.7}$$

where κ is the surface gravity, which in GR for a nonrotating BH is given by $\ddot{r} = \kappa = c^4/4GM$. In general,

$$\kappa^2 = \frac{1}{4}\frac{s(r)'^2}{s(r)p(r)}, \tag{47.8}$$

with $c = k_B = 1$, which in our MG leads to (for $B \ll 0$)

$$\kappa = \frac{1}{r_H^2}\left(1 - \frac{2B}{r_H}\right). \tag{47.9}$$

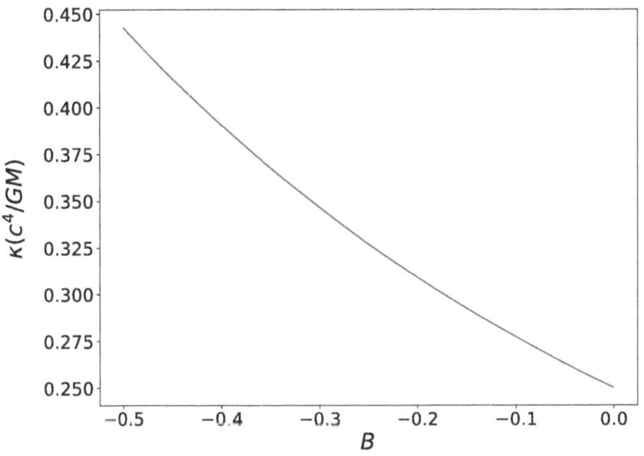

Fig. 47.2 Surface gravity as a function of B for a nonrotating BH

Therefore, while in GR, r_H and κ decrease with the increase of specific angular momentum and charge of the BH for a given mass, in MG, both of them increase with the decrease in B (i.e. magnitude of B increases which implies stronger modifications to the gravity), see Figs. 47.1 and 47.2. This implies that the temperature of the BH in MG increases.

Now we assume the BH to be a grey-body having the temperature and surface area respectively equal to the Hawking temperature and event horizon area of the BH. If the grey-body parameter is denoted by Γ, in GR the rate of evaporation of a BH is

$$Q_{GR} = -c^2 \frac{dM}{dt} = 4\pi \Gamma r_H^2 \sigma T_H^4, \tag{47.10}$$

where σ is the Stefan-Boltzmann constant. Thus, in GR the time taken for a BH of mass M to evaporate completely is

$$t_{GR} = 2.1 \times 10^{67} \text{yrs} \times \frac{1}{\Gamma} \left(\frac{M}{M_{\odot}} \right)^3. \tag{47.11}$$

Now in MG, the temperature and radius change. Therefore, we define the ratio of the temperatures in MG to GR as $\Theta = T_{MG}/T_{GR}$, and the corresponding radii ratio as $\rho = r_{\text{horizon,MG}}/r_{\text{horizon,GR}}$. Therefore, the rate of evaporation of a BH in MG can be written in terms of the same rate in GR as

$$\frac{Q_{MG}}{Q_{GR}} = \xi = \rho^2 \Theta^4. \tag{47.12}$$

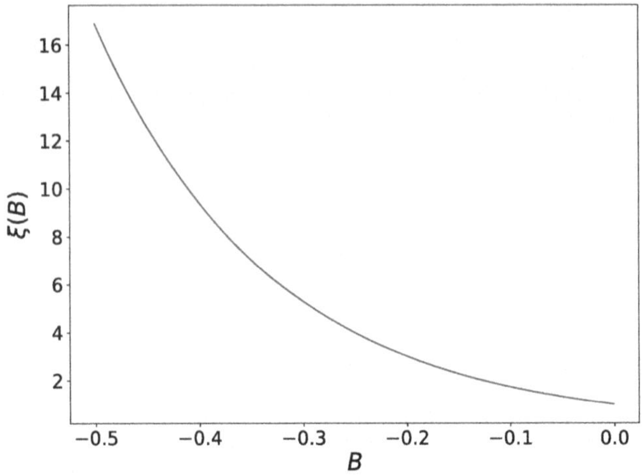

Fig. 47.3 The variation of ξ as a function of B

The variation of ξ with B is shown in Fig. 47.3. Therefore, the time of evaporation (t_{MG}) changes in MG w.r.to that in GR as

$$\frac{t_{MG}}{t_{GR}} = \frac{1}{\xi} = \frac{1}{\rho^2\Theta^4}, \tag{47.13}$$

which leads to the evaporation time of a BH of initial mass M in MG as

$$t_{MG} = 2.1 \times 10^{67}\text{yrs} \times \frac{1}{\Gamma\rho^2\Theta^4}\left(\frac{M}{M_\odot}\right)^3. \tag{47.14}$$

Since, both ρ and Θ are greater than unity, the time of evaporation of a nonrotating BH clearly decreases in MG compared to that in GR.

47.5 Minimum Survival Mass of Primordial Black Hole

In GR, the maximum initial mass of a PBH that evaporates completely in the current age of the universe can be computed easily based on the above discussion and greybody assumption, which is 2.5×10^{14} g, or $1.25 \times 10^{-19} M_\odot$ [13]. In MG, since the rate of evaporation is more than that of GR by a factor of ξ, the mass of PBH which will be evaporated by today is $\xi \times 2.5 \times 10^{14}$ g, according to the chosen MG theory. The dependence of the maximum evaporated mass of PBH on B is shown in Fig. 47.4. Depending on B, the mass of PBH that is expected to be completely evaporated by today can be even an order of magnitude higher than that predicted by GR, even in the given restricted range of B.

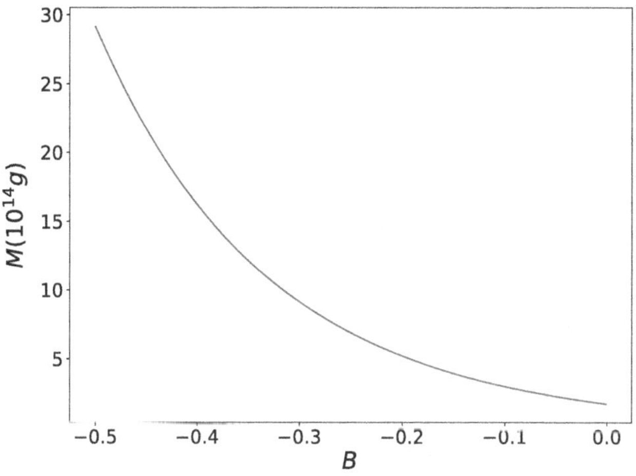

Fig. 47.4 Maximum initial PBH mass that completely evaporated today as a function of B

47.6 Implications of the Result

In GR, PBHs which have a mass much larger than 10^{15} g are practically unaffected by Hawking radiation over the age of the universe. Hence, PBHs of mass $\gtrsim 10^{15}$ g are considered to be a potential dark matter candidate which makes up to 26.8% of the critical density. On the other hand, PBHs would have formed much earlier than the end of the radiation-dominated era. Therefore, they would not be affected by the well-known BigBang Nucleosynthesis (BBN) constraint that the baryons can have at most 5% of the critical density, hence PBH contribution to critical density and dark matter would be non-baryonic. As Hawking radiation affects the survival of PBHs in MG, it plays an important role in constraining the contribution of PBHs to the dark matter, because massive PBHs are more constrained to contribute to dark matter [13]. For $B = -0.47$, the minimum mass of PBH that survives evaporation through Hawking radiation increases 10 folds, indicating a strong dependence of the minimal survival mass on the MG parameter B.

According to MG, a PBH can survive a lesser time. Therefore, we can expect that the faster evaporation in MG constrains on PBH as a potential candidate for the dark matter more strictly. This offers the scope of alternative theories, including axions and Weakly Interacting Massive Particles (WIMPs), to be better in a MG model, even though they are not yet discovered. Hence, the present work further widens the debate on the origin of dark matter. Although in these calculations, we have assumed that the age of the universe is independent of the gravity theory, this needs to be verified. Further, the rotating BHs would add up constraints, what needs to be explored in future.

References

1. Bird, S., et al. (2023). *Physics of the Dark Universe, 41*, 101231.
2. Hütsi, G., Raidal, M., Urrutia, J., Vaskonen, V., & Veermäe, H. (2023). *Physical Review D, 107*, 043502.
3. Ade, P. A. R., et al. (2016). *Astronomy & Astrophysics, 594*(13), A13.
4. Cooney, A., DeDeo, S., & Psaltis, D. (2010). *Physical Review D, 82*, 064033.
5. Das, U., & Mukhopadhyay, B. (2015). *Journal of Cosmology and Astroparticle Physics, 05*, 045.
6. Starobinsky, A. A. (1980). *Physics Letters B, 91*, 99.
7. Damour, T., & Esposito-Farése, G. (1993). *Physical Review Letters, 70*, 2220.
8. Arapoğlu, S., Deliduman, C., & Ekşi, K.Y. (2011). *Journal of Cosmology and Astroparticle Physics, 7*, 020.
9. Guerrero, M., Rubiera-Garcia, D., & Wojnar, A. (2022). *European Physical Journal C, 82*, 707.
10. Kalita, S., & Mukhopadhyay, B. (2019). *European Physical Journal C, 79*, 877.
11. Bhaumik, N., & Jain, R. K. (2020). *Journal of Cosmology and Astroparticle Physics, 01*, 037.
12. Whitt, B. (1984). *Physics Letters B, 145*, 176.
13. Carr, B., et al. (2021). *Reports on Progress in Physics, 84*, 116902.

Part XII
Astrochemistry and Astrobiology

Chapter 48
Deciphering Interstellar Ice Morphology: Atomistic Simulations Reveal the Complex Behavior of Ethanethiol

Jeet Majumdar, Shubhadeep Nag, Tejender S. Thakur, Subramanian Yashonath, Bhalamurugan Sivaraman, and Prabal K. Maiti

Abstract Ethanethiol (C_2H_5SH), a molecule detected in the interstellar medium (ISM), indicates the rich chemistry involving sulfur atoms. However, its behavior at low temperatures remains elusive, particularly the reported transition from an amorphous phase to a crystal. This study employs classical molecular dynamics (MD) simulations to reproduce the liquid-state properties of ethanethiol and to simulate the initial amorphous state of ethanethiol films deposited on a KBr substrate. The amorphous ethanethiol did not show spontaneous crystallization upon increasing temperature. Also, ethanethiol ice crystals exhibit melting behavior on KBr substrate at elevated temperatures. Our MD simulations of thin ice samples do not show any signature of reversible phase change. It will be interesting to continue this study with a thicker sample, which is beyond our current computational means. These findings underscore the complexity of icy mantle morphology on cold ISM dust grains.

J. Majumdar · S. Nag · P. K. Maiti (✉)
Department of Physics, Centre for Condensed Matter Theory, Indian Institute of Science, Bangalore, India
e-mail: jeetmajumdar@iisc.ac.in; maiti@iisc.ac.in

T. S. Thakur
Biochemistry and Structural Biology Division, CSIR—Central Research Drug Institute, Lucknow, India

S. Yashonath
Solid State and Structural Chemistry Unit, Indian Institute of Science, Bangalore, India

B. Sivaraman
Physical Research Laboratory, Navrangpura, Ahmedabad, India
e-mail: bhala@prl.res.in

S. Ghosh, A. R. Rao (eds.), *The Relativistic Universe: From Classical to Quantum*, Astrophysics and Space Science Proceedings 61,
https://doi.org/10.1007/978-3-031-90186-7_48

48.1 Introduction

The formation of ethanethiol in the ISM likely occurs through a combination of two main processes. In frigid temperatures, the icy dust grain surfaces provide a platform for gas-phase molecules to stick and react. Here, atomic hydrogen (H) could interact with ethanethiol's precursor, C_2H_5S radical, to form the molecule [1–3]. However, the efficiency of such reactions is bounded by the availability of free radicals that can react to form ethanethiol [3, 4]. The major supply of these free radicals comes from another process that occurs in high-energy environments, such as shocks caused by supernovae or stellar winds where the gas-phase reactions become more prominent. Several such reaction mechanisms have been proposed so far by Gorai et al. [3]. A potential route in this scenario involves the interaction of methanethiol with free-floating carbon atoms (C) and radicals, leading to the formation of ethanethiol [5]. Observations also suggest a significant depletion of sulfur on interstellar dust grains, potentially bound as S_8 or other allotropes [6]. Thus, shock-induced sputtering of these grains has the potential to liberate a substantial portion of sequestered sulfur, making it available for gas-phase reactions that could lead to the formation of ethanethiol and other sulfur-bearing species [5]. Both processes highlight the diverse chemical nature within the interstellar medium.

Studies suggest that ethanethiol can undergo reversible phase transitions between amorphous and crystalline phases with temperature [7]. This highlights the dynamic nature of interstellar ices and their potential role in complex molecule formation. The result of this laboratory-based experiment was more astonishing as it also showed the dependence on the thickness and deposition temperature on the amorphous to crystalline phase transition. The adsorbed ethanethiol layer on a potassium-bromide (KBr) substrate remains amorphous as the temperature was increased from 4 to 100 K. However, upon further heating to a temperature range between 110 and 120 K, the adsorbed phase undergoes a transition to a crystalline state. Beyond 125 K and up to 130 K, the study suggests a potential return to an amorphous state for the adsorbed phase. This entire observation was based on the IR spectra of the ethanethiol molecules.

Prior computational investigations of ethanethiol have primarily concentrated on the liquid phase and identification of its various vibrational modes [8–10]. The potential formation of ethanethiol within the icy mantles of cold interstellar regions suggests a need for further evaluation of current simulation studies to ensure their applicability to such environments. The recent detection of ethanethiol in the interstellar medium (ISM) highlights the paucity of readily available data on its crystal structure in the scientific literature.

Therefore, the main aim of this work is to have a microscopic understanding of the reversible phase changes as seen in the experiments. This first requires the prediction of the possible crystal structure, which is currently missing in the literature. We have first predicted a possible crystal structure of ethanethiol ice and used it to understand the thermodynamics of pure ethanethiol in the temperature range of 10–

140 K. Then, we used this model to study the properties of deposited ethanethiol molecules on KBr substrate as done in the experiment of Pavithraa et al. [7].

48.2 Method

In the absence of any reported crystal structure of ethanethiol, we performed a crystal structure prediction (CSP) for this molecule to use it as the initial configuration for our simulation. At first, a rigid CSP is performed with COMPASS force field (FF) [11]. The top 10 predicted structures were then taken for subsequent dispersion corrected (B3LYP-D) periodic-DFT-based energy re-ranking [12]. The final top-ranked structure was considered as the unit cell of the ethanethiol ice structure.

The CSP crystal structure has a triclinic (P-1) unit cell, as shown in Fig. 48.1a. The unit cell lengths are $a = 6.61$ Å, $b = 6.09$ Å, $c = 4.43$ Å, and the associated angles are $\alpha = 87.31°$, $\beta = 85.44°$, $\gamma = 74.18°$. To study the properties of bulk ethanethiol, an initial configuration of bulk ethanethiol ice was generated by replicating this unit cell in all three directions to construct a supercell ($5 \times 5 \times 5$) that consists of 250 molecules. A supercell of ($n_x \times n_y \times n_z$) denotes the number of replicas of unit cells placed along the x, y, and z directions respectively; thereby denoting different sizes of crystal. We used OPLS-AA FF parameters to describe the bonded interactions and non-bonded Lennard-Jones interactions [13]. The ESP charges from CSP calculation were reused for electrostatics. The initial bulk crystal was energy-minimized using the steepest descent method. This was followed by heating the system at a rate of 0.05 K ps^{-1}, and then performing NPT equilibration at temperatures 10–298 K at atmospheric pressure. To maintain constant temperature (T) and pressure (P) throughout the simulation, Berendsen thermostat and barostat were employed with temperature coupling constant of 1 ps and pressure coupling constant of 0.1 ps respectively [14, 15]. We used an integration time-step of 1 fs and used periodic boundary conditions (PBC) in all three directions. Short-range Lennard-Jones (LJ) and electrostatic interactions were treated with a cutoff distance of 10 Å with smooth switching from 8 Å. Long-range electrostatics were handled using the Particle Mesh Ewald (PME) method with fourth-order interpolation [16, 17]. The neighbour list was updated every 10 time-steps using the velocity Verlet algorithm. All the simulations were performed using GROMACS MD engine [18].

To study the reversible phase transition of ethanethiol as reported by Pavithraa et al. [7], we conducted deposition simulation, and also adsorption simulation of ethanethiol ice over KBr substrate. The 'vacuum deposition' of ethanethiol over KBr was carried out using PYTHINFILM—a python package that works in tandem with GROMACS [18, 19]. The deposition protocol is similar to a previous work from our group further details of the procedure can be found in reference [19, 20]. The potassium (K) and bromine (Br) atoms of KBr were parameterized using FF parameters developed by Barbosa et al. [21, 22].

Our study of adsorbed ethanethiol ice over KBr substrate was done by placing ethanethiol ice crystal of supercell size (6×6×6), (6×6×12), (6×6×18), (6×6×24) on the substrate. To place the ice at an appropriate distance, minimum energy criteria in orientation and distance were used. The minimum energy orientation is the same as that of the initial unit cell found by CSP, and the energetically minimum distance was 2.546 Å between the surface of the substrate and ice.

Density of states (DoS) was calculated using Two-phase thermodynamics (2PT) [23, 24]. For the 2PT calculations, NVT run of 20 ps was performed with high-frequency dumping of position, velocities, and energies of atoms at every 2 fs.

48.3 Results and Discussion

48.3.1 Bulk Ethanethiol

The low temperature stability of ethanethiol crystal structure was studied on a supercell of (5 × 5 × 5) dimension. A snapshot of the initial configuration system is given in Fig. 48.1b. As the temperature increases, the density of the system changes. The time evolution of this system density is plotted in Fig. 48.1c. It could be noted that as temperature crosses 120 K, the density drops from 1.178 g cm^{-3} to 1.139 g cm^{-3}, implying a phase change. The density of the new phase is lower

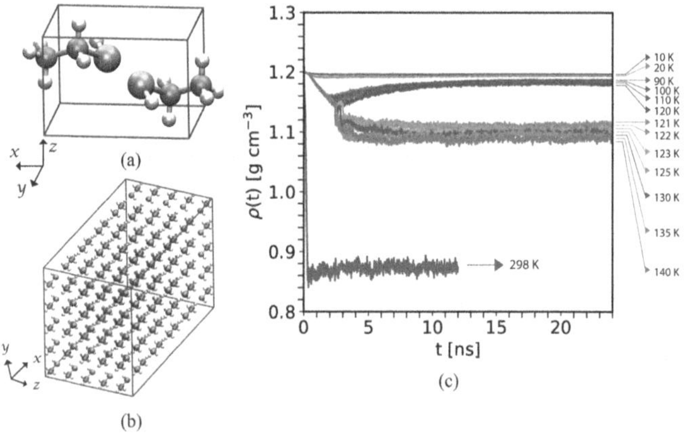

Fig. 48.1 (**a**) Predicted crystal structure for ethanethiol. Sulfur, carbon, and hydrogen atoms are shown as larger spheres, medium-sized spheres, and smaller spheres, respectively. Arrows pointing along the *x*-, *y*-, and *z*-axis indicate orientation. (**b**) Crystal supercell (5 × 5 × 5) consisting of 250 ethanethiol molecules after 24 ns simulation at 10 K. (**c**) The time evolution of the density of bulk ethanethiol crystal, heated from 10 to 140 K at a rate of 0.05 K ps^{-1} in ambient pressure. The density equilibration shows that ethanethiol crystal undergoes a phase transition at 120 K. For reference, the density of ethanethiol upon direct heating to room temperature (298 K) is also shown

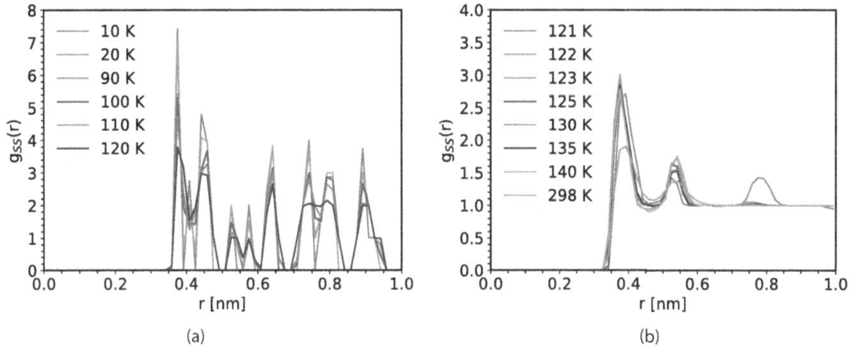

Fig. 48.2 RDF of bulk ethanethiol at various temperatures as the crystal sample was heated to higher temperatures to the range where it shows the first change in phase. The first occurrence of phase change happens at 121 K when the RDF reduces to only 3 peaks of smaller amplitudes

than the phase of the initial crystal, but it is much higher than the density of liquid ethanethiol at 298 K (shown in red in Fig. 48.1c). Experimentally, the density of ethanethiol at 298 K is known to be 0.833 g cm^{-3} [10, 25]. In our simulation, we get this density to be 0.872 g cm^{-3}. The phase change is also evident from the radial distribution function (RDF) as plotted in Fig. 48.2.

Another important quantity that is often used as a signature of crystal ethanethiol is the InfraRed (IR) spectra of S-H stretching vibration [7, 26, 27]. Literature suggests that, in the crystalline phase, S-H vibration mode gives rise to a sharp peak, which effectively captures the phase of the sample. We calculated DoS that can reveal the modes of vibrations in the IR frequency domain of ethanethiol spectra.

Figure 48.3 shows the full DoS of ethanethiol at 10 and 298 K. The major peaks in the DoS plot can be mapped to different modes of vibration and are marked in the liquid DoS plot of 298 K. The S-H peak can be identified as the highest non-overlapping peak at 2570 cm^{-1}. The amplitude of this peak can be visually distinguished between the solid and liquid phases. In the crystal phase at 10 K, the S-H peak amplitude is approximately an order of magnitude higher than its value in the liquid phase at 298 K.

Figure 48.4a shows this peak (enlarged view) as a function of temperature. Fitting with a Gaussian function allows us to measure the peak details. In the experiment, this S-H peak position is at 2528 cm^{-1} [7]. In our simulation, we get this peak at 2570 cm^{-1}. Therefore, the full width at half maxima (FWHM) of this peak is plotted in Fig. 48.4b which shows that at 121 K the FWHM shows a sudden increase by around 7 cm^{-1}. This quantity is related to experimentally obtained IR spectra with some parameter rescaling. Therefore, this confirms the phase change of ethanethiol from crystal to non-crystalline form at 121 K in our simulation.

In Fig. 48.4c, we have plotted the temperature-weighted entropy (TS) and enthalpy (H) of the system with temperature. The discontinuity in both thermodynamic quantities at 121 K implies a phase change. The rate of change of H and TS differs characteristically before and after the change of phase. At 121 K,

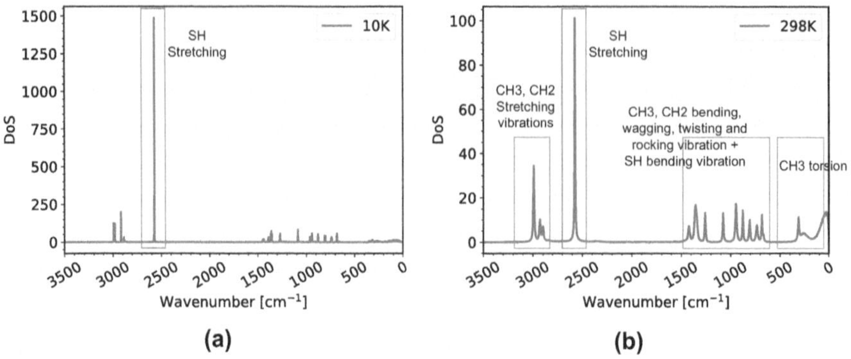

Fig. 48.3 DoS of ethanethiol molecules in (**a**) crystal phase at 10 K, and in (**b**) liquid phase at 298 K. The vibration corresponding to the major peaks are marked where the tallest peak corresponding to $2570\,cm^{-1}$ indicates the S-H mode of stretching vibration

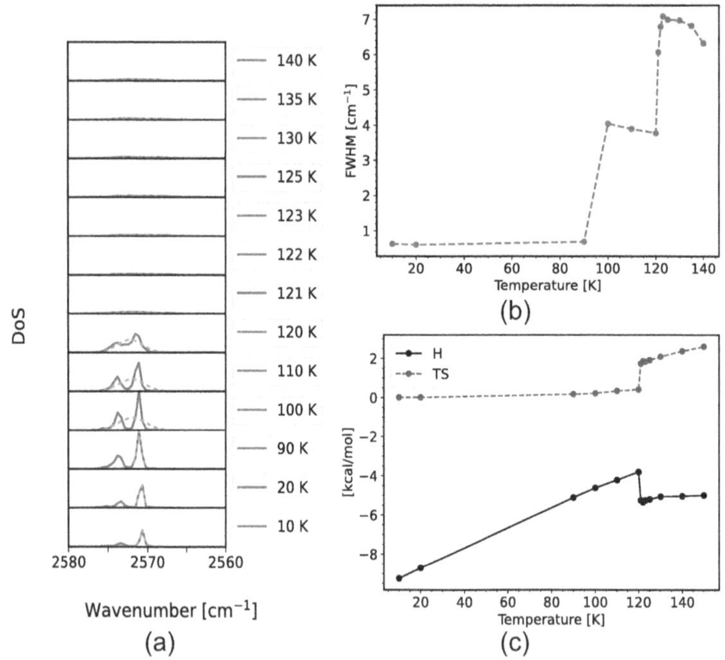

Fig. 48.4 (**a**) The spectra of S-H stretching vibration mode in the DoS of bulk ethanethiol as system was heated from low temperature crystal configuration at 10 K to higher temperatures. (**b**) The solid line shows the actual DoS plot and the broken line shows the Gaussian fitting. (**b**) The measured FWHM of the peak. (**c**) The enthalpy (solid line) and the temperature-weighted entropy (broken line) with temperature. All the plot captures the abrupt phase change of ethanethiol at 121 K

Table 48.1 Comparison of experimental properties of pure ethanethiol in literature vs. our simulations

Property	Exp.	Ref	This work
Density [g cm^{-3}]	0.833 (298 K)	[10]	0.872
Heat of fusion [kcal $mole^{-1}$]	1.19 (125.26 K)	[28]	1.44 (121 K)
Heat of vaporization [kcal $mole^{-1}$]	6.35 (298.15 K)	[10]	8.09 (298 K)
S-H stretching mode [cm^{-1}]	2528	[7]	2570

ΔH is 1.44 kcal mol^{-1} which can be compared to its experimental counterpart of 1.19 kcal mol^{-1} [28].

At 121 K, we calculate the jump in enthalpy, which gives us the heat of fusion as 1.44 kcal mol^{-1}. Again, at the liquid-gas phase boundary, the heat of vaporization can be computed using Eq. (48.1) as prescribed in [29]:

$$\Delta H_{vap}(T) = E_{gas}^{potential}(T) - E_{liquid}^{potential}(T) + RT \ . \tag{48.1}$$

Using Eq. (48.1) we get the heat of vaporization from our simulation at 298 K to be 8.09 kcal mol^{-1}. Both the heat of fusion and heat of vaporization can be compared with experimentally known values and are summarized in Table 48.1.

This study on bulk ethanethiol and its first phase transition temperature helps us understand the temperature range over which it can sustain crystallinity in our simulation. In the next sections, we use these results to understand the experimental observation of reversible phase transition in ethanethiol on substrate from our simulation perspective.

48.3.2 Deposited Ethanethiol on KBr Substrate

In the experiment, ethanethiol deposited on the KBr substrate showed reversible phase change with temperature. However, this phase change was observed only for the 'thick' samples ($\sim$200 μm) and not for the 'thin' samples ($\sim$200 nm). The thickness of the 'thick' sample is prohibitively large; therefore, through MD simulation, we probe into the 'thin' sample range.

Here, we deposit 5000 ethanethiol molecules on KBr substrate at 10 K. This substrate, of $4 \times 4\,nm^2$ area, has its atoms restrained to their initial coordinates with a force constant 1000 kJ mol^{-1} as shown in Fig. 48.5a. Molecules were released from a height of 6 nm and with an average velocity of 0.6 nm ps^{-1} towards the KBr surface. A vacuum length of 20 nm was kept from the insertion distance to the box height to prevent periodic interaction along the z-direction. The rate of heating, and equilibrium simulation settings were kept the same as that of pure bulk ethanethiol simulations described before. At each temperature, simulations were carried out for 100 ns.

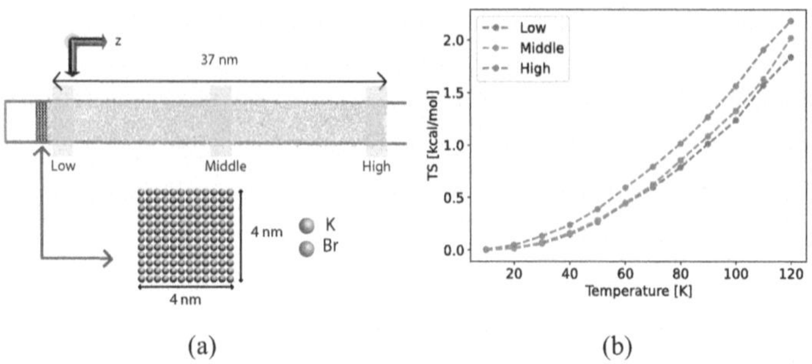

Fig. 48.5 (**a**) Deposited ethanethiol molecules over KBr substrate with marked regions that were analyzed independently. Deposited sample measure to 37 nm, and each of these regions are of 3 nm. There is also a 20 nm vacuum between the deposited sample and the box ceiling along z. (**b**) Temperature-weighted entropy of the deposited molecules in the respective regions as a function of temperature. Compared to 'Low' region molecules, the 'High' region molecules show a greater rate in the increase of disorderliness with temperature

These deposited molecules on the KBr substrate were initially found to be in an amorphous phase, as shown in Fig. 48.5a. This system, thus created, was heated and simulated at higher temperatures in NVT ensemble (till 120 K, the simulation temperature of crystal to non-crystalline transformation), which we then analyzed based on regions that are of equal size (3 nm): the 'Low', 'Middle', and 'High'. Figure 48.5a marks these regions of the system. The 'Low' region stands for the lowermost part of the deposited sample, which is in close proximity to the substrate atoms. The 'Middle' region is defined with a sufficient buffer zone separating it from the top vacuum and the bottom substrate. The 'High' region is at the top of the deposited sample, which is at the boundary of the vacuum region. In these individual regions, we analyze the system in tandem with experiments where crystallinity is detected through IR spectra.

We perform DoS calculations in these regions. The representative spectra for the 'Middle' and 'High' are plotted in Fig. 48.6a, b; the 'Low' region which showed similar behavior. Fitting Gaussian on the S-H peak we monitor the FWHM of this peak in the individual regions that is plotted in Fig. 48.6c, d. Compared with bulk FWHMs (given in Sect. 48.3.1), it can be confirmed that the deposited sample (at 10 K) is in amorphous phase. With increasing temperature, there is a declining trend of FWHM in all the regions of the sample. Therefore, deposited amorphous ethanethiol tends to achieve towards crystallinity with increasing temperature from 10 to 120 K. However, the total reduction of FWHM from 10 to 120 K is less than $\sim$20% which is far from a crystalline sample.

As a representative thermodynamic property, we calculated the TS of the molecules in these regions, and it is plotted in Fig. 48.5b. The plots show that for all the regions in the sample, entropy rises with temperature, and there is no abrupt jump that could be associated with phase change. We also see that the rate of increase in disorderliness with rising temperature is higher for the molecules in the 'High' region than the 'Low' region. Thus effectively, the thin sample does

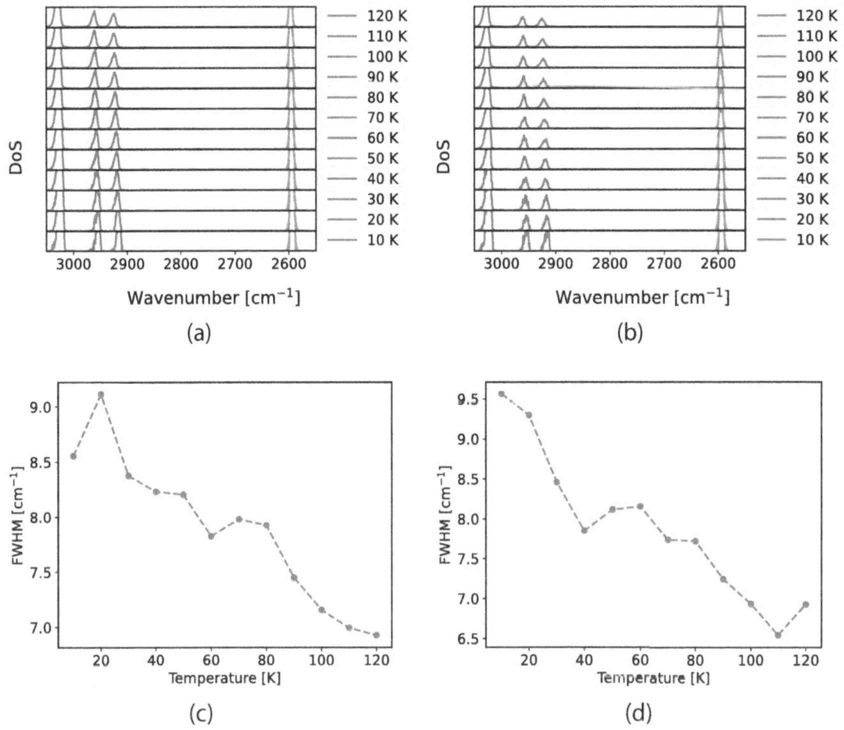

Fig. 48.6 Spectra of ethanethiol in (**a**) 'Middle' (**b**) 'High' region of the deposited sample as a function of temperature. The FWHM of the S-H peak as a function of temperature is given in (**c**) 'Middle' (**d**) 'High'

not crystallize under increasing temperature even below its first phase transition point. The results could also mean that an extremely slow crystallization process is taking place in our sample, a common bottleneck in most equilibrium simulations. Therefore, in the next section, we probe deeper with adsorbed ethanethiol ice over the KBr substrate.

48.3.3 Adsorbed Ethanethiol Ice on KBr Substrate

To understand the experimental observation in the case of 'thin' ice, here we adopted a different perspective. To facilitate this study, adsorbed ethanethiol crystals of varying thicknesses were placed on a KBr substrate to test its crystalline stability with time.

It consists of ice crystals of different supercells: $(6 \times 6 \times 6)$, $(6 \times 6 \times 12)$, $(6 \times 6 \times 18)$, and $(6 \times 6 \times 24)$ as described in the method section. Thus the initial height

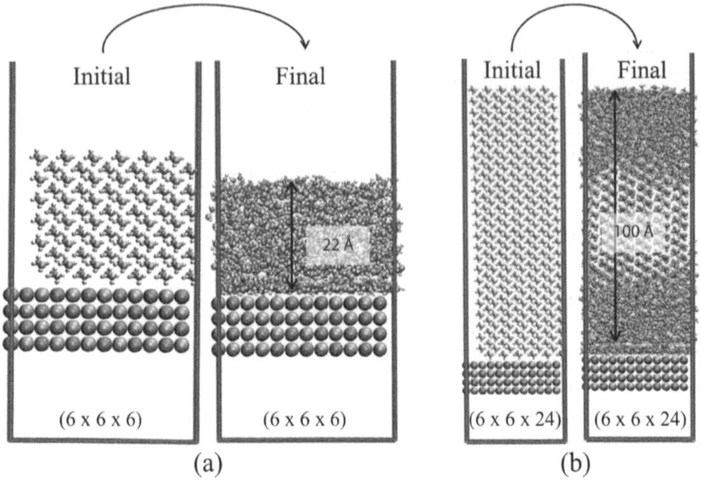

Fig. 48.7 Initial and final configuration of adsorbed ethanethiol on KBr substrate for sample size of 22 and 100 Å sample height. The final configuration is represents the system after 200 ns simulation run

of adsorbed ethanethiol crystal to begin with varied from a thickness of 26.58 Å for the smallest $(6 \times 6 \times 6)$ system to 106.32 Å for the largest $(6 \times 6 \times 24)$ system. Here, since our objective is to check the crystal stability close to our bulk phase-change temperature (121 K), we slowly heated (0.05 K ps^{-1}) all the systems to 110 K and monitored their structure with time.

Our simulations revealed that the smallest system structure $(6 \times 6 \times 6)$ exhibited melting within 5 ns. The $(6 \times 6 \times 12)$ system underwent melting after a simulation time of 150 ns. Interestingly, even after 200 ns of simulation, the $6 \times 6 \times 18$ and $(6 \times 6 \times 24)$ systems showed remarkable resilience against complete melting. The initial and final configurations of the smallest $(6 \times 6 \times 6)$ and largest $(6 \times 6 \times 24)$ systems are given in Fig. 48.7. The larger system exhibited a gradual reduction in their crystal regions, starting from the exposed surfaces of both the vacuum and substrate. These observations are evident from the snapshots in Fig. 48.7. The nature of melting for the remaining intermediate sample heights is similar.

48.4 Conclusion

In this study, we addressed the knowledge gap in low temperature bulk ethanethiol properties using MD simulation. Starting from bulk ethanethiol crystal structure prediction, we successfully reproduce the liquid-phase density of ethanethiol and deliver the heat-of-fusion and vaporization properties comparable to those of the experimental data. Furthermore, we used the S-H stretching vibration peak from the DoS spectra as a fingerprint to identify crystalline ethanethiol ice. It exhibited a significantly higher intensity in ordered crystals compared to room-temperature liquid molecules.

To understand the observed reversible phase change in thin ethanethiol films on a KBr substrate, we modeled deposited ethanethiol molecules. Consistent with experiments, the initial state resembled an amorphous phase, and no spontaneous crystallization occurred upon temperature increase. To address the possibility of a slow crystallization process, we investigated the stability of pre-formed ethanethiol ice films on KBr at elevated temperatures (where the amorphous-to-crystal transition was observed experimentally). For all simulated 'thin' film thicknesses (comparable to those used in the lab), the initial crystal structure degraded over time. The breakdown of crystallinity initiated at both the KBr interface and the vacuum-facing surface, with the propagation rate likely dependent on the distance from the edges.

This behavior highlights the possible role of hydrogen bonds in crystal formation, which is absent in ethanethiol due to the lack of highly electronegative atoms. In conclusion, our simulations support the absence of the reported amorphous-to-crystal transition in 'thin' ethanethiol films, aligning with the experimental observations.

Acknowledgments We acknowledge funding through IISc STC project, Grant No. ISTC/PHY/PKM/500. J.M. thanks MHRD, India for financial support.

References

1. Müller, H. S., Belloche, A., Xu, L.-H., Lees, R. M., Garrod, R. T., Walters, A., van Wijngaarden, J., Lewen, F., Schlemmer, S., & Menten, K. M. (2016). Exploring molecular complexity with alma (EMoCA): Alkanethiols and alkanols in sagittarius b2 (n2). *Astronomy & Astrophysics, 587*, A92.
2. Cuppen, H., Walsh, C., Lamberts, T., Semenov, D., Garrod, R., Penteado, E., & Ioppolo, S. (2017). Grain surface models and data for astrochemistry. *Space Science Reviews, 212*, 1–58.
3. Gorai, P., Das, A., Das, A., Sivaraman, B., Etim, E. E., & Chakrabarti, S. K. (2017). A search for interstellar monohydric thiols. *The Astrophysical Journal, 836*(1), 70.
4. Hudson, R. L., Gerakines, P. A., & Yarnall, Y. Y. (2024). Infrared spectroscopic and physical properties of methanol ices—reconciling the conflicting published band strengths of an important interstellar solid. *The Astrophysical Journal, 970*(2), 108.
5. Rodríguez-Almeida, L. F., Jiménez-Serra, I., Rivilla, V. M., Martín-Pintado, J., Zeng, S., Tercero, B., de Vicente, P., Colzi, L., Rico-Villas, F., Martín, S., et al. (2021). Thiols in the interstellar medium: First detection of HC (O) SH and confirmation of C2H5SH. *The Astrophysical Journal Letters, 912*(1), L11.
6. Shingledecker, C. N., Lamberts, T., Laas, J. C., Vasyunin, A., Herbst, E., Kästner, J., & Caselli, P. (2020). Efficient production of S8 in interstellar ices: The effects of cosmic-ray-driven radiation chemistry and nondiffusive bulk reactions. *The Astrophysical Journal, 888*(1), 52.
7. Pavithraa, S., Methikkalam, R., Gorai, P., Lo, J.-I., Das, A., Sekhar, B. R., Pradeep, T., Cheng, B.-M., Mason, N., & Sivaraman, B. (2017). Qualitative observation of reversible phase change in astrochemical ethanethiol ices using infrared spectroscopy. *Spectrochimica Acta Part A: Molecular and Biomolecular Spectroscopy, 178*, 166–170.

8. Miller, B. J., Yekutiel, M., Sodergren, A. H., Howard, D. L., Dunn, M. E., Vaida, V., & Kjaergaard, H. G. (2010). Overtone spectra of 2-mercaptoethanol and 1, 2-ethanedithiol. *The Journal of Physical Chemistry A, 114*(48), 12692–12700.
9. Phillips, M. D., Anderson, D. E., Nelson, J. C., Evans, H. T., Dutton, S. E., & Vasiliou, A. K. (2022). The infrared spectrum of matrix-isolated ethanethiol. In *Physical chemistry research at Undergraduate Institutions: Innovative and impactful approaches* (Vol. 1, pp. 85–98). ACS Publications.
10. Dodda, L. S., Vilseck, J. Z., Tirado-Rives, J., & Jorgensen, W. L. (2017). 1.14* cm1a-lbcc: localized bond-charge corrected cm1a charges for condensed-phase simulations. *The Journal of Physical Chemistry B, 121*(15), 3864–3870.
11. Sun, H. (1998). Compass: An ab initio force-field optimized for condensed-phase applications overview with details on alkane and benzene compounds. *The Journal of Physical Chemistry B, 102*(38), 7338–7364.
12. Grimme, S., Antony, J., Ehrlich, S., & Krieg, H. (2010). A consistent and accurate ab initio parameterization of density functional dispersion correction (DFT-D) for the 94 elements H-Pu. *The Journal of Chemical Physics, 132*(15).
13. Jorgensen, W. L., & Tirado-Rives, J. (2005). Potential energy functions for atomic-level simulations of water and organic and biomolecular systems. *Proceedings of the National Academy of Sciences, 102*(19), 6665–6670.
14. Bussi, G., Donadio, D., & Parrinello, M. (2007). Canonical sampling through velocity rescaling. *The Journal of Chemical Physics, 126*(1).
15. Berendsen, H. J., Postma, J. v., Van Gunsteren, W. F., DiNola, A., & Haak, J. R. (1984). Molecular dynamics with coupling to an external bath. *The Journal of Chemical Physics, 81*(8), 3684–3690.
16. Darden, T., York, D., & Pedersen L. (1993). Particle mesh ewald: An n log (n) method for Ewald sums in large systems. *The Journal of Chemical Physics, 98*(12), 10089–10092.
17. Essmann, U., Perera, L., Berkowitz, M. L., Darden, T., Lee, H., & Pedersen, L. G. (1995). A smooth particle mesh ewald method. *The Journal of Chemical Physics, 103*(19), 8577–8593.
18. Abraham, M. J., Murtola, T., Schulz, R., Páll, S., Smith, J. C., Hess, B., & Lindahl, E. (2015). Gromacs: High performance molecular simulations through multi-level parallelism from laptops to supercomputers. *SoftwareX, 1*, 19–25.
19. Stroet, M., Sanderson, S., Sanzogni, A. V., Nada, S., Lee, T., Caron, B., Mark, A. E., Burn, P. L. (2022). PyThinFilm: Automated molecular dynamics simulation protocols for the generation of thin film morphologies. *Journal of Chemical Information and Modeling, 63*(1), 2–8.
20. Ramachandran, R., Hazarika, A., Gupta, S., Nag, S., Meka, J., Thakur, T. S., Yashonath, S., Vishwakarma, G., Chou, S.-L., Wu, Y.-J., et al. (2024). Amorphous 1-propanol interstellar ice beyond its melting point. *Monthly Notices of the Royal Astronomical Society, 530*(1), 1027–1034.
21. Fuentes-Azcatl, R., & Barbosa, M. C. (2018). Potassium bromide, kbr/ε: New force field. *Physica A: Statistical Mechanics and Its Applications, 491*, 480–489.
22. Nag, S., Majumdar, J., Sivaraman, B., Yashonath, S., & Maiti, P. K. (2023). Influence of the substrate on the density and infrared spectra of the adsorbed methanol ice of different thicknesses using molecular dynamics simulation. *Monthly Notices of the Royal Astronomical Society, 522*(3), 3656–3664.
23. Lin, S.-T., Blanco, M., & Goddard III, W. A. (2003). The two-phase model for calculating thermodynamic properties of liquids from molecular dynamics: Validation for the phase diagram of Lennard-Jones fluids. *The Journal of Chemical Physics, 119*(22), 11792–11805.
24. Lin, S.-T., Maiti, P. K., & Goddard III, W. A. (2010). Two-phase thermodynamic model for efficient and accurate absolute entropy of water from molecular dynamics simulations. *The Journal of Physical Chemistry B, 114*(24), 8191–8198.
25. Haines, W. E., Helm, R. V., Cook, G. L., & Ball, J. S. (1956). Purification and properties of ten organic sulfur compounds-second series. *The Journal of Physical Chemistry, 60*(5), 549–555.
26. Wolff, H., & Szydlowski, J. (1985). Vibrational spectra and rotational isomerism of ethanethiol and ethanethiol-d 1. *Canadian Journal of Chemistry, 63*(7), 1708–1712.

27. Hudson, R. L., & Gerakines, P. A. (2018). Infrared spectra and interstellar sulfur: New laboratory results for H2S and four malodorous thiol ices. *The Astrophysical Journal, 867*(2), 138.
28. McCullough, J., Scott, D., Finke, H., Gross, M., Williamson, K., Pennington, R., Waddington, G., & Huffman, H. (1952). Ethanethiol (ethyl mercaptan): Thermodynamic properties in the solid, liquid and vapor states. Thermodynamic functions to 1000° k. *Journal of the American Chemical Society, 74*(11), 2801–2804.
29. Wang, J., & Hou, T. (2011). Application of molecular dynamics simulations in molecular property prediction. 1. Density and heat of vaporization. *Journal of Chemical Theory and Computation, 7*(7), 2151–2165.

Chapter 49
Representation of Continuum Equations in Physical Components for Arbitrary Curved Surfaces Appearing in Biophysical Modeling

Sujit Kumar Nath

Abstract Continuum equations are ubiquitous in physical modeling of elastic, viscous, and viscoelastic systems. The equations of continuum mechanics take nontrivial forms on curved surfaces. Although the curved surface formulation of the continuum equations are derived in many excellent references available in the literature, they are not readily usable for solving physical problems due to the covariant, contravariant or mixed nature of the stress and strain tensors in the equations. We present the continuum equations in terms of physical components for general curved surfaces. This general formulation of the continuum equations can be used readily for modeling physical problems on arbitrary curved surfaces which appear frequently in biological cells and tissues as well as in engineering systems. We demonstrate this with the help of some examples.

49.1 Introduction

Equations of continuum mechanics are essential tools for modeling a broad range of physical systems, starting from micro/nano-scale biological systems of cells [1, 2], to macro-scale ecological systems [3, 4] to large scale (parsec) astrophysical systems [5–8]. However, the physical systems under consideration in various contexts are not always planar and not suitable for representing in a Cartesian coordinate system. Even in some cases, such as fluid flow on curved surfaces, it is necessary to

S. K. Nath (✉)
Turing Center for Living Systems, Aix Marseille Univ, Université de Toulon, CNRS, CPT (UMR 7332), Marseille, France

Division of Cardiovascular Sciences, Faculty of Biology, Medicine and Health, The University of Manchester, Manchester, UK

School of Computing and Faculty of Biological Sciences, University of Leeds, Leeds, UK
e-mail: sujit-kumar.nath@univ-amu.fr

S. Ghosh, A. R. Rao (eds.), *The Relativistic Universe: From Classical to Quantum*, Astrophysics and Space Science Proceedings 61,
https://doi.org/10.1007/978-3-031-90186-7_49

represent the continuum equations in a way that takes care of the geometry of the curved surface. There is a plethora of excellent references in the literature which derives the continuum equations in curved geometry [9–12]. However, most of these literatures leave the final equations with stress, strain, displacement, velocity etc. represented in covariant, contravariant or mixed tensor components. As a result, these equations are not readily usable for physical modeling purpose as the dimensions of covariant, contravariant or mixed components of stress, strain, velocity, displacement etc. do not always match with their corresponding physical dimensions.

In this chapter we present continuum mechanical equations in physical components for arbitrary curved surfaces by consolidating previous seminal works by Sokolnikoff, Truesdell, McConnell, Altman and Oliveira etc. [9, 12–16]. There are extensive amount of references on physical coordinates and continuum equations on arbitrary curved surfaces, separately. However, references for continuum equations in physical components of arbitrary curved surfaces are very rare. As a result, presenting the correct form of continuum equations for solving physical problems on curved surfaces is very important and essential. In this chapter we present the continuum mechanical equations in terms of the physical components of vectors and tensors. The main contribution of this chapter is to summarize the general formulation of continuum equations which are immediately applicable in the modeling of physical systems in arbitrary curved geometry without any dimensional ambiguity.

49.2 The Concept of Physical Components of Vectors and Tensors

Intuitively, the physical components of any vector or tensor are the components whose dimensions are the same as their dimensions while measured during any physical/experimental measurement. The continuum mechanical equations are generally represented in covariant, contravariant or mixed tensor notations when the equations are derived for arbitrary curved surfaces. Therefore, these equations are not readily usable for physical modeling as the dimensions of the covariant, contravariant, or mixed tensors do not match with their physically measured counterparts. Hence, it is important to represent these equations in their correct physical components which can be used for physical modeling purpose, being immune to any dimensional mismatch.

49.2.1 Physical Components of Vectors

Let $\mathbf{v} = v^i \mathbf{b}_i = v_i \mathbf{b}^i$ be the representation of a vector $\mathbf{v}$ in its covariant and contravariant bases $\{\mathbf{b}_i\}$ and $\{\mathbf{b}^i\}$ with components v^i and v_i respectively. To demonstrate the problem at hand with an example, let us consider the dynamics of a particle in plane polar coordinate. The parameterization of the 2D-plane in polar coordinate is given by $\mathbf{P} = (r\cos\theta, r\sin\theta)$, where r is the radial distance of the point $\mathbf{P}$ from the origin and θ is the angular coordinate. In this coordinate system the covariant basis vectors are given by $\mathbf{b}_r = (\cos\theta, \sin\theta)$, $\mathbf{b}_\theta = (-r\sin\theta, r\cos\theta)$, and the contravariant basis vectors are $\mathbf{b}^r = (\cos\theta, \sin\theta)$, $\mathbf{b}^\theta = \left(-\frac{\sin\theta}{r}, \frac{\cos\theta}{r}\right)$. The corresponding covariant and contravariant components of the acceleration $\mathbf{a} = a^i\mathbf{b}_i = a_i\mathbf{b}^i$, $i \in \{r, \theta\}$ can be written as [13]

$$a^r = \frac{d^2r}{dt^2} - r\left(\frac{d\theta}{dt}\right)^2, \quad a^\theta = \frac{d^2\theta}{dt^2} + \frac{2}{r}\frac{dr}{dt}\frac{d\theta}{dt}, \tag{49.1}$$

$$a_r = \frac{d^2r}{dt^2} - r\left(\frac{d\theta}{dt}\right)^2, \quad a_\theta = r^2\frac{d^2\theta}{dt^2} + 2r\frac{dr}{dt}\frac{d\theta}{dt}. \tag{49.2}$$

We see in Eqs. (49.1) and (49.2) that the dimensions of a^r, a^θ, a_r and a_θ are LT^{-2}, T^{-2}, LT^{-2} and $\mathrm{L}^2\mathrm{T}^{-2}$, respectively. Clearly, not all the dimensions are same as the dimension of acceleration. Even $\dim[a^\theta]$ and $\dim[a_\theta]$ also differ from each other. It is to be noted that neither the covariant (a_θ) nor the contravariant component (a^θ), is same as the physically realizable θ-component of the acceleration, which is $r\frac{d^2\theta}{dt^2} + 2\frac{dr}{dt}\frac{d\theta}{dt}$, as derived from the first principles in many standard textbooks [17, 18]. It is also to be noted that although the geometry here is planar, one needs to distinguish the covariant, contravariant, and the physical components since the metric tensor for polar coordinate system is not constant. Therefore, the question here is "what is the link between the covariant, contravariant, and the physically realizable components of acceleration?".

Similar problems are encountered when we solve continuum mechanics problems in curvilinear coordinates. As an example, we consider the case of plane elastic problems in polar coordinate. There are three stress and three strain components, which are (rr), $(r\theta)$, and $(\theta\theta)$-components (Fig. 49.1 shows the geometrical representation of the stress components). Now, the question is whether these components are covariant, contravariant, or of any other kind, and how to write the equation of motion in correct form in terms of these physically realizable stress components $\tau(rr)$, $\tau(r\theta)$and $\tau(\theta\theta)$. One way to write the equation of motion in correct form is to derive the equations from the first principles using the force balance equation. However, this method sometimes takes enormous effort, and for complicated coordinate systems this task turns out to be extremely challenging and often unrealistic. Another way is to use the standard form of equations available in the literature in covariant or contravariant components for arbitrary coordinate systems and convert them accordingly in terms of physical components. However, if we just directly plugin the physically realizable stress components $\tau(rr)$, $\tau(r\theta)$and $\tau(\theta\theta)$ in the equation of motion, that gives rise to erroneous form of equation for plane elastic problems in polar coordinate (or generally, in any curvilinear coordinate).

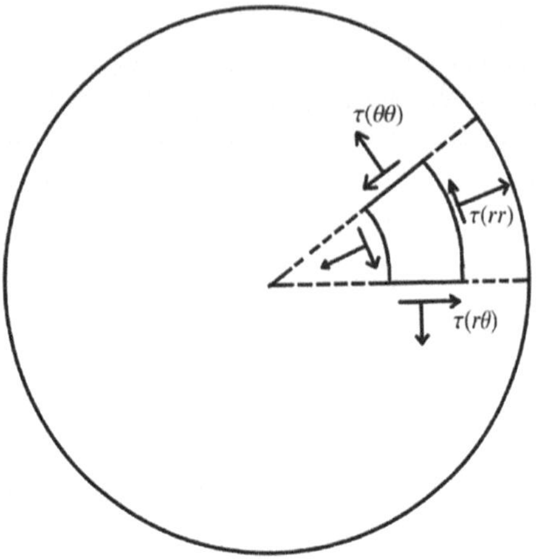

Fig. 49.1 Physical representation of plane-stress components $\tau(rr), \tau(r\theta), \tau(\theta\theta)$ in plane-polar coordinate. $\tau(rr)$ acts radially, $\tau(\theta\theta)$ acts in the positive θ-direction, and $\tau(r\theta)$ is the shear stress

To demonstrate this problem, let us consider the standard form of the equation of motion for an elastic body in contravariant components,

$$\nabla_j \tau^{ij} + F^i = \rho\, a^i, \tag{49.3}$$

which, after expanding the covariant derivative, gives

$$\frac{\partial \tau^{ij}}{\partial x^j} + \Gamma^i_{j\alpha}\tau^{\alpha j} + \Gamma^j_{j\alpha}\tau^{i\alpha} + F^i = \rho\, a^i, \tag{49.4}$$

where x^is are the coordinate variables, τ^{ij}s are the stress components, Γ^i_{jk}s are the Christoffel symbols, F^is are the force components, and a^is are the acceleration components. If we now try to obtain the equation of motion in terms of physically realizable components in plane polar coordinate, by replacing the contravariant components of stress tensor τ^{ij}, in Eq. (49.4), by $\tau(rr), \tau(r\theta), \tau(\theta\theta)$ etc., the resulting equations turn out to be erroneous (i.e., they do not match with the equation of motion derived from the first principles). Quoting from Truesdell's paper [14], he writes "In 1944 I first had the puzzling experience of getting an apparently wrong form of the equations of elasticity in polar coordinates by specializing the perfectly correct general tensor equations to the case at hand". In Sect. 49.3 we show how to obtain the correct form of the equation of motion in terms of the physically realizable components, directly from Eq. (49.4).

To circumvent this problem, it is necessary to represent the vectors and tensors in their physical components while modeling physical systems. In this purpose, it is useful to consider the unit vectors

$$\mathbf{e}_i = \frac{1}{\sqrt{g_{ii}}}\,\mathbf{b}_i, \quad \mathbf{f}^i = \frac{1}{\sqrt{g^{ii}}}\,\mathbf{b}^i, \tag{49.5}$$

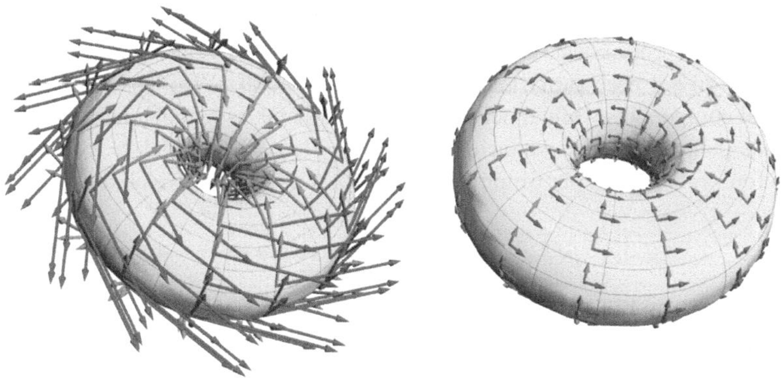

Fig. 49.2 Basis vectors on the surface of a torus parameterized by $\mathbf{r} = [(a + b\cos\theta)\cos\phi, (a + b\cos\theta)\sin\phi, b\sin\theta]$, where a and b are the radii of the major and minor circles respectively, and θ, ϕ are the angular parameters. Left panel shows the covariant basis vectors $\mathbf{b}_\phi$ (red arrows) and $\mathbf{b}_\theta$ (green arrows) without normalization. Right panel shows the normalized basis vectors $\mathbf{e}_\phi = \mathbf{b}_\phi/\sqrt{g_{\phi\phi}}$ (red) and $\mathbf{e}_\theta = \mathbf{b}_\theta/\sqrt{g_{\theta\theta}}$ (green) after normalization (Color figure online)

and their duals

$$\mathbf{e}^i = \sqrt{g_{ii}}\,\mathbf{b}^i, \quad \mathbf{f}_i = \sqrt{g^{ii}}\,\mathbf{b}_i \tag{49.6}$$

to construct the bases, instead of using the covariant or contravariant bases directly, where $g_{ij} = \mathbf{b}_i \cdot \mathbf{b}_j$ and $g^{ij} = \mathbf{b}^i \cdot \mathbf{b}^j$ for any i and j. The covariant basis vectors and their normalized counterparts are shown for an example case of a 2-dimensional toroidal surface in Fig. 49.2. Note that the dual basis vectors in Eq. (49.6) are not necessarily unit vectors in general curvilinear coordinates (although they are unit vectors in orthogonal curvilinear coordinates). Any arbitrary vector $\mathbf{v}$ can now be written as

$$\mathbf{v} = \left[v^i\right]_{\mathcal{E}} \mathbf{e}_i = \left[v_i\right]_{\mathcal{E}^*} \mathbf{e}^i = \left[v^i\right]_{\mathcal{F}} \mathbf{f}_i = \left[v_i\right]_{\mathcal{F}^*} \mathbf{f}^i, \tag{49.7}$$

where the notations $\mathcal{E}, \mathcal{E}^*, \mathcal{F}, \mathcal{F}^*$ denote the bases $\{\mathbf{e}_i\}$, $\{\mathbf{e}^i\}$, $\{\mathbf{f}_i\}$ and $\{\mathbf{f}^i\}$ respectively, and the components $\left[v^i\right]_\alpha$, $\alpha \in \{\mathcal{E}, \mathcal{F}\}$, and $\left[v_i\right]_\alpha$, $\alpha \in \{\mathcal{E}^*, \mathcal{F}^*\}$ in Eq. (49.7) are called the physical components of $\mathbf{v}$ in representative bases. Therefore, we obtain the relation between the physical components and the covariant or contravariant components as

$$\left[v^i\right]_{\mathcal{E}} = \sqrt{g_{ii}}\, v^i, \quad \left[v^i\right]_{\mathcal{F}} = \frac{1}{\sqrt{g^{ii}}}\, v^i, \quad \left[v_i\right]_{\mathcal{E}^*} = \frac{1}{\sqrt{g_{ii}}}\, v_i, \quad \left[v_i\right]_{\mathcal{F}^*} = \sqrt{g^{ii}}\, v_i. \tag{49.8}$$

Note that in the above formulae in Eq. (49.8) the right hand sides (RHS) are not summed over the repeated index i. Since for orthogonal curvilinear coordinates

$g_{ii} = 1/g^{ii}$, it is clear from Eq. (49.8) that $[v^i]_{\mathcal{E}} = [v^i]_{\mathcal{F}}$ and $[v_i]_{\mathcal{E}^*} = [v_i]_{\mathcal{F}^*}$ in orthogonal curvilinear coordinates. However, for oblique coordinates these equalities do not necessarily hold. Therefore, it is necessary to choose the correct physical components while modeling physical systems using continuum equations. In plane polar coordinate we have $g_{rr} = g^{rr} = 1$, $g_{\theta\theta} = r^2$ and $= g^{\theta\theta} = 1/r^2$. Therefore, considering our example in Eqs. (49.1) and (49.2), we obtain

$$[a^r]_{\mathcal{E}} = \sqrt{g_{rr}}\, a^r = a^r,\ [a^\theta]_{\mathcal{E}} = \sqrt{g_{\theta\theta}}\, a^\theta = r\, a^\theta, \tag{49.9}$$

$$[a_r]_{\mathcal{E}^*} = \frac{1}{\sqrt{g_{rr}}}\, a_r = a_r,\ [a_\theta]_{\mathcal{E}^*} = \frac{1}{\sqrt{g_{\theta\theta}}}\, a_\theta = \frac{a_\theta}{r}. \tag{49.10}$$

It is now evident that all the physical components $[a^r]_{\mathcal{E}}$, $[a^\theta]_{\mathcal{E}}$, $[a_r]_{\mathcal{E}^*}$, and $[a_\theta]_{\mathcal{E}^*}$ have the dimension of acceleration. Also the same is true for the physical components $[a^r]_{\mathcal{F}}$, $[a^\theta]_{\mathcal{F}}$, $[a_r]_{\mathcal{F}^*}$, and $[a_\theta]_{\mathcal{F}^*}$. In the next section we show that for a general tensor of any arbitrary order all the physical components carry the same physical dimension.

49.2.2 Physical Components of Tensors

We shall discuss here the physical components of second order tensors only. The case of tensors of any arbitrary order can easily be generalized from this discussion. Let $\mathbb{T} = \tau^{ij}\mathbf{b}_i \otimes \mathbf{b}_j = \tau_{ij}\mathbf{b}^i \otimes \mathbf{b}^j = \tau^i_{.j}\mathbf{b}_i \otimes \mathbf{b}^j = \tau_i^{.j}\mathbf{b}^i \otimes \mathbf{b}_j$ be a second order tensor, represented in the covariant, contravariant, and mixed bases. We can represent the physical components in the same way we did for vectors by representing the tensor components in unit bases. Therefore, the 16 physical components of the tensor $\mathbb{T}$ are given by

$$[\tau^{ij}]_{\mathcal{E}\otimes\mathcal{E}} = \sqrt{g_{ii}g_{jj}}\,\tau^{ij},\ [\tau^{ij}]_{\mathcal{E}\otimes\mathcal{F}} = \sqrt{\frac{g_{ii}}{g^{jj}}}\,\tau^{ij},$$

$$[\tau^{ij}]_{\mathcal{F}\otimes\mathcal{E}} = \sqrt{\frac{g_{jj}}{g^{ii}}}\,\tau^{ij},\ [\tau^{ij}]_{\mathcal{F}\otimes\mathcal{F}} = \frac{1}{\sqrt{g^{ii}g^{jj}}}\,\tau^{ij},$$

$$[\tau^i_{.j}]_{\mathcal{E}\otimes\mathcal{E}^*} = \sqrt{\frac{g_{ii}}{g_{jj}}}\,\tau^i_{.j},\ [\tau^i_{.j}]_{\mathcal{E}\otimes\mathcal{F}^*} = \sqrt{g_{ii}g^{jj}}\,\tau^i_{.j},$$

$$[\tau^i_{.j}]_{\mathcal{F}\otimes\mathcal{E}^*} = \frac{1}{\sqrt{g^{ii}g_{jj}}}\,\tau^i_{.j},\ [\tau^i_{.j}]_{\mathcal{F}\otimes\mathcal{F}^*} = \sqrt{\frac{g^{jj}}{g^{ii}}}\,\tau^i_{.j},$$

$$[\tau_i^{.j}]_{\mathcal{E}^*\otimes\mathcal{E}} = \sqrt{\frac{g_{jj}}{g_{ii}}}\,\tau_i^{.j},\ [\tau_i^{.j}]_{\mathcal{E}^*\otimes\mathcal{F}} = \frac{1}{\sqrt{g_{ii}g^{jj}}}\,\tau_i^{.j},$$

$$\left[\tau_i^{\cdot j}\right]_{\mathcal{F}^*\otimes\mathcal{E}} = \sqrt{g^{ii}g_{jj}}\;\tau_i^{\cdot j},\;\left[\tau_i^{\cdot j}\right]_{\mathcal{F}^*\otimes\mathcal{F}} = \sqrt{\frac{g^{ii}}{g^{jj}}}\;\tau_i^{\cdot j},$$

$$\left[\tau_{ij}\right]_{\mathcal{E}^*\otimes\mathcal{E}^*} = \frac{1}{\sqrt{g_{ii}g_{jj}}}\,\tau_{ij},\;\left[\tau_{ij}\right]_{\mathcal{E}^*\otimes\mathcal{F}^*} = \sqrt{\frac{g^{jj}}{g_{ii}}}\,\tau_{ij},$$

$$\left[\tau_{ij}\right]_{\mathcal{F}^*\otimes\mathcal{E}^*} = \sqrt{\frac{g^{ii}}{g_{jj}}}\,\tau_{ij},\;\left[\tau_{ij}\right]_{\mathcal{F}^*\otimes\mathcal{F}^*} = \sqrt{g^{ii}g^{jj}}\,\tau_{ij}. \tag{49.11}$$

It can be shown that all the physical components of a tensor have the same dimension which is same as their dimension in Cartesian coordinate [13]. However, the method opted in [13] is not always applicable for tensor fields on surfaces of codimension one, as there may not always exist a suitable Cartesian coordinate for representing a tensor field on such a surface. Here we present an alternative way to prove that all the physical components of a tensor have the same dimension with the help of intrinsic geometry of the surface. We again demonstrate this with the help of second order tensors, which can be easily generalized for tensors of arbitrary order.

We first derive a few results relating the physical dimensions of the covariant and contravariant metric tensor and vector components. Considering the squared arc length

$$ds^2 = g_{ij}\,dx^i dx^j = g^{ij}\,dx_i dx_j = dx^i dx_i\;, \tag{49.12}$$

we obtain

$$\dim\left[g_{ij}\right] = \frac{L^2}{\dim\left[dx^i\right]\dim\left[dx^j\right]}, \tag{49.13}$$

$$\dim\left[g^{ij}\right] = \frac{L^2}{\dim\left[dx_i\right]\dim\left[dx_j\right]}, \tag{49.14}$$

$$\dim\left[dx^i\right] = \frac{L^2}{\dim\left[dx_i\right]}, \tag{49.15}$$

where L denotes the dimension of arc length ds in Eq. (49.12). Substituting the $\dim\left[dx^i\right]$, $\dim\left[dx^j\right]$ from Eq. (49.15) to Eq. (49.13) and comparing with Eq. (49.14), we obtain

$$\dim\left[g_{ij}\right] = \frac{1}{\dim\left[g^{ij}\right]}. \tag{49.16}$$

Again, from Eq. (49.12) we see that

$$\dim\left[\sqrt{g_{ii}}\right] = \frac{L}{\dim\left[dx^i\right]}, \tag{49.17}$$

and

$$\dim\left[\sqrt{g^{ii}}\right] = \frac{L}{\dim\left[dx_i\right]}. \tag{49.18}$$

Therefore, Eqs. (49.13) and (49.14) gives

$$\dim\left[g_{ij}\right] = \dim\left[\sqrt{g_{ii}}\right] \dim\left[\sqrt{g_{jj}}\right], \tag{49.19}$$

and

$$\dim\left[g^{ij}\right] = \dim\left[\sqrt{g^{ii}}\right] \dim\left[\sqrt{g^{jj}}\right]. \tag{49.20}$$

We now use the results Eqs. (49.12)–(49.20) for proving the equality of the physical dimensions of all the physical components of a tensor. First we show that all the physical components of a tensor in a fixed basis have the same physical dimension. For this purpose, without loss of generality, we fix our basis to be $\mathcal{E} \otimes \mathcal{F}^*$. Let us now consider the scalar value $\tau^{i}_{\cdot j}\, dx_i dx^j$. For any fixed values of i and j, such as $\tau^{p}_{\cdot q}\, dx_p dx^q$, when $i = p$, $j = q$ fixed, and $\tau^{m}_{\cdot n}\, dx_m dx^n$, when $i = m$, $j = n$ fixed, we have

$$\dim\left[\tau^{p}_{\cdot q}\, dx_p dx^q\right] = \dim\left[\tau^{m}_{\cdot n}\, dx_m dx^n\right]. \tag{49.21}$$

The dimensions of the both sides of Eq. (49.21) match as they are the typical constitutive terms of the scalar value of $\mathbb{T}(\mathbf{u}, \mathbf{v})$, i.e., when $\mathbb{T}$ is applied on a pair of vectors $(\mathbf{u}, \mathbf{v})$. Substituting the values of $\dim\left[dx_p\right]$, $\dim\left[dx^q\right]$, $\dim\left[dx_m\right]$, $\dim\left[dx^n\right]$ in Eq. (49.21) using Eqs. (49.17) and (49.18), we obtain

$$\dim\left[\tau^{p}_{\cdot q}\right] \frac{L}{\dim\left[\sqrt{g^{pp}}\right]} \frac{L}{\dim\left[\sqrt{g_{qq}}\right]} = \dim\left[\tau^{m}_{\cdot n}\right] \frac{L}{\dim\left[\sqrt{g^{mm}}\right]} \frac{L}{\dim\left[\sqrt{g_{nn}}\right]}. \tag{49.22}$$

Now, using Eqs. (49.16) and (49.11), we obtain from Eq. (49.22)

$$\begin{aligned} &\dim\left[\tau^{p}_{\cdot q}\right] \dim\left[\sqrt{g_{pp}}\right] \dim\left[\sqrt{g^{qq}}\right] = \dim\left[\tau^{m}_{\cdot n}\right] \dim\left[\sqrt{g_{mm}}\right] \dim\left[\sqrt{g^{nn}}\right], \\ \text{or,}\quad &\dim\left[\left[\tau^{p}_{\cdot q}\right]_{\mathcal{E}\otimes\mathcal{F}^*}\right] = \dim\left[\left[\tau^{m}_{\cdot n}\right]_{\mathcal{E}\otimes\mathcal{F}^*}\right]. \end{aligned} \tag{49.23}$$

We now show the dimensional equality of the physical tensor components represented in different bases such as $\mathcal{E} \otimes \mathcal{F}^*$ and $\mathcal{E} \otimes \mathcal{E}$. From the relation

$$\dim\left[\tau^{p}_{\cdot q}\, dx_p dx^q\right] = \dim\left[\tau^{mn}\, dx_m dx_n\right], \tag{49.24}$$

and following the similar calculation as in Eq. (49.22), and using Eqs. (49.16) and (49.11), we obtain

$$\dim\left[\tau^{p}_{\cdot q}\right]\dim\left[\sqrt{g_{pp}}\right]\dim\left[\sqrt{g^{qq}}\right]=\dim\left[\tau^{mn}\right]\dim\left[\sqrt{g_{mm}}\right]\dim\left[\sqrt{g_{nn}}\right],$$

$$\text{or,}\quad \dim\left[\left[\tau^{p}_{\cdot q}\right]_{\mathcal{E}\otimes\mathcal{F}^*}\right]=\dim\left[\left[\tau^{mn}\right]_{\mathcal{E}\otimes\mathcal{E}}\right]. \tag{49.25}$$

Similarly, all other physical components (exhaustively listed in Eq. (49.11) for a second order tensor) can be shown to have the same physical dimension. Note that the above derivations are independent of any embedding Cartesian coordinate, unlike the derivations given in [13] (for tensors) and [14] (for vectors). Therefore, the derivations in this chapter are intrinsic in nature and can be applied to any arbitrary curved surface, such as a 2-dimensional surface of a torus, where the methods followed in [13] and [14] might not be applicable. We now demonstrate an application of the physical components of vectors and tensors in the modeling of physical systems using the equations of continuum mechanics.

49.3 Equation of Motion of an Elastic Body in Physical Components

We have seen in Sect. 49.2 that all the physical components of a tensor carry the same physical dimension irrespective of their representative bases. However, the basis vectors in $\mathcal{E}^* = \{\mathbf{e}^i\}$ and $\mathcal{F} = \{\mathbf{f}_i\}$ do not have unit norm in general, unlike the vectors in the bases $\mathcal{E} = \{\mathbf{e}_i\}$ and $\mathcal{F}^* = \{\mathbf{f}^i\}$ which have unit norm. Therefore, it is customary to use the bases $\mathcal{E} = \{\mathbf{e}_i\}$ and $\mathcal{F}^* = \{\mathbf{f}^i\}$ for defining the components of physically measurable vectors or tensors.

We now propose a method to obtain the continuum equations in physical components, directly from their general covariant or contravariant form. Here we use the bases $\mathcal{E} = \{\mathbf{e}_i\}$ and $\mathcal{F}^* = \{\mathbf{f}^i\}$ (with basis vectors of unit norm) to obtain the physical components. Our formalism here is motivated by the previous work due to Altman and Oliveira [12]. We know from standard elasticity theory that the equation relating the strain and displacement (for small displacements, neglecting the nonlinear terms) is given by

$$\epsilon_{ij} = \frac{1}{2}\left(\frac{\partial u_i}{\partial x^j} + \frac{\partial u_j}{\partial x^i}\right) - \Gamma^{\alpha}_{ij} u_\alpha\,, \tag{49.26}$$

where ϵ_{ij} is the strain component, u_i is the displacement component, and Γ^{α}_{ij} is the Christoffel symbol of second kind. We also have the equation of motion (EOM) in a general coordinate as

$$\nabla_j \tau^{ij} + F^i = \rho\, a^i, \tag{49.27}$$

where τ^{ij} is the stress, F^i is the force per unit volume, a^i is the acceleration, and ∇ denotes the covariant derivative. Finally, we also have the constitutive relation between stress and strain which, in case of linear elasticity of homogeneous and isotropic media, can be written as

$$\tau^{i}_{.j} = \lambda\,\Theta\,\delta^{i}_{.j} + 2\mu\epsilon^{i}_{.j}\,, \tag{49.28}$$

where λ, μ are the Lamé's constants and $\Theta = \epsilon^{i}_{.i}$ is the dilation. However, as we have discussed in Sect. 49.2.1, the dimensions of the vector or tensor quantities $u_i, a^i, \epsilon_{ij}, \tau^{ij}, \tau^{i}_{.j}$ etc. are not the same as they are realized during physical/experimental measurements. Hence, it is necessary to convert Eqs. (49.26)–(49.28) in terms of physical components so that they can be applied to model physical systems. It can be easily understood that for rectangular Cartesian coordinate this conversion is not necessary as the covariant, contravariant or mixed components of tensors coincide with their physical components automatically. However, in general, even in the cases of orthogonal curvilinear coordinates, the conversion of Eqs. (49.26)–(49.28) in physical components is necessary to avoid any dimensional mismatch.

Denoting the physical components by $u(i)$, $F(i)$, $\epsilon(ij)$, $\tau(ij)$ etc. with respect to the bases $\mathcal{E} = \{\mathbf{e}_i\}$, $\mathcal{F}^* = \{\mathbf{f}^i\}$ and using the relations in Eqs. (49.8) and (49.11) we obtain the strain-displacement relation equation (49.26) in their corresponding physical components as

$$\epsilon(ij) = \frac{\sqrt{g^{ii}g^{jj}}}{2}\left[\frac{\partial}{\partial x^j}\left(\frac{u(i)}{\sqrt{g^{ii}}}\right) + \frac{\partial}{\partial x^i}\left(\frac{u(j)}{\sqrt{g^{jj}}}\right)\right] - \sqrt{g^{ii}g^{jj}}\,\Gamma^{\alpha}_{ij}\frac{u(\alpha)}{\sqrt{g^{\alpha\alpha}}}\,, \tag{49.29}$$

where no summation is implied over the same type of indices, i.e., the metric tensor components g^{ii}, g^{jj} etc. are not summed over i or j. Next, expanding the covariant derivative in Eq. (49.27) we obtain

$$\frac{\partial \tau^{ij}}{\partial x^j} + \Gamma^{i}_{j\alpha}\tau^{\alpha j} + \Gamma^{j}_{j\alpha}\tau^{i\alpha} + F^i = \rho\, a^i. \tag{49.30}$$

Replacing the contravariant type components of the stress tensor $\tau^{ij}, \tau^{i\alpha}$ etc. in terms of physical components (using the relations in Eq. (49.11)) we obtain the equation of motion as

$$\frac{\partial}{\partial x^j}\left(\frac{\tau(ij)}{\sqrt{g_{ii}g_{jj}}}\right) + \Gamma^{i}_{j\alpha}\left(\frac{\tau(\alpha j)}{\sqrt{g_{\alpha\alpha}g_{jj}}}\right) + \Gamma^{j}_{j\alpha}\left(\frac{\tau(i\alpha)}{\sqrt{g_{ii}g_{\alpha\alpha}}}\right) + \frac{F(i)}{\sqrt{g_{ii}}} = \rho\,\frac{a(i)}{\sqrt{g_{ii}}}, \tag{49.31}$$

where the summations are applicable only over common indices of different type (covariant or contravariant as determined by Eq. (49.30). Similarly, we obtain the constitutive relation between stress and strain in terms of physical components as

$$\tau(ij) = \lambda\left[\sum_{i=1}^{n}\frac{\epsilon(ii)}{\sqrt{g_{ii}g^{ii}}}\right]\delta(ij) + 2\,\mu\,\epsilon(ij), \tag{49.32}$$

where n is the dimension of the manifold. It is to be noted that for orthogonal coordinates the constitutive relation in physical components retains the same form as its mixed tensor form as in Eq. (49.28). Equations (49.29), (49.31) and (49.32) are the general form of the strain-displacement relation, EOM, and constitutive relation of an elastic body in physical components in any arbitrary curvilinear coordinate, which are not only limited to orthogonal coordinates.

We next demonstrate, with an example of plane polar coordinate, how Eqs. (49.29), (49.31) and (49.32) can be used to readily obtain the EOM of an elastic body in physical components. For a plane polar coordinate we have the Parameterization of a 2-dimensional plane as $\mathbf{P}(r, \theta) = (r\cos\theta, r\sin\theta)$, where $\mathbf{P}$ is the position vector, r is the radial coordinate, and θ is the angular coordinate. The covariant basis vectors are

$$\mathbf{b}_r = (\cos\theta, \sin\theta) \quad \text{and} \quad \mathbf{b}_\theta = (-r\sin\theta, r\cos\theta)\,. \tag{49.33}$$

The basis vectors $\mathbf{b}_r$, $\mathbf{b}_\theta$ and their normalized counterparts are shown in Fig. 49.3. The metric tensor (g_{ij}) and its inverse (g^{ij}), where $i, j \in \{r, \theta\}$, are given by

$$\begin{pmatrix} g_{rr} & g_{r\theta} \\ g_{\theta r} & g_{\theta\theta} \end{pmatrix} = \begin{pmatrix} 1 & 0 \\ 0 & r^2 \end{pmatrix} \quad \text{and} \quad \begin{pmatrix} g^{rr} & g^{r\theta} \\ g^{\theta r} & g^{\theta\theta} \end{pmatrix} = \begin{pmatrix} 1 & 0 \\ 0 & \frac{1}{r^2} \end{pmatrix}. \tag{49.34}$$

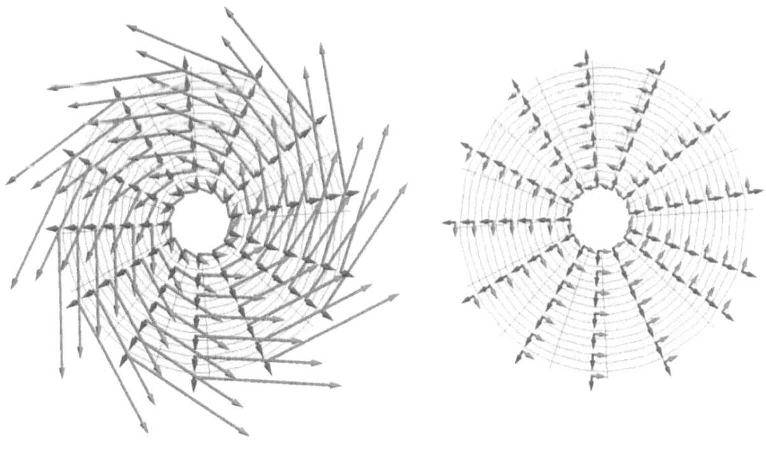

Fig. 49.3 Basis vectors in the plane polar coordinate parameterized by $\mathbf{P} = (r\cos\theta, r\sin\theta)$, where r and θ are the radial and angular coordinates respectively. Left panel shows the covariant basis vectors $\mathbf{b}_r$ (red arrows) and $\mathbf{b}_\theta$ (green arrows) without normalization. Right panel shows the normalized basis vectors $\mathbf{e}_r = \mathbf{b}_r/\sqrt{g_{rr}}$ (red) and $\mathbf{e}_\theta = \mathbf{b}_\theta/\sqrt{g_{\theta\theta}}$ (green) (Color figure online)

Therefore, the dual basis vectors are given by

$$\mathbf{b}^r = g^{rj}\mathbf{b}_j = (\cos\theta, \sin\theta) \;\text{ and }\; \mathbf{b}^\theta = g^{\theta j}\mathbf{b}_j = \left(-\frac{\sin\theta}{r}, \frac{\cos\theta}{r}\right). \tag{49.35}$$

The Christoffel symbols are calculated to be

$$\Gamma^r_{\theta\theta} = -r, \Gamma^\theta_{r\theta} = \frac{1}{r} \;\text{ and }\; \Gamma^r_{rr} = \Gamma^r_{r\theta} = \Gamma^\theta_{\theta\theta} = \Gamma^\theta_{rr} = 0. \tag{49.36}$$

Therefore, using Eq. (49.29), we obtain the strain-displacement relations in physical components for plane polar coordinate as

$$\epsilon(rr) = \frac{\partial u(r)}{\partial r}, \tag{49.37}$$

$$\epsilon(r\theta) = \frac{1}{2r}\left[\frac{\partial u(r)}{\partial \theta} - u(\theta) + r\frac{\partial u(\theta)}{\partial r}\right], \tag{49.38}$$

$$\epsilon(\theta\theta) = \frac{1}{r}\left[\frac{\partial u(\theta)}{\partial \theta} - u(r)\right]. \tag{49.39}$$

Similarly, we obtain the EOM in physical components directly from Eq. (49.31) as

$$\frac{\partial\tau(rr)}{\partial r} + \frac{1}{r}\frac{\partial\tau(r\theta)}{\partial\theta} + \frac{\tau(rr) - \tau(\theta\theta)}{r} + F(r) = \rho a(r), \tag{49.40}$$

$$\frac{\partial\tau(r\theta)}{\partial r} + \frac{1}{r}\frac{\partial\tau(\theta\theta)}{\partial\theta} + \frac{2\tau(r\theta)}{r} + F(\theta) = \rho a(\theta). \tag{49.41}$$

We obtain the constitutive relations in physical components from Eq. (49.32) as

$$\tau(rr) = \lambda\, [\epsilon(rr) + \epsilon(\theta\theta)] + 2\,\mu\,\epsilon(rr), \tag{49.42}$$

$$\tau(r\theta) = 2\,\mu\,\epsilon(r\theta), \tag{49.43}$$

$$\tau(\theta\theta) = \lambda\, [\epsilon(rr) + \epsilon(\theta\theta)] + 2\,\mu\,\epsilon(\theta\theta). \tag{49.44}$$

We see from Eqs. (49.37)–(49.44) that the strain-displacement relations, EOM, and the constitutive relations in plane polar coordinate obtained from Eqs. (49.29), (49.31) and (49.32) clearly match with the results in the literature [19] but obtained via different procedures. The advantage of having Eqs. (49.29), (49.31) and (49.32) is that they can be applied to any curvilinear coordinate or curved surfaces readily, which are also not limited to the orthogonal coordinates. The strain-displacement relationship and EOM have been derived for orthogonal curvilinear coordinates via different methods in seminal works by Sokolnikoff, Timoshenko etc. [19, 20]. However, the general form of strain-displacement relationship, EOM, and the constitutive relation in physical components as given in Eqs. (49.29), (49.31) and (49.32) are very scarce in the previous literature.

49.4 Conclusion

In this chapter we have summarized different types of physical components of vectors and tensors, scattered in earlier works in the literature [13–16]. We have provided an intrinsic proof to show that all the physical components of a vector or tensor have the same physical dimension. Finally, we use these physical components to derive the general form of the equations of elasticity which are readily applicable for physical modeling purpose. The extension of this study to derive the hydrodynamic equations in physical components is straightforward.

Acknowledgments I thank the organizers of ISRA 2023. My special thanks are due to Shubhrangshu Ghosh and Banibrata Mukhopadhyay. The project leading to this publication has received funding from France 2030, the French Government program managed by the French National Research Agency (ANR-16-CONV-0001) and from Excellence Initiative of Aix-Marseille University - A*MIDEX.

References

1. Picioreanu, C., Van Loosdrecht, M. C., & Heijnen, J. J. (2001). Two-dimensional model of biofilm detachment caused by internal stress from liquid flow. *Biotechnology and Bioengineering, 72*(2), 205 218.
2. Gaffney, E. A., Gadêlha, H., Smith, D. J., Blake, J. R., & Kirkman-Brown, J. C. (2011). Mammalian sperm motility: Observation and theory. *Annual Review of Fluid Mechanics, 43*(1), 501 528.
3. Toner, J., & Tu, Y. (1995). Long-range order in a two-dimensional dynamical XY model: How birds fly together. *Physical Review Letters, 75*(23), 4326.
4. Toner, J., Tu, Y., & Ramaswamy, S. (2005). Hydrodynamics and phases of flocks. *Annals of Physics, 318*(1), 170–244.
5. Nath, S. K., & Mukhopadhyay, B. (2015). Origin of nonlinearity and plausible turbulence by hydromagnetic transient growth in accretion disks: Faster growth rate than magnetorotational instability. *Physical Review E, 92*(2), 023005.
6. Nath, S. K., & Mukhopadhyay, B. (2016). A pure hydrodynamic instability in shear flows and its application to astrophysical accretion disks. *The Astrophysical Journal, 830*(2), 86.
7. Nath, S. K., Mukhopadhyay, B., & Chattopadhyay, A. K. (2013). Magnetohydrodynamic stability of stochastically driven accretion flows. *Physical Review E—Statistical, Nonlinear, and Soft Matter Physics, 88*(1), 013010.
8. Nath, S. K., & Chattopadhyay, A. K. (2014). Cross-correlation-aided transport in stochastically driven accretion flows. *Physical Review E, 90*(6), 063014.
9. Sokolnikoff, I. S. (1964). *Tensor analysis theory and applications to geometry and mechanics of continua* (2nd ed.). John Wiley & Sons.
10. Lebedev, L. P., Cloud, M. J., & Eremeyev, V. A. (2010). *Tensor analysis with applications in mechanics*. World Scientific.

11. Audoly, B., & Pomeau, Y. (2010). *Elasticity and geometry. From hair curls to the nonlinear response of shells*. Oxford University Press.
12. Altman, W., & Oliveira, A. M. D. (1993). Nonlinear equations of the theories of elasticity and shells in terms of anholonomic components. *Applied Mechanics Reviews, 46*(11S), S92.
13. Altman, W., & de Oliveira, A. M. (1995). Physical and anholonomic components of tensors of elasticity theory. *International Journal of Non-linear Mechanics, 30*(3), 341–358.
14. Truesdell, C. (1953). The physical components of vectors and tensors. *ZAMM-Journal of Applied Mathematics and Mechanics/Zeitschrift für Angewandte Mathematik und Mechanik, 33*(10–11), 345–356.
15. McConnell, A. J. (2014). *Applications of tensor analysis*. Courier Corporation.
16. Frederick, D. (1956). Physical interpretation of physical components of stress and strain. *Quarterly of Applied Mathematics, 14*(3), 323–327.
17. Simmons, G. F. (2016). *Differential equations with applications and historical notes*. CRC Press.
18. Goldstein, H. (2011). *Classical mechanics*. Pearson Education India.
19. Sokolnikoff, I. S. (1946). Mathematical theory of elasticity. Dover Publications Inc., New York.
20. Timoshenko, S. (1941). *Strength of materials, part II. Advanced theory and problems*. D. Van Nostrand Company Inc., New York.

Part XIII
Special Contribution

Chapter 50
Black Holes, White Dwarfs, Neutron Stars, Related Fluid Mechanics and Nuclear Astrophysics: My Research Journey from 1997 to ...

Banibrata Mukhopadhyay

Abstract I encapsulate here some of my main research dividing roughly into six fields. Since the beginning of my research in 1997 (when I published my first paper in 1999 in a refereed journal), while I explored many fields, mostly have been centering around compact object physics, underlying (magneto)fluid physics and nuclear astrophysics, gravitation: Einstein's and its modification, first and second quantizations in curved spacetime and their implications in astrophysics, astroparticle physics and cosmology. Here I touch upon exciting developments done by me, and my group in some of these fields. Many more, however, are reported by many of my present and past mentees in the same volume.

50.1 Introduction

The topics of black holes, white dwarfs and neutron stars are the central part of the modern astrophysics. In order to understand them and their effects in the universe successfully, one needs to look at their environments which mostly are the magnetohydrodynamical flows. In addition, to understand the formation and inner physics of neutron stars and white dwarfs, nuclear physics is indispensable. All these physics and astrophysics are often deeply involved with general relativity.

Over more than a quarter of a century, I have been deeply working on all the above topics. In fact, all of them are my core research fields. In this special chapter, I attempt to highlight some of my important research works in the journey from 1997 till date, when my first journal paper was published in 1999.

Over the last 27 years, I am fortunate to have chances to work with many mentees, mentors, and friends/collaborators (including stalwarts). Among the

B. Mukhopadhyay (✉)
Department of Physics, Indian Institute of Science, Bangalore, India
e-mail: bm@iisc.ac.in

S. Ghosh, A. R. Rao (eds.), *The Relativistic Universe: From Classical to Quantum*,
Astrophysics and Space Science Proceedings 61,
https://doi.org/10.1007/978-3-031-90186-7_50

mentees, majority are Ph.D. students (as of this writing, 17) and postdoctoral fellows (as of this writing, 12), also there are a few (as of this writing, 12) undergraduate students (counting only the ones who have written at least one paper with me or finished a project completely with me) as well. Overall, it has been a great experience so far in my research journey.

My research has mostly been in the following topics: Astrophysical Fluid Dynamics and Accretion Phenomena; Stellar Physics and Gravitation: Black Holes, White Dwarfs and Neutron Stars; Nuclear Astrophysics; Modified Gravity; Investigation of Spinor and Electromagnetic Fields in Curved Space-time; Field Theory in Curved Space-time and Related Particle Physics; Astrophysical Data Analysis. Many of my past and present mentees, particularly, have been contributing to the present volume, covering some of my research. Hence, I plan to touch upon some of the remaining best, in my view, research results of mine. They are: modeling accretion disks, particularly in their hard-state, by tracking underlying nuclear physics and with magnetic fields, by proposing a pseudo-Newtonian prescription; measuring spin of black holes by proposing a unified QPO model; establishing a new super-Chandrasekhar mass-limit of white dwarfs; uncovering the origin of hydrodynamical instability in rotating shear flows and accretion disks; enlightening spinors in gravitational background and the origin of baryogenesis. In the next sections, one by one I touch upon some of my research, on the aforementioned topics.

50.2 Nuclear Astrophysics

The gravitational field of a black hole is so powerful that even electromagnetic radiation (e.g. visible light) cannot escape it. This leads to energy released per unit mass infalling/accreting, $GM/R_{in} \sim c^2/6 \sim 10^{20}$ erg/gm, where R_{in} is the inner edge of the flow (accretion disk), often the same as the innermost stable circular orbit, c the speed of light, G the Newton's gravitation constant, and M the mass of the black hole. This energy is even quite greater than the nuclear energy in several astrophysical systems, confirming the high efficiency of an accreting system. Depending on the process of accretion (as of current understanding), whether by Roche-lobe overflow or wind-accretion, an accretion disk can have highest temperature (T), respectively, either $\sim 10^7$ K emitting (multicolor) blackbody radiation or much higher $T \sim 10^9$–10^{11} K emitting nonthermal radiation. The latter class of accretion is expected to have enough scope to produce significant thermonuclear reactions.

During my Ph.D., I explored the old but fascinating field of nuclear astrophysics, particularly around black holes. One of my aims during Ph.D., in 1997–2000, was to investigate the thermonuclear reactions in the hot ($T \geq 10^9$ K), mostly optically thin, sub-Keplerian accretion disks around a black hole [1]. Also I explored how nuclear energy generation and absorption affect or control the accretion dynamics.

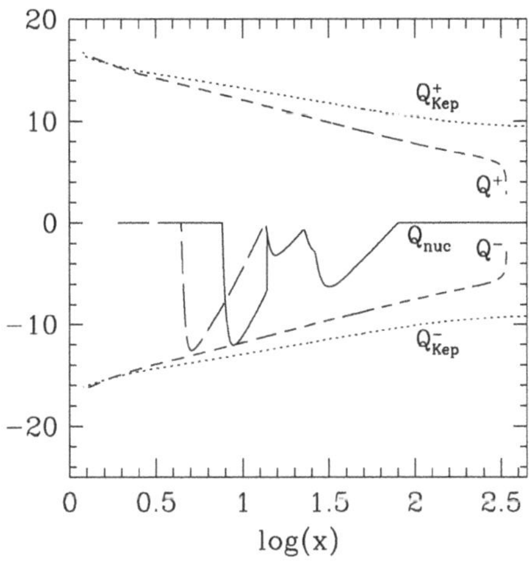

Fig. 50.1 Variations of viscous dissipation (Q^+) and radiation energy (Q^-), along with the corresponding values in a Keplerian accretion disks ($Q^+_{\rm Kep}$ and $Q^-_{\rm Kep}$, respectively), and nuclear energy ($Q_{\rm nuc}$) per unit area per unit time as functions of radial distance from a black hole. Here $M = 10M_\odot$ with $M_\odot$ being the mass of Sun, accretion rate $\dot{M} = 0.01$ Eddington rate, Shakura-Sunyaev $\alpha = 0.05$. See Figure 5(d) of [1] and related description therein for details

Figure 50.1 shows that nuclear energy is not negligible compared to the viscous energy.

Of late, as a continuation of above work, I explored the thermonuclear reactions in the outflows/jets launched from hot accretion disks around stellar mass and supermassive black holes [2]. Such studies help us understand the origin of super-solar abundance of elements, e.g. Mg, Si, S, Ar, Ca, Cr, Mn, and Co, as one detects. See Fig. 50.2 for Active Galactic Nucleus (AGN). For a stellar mass black hole, similar results are expected from the outflow launched from a farther radius, as a similar density would have been acquired by the disk fluid at an outer radius.

I also explored my expertise of nucleosynthesis in accretion disks to uncover nucleosynthesis in the supernova (SN) disks, also called Gamma-Ray burst (GRB) accretion disks. There are many SN-GRB associations, e.g., GRB 980425, SN 1998bw; SN 030329, SN 2003dh; GRB 0312003, SN 2003lw; GRB 021211, SN 2002lt; etc.

During core-collapse of a massive rotating star, an accretion disk with high density and temperature forms (sometimes via fall back materials of supernova ejecta). Such a disk, survived for 2–1000 s, is very favorable for neutrino emission, leading to a huge neutrino flux/luminosity. It is argued that [3] the jet produced from this disk is essentially the GRB and, hence, GRBs are often argued to be the birth-cry of a black hole. Such a disk is very useful for probing the black hole properties during its formation stage. Nucleosynthesis in SN disks and underlying outflows is useful to pinpoint abundances of many semi-heavy to heavy element isotopes and their contamination to the galaxies [4].

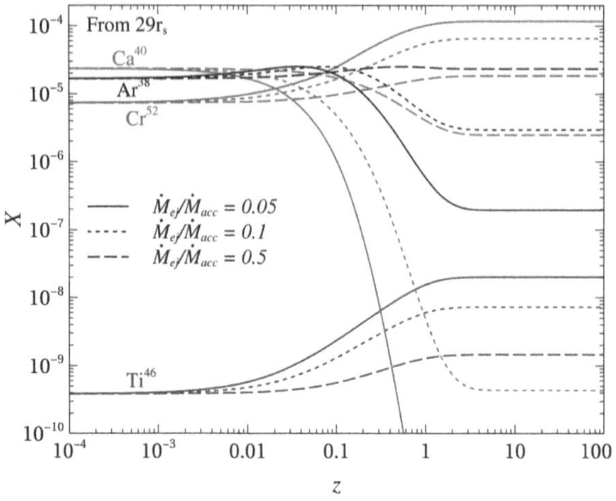

Fig. 50.2 Evolution of a few representative abundances in an outflow launched from 29 Schwarzschild radius (r_s) in AGN around a black hole of $M = 10^7 M_\odot$, with different $\dot{M}_{ej}/\dot{M}_{acc} = 0.05, 0.1, 0.5$. Here z is the distance from accretion disk and $\dot{M}_{ej}$ and $\dot{M}_{acc}$ are respectively outflow and inflow rates. See Figure 14 of [4] and related description therein for details

Fig. 50.3 Variation of black hole mass in units of $M_\odot$ with its spin (Kerr parameter), when the formation stage accretion rate in units of $M_\odot$/s is a parameter. See [5]

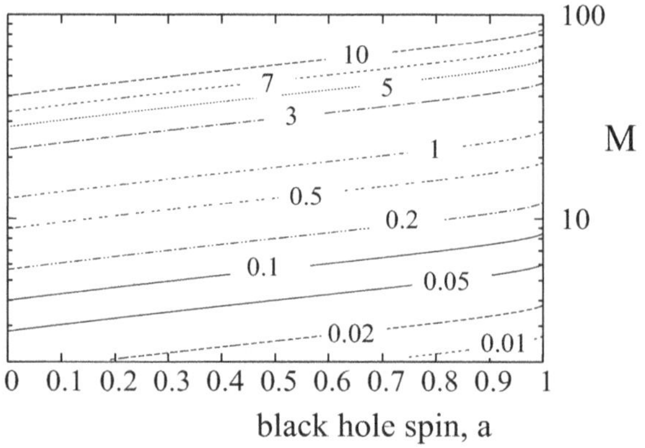

As mentioned above, the SN disk is very useful to probe black hole properties. If we constrain the luminosities of such disk flows, e.g. if they are more of highly luminous neutrino dominated optically thick or lowly luminous advective optically thin disks, black hole spin and mass could be coupled [5]. Therefore, given the black hole mass independently, spin can be inferred, as shown in Fig. 50.3. This also effectively reduces the number of black hole hairs.

50.3 Relativistic Astrophysics and Accretion Disks

About one-third of my research papers center around accretion disk phenomena, primarily around a black hole. An accretion disk around a black hole is the general relativistic magnetohydrodynamic (GRMHD) flow. Other than black hole (also neutron star), however, the accretion disk could be explained by Newtonian mechanics, at best we then need special relativity, e.g. protoplanetary disk, star forming disk.

GRMHD formalism is very complicated, in fact the set of hydrodynamical disk flows (without magnetic fields) itself is cumbersome in the GR framework around a rotating black hole. In order to avoid complicacy, often researchers use Newtonian formalism but modifying the underlying gravitational potential/force, called pseudo-Newtonian potential or just pseudo-potential, to solve the problem. For many practical purposes, this does not lose the information of important observables. Pseudo-potential is very useful to study accretion dynamics approximately without going through the detailed complicated GR equations. I proposed a method to obtain pseudo-potential for any metric. Starting from the Lagrangian of a spacetime metric, I showed how a pseudo-potential can be obtained [6]. I further established the pseudo-potential of the Keplerian accretion disk around a Kerr black hole. The corresponding magnitude of force is given by

$$F = \frac{(r^2 - 2a\sqrt{r} + a^2)^2}{r^3[\sqrt{r}(r-2) + a]^2}, \tag{50.1}$$

where r is the distance from the black hole in units of GM/c^2, a is the angular momentum per unit mass of the black hole in units of GM/c. Note that far away from the black hole, the force reduces to the Newtonian force. This pseudo-potential reproduces the marginally stable orbit radii exactly as of GR. It also reproduces the marginally bound orbit radii and the energy at marginally stable orbits with very small errors with respect to GR, which are within observational limits.

Many subsequent works followed my approach or potential for modeling accretion disks (e.g., [7–10], to mention very few). I myself used it in modeling accretion disks, from hydrodynamic to magnetohydrodynamic, around stellar mass to supermassive black holes. In one of my models of the accretion disk explaining hard luminous states, hence advective and sub-Keplerian yet luminous, I used my pseudo-potential to describe magnetically arrested advection accretion flow (MA-AAF) [11]. I used MHD momentum balance equations with appropriate boundary conditions for geometrically thick accretion disks and my Kerr pseudo-potential (see [12], for the detailed set of of equations). Figure 50.4 shows that such a disk around a highly spinning black hole exhibits luminosity $\sim 10^{40}$ erg/s even at a sub-Eddington accretion rate. Here, the magnetic energy, which is super-Eddington at the supply far away from the black hole, dominates the matter energy which is sub-Eddington. This highly luminous hard state, i.e. MA-AAF, has many implications, e.g. hard states of ultra-luminous X-ray sources (ULXs), resolving

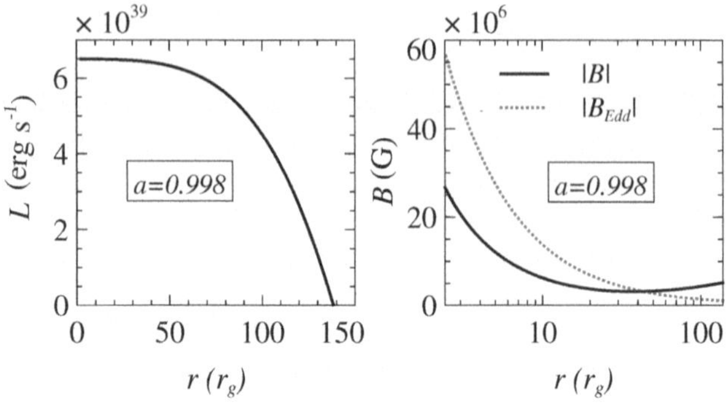

Fig. 50.4 Luminosity (left) and magnetic field (right) of MA-AAF for a black hole of $M = 20M_{\odot}$ and spin (Kerr parameter) $a = 0.998$, with accretion rate 0.05 Eddington rate. Here B_{Edd} is the Eddington magnetic field and $r_g = GM/c^2$. See Figure 7(c–d) of [12] and related description therein for details

dichotomy between various blazar classes, throughout almost similar luminosity in microquasar GRS 1915+105 independent of spectral states (at most a factor of 4 difference between hard and soft states).

Later on, by GRMHD numerical simulations, my results have been verified even at sub-Eddington magnetic field, reported in this volume itself by two of my current Ph.D. students. See [13, 14].

50.4 High Frequency QPOs and Black Hole Spin

Quasi-periodic oscillation (QPO) is a phenomenon being observed in many black hole and neutron star sources (e.g. [15–20]). Their frequencies span from a fraction of Hz to kHz. However, the exact origin of QPOs is not yet well-understood. Are QPOs originating from a unique process, e.g. in accretion disks around the compact object, independent of the nature of compact object and the underlying frequency? Over the years many models have been proposed, e.g. [21–23] to mention very few, however these models are neither beyond the caveats nor universal.

On the other hand, the spin of black holes in various astrophysical sources has not been confirmed yet. There are attempts to measure them based on a few methods, e.g. continuum fitting [24], iron line [25], but often they give nonconvergent results. A QPO also could be a very good precursor for the spin of black hole, provided the QPO is modeled accurately. This also would settle the apparent dichotomy of spin measurements by other methods. Over the years, I have been looking at the issue of QPO origin, particularly emphasizing on a unified model. Based on which, it is possible to measure the spin of black holes [26, 27]. I showed that QPOs form by the nonlinear resonance of fundamental modes present in accretion disks, which appear as the forced oscillator. In accretion disks, the fundamental modes are not

Table 50.1 Physical parameters for black holes with twin QPOs and their spin measured by me, where ν_h and ν_l are respectively higher and lower QPO frequencies, and m is associated to modes. See [27]

Black hole	M ($M_\odot$)	Estimated a	ν_h (Hz)	ν_l (Hz)	m
GRO J1655-40	$7.02^{+0.22}_{-0.2}$	0.91 ∓ 0.02	450	300	−1
GRO J1655-40	$5.4^{+0.08}_{-0.3}$	$0.8^{-0.19}_{+0.06}$	450	300	−2
XTE J1550-564	10.92–11.07	0.57 ± 0.01	276	184	−1
H1743-322	9.97–10.91	0.61–0.7	240	163	−2
H1743-322	9.25–10.86	0.87–0.71	240	163	−2
XTE J1859+226	5.9–9.69	0.57–0.61	227.5	128.6	−2
IGR J17091-3624	11.04–15.46	0.6–0.67	164	66	−2
IGR J17091-3624	8.89–10.63	0.66–0.99	164	66	−3
GRS 1915+105	$12.4^{+2.0}_{-0.75}$	$0.94^{-0.14}_{+0.05}$	168	113	−2
			69	41	−6
			69	41	−6

just those of test particles given by GR. Rather, they are the specific combinations of test particles' fundamental modes in GR and are involved with the disk dynamical properties, e.g. velocity. The forcing arises by the spin of the underlying compact object and any other external modes. For details, see [28] appeared in this volume. In order to reproduce observed QPOs, particularly high frequency ones and their properties, with minimum χ^2, I measured the spin of the black hole which is a best-fit parameter. Table 50.1 shows the measured spin parameter for some black holes, with twin high frequency QPOs, exhibiting a range, though $a > 0.5$. For XTE J1550+564, the reported measured spin tallies with that measured by the iron-line. However, for the mass range 10.26–11.58$M_\odot$ the spin would turn out to be 0.99–0.89. See [27] and [28] in this volume for details. It would be very interesting to measure the spin of the same black holes from their high and low frequency QPOs if they have both, e.g. for GRS 1915+105, by the same method. This will confirm the robustness of the technique.

50.5 Hydrodynamic Instability

There is a century old issue on the origin of hydrodynamic instability in shear flows. There are laboratory flows, e.g. plane Couette flow, where flows are exhibited to be turbulent at a finite (or lower) Reynolds number (Re), but according to the linear theory they are stable for all (or much higher) Re. In astrophysics also, e.g. accretion disks, often the flows can not be explained to be unstable and turbulent in the linear hydrodynamical theory, which are observationally understood to be turbulent. Although the presence of magnetic field unfolds the issue, there are many accreting systems which are cold and practically neutral in charge. Hence, there is no role in magnetic fields, and any instability and turbulence therein has to be of hydrodynamical origin. I am involved towards the resolution of this issue over the decades.

I initiated looking at this issue taking the stoke of nonnormality of the underlying perturbation modes [29–31]. Such nonnormal modes, while linearly stable, produce transient growth of perturbation whose peak ($G_{\max}$) could be very high, depending on Re and initial conditions, given by

$$G_{\max} \sim Re^{2/3} \text{ for a Keplerian disk, } G_{\max} \sim Re^2 \text{ for plane Couette flow.} \quad (50.2)$$

The proportionality constants therein depend on the initial conditions, which could be optimized. In an astrophysical system, Re could be indeed very high [32], so does $G_{\max}$, which is expected to lead to nonlinearity and turbulence.

Moreover, such a flow is expected to be amended with a force, which is expected to be stochastic in nature hence behaving as noise. This force could be the result of thermal fluctuations (however small be) in the flow, Brownian ratchet, grain-fluid interactions, feedback from outflows in astrophysical disks, etc. I explored the section of rotating linear shear flow or plane Couette flow with the Coriolis effect in the presence of external force. Such a flow mimics a small section of an accretion disk. I found that based on the strength of force and boundary conditions appropriate for the systems under consideration, e.g. accretion disk, the background velocity profile is modified to parabolic flow. This leads the flow practically to a plane Poiseuille flow or Couette-Poiseuille flows in the presence of Coriolis force, depending on the frame of reference. We already know the classic result that plane Poiseuille flow becomes linearly unstable at a $Re \sim 5772$, though laboratory experiments exhibit the emergence of turbulence at a much lower Re. I showed that a rotating Poiseuille flow, which mimics a small section of the accretion disk with external force, becomes linearly unstable and plausibly turbulent at a lower $Re < 5772$. This explains the transition to turbulence of accretion disks hydrodynamically.

Following previous work [29, 33], I linearly perturb the Navier-Stokes equation with background velocity $\mathbf{U} = (0, \xi(1 - x^2) - x, 0)$, where x is the dimensionless local radial coordinate (in a local box, the x-component of the Cartesian coordinate) and ξ indicates the measure of the deviation from plane shear flow due to forcing. This practically describes a small portion of an accretion disk, which further rotates with the angular velocity $\Omega \sim r^{-q}$, where r is the distance from the central source of gravity. The solution of the perturbation equation is given by

$$Q = \sum_{m=1}^{\infty} C_m Q_{x,m}(x) \exp[-i\sigma_m t], \quad (50.3)$$

where σ_m is the eigenvalue of the mth mode, with the eigenvector $Q_{x,m}$, determining the growth/decay rate of the perturbation. Figure 50.5 shows the maximum growth

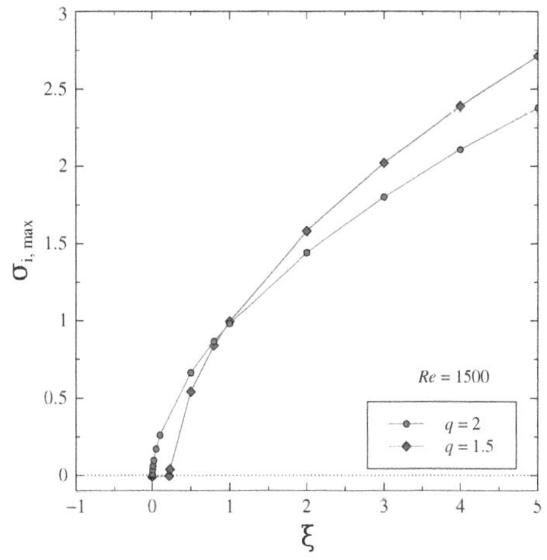

Fig. 50.5 Maximum growth rate as a function of external force parameter for different rotation profiles with $Re = 1500$

rate of perturbation with external force. It is clear that even with a tiny force, a non-growing mode (in a linear shear flow) acquires a positive imaginary eigenvalue leading to an exponential growth of perturbation, leading to nonlinearity and turbulence. Note that a constant angular momentum flow appears to be more sensitive to the tiny forcing than the Keplerian flow.

50.6 Super-Chandrasekhar White Dwarfs

I initiated this topic about a dozen years back from this writing. Subsequently, the rest of the community took it over. The observational motivation lies with peculiar type Ia supernovae (SNeIa) whose high peak luminosity apparently requires a significantly super-Chandrasekhar white dwarf progenitor. The peculiarity in such SNeIa lies also with its peak-stretch relationship (Philip's law) [34], see Fig. 50.6 borrowed from [35]. In 2012, I proposed such peculiar SNeIa to be originating from highly magnetized white dwarfs or B-WDs. Subsequently, ourselves and other groups have verified it by introducing newer and newer physics based on semi-analytical to numerical calculations to numerical simulation, e.g. see [36] in this volume.

Our venture has been the following:

(1) Assuming constant magnetic field or highly fluctuating magnetic field so that Lorentz force is negligible throughout the star [37, 38]. This leads to quantum effect by Landau quantization affecting electron degenerate gas equation of state (EoS) of white dwarfs, here B-WDs, and hence mass.
(2) Introducing varying magnetic field, hence Lorentz force, affecting the mass. This is purely a classical effect [39].

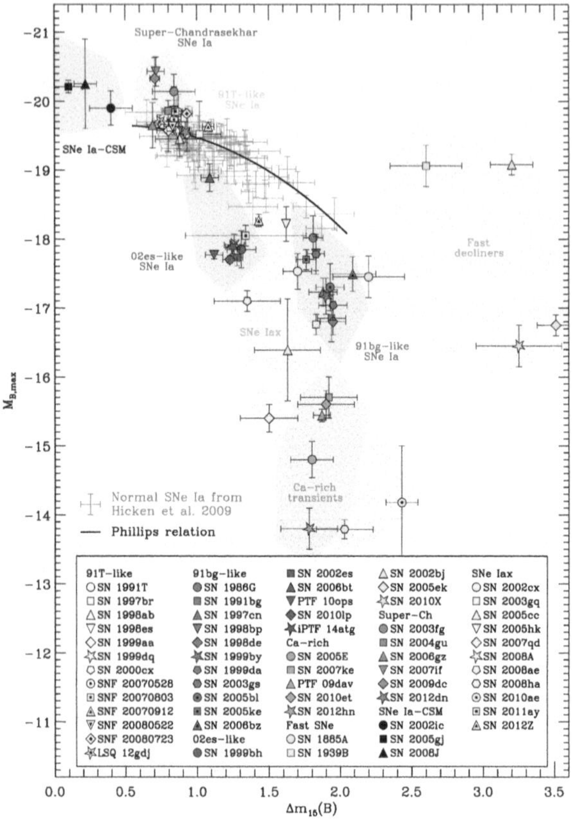

Fig. 50.6 Variation of peak luminosity in absolute B-band magnitude with the 15 days light-curve decline rate from the peak in B-band of several SNeIa lightcurves. Note that many SNeIa do not obey the Philips relation, hence peculiar in many folds. Super-Chandrasekhar white dwarfs are indicated at the top with very high luminosity. See Figure 1 of [35], which is borrowed here

(3) Numerical simulation of B-WDs based on the Einstein-Maxwell equation solution in two dimensions with different magnetic field geometries [40, 41].
(4) White dwarfs in a modified gravity [42–44].
(5) Continuous gravitational waves from rotating magnetized white dwarfs or B-WDs [45].
(6) Virial theorem of highly magnetized stars and estimates of the mass and magnetic field of B-WDs based on magnetized virial theorem [46].
(7) Exploring B-WDs and their luminosity at finite temperature [47].
(8) Cooling and magnetic field delay of B-WDs [48].
(9) White dwarf in noncommutative space (spacetime magnetic field) and its super-Chandrasekhar mass-limit [49].
(10) Anisotropic magnetized white dwarfs with different magnetic field profiles leading to different mass–radius relations [50].

Let me describe briefly one of the basic results. The equations governing a white dwarf with Fermi energy $E_F >> k_B T$, with k_B being the Boltzmann constant, are

$$\frac{1}{\rho}\left(\frac{dP}{dr}+\frac{B^2}{8\pi}\right)=F_r+\frac{\mathbf{B}\cdot\nabla\mathbf{B}|_r}{4\pi\rho},\quad \frac{dm}{dr}=4\pi r^2\rho, \tag{50.4}$$

with EoS $P = K\rho^\Gamma$, where P is the pressure, ρ the density, B the magnitude of the magnetic field vector $\mathbf{B}$ which has to be prescribed for a given star unless one solves Maxwell's equations simultaneously, F_r the radial component of the gravitational force, r the radial coordinate at an arbitrary point of the star, m the mass of the sphere of radius r, and K and Γ are the polytropic constants.

Following the Lane-Emden formalism [37] with constant magnetic field leading to Landau quantization, one can easily show that

$$M \propto K^{3/2}\rho_c^{\frac{3-n}{2n}},\quad R \propto K^{1/2}\rho_c^{\frac{1-n}{2n}}, \tag{50.5}$$

where M and R are the mass and radius respectively, ρ_c is the central density, of the white dwarf and $n = 1/(\Gamma - 1)$.

In the absence of magnetic fields (or for weak magnetic fields), $\Gamma = 4/3$ (hence $n = 3$) at high density ($\geq 10^9$ gm/cc) with relativistic EoS. Hence, the white dwarf with high central density leads to M independent of ρ_c, hence limiting mass. According to the Newtonian calculation, this corresponds to $R \to 0$, though GR treatment explains the finite radius of white dwarf at its mass-limit.

In the presence of strong magnetic field, $K = K_m \propto B^{-1} \propto \rho_c^{-2/3}$ at high density. As a result, M becomes independent of ρ_c for $\Gamma = 2$ (hence $n = 1$), leading to the mass-limit of magnetized white dwarfs. Note that in the presence of strong magnetic field, $P \propto \rho^2$ at high density unlike the nonmagnetic case when $P \propto \rho^{4/3}$.

Thus the mass-limits for nonmagnetic and magnetic white dwarfs respectively are

$$M_{\text{Chandrasekhar}} = \frac{\sqrt{6}}{32\pi}\left(\frac{hc}{G}\right)^{3/2}\left(\frac{2}{\mu_e}\right)^2\frac{2.018}{m_H^2} \approx \frac{5.9}{\mu_e^2}M_\odot, \tag{50.6}$$

$$M_{\text{Mukhopadhyay}} = \left(\frac{hc}{2G}\right)^{3/2}\frac{1}{(\mu_e m_H)^2} \approx \frac{10.312}{\mu_e^2}M_\odot, \tag{50.7}$$

where h is the Planck's constant, μ_e the mean molecular weight per electron and m_H the mass of hydrogen atom.

Therefore, for $\mu_e = 2$ which corresponds to carbon-oxygen white dwarfs, $M_{\text{Chandrasekhar}} = 1.47M_\odot$ and $M_{\text{Mukhopadhyay}} = 2.58M_\odot$. Hence, the magnetic field can easily increase the white dwarf mass-limit to 75%. For detailed explorations with varying magnetic fields, see the respective references mentioned above. In fact, depending on the field profile, white dwarfs can have different mass-limit, as is the case for neutron stars w.r.to the uncertainty in EoS. However, they are still expected to have a mass–radius curve exhibiting ultimate mass-limit considering various instabilities.

50.7 Spinors in Gravity Background

Many astrophysical systems are involved with fermions. For example, stars, compact stars, accretion disks etc. consist of baryons and leptons, apart from photons, plausible axions etc. In order to describe them successfully, one has to understand first the behavior of the underlying particles in curved spacetime, i.e. under gravitational field background.

Over the decades I have been exploring this, first understanding the solution of Dirac equation in various curved backgrounds, fully on academic interest. Subsequently, I looked at applications of said solutions in various astrophysical and cosmological problems, e.g. baryogenesis, neutrino oscillations, geometric phases etc., some of them are reported in this volume by my present/past mentees. In this chapter, I touch upon leptogenesis and baryogenesis.

The Dirac equation in curved spacetime in local coordinates can be written as (see, e.g., [51])

$$\left(i\gamma^\mu \partial_\mu - m + \gamma^\mu \gamma^5 B_\mu\right)\psi = 0, \tag{50.8}$$

where the γ-matrices have their usual meaning as in the Minkowski space-time, m is the mass of the fermion represented by the wave function ψ, and gravitational 4-vector potential (not to confuse with the magnetic field vector) B_μ is given by

$$B^\mu = \epsilon^{abc\mu}\Lambda_{b\lambda}\left(\partial_a \Lambda_c^\lambda + \Gamma^\lambda_{\alpha\beta}\Lambda_c^\alpha \Lambda_a^\beta\right), \tag{50.9}$$

where Λ-s are vierbein and $c = \hbar = k_B = 1$, with $\hbar = h/2\pi$. Note that following Schwinger [52], we remove the vector term from Eq. (50.8) which is apparently anti-Hermitian.

Based on plane wave assumption, when locally B^μ behaves as a constant background field, the dispersion energy of a fermion can be written as

$$E_{\uparrow\downarrow} = \sqrt{(\mathbf{p} \mp \mathbf{B})^2 + m^2} \pm B_0, \tag{50.10}$$

where $\uparrow$ and $\downarrow$ indicate respectively spin-up and spin-down fermions. Hence, spacetime curvature, more precisely the classical spin of spacetime, couples differently between the quantum up-spin and down-spin. This semi-classical effect has many consequences, apart from particle genesis, e.g. gravitational Berry curvature/phase and Aharonov-Bohm effect [53], neutrino oscillation [51, 54].

Now neutrinos are purely left-handed (according to Standard Model) or almost (due to finite mass) left-handed and anti-neutrinos are purely right-handed (according to Standard Model) or almost (due to finite mass) right-handed. Hence $E_\uparrow$ and $E_\downarrow$ are respectively energies for neutrinos and anti-neutrinos. Due to this split in energy under gravity, they are expected to lead to difference in the number density by simple statistical mechanics, given by

$$\Delta n = \int \frac{d^3 p}{8\pi^3}\left(\frac{1}{e^{E_\uparrow/T}+1} - \frac{1}{e^{E_\downarrow/T}+1}\right). \tag{50.11}$$

For ultra-relativistic neutrinos $\Delta n \approx T^2(B_0 - |\mathbf{B}|)$. Therefore, in a strong gravity, neutrino asymmetry hence leptogenesis takes place. From the conservation of total baryon and lepton number, this leptogenesis will induce baryogenesis. Therefore, effectively baryogenesis is possible in principle even in today's universe due to, e.g., the presence of abundant black holes in the universe.

50.8 Summary

Since 1997, I have been involved in various topics of theoretical physics focusing particularly on astrophysics, astroparticle physics, astronomy and a bit of cosmology. In the present chapter, I have touched upon some of my research over slightly more than a quarter of a century, as of this writing. Prior to 1997, I had a very brief research stint, during my master's degree and early post-master's phase, on particle phenomenology and related lie algebra, mostly via brief projects.

I am successfully working in a variety of topics under the astrophysics umbrella because of the continuous support from my mentees/advisees, collaborators and mentors. Some of my advisees who have contributed to this special Springer volume are: Shubhrangshu Ghosh, Sujit Kumar Nath, Surajit Kalita, Srishty Aggarwal, Mayusree Das, Rohan Raha, Zenia Zuraiq, Mayank Pathak, Mriganka Dutta, Prasad Basu, Debbijoy Bhattacharya, Drisya Karinkuzhi, Debabrata Deb, Chandrachur Chakraborty, Abhishek Kumar Jha, Koushik Chatterjee, Achal Kumar, Arghya Ranjan Das, Panchajanya Dey, Anindya Guria, Soumya Kanti Ganguly. My present/past collaborators, some of them are friends and some other are/were mentor as well, contributing to this volume are: Vinod Krishan, A. R. Rao, T. R. Govindarajan, Sandip Kumar Chakrabarti, Tomasz Bulik, Kaushik Ghosh, Amit Kumar Chattopadhyay, Subhashish Banerjee, Niayesh Afshordi, M. C. Ramadevi. I am grateful to all of them, as well as other colleagues who made this special Springer volume successful by contributing their work.

My specific interests are in fluid mechanics and its astrophysics applications, nuclear astrophysics, relativistic astrophysics, stellar physics, gravitational physics and its applications to astrophysics and cosmology, field theory in curved spacetime in connection to astroparticle physics, a bit of observational astrophysics or astronomy. I am lucky enough to have been successfully working in all these topics of my interest. While they may apparently seem to be a bit unrelated, broadly they are connected and in the present chapter I have attempted to describe the same. All of them are related to gravitational physics, often of black hole, and its interplay with underlying nuclear physics and particles, which collectively lead to fluids, often magnetized fluids due to high temperature.

Acknowledgments I am grateful to Professor Sandip Kumar Chakrabarti who considered me to supervise for my Ph.D. and introduced me to the field of relativistic astrophysics. I further gratefully thank Professor Ramesh Narayan for supervising during my postdoctoral stint in the Harvard University, and Professor A. R. Rao who inspired me a lot during various stages and in various affairs of my life.

Most importantly, I owe my all mentees, who have chosen to work with me and trusted me to allow supervising them. I am specially indebted to Shubhrangshu Ghosh, my first research/Ph.D. student, who organized the conference centering around my 50th birth anniversary, based on which this volume is. He did a tremendous job during the conference in organizing it elegantly and subsequently editing this special volume. Special thanks go to my current advisees (in the reverse alphabetic order): Zenia, Srishty, Rohan, Panchajanya, Mriganka, Mayusree, Mayank, Anindya, Abhishek, who strongly supported the idea of Shubhrangshu to organize the conference in Sikkim and joined me throughout.

Finally, I thank my all collaborators, friends, colleagues and other mentors, who made me what I am.

References

1. Mukhopadhyay, B., & Chakrabarti, S. K. (2000). *Astronomy & Astrophysics, 353*, 1029.
2. Datta, S. R., & Mukhopadhyay, B. (2019). *Monthly Notices of the Royal Astronomical Society, 486*, 1641. https://doi.org/10.1093/mnras/stac835
3. MacFadyen, A. I., & Woosley, S. E. (1999). *Astrophysical Journal, 524*, 262. https://doi.org/10.1086/307790
4. Banerjee, I., & Mukhopadhyay, B. (2013). *Astrophysical Journal, 778*, 8. https://doi.org/10.1088/0004-637X/778/1/8
5. Banerjee, I., & Mukhopadhyay, B. (2013). *Physical Review Letters, 111*, 061101. https://doi.org/10.1103/PhysRevLett.111.061101
6. Mukhopadhyay, B. (2002). *Astrophysical Journal, 581*, 427. https://doi.org/10.1086/344227
7. Ivanov, R. I., & Prodanov, E. M. (2005). *Physics Letters B, 611*, 34. https://doi.org/10.1016/j.physletb.2005.02.047
8. Shafee, R., Narayan, R., & McClintock, J. E. (2008). *Astrophysical Journal, 676*, 549. https://doi.org/10.1086/527346
9. Sarkar, T., Ghosh, S., & Bhadra, A. (2014). *Physical Review D, 90*, 063008. https://doi.org/10.1086/527346
10. Papavasileiou, Th. V., Kosmas, O., & Kosmas, T. S. (2024) arXiv:2408.02415v1. https://doi.org/10.48550/arXiv.2408.02415
11. Mondal, T., & Mukhopadhyay, B. (2019). *Monthly Notices of the Royal Astronomical Society, 482*, 24. https://doi.org/10.1093/mnrasl/sly165
12. Mondal, T., & Mukhopadhyay, B. (2020). *Monthly Notices of the Royal Astronomical Society, 495*, 350. https://doi.org/10.1093/mnras/staa1161
13. Raha, R., Mukhopadhyay, B., & Chatterjee, K. (2024). Magnetically arrested advective accretion flows and jets/outflows around stellar mass black holes: Explaining hard state ULXs with GRMHD simulations. In *The relativistic universe: From classical to quantum*. Springer.

14. Pathak, M., & Mukhopadhyay, B. (2024). Accretion disk-outflow/jet and hard state ULXs. In *The relativistic universe: From classical to quantum*. Springer.
15. Motch, C., Ricketts, M. J., Page, C. G., Ilovaisky, S. A., & Chevalier, C. (1983). *Astronomy & Astrophysics, 119*, 171.
16. van der Klis, M., Jansen, F., Paradijs, J., et al. (1985). *Nature, 316*, 225. https://doi.org/10.1038/316225a0
17. Mouche, C. W. (2002). *Astrophysical Journal, 580*, 423. https://doi.org/10.1086/343095
18. Titarchuk, L., & Wood, K. (2002). *Astrophysical Journal, 577*, L23. https://doi.org/10.1086/344147
19. Belloni, T. M., Sanna, A., & Méndez, M. (2012). *Monthly Notices of the Royal Astronomical Society, 426*, 1701. https://doi.org/10.1111/j.1365-2966.2012.21634.x
20. Franchini, A., Motta, & S. E., Lodato, G. (2016). *Monthly Notices of the Royal Astronomical Society, 467*, 145. https://doi.org/10.1093/mnras/stw3363
21. Chakrabarti, S. K., & Manickam, S. G. (2000). *Astrophysical Journal, 531*, L41. https://doi.org/10.1086/312512
22. Abramowicz, M. A., Karas, V., Kluźniak, W., Lee, W. H., & Rebusco, P. (2003). *Publications of the Astronomical Society of Japan, 55*, 467. https://doi.org/10.1093/pasj/55.2.467
23. Titarchuk, L., & Osherovich, V. (2000). *Astrophysical Journal, 542*, L111. https://doi.org/10.1086/312935
24. Li, L.-X., Zimmerman, E. R., Narayan, R., & McClintock, J. E. (2005). *Astrophysical Journal S, 157*, 335. https://doi.org/10.1086/428089
25. Reynolds, C. S., & Fabian, A. C. (2008). *Astrophysical Journal, 675*, 1048. https://doi.org/10.1086/527344
26. Mukhopadhyay, B. (2009). *Astrophysical Journal, 694*, 387. https://doi.org/10.1088/0004-637X/694/1/387
27. Das, A. R., & Mukhopadhyay, B. (2023). *Astrophysical Journal, 955*, 86. https://doi.org/10.3847/1538-4357/acf1fb
28. Das, A. R., & Mukhopadhyay, B. (2024). QPOs in compact sources as a non-linear hydrodynamical resonance: Determining spin of compact objects. In *The relativistic universe: From classical to quantum*. Springer.
29. Mukhopadhyay, B., Afshordi, N., & Narayan, R. (2005). *Astrophysical Journal, 629*, 383. https://doi.org/10.1086/431419
30. Afshordi, N., Mukhopadhyay, B., & Narayan, R. (2005). *Astrophysical Journal, 629*, 373 . https://doi.org/10.1086/431418
31. Nath, S. K., & Mukhopadhyay, B. (2015). *Physical Review E, 92*, 023005. https://doi.org/10.1103/PhysRevE.92.023005
32. Mukhopadhyay, B. (2013). *Physics Letters B, 721*, 151. https://doi.org/10.1016/j.physletb.2013.02.056
33. Ghosh, S., & Mukhopadhyay, B. (2021). *Astrophysical Journal, 922*, 161. https://doi.org/10.3847/1538-4357/ac1118
34. Phillips, M. M. (1993). *Astrophysical Journal, 413*, L105. https://doi.org/10.1086/186970
35. Taubenberger, S. (2017). The extremes of thermonuclear supernovae. In A. Alsabti & P. Murdin (Eds.), *Handbook of supernovae* (p. 317). Springer. https://doi.org/10.1007/978-3-319-21846-5_37
36. Zuraiq, Z., Kumar, A., Hackett, A. J., Bhattarai, S., Tout, C. A., & Mukhopadhyay, B. (2024). Simulating super-Chandrasekhar white dwarfs. In *The relativistic universe: From classical to quantum*. Springer
37. Das, U., & Mukhopadhyay, B. (2013). *Physical Review Letters, 110*, 071102. https://doi.org/10.1103/PhysRevLett.110.071102
38. Das, U., & Mukhopadhyay, B. (2014). *Modern Physics Letters A, 29*, 1450035. https://doi.org/10.1142/S0217732314500357
39. Das, U., & Mukhopadhyay, B. (2014). *Journal of Cosmology and Astroparticle Physics, 06*, 050. https://doi.org/10.1088/1475-7516/2014/06/050

40. Das, U., & Mukhopadhyay, B. (2015). *Journal of Cosmology and Astroparticle Physics, 05*, 016. https://doi.org/10.1088/1475-7516/2015/05/016
41. Subramanian, S., & Mukhopadhyay, B. (2015). *Monthly Notices of the Royal Astronomical Society, 454*, 752. https://doi.org/10.1093/mnras/stv1983
42. Das, U., & Mukhopadhyay, B. (2015). *Journal of Cosmology and Astroparticle Physics, 05*, 045. https://doi.org/10.1088/1475-7516/2015/05/045
43. Mukhopadhyay, B. (2015). *Current Science, 109*, 2250. https://doi.org/10.48550/arXiv.1609.01862
44. Kalita, S., & Mukhopadhyay, B. (2018). *Journal of Cosmology and Astroparticle Physics, 09*, 007. https://doi.org/10.1088/1475-7516/2018/09/007
45. Kalita, S., & Mukhopadhyay, B. (2019). *Monthly Notices of the Royal Astronomical Society, 490*, 2692. https://doi.org/10.1093/mnras/stz3383
46. Mukhopadhyay, B., Sarkar, A., & Tout, C. A. (2021). *Monthly Notices of the Royal Astronomical Society, 500*, 763. https://doi.org/10.1093/mnras/staa3136
47. Gupta, A., Mukhopadhyay, B., & Tout, C. A. (2020). *Monthly Notices of the Royal Astronomical Society, 496*, 894. https://doi.org/10.1093/mnras/staa1575
48. Bhattacharya, M., Hackett, A. J., Gupta, A., Tout, C. A., & Mukhopadhyay, B. (2022). *Astrophysical Journal, 925*, 133. https://doi.org/10.3847/1538-4357/ac450b
49. Kalita, S., Govindarajan, T. R., & Mukhopadhyay, B. *International Journal of Modern Physics D, 30*, 2150101 (2021). https://doi.org/10.1142/S0218271821501017
50. Deb, D., & Mukhopadhyay, B. (2022). Weber, F. *Astrophysical Journal, 926*, 66. https://doi.org/10.3847/1538-4357/ac410b
51. Mukhopadhyay, B. (2007). *Classical and Quantum Gravity, 24*, 1433. https://doi.org/10.1088/0264-9381/24/6/004
52. Schwinger, J. (1970). *Particles, sources and fields* (Vol. 1, p. 401). Addison-Wesley.
53. Ghosh, T., & Mukhopadhyay, B. (2021). *International Journal of Modern Physics D, 30*, 2150090. https://doi.org/10.1142/S0218271821500905
54. Sinha, M., & Mukhopadhyay, B. (2008). *Physical Review D, 77*, 025003. https://doi.org/10.1103/PhysRevD.77.025003

The manufacturer's authorised representative in the EU is Springer Nature Customer Service Centre GmbH, Europaplatz 3, 69115 Heidelberg, Germany. If you have any concerns regarding our products, please contact ProductSafety@springernature.com

Printed and bound by CPI Group (UK) Ltd, Croydon, CR0 4YY
01/10/2026
02225954-0020